THE INTERNATIONAL SERIES OF
MONOGRAPHS ON CHEMISTRY

FOUNDING EDITOR

J. S. ROWLINSON, FRS

GENERAL EDITORS

PROFESSOR M. L. H. GREEN, FRS
PROFESSOR J. HALPERN, FRS
PROFESSOR T. MUKAIYAMA
PROFESSOR R. L. SCHOWEN
PROFESSOR J. M. THOMAS, FRS
PROFESSOR A. H. ZEWAIL

THE INTERNATIONAL SERIES OF MONOGRAPHS ON CHEMISTRY

1. J. D. Lambert: *Vibrational and rotational relaxation in gases*
2. N. G. Parsonage and L. A. K. Staveley: *Disorder in crystals*
3. G. C. Maitland, M. Rigby, E. B. Smith, and W. A. Wakeham: *Intermolecular forces: their origin and determination*
4. W. G. Richards, H. P. Trivedi, and D. L. Cooper: *Spin-orbit coupling in molecules*
5. C. F. Cullis and M. M. Hirschler: *The combustion of organic polymers*
6. R. T. Bailey, A. M. North, and R. A. Pethrick: *Molecular motion in high polymers*
7. Atta-ur-Rahman and A. Basha: *Biosynthesis of indole alkaloids*
8. J. S. Rowlinson and B. Widom: *Molecular theory of capillarity*
9. C. G. Gray and K. E. Gubbins: *Theory of molecular fluids, volume 1: Fundamentals*
10. C. G. Gray and K. E. Gubbins: *Theory of molecular fluids, volume 2: Application* (in preparation)
11. S. Wilson: *Electron correlations in molecules*
12. E. Haslam: *Metabolites and metabolism: a commentary on secondary metabolism*
13. G. R. Fleming: *Chemical applications of ultrafast spectroscopy*
14. R. R. Ernst, G. Bodenhausen, and A. Wokaun: *Principles of nuclear magnetic resonance in one and two dimensions*
15. M. Goldman: *Quantum description of high-resolution NMR in liquids*
16. R. G. Parr and W. Yang: *Density-functional theory of atoms and molecules*
17. J. C. Vickerman, A. Brown, and N. M. Reed (editors): *Secondary ion mass spectrometry: principles and applications*
18. F. R. McCourt, J. Beenakker, W. E. Köhler, and I. Kuščer: *Nonequilibrium phenomena in polyatomic gases, volume 1: Dilute gases*
19. F. R. McCourt, J. Beenakker, W. E. Köhler, and I. Kuščer: *Nonequilibrium phenomena in polyatomic gases, volume 2: Cross-sections, scattering, and rarefied gases*
20. T. Mukaiyama: *Challenges in synthetic organic chemistry*
21. P. Gray and S. K. Scott: *Chemical oscillations and instabilities: non-linear chemical kinetics*
22. R. F. W. Bader: *Atoms in molecules: a quantum theory*
23. J. H. Jones: *The chemical synthesis of peptides*
24. S. K. Scott: *Chemical chaos*
25. M. S. Child: *Semiclassical mechanics with molecular applications*
26. D. T. Sawyer: *Oxygen chemistry*
27. P. A. Cox: *Transition metal oxides: an introduction to their electronic structure and properties*

Oxford University Press, Walton Street, Oxford OX2 6DP
Oxford New York Toronto
Delhi Bombay Calcutta Madras Karachi
Kuala Lumpur Singapore Hong Kong Tokyo
Nairobi Dar es Salaam Cape Town
Melbourne Auckland Madrid
and associated companies in
Berlin Ibadan

Oxford is a trade mark of Oxford University Press

Published in the United States
by Oxford University Press Inc. New York

© Richard R. Ernst, Geoffrey Bodenhausen, and Alexander Wokaun, 1987

First published 1987
First published in paperback (with corrections) 1990
Reprinted (with further corrections) 1991 (twice), 1992, 1994

All rights reserved. No part of this publication may be
reproduced, stored in a retrieval system, or transmitted, in any
form or by any means, without the prior permission in writing of Oxford
University Press. Within the UK, exceptions are allowed in respect of any
fair dealing for the purpose of research or private study, or criticism or
review, as permitted under the Copyright, Designs and Patents Act, 1988, or
in the case of reprographic reproduction in accordance with the terms of
licences issued by the Copyright Licensing Agency. Enquiries concerning
reproduction outside those terms and in other countries should be sent to
the Rights Department, Oxford University Press, at the address above.

This book is sold subject to the condition that it shall not, by way
of trade or otherwise, be lent, re-sold, hired out, or otherwise circulated
without the publisher's prior consent in any form of binding or cover
other than that in which it is published and without a similar condition
including this condition being imposed on the subsequent purchaser.

British Library Cataloguing in Publication Data
Ernst, Richard R.
Principles of nuclear magnetic resonance in one
and two dimensions.— (The international series
of monographs on chemistry; 14)
1. Nuclear magnetic resonance spectroscopy
I. Title II. Bodenhausen, Geoffrey III. Wokaun,
Alexander IV. Series
538'.362 QC762

Library of Congress Cataloging in Publication Data
Ernst, Richard R.
Principles of nuclear magnetic resonance in one
and two dimensions.— (The international series
of monographs on chemistry; 14)
Bibliography: p.
Includes index.
1. Nuclear magnetic resonance spectroscopy
II. Bodenhausen, Geoffrey. II. Wokaun, Alexander.
III. Title. IV. Series.
QD96.N8E76 1986 543'.0877 85-26014
ISBN 0 19 855647 0 (Pbk)

Printed in the United Kingdom by
The Universities Press, (Belfast) Ltd, Northern Ireland

Principles of Nuclear Magnetic Resonance in One and Two Dimensions

Richard R. Ernst, Geoffrey Bodenhausen,
and Alexander Wokaun

*Laboratorium für Physikalische Chemie
Eidgenössische Technische Hochschule
Zürich*

CLARENDON PRESS · OXFORD

PREFACE TO THE PAPERBACK EDITION

Since the completion of the first edition, important developments have been going on in virtually all fields of magnetic resonance, and the astounding wealth of useful NMR techniques has continued to grow, opening up further new and exciting fields of application. To name just a few particularly active domains, new experiments have been developed that involve manipulations of the magnetization in the rotating frame, three-dimensional spectroscopy, computer-supported data processing, two-dimensional spectroscopy of partially ordered phases and of solid materials with slow motion, sample rotations about one or two angles, and, last but not least, magnetic resonance imaging and *in vivo* spectroscopy, where a completely new discipline of medical diagnosis has emerged. Nuclear magnetic resonance still appears to be as lively as ever, and is likely to hold many further exciting surprises in store.

We have resisted the temptation to bring the text and references up to date. This would have compelled us to add further chapters, a dangerous enterprise in this day and age, for it is uncertain that the speed of writing can match the rate at which further developments appear in the literature. We have therefore limited the changes to the correction of errors and inaccurate formulations.

We are greatly indebted to a group of scientists in Kazan, USSR, who, under the guidance of Professor Kev Minullinovich Salikhov, not only undertook the task of translating this volume into Russian, but also made the effort to check many of the equations, and pointed out a number of errors in the original text. This team consisted of Nikolay Kuz'mich Andreev, Aleksandr Vasil'evich Egorov, Vladimir L'vovich Ermakov, Kamil' Akhatovich Il'yasov, Il'dus El'brusovich Ismaev, Dmitry Yakovlevich Osokin, and Vladimir Nikolaevich Zinin.

We hope that the publication in paperback form enables the book to reach a wider readership, particularly among younger scientists.

Zurich   R. R. E.
Lausanne   G. B.
Bayreuth   A. W.
September 1989

PREFACE

This monograph has a long history, and it was never intended to become what it finally became. Initially, we set out to compile a brief, timely review on two-dimensional spectroscopy. However, the undertaking soon outgrew its original setting. We found what had to be found: the same principles govern magnetic resonance phenomena irrespective of whether they are investigated by one- or two-dimensional spectroscopy.

The monograph has therefore grown naturally into a general account of advanced NMR spectroscopy including one- and two-dimensional techniques. The main emphasis is on principles rather than on practical aspects, although the reader may find sections where practical considerations dominate. Fourier spectroscopy has unified solid- and liquid-state NMR in an unprecedented manner. We have tried therefore to discuss principles that are relevant to both subjects.

Chapter 1 is intended to be introductory, and presents a rather cursory survey of some historical developments which led to the present state of the art.

Chapter 2 treats the dynamics of nuclear spin systems, starting with the equation of motion and providing the appropriate tools to handle density operators. The driving terms of the equation of motion, i.e. the Hamiltonian, the relaxation superoperator, and the chemical exchange superoperator are discussed in view of a general formulation of Fourier spectroscopy.

The fact that nuclear spin Hamiltonians can be manipulated easily has contributed enormously to the success of NMR spectroscopy. Chapter 3 briefly discusses various tools that are available to the NMR 'alchemist', such as double resonance, multiple-pulse sequences, etc. These procedures are best described in terms of average Hamiltonian theory, which is treated in Chapter 3, both for the usual periodic perturbations as well as for aperiodic perturbations, an aspect of particular relevance in the context of two-dimensional spectroscopy.

Chapter 4 covers one-dimensional Fourier spectroscopy and begins with a section which should help to clarify the relationship between spectroscopy and general response theory, developed originally in the context of electrical engineering. Linear and non-linear response theory are first discussed for classical systems, and then generalized to quantum systems. The basic features of one-dimensional Fourier spectroscopy can be discussed in terms of classical Bloch equations, but specific aspects of coupled spin systems require a quantum mechanical treatment.

In Chapter 5, the basic properties of multiple-quantum transitions are discussed, including a short account of continuous-wave detection, in order to allow a comparison with the more powerful two-dimensional techniques. The excitation and evolution of multiple-quantum coherence are also discussed in Chapter 5, while practical applications of two-dimensional multiple-quantum spectra are deferred to Chapter 8.

In Chapter 6, the principles of two-dimensional spectroscopy are developed in general terms. Various methods for the separation of interactions, such as chemical shifts and couplings, are reviewed in Chapter 7. Two-dimensional correlation methods based on coherence transfer are discussed in Chapter 8, while the study of dynamic processes, such as chemical exchange and cross-relaxation, is reviewed in Chapter 9. Finally, Chapter 10 gives a brief account of the principles of NMR imaging. Since many imaging methods employ concepts of two-dimensional spectroscopy, it seems appropriate to discuss the principles in this monograph.

We hope that this volume helps to fulfil the needs of NMR spectroscopists who wish to deepen their understanding of the basic features of time-domain spectroscopy and who intend to apply the recently developed tools of spectroscopy in one and two dimensions.

We have attempted to present the material in an intuitive manner as well as in a rigorous mathematical framework to satisfy the tastes of a wide spectrum of readers. Obviously this was not possible without numerous compromises.

We would like to express our deep gratitude to those who were instrumental in guiding our interests towards time domain spectroscopy. Professor Hans Primas has contributed much in terms of mathematical tools and basic concepts. Dr Weston A. Anderson inspired the search for more sensitive NMR methods which led to Fourier spectroscopy and Professor Jean Jeener first expressed the idea of performing spectroscopy in two dimensions.

We are much obliged to our colleagues and co-workers who contributed many ideas, corrected misconceptions, and performed numerous elegant experiments. Special thanks are due to Walter P. Aue, Peter Bachmann, Enrico Bartholdi, Lukas Braunschweiler, Peter Brunner, Douglas P. Burum, Pablo Caravatti, Christopher J. R. Counsell, Gerhard W. Eich, Christian Griesinger, Alfred Höhener, Yong-Ren Huang, Jiri Karhan, Herbert Kogler, Roland Kreis, René O. Kühne, Anil Kumar, Malcolm H. Levitt, Max Linder, Andrew A. Maudsley, Slobodan Macura, Beat H. Meier, Beat U. Meier, Anita Minoretti, Luciano Müller, Norbert Müller, Kuniaki Nagayama, Peter Pfändler, Umberto Piantini, Christian Radloff, Mark Rance, Michael Reinhold, Thierry Schaffhauser, Günther Schatz, Stephan Schäublin, Christian Schönenberger, Ole W. Sørensen, Dieter Suter and Stephen C. Wimperis. We should also like to mention the very fruitful collaboration with Kurt Wüthrich and Gerhard Wagner.

We are indebted to many authors around the world, not merely for their kind permission to reproduce figures from their work, but also for stimulating discussions and the many clarifying concepts that they have contributed. Particularly stimulating were contacts with Ad Bax, Philip H. Bolton, Ray Freeman, Robert G. Griffin, Jean Jeener, Horst Kessler, Alex Pines, Robert L. Vold, Regitse R. Vold, John S. Waugh, and members of their research groups.

This monograph could not have been written without the patience of Miss Irène Müller and Mrs Dorothea Spörri who carefully typed the numerous incarnations of the manuscript.

Zürich R. R. E.
March 1985 G. B.
A. W.

ACKNOWLEDGEMENTS

We are indebted to the following authors and publishers for permission to reproduce figures:

A. Bax: Figs. 6.5.14, 8.5.5, 8.5.10
P. H. Bolton: Fig. 6.5.10
R. Freeman: Figs. 4.1.4, 4.1.6, 4.2.9, 4.2.10, 4.2.13, 4.2.14, 4.2.15, 4.2.18, 4.2.19, 4.7.7, 5.2.5, 6.5.13, 7.2.12, 7.2.14, 7.2.15, 7.2.16, 8.2.11, 8.4.4
R. G. Griffin: Figs. 7.3.9, 7.3.10
R. N. Grimes: Fig. 8.2.7
U. Haeberlen: Fig. 7.3.2
W. S. Hinshaw: Fig. 10.3.2
H. Kessler: Fig. 8.5.4
M. H. Levitt: Fig. 4.2.12
G. E. Maciel: Fig. 9.10.2
P. Meakin: Fig. 4.2.4
S. Ogawa: Fig. 4.6.2
A. Pines: Figs. 5.3.2, 8.4.10, 8.4.11
D. J. Ruben: Fig. 6.5.9
V. Rutar: Fig. 7.2.10
O. W. Sørensen: Fig. 4.5.6
R. L. Vold: Figs. 5.4.2, 5.4.3
G. Wagner: Fig. 8.3.11
J. S. Waugh: Fig. 7.3.4
K. Wüthrich: Figs. 8.2.4, 8.2.5, 8.2.6, 8.3.3, 8.3.7, 9.7.4, 9.7.6, 9.7.8
D. Ziessow: Fig. 6.5.12

Academic Press:
from *Advances in Magnetic Resonance*: Fig. 4.7.7;
from *Biochemical and Biophysical Research Communications*: Figs. 8.2.6, 8.3.7;
from the *Journal of Molecular Biology*: Figs. 8.2.4, 8.2.5, 9.7.4, 9.7.8;
from the *Journal of Magnetic Resonance*: Figs. 4.1.9, 4.2.4, 4.2.9, 4.2.10, 4.2.12, 4.2.13, 4.2.14, 4.2.15, 4.2.17, 4.2.18, 4.2.19, 4.4.3, 4.5.4, 4.5.6, 4.5.7, 4.6.5, 4.6.6, 4.6.7, 4.6.8, 4.6.9, 4.6.10, 5.4.2, 5.4.3, 6.3.2, 6.3.3, 6.5.4, 6.5.9, 6.5.11, 6.5.12, 6.5.13, 6.6.5, 6.6.6, 6.8.1, 6.8.2, 7.2.2, 7.2.4, 7.2.10, 7.2.12, 7.2.13, 7.2.14, 7.2.15, 7.2.16, 7.3.2, 8.2.11, 8.3.3, 8.3.11, 8.3.12, 8.4.4, 8.4.9, 8.5.5, 8.5.8, 8.5.10, 9.4.1, 9.6.3, 9.10.2, 10.1.1, 10.4.5, 10.4.9, 10.5.1, 10.5.2.

ACKNOWLEDGEMENTS

American Chemical Society:
from the *Journal of the American Chemical Society*: Figs. 8.2.7, 8.3.6, 8.5.4, 9.6.1, 9.6.2, 9.6.4, 9.7.5, 9.7.7, 9.8.2, 9.9.1;
from *Macromolecules*: Fig. 9.10.4.

American Institute of Physics:
from the *Journal of Chemical Physics*: Figs. 4.2.11, 5.2.5, 7.2.7, 7.3.4, 7.3.5, 7.3.6, 7.3.7, 7.3.9, 7.3.10, 8.2.9, 8.5.13, 9.8.1;
from the *Review of Scientific Instruments*: Fig. 4.3.3.

American Physical Society:
from *Physical Review Letters*: Figs. 5.3.2, 8.4.11.

Blackwell Scientific Publishers:
from *Pure and Applied Chemistry*: Fig. 4.7.4.

Chimia:
from *Chimia*: Fig. 4.1.3.

Elsevier Biomedical Press:
from *Biochimica Biophysica Acta*: Fig. 9.7.6.

The Institute of Petroleum:
Fig. 4.3.4.

National Academy of Sciences:
Fig. 4.6.2

Macmillan Journals:
from *Nature*: Fig. 10.3.2.

North Holland Publishing Company:
from *Chemical Physics Letters*: Figs. 5.3.6, 8.4.10, 8.5.6, 8.5.11, 8.5.12.

Pergamon Press:
from *Progress in NMR Spectroscopy*: Figs. 2.1.4, 2.1.5, 2.1.6, 2.1.7, 4.4.4, 4.4.5, 4.4.6.

Taylor and Francis:
from *Molecular Physics*: Figs. 4.7.1, 5.4.1, 8.4.3, 8.4.6, 8.4.7, 8.4.8, 9.5.1, 9.7.2, 9.7.3, 9.9.2.

CONTENTS

SYMBOLS, TRANSFORMATIONS, AND ABBREVIATIONS — xix

1. INTRODUCTION — 1
2. THE DYNAMICS OF NUCLEAR SPIN SYSTEMS — 9
   2.1. Equation of motion — 9
      2.1.1. Density operator — 9
         2.1.1.1. Density operator equation — 12
         2.1.1.2. Expectation values — 13
         2.1.1.3. Schrödinger and Heisenberg representation — 13
         2.1.1.4. Reduced spin density operator — 14
      2.1.2. Explicit matrix representation of the master equations — 15
      2.1.3. Liouville operator space — 17
      2.1.4. Superoperators — 19
         2.1.4.1. Commutator superoperators — 20
         2.1.4.2. Unitary transformation superoperators — 20
         2.1.4.3. Projection superoperators — 21
         2.1.4.4. General representation of superoperators — 23
         2.1.4.5. Matrix representation of a superoperator — 23
         2.1.4.6. Eigenvalues and eigenoperators of superoperators — 24
         2.1.4.7. Superoperator algebra — 25
      2.1.5. Products of Cartesian spin operators — 25
         2.1.5.1 Systems of spins $I = \frac{1}{2}$ — 26
         2.1.5.2 Systems containing spins $S > \frac{1}{2}$ — 29
      2.1.6. Products involving shift operators — 32
      2.1.7. Polarization operators — 33
      2.1.8. Cartesian single-transition operators — 34
      2.1.9. Single-transition shift operators — 38
      2.1.10. Irreducible tensor operators — 40
      2.1.11. Coherence transfer — 43
   2.2. The nuclear spin Hamiltonian — 44
      2.2.1. Interactions of nuclear spins — 44
         2.2.1.1. Interactions linear in the spin operators — 44
         2.2.1.2. Interactions bilinear in the spin operators — 46
         2.2.1.3. Interactions quadratic in the spin operators — 48
   2.3. The relaxation superoperator — 49
      2.3.1. Semi-classical relaxation theory — 50
         2.3.1.1. Restriction to secular contributions — 52
         2.3.1.2. Extreme narrowing — 52
      2.3.2. Matrix representation of the relaxation superoperator — 53
      2.3.3. Specific relaxation mechanisms — 55
         2.3.3.1. First-rank relaxation mechanisms — 55
         2.3.3.2. Second-rank relaxation mechanisms — 56
   2.4. Spin dynamics due to chemical reactions — 57
      2.4.1. Description of reaction networks in classical kinetics — 58

	2.4.2.	Exchange in systems without spin–spin couplings	60
		2.4.2.1. Modified Bloch equations for first-order reactions	61
		2.4.2.2. Higher-order reactions for spin systems without spin–spin couplings	63
	2.4.3.	Density operator description of exchanging systems with spin–spin couplings	64
	2.4.4.	Density operator equation and exchange superoperator for first-order reactions	68

3. MANIPULATION OF NUCLEAR SPIN HAMILTONIANS — 70

3.1. The tools for manipulations — 70
3.2. Average Hamiltonian theory — 71
    3.2.1. Exact calculation of $\bar{\mathcal{H}}$ — 72
    3.2.2. Cumulant expansion of the propagator — 72
    3.2.3. Averaging by time-dependent perturbations — 75
    3.2.4. Truncation of internal Hamiltonians — 79
    3.2.5. Floquet theory — 81
3.3. Average Hamiltonian due to aperiodic perturbations — 83
    3.3.1. General condition for an average Hamiltonian — 84
    3.3.2. Average Hamiltonian in spin–echo experiments — 86
    3.3.3. Cancellation of irrelevant terms — 88

4. ONE-DIMENSIONAL FOURIER SPECTROSCOPY — 91

4.1. Response theory — 92
    4.1.1. Linear response theory — 93
    4.1.2. Time and frequency domains — 96
    4.1.3. Linear data-processing — 99
        4.1.3.1. Apodization — 101
        4.1.3.2. Resolution enhancement — 106
    4.1.4. Non-linear response theory — 109
    4.1.5. Quantum-mechanical response theory — 111
    4.1.6. Stochastic response theory — 112
4.2. Classical description of Fourier spectroscopy — 115
    4.2.1. Bloch equations in the rotating frame — 116
    4.2.2. Ideal pulse experiment — 118
    4.2.3. Off-resonance effects due to finite pulse amplitude — 119
    4.2.4. Longitudinal interference in repetitive pulse experiments — 124
    4.2.5. Transverse interference in repetitive pulse experiments — 125
    4.2.6. Remedies for phase and intensity anomalies due to transverse interference — 131
        4.2.6.1. Quenching of transverse magnetization by magnetic field gradient pulses — 131
        4.2.6.2. Scrambling of transverse interference by random pulse intervals — 132
        4.2.6.3. Quadriga Fourier spectroscopy — 132
        4.2.6.4. Four-phase Fourier spectroscopy — 133
        4.2.6.5. Phase-alternated pulse sequences — 133

	4.2.7.	Remedies for anomalies due to non-ideal pulses: composite pulses	133
		4.2.7.1. Minimum residual M_z component after a $\pi/2$-pulse	136
		4.2.7.2. Minimum phase dispersion of transverse magnetization after a $\pi/2$-pulse	137
		4.2.7.3. Accurate inversion: sequences optimized by calculation of trajectories	137
		4.2.7.4. Recursive expansion procedures	140
		4.2.7.5. Accurate refocusing	144
		4.2.7.6. Composite z-pulses	145
		4.2.7.7. Cyclic composite pulses	146
4.3	Sensitivity of Fourier spectroscopy		148
	4.3.1.	Signal-to-noise ratio of Fourier spectra	148
		4.3.1.1. The signal	148
		4.3.1.2. The noise	150
		4.3.1.3. The sensitivity	151
		4.3.1.4. Optimization of the weighting function	152
		4.3.1.5. Optimization of the signal energy	153
	4.3.2.	Signal-to-noise ratio in slow-passage spectra	155
	4.3.3.	Sensitivity comparison of Fourier and slow-passage experiments	156
	4.3.4.	Sensitivity enhancement by recycling of magnetization	157
4.4.	Quantum-mechanical description of Fourier spectroscopy		158
	4.4.1.	Density operator formalism applied to Fourier spectroscopy	159
	4.4.2.	Equivalence of slow-passage and Fourier spectroscopy	162
		4.4.2.1. Fourier spectroscopy	163
		4.4.2.2. Slow-passage spectroscopy	163
		4.4.2.3. Comparison of Fourier and slow-passage spectra	164
	4.4.3.	Fourier spectroscopy of non-equilibrium systems	165
	4.4.4.	Selective and semi-selective pulses	171
	4.4.5.	Identification of the terms in the density operator	173
	4.4.6.	Composite rotations	177
		4.4.6.1. Intervals with a central refocusing pulse	177
		4.4.6.2. Bilinear rotations with transverse components	178
		4.4.6.3. Sequences without 'sandwich symmetry'	178
		4.4.6.4. Phase cycles	179
		4.4.6.5. Phase shifts and r.f. rotation angles	179
		4.4.6.6. Heteronuclear systems	180
4.5.	Heteronuclear polarization transfer		180
	4.5.1.	Transfer of spin order	182
	4.5.2.	Polarization transfer by nuclear Overhauser effect	184
	4.5.3.	Cross-polarization in the rotating frame	185
	4.5.4.	Adiabatic polarization transfer	191
	4.5.5.	Polarization transfer by radio-frequency pulses	192
	4.5.6.	Editing procedures based on polarization transfer	198
4.6.	Investigation of dynamic processes, relaxation, and chemical exchange		201

		4.6.1.	Longitudinal relaxation	202
			4.6.1.1. Inversion recovery methods	203
			4.6.1.2. Saturation recovery methods	204
			4.6.1.3. Progressive saturation	205
			4.6.1.4. Selective perturbations	205
		4.6.2.	Transverse relaxation	206
		4.6.3.	Chemical reactions and exchange processes	209
			4.6.3.1. One-sided first-order reaction	211
			4.6.3.2. Two-sided first-order reaction	215
			4.6.3.3. Transient chemical reactions with coupled spin systems	217
			4.6.3.4. Experimental preparation of chemical non-equilibrium states	219
	4.7.	Fourier double resonance		220
		4.7.1.	Theoretical formulation of Fourier double resonance	222
			4.7.1.1. Double-resonance irradiation during detection	223
			4.7.1.2. Continuous Fourier double resonance	225
		4.7.2.	Fourier double resonance of two coupled spins $I = 1/2$	226
			4.7.2.1. Strong coupling	226
			4.7.2.2. Weak coupling	227
		4.7.3.	Spin tickling	230
		4.7.4.	Spin decoupling treated by average Hamiltonian theory	232
			4.7.4.1. Heteronuclear spin decoupling	233
			4.7.4.2. Off-resonance decoupling	234
		4.7.5.	Time-shared decoupling	236
		4.7.6.	Broadband decoupling and scaling of heteronuclear interactions	237
			4.7.6.1. Multiple-pulse decoupling techniques	238
			4.7.6.2. Scaling of heteronuclear couplings	238
		4.7.7.	Illusions of decoupling	239

5. MULTIPLE-QUANTUM TRANSITIONS 242

	5.1.	Number of transitions		244
	5.2.	Detection of multiple-quantum transitions by continuous-wave NMR		247
		5.2.1.	Intensity of multiple-quantum transitions	250
		5.2.2.	Saturation of multiple-quantum transitions	251
		5.2.3.	Level shift of multiple-quantum transitions	252
		5.2.4.	Line-widths of multiple-quantum transitions	254
		5.2.5.	Applications of CW multiple-quantum NMR	255
	5.3.	Time-domain multiple-quantum spectroscopy		256
		5.3.1.	Excitation and detection of multiple-quantum coherence	257
			5.3.1.1. Non-selective pulses	258
			5.3.1.2. Selective single-quantum pulses	261
			5.3.1.3. Selective multiple-quantum pulses	261
			5.3.1.4. Spin-connectivity selective excitation	263
			5.3.1.5. Selective excitation of specific orders	265
		5.3.2.	Offset-dependence of multiple-quantum frequencies and separation of orders	265
		5.3.3.	Structure of multiple-quantum spectra	270
		5.3.4.	Multiple-quantum double resonance	273

5.4.	Relaxation of multiple-quantum coherence	274
	5.4.1. Correlated external random fields	274
	5.4.2. Quadrupolar relaxation	276
	5.4.3. Measurement of multiple-quantum relaxation rates and effects of magnetic field inhomogeneity	278
6.	**TWO-DIMENSIONAL FOURIER SPECTROSCOPY**	**283**
6.1.	Basic principles	283
6.2.	Formal theory of two-dimensional spectroscopy	287
	6.2.1. Explicit matrix representation	289
	6.2.2. Expansion of the density operator in single-transition operators	291
6.3.	Coherence transfer pathways	292
	6.3.1. Selection of pathways	294
	6.3.2. Multiple transfer	298
6.4.	Two-dimensional Fourier transformation	301
	6.4.1. Properties of the complex two-dimensional Fourier transformation	302
	6.4.1.1. Vector notation	302
	6.4.1.2. Similarity theorem	303
	6.4.1.3. Convolution theorem	303
	6.4.1.4. Power theorem	304
	6.4.1.5. Projection cross-section theorem	304
	6.4.1.6. Kramers–Kronig relations in two dimensions	306
	6.4.2. Hypercomplex two-dimensional Fourier transformation	307
6.5.	Peak shapes in two-dimensional spectra	309
	6.5.1. Basic peakshapes	310
	6.5.2. Inhomogeneous broadening and interference of neighbouring peaks	312
	6.5.3. Techniques for obtaining pure two-dimensional absorption peaks	317
	6.5.3.1. Real Fourier transformation in t_1	317
	6.5.3.2. Time-reversal in a complementary experiment	323
	6.5.3.3. Combination of two experiments in quadrature	323
	6.5.4. Absolute-value spectra	326
	6.5.5. Projections of two-dimensional spectra	327
	6.5.6. Two-dimensional filtering	329
	6.5.6.1. Matched filter	331
	6.5.6.2. Lorentz–Gauss transformation	331
	6.5.6.3. Pseudo-echo transformation	333
6.6.	Manipulations of two-dimensional spectra	336
	6.6.1. Shearing transformations	336
	6.6.2. Delayed acquisition	338
	6.6.3. Time-proportional phase increments	340
	6.6.4. Symmetrization	341
	6.6.5. Pattern recognition	343
	6.6.6. Single-channel detection	346
6.7.	Operator terms and multiplet structures in two-dimensional spectra	347
6.8.	Sensitivity of two-dimensional spectra	349
	6.8.1. The signal envelope	349

		6.8.2.	Thermal noise and t_1-noise	351
		6.8.3.	Sensitivity	352
		6.8.4.	Comparison of the sensitivities of one- and two-dimensional experiments	353
		6.8.5.	Optimization of two-dimensional experiments	354
			6.8.5.1. Low resolution in the ω_1-domain	355
			6.8.5.2. High resolution in the ω_1-domain	355
			6.8.5.3. Practical recommendations	357
7.	TWO-DIMENSIONAL SEPARATION OF INTERACTIONS			358
	7.1.	Basic principles		358
	7.2.	Separation of chemical shifts and scalar couplings in isotropic phase		360
		7.2.1.	Homonuclear systems	360
		7.2.2.	Two-dimensional separation in heteronuclear systems	366
			7.2.2.1. Separation of S-spin multiplets and S-spin chemical shifts	367
			7.2.2.2. Removal of the chemical shifts from the ω_1-domain by shearing of the two-dimensional spectrum	368
			7.2.2.3. Separation of S-spin multiplets vs. S-spin chemical shifts by refocusing and gated coupling	369
			7.2.2.4. Separation of S-spin multiplets vs. S-spin chemical shifts by refocusing and inversion of I nuclei	370
			7.2.2.5. Separation of S-spin multiplets due to coupling to a selected I-spin vs. S-spin chemical shifts	370
			7.2.2.6. Separation of long-range IS couplings vs. S-spin chemical shifts	370
			7.2.2.7. Separation of one-bond IS couplings vs. S-spin chemical shifts	372
			7.2.2.8. Artefacts	373
			7.2.2.9. Pure two-dimensional absorption	374
		7.2.3.	Strong coupling effects in refocusing experiments	374
		7.2.4.	Echo modulation by non-resonant nuclei	381
	7.3.	Separation of chemical shifts and dipolar couplings in oriented phase		383
		7.3.1.	Homonuclear separated local field spectra	383
		7.3.2.	Heteronuclear separated local field spectra	384
		7.3.3.	Correlation of chemical shielding and dipolar coupling tensors in static powders	387
			7.3.3.1. Full simulation of powder spectra	390
			7.3.3.2. Ridge plots	391
		7.3.4.	Separation of $\mathcal{H}_{IS}$ and $\mathcal{H}_{ZS}$ under magic-angle spinning conditions	393
	7.4.	Separation of isotropic and anisotropic chemical shifts		396
		7.4.1.	Rotation-synchronized pulses	397
		7.4.2.	Synchronous sampling with scaling	397
		7.4.3.	Magic-angle flipping	398
		7.4.4.	Magic-angle hopping	399

8. TWO-DIMENSIONAL CORRELATION METHODS BASED ON COHERENCE TRANSFER — 400

- 8.1. Coherence transfer in two-dimensional correlation spectroscopy: amplitudes and selection rules — 402
- 8.2. Homonuclear two-dimensional correlation spectroscopy — 405
  - 8.2.1. Weakly coupled two-spin systems — 406
  - 8.2.2. Applications to complicated spectra — 411
  - 8.2.3. Connectivity and multiplet effects in weakly coupled systems — 414
  - 8.2.4. Strong coupling in two-dimensional correlation spectroscopy — 422
  - 8.2.5. Magnetic equivalence — 426
- 8.3. Modified two-dimensional correlation experiments — 427
  - 8.3.1. Delayed acquisition: spin–echo correlation spectroscopy — 428
  - 8.3.2. Constant-time correlation spectroscopy: ω_1-decoupling — 429
  - 8.3.3. Filtering and editing — 431
    - 8.3.3.1. Multiple-quantum filtering — 434
    - 8.3.3.2. p-Spin-filters — 438
    - 8.3.3.3. Filtering according to coupling network connectivity — 438
  - 8.3.4. Relayed coherence transfer — 440
  - 8.3.5. Coherence transfer by an average Hamiltonian in total correlation spectroscopy — 444
- 8.4. Homonuclear two-dimensional multiple-quantum spectroscopy — 449
  - 8.4.1. Excitation and detection of multiple-quantum coherence — 449
  - 8.4.2. Double-quantum spectra of two-spin systems — 451
  - 8.4.3. Multiple-quantum spectra of scalar-coupled networks in isotropic phase — 456
  - 8.4.4. Multiple-quantum spectra of dipole-coupled nuclei in anisotropic phase — 463
  - 8.4.5. Double-quantum spectra of quadrupolar nuclei with $S = 1$ in anisotropic phase — 465
- 8.5. Heteronuclear coherence transfer — 467
  - 8.5.1. Sensitivity considerations — 468
  - 8.5.2. Coherence transfer pathways — 470
  - 8.5.3. Heteronuclear two-dimensional correlation spectroscopy in isotropic phase — 471
    - 8.5.3.1. Transfer of in-phase magnetization — 473
    - 8.5.3.2. Broadband decoupling — 474
    - 8.5.3.3. Decoupling by refocusing pulses — 475
    - 8.5.3.4. Bilinear rotation decoupling — 475
    - 8.5.3.5. Editing of heteronuclear correlation spectra — 477
  - 8.5.4. Relayed heteronuclear correlation spectroscopy — 479
  - 8.5.5. Heteronuclear correlation experiments involving double transfer — 482
  - 8.5.6. Heteronuclear correlation in solids — 485

9. DYNAMIC PROCESSES STUDIED BY TWO-DIMENSIONAL EXCHANGE SPECTROSCOPY — 490

- 9.1. Polarization transfer in one- and two-dimensional methods — 490
- 9.2. Selection of coherence transfer pathways — 494

CONTENTS

9.3. Cross-relaxation and exchange in systems without resolved couplings ... 495
    9.3.1. Slow exchange ... 498
    9.3.2. Two-site systems ... 499
    9.3.3. Multiple-site exchange ... 500

9.4. Two-dimensional exchange spectroscopy in coupled spin systems ... 501
    9.4.1. Density operator treatment ... 502
    9.4.2. Zero-quantum interference ... 503
    9.4.3. Longitudinal scalar or dipolar order ... 506
    9.4.4. Suppression of J cross-peaks ... 506

9.5. Two-dimensional exchange difference spectroscopy ... 508

9.6. Determination of rate constants by 'accordion' spectroscopy ... 510

9.7. Cross-relaxation and nuclear Overhauser effects ... 516
    9.7.1. Intramolecular cross-relaxation ... 517
    9.7.2. Intramolecular cross-relaxation in a two-spin system ... 519
    9.7.3. Intramolecular cross-relaxation in a system with equivalent spins ... 521
    9.7.4. Intermolecular cross-relaxation ... 522
    9.7.5. Cross-relaxation in the slow-motion limit: applications to macromolecules ... 523

9.8. Chemical exchange ... 528

9.9. Indirect detection of longitudinal relaxation in multilevel spin systems ... 530

9.10. Dynamic processes in solids ... 533

10. NUCLEAR MAGNETIC RESONANCE IMAGING ... 539

10.1. Classification of imaging techniques ... 541

10.2. Sequential point techniques ... 542
    10.2.1. Sensitive point technique ... 542
    10.2.2. Field focusing NMR (FONAR) and topical NMR ... 544

10.3. Sequential line technique ... 545
    10.3.1. Sensitive line or multiple sensitive point method ... 545
    10.3.2. Line scan technique ... 547
    10.3.3. Echo line imaging ... 548

10.4. Sequential plane technique ... 548
    10.4.1. Projection–reconstruction technique ... 549
    10.4.2. Fourier imaging ... 553
    10.4.3. Spin-warp imaging ... 554
    10.4.4. Rotating-frame imaging ... 556
    10.4.5. Planar and multiplanar imaging ... 557
    10.4.6. Echo planar imaging ... 559

10.5. Comparison of sensitivity and performance time of various imaging techniques ... 560
    10.5.1. Sensitivity ... 560
    10.5.2. Performance time ... 563

REFERENCES ... 565

INDEX ... 597

SYMBOLS, TRANSFORMATIONS, AND ABBREVIATIONS

Symbols

A	(boldface): vector
A	(bold sans serif): matrix
a_{kl}	normalized integrated cross-peak intensity, eqn (9.3.17)
$a(\omega)$	Lorentzian absorption signal, eqn (1.2.19)
$a_{rs}(\omega_1), a_{rs}(\omega_2)$	Lorentzian absorption signal in ω_1- and ω_2-domains associated with transition $\|r\rangle \to \|s\rangle$, eqn (6.5.5).
B	amplitude of magnetic induction, called 'magnetic field' for simplicity
B	magnetic field vector
B_0	static magnetic field, eqn (4.2.3)
$B_1, B_{\text{r.f.}}$	radio-frequency (r.f.) field, eqn (4.2.5)
ΔB_0	offset field, eqn (4.2.21)
B_{eff}	effective field, eqn (4.2.22)
b_{kl}	spatial part of dipolar coupling, eqn (2.2.16)
$\mathbf{D}_{kl}$	dipolar coupling tensor, eqn (2.2.15)
$\mathcal{D}^l_{m'm}$	element of Wigner rotation matrix, eqn (2.1.146)
$d(\omega)$	Lorentzian dispersion signal, eqn (4.2.19)
$d_{rs}(\omega_1), d_{rs}(\omega_2)$	Lorentzian dispersion signal in ω_1- and ω_2-domains associated with transition $\|r\rangle \to \|s\rangle$, eqn (6.5.5)
F_x, F_y, F_z	Cartesian components of total angular momentum spin operator, eqn (2.1.44)
$\mathcal{F}$	Fourier transformation
$\mathcal{F}^c, \mathcal{F}^s$	cosine and sine Fourier transformations
$g^{(q,q')}(\tau), g^{(q)}(\tau)$	correlation functions, eqns (2.3.9) and (2.3.14a)
$H(\omega)$	frequency response function, eqn (4.1.14)
$H'(p)$	transfer function, eqn (4.1.13)
$\mathcal{H}_{rs}$	element of $\mathcal{H}$
$\mathcal{H}_{rs\,tu}$	element of matrix representation of Hamiltonian superoperator, eqn (2.1.37)
$\hat{\mathcal{H}}$	Hamiltonian superoperator (Liouville superoperator), eqn (2.1.54)
$h(t)$	impulse response function eqn (4.1.7)
$h(t), h(t_1, t_2)$	time-domain weighting functions, eqns (4.3.2) and (6.8.3)
$\hbar = h/(2\pi)$	Planck constant
$\mathcal{H}$	Hilbert transformation
$\mathcal{H}$	Hilbert space
$\mathcal{H}$	Hamilton operator
$\mathbf{\mathcal{H}}$	matrix representation of Hamilton operator
$I_0(\theta)$	zero-order modified Bessel function, eqn (4.1.39)
I_{kl}	integrated cross-peak intensity, eqn (9.3.16)

SYMBOLS, TRANSFORMATIONS, AND ABBREVIATIONS

I_{kx}, I_{ky}, I_{kz}	Cartesian components of angular momentum spin operators
J_{kl}	isotropic scalar spin–spin coupling constant, in Hz
$J(\omega), J^{(q,q')}(\omega),$ $J^{(q)}(\omega)$	power spectral density function, eqns (2.3.12) and (2.3.14b)
$J_{kl}(\omega)$	power spectral density function for k, l-interaction, eqn (9.4.9)
$J_{\kappa\lambda\,\mu\nu}(\omega)$	power spectral density function, eqn (2.3.22)
$\mathbf{J}_{kl}$	indirect spin–spin coupling tensor, eqn (2.2.11)
$K_{ji} = k_{ij}$	first-order rate constant of reaction $[A_i] \to [A_j]$, eqn (2.4.12)
$\mathbf{K}$	exchange matrix
k	Boltzmann constant
L	total spin quantum number, eqn (5.1.3)
$\mathbf{L} = \mathbf{K} - \mathbf{R}$	see eqn (2.4.23)
$\mathbf{L}^+ = i\mathbf{\Omega} - \mathbf{\Lambda} + \mathbf{K},$	see eqn (2.4.22)
$\hat{L}$	Liouville superoperator
$\mathscr{L}$	Liouville space
$\mathscr{L}^c$	composite Liouville space, eqn (2.4.33)
M_k	magnetic quantum number of spin I_k
M_a	magnetic quantum number of state $|a\rangle$
$M^+ = M_x + iM_y$	complex magnetization, eqn (4.2.16)
M_0	equilibrium magnetization (longitudinal component)
M_∞	steady-state magnetization (longitudinal component), eqns (4.2.34) and (4.6.5)
$\Delta M = M_z - M_0$	deviation from equilibrium magnetization
$\Delta M_{ab} = M_a - M_b$	difference in quantum numbers of states $|a\rangle$ and $|b\rangle$
$\mathbf{M}$	vector (M_x, M_y, M_z) of magnetization, eqn (2.3.1)
$\mathbf{M}^+$	vector $(M_1^+, M_2^+, \ldots, M_n^+)$ of complex magnetizations, eqns (2.4.20) and (9.3.1)
$\mathbf{M}_z$	vector $(M_{1z}, M_{2z}, \ldots, M_{nz})$ of longitudinal magnetization components, eqn (2.4.21)
$\mathbf{M}_0$	vector $(M_{10}, M_{20}, \ldots, M_{n0})$ of longitudinal magnetization components in equilibrium, eqn (2.4.21)
$\Delta \mathbf{M} = \mathbf{M}_z - \mathbf{M}_0$	see eqn (2.4.25)
N	number of spins in molecule
$\mathbf{N}, \mathbf{N}^+, \mathbf{N}^-$	stoichiometric matrices, eqn (2.4.7)
P_a	population of state $|a\rangle$, eqn (2.1.10)
P_i	projection operator, eqn (2.1.69)
$\mathbf{P}$	vector of populations
$\mathbf{P}_0$	vector of equilibrium populations
p	order of coherence, eqn (2.1.11)
p_{ab}	order of coherence $|a\rangle\langle b|$
Δp_i	change in coherence order under propagator U_i, eqn (6.3.8)
$\Delta \mathbf{p}$	vector of $\{\Delta p_i\}$, eqn (6.3.18)
$\mathbf{Q}_k$	quadrupole coupling tensor of spin I_k, eqn (2.2.20)

SYMBOLS, TRANSFORMATIONS, AND ABBREVIATIONS

q	number of spins involved in coherence, eqn (5.3.31)
q	number of terms in product operator, eqn (2.1.87)
R	rotation operator, unitary transformation operator
$R_{\alpha\alpha'\beta\beta'}$	element of Redfield relaxation matrix, eqn (2.3.21)
R_C	cross-relaxation rate constant, eqn (9.3.21)
R_L	leakage relaxation rate constant, eqn (9.3.21)
R	matrix representation of a rotation operator R
R	relaxation matrix, eqn (2.3.2)
$\mathbf{R}_x, \mathbf{R}_y, \mathbf{R}_z$	rotation matrices, eqns (4.2.25) and (4.2.49)
$\hat{\hat{R}}$	unitary transformation superoperator, eqn (2.1.62)
T_1	longitudinal relaxation time
T_2	transverse relaxation time (homogeneous contribution)
$1/T_2$	homogeneous relaxation rate constant
$1/T_2^+$	inhomogeneous decay rate constant
$1/T_2^*$	total decay rate constant of transverse magnetization
$1/T_2^a$	adiabatic relaxation rate constant, eqn (2.3.28)
$1/T_2^{na}$	non-adiabatic relaxation rate constant, eqn (2.3.29)
T_{lm}	component of irreducible tensor operator, eqn (2.1.146)
U	unitary propagator
$u(\omega)$	dispersive component of complex spectrum, eqn (4.2.18)
$v(\omega)$	absorptive component of complex spectrum, eqn (4.2.18)
W_{ij}	element of **W**, transition probability between states $|i\rangle$ and $|j\rangle$
W_{ij}^{AB}	contribution to W_{ij} due to dipolar AB interaction
W_{ij}^{RF}	contribution to W_{ij} due to external random fields
W_0, W_1, W_2	zero-, single-, and double-quantum transition probabilities, i.e. between states differing by $\Delta M_{ij} = 0$, 1, and 2, respectively
$W_0^{AB}, W_1^{AB}, W_2^{AB}$	contribution to W_0, W_1, and W_2 due to dipolar AB interaction
W_1^{RF}	contribution to W_1 due to external random fields
W	matrix of transition probabilities, eqn (2.3.3)
α	spin state $|M = +1/2\rangle$
β	spin state $|M = -1/2\rangle$
β	rotation angle of r.f. pulse ('flip angle'), eqn (4.2.13)
β_{eff}	effective β, eqns (4.2.24), (5.3.12)–(5.3.15)
β_{opt}	optimum β, eqns (4.2.36) and (4.2.41)
β_I, β_S	inverse spin temperature, eqn (4.5.8)
β_L	inverse lattice temperature
$\Gamma_{rs\,tu}$	element of matrix representation of $\hat{\hat{\Gamma}}$, eqn (2.1.36)
$\hat{\hat{\Gamma}}$	relaxation superoperator, eqn (2.1.34)
γ	gyromagnetic ratio
$\Delta_{rs\,tu}$	spin flip number, eqn (4.4.51)
η	enhancement factor, eqns (4.5.13) and (4.5.25)
η_k	asymmetry parameter, eqn (2.2.23)
θ	tilt angle, eqn (4.2.23)
κ	see eqns (6.5.11) and (6.5.13)

Λ_M	number of states with quantum number M, eqn (5.1.4)
$\boldsymbol{\Lambda}$	transverse relaxation matrix, eqn (2.4.22)
$\lambda = 1/T_2$	homogeneous contribution to half-width at half-height
$\lambda^+ = 1/T_2^+$	inhomogeneous contribution to half-width at half-height
λ^*	total half-width at half-height
$\lambda_{rs}^{(e)}, \lambda_{rs}^{(d)}$	homogeneous half-widths of transition $|r\rangle \to |s\rangle$ in evolution and detection periods, eqn (6.2.8)
λ_+, λ_-	eigenvalues of dynamic matrix $\mathbf{L}$, eqn (9.3.19)
μ_k	$+, -, \alpha,$ or β, see eqn (5.3.31)
$\nu_{jl}, \nu_{jl}^+, \nu_{lj}^-$	elements of stoichiometric matrices, eqns (2.4.2) and (2.4.9)
$\hat{\hat{\Xi}}$	exchange superoperator, eqn (2.4.34)
$\Xi_{j\alpha\alpha', s\beta\beta'}$	element of matrix representation of $\hat{\hat{\Xi}}$, eqn (2.4.39)
ξ_l	reaction number of reaction l, eqn (2.4.5)
$\boldsymbol{\xi}$	vector of reaction numbers
ρ	density operator of a system including lattice
σ	density operator of spin system, eqn (2.1.32)
σ_n	r.m.s. noise amplitude in time domain, eqn (4.3.7)
σ_N	r.m.s. noise amplitude in frequency domain, eqn (4.3.9)
σ_0	equilibrium density operator
σ^c	composite density operator, eqn (2.4.32)
$\sigma^{\square}$	concentration-dependent density operator, eqn (2.4.41)
$\sigma(\tau_-), \sigma(\tau_+)$	density operator just before and just after perturbation at time τ
$\sigma(t_i^-), \sigma(t_i^+)$	density operator just before and just after transformation by propagator U_i, eqn (6.3.4)
$\boldsymbol{\sigma}_k$	chemical shielding tensor, eqn (2.2.1)
τ_c	correlation time
τ_c^{kl}	correlation time of k, l interaction
φ, ϕ	phase
$\boldsymbol{\varphi}$	vector of r.f. phases, eqn (6.3.20)
χ	proportionality factor, see eqns (9.4.14) and (9.6.1)
Ω	frequency offset of resonance with respect to carrier frequency
Ω_k	frequency offset (chemical shift) of spin I_k with respect to carrier frequency, eqn (2.2.10)
Ω_{tot}	total spectral range, eqn (4.3.32)
$\boldsymbol{\Omega} = (\alpha, \beta, \gamma)$	Euler angles
$\omega_0 = -\gamma B_0$	Larmor frequency in laboratory frame
$\omega_{0k} = -\gamma_k(1-\sigma_k)B_0$	Larmor frequency of spin I_k in laboratory frame, eqns (2.2.2)
$\omega_{\alpha\beta}$	frequency of transition $|\alpha\rangle \to |\beta\rangle$ in laboratory frame, eqn (2.3.4)
ω_{rs}	frequency offset of transition $|r\rangle \to |s\rangle$ with respect to carrier frequency
$\omega_p^{(q)}$	see eqn (2.3.10)
$\omega_{rs}^{(e)}, \omega_{rs}^{(d)}$	frequency offset in evolution and detection periods of

SYMBOLS, TRANSFORMATIONS, AND ABBREVIATIONS xxiii

	transition $|r\rangle \leftrightarrow |s\rangle$ with respect to carrier frequency, eqn (6.2.9)
$\Delta\omega_{rs}^{(e)}, \Delta\omega_{rs}^{(d)}$	frequency offset in evolution and detection periods with respect to resonance $\omega_{rs}^{(e)}, \omega_{rs}^{(d)}$, eqn (6.5.3)
$\Delta\omega = \omega - \Omega$	frequency offset with respect to resonance frequency
$\Delta\omega_k = \omega - \Omega_k$	frequency offset with respect to resonance frequency of spin I_k
$\Delta\omega_1 = \omega_1 - \omega_{rs}^{(e)}$	frequency offset with respect to resonance frequency of transition $|r\rangle \to |s\rangle$ in evolution period
$\Delta\omega_2 = \omega_2 - \omega_{rs}^{(d)}$	frequency offset with respect to resonance frequency of transition $|r\rangle \to |s\rangle$ in detection period
$\mathbb{1}$	unity operator
a^*	complex conjugate of a
$A^\dagger$	adjoint of operator A ($A_{rs}^\dagger = A_{sr}^*$)
$\tilde{\mathbf{A}}$	transposed form of $\mathbf{A}$
$\xrightarrow{\mathcal{H}\tau}$	arrow representation of unitary transformation $U = \exp\{-i\mathcal{H}\tau\}$, see eqn (2.1.65)
$\oplus$	direct sum
$\otimes$	direct product
$\mathbf{a} \times \mathbf{b}$	vector product
$\mathbf{a} \cdot \mathbf{b}$	scalar product
$\in$	belongs to
$\propto$	proportional to

Conventions on rotations and basis transformations

(a) Rotations induced by a unitary transformation R

Transformation of wavefunctions: $\psi' = R\psi$.
Transformation of operators: $A' = RAR^{-1}$.
For a positive rotation about the x-axis through an angle β, R is given by $R = \exp\{-i\beta F_x\}$, hence $A' = \exp\{-i\beta F_x\} A \exp\{i\beta F_x\}$.

(b) Transformations induced by a unitary basis transformation T

Original basis functions: $(\widetilde{\phi}) = (\phi_1, \phi_2, \ldots, \phi_n)$.
Transformed basis functions $(\widetilde{\phi^T}) = (\phi_1^T, \phi_2^T, \ldots, \phi_n^T)$
with the relation

$$(\widetilde{\phi^T}) = (\widetilde{\phi})T.$$

The expansion coefficients of an arbitrary wavefunction ψ,

$$\psi = \sum_i \phi_i c_i = \sum \phi_i^T c_i^T$$

transform as follows
$$(c^T) = T^{-1}(c).$$
The representation of an operator A in the transformed frame is given by
$$A^T = T^{-1}AT.$$

Abbreviations

CSA	chemical shielding anisotropy
CW	continuous wave
eff	effective
FID	free induction decay
lab	laboratory frame
NOE	nuclear Overhauser effect
opt	optimum
pQT	p quantum transition
r.f.	radio-frequency
r.m.s.	root mean square
rot	rotating frame
S/N	signal-to-noise ratio
ZQT	zero-quantum transition

1
INTRODUCTION

In the course of the last two decades, nuclear magnetic resonance (NMR) has gone through a profound renaissance. Slow-passage spectroscopy techniques have largely become obsolete, and more versatile pulse techniques have begun to dominate the NMR scene. The rediscovery of the time domain has revived interest and stimulated creativity for new methodological developments. A surprising diversity of novel approaches and ingenious procedures has been invented and has radically changed the scope of NMR spectroscopy.

The basic principles of time-domain spectroscopy were known, of course, decades before. However, the advent of digital computers and more sophisticated electronics made it possible to put into practice ideas that had previously only existed in theory.

The plentiful arsenal of powerful time-domain techniques has allowed the spectroscopist to conquer new and exciting fields of application. Nuclear magnetic resonance has become one of the most successful analytical techniques for an extremely wide palette of applications ranging from solid state physics to all branches of chemistry, molecular biology, and medical diagnosis.

In the beginning was slow passage

The history of NMR began with the application of traditional methods of spectroscopy to nuclear magnetic resonance (1.1–1.6). The purpose of spectroscopy is the characterization of a molecular system by a 'spectrum'. The spectrum presents in a most direct way the resonant properties of a system and can provide insight into its quantum mechanical structure.

Spectroscopy stands in close analogy to techniques which measure the transfer function of an electronic device. Indeed, we may identify the complex spectrum (combining absorption and dispersion spectra into a complex function) with the transfer function of a system. It is well known that the transfer function completely characterizes a linear, time-independent system (1.7–1.10). Many of the concepts of spectroscopy stem from the consideration of linear or approximately linear systems for which a simple and elegant mathematical treatment is possible and for which we have some intuitive understanding. However, most physical systems are inherently non-linear, and the application of the concepts of linear response theory requires some caution.

The pedestrian way to measure a spectrum or a transfer function consists in applying a monochromatic perturbation to the system and to determine the (complex) amplitude of the response. A time-consuming, point-by-point measurement allows one to trace out the complete spectral function. In practice, a slow frequency-sweep is applied to obtain a continuous spectrum. We call this a slow-passage or a 'continuous-wave' (CW) spectrum. This traditional technique of spectroscopy dominated the first 25 years (1945–70) of high-resolution NMR spectroscopy, while pulsed excitation remained confined for the most part to relaxation-time measurements.

It was soon recognized that the inherent non-linearities of spin systems lead to distortions of the recorded spectra. The non-linearities cause saturation effects, including intensity distortions and line broadening (1.2), while a rapid frequency sweep results in transient oscillations after passing through resonance, called 'wiggles' (1.11–1.14). Very slow sweep rates and sufficiently weak radio-frequency irradiation are therefore required to obtain undistorted CW spectra.

The powers of evil in NMR

The widespread application of NMR spectroscopy has been seriously impeded by its inherently low sensitivity. The powers of evil have their origin in the extremely small energy quanta ($\sim 10^{-6}$ eV) involved in NMR transitions. Continuous-wave spectroscopy is quite inefficient with regard to measurement time. A low information flux is typical for sequential measurement techniques (1.14). The optimization of sensitivity has long been one of the prime concerns of NMR methodology. Numerous approaches have been proposed and tested, the most successful of which are:

1. High-field spectroscopy: the sensitivity is approximately proportional to $\gamma^{\frac{5}{2}} B_0^{\frac{3}{2}}$ (1.15–1.17);
2. Use of large sample volumes;
3. Cooling of the sample to increase the Curie magnetization (proportional to $1/T$);
4. Enhancement of the magnetization by
   (a) Heteronuclear Overhauser effect (1.18, 1.19);
   (b) Cross-polarization in solids (1.20–1.22) and in liquids (1.23–1.27);
   (c) Transfer of coherence by r.f. pulses (1.23, 1.28–1.30);
   (d) Electron nuclear Overhauser effect (1.31, 1.32);
5. Indirect detection of resonance by
   (a) Electron nuclear double resonance (ENDOR) (1.33);

(b) Internuclear double resonance (INDOR) (1.34);
(c) Cross-polarization in solids (1.20, 1.35–1.37) and in liquids (1.23, 1.24);
(d) Transfer of coherence by r.f. pulses (1.28–1.30).
6. Shortening of the longitudinal relaxation time by addition of paramagnetic materials (1.38);
7. Flowing sample techniques to circumvent saturation effects (1.39, 1.40).

In addition to these procedures, which often require modified instrumentation and favourable systems, the sensitivity can also be optimized by a judicious design of the measurement and data-processing procedure.

Trading time for sensitivity

In the course of the 1960s, the basic relation between measurement time and sensitivity (1.14) became common knowledge among NMR spectroscopists. The achievable sensitivity is proportional to the square root of the available measuring time. The recognition of this fact led to the use of signal averaging as an alternative to long-term single-scan measurements (1.41–1.45). The CAT (computer-averaged transients) became a familiar pet in NMR laboratories. In particular, it was realized that averaging of rapid scans has major advantages over a single slow-passage experiment performed in the same total time for the following two reasons:

1. Signal averaging allows the suppression of low-frequency noise (1.45). The scheme is in this respect equivalent to a modulation method which shifts the relevant signals to higher frequencies. It is therefore insensitive to slow instrumental drifts and low-frequency instabilities.

2. Averaging of rapid scans can lead to a stronger average magnetization (1.46). The optimum repetition time is of the order of the recovery time T_1 of the magnetization.

However, a very serious drawback of rapid scanning is the resulting distortion of the signal which leads to a significant reduction of resolution (1.14, 1.46). These line-shape distortions can be corrected by suitable deconvolution procedures, which are used in correlation spectroscopy and rapid-scan Fourier spectroscopy (1.47–1.50). Rapid scan techniques can approach the sensitivity of Fourier methods, and require a similar computational effort. They have the advantage that selected regions of the NMR spectra can be scanned.

Exorcising the evil from NMR by r.f. pulses

The 1970s were dominated by the revolutionary development of pulse Fourier spectroscopy, which paved the way to modern NMR spectroscopy and to an unprecedented expansion of NMR applications. The starting point was the idea to design a multichannel spectrometer which would allow the simultaneous measurement of numerous points of a spectrum. The performance time is reduced in proportion to the number of available channels. Attempts were indeed undertaken by W. A. Anderson to construct such spectrometers, but it was soon recognized that the instrumental effort becomes exorbitant when the number of channels is increased.

The fact that a short delta-function-like pulse can be considered as a multifrequency source which allows simultaneous excitation of all resonance frequencies is quite well established (1.7–1.10). According to the superposition principle, which is valid in linear systems, the response to a delta function pulse, known as impulse response, is a linear superposition of the responses of all frequency components. The transfer function can be obtained from the impulse response simply by frequency analysis, i.e. by a Fourier transformation.

It had also been known for many years (1.51) that the free induction decay, which is equivalent to the impulse response in linear system theory, and the complex spectrum (equivalent to the transfer function) form a Fourier transform pair. Knowledge of these facts leads almost naturally to the recognition that a dramatic sensitivity improvement can be realized by recording the response to a short r.f. pulse in the form of a free induction decay and subsequently computing the desired spectrum by means of a numerical Fourier transformation (1.14, 1.52). The practical realization of pulse Fourier spectroscopy was made possible by the introduction of inexpensive computers in the late-1960s and by the development of the fast Fourier transform (FFT) algorithm.

Nowadays, traditional CW spectrometers have almost completely been replaced by pulse Fourier instruments, although these have some peculiar drawbacks, such as problems of dynamic range, baseline artefacts, and frequency folding. Nevertheless, the inherent advantages of greater sensitivity, high resolution, and the absence of line-shape distortions contributed to make Fourier spectroscopy the preferred experimental technique in NMR. Several monographs are available which describe the practical aspects of Fourier spectroscopy (1.53–1.55).

We should like to emphasize that the principles of Fourier spectroscopy are not restricted to nuclear magnetic resonance. The same principles can be applied to many other fields of spectroscopy, including electron spin resonance and rotational spectroscopy in the microwave region (1.56), nuclear quadrupole resonance (1.57), pulsed optical transient spectroscopy, and ion cyclotron resonance (1.58).

A different, although related type of Fourier spectroscopy was introduced as early as 1951 in the field of infra-red spectroscopy (1.59–1.61). Infra-red Fourier spectroscopy does not involve a time-domain experiment in the same sense as Fourier NMR, but relies on the observation of interference as a function of a path-length difference which leads to an interferogram that resembles a symmetrized free induction decay.

Unfortunately, the cross-fertilization between the various fields of Fourier spectroscopy has been quite meagre and many of the basic principles, particularly in the context of data-processing, have been developed independently in several fields (1.62). It can be quite revealing to consult papers on the same subject dressed up in a slightly different disguise. We shall occasionally include cross-references to applications in other fields.

The universality of pulse experiments

The widespread application of pulse experiments has unified NMR technology in an unprecedented manner. The same techniques can be used nowadays for solid-state NMR (1.22), high-resolution NMR in liquids (1.53–1.55), and for nuclear quadrupole resonance (1.57). Multipurpose spectrometers are currently available which are suitable for all these applications. Pulse experiments are predestinate for measuring transient properties of spin systems. In particular, they are commonly used for the measurement of relaxation times (1.63–1.67), for the study of diffusion processes (1.68), and for the investigation of chemical reactions (1.69–1.71).

Putting noise to good use

Pulse Fourier spectroscopy represents only one particular realization of the multichannel concept for improving sensitivity. Instead of using a pulse as a wide-band frequency source, it was suggested many years ago that random noise could be used for wideband stimulation of linear and non-linear systems (1.72). This technique has been used for testing electronic systems and for describing hydrodynamic processes (1.73, 1.74) and biological systems (1.75). Under the name of stochastic resonance, it has also been applied to NMR (1.76–1.82). The technique has a number of interesting properties in comparison with pulse Fourier spectroscopy. However, it appears to be less versatile for the realization of more sophisticated experiments and it has not yet attained the status of a general tool in NMR.

Spin alchemy: sorcery in NMR

Many of the new NMR methods rely on the fact that it is possible to modify the Hamiltonian almost at will to extract desired information. On the one hand, the spectra may be simplified by eliminating or scaling selected interactions such as homo- or heteronuclear dipolar couplings. On the other hand, the information content can also be enhanced by introducing additional perturbations. The Hamiltonian can be manipulated to such an extent that some experiments border on sorcery. The range of manipulations includes double resonance which may be used for spin decoupling (1.83–1.85) and spin tickling (1.84, 1.86), multiple-pulse methods for removing dipolar couplings between abundant spins in solids (1.22, 1.87–1.90), magic-angle sample spinning for eliminating the anisotropic part of chemical shifts (1.91–1.94), etc. Other specialized techniques will be discussed in Chapters 4, 7, 8, and 9. Most aspects of spin alchemy involve transient phenomena in a vital way, and can only be fully exploited in time-domain spectroscopy. This applies in particular to the creation of effective Hamiltonians based on aperiodic perturbations, which are employed in many two-dimensional experiments.

The discovery of a new dimension

In conventional magnetic resonance, the spectrum or frequency response $S(\omega)$ is a complex function of a single-frequency variable. In double resonance, a response function $S(\omega_1, \omega_2)$ is measured as a function of two variables, one of which is usually a parameter rather than a continuous variable, and the existence of a second dimension is often not fully appreciated.

The first proposal to introduce a true second frequency dimension was made by Jeener in 1971 (1.95). He proposed a two-pulse experiment in time domain, which became the germ cell of two-dimensional spectroscopy (1.96). The basic secret of two-dimensional (2D) time-domain spectroscopy is the use of two independent precession periods during which the coherence can evolve. The precession frequency of the coherence is suddenly changed between the evolution and detection periods, either because the effective Hamiltonian is switched by one of the tricks of spin alchemy, or alternatively because the coherence is transferred from one transition to another. It is important to note that the coherence is only observed in the detection period. The evolution during the preceding time period is monitored indirectly through the phase and amplitude of the magnetization at the beginning of the detection period. This scheme has many decisive advantages, making it possible, for example, to observe multiple-quantum coherence indirectly. Four major classes of 2D spectroscopy methods can be distinguished.

THE DISCOVERY OF A NEW DIMENSION

1. Shedding light on the spectral chaos

One-dimensional spectra that are rendered inscrutable because of severe overlap may be unravelled by separating interactions of different physical origin, e.g. chemical shifts and couplings, thus making it possible to spread the signals in a second frequency dimension much like opening a Venetian blind. In many circumstances, resonances that overlap in conventional one-dimensional spectra can be unravelled in this manner. A wide variety of applications concerned with isotropic liquids, liquid crystals, and solids will be discussed in Chapter 7.

2. Two-dimensional correlation: the tree of knowledge

The phenomena of coherence transfer can be put to use in two-dimensional time-domain spectroscopy to identify pairs of nuclei that are connected through scalar or dipolar couplings. The techniques of homo- and heteronuclear correlation spectroscopy (Chapter 8) allow one to identify the topology of the coupling network by inspection of the two-dimensional spectra. Coherence transfer gives a detailed picture of the coupling networks and the relationship between spectra and energy-level diagrams.

3. The forbidden fruits of spectroscopy

The observation of forbidden multiple-quantum transitions by CW techniques (1.97–1.99) is hampered by the difficulty of separating various orders and by line-broadening effects. The introduction of two-dimensional time-domain spectroscopy turned out to yield decisive advantages in this area (1.100, 1.101), since undistorted lineshapes can be readily obtained, and multiple-quantum transitions of various orders can be cleanly separated. Since the coherence that precesses in the evolution period of a two-dimensional experiment need not be detected, the usual selection rules can be circumvented. It has now become possible to taste the forbidden fruits of spectroscopy with impunity.

4. Mapping slow dissipative processes

Magnetic resonance has proven unique for probing dissipative dynamic processes such as chemical exchange and cross-relaxation (1.69–1.71). Two-dimensional spectroscopy has given a fresh impetus to this area, and has been particularly successful for mapping cross-relaxation pathways, nuclear Overhauser effects, spin diffusion, and slow chemical exchange (1.102–1.104).

INTRODUCTION

After this rather cursory introduction, which traced some of the historical developments, we present in the following two chapters some of the basic aspects of nuclear magnetic resonance to provide a foundation for the subsequent chapters on one- and two-dimensional Fourier spectroscopy.

2
THE DYNAMICS OF NUCLEAR SPIN SYSTEMS

The dynamics of isolated spins can be understood in terms of the motion of classical magnetization vectors (see §4.2). To describe coupled spins, however, it is necessary to have recourse to a quantum mechanical formalism where the state of the system is expressed by a state function or, more generally, by a density operator.

In this chapter, we give a condensed account of the density operator treatment, in particular of those aspects which are relevant to pulse experiments in liquids and solids. The equation of motion of the density operator is introduced in §2.1. The properties of the system are represented by the total Hamiltonian $\mathcal{H}$ which forms the *spiritus rector* of the motion of the entire molecular system. For magnetic resonance, however, it is sufficient to derive a reduced *spin Hamiltonian* $\mathcal{H}^s$ which reigns over the nuclear spin assembly (§2.2). The spin Hamiltonian does not take into account possible time-dependent random interactions between the spin system and the environment. The effects of these interactions can, however, be represented by the relaxation superoperator described in §2.3. Finally, the implications of chemical exchange are discussed in §2.4.

2.1. Equation of motion

2.1.1. Density operator

The theoretical discussion given in the following sections is based on the density operator, which permits the most convenient description of quantum mechanical system dynamics. It seems appropriate to recall some basic properties. More extensive treatments of the density operator theory are given by Fano (2.1), Weissbluth (2.2), Böhm (2.3), Blum (2.4), and Slichter (2.5). An excellent introduction to the basic mathematical tools relevant to NMR spectroscopy is provided by a recent monograph of Goldmann (2.70).

In order to define the density operator ρ of the entire quantum mechanical system (including the lattice) and to derive its equation of motion, we start with the time-dependent Schrödinger equation for the evolution of a state function $|\psi(t)\rangle$,

$$\frac{\mathrm{d}}{\mathrm{d}t}|\psi(t)\rangle = -\mathrm{i}\mathcal{H}(t)|\psi(t)\rangle. \qquad (2.1.1)$$

$\mathcal{H}(t)$ is the Hamiltonian or total energy operator of the system, which may itself be time-dependent. In this monograph, we invariably measure energy eigenvalues in angular frequency units, and $\hbar$ does not appear in eqn (2.1.1). We may expand the state function $\psi(t)$ in terms of a complete orthonormal base $\{|i\rangle, i = 1, 2, \ldots, n\}$

$$|\psi(t)\rangle = \sum_{i=1}^{n} c_i(t) |i\rangle \qquad (2.1.2)$$

where the time-dependence of $|\psi(t)\rangle$ is expressed by the time-dependent coefficients $c_i(t)$ and n is the dimension of the vector space of all admissible state functions, called Hilbert space. The scalar product in this space is defined by

$$\langle \chi | \psi \rangle = \sum \int d\tau \, \chi^* \psi \qquad (2.1.3)$$

where the symbol $\sum \int d\tau$ indicates summation over all values of the discrete variables, and integration over the complete domain of the continuous variables of the state functions.

With regard to the definition of the density operator, two cases can be distinguished.

1. In an idealized *pure state*, all spin systems of the ensemble are in the same state and can be described by the same normalized state function $|\psi(t)\rangle$ with $\langle \psi(t) | \psi(t) \rangle = 1$. The corresponding *density operator* ρ is defined by the product of the ket $|\psi(t)\rangle$ and the bra $\langle \psi(t)|$,

$$\rho(t) = |\psi(t)\rangle\langle\psi(t)| = \sum_i \sum_j c_i(t) c_j^*(t) |i\rangle\langle j|. \qquad (2.1.4)$$

2. The situation is different for an ensemble in a *mixed state*, e.g. for an ensemble in thermal equilibrium. Here we can only indicate probabilities p^k that a spin system of the ensemble is in one of several possible states $|\psi^k(t)\rangle$. The density operator is then understood as an average over the ensemble,

$$\rho(t) = \sum_k p^k |\psi^k(t)\rangle\langle\psi^k(t)| \qquad (2.1.5)$$

$$= \sum_k p^k \sum_i \sum_j c_i^k(t) c_j^{k*}(t) |i\rangle\langle j| = \sum_i \sum_j \overline{c_i(t) c_j^*(t)} |i\rangle\langle j|. \qquad (2.1.6)$$

where $\sum_k p^k = 1$, and where the bar denotes the ensemble average.

The physical significance of the density operator may be appreciated by considering its matrix elements in the orthonormal basis $\{|j\rangle\}$. For a pure state we obtain

$$\langle r| \rho(t) |s\rangle = \sum_i \sum_j c_i(t)c_j^*(t)\langle r|i\rangle\langle j|s\rangle$$
$$= c_r(t)c_s^*(t), \qquad (2.1.7)$$

whereas for mixed states we find

$$\langle r| \rho(t) |s\rangle = \sum_k p^k c_r^k(t)c_s^{k*}(t) = \overline{c_r(t)c_s^*(t)}. \qquad (2.1.8)$$

Thus the matrix elements are simply products of the expansion coefficients of the state function $\psi(t)$. Clearly, $\rho(t)$ is a Hermitian operator

$$\langle r| \rho(t) |s\rangle = \langle s| \rho(t) |r\rangle^*. \qquad (2.1.9)$$

A particularly simple interpretation of the density matrix is possible in the eigenbasis of the Hamiltonian $\mathcal{H}$. The diagonal element

$$\rho_{rr} = \langle r| \rho(t) |r\rangle = \overline{|c_r(t)|^2} = P_r \qquad (2.1.10)$$

is equal to the probability that the spin system is found in the eigenstate $|r\rangle$. P_r is the population of state $|r\rangle$.

The off-diagonal element

$$\rho_{rs} = \langle r| \rho(t) |s\rangle = \overline{c_r(t)c_s^*(t)}, \qquad (2.1.11)$$

on the other hand, indicates a 'coherent superposition' of eigenstates $c_r(t)|r\rangle + c_s(t)|s\rangle$ in $\psi(t)$ of eqn (2.1.2) in the sense that the time dependence and the phase of the various members of the ensemble are correlated with respect to $|r\rangle$ and $|s\rangle$. Such a coherent superposition is simply called 'coherence'. The matrix element $\rho_{rs}(t)$ is the complex amplitude of the coherence expressed by the operator $|r\rangle\langle s|$. A coherence can be associated with a transition between two eigenstates $|r\rangle$ and $|s\rangle$. If the two states span an allowed transition with a difference in magnetic quantum numbers $\Delta M_{rs} = M_r - M_s = \pm 1$, the coherence ρ_{rs} is related to the transverse magnetization components $M^{\pm(rs)} = M_x^{(rs)} \pm iM_y^{(rs)}$. In general, a matrix element ρ_{rs} represents p-quantum coherence ($p = M_r - M_s$) which, for $p \neq \pm 1$, does not lead to observable magnetization and can only be detected indirectly.

Because the state functions have been assumed to be normalized

$$\langle \psi | \psi \rangle = \sum_{r=1}^n |c_r(t)|^2 = 1, \qquad (2.1.12)$$

the trace of the density operator is equal to unity

$$\text{tr}\{\rho\} = \sum_{r=1}^n \rho_{rr} = \sum_{r=1}^n P_r = 1. \qquad (2.1.13)$$

Pure states are often referred to as 'states of maximum information'. Since all subsystems behave equally, one has complete knowledge of all constituent parts of the system. Pure states can be recognized by the fact that ρ is idempotent

$$\rho^2 = |\psi\rangle\langle\psi|\psi\rangle\langle\psi| = |\psi\rangle\langle\psi| = \rho \qquad (2.1.14)$$

and that

$$\text{tr}\{\rho^2\} = \text{tr}\{\rho\} = 1. \qquad (2.1.15)$$

It is worth noting that the density operator ρ of a pure state is a projection operator (compare eqns (2.1.4) and (2.1.69)).

Mixed states are sometimes called 'states of less than maximum information'. It can be shown that for mixed states

$$\text{tr}\{\rho^2\} < 1. \qquad (2.1.16)$$

Our knowledge of the individual system is less complete in a mixed state, and $\text{tr}\{\rho^2\}$ may be considered as a measure of information. The density operator for a mixed state is no longer a projection operator.

2.1.1.1. Density operator equation

From the time-dependent Schrödinger equation (2.1.1), one easily derives the equation of motion for the density operator

$$\frac{d}{dt}\rho(t) = -i[\mathcal{H}(t), \rho(t)]. \qquad (2.1.17)$$

This differential equation, called Liouville–von Neumann equation or simply density operator equation, is of central importance for calculating the dynamics of quantum mechanical systems. Its formal solution may be written

$$\rho(t) = U(t)\rho(0)U(t)^{-1}; \quad U(t) = T \exp\left\{-i\int_0^t \mathcal{H}(t')\,dt'\right\} \qquad (2.1.18)$$

where the Dyson time-ordering operator T (2.6) defines a prescription for evaluating the exponential functions in cases where the Hamiltonians at different times do not commute, $[\mathcal{H}(t'), \mathcal{H}(t'')] \neq 0$ (see eqn (3.2.12)). The use of this equation will be illustrated by numerous examples throughout this book. By selecting a suitable rotating frame, the Hamiltonian can often be made time-independent within finite segments of time, and the evolution can be expressed by a sequence of unitary transformations of the type

$$\rho(t + \tau_1 + \tau_2) = \exp\{-i\mathcal{H}_2\tau_2\}\exp\{-i\mathcal{H}_1\tau_1\}\rho(t)\exp\{+i\mathcal{H}_1\tau_1\}\exp\{+i\mathcal{H}_2\tau_2\}$$
$$(2.1.19)$$

with the propagators $\exp\{-i\mathcal{H}_k\tau_k\}$. This equation applies to any sequence of intervals τ_k in which time-independent average Hamiltonians $\mathcal{H}_k$ can be defined.

2.1.1.2. Expectation values

For normalized state functions, the expectation value $\langle A \rangle$ of an arbitrary observable operator A

$$\langle A \rangle = \sum_k p^k(t) \langle \psi^k(t) | A | \psi^k(t) \rangle \tag{2.1.20}$$

can be expressed by $\rho(t)$

$$\langle A \rangle = \sum_k p^k(t) \sum_r \sum_s c_r^{k*}(t) c_s^k(t) \langle r | A | s \rangle$$

$$= \sum_r \sum_s \rho_{sr}(t) A_{rs} \tag{2.1.21}$$

leading to the expression

$$\langle A \rangle = \text{tr}\{A\rho(t)\}. \tag{2.1.22}$$

Thus the expectation value is found by evaluating the trace of the product of the observable operator and the density operator. The trace can be calculated from the product of the matrix representations of A and $\rho(t)$ in an arbitrary basis $\{|r\rangle\}$ or via an expansion of the two operators A and $\rho(t)$ in terms of orthogonal basis operators.

2.1.1.3. Schrödinger and Heisenberg representation

Equation (2.1.22) refers to the so-called 'Schrödinger representation' where the time dependence of the system is associated with the state functions or with the density operator $\rho(t)$, while the observable operator A remains time-independent. Sometimes, it is more convenient to transfer the time dependence to the observable operator A

$$\langle A \rangle = \text{tr}\{A\rho(t)\}$$

$$= \text{tr}\left\{AT\exp\left\{-i\int_0^t \mathcal{H}(t')\,dt'\right\}\rho(0)\exp\left\{i\int_0^t \mathcal{H}(t')\,dt'\right\}\right\}$$

$$= \text{tr}\left\{T\exp\left\{i\int_0^t \mathcal{H}(t')\,dt'\right\}A\exp\left\{-i\int_0^t \mathcal{H}(t')\,dt'\right\}\rho(0)\right\}$$

$$= \text{tr}\{A(t)\rho(0)\} \tag{2.1.23}$$

where the Heisenberg operator $A(t)$ is the solution of the differential equation

$$\frac{d}{dt}A(t) = -i[A(t), \mathcal{H}(t)] \tag{2.1.24}$$

with the initial condition $A(0) = A$. Attributing the time-dependence to the observable operator amounts to using the so-called 'Heisenberg representation'. Although it provides less intuitive insight, this representation is of advantage when the evolution under a given Hamiltonian should be discussed for various initial conditions $\rho(0)$. As long as relaxation is not considered, both representations are only trivially different. However, for proper consideration of relaxation and exchange processes it is preferable to work in the Schrödinger representation. In the following, we use only the more 'natural' Schrödinger representation although most of the results can also be expressed in the Heisenberg representation.

The density operator in thermal equilibrium at temperature T is given by

$$\rho_0 = \frac{1}{Z} \exp\{-\mathcal{H}\hbar/kT\} \qquad (2.1.25)$$

where

$$Z = \text{tr}\{\exp\{-\mathcal{H}\hbar/kT\}\} \qquad (2.1.26)$$

is the partition function of the system. By evaluating ρ_0 in the eigenbasis $\{|r\rangle\}$ of the Hamiltonian, one easily verified that the probability distribution of the eigenstates $|r\rangle$, $P_r = \rho_{rr}$, correctly describes a Boltzmann distribution

$$P_r = \frac{1}{Z} \exp\{-E_r\hbar/kT\}. \qquad (2.1.27)$$

2.1.1.4. Reduced spin density operator

So, far, the density operator $\rho(t)$ has been formulated for the entire quantum mechanical system. The basis functions of the full Hilbert space depend on both space and spin coordinates of all electrons and nuclei that constitute the system. However, for magnetic resonance applications it is usually sufficient to calculate expectation values of a restricted set of operators $\{Q\}$ which act exclusively on nuclear or electronic spin variables. The remaining degrees of freedom are referred to as the 'lattice'.

For the calculation of such expectation values $\langle Q \rangle$, knowledge of the complete density operator $\rho(t)$ is, fortunately, not required. It is sufficient to define a reduced *spin density operator* $\sigma(t)$, which is obtained from $\rho(t)$ by trace formation over all degrees of freedom of the lattice.

The basis function $|\alpha\rangle$ of the entire system can be written as products of functions $|f\rangle$, depending on the lattice variables only, and functions $|s\rangle$, depending exclusively on the spin coordinates of the restricted system of interest

$$|\alpha\rangle = |fs\rangle. \qquad (2.1.28)$$

Evaluating eqn (2.1.22), we then find

$$\langle Q \rangle(t) = \sum_{s,s'} \sum_{f,f'} \langle sf| Q |f's' \rangle \langle s'f'| \rho(t) |fs \rangle. \qquad (2.1.29)$$

Since Q acts on the spin variables only, the matrix representation of Q is diagonal in the lattice variables,

$$\langle sf| Q |f's' \rangle = \langle s| Q |s' \rangle \delta_{ff'}, \qquad (2.1.30)$$

assuming orthonormal lattice functions $|f\rangle$. Defining a reduced density operator $\sigma(t)$ by

$$\langle s'| \sigma(t) |s \rangle = \sum_{f} \langle s'f| \rho(t) |fs \rangle \qquad (2.1.31)$$

or, equivalently, by

$$\sigma(t) = \text{tr}_f \{\rho(t)\} \qquad (2.1.32)$$

where tr_f denotes a partial trace over the lattice variables, we obtain for the expectation value of the operator Q

$$\langle Q \rangle(t) = \text{tr}_s \{Q\sigma(t)\}. \qquad (2.1.33)$$

The dynamics of the reduced density operator proceed according to the equation

$$\frac{d}{dt} \sigma(t) = -i[\mathcal{H}^s, \sigma(t)] - \hat{\Gamma}\{\sigma(t) - \sigma_0\} \qquad (2.1.34)$$

which replaces eqn (2.1.17), and is often called the 'quantum mechanical master equation'. Here, $\mathcal{H}^s$ is the spin Hamiltonian acting only on the spin variables, and is obtained by averaging the full Hamiltonian over the lattice coordinates,

$$\mathcal{H}^s = \sum_f \langle f| \mathcal{H} |f \rangle = \text{tr}_f \{\mathcal{H}\}. \qquad (2.1.35)$$

Its various terms will be summarized in §2.2. The relaxation superoperator $\hat{\Gamma}$ in eqn (2.1.34) accounts for the dissipative interactions between the spin system and the lattice and drives the density operator towards its equilibrium value σ_0 (§2.3).

The integration of the master equation eqn (2.1.34) is in general a formidable task, particularly for an elaborate relaxation superoperator $\hat{\Gamma}$. We shall describe several approaches which can lead to manageable equations by an adequate choice of the basis in which the density operator is expressed.

2.1.2. Explicit matrix representation of the master equations

The most direct brute force approach to solve the master equation, eqn (2.1.34) proceeds via explicit matrix representations of the operators. Let us assume an arbitrary set of basis functions $\{|r\rangle\}$ and evaluate the matrix elements $\sigma_{re} = \langle r| \sigma |s\rangle$. The relaxation superoperator $\hat{\Gamma}$ may transform any matrix element σ_{tu} into σ_{rs}, thus requiring a representation with two pairs of indices

$$\frac{d}{dt}\sigma_{rs} = -i\sum_{k}(\mathcal{H}_{rk}\sigma_{ks} - \sigma_{rk}\mathcal{H}_{ks}) - \sum_{tu}\Gamma_{rs\,tu}\{\sigma_{tu} - \sigma_{0\,tu}\}. \quad (2.1.36)$$

The terms originating from the commutator with the Hamiltonian $\mathcal{H}$ may be expressed by matrix elements $\mathcal{H}_{rs\,tu}$ of the commutator superoperator $\hat{\mathcal{H}}$ (see § 2.1.4)

$$\frac{d}{dt}\sigma_{rs} = -i\sum_{tu}\mathcal{H}_{rs\,tu}\sigma_{tu} - \sum_{tu}\Gamma_{rs\,tu}\{\sigma_{tu} - \sigma_{0\,tu}\} \quad (2.1.37)$$

with the supermatrix elements $\mathcal{H}_{rs\,tu} = \mathcal{H}_{rt}\delta_{us} - \delta_{rt}\mathcal{H}_{us}$. This represents a system of n^2 coupled linear differential equations, n being the dimension of the spin space. If the n^2 elements σ_{rs} are arranged in the form of a column vector $\boldsymbol{\sigma}$, eqn (2.1.37) can be expressed in matrix form

$$\frac{d}{dt}\boldsymbol{\sigma} = -i\hat{\mathcal{H}}\boldsymbol{\sigma} - \boldsymbol{\Gamma}(\boldsymbol{\sigma} - \boldsymbol{\sigma}_0). \quad (2.1.38)$$

If we assume a time-independent Hamiltonian we obtain the formal solution

$$\boldsymbol{\sigma}(t) = \boldsymbol{\sigma}^{ss} + \exp\{(-i\hat{\mathcal{H}} - \boldsymbol{\Gamma})t\}\{\boldsymbol{\sigma}(0) - \boldsymbol{\sigma}^{ss}\} \quad (2.1.39)$$

where $\boldsymbol{\sigma}^{ss}$ is the density operator describing the steady-state, which may be obtained by setting the time-derivative of eqn (2.1.38) equal to zero. By neglecting relaxation, we immediately obtain the solution

$$\sigma(t) = R(t)\sigma(0)R^{-1}(t) \quad (2.1.40)$$

with $R(t) = \exp\{-i\mathcal{H}t\}$. This unitary transformation is shown schematically in Fig. 2.1.1. The same transformation can also be written as a multiplication of a matrix of dimension $n^2 \times n^2$ with the column vector $\boldsymbol{\sigma}(0)$, as depicted schematically in Fig. 2.1.1

$$\boldsymbol{\sigma}(t) = \mathbf{R}(t)\boldsymbol{\sigma}(0). \quad (2.1.41)$$

The elements transform according to

$$\sigma_{rs}(t) = \sum_{tu} R_{rs\,tu}(t)\sigma_{tu}(0) \quad (2.1.42)$$

with

$$R_{rs\,tu}(t) = R_{rt}(t)R_{us}^{-1}(t). \quad (2.1.43)$$

The transformation of σ_{tu} into σ_{rs} expressed in eqn (2.1.42) is often referred to as 'coherence transfer', a concept that is of central importance

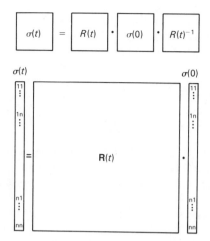

FIG. 2.1.1. Evolution of the density operator in the absence of relaxation. The density operator $\sigma(t)$ is obtained by forming the matrix product $R(t)\,\sigma(0)\,R^{-1}(t)$ (eqn (2.1.40)). If the elements of $\sigma(t)$ are arranged in a column vector $\boldsymbol{\sigma}(t)$, the evolution can be expressed by a matrix $\mathbf{R}(t)$ of dimension $n^2 \times n^2$.

in the context of pulse experiments and particularly in two-dimensional NMR spectroscopy.

2.1.3. Liouville operator space

In many situations, it is not necessary to consider the entire density matrix, either because certain elements are irrelevant for the selected observable, or because the number of the actual degrees of freedom is smaller than the total number of matrix elements. In such cases, it is worthwhile to expand the density operator in terms of a suitably chosen set of basis operators and to derive equations for their time-dependent coefficients.

For illustration, consider the simple case of N spins of arbitrary spin quantum number I_k, all with the same Zeeman frequency Ω and with the initial state $\sigma(0) = F_x = \sum I_{kx}$.[1] The time evolution under the Zeeman Hamiltonian corresponds to a pure rotation about the z-axis

$$\sigma(t) = F_x \cos \Omega t + F_y \sin \Omega t \qquad (2.1.44)$$

---

[1] In this simplified notation, which will often be used in the following, only relevant terms of the density operator σ are retained, disregarding the normalization required in eqn (2.1.13). The full density operator corresponding to the abbreviated form $\sigma(0) = F_x$ would be $\sigma(0) = (\mathbb{1} + cF_x)/\mathrm{tr}\{\mathbb{1}\}$, where $\mathrm{tr}\{\mathbb{1}\} = (2I+1)^N$ for N spins with quantum number I.

Although the dimension of the system may be quite large, only one degree of freedom is involved in the time evolution, a simplification which suggests the use of an operator representation of σ, using F_x and F_y as basis operators. To treat more general situations, we have to expand the density operator in a complete set of basis operators $\{B_s\}$

$$\sigma(t) = \sum_{s=1}^{n^2} b_s(t) B_s. \qquad (2.1.45)$$

If we assume a Hilbert space of dimension n spanned by n independent functions, there are n^2 independent operators. This can be easily verified by considering the $n \times n$ matrix representations of operators acting on the Hilbert space. Each of the n^2 matrix elements may represent an independent operator. Various useful sets of basis operators $\{B_s\}$ will be treated in §§ 2.1.5–2.1.10.

The basis operators B_s span an operator space of dimension n^2, called *Liouville space*. The trace metric defined by the scalar product,

$$\langle A | B \rangle = \text{tr}\{A^\dagger B\}, \qquad (2.1.46)$$

fulfils all properties required for a Hermitian scalar product. $A^\dagger$ is the adjoint operator with the matrix elements $(A^\dagger)_{kl} = A_{lk}^*$. The scalar product of two operators is equivalent to the scalar product of two n^2-dimensional vectors consisting of the sets of elements $\{A_{lk}\}$ and $\{B_{lk}\}$

$$\langle A | B \rangle = \sum_{kl} A_{kl}^\dagger B_{lk} = \sum_{kl} A_{lk}^* B_{lk}. \qquad (2.1.47)$$

Considering eqn (2.1.22), it is evident that there is a relationship between the expectation value of an operator A and the scalar product $\langle A | \sigma \rangle$. It is important to note, however, that the scalar product involves the adjoint operator

$$\langle A \rangle = \text{tr}\{A \sigma(t)\}, \qquad (2.1.48)$$

$$\langle A | \sigma(t) \rangle = \text{tr}\{A^\dagger \sigma(t)\}. \qquad (2.1.49)$$

For a Hermitian operator $A = A^\dagger$, this distinction is obviously irrelevant.

There is a close analogy between the Hilbert space $\mathcal{H}$ spanned by the state functions and the Liouville space $\mathcal{L}$ spanned by the corresponding linear operators. However, in addition to having the properties of a unitary vector space, $\mathcal{L}$ forms an operator algebra where the product of two operators is defined as well. For single-transition shift operators (see § 2.1.9) one finds, for example, the relation

$$I^{+(rs)} I^{+(tu)} = I^{+(ru)} \delta_{st}. \qquad (2.1.50)$$

No similar product relation exists for the state functions $|r\rangle$.

The coefficients a_s in the expansion of an arbitrary operator A in terms of a set of orthogonal basis operators $\{B_s\}$

$$A = \sum_{s=1}^{n^2} a_s B_s, \tag{2.1.51}$$

with $\langle B_s | B_t \rangle = \mathrm{tr}\{B_s^\dagger B_t\} = 0$ for $s \neq t$, can be determined by the scalar product

$$a_s = \frac{\langle B_s | A \rangle}{\langle B_s | B_s \rangle} = \frac{\mathrm{tr}\{B_s^\dagger A\}}{\mathrm{tr}\{B_s^\dagger B_s\}}. \tag{2.1.52}$$

For normalized basis operators, the denominator is unity.

2.1.4. Superoperators

The analogy between Hilbert and Liouville space suggests the introduction of superoperators which define operator relations in Liouville space $\mathscr{L}$ (2.7–2.9). An example of such an operator relation is the commutator in eqn (2.1.17)

$$[\mathscr{H}, \sigma] = \mathscr{H}\sigma - \sigma\mathscr{H}. \tag{2.1.53}$$

It may be written in the abbreviated superoperator form

$$\hat{\hat{\mathscr{H}}}\sigma \equiv [\mathscr{H}, \sigma]. \tag{2.1.54}$$

An operator $\hat{\hat{S}}$ acting on operators is called a linear superoperator if

1. $\hat{\hat{S}}A \in \mathscr{L}$ is defined for all operators $A \in \mathscr{L}$;
2. $\hat{\hat{S}}(aA + bB) = a\hat{\hat{S}}A + b\hat{\hat{S}}B$. \hfill (2.1.55)

We denote superoperators by capital letters with a double circumflex, by analogy with the single circumflex used in many textbooks on quantum mechanics to identify operators $\hat{A}$ acting on state functions $|\psi\rangle$. However, we omit the single circumflex on operators and use the same capital letters to denote operators as well as their matrix representations.

In complete analogy to operators, superoperators $\hat{\hat{S}}$ may be represented by (super)matrices in any suitable orthonormal operator basis $\{B_s\}$. The matrix representation of the superoperator $\hat{\hat{S}}$ is of dimension $n^2 \times n^2$ and consists of the matrix elements

$$S_{rs} = \langle B_r | \hat{\hat{S}} B_s \rangle = \mathrm{tr}\{B_r^\dagger \hat{\hat{S}} B_s\}. \tag{2.1.56}$$

Superoperators can be classified according to the same criteria as operators. The adjoint superoperator $\hat{\hat{S}}^\dagger$ is defined by

$$\langle A | \hat{\hat{S}} B \rangle = \langle \hat{\hat{S}}^\dagger A | B \rangle. \tag{2.1.57}$$

For a Hermitian superoperator $\hat{S}^{\dagger} = \hat{S}$. A unitary superoperator is defined by the relation $\hat{S}^{-1} = \hat{S}^{\dagger}$. We shall briefly describe a few particularly important classes of superoperators.

2.1.4.1. Commutator superoperators

For each operator C, a commutator superoperator $\hat{C}$ may be defined by

$$\hat{C}A = CA - AC = [C, A]. \qquad (2.1.58)$$

Commutator superoperators are also called *derivative* superoperators and can be represented as the difference

$$\hat{C} = \hat{C}^{L} - \hat{C}^{R} \qquad (2.1.59)$$

where the left- and right-translation superoperators $\hat{C}^{L}$ and $\hat{C}^{R}$, respectively, are defined by

$$\hat{C}^{L}A = CA \qquad (2.1.60)$$

and

$$\hat{C}^{R}A = AC. \qquad (2.1.61)$$

When C is a Hermitian operator, then $\hat{C}$ is a Hermitian commutator superoperator.

We use the convention that $\hat{\mathcal{H}}$, $\hat{I}_{k\alpha}$, $\hat{S}_{k\alpha}$, and $\hat{F}_{\alpha}$, associated with the Hamiltonian $\mathcal{H}$ and with the spin operators $I_{k\alpha}$, $S_{k\alpha}$, and F_{α} respectively, denote commutator superoperators. The superoperator $\hat{\mathcal{H}}$, which is the driving term in the density operator equation eqn (2.1.17), is specifically called the *Liouville superoperator*. Superoperators designated by other letters will be defined upon usage.

2.1.4.2. Unitary transformation superoperators

The unitary transformation RAR^{-1} with $R = \exp\{-iG\}$ may be expressed by the unitary superoperator $\hat{R}$

$$\hat{R}A = \exp\{-i\hat{G}\}A = \exp\{-iG\}A\exp\{iG\}, \text{ with } \hat{R}^{-1} = \hat{R}^{\dagger}, \quad (2.1.62)$$

where $\hat{G}$ is the commutator superoperator associated with the Hermitian operator G. This identity can easily be proven by noting that the corresponding translation superoperators $\hat{G}^{L}$ and $\hat{G}^{R}$ commute (2.8). For $G = \mathcal{H}\tau$ with a Hamiltonian that is assumed to be time-independent in the interval τ, $\hat{R}$ is the time evolution superoperator

$$\hat{R}(\tau) = \exp\{-i\hat{\mathcal{H}}\tau\}. \qquad (2.1.63)$$

Hence

$$\begin{aligned}\hat{R}(\tau)\sigma(t) &= \exp\{-i\hat{\mathcal{H}}\tau\}\sigma(t) \\ &= \exp\{-i\mathcal{H}\tau\}\sigma(t)\exp\{i\mathcal{H}\tau\} \\ &= \sigma(t+\tau). \qquad (2.1.64)\end{aligned}$$

For the sake of economy of notation, we shall sometimes use a schematic notation for unitary transformation

$$\sigma(t) \xrightarrow{\mathcal{H}\tau} \sigma(t+\tau). \tag{2.1.65}$$

This 'arrow representation' is equivalent to eqn (2.1.64).

The arrow notation has the advantage that a series of transformations can be written chronologically. Thus two consecutive unitary transformations expressed by

$$\hat{\hat{R}}_2 \hat{\hat{R}}_1 \sigma(t) = \exp\{-i\hat{\mathcal{H}}_2 \tau_2\} \exp\{-i\hat{\mathcal{H}}_1 \tau_1\} \sigma(t) \tag{2.1.66}$$

$$= \exp\{-i\mathcal{H}_2 \tau_2\} \exp\{-i\mathcal{H}_1 \tau_1\} \sigma(t)$$

$$\times \exp\{i\mathcal{H}_1 \tau_1\} \exp\{i\mathcal{H}_2 \tau_2\} \tag{2.1.67}$$

can be represented by

$$\sigma(t) \xrightarrow{\mathcal{H}_1 \tau_1} \sigma(t+\tau_1) \xrightarrow{\mathcal{H}_2 \tau_2} \sigma(t_1+\tau_1+\tau_2). \tag{2.1.68}$$

Note that the algebraic signs and the chronological sequence of the terms written above the arrows correspond to the arguments of the time-ordered exponential operators on the *right-hand side* of eqn (2.1.67).

2.1.4.3. Projection superoperators

In Hilbert space $\mathcal{H}$, a projection operator P_j, which projects an arbitrary state function $|\psi\rangle$ on to the function $|j\rangle$, may be represented (2.10–2.12) by

$$P_j = \frac{|j\rangle\langle j|}{\langle j|j\rangle}. \tag{2.1.69}$$

Let us assume, for example, that we can represent $|\psi\rangle$ as a linear combination of orthogonal functions $|i\rangle$,

$$|\psi\rangle = \sum_i c_i |i\rangle. \tag{2.1.70}$$

Then, we obtain the projection

$$P_j |\psi\rangle = \sum_i c_i |j\rangle \frac{\langle j|i\rangle}{\langle j|j\rangle} = c_j |j\rangle \tag{2.1.71}$$

which is just the 'amount' of $|j\rangle$ contained in $|\psi\rangle$.

Projection operators may be used for the spectral resolution of an operator A. The spectrum of an operator is defined as the complete set of its eigenvalues $\{a_j, j=1,\ldots,n\}$. If $|j\rangle$ are the corresponding eigenfunctions and P_j the associated projection operators defined in eqn (2.1.69), it is possible to represent A by its eigenvalues ('spectral resolution' of A)

$$A = \sum_j a_j P_j. \tag{2.1.72}$$

The action of A on an arbitrary function $\psi(t) = \sum c_j(t) |j\rangle$ is then simply given by

$$A\psi(t) = \sum_j a_j c_j(t) |j\rangle. \tag{2.1.73}$$

This particular set of projection operators $\{P_j\}$ is called the *spectral set* of the operator A. It has the property

$$\sum_{j=1}^{n} P_j = \mathbb{1}. \tag{2.1.74}$$

In analogy to the projection operators P_j, it is possible to define *projection superoperators* on the Liouville space $\mathscr{L}$ which project an arbitrary operator A on to an operator B

$$\hat{\hat{P}}_B = \frac{|B\rangle\langle B|}{\langle B|B\rangle} \tag{2.1.75}$$

with

$$\hat{\hat{P}}_B A = |B\rangle \frac{\langle B|A\rangle}{\langle B|B\rangle} = B \frac{\text{tr}\{B^\dagger A\}}{\text{tr}\{B^\dagger B\}}. \tag{2.1.76}$$

Projection operators are idempotent, $\hat{\hat{P}}_B \hat{\hat{P}}_B = \hat{\hat{P}}_B$.

A projection superoperator that is of importance in the context of two-dimensional spectroscopy is the p-quantum projection superoperator $\hat{\hat{P}}^{(p)}$. When applied to the density operator σ it selects only operators $I^{+(rs)}$ associated with a given change in quantum number $p = \Delta M_{rs} = M_r - M_s$, i.e.

$$\hat{\hat{P}}^{(p)} = \frac{1}{N} \sum_{k=0}^{N-1} \hat{\hat{R}}_z(k2\pi/N) \exp\{ipk2\pi/N\}, \tag{2.1.77}$$

where N is the number of orders p to be distinguished. Similarly,

$$\hat{\hat{P}}^{(|p|)} = \hat{\hat{P}}^{(p)} + \hat{\hat{P}}^{(-p)} = \frac{2}{N} \sum_{k=0}^{N-1} \hat{\hat{R}}_z(k2\pi/N) \cos(pk2\pi/N)$$

where $\hat{\hat{R}}_z(\phi)$ represents a rotation through an angle ϕ about the z-axis, which defines the axis of quantization. We shall see in § 6.3 that such projections can actually be achieved experimentally in an elegant manner.

2.1.4.4. General representation of superoperators

Given a complete operator basis $\{B_s; s = 1, \ldots, n^2\}$, it is possible to represent any superoperator $\hat{S}$ in the form

$$\hat{S} = \sum_{jk} s_{jk} \hat{B}_j^L \hat{B}_k^R \tag{2.1.77a}$$

with the corresponding left- and right-translation superoperators $\hat{B}_j^L$ and $\hat{B}_k^R$, or by

$$\hat{S}A = \sum_{jk} s_{jk} B_j A B_k. \tag{2.1.77b}$$

It is obvious that there are n^4 linearly independent superoperators.

2.1.4.5. Matrix representation of a superoperator

Equation (2.1.77b) may be evaluated by expressing all operators by matrices in an arbitrary basis

$$(\hat{S}A)_{pq} = \sum_l \sum_m \sum_{jk} s_{jk} B_{j,pl} A_{lm} B_{k,mq}$$

$$= \sum_l \sum_m S_{pq,lm} A_{lm} \tag{2.1.78}$$

with the supermatrix elements

$$S_{pq,lm} = \sum_{jk} s_{jk} B_{j,pl} B_{k,mq} \tag{2.1.79}$$

which may be expressed by elements of the direct matrix products ($\otimes$)

$$S_{pq,lm} = \sum_{jk} s_{jk} (B_j \otimes \tilde{B}_k)_{pq,lm}, \tag{2.1.80}$$

where $\tilde{B}_k$ represents the transposed matrix B_k. $(S_{pq,lm})$ is the matrix representation of the superoperator $\hat{S}$. This shows that the matrix representation of a superoperator can be obtained as a sum of direct products of the matrix representations of the constituent operators.

The representation of supermatrices by direct products of conventional matrices can be applied to compute the supermatrix representations of commutators and unitary transformations. For the commutator superoperator $\hat{C}$, we obtain

$$\hat{C}A = CAE - EAC \tag{2.1.81}$$

where E represents the unity operator. This leads to the matrix representation

$$(\hat{C}) = (C) \otimes (\tilde{E}) - (E) \otimes (\tilde{C}). \tag{2.1.82}$$

$$(\hat{\tilde{I}}_x) = (I_x) \otimes (\tilde{E}) - (E) \otimes (\tilde{I}_x)$$

$$= \tfrac{1}{2} \begin{bmatrix} 0 & 1 \\ 1 & 0 \end{bmatrix} \otimes \begin{bmatrix} 1 & 0 \\ 0 & 1 \end{bmatrix} - \begin{bmatrix} 1 & 0 \\ 0 & 1 \end{bmatrix} \otimes \tfrac{1}{2} \begin{bmatrix} 0 & 1 \\ 1 & 0 \end{bmatrix}$$

$$= \tfrac{1}{2} \begin{bmatrix} 0 & 0 & 1 & 0 \\ 0 & 0 & 0 & 1 \\ 1 & 0 & 0 & 0 \\ 0 & 1 & 0 & 0 \end{bmatrix} - \tfrac{1}{2} \begin{bmatrix} 0 & 1 & 0 & 0 \\ 1 & 0 & 0 & 0 \\ 0 & 0 & 0 & 1 \\ 0 & 0 & 1 & 0 \end{bmatrix}$$

$$= \tfrac{1}{2} \begin{bmatrix} 0 & -1 & 1 & 0 \\ -1 & 0 & 0 & 1 \\ 1 & 0 & 0 & -1 \\ 0 & 1 & -1 & 0 \end{bmatrix}$$

FIG. 2.1.2. Construction of the matrix representation of the commutation superoperator $\hat{\tilde{I}}_x$ for an isolated spin $\tfrac{1}{2}$.

As an example, the matrix representation of the commutator superoperator $\hat{\tilde{I}}_x$ for a single spin $\tfrac{1}{2}$ is constructed in Fig. 2.1.2 according to eqn (2.1.82).

The matrix representation of a unitary transformation superoperator $\hat{R}A = RAR^{-1}$, on the other hand, is obtained in the form

$$(\hat{R}) = (R) \otimes (R^*) \tag{2.1.83}$$

where (R^*) is the complex conjugate of the matrix (R).

2.1.4.6. Eigenvalues and eigenoperators of superoperators

The eigenvalue problem of a superoperator $\hat{S}$ is given by

$$\hat{S} Q_j = s_j Q_j \qquad j = 1, \ldots, n^2 \tag{2.1.84}$$

with n^2 eigenoperators Q_j and the corresponding eigenvalues s_j.

Simple relations can be obtained for the eigenvalues of a commutator superoperator $\hat{\mathcal{H}}$. If we denote the eigenvalues of $\mathcal{H}$ by ε_r, $r = 1, \ldots, n$, we obtain for the eigenvalues ω_{rs} of $\hat{\mathcal{H}}$,

$$\omega_{rs} = \varepsilon_r - \varepsilon_s, \qquad r, s = 1, \ldots, n. \tag{2.1.85}$$

This shows that the set of eigenvalues $\{\omega_{rs}\}$ of the commutator superoperator consists of the entire set of differences of eigenvalues of $\mathcal{H}$.

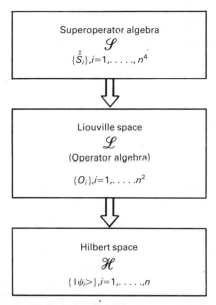

FIG. 2.1.3. The hierarchy of linear spaces in quantum mechanics. Superoperators cause a linear mapping of the operator algebra while operators effect a linear mapping of the Hilbert space.

2.1.4.7. Superoperator algebra

Just as the linear operators $\{B_s\}$ which span the Liouville space $\mathscr{L}$ form an operator algebra, the superoperators in turn form an algebra, since they span a vector space (of dimension $n^2 \times n^2$) in which products are defined. The hierarchy of linear spaces is shown in Fig. 2.1.3.

In the following sections, we describe sets of basis operators which have proven to be particularly useful for the expansion of the density operator in the context of pulse Fourier spectroscopy.

2.1.5. Products of Cartesian spin operators

Numerous choices are possible for the expansion of the density operator in terms of a complete set of orthogonal basis operators $\{B_s\}$ according to eqn (2.1.45). A proper selection is often essential to ease the solution of a particular problem. In §§ 2.1.5–2.1.10 we present different sets which have proven convenient in the treatment of pulse experiments.

The first and perhaps most straightforward choice is based on the angular momentum operators I_{kx}, I_{ky}, and I_{kz} of the individual spins which obey the usual cyclic commutation rules

$$[I_{k\alpha}, I_{k\beta}] = iI_{k\gamma} \qquad (2.1.86)$$

with α, β, $\gamma = x$, y, z and cyclic permutations. They can be considered as the generating operators of the operator algebra, and it is possible to span the entire Liouville space of N spins by products (including powers if $I_k > \frac{1}{2}$) of the $3N$ generating operators.

2.1.5.1. Systems of spins $I = \frac{1}{2}$

For systems with spins $I_k = \frac{1}{2}$, the basis operators B_s can be defined by the the products (2.13)

$$B_s = 2^{(q-1)} \prod_{k=1}^{N} (I_{k\alpha})^{a_{ks}} \qquad (2.1.87)$$

where N = total number of $I = \frac{1}{2}$ nuclei in the spin system, k = index of nucleus, $\alpha = x$, y, or z, q = number of operators in the product, $a_{ks} = 1$ for q of the spins, and $a_{ks} = 0$ for the $N - q$ remaining spins.

The product operators defined in eqn (2.1.87) for spin $\frac{1}{2}$ nuclei are orthogonal with respect to formation of the trace; however the normalization depends on the total number N of spins in the system

$$\text{tr}\{B_r B_s\} = \delta_{r,s} 2^{N-2}. \qquad (2.1.88)$$

The complete basis set $\{B_s\}$ for a system with N spins $\frac{1}{2}$ consists of 4^N product operators B_s.

As an example, we list the complete set of 16 product operators B_s for a two-spin system with $I = \frac{1}{2}$

$$\begin{aligned}
q = 0 \quad & \tfrac{1}{2}E \quad (E = \text{unity operator}), \\
q = 1 \quad & I_{1x}, I_{1y}, I_{1z}, I_{2x}, I_{2y}, I_{2z}, \\
q = 2 \quad & 2I_{1x}I_{2x}, 2I_{1x}I_{2y}, 2I_{1x}I_{2z}, \\
& 2I_{1y}I_{2x}, 2I_{1y}I_{2y}, 2I_{1y}I_{2z}, \\
& 2I_{1z}I_{2x}, 2I_{1z}I_{2y}, 2I_{1z}I_{2z}.
\end{aligned} \qquad (2.1.89)$$

The spectroscopic meaning of these operators in terms of the signals that can be observed in Fourier spectroscopy will be discussed in § 4.4.5.

Products of Cartesian spin operators are particularly useful for calculating the evolution of the density operator for a weakly coupled spin system where all terms in the Hamiltonian commute (eqn (2.2.14)). The effect of these terms may be evaluated in a cascade of transformations

$$\sigma(t + \tau) = \prod_k \exp(-i\Omega_k \tau I_{kz}) \prod_{k<l} \exp(-i\pi J_{kl} \tau 2I_{kz}I_{lz}) \sigma(t)$$
$$\times \prod_{k<l} \exp(i\pi J_{kl} \tau 2I_{kz}I_{lz}) \prod_k \exp(i\Omega_k \tau I_{kz}) \qquad (2.1.90)$$

or, symbolically, in the arrow notation of eqn (2.1.65)

$$\sigma(t) \xrightarrow{\Omega_1 \tau I_{1z}} \xrightarrow{\Omega_2 \tau I_{2z}} \cdots \xrightarrow{\pi J_{12}\tau 2 I_{1z}I_{2z}} \xrightarrow{\pi J_{13}\tau 2 I_{1z}I_{3z}} \cdots \sigma(t+\tau). \quad (2.1.91)$$

Each of these transformations corresponds to a rotation in a three-dimensional operator subspace. The evolution under the chemical shift terms and under r.f. pulses induces rotations in the single-spin subspaces spanned by the angular momentum operators (I_{kx}, I_{ky}, I_{kz}). Thus the transformation

$$I_{k\beta} \xrightarrow{\phi I_{k\alpha}} I_{k\beta} \cos\phi + I_{k\gamma} \sin\phi, \quad (2.1.92)$$

where $\alpha, \beta, \gamma = x, y, z$ and cyclic permutations, expresses a rotation through an angle $\phi = -\gamma B_\alpha \tau$ in physical space about the α-axis.

We should note at this point that we consistently define positive rotations (frequencies and angles) in the right-handed sense (clockwise rotations if looking in the direction of the rotation vector). A positive rotation about the z-axis leads to the transformations $x \to y \to -x \to -y$. For a positive gyromagnetic ratio γ_k, the Larmor frequency vectors $\Omega_k = -\gamma_k \mathbf{B}_0$ point to the negative z-axis if $\mathbf{B}_0$ is oriented along the positive z-axis. The magnetic field vector $\mathbf{B}_0$ and the rotation frequency vectors Ω_k are antiparallel both in the laboratory and in the rotating frame for positive γ_k (see Fig. 4.2.2). This consistent definition is in contrast to common usage in many NMR papers and monographs which define $\Omega_k = \gamma_k B_0$ for convenience.

As a consequence, we define a β_x pulse as a pulse with rotation angle β about the positive x-axis (leading to transformations $z \to -y \to -z \to y$). The corresponding magnetic field vector $\mathbf{B}_1$ would point along the negative x-axis for positive γ_k.

Scalar and dipolar coupling terms induce rotations within the following four operator subspaces

$$\begin{aligned} &(I_{kx}, 2I_{ky}I_{lz}, 2I_{kz}I_{lz}), \\ &(2I_{kx}I_{lz}, I_{ky}, 2I_{kz}I_{lz}). \\ &(I_{lx}, 2I_{kz}I_{ly}, 2I_{kz}I_{lz}), \\ &(2I_{kz}I_{lx}, I_{ly}, 2I_{kz}I_{lz}). \end{aligned} \quad (2.1.93)$$

These subspaces are isomorphous to the Cartesian subspace (I_{kx}, I_{ky}, I_{kz}). The rotations in these spaces are shown schematically in Fig. 2.1.4. By way of example, consider the transformation

$$I_{kx} \xrightarrow{\pi J_{kl}\tau 2 I_{kz}I_{lz}} I_{kx} \cos(\pi J_{kl}\tau) + 2I_{ky}I_{lz} \sin(\pi J_{kl}\tau). \quad (2.1.94)$$

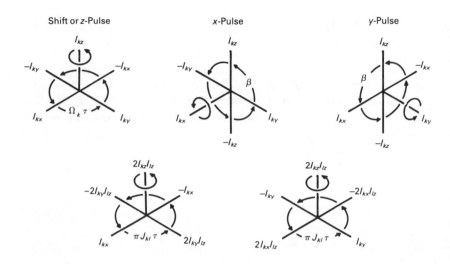

FIG. 2.1.4. Rotations in subspaces spanned by products of Cartesian spin operators, corresponding to chemical shifts (rotation about I_{kz}), r.f. pulses (rotations about I_{kx} and I_{ky}), and weak scalar couplings (rotations about $2 I_{kz}I_{lz}$). Note that we consistently use positive rotations (frequencies and angles) in the right-handed sense (clockwise when looking along the rotation vector). (Reproduced from Ref. 2.13.)

which corresponds to the conversion of in-phase coherence of spin k into antiphase coherence with respect to the coupling partner l. The nomenclature of these operator terms is described in § 4.4.5.

If the density operator contains products of operators belonging to different spins (k and l), each constituent operator $I_{k\alpha}$ transforms individually, e.g. for a propagator with linear terms we obtain

$$2I_{k\beta}I_{l\beta'} \xrightarrow{\phi I_{k\alpha}+\phi' I_{l\alpha'}} 2(I_{k\beta}\cos\phi + I_{ky}\sin\phi)$$

$$\times (I_{l\beta'}\cos\phi' + I_{ly'}\sin\phi') \quad (2.1.95)$$

and for Hamiltonians with bilinear terms we have

$$2I_{k\beta}I_{l\beta'} \xrightarrow{\phi 2I_{k\alpha}I_{l\alpha'}} 2(I_{k\beta}\cos\phi + 2I_{ky}I_{l\alpha'}\sin\phi)$$

$$\times (I_{l\beta'}\cos\phi + 2I_{k\alpha}I_{ly'}\sin\phi). \quad (2.1.96a)$$

Thus one obtains, for example,

$$2I_{kx}I_{lz} \xrightarrow{\pi J_{kl}\tau 2I_{kz}I_{lz}} 2I_{kx}I_{lz}\cos(\pi J_{kl}\tau) + I_{ky}\sin(\pi J_{kl}\tau), \quad (2.1.96b)$$

$$2I_{ky}I_{lz} \xrightarrow{\pi J_{kl}\tau 2I_{kz}I_{lz}} 2I_{ky}I_{lz}\cos(\pi J_{kl}\tau) - I_{kx}\sin(\pi J_{kl}\tau) \quad (2.1.96c)$$

as illustrated in Fig. 2.1.4. For the evaluation of bilinear rotations it is important to note that

$$[2I_{k\alpha}I_{l\alpha'}, 2I_{k\beta}I_{l\beta'}] = 0 \tag{2.1.97}$$

if $\alpha \neq \beta$ and $\alpha' \neq \beta'$ simultaneously. This implies invariance of terms under corresponding transformations.

Rotations about tilted axes can be decomposed: for example, the effect of an r.f. pulse with arbitrary phase φ (defined as excursion from the x- towards the y-axis of the rotating frame) is described by the three steps

$$\sigma(t_-) \xrightarrow{\varphi \sum_k I_{kz}} \xrightarrow{\beta \sum_k I_{kx}} \xrightarrow{\psi \sum_k I_{kz}} \sigma(t_+). \tag{2.1.98}$$

By constructing special pulse sequences, it is possible to generate effective Hamiltonians which contain coupling terms with transverse spin operators (see § 4.4.6.2). Thus the effect of the pulse sandwich $[(\pi/2)_x - \tau - (\pi)_x - \tau - (\pi/2)_x]$ may be written (2.13)

$$\sigma(t) \xrightarrow{\sum \pi J_{kl} 2\tau 2 I_{ky} I_{ly}} \sigma(t + 2\tau). \tag{2.1.99}$$

Transformations of this type correspond to rotations in eight additional subspaces spanned by the operators

$$(2I_{kx}I_{lx}, 2I_{ky}I_{lx}, I_{kz}),$$
$$(2I_{kx}I_{lx}, I_{ky}, 2I_{kz}I_{lx}),$$
$$(2I_{kx}I_{ly}, 2I_{ky}I_{ly}, I_{kz}),$$
$$(I_{kx}, 2I_{ky}I_{ly}, 2I_{kz}I_{ly}),$$
$$(2I_{kx}I_{lx}, 2I_{kx}I_{ly}, I_{lz}),$$
$$(2I_{kx}I_{lx}, I_{ly}, 2I_{kx}I_{lz}),$$
$$(2I_{ky}I_{lx}, 2I_{ky}I_{ly}, I_{lz}),$$
$$(I_{lx}, 2I_{ky}I_{ly}, 2I_{ky}I_{lz}), \tag{2.1.100}$$

which are all isomorphous to the subspace (I_{kx}, I_{ky}, I_{kz}) and can be obtained from eqn (2.1.93) by cyclic index permutation. Rotations in the first four of these subspaces are illustrated in Fig. 2.1.5.

2.1.5.2. Systems containing spins $S > \frac{1}{2}$

In systems with spin quantum numbers $I \geq 1$ or $S \geq 1$, the complete set of basis operators $\{B_s\}$ includes quadratic and higher terms of the type I_{kz}^2, S_{lz}^2, $I_{kx}S_{kz}^2$, etc. The relevance of such terms for the resulting spectrum will be discussed in § 4.4.5. At this point it suffices to note that the spectrum of a nucleus I_k with weak scalar or dipolar coupling to $S_l = 1$

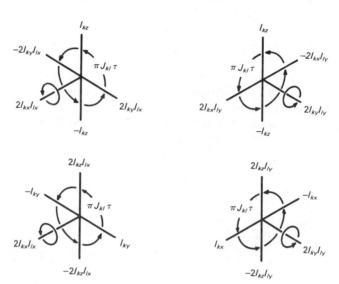

FIG. 2.1.5. Bilinear rotations in subspaces spanned by products of Cartesian spin operators, induced by transformations that can be brought about by pulse sequences with propagators of the type $\exp\{-i\varphi 2I_{kx}I_{lx}\}$ and $\exp\{-i\varphi 2I_{ky}I_{ly}\}$; see eqn (2.1.100). (Reproduced from Ref. 2.13.)

consists of three lines of equal intensity. We decompose the I_k triplet coherence into a coherence associated with the central line and in-phase and antiphase coherence of the outer lines

$$I_{kx}[0, 1, 0] = I_{kx}(\mathbb{1}_l - S_{lz}^2),$$
$$I_{kx}[1, 0, 1] = I_{kx}S_{lz}^2,$$
$$I_{kx}[-1, 0, 1] = I_{kx}S_{lz}. \qquad (2.1.101)$$

The corresponding y-terms are defined by analogy. The values in brackets indicate the intensities of the corresponding multiplet lines in the I-spin spectrum (compare Fig. 4.4.6). In mathematical terms, they represent the coefficients of three polarization operators of spin S_l: $S_l^{[-1]} = \frac{1}{2}(S_{lz}^2 - S_{lz})$, $S_l^{[0]} = \mathbb{1} - S_{lz}^2$, and $S_l^{[1]} = \frac{1}{2}(S_{lz}^2 + S_{lz})$ (see § 2.1.7).

The central component of the triplet is invariant to J_{kl}-coupling, while the outer lines evolve in analogy to doublets involving $S_l = \frac{1}{2}$ nuclei, but with twice the usual coupling constant. These transformations correspond to rotations in the two subspaces shown in Fig. 2.1.6, which are spanned by the terms

$$(I_{kx}S_{lz}^2, I_{ky}S_{lz}, I_{kz}S_{lz}),$$
$$(I_{kx}S_{lz}, I_{ky}S_{lz}^2, I_{kz}S_{lz}). \qquad (2.1.102)$$

If an arbitrary spin I_k is coupled to a spin $S_l = \frac{3}{2}$ (which may be a group spin if AX_3 groups are treated in terms of symmetrized eigenfunctions),

Scalar coupling to S=1

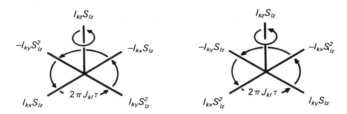

FIG. 2.1.6. Rotations in operator subspaces which describe the evolution of the triplet of a spin I under weak scalar coupling to a spin $S = 1$. (Reproduced from Ref. 2.13.)

the I-spin quartet can be decomposed into in-phase and antiphase magnetization with the inner and outer transitions of the quartet

$$I_{kx}[0, 1, 1, 0] = I_{kx}\tfrac{1}{2}(\tfrac{2}{4}\mathbb{1}_l - S_{lz}^2),$$
$$I_{kx}[0, -1, 1, 0] = -I_{kx}S_{lz}(\tfrac{2}{4}\mathbb{1}_l - S_{lz}^2),$$
$$I_{kx}[1, 0, 0, 1] = -I_{kx}\tfrac{1}{2}(\tfrac{1}{4}\mathbb{1}_l - S_{lz}^2),$$
$$I_{kx}[-1, 0, 0, 1] = I_{kx}S_{lz}\tfrac{1}{3}(\tfrac{1}{4}\mathbb{1}_l - S_{lz}^2) \quad (2.1.103)$$

and corresponding y-components.

The evolution under the scalar coupling can be described by rotations in four three-dimensional subspaces analogous to those depicted in Fig. 2.1.4. For the inner components, $\phi = \pi J_{kl}\tau$, while the outer components precess with $\phi = 3\pi J_{kl}\tau$.

In oriented systems, the quadrupolar interaction for $S = 1$ and $\eta = 0$ (axially symmetric quadrupole tensor, see § 2.2.1.3) has the form

$$\mathcal{H}_Q = \omega_Q(S_z^2 - \tfrac{1}{3}\mathbf{S}^2) \quad (2.1.104)$$

where ω_Q corresponds to one-half of the quadrupolar splitting. The magnetization of the doublet may be decomposed into in-phase components S_x and S_y, and antiphase components $\{S_xS_z + S_zS_x\}$ and $\{S_yS_z + S_zS_y\}$. The effect of the quadrupolar coupling corresponds to rotations in the spaces shown in Fig. 2.1.7.

Quadrupolar coupling S=1

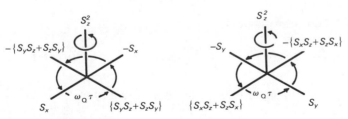

FIG. 2.1.7. Rotations in operator subspaces for $S = 1$ spins, induced by the quadrupole coupling ($\eta = 0$) in oriented systems. (Reproduced from Ref. 2.13.)

In contrast to the operator products used for $I=\frac{1}{2}$, products of the type $S_x S_z$ which involve operators of the *same* spin $S>\frac{1}{2}$ are non-Hermitian and do not represent convenient basis operators B_s to expand the density operator. The expressions in curly brackets in Fig. 2.1.7 (anticommutators), however, are Hermitian.

2.1.6. Products involving shift operators

Although Cartesian angular momentum operators and their products often yield elegant descriptions of pulse experiments, there are situations, in particular in multiple-quantum spectroscopy, where raising and lowering operators are more straightforward to use

$$I_k^+ = I_{kx} + iI_{ky}, \qquad (2.1.105)$$

$$I_k^- = I_{kx} - iI_{ky}. \qquad (2.1.106)$$

Together with the operators I_{kz} and $\mathbb{1}_k$, the products of these operators also form a complete set $\{B_s\}$ of 4^N operators for a system with N spins $\frac{1}{2}$.

In a two-spin system we have the identities

$$I_k^+ I_l^+ = \tfrac{1}{2}[2I_{kx}I_{lx} - 2I_{ky}I_{ly} + i2I_{kx}I_{ly} + i2I_{ky}I_{lx}],$$
$$I_k^- I_l^- = \tfrac{1}{2}[2I_{kx}I_{lx} - 2I_{ky}I_{ly} - i2I_{kx}I_{ly} - i2I_{ky}I_{lx}],$$
$$I_k^+ I_l^- = \tfrac{1}{2}[2I_{kx}I_{lx} + 2I_{ky}I_{ly} - i2I_{kx}I_{ly} + i2I_{ky}I_{lx}],$$
$$I_k^- I_l^+ = \tfrac{1}{2}[2I_{kx}I_{lx} + 2I_{ky}I_{ly} + i2I_{kx}I_{ly} - i2I_{ky}I_{lx}]. \qquad (2.1.107)$$

These operators are useful to express ±2-quantum and zero quantum coherence, as shown in §§ 4.4.5 and 5.3.3.

We can easily derive the transformation relations

$$I_k^\pm \xrightarrow{\phi I_{kz}} I_k^\pm \exp\{\mp i\phi\}, \qquad (2.1.108)$$

$$I_k^\pm \xrightarrow{\phi I_{kx}} I_k^\pm \cos^2\frac{\phi}{2} + I_k^\mp \sin^2\frac{\phi}{2} \pm iI_{kz}\sin\phi. \qquad (2.1.109)$$

For a phase-shifted r.f. pulse with the Hamiltonian

$$\mathcal{H}\tau = \phi[I_{kx}\cos\varphi + I_{ky}\sin\varphi] \qquad (2.1.110)$$

we find the transformation

$$I_k^\pm \xrightarrow{\mathcal{H}\tau} I_k^\pm \cos^2\frac{\phi}{2} + I_k^\mp \sin^2\frac{\phi}{2}\exp\{\pm i2\varphi\} \pm iI_{kz}\sin\phi\exp\{\pm i\varphi\}. \qquad (2.1.111)$$

Transformations under bilinear terms have the form

$$I_k^\pm \xrightarrow{\phi 2I_{kz}I_{l\alpha}} I_k^\pm \cos\phi \mp i2I_k^\pm I_{l\alpha}\sin\phi, \qquad (2.1.112)$$

$$I_k^\pm \xrightarrow{\phi 2I_{kx}I_{l\alpha}} I_k^\pm \cos^2\frac{\phi}{2} + I_k^\mp \sin^2\frac{\phi}{2} \pm i2I_{kz}I_{l\alpha}\sin\phi \quad (2.1.113a)$$

$$I_k^\pm \xrightarrow{\phi 2I_{ky}I_{l\alpha}} I_k^\pm \cos^2\frac{\phi}{2} - I_k^\mp \sin^2\frac{\phi}{2} - 2I_{kz}I_{l\alpha}\sin\phi \quad (2.1.113b)$$

with $\alpha = x, y,$ or z.

2.1.7. Polarization operators

To provide a link to single transition operators, it is useful to introduce polarization operators (2.14, 2.15). For spins $I = \frac{1}{2}$ we define

$$I_k^\alpha = \tfrac{1}{2}\mathbb{1}_k + I_{kz},$$
$$I_k^\beta = \tfrac{1}{2}\mathbb{1}_k - I_{kz}. \quad (2.1.114)$$

Hence

$$I_{kz} = \tfrac{1}{2}(I_k^\alpha - I_k^\beta).$$

The definitions in terms of kets and bras and the matrix representations for a spin $I = \frac{1}{2}$ show that these single-element operators are the diagonal counterparts of the shift operators

$$I_k^\alpha = |\alpha\rangle\langle\alpha| = \begin{pmatrix}1 & 0\\ 0 & 0\end{pmatrix}, \quad I_k^\beta = |\beta\rangle\langle\beta| = \begin{pmatrix}0 & 0\\ 0 & 1\end{pmatrix},$$

$$I_k^+ = |\alpha\rangle\langle\beta| = \begin{pmatrix}0 & 1\\ 0 & 0\end{pmatrix}, \quad I_k^- = |\beta\rangle\langle\alpha| = \begin{pmatrix}0 & 0\\ 1 & 0\end{pmatrix}. \quad (2.1.115)$$

For spins with $I_k > \frac{1}{2}$, the polarization operators corresponding to each of the $(2I + 1)$ states may be specified by their quantum number M_k. Thus, for $I_k = 1$, one must distinguish the polarization operators $I_k^{[+1]}$, $I_k^{[0]}$, and $I_k^{[-1]}$. For $I_k = \frac{1}{2}$, we have $I_k^{[+\frac{1}{2}]} = I_k^\alpha$ and $I_k^{[-\frac{1}{2}]} = I_k^\beta$.

We may now construct a complete set of 4^N single-element operators for an N spin-$\frac{1}{2}$ system by products of the form

$$B_s = \prod_{k=1}^{N} I_k^{\mu_{ks}} \quad (2.1.116)$$

where $\mu_{ks} = +, -, \alpha,$ or β. If the product extends over all spins k, the unity operator is no longer required. The matrix representation of each of these product operators contains a single matrix element with unit value.

For the representation of a weakly coupled system with non-equilibrium populations (so-called 'incoherent non-equilibrium state'; see § 4.4.3), only products of polarization operators ($\mu_{ks} = \alpha, \beta$) are required, each product identifying a particular eigenstate. For a two-spin system, for example, we have

$$\sigma = P_{\alpha\alpha}I_1^\alpha I_2^\alpha + P_{\alpha\beta}I_1^\alpha I_2^\beta + P_{\beta\alpha}I_1^\beta I_2^\alpha + P_{\beta\beta}I_1^\beta I_2^\beta \qquad (2.1.117)$$

where $P_{\alpha\alpha}$, $P_{\alpha\beta}$, ... are real numbers that express the populations of the eigenstates $|\alpha\alpha\rangle$, $|\alpha\beta\rangle$, ... respectively.

If polarization operators are used to expand the density operator, the following transformations are required to describe the effect of r.f. pulses

$$I_k^\alpha \xrightarrow{\phi[I_{kx}\cos\varphi + I_{ky}\sin\varphi]} I_k^\alpha \cos^2\phi/2 + I_k^\beta \sin^2\phi/2$$

$$+ \frac{i}{2}\sin\phi[I_k^+ \exp\{-i\varphi\} - I_k^- \exp\{+i\varphi\}], \qquad (2.1.118)$$

$$I_k^\beta \xrightarrow{\phi[I_{kx}\cos\varphi + I_{ky}\sin\varphi]} I_k^\beta \cos^2\phi/2 + I_k^\alpha \sin^2\phi/2$$

$$- \frac{i}{2}\sin\phi[I_k^+ \exp\{-i\varphi\} - I_k^- \exp\{+i\varphi\}]. \qquad (2.1.119)$$

Polarization operators are obviously invariant under I_{kz}.

2.1.8. Cartesian single-transition operators

Single-transition operators are particularly useful for the description of selective excitation of a single transition in a complicated spin system (2.16, 2.17).

The single-transition operators $I_x^{(rs)}$, $I_y^{(rs)}$, $I_z^{(rs)}$, $I^{+(rs)}$, and $I^{-(rs)}$ refer to the transition between two energy levels $|r\rangle$ and $|s\rangle$. All other energy levels are disregarded, and the subsystem is treated as a virtual two-level system. The single-transition operators associated with the transition between $|r\rangle$ and $|s\rangle$, which may represent a zero-, single-, or multiple-quantum transition, are defined in the following way

$$\langle i|I_x^{(rs)}|j\rangle = \frac{1}{2}(\delta_{ir}\delta_{js} + \delta_{is}\delta_{jr}),$$

$$\langle i|I_y^{(rs)}|j\rangle = \frac{i}{2}(-\delta_{ir}\delta_{js} + \delta_{is}\delta_{jr}),$$

$$\langle i|I_z^{(rs)}|j\rangle = \frac{1}{2}(\delta_{ir}\delta_{jr} - \delta_{is}\delta_{js}). \qquad (2.1.120)$$

The matrix representations of these three operators (Fig. 2.1.8) lead, after elimination of rows and columns containing only zeroes, to the Pauli matrices. It is easily seen that reversing the order of indices leads to the

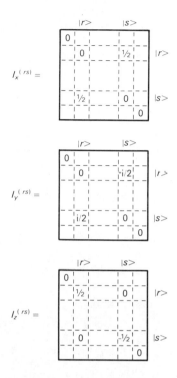

FIG. 2.1.8. Matrix representations of the three single-transition operators $I_x^{(rs)}$, $I_y^{(rs)}$, and $I_z^{(rs)}$. They are represented in the eigenbasis of the corresponding Hamiltonian.

relations

$$I_x^{(sr)} = I_x^{(rs)},$$
$$I_y^{(sr)} = -I_y^{(rs)},$$
$$I_z^{(sr)} = -I_z^{(rs)}. \quad (2.1.121)$$

These sign changes have to be taken into account when constructing single-transition operators for arbitrary eigenstates $|r\rangle$ and $|s\rangle$.

The three operators belonging to a particular transition between $|r\rangle$ and $|s\rangle$ obey the standard commutation relations, eqn (2.1.86)

$$[I_\alpha^{(rs)}, I_\beta^{(rs)}] = iI_\gamma^{(rs)} \quad (2.1.122)$$

where $\alpha, \beta, \gamma = x, y, z$, and cyclic permutations.

For operators describing two connected transitions that involve three distinct states $|r\rangle$, $|s\rangle$, and $|t\rangle$, the following commutation rules hold:

$$[I_x^{(rt)}, I_x^{(st)}] = [I_y^{(rt)}, I_y^{(st)}] = \frac{i}{2} I_y^{(rs)},$$

$$[I_z^{(rt)}, I_z^{(st)}] = 0,$$

$$[I_x^{(rt)}, I_y^{(st)}] = \frac{i}{2} I_x^{(rs)},$$

$$[I_x^{(rt)}, I_z^{(st)}] = \frac{-i}{2} I_y^{(rt)},$$

$$[I_y^{(rt)}, I_z^{(st)}] = \frac{i}{2} I_x^{(rt)}. \tag{2.1.123}$$

Changing the order of indices results in sign changes according to eqn (2.1.121), e.g.

$$[I_x^{(rt)}, I_y^{(ts)}] = \frac{-i}{2} I_x^{(rs)}. \tag{2.1.124}$$

Operators belonging to non-connected transitions always commute

$$[I_\alpha^{(rs)}, I_\beta^{(tu)}] = 0. \tag{2.1.125}$$

It should be noted that there are linear dependences among the z-components:

$$I_z^{(rs)} + I_z^{(st)} + I_z^{(tr)} = 0. \tag{2.1.126}$$

Single-transition operators are normally defined in the eigenbasis of the Hamiltonian, while the products of shift and polarization operators, defined in the previous section, refer to the product basis. For weakly coupled spin systems, the two bases are identical and it is possible to identify the operators. For a two-spin system with $I = \frac{1}{2}$ we find

$$I_x^{(1,2)} = I_1^\alpha I_{2x}, \qquad I_y^{(1,2)} = I_1^\alpha I_{2y},$$

$$I_x^{(3,4)} = I_1^\beta I_{2x}, \qquad I_y^{(3,4)} = I_1^\beta I_{2y},$$

$$I_x^{(1,3)} = I_{1x} I_2^\alpha, \qquad I_y^{(1,3)} = I_{1y} I_2^\alpha,$$

$$I_x^{(2,4)} = I_{1x} I_2^\beta, \qquad I_y^{(2,4)} = I_{1y} I_2^\beta,$$

$$I_x^{(1,4)} = \tfrac{1}{2}(2I_{kx}I_{lx} - 2I_{ky}I_{ly}) \qquad I_y^{(1,4)} = \tfrac{1}{2}(2I_{kx}I_{ly} + 2I_{ky}I_{lx})$$

$$= \tfrac{1}{2}(I_k^+ I_l^+ + I_k^- I_l^-), \qquad = \frac{i}{2}(-I_k^+ I_l^+ + I_k^- I_l^-),$$

$$I_x^{(2,3)} = \tfrac{1}{2}(2I_{kx}I_{lx} + 2I_{ky}I_{ly}) \qquad I_y^{(2,3)} = \tfrac{1}{2}(2I_{ky}I_{lx} - 2I_{kx}I_{ly})$$

$$= \tfrac{1}{2}(I_k^+ I_l^- + I_l^- I_k^+), \qquad = \frac{i}{2}(-I_k^+ I_l^- + I_k^- I_l^+)$$

$$\tag{2.1.127}$$

where the states are numbered $|1\rangle = |\alpha\alpha\rangle$, $|2\rangle = |\alpha\beta\rangle$, $|3\rangle = |\beta\alpha\rangle$, and $|4\rangle = |\beta\beta\rangle$.

The transformation properties of single-transition operators under selective pulses can be described in three-dimensional subspaces. If the coherence and the r.f. pulse involve one and the same transition between a pair of eigenstates $|r\rangle$ and $|s\rangle$, we have the usual transformations

$$I_\beta^{(rs)} \xrightarrow{\phi I_\alpha^{(rs)}} I_\beta^{(rs)} \cos \phi + I_\gamma^{(rs)} \sin \phi, \qquad (2.1.128)$$

with $\alpha, \beta, \gamma = x, y, z$, and cyclic permutations. However, if a pulse is applied to a different transition that is connected through a common eigenstate, the coherence transforms according to

$$I_x^{(st)} \xrightarrow{\phi I_x^{(rs)}} I_x^{(st)} \cos \phi/2 + I_y^{(rt)} \sin \phi/2. \qquad (2.1.129)$$

It should be emphasized that the angle of rotation is apparently halved whenever the transformed operator and the rotation operator have only one index in common, implying that the coherence is completely transferred for $\phi = \pi$. For a three-level system, there are seven subspaces spanned by three orthogonal operators each, which form a basis for the representation of three-dimensional rotations of the type of eqns (2.1.128) and (2.1.129)

$$(I_x^{(1,2)}, I_y^{(1,2)}, I_z^{(1,2)}), (I_x^{(1,3)}, I_y^{(1,3)}, I_z^{(1,3)}), (I_x^{(2,3)}, I_y^{(2,3)}, I_z^{(2,3)}),$$
$$(2I_x^{(1,2)}, 2I_x^{(2,3)}, 2I_x^{(1,3)}), (2I_x^{(1,2)}, 2I_y^{(1,3)}, 2I_y^{(2,3)}),$$
$$(2I_y^{(1,2)}, 2I_x^{(1,3)}, 2I_x^{(2,3)}), (2I_y^{(1,2)}, 2I_y^{(1,3)}, 2I_y^{(2,3)}). \qquad (2.1.130)$$

The rotations in the latter four operator spaces are shown schematically in Fig. 2.1.9.

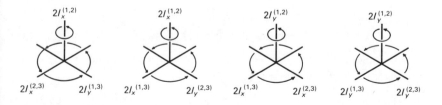

FIG. 2.1.9. Rotations in subspaces spanned by single-transition operators (eqn (2.1.130)).

2.1.9. Single-transition shift operators

Single-transition shift operators can be defined in terms of the Cartesian components or as products of kets and bras

$$I^{+(rs)} = I_x^{(rs)} + iI_y^{(rs)} = |r\rangle\langle s|,$$
$$I^{-(rs)} = I_x^{(rs)} - iI_y^{(rs)} = |s\rangle\langle r|. \qquad (2.1.131)$$

Because the y-component is reversed in sign if we permute the indices, we have formally

$$I^{-(rs)} = I^{+(sr)}. \qquad (2.1.132)$$

However, it is desirable to use ordered indices such that $M_r > M_s$, to ensure that the raising and lowering operators increase and decrease the magnetic quantum numbers

$$I^{+(rs)}|s\rangle = |r\rangle,$$
$$I^{-(rs)}|r\rangle = |s\rangle. \qquad (2.1.133)$$

The difference in magnetic quantum numbers $\Delta M_{rs} = M_r - M_s$ determines the *order of coherence* p: for $I^{+(rs)}$, $p = \Delta M_{rs}$ and for $I^{-(rs)}$, $p = -\Delta M_{rs}$.

In a system of two weakly-coupled nuclei with $I = \frac{1}{2}$, we find the following identities

$$I^{+(1,2)} = |1\rangle\langle 2| = |\alpha\alpha\rangle\langle\alpha\beta| = I_1^\alpha I_2^+,$$
$$I^{+(3,4)} = |3\rangle\langle 4| = |\beta\alpha\rangle\langle\beta\beta| = I_1^\beta I_2^+,$$
$$I^{+(1,3)} = |1\rangle\langle 3| = |\alpha\alpha\rangle\langle\beta\alpha| = I_1^+ I_2^\alpha,$$
$$I^{+(2,4)} = |2\rangle\langle 4| = |\alpha\beta\rangle\langle\beta\beta| = I_1^+ I_2^\beta,$$
$$I^{+(1,4)} = |1\rangle\langle 4| = |\alpha\alpha\rangle\langle\beta\beta| = I_1^+ I_2^+,$$
$$I^{+(2,3)} = |2\rangle\langle 3| = |\alpha\beta\rangle\langle\beta\alpha| = I_1^+ I_2^-,$$
$$I^{-(1,2)} = |2\rangle\langle 1| = |\alpha\beta\rangle\langle\alpha\alpha| = I_1^\alpha I_2^-,$$
$$I^{-(3,4)} = |4\rangle\langle 3| = |\beta\beta\rangle\langle\beta\alpha| = I_1^\beta I_2^-,$$
$$I^{-(1,3)} = |3\rangle\langle 1| = |\beta\alpha\rangle\langle\alpha\alpha| = I_1^- I_2^\alpha,$$
$$I^{-(2,4)} = |4\rangle\langle 2| = |\beta\beta\rangle\langle\alpha\beta| = I_1^- I_2^\beta,$$
$$I^{-(1,4)} = |4\rangle\langle 1| = |\beta\beta\rangle\langle\alpha\alpha| = I_1^- I_2^-,$$
$$I^{-(2,3)} = |3\rangle\langle 2| = |\beta\alpha\rangle\langle\alpha\beta| = I_1^- I_2^+. \qquad (2.1.134)$$

To complete the operator set, it is convenient to introduce polarization operators $I^{(rr)}$ defined by (see § 2.1.7)

$$I^{(rr)} = |r\rangle\langle r|. \qquad (2.1.135)$$

In the case of a weakly-coupled two-spin system, we immediately find the identities

$$I^{(1,1)} = I_1^\alpha I_2^\alpha, \qquad I^{(3,3)} = I_1^\beta I_2^\alpha,$$
$$I^{(2,2)} = I_1^\alpha I_2^\beta, \qquad I^{(4,4)} = I_1^\beta I_2^\beta. \qquad (2.1.136)$$

The operators $I^{+(rs)}$, $I^{-(rs)}$, and $I^{(rr)}$ each contain a single matrix element different from zero in the eigenbasis of the Hamiltonian. In a system with n eigenstates there are n^2 orthogonal operators which span the full Liouville operator space.

Single-transition shift operators are eigenoperators of the Hamiltonian superoperator

$$I^{\pm(rs)} \xrightarrow{\mathcal{H}\tau} I^{\pm(rs)} \exp\{\mp i\omega_{rs}\tau\} \qquad (2.1.137)$$

where $\omega_{rs} = \varepsilon_r - \varepsilon_s = \langle r|\mathcal{H}|r\rangle - \langle s|\mathcal{H}|s\rangle$. Note that if we are dealing with nuclei with $\gamma > 0$, the states with the highest magnetic quantum number have the lowest energy. Of particular importance is the transformation under z-rotations

$$I^{\pm(rs)} \xrightarrow{F_z\phi} I^{\pm(rs)} \exp\{\mp i\phi p\} \qquad (2.1.138)$$

where $F_z = \sum_k I_{kz}$ and $p = \Delta M_{rs} = M_r - M_s$ is the order of coherence.

Since single-transition operators are defined in the eigenbasis of the Hamiltonian, they are also suitable for describing strongly coupled systems. Consider the transformation U that diagonalizes the Hamiltonian and transforms the product basis $\{|\phi_i\rangle\}$ into the eigenbasis $\{|\psi_i\rangle\}$

$$|\psi_i\rangle = \sum_j |\phi_j\rangle U_{ji}. \qquad (2.1.139)$$

The corresponding transformation of the single-transition operators from the weak coupling limit into the strong coupling eigenbasis is given by

$$I^{+(rs)}_{(\psi)} = \sum_{tu} I^{+(tu)}_{(\phi)} U_{tu\,rs} \qquad (2.1.140)$$

or, equivalently,

$$|\psi_r\rangle\langle\psi_s| = \sum_{tu} |\phi_t\rangle\langle\phi_u| U_{tu\,rs} \qquad (2.1.141)$$

where the elements of the supermatrix are

$$U_{tu\,rs} = U_{tr} U_{us}^*. \qquad (2.1.142)$$

Consider by way of example a strongly coupled two-spin system, which will often be referred to when discussing characteristic features of strong coupling. The diagonalization of the Hamiltonian is achieved by the transformation

$$(\alpha\alpha, [\alpha\beta c + \beta\alpha s], [\beta\alpha c - \alpha\beta s], \beta\beta) = (\alpha\alpha, \alpha\beta, \beta\alpha, \beta\beta)\mathbf{U} \quad (2.1.143)$$

with

$$\mathbf{U} = \begin{pmatrix} 1 & 0 & 0 & 0 \\ 0 & c & -s & 0 \\ 0 & s & c & 0 \\ 0 & 0 & 0 & 1 \end{pmatrix}$$

where $c = \cos\theta$, $s = \sin\theta$, and $\theta = \frac{1}{2}\tan^{-1}\{2\pi J/(\Omega_A - \Omega_X)\}$.

In a two-spin system, the 16×16 (super)matrix $\mathbf{U}$ has four blocks of dimension 1×1 (for example $I_{(\psi)}^{+(1,4)} = I_{(\phi)}^{+(1,4)}$ remains invariant), four 2×2 blocks for the eight single-transition operators $I^{\pm(rs)}$, and one 4×4 block containing the zero-quantum coherences and the states with $M_r = 0$. The single-quantum transitions mix according to

$$\begin{pmatrix} I_{(\psi)}^{\pm(1,2)} \\ I_{(\psi)}^{\pm(1,3)} \\ I_{(\psi)}^{\pm(2,4)} \\ I_{(\psi)}^{\pm(3,4)} \end{pmatrix} = \begin{pmatrix} c & s & 0 & 0 \\ -s & c & 0 & 0 \\ 0 & 0 & c & s \\ 0 & 0 & -s & c \end{pmatrix} \begin{pmatrix} I_1^\alpha I_2^\pm \\ I_1^\pm I_2^\alpha \\ I_1^\pm I_2^\beta \\ I_1^\beta I_2^\pm \end{pmatrix} \quad (2.1.144)$$

while the zero-quantum terms mix according to

$$\begin{pmatrix} I_{(\psi)}^{+(2,3)} \\ I_{(\psi)}^{-(2,3)} \\ I_{(\psi)}^{(2,2)} \\ I_{(\psi)}^{(3,3)} \end{pmatrix} = \begin{pmatrix} c^2 & -s^2 & -sc & +sc \\ -s^2 & +c^2 & -sc & +sc \\ +cs & +cs & +c^2 & +s^2 \\ -cs & -cs & +s^2 & +c^2 \end{pmatrix} \begin{pmatrix} I_1^+ I_2^- \\ I_1^- I_2^+ \\ I_1^\alpha I_2^\beta \\ I_1^\beta I_2^\alpha \end{pmatrix} \quad (2.1.145)$$

with the polarization operators defined in eqn (2.1.135). Note that the effect of r.f. pulses can be calculated most easily in the product basis, while free evolution is expressed more conveniently in the eigenbasis. We shall see in Chapters 7 and 8 how the combination of both basis sets can lead to simple expressions for coherence transfer in strongly coupled systems.

2.1.10. Irreducible tensor operators

For the description of three-dimensional rotations it is sometimes of advantage to expand the density operator in terms of irreducible tensor operators T_{lm}. These are of particular importance for the description of isolated spins $I > \frac{1}{2}$, although it is also possible to describe coupled spins with irreducible tensor operators. The mathematical convenience arises from the fact that irreducible spherical tensors transform according to the irreducible representations of the three-dimensional rotation group. In the context of spin dynamics, they have been used extensively by Sanctuary (2.18).

A general rotation $R(\alpha, \beta, \gamma)$ expressed by the three Euler angles α, β, γ leads to the transformation

$$\hat{R}(\alpha, \beta, \gamma)T_{lm} = R(\alpha, \beta, \gamma)T_{lm}R^{-1}(\alpha, \beta, \gamma)$$
$$= \sum_{m'} T_{lm'}\mathcal{D}^l_{m'm}(\alpha, \beta, \gamma) \quad (2.1.146)$$

with the elements of the Wigner rotation matrix $\mathcal{D}^l_{m'm}(\alpha, \beta, \gamma)$ of order l (2.19–2.21). The rotation operator can be decomposed in terms of individual rotations

$$R(\alpha, \beta, \gamma) = e^{-i\alpha I_z} e^{-i\beta I_y} e^{-i\gamma I_z}. \quad (2.1.147)$$

This implies first a rotation by γ about the z-axis, followed by a rotation β about the y-axis and finally by a rotation α again about the z-axis, in agreement with the conventions of Brink and Satchler (2.21).

In schematic notation we may write

$$T_{lm} \xrightarrow{\gamma I_z} \xrightarrow{\beta I_y} \xrightarrow{\alpha I_z} \sum_{m'} T_{lm'}\mathcal{D}^{(l)}_{m'm}(\alpha, \beta, \gamma). \quad (2.1.148)$$

The rank l of the tensor operator does not change under a rotation. It denotes the irreducible representation of the rotation group which governs the transformation.

For a single spin $I_k = \frac{1}{2}$, we readily find the relations

$$\begin{aligned}
T^{(k)}_{00} &= \frac{1}{\sqrt{2}}\mathbb{1}_k & &= \frac{1}{\sqrt{2}}(I^\alpha_k + I^\beta_k), \\
T^{(k)}_{11} &= -(I_{kx} + iI_{ky}) & &= -I^+_k, \\
T^{(k)}_{10} &= \sqrt{2}\,I_{kz} & &= \frac{1}{\sqrt{2}}(I^\alpha_k - I^\beta_k), \\
T^{(k)}_{1,-1} &= (I_{kx} - iI_{ky}) & &= I^-_k.
\end{aligned}$$
$$(2.1.149)$$

Note in particular the sign in the definition of T_{11}. The tensor operators are orthogonal and have been normalized

$$\langle T^{(k)}_{lp} | T^{(k')}_{l'p'} \rangle = \delta_{ll'}\delta_{pp'}\delta_{kk'} \quad (2.1.150)$$

In systems with several spins, the tensor operators can be expressed as linear combinations of products of single-spin tensor operators. The corresponding coefficients are given by the Clebsch–Gordon coefficients $(l_1 l_2 m_1 m_2 | lm)$ (2.19–2.21).

For a two-spin system, for example, we may form the tensor operators T_{lm} from the tensor operators $T^{(1)}_{l_1 m_1}$ and $T^{(2)}_{l_2 m_2}$ of the two individual spins, using the notation of Ref. 2.21,

$$T^{(12)}_{lm} = \sum_{m_1} (l_1 l_2 m_1, m - m_1 | lm) T^{(1)}_{l_1 m_1} T^{(2)}_{l_2, m-m_1}. \quad (2.1.151)$$

The operators $T^{(12)}_{lm}$ are again irreducible tensor operators with rank l expressed now as a sum of products of the tensor operators $T^{(1)}_{l_1 m_1}$ and $T^{(2)}_{l_2 m_2}$.

An alternative formulation is in terms of the so-called 3j symbols (2.19–2.21) which have more symmetric properties

$$T^{(12)}_{lm} = (-1)^{l_2 - l_1 - m}\sqrt{(2l+1)} \sum_{m_1} \begin{pmatrix} l_1 & l_2 & l \\ m_1 & m-m_1 & -m \end{pmatrix} T^{(1)}_{l_1 m_1} T^{(2)}_{l_2 m - m_1}. \quad (2.1.152)$$

The operators obtained from eqns (2.1.151) and (2.1.152) are orthogonal but not normalized. For a two-spin system we obtain, after normalization, the following irreducible tensor operators in addition to the single-spin operators of eqn (2.1.149)

$$T^{(12)}_{00} = -\frac{2}{\sqrt{3}}(\tfrac{1}{2}I_1^+ I_2^- + \tfrac{1}{2}I_1^- I_2^+ + I_{1z}I_{2z})$$

$$T^{(12)}_{10} = \frac{1}{\sqrt{2}}(-I_1^+ I_2^- + I_1^- I_2^+),$$

$$T^{(12)}_{1\pm 1} = (-I_1^\pm I_{2z} + I_{1z}I_2^\pm),$$

$$T^{(12)}_{20} = \sqrt{\tfrac{2}{3}}(3I_{1z}I_{2z} - \mathbf{I}_1 \mathbf{I}_2),$$

$$T^{(12)}_{2\pm 1} = \mp (I_1^\pm I_{2z} + I_{1z}I_2^\pm),$$

$$T^{(12)}_{2\pm 2} = I_1^\pm I_2^\pm. \quad (2.1.153)$$

Under free precession, the rank l changes while m is conserved, whereas rotations under r.f. pulses conserve l and change m. The quantum number m corresponds to the order p of single- or multiple-quantum coherence. Although the effect of r.f. pulses is described in an elegant manner, it is more cumbersome to describe free precession under an arbitrary Hamiltonian with chemical shifts and scalar or dipolar couplings in terms of irreducible tensor operators. Free precession can be evaluated conveniently in terms of $I^{\pm(rs)}$ operators, and the density operator is transformed into the T_{lm} basis just before the pulse, to be reconverted into $I^{\pm(rs)}$ to describe subsequent evolution.

2.1.11. Coherence transfer

The concept of coherence is an extension of the notion of 'transverse magnetization'. While the latter is necessarily associated with allowed transitions $|r\rangle \leftrightarrow |s\rangle$ with $M_r - M_s = \pm 1$, the idea of coherence is more general, since it applies to any arbitrary pair of states (eqn (2.1.11)). If we consider a matrix representation of the density operator in the eigenbasis, a non-vanishing off-diagonal element σ_{rs} represents coherence between states $|r\rangle$ and $|s\rangle$.

The origin of off-diagonal density matrix elements becomes clear from eqn (2.1.11). A non-zero element σ_{rs} indicates that the state function $|\psi(t)\rangle$ in eqn (2.1.2) consists of a superposition of the eigenfunctions $|r\rangle$ and $|s\rangle$, possibly combined with additional eigenfunctions functions

$$|\psi(t)\rangle = c_r(t)|r\rangle + c_s(t)|s\rangle + \ldots . \qquad (2.1.154)$$

A coherent state is invariably a non-eigenstate of the Hamiltonian which evolves in time. The evolution is coherent insofar as the members of the molecular ensemble have the same time-dependence $c_r(t)$ and $c_s(t)$. A coherent state must clearly be distinguished from a statistical ensemble of spins in either of the two eigenstates $|r\rangle$ or $|s\rangle$ which would not lead to coherence: the off-diagonal elements in the density operator would vanish as can be inferred from eqn (2.1.6).

Magnetic resonance experiments are insensitive to higher-order coherences where more than two eigenstates are involved, and it is sufficient to consider coherence between pairs of states. The difference in magnetic quantum numbers $\Delta M_{rs} = p_{rs}$ is referred to as the *order of coherence*, which in systems with N coupled spins with spin quantum numbers I can take the values $-N(2I+1), \ldots, +N(2I+1)$. We can distinguish zero-quantum coherence ($p_{rs} = 0$), single-quantum coherence ($p_{rs} = \pm 1$), which corresponds to observable transverse magnetization or to single-quantum combination lines, and generally p-quantum coherence.

In some cases, it is convenient to consider individual coherence components $I^{+(rs)}$ and $I^{-(rs)}$ associated with two specific energy levels. In other cases, groups of related coherence components may be represented by a single operator, e.g. in-phase coherence I_{kx}, or antiphase coherence $2I_{kx}I_{lz}$, etc. (see § 4.4).

Coherence transfer describes the transformation of coherence from one transition to another. For example, a selective π-pulse applied to transition (rs) transfers the coherence from transition (st) to the transition (rt) as shown in eqn (2.1.129)

$$|\psi_{st}(t_0)\rangle = c_s(t_0)|s\rangle + c_t(t_0)|t\rangle,$$
$$\downarrow \pi_x^{(rs)}$$
$$|\psi_{rt}(t_0)\rangle = -ic_s(t_0)|r\rangle + c_t(t_0)|t\rangle. \qquad (2.1.155)$$

Coherence transfer involves the exchange of off-diagonal matrix elements in the density matrix.

Coherence can be transferred between transitions belonging to the same spin or to different spins. By suitable design of the propagator it is possible, for example, to transfer in-phase coherence from spin k to spin l, e.g.

$$I_{kx} \xrightarrow{\text{precession}} 2I_{ky}I_{lz} \xrightarrow{\text{r.f. pulse}} -2I_{kz}I_{ly} \xrightarrow{\text{precession}} I_{lx}. \quad (2.1.156)$$

We shall see later that coherence transfer is of central importance in many advanced pulse experiments. Coherence transfer is possible only between certain pairs of transitions, subject to the 'coherence transfer selection rules' that will be discussed in § 8.1.

2.2. The nuclear spin Hamiltonian

The complete Hamiltonian $\mathcal{H}$ of a molecular system is in most cases enormously complex, and the determination of exact solutions of the equations of motion for the entire quantum mechanical system is a very ambitious undertaking. The fact that magnetic resonance experiments can be described by a drastically simplified *spin* Hamiltonian $\mathcal{H}^s$ has proven to be one of the most valuable assets of magnetic resonance. There is good reason for the envy of optical spectroscopists toward magnetic resonance: The reduced Hilbert space of the spin Hamiltonian is of finite dimension and permits closed solutions for the analysis of sophisticated experiments on rather complex systems.

The nuclear spin Hamiltonian contains only nuclear spin operators and a few phenomenological constants that originate from the reduction process of the complete Hamiltonian (eqn (2.1.35)) which can, at least in principle, be deduced by quantum chemical calculations (2.23). It is not the purpose of this monograph to give an account of the physical significance of these parameters. We shall merely summarize in this section the terms which will be needed in the later discussions.

2.2.1. Interactions of nuclear spins

Nuclear spins are subject to three types of interactions.

2.2.1.1. Interactions linear in the spin operators

Linear terms in the Hamiltonian comprise the Zeeman interaction with the static field $\mathbf{B}_0$ and the interaction with the radio-frequency field $\mathbf{B}_{\text{r.f.}}(t)$. The Zeeman interaction is modified by the chemical shielding, expressed by the tensor $\boldsymbol{\sigma}_k$

$$\mathcal{H}_Z = -\sum_{k=1}^{N} \gamma_k \mathbf{I}_k (1 - \boldsymbol{\sigma}_k) \mathbf{B}_0. \qquad (2.2.1)$$

If the chemical shielding is weak, $\|\sigma_k\| \ll 1$, and if $\mathbf{B}_0$ is parallel to the z-axis, we can write the Zeeman interaction in the form

$$\mathcal{H}_Z = \sum_{k=1}^{N} \omega_{0k} I_{kz} \qquad (2.2.2a)$$

with the Larmor frequency

$$\omega_{0k} = -\gamma_k (1 - \sigma_{zz}^k(\theta, \phi)) B_0 \qquad (2.2.2b)$$

and the chemical shift

$$\sigma_{zz}^k(\theta, \phi) = \sigma_{11}^k \sin^2\theta \cos^2\phi + \sigma_{22}^k \sin^2\theta \sin^2\phi + \sigma_{33}^k \cos^2\theta \quad (2.2.2c)$$

where σ_{11}^k, σ_{22}^k, and σ_{33}^k are the principal values of the chemical shielding tensor $\boldsymbol{\sigma}_k$. The polar angle θ and the azimuthal angle ϕ describe the orientation of the magnetic field $\mathbf{B}_0$ in the principal axis system of the chemical shielding tensor. A positive frequency ω_{0k} leads to a positive rotation $(x \to y \to -x \to -y)$.

The interaction with the radio-frequency field has the same form as the Zeeman interaction

$$\mathcal{H}_{\text{r.f.}}(t) = -\sum_{k=1}^{N} \gamma_k \mathbf{I}_k \mathbf{B}_{\text{r.f.}}(t). \qquad (2.2.3)$$

The applied r.f. field with frequency $\omega_{\text{r.f.}}$ is normally linearly polarized, with a phase φ

$$\mathbf{B}_{\text{r.f.}}(t) = 2B_1 \cos \omega_{\text{r.f.}} t [\mathbf{e}_x^L \cos \varphi + \mathbf{e}_y^L \sin \varphi]. \qquad (2.2.4)$$

It is well known that a linearly oscillating r.f. field can be decomposed into two counter-rotating components, one of which can be neglected to an excellent approximation at high static field $\mathbf{B}_0$. One obtains the Hamiltonian

$$\mathcal{H}_{\text{r.f.}}(t) = -B_1 \sum_{k=1}^{N} \gamma_k \{I_{kx} \cos(\omega_{\text{r.f.}} t + \varphi) + I_{ky} \sin(\omega_{\text{r.f.}} t + \varphi)\}. \quad (2.2.5)$$

Before attempting to solve the density operator equation, eqn (2.1.34), for the Hamiltonian

$$\mathcal{H}(t) = \mathcal{H}_0 + \mathcal{H}_{\text{r.f.}}(t), \qquad (2.2.6)$$

it is advisable to render it time-independent by a transformation into a frame rotating with the radio-frequency $\omega_{\text{r.f.}}$ about the z-axis,

$$\begin{aligned}\mathcal{H}^r &= R^{-1}\mathcal{H}(t)R \\ &= \mathcal{H}_0 + \mathcal{H}_{\text{r.f.}}^r\end{aligned} \qquad (2.2.7)$$

with $R = \exp\{-i\omega_{r.f.}F_z t\}$ and

$$\mathcal{H}^r_{r.f.} = -B_1 \sum_{k=1}^{N} \gamma_k \{I_{kx} \cos \varphi + I_{ky} \sin \varphi\}. \quad (2.2.8)$$

The high-field Hamiltonian $\mathcal{H}_0$ is invariant under a z-rotation. Taking into account that both the high-field relaxation superoperator $\hat{\Gamma}$ and the equilibrium density operator σ_0 are also invariant under a z-rotation, the following differential equation is found for the transformed density operator in the rotating frame

$$\dot{\sigma}^r = -i[\mathcal{H}^r - \omega_{r.f.}F_z, \sigma^r] - \hat{\Gamma}\{\sigma^r - \sigma_0\} \quad (2.2.9)$$

with $\sigma^r = R^{-1}\sigma R$. The term $-\omega_{r.f.}F_z$ in the commutator expresses a shift of all Larmor frequencies by $-\omega_{r.f.}$ and leads to a new frequency origin at $\omega = \omega_{r.f.}$. The Larmor frequency of spin k in the rotating frame, i.e. the offset with respect to the carrier, is denoted by

$$\Omega_k = \omega_{0k} - \omega_{r.f.}. \quad (2.2.10)$$

2.2.1.2. Interactions bilinear in the spin operators

Bilinear terms in the Hamiltonian result from interactions between nuclear spin moments. It is customary to divide them into two categories.

1. *Indirect, electron-mediated interactions* which result from electron–nuclear interactions have the form

$$\mathcal{H}_J = 2\pi \sum_{k<l} \mathbf{I}_k \mathbf{J}_{kl} \mathbf{I}_l \quad (2.2.11)$$

with the indirect spin–spin coupling tensor $\mathbf{J}_{kl}$. The anisotropic part can be separated,

$$\mathbf{J}_{kl} = \mathbf{J}^{iso}_{kl} + \mathbf{J}^{aniso}_{kl} = J_{kl}\mathbb{1} + \mathbf{J}^{aniso}_{kl} \quad (2.2.12a)$$

with $\mathrm{tr}\{\mathbf{J}^{aniso}_{kl}\} = 0$ and $J_{kl} = \frac{1}{3}\mathrm{tr}\{\mathbf{J}_{kl}\}$, leading to

$$\mathcal{H}^{iso}_J = 2\pi \sum_{k<l} J_{kl}\mathbf{I}_k\mathbf{I}_l,$$

$$\mathcal{H}^{aniso}_J = 2\pi \sum_{k<l} \mathbf{I}_k\mathbf{J}^{aniso}_{kl}\mathbf{I}_l. \quad (2.2.12b)$$

It should be noted that the anisotropic part cannot easily be distinguished from direct (dipolar) contributions.

In the context of high-resolution NMR in liquid phase, we shall often encounter the spin Hamiltonian

$$\mathcal{H}_0 = \mathcal{H}_Z + \mathcal{H}_J^{\text{iso}},$$

$$\mathcal{H}_0 = -\sum_k \gamma_k(1-\sigma_k^{\text{iso}})B_0 I_{kz} + \sum_{k<l} 2\pi J_{kl}\mathbf{I}_k\mathbf{I}_l$$

$$= \sum_k \omega_{0k} I_{kz} + \sum_{k<l} 2\pi J_{kl}\mathbf{I}_k\mathbf{I}_l. \quad (2.2.13)$$

For weak coupling, i.e. if $2\pi|J_{kl}| \ll |\omega_{0k}-\omega_{0l}|$, one may retain only the secular components of the scalar coupling Hamiltonian

$$\mathcal{H}_0 = \sum_k \omega_{0k} I_{kz} + \sum_{k<l} 2\pi J_{kl} I_{kz} I_{lz}. \quad (2.2.14)$$

2. *Direct dipolar interactions* between the nuclear magnetic moments provide structural information. The contributions to the Hamiltonian have the form

$$\mathcal{H}_D = \sum_{k<l} \mathbf{I}_k \mathbf{D}_{kl} \mathbf{I}_l \quad (2.2.15)$$

or, explicitly,

$$\mathcal{H}_D = \sum_{k<l} b_{kl}\left\{ \mathbf{I}_k \mathbf{I}_l - 3\frac{1}{r_{kl}^2}(\mathbf{I}_k \mathbf{r}_{kl})(\mathbf{I}_l \mathbf{r}_{kl})\right\} \quad (2.2.16)$$

with $b_{kl} = \mu_0\gamma_k\gamma_l\hbar/(4\pi r_{kl}^3)$ in SI units. The internuclear unit vector $\mathbf{r}_{kl}/|\mathbf{r}_{kl}|$ can be expressed in polar coordinates θ_{kl}, ϕ_{kl} to obtain a representation of the dipolar Hamiltonian in terms of irreducible tensor operators (2.27)

$$\mathcal{H}_D = \sum_{k<l}\sum_{q=-2}^{2} F_{kl}^{(q)} A_{kl}^{(q)}. \quad (2.2.17)$$

The functions $F_{kl}^{(q)}$ describe the orientation and $A_{kl}^{(q)}$ contain the spin operators

$$A_{kl}^{(0)} = b_{kl}\{I_{kz}I_{lz} - \tfrac{1}{4}(I_k^+ I_l^- + I_k^- I_l^+)\}, \quad F_{kl}^{(0)} = 1 - 3\cos^2\theta_{kl},$$
$$A_{kl}^{(1)} = -\tfrac{3}{2}b_{kl}(I_{kz}I_l^+ + I_k^+ I_{lz}), \quad F_{kl}^{(1)} = \sin\theta_{kl}\cos\theta_{kl}\exp\{-i\phi_{kl}\},$$
$$A_{kl}^{(-1)} = -\tfrac{3}{2}b_{kl}(I_{kz}I_l^- + I_k^- I_{lz}), \quad F_{kl}^{(-1)} = \sin\theta_{kl}\cos\theta_{kl}\exp\{+i\phi_{kl}\},$$
$$A_{kl}^{(2)} = -\tfrac{3}{4}b_{kl}I_k^+ I_l^+, \quad F_{kl}^{(2)} = \sin^2\theta_{kl}\exp\{-2i\phi_{kl}\},$$
$$A_{kl}^{(-2)} = -\tfrac{3}{4}b_{kl}I_k^- I_l^-, \quad F_{kl}^{(-2)} = \sin^2\theta_{kl}\exp\{+2i\phi_{kl}\}$$

where θ_{kl} is the angle between the magnetic field $\mathbf{B}_0$ and the internuclear vector $\mathbf{r}_{kl}$, and ϕ_{kl} is the azimuthal angle with respect to the x-axis. Comparing these terms with the conventional 'dipolar alphabet' (2.5,

2.25, 2.27), we can make the identifications

$$A + B = \frac{1}{b_{kl}} F_{kl}^{(0)} A_{kl}^{(0)},$$

$$C = \frac{1}{b_{kl}} F_{kl}^{(1)} A_{kl}^{(1)},$$

$$D = \frac{1}{b_{kl}} F_{kl}^{(-1)} A_{kl}^{(-1)},$$

$$E = \frac{1}{b_{kl}} F_{kl}^{(2)} A_{kl}^{(2)},$$

$$F = \frac{1}{b_{kl}} F_{kl}^{(-2)} A_{kl}^{(-2)}.$$

In the high-field approximation, it is normally possible to neglect non-secular contributions and retain only the term with $q = 0$

$$\mathcal{H}_D^{trunc} = \sum_{k<l} b_{kl} \tfrac{1}{2}(1 - 3\cos^2\theta_{kl})[3I_{kz}I_{lz} - \mathbf{I}_k\mathbf{I}_l]. \qquad (2.2.18)$$

In heteronuclear spin systems (e.g. I_k = proton, S_l = carbon-13) further simplification can be obtained due to weak coupling by dropping all terms that involve transverse spin operators,

$$\mathcal{H}_D^{IS} = b_{kl}(1 - 3\cos^2\theta_{kl})I_{kz}S_{lz}. \qquad (2.2.19)$$

2.2.1.3. Interactions quadratic in the spin operators

Quadratic terms arise from the electric quadrupole interaction and may be interpreted as nuclear interactions with electric field gradients. The contributions to the Hamiltonian (which vanish for $I_k = \tfrac{1}{2}$ nuclei) have the form

$$\mathcal{H}_Q = \sum_{k=1}^{N} \mathbf{I}_k \mathbf{Q}_k \mathbf{I}_k, \qquad (2.2.20)$$

with the quadrupole coupling tensor $\mathbf{Q}_k$, which may be expressed in terms of the electric field gradient tensor $\mathbf{V}_k$ at the site of nucleus k

$$\mathbf{Q}_k = \frac{eQ_k}{2I_k(2I_k - 1)\hbar} \mathbf{V}_k. \qquad (2.2.21)$$

Q_k is the nuclear quadrupole moment of nucleus k.

In terms of the quadrupolar frequency ω_{Qk},

$$\omega_{Qk} = \frac{3e^2 q_k Q_k}{4I_k(2I_k - 1)\hbar}, \qquad eq_k = V_{kzz} \qquad (2.2.22)$$

and the asymmetry parameter η_k,

$$\eta_k = (V_{kxx} - V_{kyy})/V_{kzz}, \tag{2.2.23}$$

the quadrupolar Hamiltonian of nucleus k can be written in its principal-axis coordinate system in the convenient form

$$\mathcal{H}_{Qk} = \omega_{Qk}\left\{(I_{kz}^2 - \tfrac{1}{3}\mathbf{I}_k^2) + \frac{\eta}{3}(I_{kx}^2 - I_{ky}^2)\right\}. \tag{2.2.24}$$

The principal axis values of the electric field gradient tensor $\mathbf{V}_k$ are arranged in the order $|V_{kxx}| \leq |V_{kyy}| \leq |V_{kzz}|$. Then, the asymmetry parameter is within the range $0 \leq \eta \leq 1$. Occasionally, different definitions of η are used (e.g. in ref. 2.24).

Note that, in general, different nuclei in a molecule or crystal will have different principal-axis systems. For a more detailed discussion of the various terms of the Hamiltonian in terms of irreducible tensor operators we refer to Refs. 2.24–2.26.

2.3. The relaxation superoperator

A detailed discussion of relaxation phenomena is beyond the scope of this volume. We shall limit this chapter to a brief survey, and refer the reader to the treatises of Abragam (2.27), Redfield (2.28), and Wolf (2.29), as well as to the original papers by Wangsness and Bloch (2.30), Hubbard (2.31), Redfield (2.32), and Argyres and Kelley (2.33), for a more complete discussion.

Relaxation in NMR can be described at four increasingly fundamental levels of physical significance.

1. *Phenomenological Bloch equations.* The longitudinal and transverse relaxation times T_1 and T_2 have been introduced (2.34) on purely phenomenological grounds, leading to the Bloch equations for the magnetization vector $\mathbf{M}(t)$ which are conveniently written in vector notation

$$\frac{d}{dt}\mathbf{M}(t) = \gamma \mathbf{M}(t) \times \mathbf{B}(t) - \mathbf{R}\{\mathbf{M}(t) - \mathbf{M}_0\}. \tag{2.3.1}$$

The relaxation matrix $\mathbf{R}$ has the form

$$\mathbf{R} = \begin{pmatrix} 1/T_2 & 0 & 0 \\ 0 & 1/T_2 & 0 \\ 0 & 0 & 1/T_1 \end{pmatrix}. \tag{2.3.2}$$

The validity of this simple form of a relaxation (super)operator is restricted to systems without spin coupling.

2. *Transition probabilities.* Longitudinal spin–lattice relaxation rates can be computed easily by second-order perturbation theory in terms of transition probabilities $W_{\alpha\beta}$ between the energy levels α and β in spin

systems with n eigenstates (2.5, 2.27). The transition probabilities can be combined into an $n \times n$ relaxation matrix $\mathbf{W}$ which describes the evolution of the populations P_α collected in a vector $\mathbf{P}$

$$\frac{d}{dt}\mathbf{P}(t) = \mathbf{W}\{\mathbf{P}(t) - \mathbf{P}_0\}. \tag{2.3.3}$$

This is the master equation of populations. The transition probabilities $W_{\alpha\beta}$ are given by

$$W_{\alpha\beta} = J_{\alpha\beta\alpha\beta}(\omega_{\alpha\beta}) \quad \text{for} \quad \alpha \neq \beta,$$
$$W_{\alpha\alpha} = -\sum_{\beta \neq \alpha} W_{\alpha\beta} \tag{2.3.4}$$

with the power spectral density $J_{\alpha\beta\alpha\beta}(\omega_{\alpha\beta})$ at the transition frequency $\omega_{\alpha\beta} = \langle \alpha | \mathcal{H}_0 | \alpha \rangle - \langle \beta | \mathcal{H}_0 | \beta \rangle$

$$J_{\alpha\beta\alpha\beta}(\omega) = \int_{-\infty}^{\infty} d\tau \, e^{-i\omega\tau} \overline{\mathcal{H}_1(t)_{\alpha\beta} \mathcal{H}_1^*(t-\tau)_{\alpha\beta}} \tag{2.3.5}$$

where $\mathcal{H}_1(t)$ is the Hamiltonian of the random perturbations due to the thermal lattice motions.

Transverse relaxation cannot be described by this approach. It requires a more fundamental density operator treatment.

3. *Semi-classical relaxation theory.* The semi-classical relaxation theory is, with regard to applications, the most useful approach. The evolution of the spin system is described quantum mechanically by a density operator, while the influence of the surroundings is represented by fluctuating random processes (2.27, 2.28, 2.31, 2.32). The results of this treatment are summarized in § 2.3.1.

4. *Quantum-mechanical relaxation theory.* On the most fundamental level of relaxation theory, both spin system and surroundings are described quantum mechanically. Starting with the Liouville–von Neumann equation for the complete density operator $\rho(t)$, a master equation is derived for the reduced density operator $\sigma(t)$ which describes the spin subsystem. This treatment can lead to a detailed understanding of the origins of the random perturbations acting on the spin system (2.27, 2.30, 2.33).

2.3.1. Semi-classical relaxation theory

Relaxation is induced by the influence of the lattice on the spin system. The interaction Hamiltonian can be written in the form

$$\mathcal{H}_1(t) = \sum_q F^{(q)}(t) A^{(q)} \tag{2.3.6}$$

where $A^{(q)}$ are operators acting on the spin system only, and $F^{(q)}(t)$ are random processes representing the lattice dynamics with $A^{(-q)} = A^{(q)\dagger}$ and $F^{(-q)}(t) = F^{(q)*}(t)$. The index q is used here at the same time to distinguish irreducible tensor components as well as different types of interactions, in contrast to eqn (2.2.17). While in quantum mechanical relaxation theory the terms $F^{(q)}(t)$ are lattice operators, these quantities are taken as classical random processes in the semi-classical theory. By second-order perturbation theory in the interaction representation, it is possible to derive a quantum-mechanical master equation (2.27)

$$\dot{\sigma}^T(t) = -\int_0^\infty d\tau \overline{[\mathcal{H}_1^T(t), [\mathcal{H}_1^T(t-\tau), \sigma^T(t) - \sigma_0]]} \qquad (2.3.7)$$

with

$$\mathcal{H}_1^T(t) = \exp\{i\mathcal{H}_0 t\}\mathcal{H}_1(t)\exp\{-i\mathcal{H}_0 t\} \qquad (2.3.8)$$

where the superscript T indicates the interaction representation. This treatment is valid provided $\tau_c \ll t, T_1, T_2$, where τ_c is the correlation time of the random processes. The random processes $F^{(q)}(t)$ are characterized by the correlation functions

$$g^{(q,q')}(\tau) = \overline{F^{(q)}(t)F^{(q')*}(t+\tau)}. \qquad (2.3.9)$$

The time-dependence of $\mathcal{H}_1^T(t)$ is most easily expressed by expanding the operators $A^{(q)}$ into eigenoperators $A_p^{(q)}$ of the Hamiltonian superoperator $\mathcal{H}_0$,

$$A^{(q)T}(t) = \exp\{i\mathcal{H}_0 t\}A^{(q)}\exp\{-i\mathcal{H}_0 t\} = \sum_p A_p^{(q)}\exp\{i\omega_p^{(q)}t\} \qquad (2.3.10)$$

where $\omega_p^{(q)}$ are differences of eigenvalues of $\mathcal{H}_0$. The time evolution of $\sigma^T(t)$ may now be written

$$\dot{\sigma}^T(t) = -\sum_{q,q'}\sum_{p,p'} \exp\{i(\omega_p^{(q)} + \omega_{p'}^{(q')})t\}[A_{p'}^{(q')}, [A_p^{(q)}, \sigma^T(t) - \sigma_0]]$$
$$\times \int_0^\infty g^{(q,-q')}(\tau)\exp\{-i\omega_p^{(q)}\tau\}\,d\tau. \qquad (2.3.11)$$

The imaginary part of the integral gives rise to small second-order shifts of the lines, which may be neglected or included in a modified Hamiltonian $\mathcal{H}_0$ (2.5, 2.27). The real part equals one-half of the spectral density

$$J^{(q,-q')}(\omega) = \int_{-\infty}^\infty d\tau\, g^{(q,-q')}(\tau)\exp\{-i\omega\tau\}. \qquad (2.3.12)$$

This leads to a general, explicit form of the relaxation superoperator

$$\dot{\sigma}^T(t) = -\tfrac{1}{2}\sum_{q,q'}\sum_{p,p'} J^{(q,-q')}(\omega_p^{(q)})\exp\{i(\omega_p^{(q)} + \omega_{p'}^{(q')})t\}$$
$$\times [A_{p'}^{(q')}, [A_p^{(q)}, \sigma^T(t) - \sigma_0]]. \qquad (2.3.13)$$

A transformation into the laboratory frame of eqn (2.3.13) leads to the result

$$\dot{\sigma}(t) = -i[\mathcal{H}_0, \sigma(t)] - \tfrac{1}{2} \sum_{q,q'} \sum_{p,p'} J^{(q,-q')}(\omega_p^{(q)})$$
$$\times [A_p^{(q')}, [A_p^{(q)}, \sigma(t) - \sigma_0]]. \quad (2.3.13a)$$

It should be noted that in this general expression the relaxation superoperator contains all non-secular contributions, many of which are usually ineffective.

In many cases, the random processes $F^{(q)}(t)$ are statistically independent if the Hamiltonian is properly partitioned, i.e.

$$g^{(q,-q')}(\tau) = \delta_{q,-q'} g^{(q)}(\tau), \quad (2.3.14a)$$
$$J^{(q,-q')}(\omega) = \delta_{q,-q'} J^{(q)}(\omega). \quad (2.3.14b)$$

This allows the elimination of one summation in eqn (2.3.13). Further simplification can be achieved in two different ways by retaining only secular terms as shown in the next two sections.

2.3.1.1. Restriction to secular contributions

The non-secular terms, for which $\omega_p^{(q)} + \omega_{p'}^{(-q)} \neq 0$, do not influence the long-term development of $\sigma^T(t)$ due to rapidly oscillating factors $\exp\{i(\omega_p^{(q)} + \omega_{p'}^{(-q)})t\}$ in eqn (2.3.13). In the absence of degenerate transition frequencies, the condition $\omega_p^{(q)} + \omega_{p'}^{(-q)} = 0$ can be fulfilled only for $p = p'$ and eqn (2.3.13) simplifies to

$$\dot{\sigma}^T(t) = -\tfrac{1}{2} \sum_q \sum_p J^{(q)}(\omega_p^{(q)})[A_p^{(-q)}, [A_p^{(q)}, \sigma^T(t) - \sigma_0]]. \quad (2.3.15)$$

The relaxation superoperator expressed by eqn (2.3.15) is invariant under the transformation to the laboratory frame. We obtain the important result for the density operator

$$\dot{\sigma}(t) = -i[\mathcal{H}_0, \sigma(t)] - \hat{\Gamma}\{\sigma(t) - \sigma_0\} \quad (2.3.16)$$

with the relaxation superoperator

$$\hat{\Gamma}\{\sigma\} = \tfrac{1}{2} \sum_q \sum_p J^{(q)}(\omega_p^{(q)})[A_p^{(-q)}, [A_p^{(q)}, \sigma]]. \quad (2.3.17)$$

The power spectral densities $J_q(\omega_p^{(q)})$ at the (allowed and forbidden) transition frequencies are the coefficients of double commutators involving the corresponding component operators of the interaction Hamiltonian $\mathcal{H}_1(t)$.

The density operator equation, eqn (2.3.16), forms the basis for the quantum mechanical treatment of one- and two-dimensional Fourier spectroscopy.

2.3.1.2. Extreme narrowing

Extreme narrowing characterizes situations with extremely short correlation times $\omega_p^{(q)}\tau_c \ll 1$ for all relevant transition frequencies $\omega_p^{(q)}$. The random processes take place in a time much shorter than the nuclear precession periods. In this case, the power spectral densities are independent of frequency over the range of interest

$$J^{(q)}(\omega_p^{(q)}) = J^{(q)}(0) \tag{2.3.18}$$

and we obtain from eqn (2.3.13) the relaxation superoperator

$$\hat{\Gamma}\{\sigma\} = \tfrac{1}{2} \sum_q J^{(q)}(0)[A^{(-q)}, [A^{(q)}, \sigma]]. \tag{2.3.19}$$

Each term in this sum corresponds to a term in eqn (2.3.6).

Finally, if all correlation functions $g^{(q)}(\tau)$ are exponential with the same correlation time τ_c, the relaxation superoperator takes the simple form

$$\hat{\Gamma}\{\sigma\} = \tau_c \overline{[\mathcal{H}_1(t), [\mathcal{H}_1(t), \sigma]]}. \tag{2.3.20}$$

2.3.2. Matrix representation of the relaxation superoperator

In the eigenbasis of the Hamiltonian $\mathcal{H}_0$, the matrix representation of the master equation, eqn (2.3.7), may be written

$$\dot{\sigma}^T_{\alpha\alpha'}(t) = \sum_{\beta\beta'} \exp\{i(\omega_{\alpha\alpha'} - \omega_{\beta\beta'})t\} R_{\alpha\alpha'\beta\beta'}(\sigma^T_{\beta\beta'}(t) - \sigma_{0\beta\beta'}). \tag{2.3.21}$$

The terms $R_{\alpha\alpha'\beta\beta'}$ are elements of the Redfield relaxation supermatrix (2.28). These elements can be conveniently expressed in terms of spectral density functions that correspond to pairs of matrix elements of the Hamiltonian $\mathcal{H}_1(t)$

$$J_{\kappa\lambda\mu\nu}(\omega) = \int_{-\infty}^{\infty} d\tau \exp\{-i\omega\tau\} \overline{\mathcal{H}_1(t)_{\kappa\lambda} \mathcal{H}_1^*(t+\tau)_{\mu\nu}}. \tag{2.3.22}$$

These functions can be related to the spectral densities defined in eqn (2.3.12)

$$J_{\kappa\lambda\mu\nu}(\omega) = \sum_{qq'} J^{(q,q')}(\omega) A^{(q)}_{\kappa\lambda} A^{(q')*}_{\mu\nu}. \tag{2.3.23}$$

The Redfield matrix elements $R_{\alpha\alpha'\beta\beta'}$ in eqn (2.3.21) may be written (2.28)

$$R_{\alpha\alpha'\beta\beta'} = \frac{1}{2} \Big\{ J_{\alpha\beta\alpha'\beta'}(\omega_{\beta'\alpha'}) + J_{\alpha\beta\alpha'\beta'}(\omega_{\alpha\beta}) - \delta_{\alpha'\beta'} \sum_\gamma J_{\gamma\beta\gamma\alpha}(\omega_{\gamma\alpha})$$
$$- \delta_{\alpha\beta} \sum_\gamma J_{\gamma\alpha'\gamma\beta'}(\omega_{\alpha'\gamma}) \Big\}. \tag{2.3.24}$$

This general expression is considerably simplified if all non-secular elements $R_{\alpha\alpha'\beta\beta'}$ with $\omega_{\beta'\beta} - \omega_{\alpha'\alpha} \neq 0$ are neglected. In this case, eqn (2.3.21) is time-independent. After transformation into the laboratory frame, one finds

$$\dot{\sigma}_{\alpha\alpha'}(t) = -i\omega_{\alpha\alpha'}\sigma_{\alpha\alpha'}(t) + \sum_{\beta\beta'} R_{\alpha\alpha'\beta\beta'}(\sigma_{\beta\beta'}(t) - \sigma_{0\beta\beta'}). \tag{2.3.25}$$

This equation corresponds to a matrix representation of eqn (2.3.16).

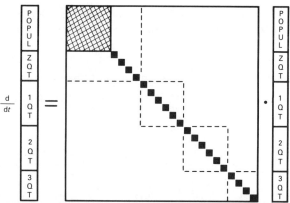

FIG. 2.3.1. Matrix representation of the relaxation superoperator. The Redfield matrix **R** has a block structure due to the neglect of non-secular contributions. The first block couples the populations and off-diagonal elements with $\Delta M = 0$ (ZQTs). Each of the following blocks couples one order ΔM of off-diagonal elements (single-quantum transitions (1QTs), double-quantum (2QTs) and higher-order transitions) among themselves. In the absence of degenerate transitions, there is no transverse cross-relaxation, and all off-diagonal elements relax independently. In this case, the Redfield matrix consists of one block coupling the populations, and a tail of diagonal elements (so-called 'Redfield kite').

The neglect of non-secular terms leads to a characteristic block structure of the Redfield matrix, as shown schematically in Fig. 2.3.1. The condition $\omega_{\alpha'\alpha} = \omega_{\beta'\beta}$ can only be fulfilled for transitions with the same order of coherence $p = \Delta M_{\alpha'\alpha} = \Delta M_{\beta'\beta}$. This implies that there can be no cross-relaxation between elements of different order.

In the absence of degenerate transitions, the structure of the relaxation matrix is further simplified, since each off-diagonal element $\sigma_{\alpha\alpha'}$ with $\alpha \neq \alpha'$ decays according to a single exponential with rate constant $R_{\alpha\alpha'\alpha\alpha'} = -(T_{2\alpha\alpha'})^{-1}$

$$\dot{\sigma}_{\alpha\alpha'}(t) = -i\omega_{\alpha\alpha'}\sigma_{\alpha\alpha'}(t) + R_{\alpha\alpha'\alpha\alpha'}\sigma_{\alpha\alpha'}(t) \quad (2.3.26a)$$

with the solution

$$\sigma_{\alpha\alpha'}(t) = \sigma_{\alpha\alpha'}(0)\exp\{-i\omega_{\alpha\alpha'}t\}\exp\{-t/T_{2\alpha\alpha'}\}. \quad (2.3.26b)$$

The transverse relaxation rate constant can be decomposed into two contributions of different physical origin (2.27)

$$T_{2\alpha\alpha'}^{-1} = (T_{2\alpha\alpha'}^{a})^{-1} + (T_{2\alpha\alpha'}^{na})^{-1} \quad (2.3.27)$$

with the adiabatic relaxation rate (often called secular contribution (2.5))

$$(T_{2\alpha\alpha'}^{a})^{-1} = \tfrac{1}{2}\{J_{\alpha\alpha\alpha\alpha}(0) - 2J_{\alpha\alpha\alpha'\alpha'}(0) + J_{\alpha'\alpha'\alpha'\alpha'}(0)\}$$

$$= \frac{1}{2}\int_{-\infty}^{\infty} d\tau \overline{[\mathcal{H}_1(t)_{\alpha\alpha} - \mathcal{H}_1(t)_{\alpha'\alpha'}][\mathcal{H}_1(t-\tau)_{\alpha\alpha} - \mathcal{H}_1(t-\tau)_{\alpha'\alpha'}]}$$

$$(2.3.28)$$

and the non-adiabatic relaxation rate (also called non-secular or lifetime contribution (2.5))

$$(T_{2\alpha\alpha'}^{na})^{-1} = \frac{1}{2}\left\{\sum_{\gamma\neq\alpha} J_{\alpha\gamma\alpha\gamma}(\omega_{\alpha\gamma}) + \sum_{\gamma\neq\alpha'} J_{\alpha'\gamma\alpha'\gamma}(\omega_{\gamma\alpha'})\right\}$$
$$= \frac{1}{2}\left\{\sum_{\gamma\neq\alpha} W_{\alpha\gamma} + \sum_{\gamma\neq\alpha'} W_{\alpha'\gamma}\right\} \quad (2.3.29)$$

with the transition probabilities given in eqn (2.3.4). The adiabatic relaxation rate is caused by fluctuations of the energy difference of states $|\alpha\rangle$ and $|\alpha'\rangle$ due to the random perturbation. It does not involve transitions and is caused by perturbations which commute with the Hamiltonian $\mathcal{H}_0$. The non-adiabatic relaxation rate, on the other hand, is a consequence of the finite lifetime of the states $|\alpha\rangle$ and $|\alpha'\rangle$.

The longitudinal relaxation is governed by a set of coupled linear differential equations. For $\alpha = \alpha'$ and $\beta = \beta'$, eqn (2.3.25) takes the form

$$\dot{\sigma}_{\alpha\alpha}(t) = +\sum_{\beta} R_{\alpha\alpha\beta\beta}(\sigma_{\beta\beta}(t) - \sigma_{0\beta\beta}). \quad (2.3.30)$$

Substituting $\sigma_{\alpha\alpha} = P_\alpha$ and $R_{\alpha\alpha\beta\beta} = W_{\alpha\beta}$, we obtain the master equation (2.3.3). The time-dependence of each population is given by a linear combination of exponential functions $\exp\{\lambda_i t\}$, where the λ_i are eigenvalues of **W**.

2.3.3. Specific relaxation mechanisms

Within the framework of a semi-classical description of relaxation, we may distinguish two principal classes of random Hamiltonians $\mathcal{H}_1(t)$: interactions which are linear and those which are bilinear in the operators of the spin system.

2.3.3.1. First-rank relaxation mechanisms

The relaxation Hamiltonian will be linear in the spin operators whenever it describes an interaction with magnetic fields which stem from sources that are external to the spin system. Examples are:

1. *Zeeman interaction.* Molecular tumbling modulates the Larmor frequency through the anisotropy of the chemical shielding.
2. *Spin–rotation interaction.* The magnetic field created by a rotating molecule is modulated by molecular reorientations caused by collisions.
3. *Dipolar interaction with 'external' spins.* The interaction with electronic or nuclear spins of solvent molecules is modulated by translation and molecular rotation.

In the semi-classical approach, all these interactions can be represented in the form

$$\mathcal{H}_1(t) = \sum_{k=1}^{N} \sum_{q=-1}^{1} F_k^{(q)}(t) I_k^{(q)} \qquad (2.3.31)$$

where the first summation runs over the nuclei k of the considered spin sytem. The terms $I_k^{(q)}$ are first-rank irreducible tensor operators of nucleus k (I_k^-, I_{kz}, and I_k^+) and $F_k^{(q)}(t)$ can be identified with spherical components of fluctuating random fields ($B_k^-(t)$, $B_{kz}(t)$, $B_k^+(t)$), acting on the nucleus k

$$\mathcal{H}_1(t) = -\sum_{k=1}^{N} \sum_{q=-1}^{1} \gamma_k B_k^{(q)}(t) I_k^{(q)}. \qquad (2.3.32)$$

This identification leads to the *random-field model*. Correlations between the fields acting on nuclei k and l are accounted for by correlation coefficients

$$C_{kl}^{(q,q')} = \frac{\overline{B_k^{(q)} B_l^{(q')}}}{[\overline{|B_k^{(q)}|^2}\, \overline{|B_l^{(q')}|^2}]^{\frac{1}{2}}}. \qquad (2.3.33)$$

Explicit results for external random-field relaxation for a weakly coupled two-spin system are given in Chapter 9: the elements W_{ij} of the matrix of transition probabilities **W** of eqn (2.3.3) consist of the terms W_{1A}^{RF} and W_{1B}^{RF} describing single-quantum transitions of spins A and B induced by the random field. The relaxation rates $1/T_2$ of zero, single, and double quantum coherence are given in eqn (9.4.10). Results for strong coupling are given in Ref. 2.69.

2.3.3.2. Second-rank relaxation mechanisms

The two most important interactions that are bilinear in the relevant spin operators are the intramolecular dipolar and the quadrupolar interactions.

1. *Intramolecular dipolar relaxation.* The dipolar Hamiltonian of eqn (2.2.17) fluctuates because of the random molecular rotation which modulates the functions $F_{kl}^{(q)}(t)$. In the case of an isotropic random motion with the correlation time τ_c, we find the corresponding power spectral densities (eqn (2.3.12) with $q = -q'$)

$$J_{kl}^{(0)}(\omega) = \tfrac{12}{15} J_{kl}(\omega),$$
$$J_{kl}^{(1)}(\omega) = \tfrac{2}{15} J_{kl}(\omega),$$
$$J_{kl}^{(2)}(\omega) = \tfrac{8}{15} J_{kl}(\omega),$$

where

$$J_{kl}(\omega) = \frac{2\tau_c^{kl}}{1 + \omega^2 (\tau_c^{kl})^2}. \qquad (2.3.34)$$

2.4 SPIN DYNAMICS DUE TO CHEMICAL REACTIONS

The evaluation of eqn (2.3.24) for a weakly coupled two-spin system leads to the transition probabilities given in eqn (9.7.4) and to the transverse relaxation rates given in eqn (9.4.7). Related expressions for a strongly coupled system are given in Ref. 2.69.

2. *Quadrupolar relaxation.* The quadrupolar relaxation rates can be derived similarly by expanding the quadrupolar Hamiltonian in irreducible tensor operators (2.26).

2.4. Spin dynamics due to chemical reactions

Chemical reactions and exchange processes bear a strong resemblance to relaxation phenomena. They are also governed by irreversible random processes. Chemical processes, like internal rotations in molecules, bond shifts, valence isomerization, chemical exchange, and chemical reactions of arbitrary complexity, can lead to exchange of nuclei between non-equivalent electronic surroundings and cause characteristic effects in the magnetic resonance spectrum.

Exchanging systems in *dynamic equilibrium* have been investigated since the early days of NMR, starting with the famous work of Gutowsky, McCall, and Slichter (2.35). In fact, it is one of the merits of NMR that kinetic information may be obtained from studies in chemical equilibrium. Much of the present knowledge on chemical exchange has been obtained by magnetic resonance studies. The extensive literature on this subject has been summarized in numerous review articles and specialized chapters in more general treatises on NMR (2.26, 2.36–2.44). The effects of 'exchange-type' reactions on the NMR spectrum are well known. They range from line-broadening via coalescence to exchange-narrowed spectra with averaged spectral parameters.

Besides equilibrium studies, it is also possible to investigate *transient* chemical reactions by NMR. Here, the system is initially brought into a chemical non-equilibrium state, and subsequently its approach to equilibrium is observed as a function of time. The non-equilibrium state can be created by stopped flow (2.45–2.52), by optically induced photoreactions in connection with chemically induced dynamic nuclear polarization (2.53–2.56), or by a sudden change of a parameter affecting the chemical equilibrium. The special merit of Fourier spectroscopy as a measuring technique for transient phenomena is obvious (2.57).

In this section we shall present a formalism sufficiently general to describe both equilibrium *and* transient exchange effects in magnetic resonance. Most earlier treatises of chemical exchange have been confined to equilibrium processes. We will stress here not so much the more traditional aspects of chemical exchange but rather emphasize the changes necessary to cover non-stationary situations and higher-order chemical reactions. At first, in § 2.4.1, the matrix formalism of classical kinetics, useful for the description of higher order reactions, is reviewed. Next, the modified Bloch equations are discussed in § 2.4.2 for the case of transient and equilibrium chemical reactions of first and higher order. Finally, a general density operator formalism is developed for the description of complex spin systems involving transient chemical reactions of arbitrary order (§ 2.4.3).

2.4.1. Description of reaction networks in classical kinetics

We assume that J molecular species A_j are participating in L two-sided reactions. Forward and backward reactions are considered as separate reactions because their rates are different in chemical non-equilibrium. We label by $l = 1, 2, \ldots, L$ the forward reactions, and by $l = L+1, L+2, \ldots, 2L$ the corresponding backward reactions. The stoichiometry of the reaction system can then be formulated in the form of a system of $2L$ linear equations (2.57–2.60)

$$\tilde{\mathbf{A}}\mathbf{N} = 0 \tag{2.4.1}$$

where $\tilde{\mathbf{A}}$ is the row vector of the J particles A_j, and $\mathbf{N}$ represents the rectangular *stoichiometric matrix of dimension* $J \times 2L$. The element v_{jl} of the matrix $\mathbf{N}$ is the stoichiometric coefficient of A_j in the reaction l. In explicit notation, eqn (2.4.1) represents a system of $2L$ chemical reactions

$$\begin{aligned}
v_{11}A_1 + v_{21}A_2 + \ldots v_{J1}A_J &= 0 \\
v_{12}A_1 + v_{22}A_2 + \ldots v_{J2}A_J &= 0 \\
&\vdots \\
v_{1,2L}A_1 + v_{2,2L}A_2 + \ldots v_{J,2L}A_J &= 0.
\end{aligned} \tag{2.4.2}$$

As an example, consider the fast keto-enol tautomerization of acetylacetone (2,4-propanedione) in the presence of diethylamine, investigated with NMR techniques by Reeves and Schneider (2.61, 2.62). The relevant particles are $A_1 = CH_3COCH_2COCH_3$ (acetylacetone keto-form), $A_2 = CH_3C(OH)CHCOCH_3$ (acetylacetone enol-form), $A_3 = CH_3CO\overset{\ominus}{C}HCOCH_3$ (enolate), $A_4 = (CH_3CH_2)_2NH$ (diethylamine), $A_5 = (CH_3CH_2)_2\overset{\oplus}{N}H_2$ (diethylammonium ion). The following reactions have to be taken into account

$$A_1 \underset{4}{\overset{1}{\rightleftharpoons}} A_2,$$

$$A_2 + A_4 \underset{5}{\overset{2}{\rightleftharpoons}} A_3 + A_5,$$

$$A_3 + A_5 \underset{6}{\overset{3}{\rightleftharpoons}} A_1 + A_4. \tag{2.4.3}$$

2.4 SPIN DYNAMICS DUE TO CHEMICAL REACTIONS

Labelling the forward reactions ($\rightarrow$) by $l = 1\ldots 3$, and the backward reactions ($\leftarrow$) by $l = 4\ldots 6$, we obtain the system of stoichiometric equations

$$(A_1 A_2 A_3 A_4 A_5) \cdot \begin{pmatrix} -1 & 0 & 1 & 1 & 0 & -1 \\ 1 & -1 & 0 & -1 & 1 & 0 \\ 0 & 1 & -1 & 0 & -1 & 1 \\ 0 & -1 & 1 & 0 & 1 & -1 \\ 0 & 1 & -1 & 0 & -1 & 1 \end{pmatrix} = 0 \quad (2.4.4)$$

with a stoichiometric matrix $\mathbf{N}$ of dimension 5×6.

The extent to which a reaction l has proceeded can be expressed by the *reaction number* $\xi_l(t)$ of reaction l, measured in mol/litre, which is the number of formula weights per unit volume which have reacted at time t (2.60). With the vector $\xi(t)$ comprising the $2L$ reaction numbers, we can represent the time-dependent concentration vector $[\mathbf{A}](t)$ in the form

$$[\mathbf{A}](t) = [\mathbf{A}](0) + \mathbf{N}\xi(t). \quad (2.4.5)$$

The rate of change of the concentrations is then determined by the time derivative

$$[\dot{\mathbf{A}}](t) = \mathbf{N}\dot{\xi}(t). \quad (2.4.6)$$

The element $\dot{\xi}_l(t)$ of $\dot{\xi}(t)$ is the *rate of reaction* l, measured in mol l^{-1} s^{-1}.

It is sometimes convenient to distinguish reactants and products in the stoichiometric matrix. For the reactants, the stoichiometric coefficients v_{jl} are negative, while for the products they are positive. We combine all positive coefficients v_{jl}^+ into a matrix $\mathbf{N}^+$, and the magnitude $|v_{jl}^-|$ of the negative coefficients into a matrix $\mathbf{N}^-$, thus writing

$$\mathbf{N} = \mathbf{N}^+ - \mathbf{N}^-. \quad (2.4.7)$$

For example, the keto-enol tautomerization considered above leads to

$$\mathbf{N} = \begin{pmatrix} 0 & 0 & 1 & 1 & 0 & 0 \\ 1 & 0 & 0 & 0 & 1 & 0 \\ 0 & 1 & 0 & 0 & 0 & 1 \\ 0 & 0 & 1 & 0 & 1 & 0 \\ 0 & 1 & 0 & 0 & 0 & 1 \end{pmatrix} - \begin{pmatrix} 1 & 0 & 0 & 0 & 0 & 1 \\ 0 & 1 & 0 & 1 & 0 & 0 \\ 0 & 0 & 1 & 0 & 1 & 0 \\ 0 & 1 & 0 & 0 & 0 & 1 \\ 0 & 0 & 1 & 0 & 1 & 0 \end{pmatrix}$$

$$= \quad \mathbf{N}^+ \quad - \quad \mathbf{N}^-. \quad (2.4.8)$$

For each reaction l, the reaction rate $\dot{\xi}_l(t)$ depends in a characteristic manner on the concentrations $[A_j]$. In many cases, the observed reactions are composite, and the reaction rate $\dot{\xi}$ can be expressed by a so-called mass-action form. It is then determined by the concentrations of the reactants $[A_j]$ and their stoichiometric coefficients v_{jl}^-, and takes the form

$$\dot{\xi}_l = k_l \prod_{j=1}^{J} [A_j]^{-v_{jl}^-}. \tag{2.4.9}$$

The constant k_l is the *rate constant* of reaction l.

Substituting the appropriate rate law, e.g. eqn (2.4.9), into eqn (2.4.6), one obtains a system of nonlinear differential equations for the J concentrations or the $2L$ reaction numbers. Except for the simplest systems where an analytical solution is possible, one has to resort to numerical integration techniques, or to invoke simplifying assumptions suggested by chemical reasoning.

The kinetic equations can easily be solved for first-order chemical reactions. Here, the reaction rate $\dot{\xi}_{jr}$ leading from species A_j to species A_r is proportional to the reactant concentration $[A_j]$,

$$\dot{\xi}_{jr} = k_{jr}[A_j]. \tag{2.4.10}$$

The time-dependence of the concentrations $[A_j]$ is then given by

$$\frac{d}{dt}[A_j](t) = -\left(\sum_{r \neq j} k_{jr}\right)[A_j](t) + \sum_{r \neq j} k_{rj}[A_r](t). \tag{2.4.11}$$

Defining the matrix elements

$$K_{jr} = k_{rj}, \quad r \neq j$$
$$K_{jj} = -\sum_{r \neq j} k_{jr}, \tag{2.4.12}$$

of the kinetic matrix **K** of dimension $J \times J$, eqn (2.4.11) can be written in matrix form as

$$\frac{d}{dt}[\mathbf{A}] = \mathbf{K}[\mathbf{A}] \tag{2.4.13}$$

with the formal solution

$$[\mathbf{A}](t) = e^{\mathbf{K}t}[\mathbf{A}](0). \tag{2.4.14}$$

2.4.2. Exchange in systems without spin–spin couplings

Chemical exchange effects for spin systems without spin–spin couplings can be described by modified Bloch equations. These equations, often called McConnell equations, have been derived (2.35, 2.63–2.65) for first-order exchange reactions. We shall first summarize these results and

2.4 SPIN DYNAMICS DUE TO CHEMICAL REACTIONS

then generalize the modified Bloch equations to one-spin systems participating in higher-order reactions.

2.4.2.1. Modified Bloch equations for first-order reactions

In the absence of chemical reactions, the magnetization $\mathbf{M}_j$ of the chemical species j obeys the conventional Bloch equations (see eqn (2.3.1))

$$\frac{d}{dt}\mathbf{M}_j(t) = \gamma(1-\sigma_j)\mathbf{M}_j(t) \times \mathbf{B}(t) - \mathbf{R}_j\{\mathbf{M}_j(t) - \mathbf{M}_{j0}\} \quad (2.4.15)$$

with the chemical shielding constant σ_j and the relaxation matrix (see eqn (2.3.2))

$$\mathbf{R}_j = \begin{pmatrix} 1/T_{2j} & 0 & 0 \\ 0 & 1/T_{2j} & 0 \\ 0 & 0 & 1/T_{1j} \end{pmatrix}. \quad (2.4.16)$$

A chemical reaction network with J species causes magnetization transfer between the various species and couples J equations of the type of eqn (2.4.15). The chemical dynamics, governed by eqn (2.4.13), leads to the modified Bloch equations

$$\frac{d}{dt}\mathbf{M}_j(t) = \gamma(1-\sigma_j)\mathbf{M}_j(t) \times \mathbf{B}(t) - \mathbf{R}_j\{\mathbf{M}_j(t) - \mathbf{M}_{j0}(t)\}$$
$$+ \sum_r K_{jr}\mathbf{M}_r(t) \quad (2.4.17)$$

where the matrix elements K_{jr} of the kinetic matrix $\mathbf{K}$ are related through eqn (2.4.12) to the chemical rate constants k_{rj}. The z-magnetization component in *magnetic* equilibrium, M_{j0}, is proportional to the instantaneous concentration $[A_j](t)$

$$M_{j0}(t) = M_0 \frac{[A_j](t)}{\sum_k [A_k]}. \quad (2.4.18)$$

In the context of one- and two-dimensional Fourier spectroscopy, the chemical rate processes normally occur in the absence of r.f. fields in the course of one or several free precession periods: The transverse and longitudinal magnetization components, described in a frame rotating with the radio-frequency $\omega_{r.f.}$, evolve independently in these intervals

$$\frac{d}{dt}M_j^+ = \left(i\Omega_j - \frac{1}{T_{2j}}\right)M_j^+ + \sum_r K_{jr}M_r^+,$$
$$\frac{d}{dt}M_{jz} = -\frac{1}{T_{1j}}(M_{jz} - M_{j0}(t)) + \sum_r K_{jr}M_{rz} \quad (2.4.19)$$

with $M_j^+ = M_{jx} + iM_{jy}$ and the chemical shift frequency $\Omega_j = -\gamma(1 - \sigma_j)B_0$. These equations are conveniently written in matrix form

$$\frac{d}{dt}\mathbf{M}^+(t) = \mathbf{L}^+\mathbf{M}^+(t), \qquad (2.4.20)$$

$$\frac{d}{dt}\mathbf{M}_z(t) = \mathbf{L}\{\mathbf{M}_z(t) - \mathbf{M}_0(t)\} + \mathbf{K}\mathbf{M}_0(t). \qquad (2.4.21)$$

The magnetization vectors $\mathbf{M}^+$, $\mathbf{M}_z$, and $\mathbf{M}_0$ comprise the magnetization components M_j^+, M_{jz}, and M_{j0} of all J chemical species. In chemical equilibrium, the last term in eqn (2.4.21) disappears.

The dynamic matrices $\mathbf{L}^+$ and $\mathbf{L}$ describe precession, relaxation, and chemical kinetics

$$\mathbf{L}^+ = i\mathbf{\Omega} - \mathbf{\Lambda} + \mathbf{K}, \qquad (2.4.22)$$

$$\mathbf{L} = -\mathbf{R} + \mathbf{K}. \qquad (2.4.23)$$

The elements of the diagonal matrix $\mathbf{\Omega}$ correspond to the chemical shift frequencies Ω_j; the transverse relaxation matrix $\mathbf{\Lambda}$ is also diagonal (in the absence of degenerate transitions) with the elements $\lambda_{ij} = \delta_{ij}T_{2j}^{-1}$. Cross-relaxation between nuclei associated with different chemical species is represented by off-diagonal elements of the longitudinal relaxation matrix $\mathbf{R}$.

Consider by way of example a system with three chemical species exchanging by first-order reactions with $k_{ij} = K_{ji}$

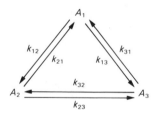

The time development of the transverse magnetization components is governed by eqn (2.4.20) which in this case has the form

$$\begin{pmatrix} \dot{M}_1^+ \\ \dot{M}_2^+ \\ \dot{M}_3^+ \end{pmatrix} = \left\{ i\begin{bmatrix} \Omega_1 & 0 & 0 \\ 0 & \Omega_2 & 0 \\ 0 & 0 & \Omega_3 \end{bmatrix} - \begin{bmatrix} T_{2(1)}^{-1} & 0 & 0 \\ 0 & T_{2(2)}^{-1} & 0 \\ 0 & 0 & T_{2(3)}^{-1} \end{bmatrix} \right. \qquad (2.4.24)$$
$$\left. + \begin{bmatrix} -k_{12}-k_{13} & k_{21} & k_{31} \\ k_{12} & -k_{21}-k_{23} & k_{32} \\ k_{13} & k_{23} & -k_{31}-k_{32} \end{bmatrix} \right\} \begin{pmatrix} M_1^+ \\ M_2^+ \\ M_3^+ \end{pmatrix}.$$

The evolution of the longitudinal components is described in close analogy with eqns (2.4.21) and (2.4.23).

In systems which are not in chemical equilibrium, the equilibrium magnetization M_{j0} associated with a species A_j is time-dependent, since it is proportional to the concentration $[A_j]$ (eqn (2.4.18)). In dynamic chemical equilibrium, on the other hand, microscopic reversibility implies that the net exchange of equilibrium magnetization $\mathbf{M}_0$ is zero, and eqn (2.4.21) may be simplified to

$$\frac{d}{dt}\Delta\mathbf{M} = \mathbf{L}\,\Delta\mathbf{M} \qquad (2.4.25)$$

where $\Delta\mathbf{M} = \mathbf{M}_z - \mathbf{M}_0$ represents the deviations from the Boltzmann distribution of the nuclear polarizations in a system with stationary concentrations.

2.4.2.2. Higher-order reactions for spin systems without spin–spin coupling

Uncoupled spins involved in higher-order reactions may be considered as 'tracers' that are carried through various molecular environments. Let us consider one particular nuclear spin. Its pathway shall be described by J molecular environments, labelled $A_1 \ldots A_J$, and by reaction rates $\dot{\xi}_{rj}$ which lead from environment A_r to A_j. Such a scheme may be represented symbolically as follows

$$A_1 + \bigcirc \underset{\dot{\xi}_{21}}{\overset{\dot{\xi}_{12}}{\rightleftarrows}} \triangle + A_2$$

$$\bigcirc + A_2 \underset{\dot{\xi}_{32}}{\overset{\dot{\xi}_{23}}{\rightleftarrows}} A_3$$

$$A_3 \underset{\dot{\xi}_{13}}{\overset{\dot{\xi}_{31}}{\rightleftarrows}} A_1 + \square$$

From the point of view of the tracer nucleus, the situation is very similar to a system of first-order reactions except that the reaction rates, expressed by the derivatives $\dot{\xi}_{rj}(t)$ of the reaction numbers, depend (linearly or non-linearly) on the concentrations of all particles in solution and may change with time. They can be calculated explicitly by solving the kinetic system of equations according to the formalism indicated in § 2.4.1.

To obtain equations equivalent to eqns (2.4.20) and (2.4.21), we introduce rate 'constants' $k_{rj}(t)$ by dividing the reaction rates $\dot{\xi}_{rj}(t)$ by the concentration of the reactant molecule $[A_r](t)$

$$k_{rj}(t) = \frac{\dot{\xi}_{rj}(t)}{[A_r](t)}. \quad (2.4.26)$$

We can now define a time-dependent pseudo first-order kinetic matrix $\mathbf{K}(t)$ by eqn (2.4.12) and achieve a *formal* equivalence with the situation of true first-order reactions. We obtain differential equations

$$\frac{d}{dt}\mathbf{M}^+(t) = \mathbf{L}^+(t)\mathbf{M}^+(t), \quad (2.4.27)$$

$$\frac{d}{dt}\mathbf{M}_z(t) = \mathbf{L}(t)\{\mathbf{M}_z(t) - \mathbf{M}_0(t)\} + \mathbf{K}(t)\mathbf{M}_0(t). \quad (2.4.28)$$

These equations are analogous to eqns (2.4.20) and (2.4.21), except for the fact that $\mathbf{L}^+$ and $\mathbf{L}$ are now time-dependent.

In *chemical equilibrium*, the concentrations $[A_r]$ and reaction rates $\dot{\xi}_{jr}$ become time-independent and a time-independent kinetic matrix $\mathbf{K}$ is obtained. Magnetic resonance of higher-order reaction networks in chemical equilibrium is thus completely equivalent to the case of first-order reactions as long as only one-spin systems are involved.

2.4.3. Density operator description of exchanging systems with spin–spin couplings

In systems containing several nuclear spins that interact through scalar couplings, the classical treatment based on modified Bloch equations is no longer valid. We have to resort to a full density-matrix treatment, which is complicated by the presence of several molecular species and their chemical interconversion.

A general density operator equation for J molecular species involved in a system of L coupled chemical reactions has been derived by Kühne *et al.* (2.57, eqn (11a))

$$\dot{\sigma}_j = -\mathrm{i}[\mathcal{H}_j, \sigma_j] - \hat{\Gamma}_j\{\sigma_j - \sigma_{j0}\} - \frac{\sigma_j}{[A_j]}\sum_{l=1}^{2L} v_{jl}^+ \dot{\xi}_l$$

$$+ \frac{1}{[A_j]}\sum_{l=1}^{2L} v_{jl}^+ \dot{\xi}_l \, \mathrm{tr}^{(j)}\left\{R_l \bigotimes_{k=1}^{J}(\sigma_k)^{\otimes \bar{v}_{kl}} R_l^{-1}\right\}. \quad (2.4.29)$$

This equation is capable of describing very complicated situations; its concise notation requires some explanatory comments. The density operator of the molecular species j is denoted by σ_j. $\mathcal{H}_j$ is the

2.4 SPIN DYNAMICS DUE TO CHEMICAL REACTIONS

corresponding Hamiltonian and $\hat{\Gamma}_j$ the relaxation superoperator. The third term in eqn (2.4.29) represents the loss in σ_j due to the chemical reactions in which the molecule A_j participates as a reactant. Because σ_j is as usual a normalized density operator (tr $\sigma = 1$), it is necessary to divide the rates of reaction $\dot{\xi}_l$ by the concentration $[A_j]$ to obtain the change in σ_j. Furthermore, the stoichiometric coefficient v_{jl}^+ (rather than v_{jl}^-) appears in the third term of eqn (2.4.29).

The last, most complicated term gives the increase in σ_j due to the chemical reactions which produce the molecule A_j. The expression

$$\bigotimes_{k=1}^{J} (\sigma_k)^{\otimes v_{kl}^-} = \sigma_l^\ddagger,$$

where $v_{kl}^- = \frac{1}{2}\{|v_{kl}| - v_{kl}\}$ is different from zero only for reactants, i.e. for species with negative stoichiometric coefficients, represents the density operator of the transition complex of reaction l expressed in an 'interaction Liouville space' which is formed by the direct product of the Liouville spaces of all particles k participating in the reaction as reactants with coefficients v_{kl}^- (Fig. 2.4.1). The resulting direct product of the density operators σ_k is indicated by $\otimes$. If $v_{kl}^- > 1$, the same density operator σ_k appears several times in the direct product (indicated by the exponent $\otimes v_{kl}^-$). R_l is the rearrangement operator which transforms the reactant density operator of reaction l into the density operator of the product. To obtain the contribution to σ_j, it is necessary to form the trace over all spaces except for the space of one molecule A_j. The multiplication with $v_{jl}^+ \dot{\xi}_l$ and the division by $[A_j]$ is analogous to that in the third term.

The non-linear density operator equation (2.4.29) is fully general and does not yet take into account the high-temperature approximation valid

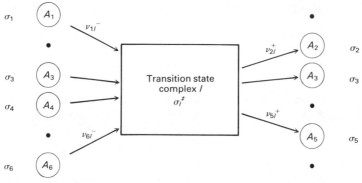

FIG. 2.4.1. Schematic representation of the transformation of the density operator during a chemical reaction l. $\sigma_1, \ldots, \sigma_k$ are the density operators of the individual reactants, while $\sigma_l^\ddagger$ is the composite density operator of the transition state complex of reaction l.

under almost all situations in magnetic resonance. If we now invoke this approximation, the equilibrium density matrices of the spin systems never deviate appreciably from unity, and the transverse magnetization components generated in the course of the experiment are correspondingly small. Consequently, eqn (2.4.29) can be linearized to a good approximation (2.66, 2.67). We introduce the deviations $\sigma'_k(t)$ from unity by defining

$$\sigma_k(t) = \frac{\mathbb{1}_k}{\text{tr}\{\mathbb{1}_k\}} + \sigma'_k(t) \tag{2.4.30}$$

and insert this expression in eqn (2.4.29). We neglect all terms of higher order in the small deviations $\sigma'_k(t)$ and obtain the equation

$$\dot{\sigma}'_j = -i[\mathcal{H}_j, \sigma'_j] - \hat{\Gamma}_j\{\sigma'_j - \sigma'_{j0}\} - \frac{\sigma'_j}{[A_j]} \sum_{l=1}^{2L} v_{jl}^+ \dot{\xi}_l(t)$$
$$+ \frac{1}{[A_j]} \sum_{l=1}^{2L} v_{jl}^+ \dot{\xi}_l(t) \text{tr}^{(j)} \left\{ R_l \bigoplus_{k=1}^{J} \sigma'_k(t)^{\oplus v_{kl}^-} R_l^{-1} \right\}. \tag{2.4.31}$$

In the following, we omit the prime on the density operator σ'. The notation used in the last term again deserves some comment. The symbol $\sigma_k(t)^{\oplus v_{kl}^-}$ implies a prescription to form a direct sum of as many replicas of the matrix $\sigma_k(t)$ as required by the stoichiometric coefficient v_{kl}^-. The resulting reactant matrices are then combined by the direct summation $\bigoplus_{k=1}^{J}$, rearranged by the reaction operator R_l, and a partial trace $\text{tr}^{(j)}$ is taken to obtain the increase to $\sigma_j(t)$ due to the reaction l.

The direct summation used above corresponds to the introduction of the so-called composite Liouville space representation. In practical situations, we can safely assume that nuclear spin functions centred on different molecules are not correlated. The entire system can then be characterized completely by the J individual component density operators σ_j. They can be combined in the form of a direct sum into a *composite density operator* σ^c,

$$\sigma^c(t) = \bigoplus_{j=1}^{J} \sigma_j(t), \tag{2.4.32}$$

that can be considered as a vector in the *composite Liouville space* $\mathcal{L}^c$ which is defined as the direct sum of the molecular Liouville spaces

$$\mathcal{L}^c = \bigoplus_{j=1}^{J} \mathcal{L}_j \tag{2.4.33}$$

with the dimension $d^c = \sum_{j=1}^{J} d_j$. The drastically reduced dimension of $\mathcal{L}^c$ in comparison with the direct product Liouville space $\mathcal{L}$ is essential for numerical solutions of the density operator equation (2.68). The trace

2.4 SPIN DYNAMICS DUE TO CHEMICAL REACTIONS

formation $\mathrm{tr}^{(j)}$, for example, is performed simply by discarding all of the components except the ones lying in subspace $\mathcal{L}_j$.

The final aim is the derivation of a master equation for the composite density operator σ^c. It is convenient to represent it in (super)matrix form. The composite density operator is then cast into a column vector σ^c formed by arranging the column vectors σ_j, representing the component density operators (see Fig. 2.1.1), into a single column.

The desired master equation can then be written in the form

$$\frac{d}{dt}\sigma^c = \{-i\hat{\mathcal{H}}_0^c - \hat{\hat{\Gamma}}^c + \hat{\hat{\Xi}}^c\}\sigma^c + \hat{\hat{\Gamma}}^c\sigma_0^c. \qquad (2.4.34)$$

The structure of this equation is shown schematically in Fig. 2.4.2. The Hamiltonian supermatrix $\hat{\mathcal{H}}_0^c$, and the relaxation supermatrix $\hat{\hat{\Gamma}}^c$ are formed from the corresponding supermatrices of the component molecules by direct summation. They have a diagonal block structure. The exchange superoperator $\hat{\hat{\Xi}}^c$, which describes all effects of the chemical transformations, has no block structure since it mediates between the density operators σ_j.

It is now straightforward to give an explicit expression for the superoperator $\hat{\hat{\Xi}}$. With the composite density operator we can rewrite eqn (2.4.31) in the form

$$\dot{\sigma}_j(t) = -i\hat{\mathcal{H}}_j\sigma_j(t) - \hat{\hat{\Gamma}}_j\{\sigma_j(t) - \sigma_{j0}\} + \hat{\hat{\Xi}}_j(t)\sigma^c(t) \qquad (2.4.35)$$

where

$$\hat{\hat{\Xi}}_j(t)\sigma^c(t) = -\frac{1}{[A_j](t)}\sum_{l=1}^{2L} v_{jl}^+\dot{\xi}_l(t)\sigma_j(t)$$

$$+ \frac{1}{[A_j](t)}\sum_{l=1}^{2L} v_{jl}^+\dot{\xi}_l(t)\mathrm{tr}^{(j)}\left\{R_l\bigoplus_{k=1}^{J}\sigma_k(t)^{\oplus v_{kl}}R_l^{-1}\right\}. \qquad (2.4.36)$$

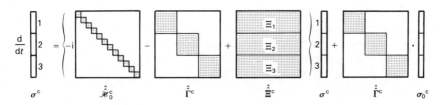

FIG. 2.4.2. Pictorial representation of the density operator equation for a system with chemical exchange (eqn (2.4.34)). The commutator $\hat{\mathcal{H}}_0^c$ is diagonal when represented in the respective eigenbases of the molecules A_j. Disregarding intermolecular cross-relaxation, the relaxation supermatrix $\hat{\hat{\Gamma}}^c$ has the block structure shown. The exchange superoperator $\hat{\hat{\Xi}}^c$, mediating the chemical transformations between the molecules, is represented by a supermatrix with diagonal and off-diagonal blocks different from zero.

The total exchange superoperator $\hat{\Xi}^c$ is finally obtained as the sum

$$\hat{\Xi}^c = \sum_{j=1}^{J} \hat{\Xi}_j. \tag{2.4.37}$$

Equation (2.4.37) provides an explicit expression for the exchange superoperator in the high-temperature limit. For *chemical non-equilibrium* systems, the exchange superoperator $\hat{\Xi}^c(t)$ is time-dependent, via the concentrations $[A_j](t)$ and rates $\dot{\xi}_l(t)$.

2.4.4. Density operator equation and exchange superoperator for first-order reactions

For reactions of first-order (or pseudo first-order), it is possible to derive simplified expressions for the exchange superoperator $\hat{\Xi}$. Following eqn (2.4.10), we describe the reaction network by rate constants k_{jr}. We can then rewrite eqn (2.4.31) for the concentration-independent density operator in the form

$$\dot{\sigma}_j = -i[\mathcal{H}_j, \sigma_j] - \hat{\Gamma}_j\{\sigma_j - \sigma_{j0}\} + \sum_{r \neq j} \frac{[A_r](t)}{[A_j](t)} k_{rj}\{R_{rj}\sigma_r R_{rj}^{-1} - \sigma_j\}. \tag{2.4.38}$$

The elements of the exchange superoperator $\hat{\Xi}$ of eqn (2.4.36) are now explicitly given by the expression

$$\Xi(t)_{j\alpha\alpha', s\beta\beta'} = (1 - \delta_{js})k_{sj}\frac{[A_s](t)}{[A_j](t)}(R_{sj})_{\alpha\beta}(R_{sj}^{-1})_{\beta'\alpha'}$$

$$- \delta_{js}\delta_{\alpha\beta}\delta_{\alpha'\beta'}\left(\sum_{r \neq j} k_{rj}\frac{[A_r](t)}{[A_j](t)}\right). \tag{2.4.39}$$

The index triple $j\alpha\alpha'$ refers to the $\alpha\alpha'$ matrix element of the density operator σ_j of the product species j, and $\Xi(t)_{j\alpha\alpha', s\beta\beta'}$ describes the rate of transfer from the matrix element $\sigma_{s\beta\beta'}$ to the element $\sigma_{j\alpha\alpha'}$. For non-equilibrium reactions, the exchange superoperator remains time-dependent even for first-order reactions as is apparent from eqn (2.4.38).

For *chemical equilibrium*, it is possible to simplify eqn (2.4.38) further by using the equilibrium condition $[A_r]k_{rj} = [A_j]k_{jr}$

$$\dot{\sigma}_j = -i[\mathcal{H}_j, \sigma_j] - \hat{\Gamma}_j\{\sigma_j - \sigma_{j0}\} + \sum_{r \neq j} k_{jr}\{R_{rj}\sigma_r R_{rj}^{-1} - \sigma_j\}. \tag{2.4.40}$$

This is a useful equation for a wide variety of equilibrium exchange phenomena in NMR. The unexpected appearance of the rate constant k_{jr}, describing the reaction from molecule A_j to molecule A_r, instead of k_{rj}, should be noted. It has its origin in the use of a concentration-independent density operator.

2.4 SPIN DYNAMICS DUE TO CHEMICAL REACTIONS

It is possible to obtain a time-independent exchange superoperator even for *non-equilibrium reactions*, by defining a concentration-dependent density operator

$$\sigma_j^\square = [A_j]\sigma_j \tag{2.4.41}$$

which is proportional to the concentration $[A_j]$.

The equations of motion differ for σ_j and $\sigma_j^\square$. This is evidenced by the relation between their time derivatives

$$\dot{\sigma}_j^\square = [A_j]\dot{\sigma}_j + [\dot{A}_j]\sigma_j \tag{2.4.42}$$

which becomes essential for time-dependent concentrations in non-equilibrium reactions. We find

$$\dot{\sigma}_j^\square = -i[\mathcal{H}_j, \sigma_j^\square] - \hat{\Gamma}_j\{\sigma_j^\square - \sigma_{j0}^\square\} + \sum_{r \neq j} \{k_{rj}R_{rj}\sigma_r^\square R_{rj}^{-1} - k_{jr}\sigma_j^\square\}. \tag{2.4.43}$$

This equation is completely analogous to the classical modified Bloch equations of eqn (2.4.17). It is the most convenient equation for handling first-order non-equilibrium chemical reactions. It can be integrated without difficulties, in contrast to eqn (2.4.38).

3
MANIPULATION OF NUCLEAR SPIN HAMILTONIANS

A major advantage of NMR in comparison with other forms of spectroscopy is the possibility of manipulating and modifying the nuclear spin Hamiltonian at will, almost without any restriction, and to adapt it to the special needs of the problem to be solved. Many infra-red and ultraviolet spectra cannot be interpreted because of the enormous complexity of incompletely resolved signal patterns. In NMR, however, it is very often possible to simplify complex spectra by modifying the Hamiltonian to an extent which permits a successful analysis.

The ease of manipulation of nuclear spin Hamiltonians is unique for purely practical reasons. Because nuclear interactions are weak, it is possible to introduce competitive perturbations of sufficient strength to override certain interactions. In optical spectroscopy, the relevant interactions are of much higher energy, and similar manipulations are virtually impossible.

In many applications of one-dimensional NMR spectroscopy, the modification of the spin Hamiltonian plays an essential role. Spin decoupling, coherent averaging by multiple-pulse techniques, sample spinning, and partial orientation in liquid crystalline solvents are nowadays standard techniques, either for the simplification of spectra or for the enhancement of their information content. In the context of two-dimensional spectroscopy, the manipulation of spin Hamiltonians turns out to be of even greater value, since it is possible to use several different average Hamiltonians in the course of a single experiment.

3.1. The tools for manipulations

Before describing in the following sections the mathematical formalism to compute manipulated Hamiltonians, we briefly present the tools which are available for modifying a Hamiltonian. Manipulation requires an external perturbation of the system, which may either be time-independent or time-dependent.

Time-independent perturbations change the parameters which govern the Hamiltonian and lead to corresponding modifications of the spectrum. Changes of temperature, pressure, solvents, and of the static magnetic field can be used to achieve the desired alteration. Many of these perturbations cannot be applied or removed sufficiently quickly to be incorporated in a two-dimensional experiment. A noticeable exception are field-cycling experiments where the sample is transported between evolution and detection periods from one magnetic field strength to another (3.1). A particularly intriguing application is time domain zero-field magnetic resonance, which can be used to measure dipolar and quadrupolar couplings in powders (3.2, 3.3).

By far the most important *time-dependent* perturbations are mechanical sample spinning and stationary or pulsed radio-frequency fields. Fast spinning leads to a spatial averaging of the inhomogeneous or anisotropic parameters of the Hamiltonian. Magnetic field inhomogeneities which cause a distribution of Larmor frequencies can be averaged out, and anisotropic interactions, like dipolar or quadrupolar couplings and the anisotropic part of chemical shifts, may be eliminated by sufficiently rapid spinning about a suitable rotation axis. The resulting spectra may be described by a modified Hamiltonian where the time-dependent terms have been removed. For slow spinning, however, families of spinning sidebands appear which can no longer be accounted for by a modified time-independent Hamiltonian alone. Here the Floquet theory may be utilized for a concise description (3.4–3.6).

A great variety of techniques that employ r.f. fields to modify the Hamiltonian have been proposed so far. Radio-frequency fields can be applied continuously, in the form of a periodic pulse train, or as aperiodic pulse sequences. The application of a continuous r.f. field leads to well-known double-resonance effects: with increasing field strength, one obtains first perturbed populations, then tickling effects, and, finally, spin decoupling (see § 4.7).

Periodic multiple-pulse sequences are astonishingly versatile for the suppression or scaling of selected interactions. For homonuclear dipolar decoupling in solids, one may use sequences like WHH-4 (3.7–3.9, 3.31), MREV-8 (3.10–3.12), BR-24 (3.13), and BLEW-48 (3.14). For heteronuclear spin decoupling in liquids, multipulse sequences such as MLEV (3.15–3.18) or WALTZ (3.19, 3.20) are quite efficient. Multiple-pulse sequences can also be designed for *scaling* homo- and heteronuclear interactions in liquids (3.21, 3.22) and solids (3.23, 3.24). Finally, external perturbations such as magnetic field inhomogeneity can be removed by refocusing in order to measure transverse relaxation (3.25).

In those cases where the net effect of a periodic perturbation can be described by a modified time-independent Hamiltonian, it is convenient to base the theoretical description on average Hamiltonian theory. This description leads to simple analytical results and is particularly useful for the analysis of cyclic multiple-pulse sequences. It can also be applied to double resonance with strong r.f. fields (§ 4.7).

However, not all of the situations mentioned above can be described by an average Hamiltonian. In many cases, more resonance lines result than a Hamiltonian of given dimension can ever account for. In the presence of a periodic perturbation, it is always possible to apply the Floquet theory (3.6). It can be used to describe sample spinning as well as multiple-pulse experiments.

In the context of two-dimensional spectroscopy, it is also possible to apply aperiodic pulse sequences in the course of the evolution or mixing period to obtain a suitably modified Hamiltonian. In the case of such aperiodic perturbations, special conditions must be fulfilled if the net effect is to be described in terms of an average Hamiltonian. If these conditions are violated, an explicit calculation of the time evolution is in order. In the following sections, a brief account of these theoretical concepts is given.

3.2. Average Hamiltonian theory

The notion of an average Hamiltonian, which represents the 'average' motion of the spin system, provides an elegant description of the effects of a time-dependent perturbation applied to the system. It has initially been introduced into magnetic resonance by Waugh to explain the effects of multiple-pulse sequences (3.7, 3.26).

The basic question underlying average Hamiltonian theory is quite simple: Let us assume that the evolution of a system is governed by a time-dependent Hamiltonian $\mathcal{H}(t)$. We may then ask whether it is possible to describe the effective evolution within a time interval t_c by an average Hamiltonian $\bar{\mathcal{H}}$.

We shall find that it is always possible to describe the overall motion during an interval $t_1 < t < t_2$ in terms of an average Hamiltonian $\bar{\mathcal{H}}(t_1, t_2)$ which, however, depends on the beginning and the end of the time interval. A time-independent average Hamiltonian $\bar{\mathcal{H}}$ for repeated observations is obtained only if

1. The Hamiltonian $\mathcal{H}(t)$ is periodic;
2. The observation is stroboscopic and synchronized with the period of the Hamiltonian.

The average Hamiltonian $\bar{\mathcal{H}}$ can be defined either by an exact calculation involving the diagonalization of the time evolution operator or by means of an expansion known as Baker–Campbell–Hausdorff expansion or Magnus expansion.

3.2.1. Exact calculation of $\bar{\mathcal{H}}$

Let us assume that the Hamiltonian $\mathcal{H}(t)$ is piecewise constant in successive time intervals

$$\mathcal{H}(t) = \mathcal{H}_k \quad \text{for} \quad (\tau_1 + \tau_2 + \ldots + \tau_{k-1}) < t < (\tau_1 + \tau_2 + \ldots + \tau_k). \quad (3.2.1)$$

In practice, $\mathcal{H}(t)$ often fulfils this condition in a suitable rotating frame. Then the density operator equation

$$\dot{\sigma} = -i[\mathcal{H}(t), \sigma] \quad (3.2.2)$$

can readily be integrated

$$\sigma(t_c) = U(t_c)\sigma(0)U(t_c)^{-1} \quad (3.2.3)$$

with

$$U(t_c) = \exp(-i\mathcal{H}_n \tau_n) \ldots \exp(-i\mathcal{H}_1 \tau_1)$$

and

$$t_c = \sum_{k=1}^{n} \tau_k.$$

A product of unitary transformations is again a unitary transformation, hence it is possible to express the entire sequence by a single transformation under an average Hamiltonian $\bar{\mathcal{H}}(t_c)$

$$U(t_c) = \exp\{-i\bar{\mathcal{H}}(t_c)t_c\}. \quad (3.2.4)$$

$\bar{\mathcal{H}}(t_c)$ can be determined by diagonalizing the explicit matrix product of the n transformations and by taking the logarithm of the resulting eigenvalues. Note that the average Hamiltonian $\bar{\mathcal{H}}(t_c)$ is only applicable for a fixed interval $t = t_c$. If, however, t_c corresponds to the period of a periodic Hamiltonian $\mathcal{H}(t)$, $\bar{\mathcal{H}}(t_c)$ also describes the motion over extended time periods, provided the observation is restricted to stroboscopic and synchronized sampling,

$$U(nt_c) = U(t_c)^n = \exp\{-i\bar{\mathcal{H}}(t_c)nt_c\}. \quad (3.2.5)$$

3.2.2. Cumulant expansion of the propagator

In many cases, it is more convenient to express $\bar{\mathcal{H}}(t_c)$ by a cumulant expansion of the exponential operator product. The Baker–Campbell–Hausdorff relation

$$e^B e^A = \exp\{A + B + \tfrac{1}{2}[B, A] + \tfrac{1}{12}([B, [B, A]] + [[B, A,]A]) + \ldots\} \quad (3.2.6)$$

yields for two sequential time intervals τ_1, τ_2 an explicit expression for $\bar{\mathcal{H}}(t_c)$

$$\bar{\mathcal{H}}(t_c) = \frac{i}{t_c} \{ -i(\mathcal{H}_1 \tau_1 + \mathcal{H}_2 \tau_2) - \tfrac{1}{2}[\mathcal{H}_2 \tau_2, \mathcal{H}_1 \tau_1]$$
$$+ \tfrac{1}{12}(i[\mathcal{H}_2 \tau_2, [\mathcal{H}_2 \tau_2, \mathcal{H}_1 \tau_1]]$$
$$+ i[[\mathcal{H}_2 \tau_2, \mathcal{H}_1 \tau_1], \mathcal{H}_1 \tau_1])$$
$$+ \ldots \}. \quad (3.2.7)$$

For commuting Hamiltonians $[\mathcal{H}_1, \mathcal{H}_2] = 0$, we obtain the exact (and obvious) result

$$\bar{\mathcal{H}}(t_c) = \frac{1}{t_c}(\mathcal{H}_1 \tau_1 + \mathcal{H}_2 \tau_2). \quad (3.2.8)$$

Analogous expressions can also be obtained for a piecewise constant Hamiltonian changing $n-1$ times in the interval $t_c = \tau_1 + \tau_2 + \tau_3 + \ldots + \tau_n$, $\mathcal{H} = \{\mathcal{H}_1, \tau_1; \mathcal{H}_2, \tau_2; \mathcal{H}_3, \tau_3; \ldots\}$. The average Hamiltonian can be divided into contributions of different orders (3.8)

$$\bar{\mathcal{H}}(t_c) = \bar{\mathcal{H}}^{(0)} + \bar{\mathcal{H}}^{(1)} + \bar{\mathcal{H}}^{(2)} + \ldots. \quad (3.2.9)$$

with

$$\bar{\mathcal{H}}^{(0)} = \frac{1}{t_c} \{ \mathcal{H}_1 \tau_1 + \mathcal{H}_2 \tau_2 + \ldots + \mathcal{H}_n \tau_n \},$$

$$\bar{\mathcal{H}}^{(1)} = -\frac{i}{2t_c} \{ [\mathcal{H}_2 \tau_2, \mathcal{H}_1 \tau_1] + [\mathcal{H}_3 \tau_3, \mathcal{H}_1 \tau_1] + [\mathcal{H}_3 \tau_3, \mathcal{H}_2 \tau_2] + \ldots \},$$

$$\bar{\mathcal{H}}^{(2)} = -\frac{1}{6t_c} \{ [\mathcal{H}_3 \tau_3, [\mathcal{H}_2 \tau_2, \mathcal{H}_1 \tau_1]] + [[\mathcal{H}_3 \tau_3, \mathcal{H}_2 \tau_2], \mathcal{H}_1 \tau_1]$$
$$+ \tfrac{1}{2}[\mathcal{H}_2 \tau_2, [\mathcal{H}_2 \tau_2, \mathcal{H}_1 \tau_1]] + \tfrac{1}{2}[[\mathcal{H}_2 \tau_2, \mathcal{H}_1 \tau_1], \mathcal{H}_1 \tau_1] + \ldots \}. \quad (3.2.10)$$

For more formal calculations, one may express the propagator $U(t_c)$ in the simple form

$$U(t_c) = \exp\{-i\mathcal{H}_n \tau_n\} \ldots \exp\{-i\mathcal{H}_1 \tau_1\} = T \exp\left\{-i \sum_k^n \mathcal{H}_k \tau_k\right\}$$
$$= \exp\{-i\bar{\mathcal{H}} t_c\}. \quad (3.2.11)$$

Here, T is the Dyson time-ordering operator (3.27, 3.28) which time-orders operators with different time arguments in operator products in a decreasing order. The following relations define the operation of T

$$T\{\mathcal{H}(t_1)\mathcal{H}(t_2)\} = \begin{cases} \mathcal{H}(t_1)\mathcal{H}(t_2) & \text{for } t_1 > t_2 \\ \mathcal{H}(t_2)\mathcal{H}(t_1) & \text{for } t_1 < t_2. \end{cases} \quad (3.2.12)$$

The effect of the time-ordering operator T on the exponential function in eqn (3.2.11) can be explicitly represented by expanding the exponential in a power series and time-ordering the factors of the various expansion terms. In this way it is also possible to verify eqns (3.2.9) and (3.2.10).

The formal equation (3.2.11) is easily generalized to continuously changing Hamiltonians, leading to the propagator

$$U(t_c) = T \exp\left\{-i\int_0^{t_c} d\tau \mathcal{H}(\tau)\right\} = \exp\{-i\bar{\mathcal{H}}t_c\}. \tag{3.2.13}$$

Expanding this exponential and collecting terms of equal order leads finally to the following expressions for the various orders of the average Hamiltonian $\bar{\mathcal{H}}(t_c)$ in analogy to eqn (3.2.10)

$$\bar{\mathcal{H}}^{(0)} = \frac{1}{t_c}\int_0^{t_c} dt_1 \mathcal{H}(t_1), \tag{3.2.14}$$

$$\bar{\mathcal{H}}^{(1)} = \frac{-i}{2t_c}\int_0^{t_c} dt_2 \int_0^{t_2} dt_1 [\mathcal{H}(t_2), \mathcal{H}(t_1)], \tag{3.2.15}$$

$$\bar{\mathcal{H}}^{(2)} = -\frac{1}{6t_c}\int_0^{t_c} dt_3 \int_0^{t_3} dt_2 \int_0^{t_2} dt_1 \{[\mathcal{H}(t_3), [\mathcal{H}(t_2), \mathcal{H}(t_1)]]$$
$$+ [[\mathcal{H}(t_3), \mathcal{H}(t_2)], \mathcal{H}(t_1)]\}. \tag{3.2.16}$$

This expansion is known as the *Magnus expansion* (3.8, 3.29, 3.30). It forms the basis of average Hamiltonian theory.

3.2.3. Averaging by time-dependent perturbations

We discuss in this section situations where the original time-independent Hamiltonian $\mathcal{H}_0$ is modified by the introduction of a time-dependent perturbation $\mathcal{H}_1(t)$. The perturbation should be chosen such that it does not explicitly appear in the resulting average Hamiltonian. Typical examples are spin decoupling, multiple-pulse experiments, and sample spinning. In these cases, the total Hamiltonian contains a time-independent and a time-dependent term

$$\mathcal{H}(t) = \mathcal{H}_0 + \mathcal{H}_1(t) \tag{3.2.17}$$

where $\mathcal{H}_0$ is the unperturbed Hamiltonian and $\mathcal{H}_1(t)$ is the perturbation that has been purposely introduced.

The general propagator $U(t)$ can be expressed according to eqn (3.2.13) by

$$U(t) = T \exp\left\{-i\int_0^t dt_1(\mathcal{H}_0 + \mathcal{H}_1(t_1))\right\}. \tag{3.2.18}$$

We now attempt to separate the effects of $\mathcal{H}_0$ and $\mathcal{H}_1(t)$ and divide the propagator into two factors

$$U(t) = U_1(t) U_0(t) \tag{3.2.19}$$

with

$$U_1(t) = T \exp\left\{-i \int_0^t dt_1 \mathcal{H}_1(t_1)\right\} \tag{3.2.20}$$

and

$$U_0(t) = T \exp\left\{-i \int_0^t dt_1 \tilde{\mathcal{H}}_0(t_1)\right\} \tag{3.2.21}$$

where $U_1(t)$ depends only on the perturbation $\mathcal{H}_1(t)$ and expresses its direct effects. $\tilde{\mathcal{H}}_0(t)$ is the Hamiltonian in the time-dependent interaction representation with respect to $\mathcal{H}_1(t)$, often called the *toggling frame*,

$$\tilde{\mathcal{H}}_0(t) = U_1^{-1}(t) \mathcal{H}_0 U_1(t). \tag{3.2.22}$$

Of particular practical interest are situations where the perturbation $\mathcal{H}_1(t)$ is *periodic* with period t_c

$$\mathcal{H}_1(t + nt_c) = \mathcal{H}_1(t) \quad \text{for} \quad n = 0, 1, 2, \ldots. \tag{3.2.23}$$

and where, in addition, $\mathcal{H}_1(t)$ is *cyclic* in the sense that

$$U_1(t_c) = 1. \tag{3.2.24}$$

$\mathcal{H}_1(t)$ has therefore no overall direct effect in the course of one full cycle.

For these conditions, the Hamiltonian $\tilde{\mathcal{H}}_0(t)$ in the interaction representation also becomes periodic and leads to the simple propagator for one cycle

$$U(t_c) = U_0(t_c) \tag{3.2.25}$$

and for n cycles

$$U(nt_c) = U_0(t_c)^n. \tag{3.2.26}$$

For *stroboscopic observation* of the time evolution in synchronism with the periodic perturbation $\mathcal{H}_1(t)$, the propagator $U_0(t_c)$ alone describes the observed time evolution of $\sigma(t)$.

In a last step, we express the propagator $U_0(t_c)$ by an average Hamiltonian $\bar{\mathcal{H}}_0$ according to eqn (3.2.13)

$$U_0(t_c) = \exp\{-i\bar{\mathcal{H}}_0 t_c\}.$$

Using the Magnus expansion, eqns (3.2.14)–(3.2.16), we find the central result of average Hamiltonian theory

$$\bar{\mathcal{H}}_0 = \bar{\mathcal{H}}_0^{(0)} + \bar{\mathcal{H}}_0^{(1)} + \ldots \tag{3.2.27}$$

with

$$\bar{\mathcal{H}}_0^{(0)} = \frac{1}{t_c} \int_0^{t_c} dt_1 \tilde{\mathcal{H}}_0(t_1), \tag{3.2.28}$$

$$\bar{\mathcal{H}}_0^{(1)} = \frac{-i}{2t_c} \int_0^{t_c} dt_2 \int_0^{t_2} dt_1 [\tilde{\mathcal{H}}_0(t_2), \tilde{\mathcal{H}}_0(t_1)], \tag{3.2.29}$$

$$\bar{\mathcal{H}}_0^{(2)} = -\frac{1}{6t_c} \int_0^{t_c} dt_3 \int_0^{t_3} dt_2 \int_0^{t_2} dt_1 \{[\tilde{\mathcal{H}}_0(t_3), [\tilde{\mathcal{H}}_0(t_2), \tilde{\mathcal{H}}_0(t_1)]]$$

$$+ [[\tilde{\mathcal{H}}_0(t_3), \tilde{\mathcal{H}}(t_2)], \tilde{\mathcal{H}}(t_1)]\}. \tag{3.2.30}$$

where the toggling frame Hamiltonian $\tilde{\mathcal{H}}_0(t)$ is given by eqns (3.2.22) and (3.2.20).

The zero-order term $\bar{\mathcal{H}}_0^{(0)}$ has a particularly simple form. The average Hamiltonian of zero order is just the time average of the toggling frame Hamiltonian $\tilde{\mathcal{H}}_0(t)$. Clearly, the purpose of introducing the perturbation $\mathcal{H}_1(t)$ is to effect a transformation into a toggling frame (eqn (3.2.22)) such that the time averages of the undesired terms in the Hamiltonian vanish in this frame.

In most cases, the removal of certain higher-order terms $\bar{\mathcal{H}}_0^{(1)}, \ldots$ is necessary to achieve an efficient suppression of the unwanted interactions. Much of the sophistication of multiple-pulse sequences is aimed at this goal. The higher-order terms contain undesired cross-terms between the various terms of the Hamiltonian. The commutators between the Hamiltonian operators at different times involved in the higher-order terms become smaller for shorter cycle times τ_c, such that a faster multiple-pulse sequence leads in general to better averaging.

The cyclic property of the perturbation is not a prerequisite for the introduction of an average Hamiltonian. In the case of a non-cyclic perturbation, $\bar{\mathcal{H}}$ contains an explicit contribution originating from $\mathcal{H}_1(t)$. In many situations, this implies that $\bar{\mathcal{H}}$ will become susceptible to minor misadjustments and variations in $\mathcal{H}_1(t)$. A cyclic perturbation which does not appear explicitly in $\bar{\mathcal{H}}$ is therefore preferable in most cases.

We conclude this section with a simple recipe for the calculation of the zeroth-order average Hamiltonian $\bar{\mathcal{H}}_0^{(0)}$ for a perturbation $\mathcal{H}_1(t)$ consisting of periods with n infinitely narrow r.f. pulses, represented by the transformations $U_1, U_2, \ldots, U_n$, and separated by free precession periods. Each pulse rotates the toggling frame into a new position, as shown schematically in Fig. 3.2.1. The toggling frame Hamiltonian $\tilde{\mathcal{H}}_0(t)$ of eqn (3.2.22) remains constant in the interval τ_k between two pulses k and $k+1$, $\tilde{\mathcal{H}}_0(t) = \tilde{\mathcal{H}}_{0(k)}$, and can be computed by stepwise

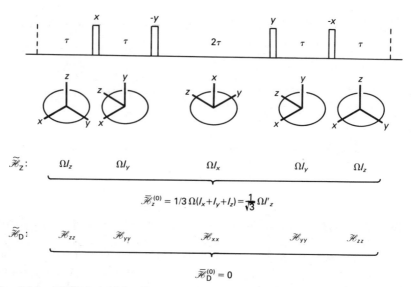

FIG. 3.2.1. WHH-4 multiple-pulse sequence for homonuclear dipolar decoupling. In each cycle of total length $\tau_c = 6\tau$, the four pulses spaced by τ and 2τ lead to rotated coordinate systems, known as toggling frame. The average Hamiltonian $\bar{\mathscr{H}}^{(0)}$ is obtained by averaging the transformed Hamiltonians $\tilde{\mathscr{H}}$ in the toggling frame. Averaging is indicated for the Zeeman interaction $\mathscr{H}_Z$ and for the dipolar interaction $\mathscr{H}_D$.

transformations

$$\tilde{\mathscr{H}}_{0(0)} = \mathscr{H}_0,$$
$$\tilde{\mathscr{H}}_{0(1)} = U_1^{-1}\mathscr{H}_0 U_1,$$
$$\tilde{\mathscr{H}}_{0(2)} = U_1^{-1} U_2^{-1} \mathscr{H}_0 U_2 U_1$$

. . . . (3.2.31)

Note the unexpected order of the transformations: all previous pulses must be arranged in reversed order and appear with reversed sense of rotation.

The average Hamiltonian is obtained from a weighted sum

$$\bar{\mathscr{H}}_0^{(0)} = \frac{1}{t_c}\sum_{k=0}^{n} \tau_k U_1^{-1} \ldots U_k^{-1} \mathscr{H}_0 U_k \ldots U_1. \quad (3.2.32)$$

For pulses of finite length, the duration of the pulses, where the toggling frame changes continuously, must also be included in the average.

Figure 3.2.1 illustrates the procedure for the first successful multiple-pulse sequence proposed for the elimination of homonuclear dipolar interactions in solids, the WHH-4 sequence (3.31). The sequence of four $\pi/2$-pulses with phases x, $-y$, y, and $-x$ has unequal intervals, $\tau_0 = \tau_1 = \tau_3 = \tau_4 = \tau$ and $\tau_2 = 2\tau$. The pulses rotate the toggling frame sequentially through the indicated orientations. The Zeeman interaction in the toggling frame can be determined by inspection of Fig. 3.2.1, since it is associated with the axis that is parallel to the laboratory frame z-axis. The average Zeeman Hamiltonian $\bar{\mathcal{H}}_Z^{(0)}$ corresponds to a new axis of quantization $z' = (1, 1, 1)$ and involves a Larmor frequency scaled by a factor $1/\sqrt{3}$. Scaling of the Zeeman interaction is typical for all dipolar decoupling sequences.

The dipolar Hamiltonian goes through the three forms $\mathcal{H}_{xx}$, $\mathcal{H}_{yy}$, and $\mathcal{H}_{zz}$ where the indices indicate the operators involved, e.g.

$$\mathcal{H}_{xx} = \sum_{k<l} b_{kl} \tfrac{1}{2}(1 - 3\cos^2\theta_{kl})[3I_{kx}I_{lx} - \mathbf{I}_k \mathbf{I}_l]. \qquad (3.2.33)$$

The resulting average Hamiltonian $\bar{\mathcal{H}}_D^{(0)}$ is zero.

For the computation of higher orders in $\bar{\mathcal{H}}_0$ it should be noted that the toggling frame Hamiltonian $\tilde{\mathcal{H}}(t)$ in Fig. 3.2.1 is symmetric in the sense that

$$\tilde{\mathcal{H}}(t) = \tilde{\mathcal{H}}(t_c - t). \qquad (3.2.34)$$

A pulse sequence with this property is called a symmetric cycle. For such cycles, it can easily be shown that all contributions of odd orders to $\bar{\mathcal{H}}_0$ vanish (3.8, 3.32),

$$\bar{\mathcal{H}}_0^{(k)} = 0 \quad \text{for} \quad k = 1, 3, 5, \ldots . \qquad (3.2.35)$$

Symmetric cycles have a significantly improved performance because the error terms are reduced. It should be noted that a symmetric cycle does not imply a pulse sequence with symmetrically arranged pulses: in fact, the symmetric cycle in Fig. 3.2.1 consists of an antisymmetric sequence of pulses.

A great number of multiple-pulse sequences with improved properties have been proposed for homonuclear dipolar decoupling. By extending the length of the sequence, it is possible to extend the suppression to higher-order terms in $\bar{\mathcal{H}}_0$. The reader is referred for more details to excellent reviews on this subject (3.8, 3.9, 3.33) and to the original literature (3.7, 3.10–3.14).

3.2.4. Truncation of internal Hamiltonians

Dominant terms of a time-independent Hamiltonian often have a truncating effect on smaller terms. This can be understood as coherent averaging in the interaction representation of the dominant term. Let us assume a Hamiltonian in the laboratory frame

$$\mathcal{H} = \mathcal{H}_0 + \mathcal{H}_1 \qquad (3.2.36)$$

where $\mathcal{H}_0$ is the dominant term. In the interaction representation of $\mathcal{H}_0$, $\mathcal{H}_1$ becomes time-dependent and we may compute an average truncated Hamiltonian $\tilde{\mathcal{H}}_1$ which includes the modifying effects of $\mathcal{H}_0$. An ubiquitous example is the truncation of nonsecular parts of the spin–spin interactions by the Zeeman Hamiltonian. $\mathcal{H}_0$ is then identified with the Zeeman interaction and $\mathcal{H}_1$ with the spin–spin coupling term.

Using the formalism of § 3.2.3, we may again split the propagator into two factors, this time with interchanged roles for $\mathcal{H}_0$ and $\mathcal{H}_1$

$$U(t) = U_0(t) U_1(t) \tag{3.2.37}$$

with

$$U_0(t) = \exp\{-i\mathcal{H}_0 t\},$$

$$U_1(t) = T \exp\left\{-i \int_0^t \tilde{\mathcal{H}}_1(t_1) \, dt_1\right\},$$

and

$$\tilde{\mathcal{H}}_1(t) = U_0(t)^{-1} \mathcal{H}_1 U_0(t).$$

The zero-order average Hamiltonian is then (eqn (3.2.28))

$$\bar{\mathcal{H}}_1^{(0)} = \frac{1}{t_c} \int_0^{t_c} \tilde{\mathcal{H}}_1(t_1) \, dt_1. \tag{3.2.38}$$

To evaluate this equation explicitly, we may express $\mathcal{H}_1$ in terms of eigenoperators Q_k of the superoperator $\hat{\mathcal{H}}_0$,

$$\mathcal{H}_1 = \sum_k a_k Q_k \tag{3.2.39}$$

with

$$\hat{\mathcal{H}}_0 Q_k = [\mathcal{H}_0, Q_k] = q_k Q_k.$$

It follows that

$$\tilde{\mathcal{H}}_1(t) = \sum_k a_k \exp\{-iq_k t\} Q_k \tag{3.2.40}$$

and

$$\bar{\mathcal{H}}_1^{(0)} = \sum_k a_k \frac{\exp\{-iq_k t_c\} - 1}{-iq_k t_c} Q_k. \tag{3.2.41}$$

Let us investigate the case that the evolution under $\mathcal{H}_0$ is cyclic for the period length t_c, i.e.

$$U_0(t_c) = 1. \tag{3.2.42}$$

This condition can only be fulfilled for special Hamiltonians: the eigenvalues of $\mathcal{H}_0$ must be multiples of the frequency $2\pi/t_c$. For example, the Zeeman Hamiltonian for a single spin species (neglecting unequal chemical shielding) fulfils this condition. For Hamiltonians $\mathcal{H}_0$ of this type, $\mathcal{H}_1(t)$ becomes periodic,

$$\tilde{\mathcal{H}}_1(t_c) = \tilde{\mathcal{H}}_1(0), \tag{3.2.43}$$

and the integration over one period eliminates all oscillating terms in eqn (3.2.40), leaving only terms containing eigenoperators Q_{k_0} with zero eigenvalues $q_{k_0} = 0$

$$\bar{\mathcal{H}}_1^{(0)} = \sum_{k_0} a_{k_0} Q_{k_0}. \tag{3.2.44}$$

This implies that $\mathcal{H}_1$ has been reduced by averaging to those parts which commute with $\mathcal{H}_0$, or in other words, $\bar{\mathcal{H}}_1^{(0)}$ is the diagonal part of $\mathcal{H}_1$ with respect to $\mathcal{H}_0$

$$[\bar{\mathcal{H}}_1^{(0)}, \mathcal{H}_0] = 0. \tag{3.2.45}$$

When $\mathcal{H}_0$ is identified with the Zeeman Hamiltonian, $\bar{\mathcal{H}}_1^{(0)}$ consists of the secular parts of $\mathcal{H}_1$ that are invariant with respect to rotations about the z-axis.

The periodicity condition on $\mathcal{H}_0$ can be dropped in practice and stroboscopic observation is not required when the period length t_c is sufficiently short, i.e. $\mathcal{H}_0$ sufficiently strong, and possible sidebands, created by $\mathcal{H}_0$, are sufficiently far removed from the spectral region of interest.

First-order truncation of internal Hamiltonians is equivalent to conventional perturbation theory. It corresponds to the suppression of the so-called non-secular parts of $\mathcal{H}_1$, i.e. of those parts which do not commute with $\mathcal{H}_0$. Examples of truncation by the neglect of non-secular parts have been given in § 2.2.1. The most important cases are the truncation of the dipolar Hamiltonian and the Hamiltonian for weak scalar coupling.

3.2.5. Floquet theory

This brief account places average Hamiltonian theory into the more general framework of Floquet theory (3.4, 3.35). The Floquet formalism aims at obtaining a general solution for the time evolution under a periodic time-dependent Hamiltonian $\mathcal{H}(t)$. Floquet theory bears similarities to average Hamiltonian theory in that it also uses an expansion of the evolution operator into 'orders' of decreasing importance. A general assessment has been given elsewhere (3.4, 3.5, 3.35) and is beyond the scope of this section, which focuses on the following features:

1. Floquet theory provides a more general approach to average

Hamiltonian theory and is useful in discussing the convergence of the expansion.

2. Floquet theory allows one to write a time evolution operator for all times, not only for integral multiples of the cycle time t_c.

3. Floquet theory can be used to discuss 'multi-photon' NMR experiments where a single-quantum transition is excited by using two or several radio-frequency quanta (3.36, 3.37).

The Floquet theorem asserts the existence of a solution for a system of linear differential equations with periodically time-dependent coefficients. Applied to the case of a periodic Hamiltonian $\mathcal{H}(t)$ with period t_c, it suggests the form of the time evolution operator $U(t)$

$$U(t) = P(t) \cdot \exp\{-i\mathcal{H}_F t\} \qquad (3.2.46)$$

where the operator $P(t)$ is periodic with the period t_c while the Floquet Hamiltonian $\mathcal{H}_F$ is time-independent.

The connection to average Hamiltonian theory becomes immediately obvious by setting $P(0) = P(nt_c) = 1$ and by assuming stroboscopic observation

$$U(nt_c) = \exp\{-i\mathcal{H}_F nt_c\}. \qquad (3.2.47)$$

A comparison with eqn (3.2.13) shows the identity $\mathcal{H}_F = \bar{\mathcal{H}}$.

Equation (3.2.46) may be evaluated for any t and is not limited to stroboscopic sampling. The two operators are expanded in the form of two interdependent series

$$\mathcal{H}_F = \sum_{k=0}^{\infty} \mathcal{H}_F^{(k)}, \qquad (3.2.48)$$

$$P(t) = \sum_{k=0}^{\infty} P^{(k)}(t) \qquad (3.2.49)$$

with the supplementary definitions $\mathcal{H}_F^{(0)} \equiv 0$, $P^{(0)} \equiv 1$. $\mathcal{H}_F^{(k)}$ and $P^{(k)}(t)$ are evaluated with the recursive relations

$$\mathcal{H}_F^{(k)} = \frac{1}{t_c} \int_0^{t_c} \left\{ \mathcal{H}(t')P^{(k-1)}(t') - \sum_{j=1}^{k-1} P^{(j)}(t')\mathcal{H}_F^{(k-j)} \right\} dt', \qquad (3.2.50)$$

$$P^{(k)}(t) = -i \int_0^t \left\{ \mathcal{H}(t')P^{(k-1)}(t') - \sum_{j=1}^{k-1} P^{(j)}(t')\mathcal{H}_F^{(k-j)} - \mathcal{H}_F^{(k)} \right\} dt'. \qquad (3.2.51)$$

It is easily seen that $\mathcal{H}_F^{(1)}$ is identical with the average Hamiltonian $\bar{\mathcal{H}}^{(0)}$, eqn (3.2.14). The higher-order terms are also equal to the corresponding terms of the average Hamiltonian series, $\mathcal{H}_F^{(k)} = \bar{\mathcal{H}}^{(k-1)}$. A proof is given in Ref. 3.35.

Maricq (3.35) discusses the convergence of the series and the practical question under which circumstances the series can be truncated after two or three terms. An often quoted criterion is $\|\mathcal{H}^2\|^{\frac{1}{2}}\tau_c < 1$, i.e. the cycle repetition rate $1/\tau_c$ should be larger than an average transition frequency of the system in the chosen reference frame. Two conclusions can be drawn.

1. The success of average Hamiltonian theory and the convergence of the expansions depends on the choice of a suitable reference frame, as expressed by the partitioning of the Hamiltonian in eqn (3.2.17).
2. For systems which exhibit a wide range of transition frequencies $\Delta\omega$, the criterion $\Delta\omega\tau_c < 1$ may be fulfilled for transition frequencies in the centre of the spectrum, but violated for transitions in the wings. This can lead to a two-time-scale behaviour, i.e. rapid establishment of a quasi-stationary state followed by a slower overall time evolution (3.6, 3.35).

The application of a periodic time-dependent Hamiltonian leads to a spectrum with a sideband structure which cannot be described by an average Hamiltonian $\bar{\mathcal{H}}$ with its finite number of transitions. The Floquet theory in the formulation of Shirley (3.4) offers a possibility for handling this situation by introducing a 'Floquet Hamiltonian' $\mathcal{H}_F$ with an infinite-dimensional matrix representation. The Floquet Hamiltonian can be represented in terms of the 'Floquet states' $|pn\rangle$ which are equivalent to dressed spin states formed by the direct product of pure spin states $|p\rangle$ and free photon states $|n\rangle$. The Floquet Hamiltonian has an infinite number of transitions which take account of the sideband families. This approach has been successfully applied to multi-photon NMR (3.36, 3.37).

3.3. Average Hamiltonian due to aperiodic perturbations

This section concerns situations where the evolution of the spin system is observed indirectly. This is typical for two-dimensional experiments: The precession during the evolution period t_1 is monitored indirectly by a systematic incrementation of t_1 in the course of a sequence of experiments, the actual observation being restricted to the detection period. Numerous experiments of this type are known, such as spin–echo spectroscopy or field cycling experiments in quadrupolar resonance and dipolar spectroscopy, which utilize the same principles of indirect detection, although they are often not classified as two-dimensional experiments.

By using indirect detection, it is feasible to apply *aperiodic* perturbations during the evolution time t_1. It is, for example, possible to switch the Hamiltonian by turning on the decoupler at a break point

$t_x = xt_1$, which moves in proportion to t_1. On the other hand, it is also possible to introduce in each experiment a refocusing pulse at a time $t_y = yt_1$. The question arises under which conditions the overall evolution during t_1 can be described by an average Hamiltonian $\bar{\mathcal{H}}$, or in mathematical terms whether the relation

$$\sigma(t_1) = \exp\{-i\bar{\mathcal{H}}t_1\}\sigma(0)\exp\{i\bar{\mathcal{H}}t_1\} \qquad (3.3.1)$$

holds for arbitrary times t_1 with a constant $\bar{\mathcal{H}}$.

3.3.1. General condition for an average Hamiltonian

To formulate conditions under which aperiodic perturbations applied during an evolution time t_1 lead to an average Hamiltonian, consider an experiment with a rather general sequence of perturbations, as shown in Fig. 3.3.1(a). The period t_1 is divided in n intervals of variable duration $\tau_j = x_j t_1$ proportional to t_1. Each interval τ_j may have a different Hamiltonian $\mathcal{H}_j$. These evolution intervals of variable duration may be separated by intervals of fixed length. In most cases of practical interest, these fixed intervals will be extremely short and correspond to non-selective r.f. pulses whose effect can be described by a unitary transformation R_j. The evolution time t_1 is defined as the sum of all variable evolution intervals

$$t_1 = \sum_{j=1}^{n} x_j t_1. \qquad (3.3.2)$$

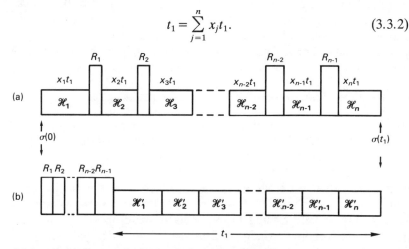

FIG. 3.3.1. Aperiodic perturbations such as may be applied in the evolution period t_1 of a two-dimensional time-domain experiment. (a) The period t_1 is divided into n intervals τ_j ($j = 1, 2, \ldots, n$) with Hamiltonians $\mathcal{H}_j$ and durations $\tau_j = x_j t_1$, separated by intervals of fixed length (typically r.f. pulses represented by unitary transformations R_j). (b) By introducing transformed Hamiltonians $\mathcal{H}_j'$ defined in eqn (3.3.4), the transformations R_j may be shifted to the beginning of the t_1 period.

Using for convenience a superoperator notation, and disregarding relaxation, the time evolution during the entire period t_1 is given by

$$\sigma(t_1) = \exp(-i\hat{\mathcal{H}}_n x_n t_1)\prod_{j=1}^{n-1} \hat{R}_j \exp(-i\hat{\mathcal{H}}_j x_j t_1)\sigma(0). \quad (3.3.3)$$

By introducing transformed Hamiltonians $\mathcal{H}'_j$ with

$$\mathcal{H}'_j = \left(\prod_{k=j}^{n-1} \hat{R}_k\right)\mathcal{H}_j$$

$$= R_{n-1}\ldots R_{j+1}R_j\mathcal{H}_j R_j^{-1}R_{j+1}^{-1}\ldots R_{n-1}^{-1}, \quad (3.3.4)$$

it is possible to shift all transformations R_j (e.g. r.f. pulses) to the beginning of the evolution period

$$\sigma(t_1) = \prod_{j=1}^{n} \exp(-i\hat{\mathcal{H}}'_j x_j t_1)\sigma'(0) \quad (3.3.5)$$

where the initial density operator $\sigma'(0)$ appears in transformed form

$$\sigma'(0) = \prod_{j=1}^{n-1} \hat{R}_j \sigma(0). \quad (3.3.6)$$

This leads to the equivalent scheme shown in Fig. 3.3.1(b) where the new initial condition $\sigma'(0)$ includes the effect of the entire sequence of transformations R_j, and where the Hamiltonians $\mathcal{H}'_j$ are modified by all transformations which *follow* the interval $x_j t_1$.

Equation (3.3.5) is equivalent to the notation

$$\sigma'(0) \xrightarrow{\mathcal{H}'_1 x_1 t_1} \xrightarrow{\mathcal{H}'_2 x_2 t_1} \ldots \xrightarrow{\mathcal{H}'_n x_n t_1} \sigma(t_1) \quad (3.3.7)$$

Using the results of the previous section, we can easily replace the sequence of evolutions in eqns (3.3.5) or (3.3.7) by an average Hamiltonian $\bar{\mathcal{H}}(t_1)$

$$\sigma'(0) \xrightarrow{\bar{\mathcal{H}}(t_1)t_1} \sigma(t_1) \quad (3.3.8)$$

where $\bar{\mathcal{H}}(t_1)$ can be expanded according to eqn (3.2.9)

$$\bar{\mathcal{H}}(t_1) = \bar{\mathcal{H}}^{(0)} + \bar{\mathcal{H}}^{(1)}(t_1) + \bar{\mathcal{H}}^{(2)}(t_1) + \ldots \quad (3.3.9)$$

with

$$\bar{\mathcal{H}}^{(0)} = \mathcal{H}'_1 x_1 + \mathcal{H}'_2 x_2 + \ldots + \mathcal{H}'_n x_n$$

$$\bar{\mathcal{H}}^{(1)}(t_1) = -\frac{i}{2}t_1\{[\mathcal{H}'_2 x_2, \mathcal{H}'_1 x_1] + [\mathcal{H}'_3 x_3, \mathcal{H}'_1 x_1]$$

$$+ [\mathcal{H}'_3 x_3, \mathcal{H}'_2 x_2] + \ldots\} \quad (3.3.10)$$

and higher terms. It is important to note that $\bar{\mathcal{H}}^{(1)}(t_1)$ is proportional to t_1. Similarly, the higher-order terms $\bar{\mathcal{H}}^{(k)}(t_1)$ are proportional to t_1^k.

This immediately leads to the basic result: A time-independent average Hamiltonian $\bar{\mathcal{H}}$ which describes the motion during the evolution time t_1 can be defined if *all* transformed Hamiltonians $\mathcal{H}_j'$, given by eqn (3.3.4), commute

$$[\mathcal{H}_j', \mathcal{H}_k'] = 0. \qquad (3.3.11)$$

In the following section, we present an example to illustrate this theorem.

3.3.2. Average Hamiltonian in spin–echo experiments

One of the most frequently used sequences involving an aperiodic perturbation is the spin–echo experiment (3.25), where a π-pulse is applied in the centre of the evolution period t_1. The Hamiltonians $\mathcal{H}_1$ and $\mathcal{H}_2$ in the first and second halves are identical to the unperturbed Hamiltonian $\mathcal{H}$.

Here the question arises again whether it is possible to describe the motion leading to the echo at $t = t_1$ by an average Hamiltonian, and it should be determined which terms of the Hamiltonian $\mathcal{H}$ contribute to the average Hamiltonian and which terms are refocused by the π-pulse.

The basic result of the previous section immediately gives the condition for the existence of an average Hamiltonian

$$[\mathcal{H}_1', \mathcal{H}_2'] = [\mathcal{H}_1', \mathcal{H}] = 0 \qquad (3.3.12)$$

where $\mathcal{H}_1'$ is the transformed Hamiltonian during the first time-period,

$$\mathcal{H}_1' = R_\pi \mathcal{H} R_\pi^{-1}.$$

We can cast the condition eqn (3.3.12) in a more convenient form by dividing $\mathcal{H}$ into

$$\mathcal{H} = \mathcal{H}^{(s)} + \mathcal{H}^{(a)}, \qquad (3.3.13)$$

the first term being symmetric and remaining invariant under a π-rotation

$$R_\pi \mathcal{H}^{(s)} R_\pi^{-1} = \mathcal{H}^{(s)}, \qquad (3.3.14)$$

and the second term being antisymmetric, such that it changes sign under a π-rotation,

$$R_\pi \mathcal{H}^{(a)} R_\pi^{-1} = -\mathcal{H}^{(a)}. \qquad (3.3.15)$$

Such a division is always possible. It is now easily verified that eqn (3.3.12) requires that $\mathcal{H}^{(s)}$ and $\mathcal{H}^{(a)}$ commute. This leads to the general requirement that the symmetric and antisymmetric parts of the Hamiltonian must commute to permit the introduction of an average Hamiltonian in a spin–echo experiment.

If $\mathcal{H}^{(s)}$ and $\mathcal{H}^{(a)}$ commute, we find the average Hamiltonian

$$\bar{\mathcal{H}} = \tfrac{1}{2}\{\mathcal{H}'_1 + \mathcal{H}'_2\} = \tfrac{1}{2}\{\mathcal{H}^{(s)} - \mathcal{H}^{(a)} + \mathcal{H}^{(s)} + \mathcal{H}^{(a)}\} = \mathcal{H}^{(s)}. \quad (3.3.16)$$

The symmetric part of $\mathcal{H}$ remains while the antisymmetric part is refocused by the π-pulse. Symmetric terms are all bilinear terms, like the homonuclear scalar and dipolar coupling and the quadrupolar interaction, while the chemical shielding and heteronuclear interactions (provided the π-pulse is applied to one of the nuclear species only) represent antisymmetric terms.

Let us consider three examples.

1. Weakly-coupled homonuclear spin system,

$$\mathcal{H} = \mathcal{H}_{ZI} + \mathcal{H}_{II} \quad (3.3.17)$$

with the Zeeman interaction $\mathcal{H}_{ZI}$ and the coupling term $\mathcal{H}_{II} = \sum 2\pi J_{kl} I_{kz} I_{lz}$. $\mathcal{H}_{ZI}$ is antisymmetric while $\mathcal{H}_{II}$ is symmetric under a π-pulse. The two terms commute. Therefore an average Hamiltonian can be formulated and

$$\bar{\mathcal{H}} = \mathcal{H}_{II}. \quad (3.3.18)$$

2. Strongly-coupled homonuclear spin system,

$$\mathcal{H} = \mathcal{H}_{ZI} + \mathcal{H}_{II} \quad (3.3.19)$$

with $\mathcal{H}_{II} = \sum 2\pi J_{kl} \mathbf{I}_k \cdot \mathbf{I}_l$. The two terms no longer commute and it is not possible to formulate an average Hamiltonian.

3. Heteronuclear spin systems with strong II and weak SS spin coupling in an experiment with pulses applied to the S spins,

$$\mathcal{H} = \mathcal{H}_{ZI} + \mathcal{H}_{II} + \mathcal{H}_{IS} + \mathcal{H}_{ZS} + \mathcal{H}_{SS}. \quad (3.3.20)$$

The terms $\mathcal{H}_{ZI}$, $\mathcal{H}_{II}$, and $\mathcal{H}_{SS}$ are symmetric, and $\mathcal{H}_{IS}$ and $\mathcal{H}_{ZS}$ are antisymmetric under an S-spin π-pulse. The key commutator is

$$[\mathcal{H}_{II}, \mathcal{H}_{IS}] \neq 0. \quad (3.3.21)$$

This implies that the symmetric and antisymmetric parts of $\mathcal{H}$ do not commute and it is not possible to define an average Hamiltonian, despite the fact that the observed S spins are weakly coupled. The practical implications are discussed in more detail in § 7.2.3.

A related question is: which terms of a Hamiltonian affect the echo amplitude in an experiment with a single or several π pulses (3.25)? This problem has been investigated (3.34). The main result can be summarized by the following theorem.

The necessary condition that a term $\mathcal{H}_k$ of the free-precession Hamiltonian affects the echo amplitude in a spin–echo experiment is that it belongs to a chain of non-commuting terms $\{\mathcal{H}_g\}$ of which at least one term does not commute with the observable F_x and at least one term is not antisymmetric under a π-rotation about a transverse axis.

If all non-commuting terms commute with the observable operator F_x, the effect of these terms will not be observable. If all terms change sign under a π-rotation, they will disappear altogether from the average Hamiltonian.

As an example consider again the case of a heteronuclear spin system with strong II interactions, eqn (3.3.20), with pulses applied exclusively to the S spins. In addition to the commutator of eqn (3.3.21), we have

$$[\mathcal{H}_{ZI}, \mathcal{H}_{II}] \neq 0. \qquad (3.3.22)$$

The three terms $\mathcal{H}_{ZI}$, $\mathcal{H}_{II}$, and $\mathcal{H}_{IS}$ form a chain of non-commuting terms. One of these does not commute with the observable

$$[\mathcal{H}_{IS}, S_x] \neq 0. \qquad (3.3.23)$$

In addition, $\mathcal{H}_{ZI}$ and $\mathcal{H}_{II}$ are symmetric under a π pulse applied to the S spins. Therefore all three terms $\mathcal{H}_{ZI}$, $\mathcal{H}_{II}$, and $\mathcal{H}_{IS}$ affect the echo amplitude. The term $\mathcal{H}_{SS}$ is symmetric and does not commute with S_x; hence it also affects the echo amplitude. The remaining term $\mathcal{H}_{ZS}$ in eqn (3.3.20) is antisymmetric and commutes with all other terms, and therefore has no effect on the echo amplitude.

3.3.3. Cancellation of irrelevant terms

There may be situations where the evolution in the presence of aperiodic perturbations can be described by an average Hamiltonian although the basic condition of eqn (3.3.11) is violated. This may happen when the Hamiltonian contains terms that are irrelevant for a particular experiment.

For simplicity, we restrict the discussion to situations where the Hamiltonian $\mathcal{H}(t)$ is switched only once at $t_x = xt_1$ from $\mathcal{H}_1$ to $\mathcal{H}_2$. Let us assume that $\mathcal{H}_1$ and $\mathcal{H}_2$ do *not* commute. The observed magnetization is then proportional to the expectation value $\langle Q \rangle$ of the observable operator Q

$$\langle Q \rangle = \mathrm{tr}\{QU_2U_1\sigma(0)U_1^{-1}U_2^{-1}\} \qquad (3.3.24)$$

with the propagators

$$U_1 = \exp\{-i\mathcal{H}_1 xt_1\}$$

and

$$U_2 = \exp\{-i\mathcal{H}_2(1-x)t_1\}. \qquad (3.3.25)$$

We briefly discuss two special cases.

1. $\mathcal{H}_1$ can be divided into two commuting parts $\mathcal{H}_{1c}$ and $\mathcal{H}_{1n}$

$$\mathcal{H}_1 = \mathcal{H}_{1c} + \mathcal{H}_{1n}, \quad [\mathcal{H}_{1c}, \mathcal{H}_{1n}] = 0 \quad (3.3.26)$$

with the properties

$$[\mathcal{H}_{1c}, \mathcal{H}_2] = 0, \quad [\mathcal{H}_{1n}, \mathcal{H}_2] \neq 0. \quad (3.3.27)$$

We obtain the expectation value

$$\langle Q \rangle = \text{tr}\{QU_2 U_{1c} U_{1n} \sigma(0) U_{1n}^{-1} U_{1c}^{-1} U_2^{-1}\}. \quad (3.3.28)$$

In situations where it is possible to prepare the system initially in a state $\sigma(0)$ which commutes with $\mathcal{H}_{1n}$

$$[\sigma(0), \mathcal{H}_{1n}] = 0, \quad (3.3.29)$$

it is obvious that U_{1n} drops out of eqn (3.3.28) and we obtain

$$\langle Q \rangle = \text{tr}\{QU_2 U_{1c} \sigma(0) U_{1c}^{-1} U_2^{-1}\}$$
$$= \text{tr}\{Q\bar{U}(t_1) \sigma(0) \bar{U}(t_1)^{-1}\} \quad (3.3.30)$$

with

$$\bar{U}(t_1) = \exp\{-i\bar{\mathcal{H}} t_1\} \quad (3.3.31)$$

and with the average Hamiltonian

$$\bar{\mathcal{H}} = x\mathcal{H}_{1c} + (1-x)\mathcal{H}_2. \quad (3.3.32)$$

For the special initial condition of eqn (3.3.29), the non-commuting part of $\mathcal{H}_1$ becomes ineffective and an average Hamiltonian, given by eqn (3.3.32) can be formulated although $\mathcal{H}_1$ and $\mathcal{H}_2$ do not commute.

2. $\mathcal{H}_2$ can be divided into two commuting parts $\mathcal{H}_{2c}$ and $\mathcal{H}_{2n}$

$$\mathcal{H}_2 = \mathcal{H}_{2c} + \mathcal{H}_{2n}, \quad [\mathcal{H}_{2c}, \mathcal{H}_{2n}] = 0 \quad (3.3.33)$$

with the properties

$$[\mathcal{H}_{2c}, \mathcal{H}_1] = 0, \quad [\mathcal{H}_{2n}, \mathcal{H}_1] \neq 0. \quad (3.3.34)$$

Then $\langle Q \rangle$ can be written in the form

$$\langle Q \rangle = \text{tr}\{QU_{2n} U_{2c} U_1 \sigma(0) U_1^{-1} U_{2c}^{-1} U_{2n}^{-1}\}$$
$$= \text{tr}\{U_{2n}^{-1} Q U_{2n} U_{2c} U_1 \sigma(0) U_1^{-1} U_{2c}^{-1}\}. \quad (3.3.35)$$

When the observable operator Q commutes with U_{2n},

$$[Q, \mathcal{H}_{2n}] = 0, \quad (3.3.36)$$

U_{2n} drops out of eqn (3.3.35) and we can again introduce an average Hamiltonian

$$\langle Q \rangle = \text{tr}\{Q\bar{U}(t_1)\sigma(0)\bar{U}(t_1)^{-1}\} \quad (3.3.37)$$

with
$$\bar{U}(t_1) = \exp\{-i\bar{\mathcal{H}}t_1\} \tag{3.3.38}$$
and
$$\bar{\mathcal{H}} = x\mathcal{H}_1 + (1-x)\mathcal{H}_{2c} \tag{3.3.39}$$

In this case, it is also possible to formulate an average Hamiltonian $\bar{\mathcal{H}}$ provided the observable operator Q commutes according to eqn (3.3.36).

This treatment can be generalized to situations with more than two periods. However, the possibility of suppressing a non-commuting part of the Hamiltonian is restricted to the first and the last period.

As an example, let us anticipate a case to be discussed more thoroughly in § 4.7.7 under the heading 'Illusions of decoupling'. We consider a heteronuclear spin system with the Hamiltonian $\mathcal{H}$ of eqn (3.3.20) and apply I spin decoupling only during the initial period xt_1. This leads to the effective Hamiltonians

$$\mathcal{H}_1 = \mathcal{H}_S + \bar{\mathcal{H}}_I$$

and

$$\mathcal{H}_2 = \mathcal{H} \tag{3.3.40}$$

with

$$\mathcal{H}_S = \mathcal{H}_{ZS} + \mathcal{H}_{SS},$$
$$\bar{\mathcal{H}}_I = \mathcal{H}_{II} + \mathcal{H}_I^{\text{r.f.}}.$$

According to eqn (3.3.26), we can make the identification

$$\mathcal{H}_{1c} = \mathcal{H}_S, \qquad \mathcal{H}_{1n} = \bar{\mathcal{H}}_I. \tag{3.3.41}$$

The special initial condition $\sigma(0) = S_{lx}$ commutes with $\mathcal{H}_{1n}$, and we find with eqn (3.3.32) the average Hamiltonian $\bar{\mathcal{H}}$ for the entire period t_1

$$\bar{\mathcal{H}} = x\{\mathcal{H}_{ZI} + \mathcal{H}_{II} + \mathcal{H}_{IS}\} + \mathcal{H}_{ZS} + \mathcal{H}_{SS} \tag{3.3.42}$$

This leads to scaling of the heteronuclear multiplet structure in the S-spin spectrum.

On the other hand, if we assume an initial condition with antiphase coherence $S_{lx}I_{kz}$, the commutation with $\mathcal{H}_{1n}$ breaks down and an average Hamiltonian can no longer be defined (3.38).

4
ONE-DIMENSIONAL FOURIER SPECTROSCOPY

The advantages of Fourier spectroscopy over conventional slow-passage methods are well known. Although sensitivity was the primary objective for the introduction of Fourier techniques in 1965 (4.1, 4.2), it is the versatility of time-domain experiments which explains the extraordinary development of modern NMR spectroscopy. On the one hand, Fourier spectroscopy allows the direct study of time-dependent phenomena such as relaxation and exchange processes. On the other hand, pulse experiments can easily be extended to incorporate polarization and coherence transfer. The separation of excitation and detection in distinct time intervals is of importance for many experiments and leads naturally to the segmentation of time in two-dimensional Fourier spectroscopy. In comparison with CW methods, Fourier spectroscopy has the additional advantage of circumventing lineshape distortions associated with fast sweeping and saturation.

Fourier spectroscopy has become the universal technique for acquiring spectroscopic data in all disciplines of NMR, including high-resolution spectroscopy of isotropic liquids (4.3, 4.4), liquid crystals, and solids (4.5, 4.6). In this sense Fourier spectroscopy has unified the various fields. This made it possible to construct multipurpose NMR spectrometers which can handle all aspects of NMR. Fourier spectroscopy helped in a sense to bridge the technological gap between NMR applications in physics, chemistry, and biology.

The fact that a dedicated computer is needed in Fourier spectrometers has beneficial side-effects: sophisticated filtering and convolution procedures can be carried out digitally, and the computer can be used to control various spectrometer functions in view of automation. On the other hand, Fourier spectroscopy is associated with a number of drawbacks, such as the limited dynamic range, the difficulty of covering wide spectral ranges, interference of subsequent scans, aliasing associated with insufficient sampling rates, and, for r.f. pulses with large rotation angles, the loss of a direct proportionality between the amplitude of a line and the population difference between the corresponding pair of eigenstates. Various means of solving these problems will be discussed in the following sections.

We shall begin the discussion of Fourier spectroscopy with a brief recapitulation of response theory, which forms the basis of Fourier transform techniques, and proceed to describe the dynamics of classical magnetization in spin systems without couplings (§ 4.2). The fundamental aspects of the relative sensitivity of Fourier and slow-passage spectroscopy are discussed in § 4.3. In coupled spin systems, Fourier spectra are not necessarily equivalent to slow-passage spectra, and non-equilibrium populations lead to deviations discussed in § 4.4. In spin systems with resolved couplings, many experimental manipulations can be carried out, both to enhance the sensitivity and to gain insight into the nature of the coupling network (§ 4.5). Various methods for studying relaxation, chemical exchange, and diffusion are reviewed in § 4.6, and § 4.7 is devoted to Fourier double resonance.

4.1. Response theory

Spectroscopy may be regarded as a subdiscipline of general response theory, the fundamental purpose being the characterization of a system by input/output relations. The methodological development of spectroscopy has been strongly influenced by the basic concepts of response theory. This is particularly evident in the context of NMR Fourier spectroscopy. But response theory is also of central importance for data-processing.

In the following, we discuss response theory in more abstract terms without reference to particular applications. The properties of the system to be investigated shall be formally represented by the system operator Φ. The system is perturbed by the input signal $x(t)$, which leads to the reaction or response signal $y(t)$ as shown in Fig. 4.1.1. The general input/output relations may be represented by

$$y(t) = \Phi\{x(t)\}. \tag{4.1.1}$$

In some situations, the input and output signals $x(t)$ and $y(t)$ can be vectors of arbitrary dimensions. We shall restrict our discussion to time-invariant systems, where the operator Φ is explicitly time-independent.

Response theory and system theoretical concepts are of central importance in electronic engineering, where they have been developed initially. They have found widespread application in subjects ranging from sociology to nuclear physics. Many textbook treatments are limited to *linear* systems, where it is possible to develop a closed theoretical

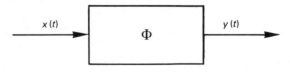

FIG. 4.1.1. The system operator Φ describes the relation between input $x(t)$ and output $y(t)$ for an arbitrary system.

framework, irrespective of the specific properties of the system under investigation (4.7–4.13).

The analysis of *non-linear* systems, where the response $y(t)$ is a non-linear function of $x(t)$, is more difficult. Few results of general applicability have been obtained so far. In most cases, the specific properties of the system under consideration become important. A unified treatment is possible, however, for 'weak' non-linearities where power series expansions still converge (4.14–4.17).

4.1.1. Linear response theory

A system represented by the operator Φ is said to be linear if the superposition principle is valid, i.e.

$$\Phi\{x_1(t) + x_2(t)\} = \Phi\{x_1(t)\} + \Phi\{x_2(t)\} = y_1(t) + y_2(t). \quad (4.1.2)$$

In this case, it is possible to represent an arbitrary input signal $x(t)$ by a linear combination of basis functions $g_k(t)$

$$x(t) = \sum_k X_k g_k(t) \quad (4.1.3)$$

or, under certain conditions, by an integral

$$x(t) = \int X(p)g(p, t)\,dp \quad (4.1.4)$$

and to consider the response of each constituent function separately

$$y(t) = \Phi\{x(t)\} = \sum_k X_k \Phi\{g_k(t)\}$$

$$= \int X(p)\Phi\{g(p, t)\}\,dp. \quad (4.1.5)$$

Of particular importance is the representation of the input signal in terms of Dirac $\delta(t)$ functions

$$x(t) = \int_{-\infty}^{\infty} x(\tau)\,\delta(t-\tau)\,d\tau. \quad (4.1.6)$$

The calculation of the response $y(t)$ only requires knowledge of the response $\Phi\{\delta(t)\}$ which is called the *impulse response* $h(t)$ of the system

$$h(t) = \Phi\{\delta(t)\}. \quad (4.1.7)$$

This immediately leads to the central result

$$y(t) = \int_{-\infty}^{\infty} h(\tau)x(t-\tau)\,d\tau$$

$$= h(t) * x(t). \quad (4.1.8)$$

Thus the response for an arbitrary input is equal to the convolution of the input signal with the impulse response of the system. The impulse response therefore completely characterizes a linear and time-independent system, and allows one to predict the response to arbitrary perturbations. Because of the causal character of any physical system, it is evident that

$$h(t) = 0 \quad \text{for} \quad t < 0 \qquad (4.1.9)$$

since the reaction can not precede its cause.

Integrating eqn (4.1.8) by parts, one finds another representation of the response of a linear system

$$y(t) = \int_{-\infty}^{\infty} \gamma(\tau) x'(t-\tau) \, d\tau$$

$$= \gamma(t) * x'(t) \qquad (4.1.10)$$

where $x'(t) = \partial x(t)/\partial t$ and $\gamma(t)$ is the *unit step response*,

$$\gamma(t) = \Phi\{u(t)\} \qquad (4.1.11)$$

with

$$u(t) = \int_{-\infty}^{t} \delta(\tau) \, d\tau = \begin{cases} 0 & \text{for} \quad t < 0 \\ 1 & \text{for} \quad t > 0. \end{cases} \qquad (4.1.12)$$

In this case, the response $y(t)$ is represented by the convolution of the derivative of the excitation $x'(t)$ with the unit step response $\gamma(t)$ of the system.

It is particularly convenient to represent the properties of a linear system in terms of the response to its eigenfunctions. It can easily be shown that exponential functions $\exp\{pt\}$ are eigenfunctions of any linear time-independent system operator for arbitrary complex p

$$y(t) = \Phi\{e^{pt}\} = H'(p) \, e^{pt}. \qquad (4.1.13)$$

The input function e^{pt} is reproduced at the output, multiplied by the complex eigenvalues $H'(p)$ which measure the change in phase and amplitude caused by the system. $H'(p)$ can be considered as a continuous function of p and is called the *transfer function* of the system.

For the special choice of a harmonic input function, $p = i\omega = i2\pi f$, one obtains the *frequency response function*

$$H(\omega) = H'(i\omega). \qquad (4.1.14)$$

Using the input signal $\exp\{i\omega t\}$ to compute the response by means of eqn (4.1.8), one finds

$$y(t) = \int_{-\infty}^{\infty} h(\tau) \, e^{i\omega t} \cdot e^{-i\omega \tau} \, d\tau = H(\omega) \, e^{i\omega t} \qquad (4.1.15)$$

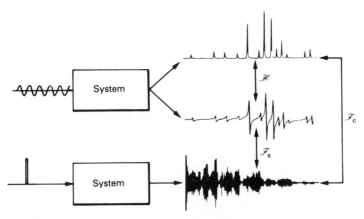

FIG. 4.1.2. The relation between frequency and impulse response for a linear system. Real and imaginary parts of the frequency response function $H(\omega)$ are Hilbert transforms of each other (denoted by $\mathcal{H}$). They are related to the impulse response via a cosine ($\mathcal{F}_c$) or a sine Fourier transformation ($\mathcal{F}_s$), respectively. (Reproduced from Ref. 4.130.)

or

$$H(\omega) = \int_{-\infty}^{\infty} h(t)\, e^{-i\omega t}\, dt \qquad (4.1.16)$$

and

$$h(t) = \frac{1}{2\pi} \int_{-\infty}^{\infty} H(\omega)\, e^{i\omega t}\, d\omega.$$

This shows that the impulse response $h(t)$ and the frequency response function $H(\omega)$ form a Fourier transform pair (see Fig. 4.1.2). Both functions fully represent the properties of any linear time-independent system.

Equation (4.1.16) is of fundamental importance in linear response theory. In the context of Fourier spectroscopy, the free induction decay can be identified with the impulse response, and the complex spectrum is equivalent to the frequency response function. For a real impulse response $(h(t) = h(t)^*)$, it follows that

$$H(-\omega) = H(\omega)^*. \qquad (4.1.17)$$

The real part of the frequency response function is even, while its imaginary part is an odd function of frequency. In the context of quadrature detection, we shall also encounter complex impulse response functions.

The causality expressed by eqn (4.1.9) leads to the so-called *dispersion relations* or *Kramers–Kronig relations* which express the fact that real and imaginary parts of the frequency response function of a linear time-invariant system (see Fig. 4.1.2) can be computed from each other by means of a Hilbert transformation (4.7, 4.10, 4.18–4.21).

$$\text{Re}\{H(\omega)\} = \frac{1}{\pi} \int_{-\infty}^{\infty} \frac{\text{Im}\{H(\omega')\}}{\omega - \omega'} d\omega'$$

$$\text{Im}\{H(\omega)\} = -\frac{1}{\pi} \int_{-\infty}^{\infty} \frac{\text{Re}\{H(\omega')\}}{\omega - \omega'} d\omega' \qquad (4.1.18)$$

or, in a more compact notation,

$$H(\omega) = \frac{-i}{\pi} \int_{-\infty}^{\infty} \frac{H(\omega')}{\omega - \omega'} d\omega'. \qquad (4.1.19)$$

Applied to Fourier spectroscopy, these relations imply that it is possible to calculate a pure phase spectrum, e.g. pure absorption, from the real part of $H(\omega)$ without knowledge of the imaginary part of the spectrum (4.21).

It is well known that a nuclear spin system is inherently non-linear in terms of its input/output relations. This is manifested by the non-linearity of the Bloch equations (see eqn (4.2.1)), as well as by the non-linearity of the density operator equation (eqn (2.1.17)) with regard to r.f. stimulations. Consequently, the linearity defined in eqn (4.1.2) does not generally apply.

It may appear astonishing that the concepts of linear response and Fourier transform theory are nevertheless applicable. This is a consequence of the fact that the non-linear effect of an r.f. pulse merely defines the initial conditions. Thus after the pulse, the transverse component is $M_x(0_+) \propto M_0 \sin(-\gamma B_1 \tau_p)$ (see eqn (4.2.14)). However, the subsequent free evolution proceeds in the absence of r.f. fields. The equations of motion for free precession are linear with respect to the magnetization vector **M** or the density operator σ. The superposition principle is indeed valid for the magnetization and a Fourier transformation of the free induction signal remains meaningful.

On the other hand, the non-linearity of the spin system has to be taken into consideration to account for the effects of strong r.f. fields, for example in multiple-pulse experiments or in stochastic resonance.

4.1.2. Time and frequency domains

The duality between the time domain and the frequency domain, which can both serve for the acquisition, manipulation, or representation of spectroscopic data, is a central concept not only in Fourier spectroscopy, but for measurement techniques in general. Both domains can have one or more dimensions. In this chapter, we limit the discussion to one dimension, while extensions to two dimensions are treated in Chapter 6. Occasionally even higher dimensions will be encountered, for example in NMR imaging (Chapter 10).

Both time and frequency domains can carry the same information, although in different forms, but it may be more convenient to consider one or the other domain in a particular situation. It is a great asset that it is possible to transform at will between the two domains to facilitate either the spectroscopic experiment, the data-processing, or the representation of the data.

The Fourier transformation (4.18, 4.22, 4.23) establishes a unique correspondence between the functions $s(t)$ in time domain and the functions $S(\omega)$ or $S(f)$ in frequency domain.

$$S(\omega) = \int_{-\infty}^{\infty} s(t)\, e^{-i\omega t}\, dt,$$

$$S(f) = \int_{-\infty}^{\infty} s(t)\, e^{-i2\pi ft}\, dt,$$

$$s(t) = \frac{1}{2\pi} \int_{-\infty}^{\infty} S(\omega)\, e^{i\omega t}\, d\omega$$

$$= \int_{-\infty}^{\infty} S(f)\, e^{i2\pi ft}\, df. \qquad (4.1.20)$$

While the frequency variable $\omega = 2\pi f$ (in units of rad/s) is convenient for formal calculations, the frequency variable f (in Hz) is often more suitable to represent spectroscopic data. Note that the functions $S(\omega)$ and $S(f)$ are not strictly the same functions, since they differ by factors 2π in the arguments.

The Fourier relations in eqn (4.1.20) demonstrate that the transformations between the two domains are almost fully symmetrical (except for the change of sign of the imaginary unit) and relations that are valid in one direction of the transformation are also valid in the opposite direction. This is important for signal-processing where the same procedures can be used for filtering, regardless of whether the signals have been recorded in time domain or in frequency domain.

In the following, some of the fundamental relations between the two domains will be summarized. Let us assume that $s(t)$ and $S(\omega)$, and $s(t)$ and $S(f)$ form Fourier transform pairs according to eqn (4.1.20). Then the following theorems hold.

1. *Similarity theorem.*

$$\mathcal{F}\{s(at)\} = \frac{1}{|a|} S(\omega/a) = \frac{1}{|a|} S(f/a). \quad (4.1.21)$$

The Fourier transform of a function with a scaled variable is scaled in a reciprocal manner, and its amplitude is multiplied by the reciprocal scaling factor, so that the integral remains constant. Broadening of a function in one domain implies narrowing of its Fourier transform and vice versa.

2. *Shift theorem.*

$$\mathcal{F}\{s(t-\tau)\} = e^{-i\omega\tau} S(\omega) = e^{-i2\pi f\tau} S(f). \quad (4.1.22)$$

A shift of a function along the time axis causes a frequency-dependent phase change in the frequency domain. For the corresponding relation with interchanged time and frequency domains, the exponent of the phase factor must be changed in sign. This theorem is the basis for understanding delayed acquisition and fold-over correction techniques (§ 6.6.2).

3. *Derivative theorem.*

$$\mathcal{F}\left\{\frac{d^k}{dt^k} s(t)\right\} = (i\omega)^k S(\omega)$$
$$= (i2\pi f)^k S(f). \quad (4.1.23)$$

The Fourier transform of a derivative of a time function appears filtered through a high-pass filter. The theorem for the reverse transformation involves a sign change of the imaginary unit.

4. *Convolution theorem.* The Fourier transform of the convolution integral of two functions $r(t)$ and $s(t)$, defined by

$$r(t) * s(t) = \int_{-\infty}^{\infty} r(\tau) s(t-\tau)\, d\tau, \quad (4.1.24)$$

can be expressed as the product of the corresponding Fourier transforms $R(\omega)$ and $S(\omega)$

$$\mathcal{F}\{r(t) * s(t)\} = R(\omega) \cdot S(\omega) = R(f) \cdot S(f),$$
$$\mathcal{F}^{-1}\{R(\omega) * S(\omega)\} = \frac{1}{2\pi} r(t) \cdot s(t),$$
$$\mathcal{F}^{-1}\{R(f) * S(f)\} = r(t) \cdot s(t). \quad (4.1.25)$$

In the context of spectroscopy, the convolution theorem plays a central role, and justifies by itself the use of the Fourier transformation. This theorem implies that any filtering process which can be expressed by a convolution according to eqn (4.1.8), can be transformed into a product in the co-domain. In most cases it is easier to calculate a Fourier transformation and compute the product in the co-domain than to evaluate the convolution integral (or the corresponding convolution sum) directly. This simplification relies on the fact that a Fourier transformation is equivalent to an expansion in terms of eigenfunctions of a linear time-invariant system, eqn (4.1.13).

5. *Power theorem.*

$$\int_{-\infty}^{\infty} |s(t)|^2 \, dt = \frac{1}{2\pi} \int_{-\infty}^{\infty} |S(\omega)|^2 \, d\omega = \int_{-\infty}^{\infty} |S(f)|^2 \, df. \qquad (4.1.26)$$

The signal energy can be computed by integration either in time or in frequency space. This is important for sensitivity considerations (§ 4.3.1.4).

4.1.3. Linear data-processing

It is very rare that a spectrum obtained by Fourier transformation of a free induction decay satisfies all demands with regard to optimum presentation. In most cases, it is desirable to subject the data to a linear filtering procedure to optimize the appearance of the spectrum. The restriction to linear processes is justified since it allows the processing of overlapping resonance lines without causing interference effects.

Linear transformation processes can always be represented by a convolution integral of the signal and the impulse response of the filter process, as inferred in § 4.1.1. In the context of Fourier spectroscopy, the spectrum $S(\omega)$ must be submitted to a filtering process characterized by a frequency-domain filter function $H(\omega)$

$$S_f(\omega) = H(\omega) * S(\omega) \qquad (4.1.27)$$

in analogy to eqn (4.1.8). The convolution integral can be evaluated directly, but one may take advantage of the convolution theorem of eqn (4.1.25) and multiply the time-domain signal $s(t)$ with the corresponding time-domain filter function $h(t)$

$$s_f(t) = h(t) \cdot s(t) \qquad (4.1.28)$$

where $h(t)$ is the Fourier transform of $H(\omega)$.

Note that the significance of $h(t)$ and $H(\omega)$ has been interchanged in comparison with § 4.1.1. We may identify $H(\omega)$ with the 'impulse response', while $h(t)$ now represents the 'frequency response' of the filter. To avoid semantic difficulties in the distinction of the two

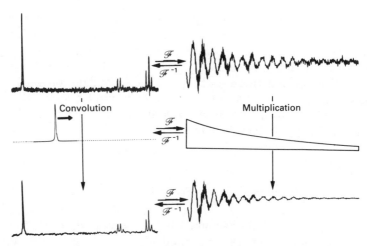

FIG. 4.1.3. A linear filtering process, which in this example enhances the sensitivity, amounts to a convolution in frequency domain (left) while the equivalent process in time domain (right) is a multiplication with a time-domain filter function. (Reproduced from Ref. 4.58.)

functions, we prefer to use the more neutral terms 'frequency-domain' and 'time-domain filter function' for $H(\omega)$ and $h(t)$, respectively.

Equation (4.1.28) demonstrates that filtering in Fourier spectroscopy boils down to the multiplication of the free induction signal with a suitable weighting function $h(t)$ prior to Fourier transformation (Fig. 4.1.3). It is one of the virtues of Fourier spectroscopy that filtering can be achieved in this extremely simple and convenient manner, perhaps with the only disadvantage that a Fourier transformation must be computed before the effect of a filter function on the spectrum can be appreciated.

The purposes of filtering may be quite diverse, and we shall mention only a few of the many possible applications.

1. Matched filtering to maximize the sensitivity (signal-to-noise ratio) in one- and in two-dimensional spectroscopy (see §§ 4.3 and 6.8).

2. Resolution enhancement by artificially narrowing the resonance lines.

3. Lineshape transformation, for example the Lorentz–Gauss transformation to eliminate the 'star effect' in two-dimensional spectroscopy (§ 6.5.6.2).

4. Apodization of free induction decays to suppress oscillating signal tails ('ripple') in the spectrum.

5. Pseudo-echo filtering to eliminate dispersive contributions to lineshapes in two-dimensional spectroscopy (§ 6.5.6.3).

6. Correction of instrumental distortions, caused for example by a finite response time.

In the following, we shall briefly discuss apodization and resolution enhancement, as they are not adequately covered in the later parts of this volume. A few remarks on resolution enhancement by zero-filling and linear prediction methods will conclude this section. More detailed treatments on filtering can be found in Refs. 4.2 and 4.24–4.26.

4.1.3.1. Apodization

In practical Fourier spectroscopy, the acquisition time t_{max} of the free induction decay is always limited and the signal $s(t)$ is known only for $0 \leq t \leq t_{max}$. This may severely limit the resolution of the spectrum, since one is restricted to calculating the Fourier transform of a truncated signal

$$s_{trunc}(t) = s(t) \quad \text{for} \quad t \leq t_{max},$$
$$= 0 \quad \text{for} \quad t > t_{max}. \quad (4.1.29)$$

The truncated signal $s_{trunc}(t)$ can be thought of as the product of the untruncated signal $s(t)$ with a rectangular weighting function,

$$s_{trunc}(t) = s(t) \cdot \Pi\left(\frac{t}{2t_{max}}\right) \quad (4.1.30)$$

with

$$\Pi(x) = 1 \quad \text{for} \quad -\tfrac{1}{2} < x < \tfrac{1}{2},$$
$$= 0 \quad \text{for} \quad |x| > \tfrac{1}{2}. \quad (4.1.31)$$

The corresponding Fourier spectrum is therefore obtained by convolution of the undistorted spectrum $S(f)$ with the Fourier transform of the rectangular weighting function

$$S_{trunc}(f) = S(f) * 2t_{max} \operatorname{sinc}(2t_{max}f) \quad (4.1.32)$$

The $\operatorname{sinc}(x)$ function, defined by

$$\operatorname{sinc}(x) = \frac{\sin \pi x}{\pi x}, \quad (4.1.33)$$

produces oscillatory signal tails ('ripple'), as shown in Fig. 4.1.4(a), which may be highly undesirable as it severely limits resolution (4.27).

The oscillations arise from the sharp cut-off of the free induction signal which introduces high frequencies. It is the purpose of *apodization* to modify the envelope of the truncated signal by multiplication with a weighting function such that these oscillations are largely suppressed. It is obvious that the envelope must tend smoothly to zero at $t = t_{max}$ to prevent such oscillations. At the same time, care should be taken to avoid excessive line-broadening.

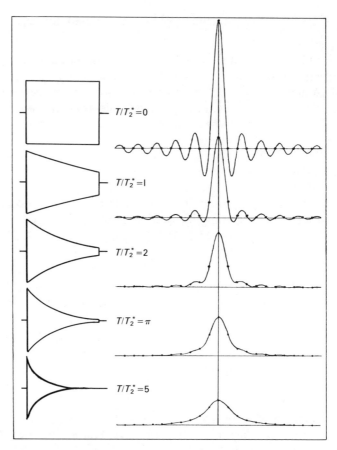

FIG. 4.1.4. Lineshapes obtained by Fourier transformation of truncated free induction decays of length $t_{max} = T$. (a) No decay of the signal ($T_2^* = \infty$): the full width at half-height of the central lobe is $\Delta f = 0.604/t_{max}$. (b) to (e): Signals with increasing decay rates $T_2^* = T$, $T/2$, T/π, and $T/5$. Note that the ripple amplitude decreases. If the free induction signal is extended by unrestricted zero-filling, the continuous curve is approached. If the signal is only extended by a factor two by zero-filling, the Fourier transformation gives the sampling points indicated by dots. A Fourier transform of the truncated signal without zero-filling only gives a sampling point for every second dot. (Adapted from Ref. 4.27.)

The selection of a suitable weighting function $h(t)$ to apodize truncated signals has been discussed in numerous papers in various fields of science, such as electrical communication, astronomy, and infra-red Fourier spectroscopy (4.28–4.37) as well as in NMR (4.2, 4.26). The approaches range from inspired guesswork to computer optimizations and purely theoretical derivations.

In the context of digital signal processing by Fourier transformation, apodization is often called 'windowing' (4.28–4.31). This term suggests that truncation errors can be minimized by properly shaping the window through which the data is observed. A certain broadening has to be admitted to minimize the amplitude of the ripple, and the larger the acceptable broadening the better the suppression of the ripple. The theoretical optimum is reached by the so-called Dolph–Chebycheff window (4.38, 4.39). This class of windows minimizes the relative ripple amplitude for any predetermined broadening B of the resonance lines.

Unfortunately there is no analytical expression for the optimum weighting function $h(t)$, but it can be obtained numerically by Fourier-transforming the corresponding frequency domain filter function $H(f)$

$$H(f) = \frac{\cos\{2P \cos^{-1}[z_0 \cos(\pi f/v_s)]\}}{\cosh\{2P \cosh^{-1}(z_0)\}} \qquad (4.1.34)$$

where $P+1$ is the number of sampling points of the free induction decay, v_s is the sampling rate, and the quantity

$$z_0 = [\cos(\pi B/2v_s)]^{-1} \approx 1 + \pi^2 B^2/(8v_s^2) \qquad (4.1.35)$$

is determined by the allowed broadening B (expressed in Hz). $H(f)$ corresponds then to the lineshape obtained for a line of infinitely narrow natural width.

In most practical applications it is not necessary to afford the trouble of adjusting the apodization function to the number of sampling points. Numerous simple approximations are known, particularly in the field of digital data-processing (4.28–4.31).

Some useful apodization (window) functions are:

1. *Cosine window*

$$h(t) = \cos(\pi t/2t_{max}), \qquad (4.1.36)$$

2. *'Hanning'-window*

$$h(t) = 0.5 + 0.5 \cos(\pi t/t_{max}) \qquad (4.1.37)$$

3. *Hamming window*

$$h(t) = 0.54 + 0.46 \cos(\pi t/t_{max}) \qquad (4.1.38)$$

4. *Kaiser window*

$$h(t) = I_0\{\theta\sqrt{(1-(t/t_{max})^2)}\}[I_0\{\theta\}]^{-1}. \qquad (4.1.39)$$

In the latter case, I_0 is the zero order modified Bessel function, and increasing θ values reduce the ripple amplitude but increase the linewidth (typical values are $\theta = \pi$, 1.5π, or 2π) (4.40).

The relevant characteristics of these windows, the resulting linewidth and the ripple amplitude are summed up in Fig. 4.1.5. The solid line

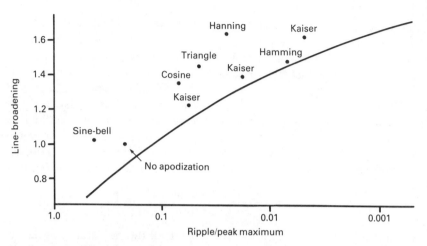

FIG. 4.1.5. Line-width versus ripple of the Fourier transform of a truncated exponentially decaying signal that has been smoothed with different types of apodization functions. The solid line indicates the optimum performance achieved with a Dolph–Chebycheff window. The full circles indicate the performance of various windows discussed in the text. The three Kaiser windows correspond to $\theta = \pi$, 1.5π, and 2π from left to right, respectively. The line-width has been normalized to unity for a truncated signal that is transformed without apodization; the amplitude of the ripple is relative to the height of the main peak.

gives the characteristics obtained with the Dolph–Chebycheff window (4.38, 4.39). Both Hamming and Kaiser windows provide good approximations to the optimum functions. The examples of Fig. 4.1.6 demonstrate that even the Hanning window performs markedly better than exponential weighting (4.27).

The sampling rate required to adequately represent the time domain signal is determined by the *sampling theorem* (4.7, 4.11, 4.18, 4.22, 4.24, 4.25). It can be stated as follows: *For a faithful representation of a signal, the sampling rate* $f_s = 1/\Delta t$ *must be at least equal to twice the highest frequency* f_{max} *contained in the signal*

$$f_s \geq 2f_{max}. \qquad (4.1.40a)$$

The highest frequency which can be retrieved after sampling with the rate f_s is called the *Nyquist frequency* f_N

$$f_N = \tfrac{1}{2}f_s. \qquad (4.1.40b)$$

Higher frequencies will appear converted to lower frequencies and are 'folded' into the frequency range $0 \leq f < f_N$. They cause an 'aliasing problem'.

If a complex time domain signal can be recorded by quadrature phase

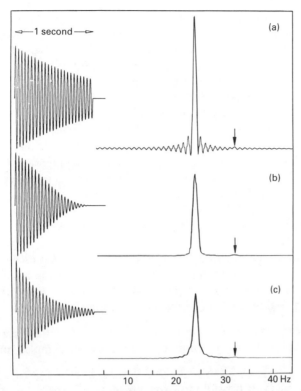

FIG. 4.1.6. (a) Exponentially decaying free induction signal of a superposition of a strong and a weak resonance and corresponding spectrum, where the weak signal is indicated by an arrow. (b) The same after apodization with a 'Hanning' function, and (c) with exponential weighting. The weak line is most clearly visible in (b). (Reproduced from Ref. 4.27.)

detection, it is possible to distinguish positive and negative frequencies and to cover the extended frequency range from $-f_N$ to $+f_N$.

In practice, the spectrum may be obtained by a *discrete* Fourier transformation of M acquired samples (4.18, 4.22)

$$S_l = \frac{1}{M} \sum_{k=0}^{M-1} s_k W^{-kl},$$

$$s_k = \sum_{l=0}^{M-1} S_l W^{kl} \qquad (4.1.41)$$

with $W = \exp\{i2\pi/M\}$. If the Fourier transformation is restricted to the M recorded samples, one does not obtain an oscillatory lineshape as discussed before (see Fig. 4.1.4), since the spacing of the frequency-domain sampling points matches the period of the oscillation.

However, the response may be sensitive to frequency shifts of the signals, and frequency components that fall between two sampling points in frequency domain may not be properly represented. To avoid this problem it is necessary to compute amplitudes at intermediate frequencies, for example by trigonometric interpolation. This can be readily achieved either by extending the sampling process beyond the limit t_{max} or by supplementing the M recorded samples by a string of zeroes prior to Fourier transformation. This is fully equivalent to a trigonometric interpolation: for example, if the length of the original time domain signal is doubled by adding M zeroes, the resulting sample spacing in frequency domain is $1/2T$, and the additional points obtained in this way appear half-way between the sampling points of the original array. Extending the *zero-filling* to infinity leads to continuous lineshapes which can be computed with a continuous Fourier transformation, eqn (4.1.20), (see Fig. 4.1.4).

It is obvious that supplementing the measured samples by zero values represents a very poor extrapolation of the signal. If the behaviour of the function $s(t)$ for $0 \leq t \leq t_{max}$ is taken into account, a better extrapolation should be possible. Several procedures have been suggested.

1. *The principle of linear prediction* (4.41–4.43) relies on representing the $(M + 1)^{st}$ (unknown) sample by the n previous samples in the form

$$x_{M+1} = a_0 x_M + a_1 x_{M-1} + \ldots + a_{n-1} x_{M-n+1}. \quad (4.1.42)$$

In this manner it is possible to extrapolate iteratively as far as desired. It has been demonstrated that such a procedure makes it possible to improve the representation of a spectrum considerably, even if only a limited set of sampling points in time domain are available.

2. *The maximum entropy method* (4.44, 4.45) reaches the same goal by making best use of the available information for the reconstruction of the most likely spectrum by maximizing its entropy. It should be noted that the computing time necessary for procedures of this type can be substantial.

4.1.3.2. Resolution enhancement

While apodization aims at a faithful representation of the spectrum, resolution enhancement attempts to achieve a transformation of the lineshape to narrow the resonance lines artificially.

In principle, it is possible to select an arbitrary desired lineshape $S_f(\omega)$ and to compute a weighting function $h(t)$ which will transform the experimental into the desired lineshape. The transformation can be obtained by multiplying the free induction decay with the function

$$h(t) = s_f^e(t)/s^e(t) \quad (4.1.43)$$

where $s^e(t)$ is the envelope of the recorded free induction signal, and $s_f^e(t)$ is the desired envelope, i.e. $s_f^e(t) = \mathscr{F}^{-1}\{S_f(\omega)\}$. In other words, the signal $s(t)$ is stripped of its 'natural' envelope $s^e(t)$ and is fitted out with an envelope $s_f^e(t)$ that produces the desired lineshape after Fourier transformation.

In practice, however, two restrictions have to be taken into account.

1. Resolution enhancement necessarily implies the enhancement of the later parts of the free induction signal, since the weighting function $h(t)$ increases with t, as shown in Fig. 4.1.7. Random noise contributions in the later parts of the signal may therefore be excessively enhanced, and the sensitivity may be deteriorated beyond an acceptable limit. A useful resolution enhancement function $h(t)$ should therefore always decay towards zero for large t as shown in Fig. 4.1.7, in order to obtain a compromise between resolution and sensitivity.

2. The achievable resolution enhancement is often restricted by the fact that the total acquisition time t_{max} is limited, and an appreciable resolution enhancement is only feasible when the sampling period is extended.

We limit the discussion to some widely used resolution enhancement functions and do not give any details on earlier methods like the convolution-difference technique (4.46).

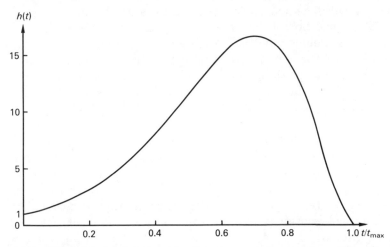

FIG. 4.1.7. Weighting function for enhancing the resolution together with a Hanning apodization window to avoid ripple

$$h(t) = 0.5[1 + \cos(\pi t/t_{max})]\exp[2\pi t/t_{max}].$$

Note that the weighting function first increases because of the exponential term, and later drops because of the apodization function.

1. *Lorentz–Gauss transformation.* Assuming that the natural decay is exponential with a time constant T_2^*, multiplication of the free induction signal with the weighting function (4.2, 4.47)

$$h(t) = \exp\{t/T_2^* - \sigma^2 t^2/2\} \qquad (4.1.44)$$

strips the line of its Lorentzian character with half-width at half-height $\omega_{\frac{1}{2}} = 1/T_2^*$ and fits it out with a Gaussian shape

$$S(\omega) = \frac{\sqrt{(2\pi)}}{\sigma} \exp\left\{\frac{-\omega^2}{2\sigma^2}\right\} \qquad (4.1.45)$$

with a half-width at half-height $\omega_{\frac{1}{2}} = 1.177\sigma$. By adjusting the parameter σ, it is in principle possible to achieve an arbitrary degree of line-narrowing, disregarding for the moment the limitations imposed by the finite acquisition time $t_{\max}$. The Gaussian shape has the advantage that the resonance lines have 'tails' that are less pronounced. At the same time, a fair apodization of the FID is achieved, thus reducing problems with truncation.

2. *Sine-bell function.* Multiplication of the free induction signal with a sine-bell function with a period equal to twice the acquisition time $t_{\max}$ (4.48)

$$h(t) = \sin(\pi t/t_{\max}) \qquad (4.1.46)$$

has the desired effect of giving the free induction signal an envelope that increases with time and that is apodized towards zero for t approaching $t_{\max}$. The application of this function is extremely simple, since it has no adjustable parameter, but the resolution enhancement is limited. The resulting lineshape is often unsatisfactory; because $h(0) = 0$, the integral of the lineshape vanishes, which implies the presence of negative signal tails that distort the baseline of the spectrum. This deficiency can be slightly reduced by shifting the phase of the sine-bell (4.49).

3. *Sensitivity-optimized resolution enhancement.* Resolution enhancement necessarily increases the high-frequency noise in the resulting spectrum and deteriorates the sensitivity. The sensitivity loss can be an order of magnitude when the resolution is enhanced by more than a factor of 2–3 (4.2). There is an exclusion principle which limits the resolution and sensitivity that are simultaneously attainable by a linear filtering process. It is therefore meaningful to impose a restriction upon the acceptable sensitivity loss and to optimize resolution under this constraint (4.2, 4.50). Calculations of this type have been performed in Ref. 4.2 and lead to the optimum class of weighting functions

$$h(t) = \frac{s^e(t)}{1 + q[s^e(t)]^2} \qquad (4.1.47)$$

with the envelope $s^e(t)$ of the experimental free induction decay. The parameter q determines the achievable resolution; the larger q, the better the resolution (at the expense of sensitivity). Although the trade-off between resolution and sensitivity is optimized with this function, the resulting lineshape is not controlled and the ripple may be worse than with a Lorentz–Gauss transformation (4.50).

4. *Ultimate ripple-free resolution enhancement.* Disregarding sensitivity, we may ask: what is the highest achievable resolution enhancement? Ultimate resolution enhancement requires complete flattening of the free induction decay (FID) by multiplication with the inverse envelope, leading to a rectangular FID of duration t_{max}. The lineshape will now exhibit excessive ripple, as shown in Fig. 4.1.4, and a full width at half-height of the central lobe of $\Delta f = 0.604/t_{max}$. This is the minimum achievable width. To suppress the ripple, a window filter must be added, ideally a Dolph–Chebycheff window (4.38, 4.39), but in practice a Hamming window is also suitable (eqn (4.1.38)), leading to the function for ultimate resolution enhancement

$$h(t) = \frac{0.54 + 0.46 \cos(\pi t/t_{max})}{s^e(t)}. \tag{4.1.48}$$

4.1.4. Non-linear response theory

Non-linear response theory is less developed than its linear counterpart, and lacks the elegance of linear response theory. Formally it is possible to extend the linear response equation (4.1.8) by including higher-order terms. This leads to a power series expansion of the response, analogous to a functional expansion originally proposed by Volterra (4.51–4.58)

$$y(t) = \Phi\{x(t)\} = \sum_{n=0}^{\infty} y_n(t)$$

with

$$y_0(t) = h_0,$$
$$y_1(t) = \int_{-\infty}^{\infty} h_1(\tau)x(t-\tau)\,d\tau,$$
$$y_2(t) = \iint_{-\infty}^{\infty} h_2(\tau_1, \tau_2)x(t-\tau_1)x(t-\tau_2)\,d\tau_1\,d\tau_2,$$
$$y_3(t) = \iiint_{-\infty}^{\infty} h_3(\tau_1, \tau_2, \tau_3)x(t-\tau_1)x(t-\tau_2)x(t-\tau_3)\,d\tau_1\,d\tau_2\,d\tau_3.$$

$$\tag{4.1.49}$$

The responsiveness of order k of the system is characterized by a kernel function $h_k(\tau_1, \ldots, \tau_k)$ which is of dimension k. In the case of a linear system, all the higher-order terms disappear, and the function $h_1(t)$ can be identified immediately with the impulse response $h(t)$ of eqn (4.1.8). For a full characterization of a non-linear system, all kernels $h_k(\tau_1, \ldots, \tau_k)$ are required, taking into account that normally the order k is not limited unless weak perturbations are applied. The causality principle implies that any kernel $h_k(\tau_1, \ldots, \tau_k)$ is zero whenever one of the arguments is negative.

The different kernels $h_k(\tau_1, \ldots, \tau_k)$ can be interpreted as impulse response functions of order k in the sense that $h_k(\tau_1, \ldots, \tau_k)$ is the response to a sequence of k delta functions applied at $t = \tau_1, \ldots, \tau_k$. For $h_1(\tau_1)$ the relation is obvious; for the higher responses the correspondence is a bit more delicate. Let us consider the quadratic response $h_2(\tau_1, \tau_2)$ and indicate a procedure for measuring it (4.54, 4.57). We apply to a quadratic system an input composed of two functions, $x(t) = x_a(t) + x_b(t)$. The quadratic response $y_2(t)$ can then be represented in the form

$$y_2(t) = y_2[x_a(t)] + y_2[x_b(t)] + 2y_2[x_a(t), x_b(t)] \quad (4.1.50)$$

where the first two terms represent the quadratic response to the individual functions, while the last term is the bilinear cross-term. In the special case that $x_a = \delta(t - t_a)$ and $x_b = \delta(t - t_b)$ we find

$$y_2[x_a(t), x_b(t)] = \iint_{-\infty}^{\infty} h_2(\tau_1, \tau_2)\delta(t - t_a - \tau_1)\delta(t - t_b - \tau_2)\,d\tau_1\,d\tau_2$$

$$= h_2(t - t_a, t - t_b) \quad (4.1.51)$$

which represents, as a function of t, a section through the two-dimensional impulse response function $h_2(t_1, t_2)$. It can be measured by a two-pulse experiment, subtracting the quadratic responses $y_2[x_a(t)]$ and $y_2[x_b(t)]$ of the single pulses. Similar procedures can be devised to measure the higher impulse response functions.

Obviously there is a close analogy between the higher-order impulse responses and multidimensional spectroscopy as described in Chapters 6 to 10. The kth order impulse response can be Fourier transformed to produce a k-dimensional frequency response function, i.e. a k-dimensional complex spectrum $H_k(\omega_1, \ldots, \omega_k)$

$$h_k(\tau_1, \ldots, \tau_k) \xrightarrow{\mathscr{F}^k} H_k(\omega_1, \ldots, \omega_k). \quad (4.1.52)$$

We must emphasize however that all even-order responses disappear in NMR; $y_k(t) = 0$ for $k = $ even. In other words, a two-pulse experiment as described above does not produce any quadratic response and does not provide a two-dimensional spectrum. It is however possible to compute two-dimensional spectra by recording a third-order impulse response and taking a two-dimensional section through its three-dimensional Fourier transform. This is relevant for stochastic multidimensional spectroscopy (see § 4.1.6).

The disappearance of even-order response functions in NMR is due to the special form of the Bloch or Liouville–von Neumann equations in the high-field approximation (4.59, 4.60). Since the response changes sign when the r.f. excitation is changed in sign, the response is an odd function of the excitation, irrespective of the amplitude of the perturbation, and the even orders disappear (see eqn (4.1.62)).

The Volterra series of eqn (4.1.49) unfortunately does not lead to an orthogonal expansion and the separation of the various terms is far from trivial. In systems of finite order where k is limited, it is possible to determine the response of maximum order, to subtract it, and progress sequentially to the lower orders. However, most systems are not of finite order and there is no maximum order. Then approximate solutions have to be considered.

In this context, the proposal of N. Wiener (4.52) to employ Gaussian noise as input provides an elegant solution, since it permits the application of orthogonal stochastic polynomials (see § 4.1.6).

4.1.5. Quantum-mechanical response theory

The quantum-mechanical response theory can be developed in close analogy to the classical response theory discussed in the preceding sections. The major contribution was that of R. Kubo (4.61, 4.62), who developed the linear response theory of quantum systems and also indicated extensions to non-linear systems. It is not astonishing that there is a close relation between the non-linear response theory of Kubo and the Volterra functional expansion (4.63).

The derivation starts with the Liouville–von Neumann equation (2.1.17)

$$\dot{\sigma} = -i[\mathcal{H}(t), \sigma] \qquad (4.1.53)$$

with

$$\mathcal{H}(t) = \mathcal{H}_0 + x(t)A \qquad (4.1.54)$$

where $\mathcal{H}_0$ is the unperturbed Hamiltonian, $x(t)$ the (classical) input function, and A the perturbation operator which couples the input to the system.

To obtain the linear response for a weak perturbation $x(t)$, the density operator is expanded

$$\sigma(t) = \sigma_0 + \Delta\sigma(t), \qquad (4.1.55)$$

which, inserted in eqn (4.1.53), immediately leads to the solution

$$\Delta\sigma(t) = -i\int_{-\infty}^{t} \exp\{-i\mathcal{H}_0(t-t')\}[A, \sigma_0]\exp\{i\mathcal{H}_0(t-t')\}x(t')\,dt'. \tag{4.1.56}$$

For the observable B, the linear response $y(t)$ is

$$y(t) = \text{tr}\{\Delta\sigma(t)B\}$$
$$= \int_0^\infty h_{AB}(\tau)x(t-\tau)\,d\tau \tag{4.1.57}$$

with the impulse response function $h_{AB}(t)$ given by

$$h_{AB}(t) = -i\,\text{tr}\{[A(t), \sigma_0]B\} \tag{4.1.58}$$

and

$$A(t) = \exp\{-i\mathcal{H}_0 t\}A\exp\{i\mathcal{H}_0 t\}. \tag{4.1.59}$$

Equation (4.1.57) allows the computation of the linear response $y(t)$ for any input $x(t)$. There is a direct analogy to the classical eqn (4.1.8).

The extension of the quantum-mechanical response theory to higher orders is straightforward (4.62) and leads to the result

$$\sigma(t) = \sigma_0 + \sum_{k=1}^{\infty} (-i)^k \int_0^\infty \int_{\tau_1}^\infty \ldots \int_{\tau_{k-1}}^\infty \hat{A}(\tau_1)\hat{A}(\tau_2)\ldots\hat{A}(\tau_k)\sigma_0$$
$$\times x(t-\tau_1)x(t-\tau_2)\ldots x(t-\tau_k)\,d\tau_1\ldots d\tau_k. \tag{4.1.60}$$

The response $y(t)$ can then be expressed by higher-order impulse response functions $h_{AB}(t_1,\ldots,t_k)$

$$y(t) = \text{tr}\{\sigma(t)B\} = \text{tr}\{\sigma_0 B\} + \sum_{k=1}^{\infty} \int_0^\infty \int_{\tau_1}^\infty \ldots \int_{\tau_{k-1}}^\infty h_{AB}(\tau_1\ldots\tau_k)$$
$$\times x(t-\tau_1)\ldots x(t-\tau_k)\,d\tau_1\ldots d\tau_k \tag{4.1.61}$$

with

$$h_{AB}(\tau_1\ldots\tau_k) = (-i)^k\,\text{tr}\{[A(\tau_1), [A(\tau_2), [\ldots[A(\tau_k), \sigma_0]\ldots]]]B\}. \tag{4.1.62}$$

In magnetic resonance, the operators A and B are mostly sums of transverse spin operators I_{lx} or I_{ly} and it is evident that the terms with k = even will disappear.

4.1.6. Stochastic response theory

In stochastic response measurements, the system Φ is excited with a Gaussian or binary random process $x(t)$ with a frequency-independent

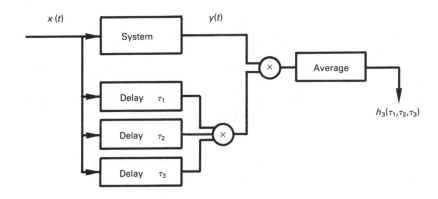

FIG. 4.1.8. Measurement of the stochastic response of a non-linear system. The operations are normally performed in a digital computer on a record of $x(t)$ and $y(t)$. The example measures the cubic response of the system.

power spectral density (white noise) and the random response $y(t)$ is analysed in terms of the properties of the system Φ (see Fig. 4.1.8). The characterization of non-linear systems by their stochastic response dates back to N. Wiener (4.52) who recognized and exploited the orthogonality properties of stochastic Hermite polynomials which allow a straightforward separation of the different orders in a Volterra expansion of the type of eqn (4.1.49).

The use of stochastic excitation in NMR has been proposed at several occasions, originally for tailored and broad-band decoupling (4.64, 4.65), and later as an alternative to one-dimensional Fourier spectroscopy (4.59, 4.66–4.69) with inherent advantages regarding the r.f. power requirements. More recently, Blümich, Ziessow, and Kaiser (4.70–4.79) have explored the use of stochastic resonance in two-dimensional spectroscopy. They have convincingly demonstrated that most of the results that can be obtained with pulse excitation (4.80) can also be attained with stochastic excitation with suitable treatment of the data.

Nevertheless stochastic resonance has remained a 'sleeping beauty' as far as practical applications are concerned, for reasons that are not too difficult to understand. First of all, a stochastic experiment produces meaningful results only after averaging, and a large number of stochastic response signals must be combined to reduce the variance of the resulting spectrum. Second, stochastic excitation is a very general procedure which extracts all the available information from the system at once. The screening is done only later in the course of data-processing. This implies a demanding experiment even if only a limited amount of information is required. It is not easily possible to adapt the experiment to obtain specific information.

In the following, we shall briefly sketch the principal aspects of stochastic NMR. We encourage the reader to consult the original literature for more details (4.70–4.79).

The separation of the various kernels or response functions $h_k(\tau_1, \ldots, \tau_k)$ to a Gaussian white random excitation $x(t)$ is based on the orthogonality of k-dimensional stochastic Hermite polynomials. The following polynomials of a white Gaussian random process $x(t)$

$$P_0 = 1,$$
$$P_1(t_1) = x(t_1),$$
$$P_2(t_1, t_2) = x(t_1)x(t_2) - \delta(t_1 - t_2),$$
$$P_3(t_1, t_2, t_3) = x(t_1)x(t_2)x(t_3) - x(t_1)\delta(t_2 - t_3)$$
$$- x(t_2)\delta(t_1 - t_3) - x(t_3)\delta(t_1 - t_2),$$
$$\text{etc.} \tag{4.1.63}$$

fulfil the orthogonality relations

$$\overline{P_k(t_1, \ldots, t_k) P_l(\tau_1, \ldots, \tau_l)} = \delta_{kl} \sum_{(mn)} \prod \delta(t_m - \tau_n) \tag{4.1.64}$$

where the summation extends over all possibilities of forming paired arrangements of the variables t_m and τ_n (4.81–4.83). The response function can be calculated by cross-correlating the output $y(t)$ with products of the delayed random input $x(t)$. For the first three response functions, we find the expressions (4.54, 4.57, 4.58, 4.73)

$$\overline{y(t)x(t - \tau_1)} = \mu_2 h_1(\tau_1),$$
$$\overline{y(t)x(t - \tau_1)x(t - \tau_2)} = 2\mu_2^2 h_2(\tau_1, \tau_2) + \mu_2 \delta(\tau_1 - \tau_2) h_0,$$
$$\overline{y(t)x(t - \tau_1)x(t - \tau_2)x(t - \tau_3)} = 6\mu_2^3 h_3(\tau_1, \tau_2, \tau_3)$$
$$+ \mu_2^2 [\delta(\tau_2 - \tau_3)h_1(\tau_1) + \delta(\tau_1 - \tau_3)h_1(\tau_2) + \delta(\tau_1 - \tau_2)h_1(\tau_3)] \tag{4.1.65}$$

with the variance per unit bandwidth μ_2 of the white Gaussian process $x(t)$. In theory, the average denotes an ensemble average, which is replaced in practice by a time average over an extended record of $y(t)$ and $x(t)$. This provides easy access to the desired response functions.

The multidimensional spectra can be computed by Fourier-transforming the response functions $h_k(\tau_1, \ldots, \tau_k)$. The spectra can however be obtained more directly by recognizing that the correlation of $y(t)$ and $x(t)$ in time domain is equivalent to a complex multiplication of the excitation spectrum $X(\omega)$ and the response spectrum $Y(\omega)$ in frequency domain. One finds the following expressions for the spectra

$H_k(\omega_1, \ldots, \omega_k)$ of dimension $k = 1, 2, 3$

$$H_1(\omega) = \frac{\langle Y(\omega)X^*(\omega)\rangle}{\langle |X(\omega)|^2\rangle},$$

$$H_2(\omega_1, \omega_2) = \frac{\langle Y(\omega_1 + \omega_2)X^*(\omega_1)X^*(\omega_2)\rangle}{2\langle |X(\omega_1)|^2 |X(\omega_2)|^2\rangle},$$

$$H_3(\omega_1, \omega_2, \omega_3) = \frac{\langle Y(\omega_1 + \omega_2 + \omega_3)X^*(\omega_1)X^*(\omega_2)X^*(\omega_3)\rangle}{6\langle |X(\omega_1)|^2 |X(\omega_2)|^2 |X(\omega_3)|^2\rangle}. \quad (4.1.66)$$

The cross-talk to the lower-order terms apparent in eqn (4.1.65) is compensated by modifying the spectra $Y(\omega)$ before multiplication (4.74). In eqn (4.1.66), the variance μ_2 has been replaced by the frequency-dependent value $\langle |X(\omega)|^2\rangle$ which is estimated from the experimental excitation process $x(t)$. Note that eqn (4.1.65) holds strictly only for Gaussian white-noise excitation. Considerably more involved expressions are applicable for coloured noise input.

In practice, extensive records of the excitation $x(t)$ and the response $y(t)$ are stored. Shorter segments are Fourier-transformed and processed according to eqn (4.1.66). An ensemble average (denoted by $\langle \ldots \rangle$) of many processed segments is necessary to obtain satisfactory statistics.

The linear frequency response $H_1(\omega)$ is equivalent to a complex one-dimensional spectrum. Recording the stochastic linear response is an alternative to pulse Fourier spectroscopy with a similar sensitivity advantage (4.59, 4.66). An example is shown in Fig. 4.1.9. To avoid problems with the statistics, it can be of advantage to use pseudo-random-noise excitation with well-determined spectral properties (4.59).

It has been mentioned before that the quadratic response disappears in high-field NMR. Two-dimensional spectra can however be obtained from the cubic frequency response $H_3(\omega_1, \omega_2, \omega_3)$ by taking two-dimensional sections through the three-dimensional function (4.70, 4.72, 4.73). We should mention that there is an intriguing analogy between a three-pulse experiment of the type discussed in Chapter 6 and the cubic stochastic response. The latter corresponds to an experiment with very weak r.f. pulses. Many forms of two-dimensional spectra have so far been obtained with stochastic excitation.

4.2. Classical description of Fourier spectroscopy

Many aspects of pulse Fourier spectroscopy can be understood by considering systems of isolated spins that can be described in terms of phenomenological Bloch equations. In particular, questions regarding the optimum design of experiments, lineshapes, and sensitivity can be

Binary pseudo-random sequence

Stochastic response

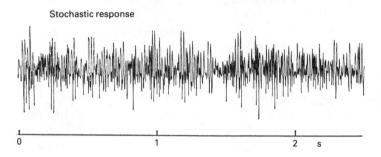

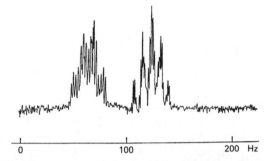

FIG. 4.1.9. Stochastic resonance of fluorine-19 in 2,4-difluorotoluene: binary pseudo-random input, stochastic response, and absorption-mode spectrum. The time-domain signal was recorded over a period of 2.5 s with 1023 sample values; the frequency-domain spectrum is represented by 512 samples for a width of 220 Hz. (Adapted from Ref. 4.59.)

discussed on classical grounds. It should however be recalled that an adequate treatment of systems with resolved scalar, dipolar, or quadrupolar couplings calls for a quantum-mechanical description in terms of density operators. In some simple cases a semi-classical picture can help to bridge the gap between the two approaches.

4.2.1. Bloch equations in the rotating frame

The response of a nuclear spin system to r.f. pulses is most conveniently calculated in a frame rotating with the applied radio-frequency. We start with the Bloch equations in the laboratory frame, which may be written in vector notation (eqn (2.3.1))

$$\dot{\mathbf{M}}(t) = \gamma \mathbf{M}(t) \times \mathbf{B}(t) - \mathbf{R}\{\mathbf{M}(t) - \mathbf{M}_0\}. \tag{4.2.1}$$

4.2 CLASSICAL DESCRIPTION

The magnetization vector **M** has the thermal equilibrium value $\mathbf{M}_0$, and **R** is the relaxation matrix

$$\mathbf{R} = \begin{pmatrix} 1/T_2 & 0 & 0 \\ 0 & 1/T_2 & 0 \\ 0 & 0 & 1/T_1 \end{pmatrix}, \quad (4.2.2)$$

with the longitudinal and transverse relaxation times T_1 and T_2. The external magnetic field $\mathbf{B}(t)$ consists of a static field $\mathbf{B}_0$ and an r.f. field $\mathbf{B}_{\text{r.f.}}(t)$

$$\mathbf{B}(t) = \mathbf{B}_0 + \mathbf{B}_{\text{r.f.}}(t). \quad (4.2.3)$$

$\mathbf{B}_{\text{r.f.}}(t)$ is normally applied in the form of a linearly oscillating field

$$\mathbf{B}_{\text{r.f.}}(t) = 2B_1 \cos(\omega_{\text{r.f.}} t + \varphi) \mathbf{e}_x \quad (4.2.4)$$

that can be decomposed into two counter-rotating components, of which only the field rotating in the same sense as the spins will be retained

$$\mathbf{B}_{\text{r.f.}}(t) = B_1 \{\cos(\omega_{\text{r.f.}} t + \varphi) \mathbf{e}_x + \sin(\omega_{\text{r.f.}} t + \varphi) \mathbf{e}_y\}. \quad (4.2.5)$$

The coefficients of the Bloch equations eqn (4.2.1) are rendered time-independent by a transformation into a frame rotating with the frequency $\omega_{\text{r.f.}}$, as shown in Fig. 4.2.1.

$$(\mathbf{e}_x, \mathbf{e}_y, \mathbf{e}_z) = (\mathbf{e}_x^r, \mathbf{e}_y^r, \mathbf{e}_z^r) \mathbf{T}(t) \quad (4.2.6)$$

$$\mathbf{T}(t) = \begin{pmatrix} \cos \omega_{\text{r.f.}} t & \sin \omega_{\text{r.f.}} t & 0 \\ -\sin \omega_{\text{r.f.}} t & \cos \omega_{\text{r.f.}} t & 0 \\ 0 & 0 & 1 \end{pmatrix}. \quad (4.2.7)$$

For the magnetization vector in the rotating frame

$$\mathbf{M}^r(t) = \mathbf{T}(t) \mathbf{M}(t), \quad (4.2.8)$$

we obtain the differential equation

$$\dot{\mathbf{M}}^r(t) = \gamma \mathbf{M}^r(t) \times \mathbf{B}^r - \mathbf{R}\{\mathbf{M}^r(t) - \mathbf{M}_0\} \quad (4.2.9)$$

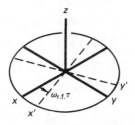

FIG. 4.2.1. Transformation from the laboratory frame with the axes x, y, and z into a frame with axes x', y', and z rotating with the frequency $\omega_{\text{r.f.}}$.

where the effective magnetic field in the rotating frame has the three components

$$B_x^r = B_1 \cos \varphi,$$
$$B_y^r = B_1 \sin \varphi,$$
$$B_z^r = B_0 + \omega_{r.f.}/\gamma = -\Omega/\gamma. \qquad (4.2.10)$$

The modified z-component of the magnetic field in the rotating frame reflects the fact that the magnetic moments apparently experience a lower field, since the effective precession frequency Ω in the rotating frame is reduced,

$$\Omega = -\gamma B_0 - \omega_{r.f.} = \omega_0 - \omega_{r.f.} \qquad (4.2.11)$$

where $\omega_0 = -\gamma B_0$ is the Larmor frequency in the laboratory frame. The r.f. field in the rotating frame is described by the amplitude B_1 and phase φ (excursion from the x-axis towards the y-axis). Both can become time-dependent for a pulsed field of variable phase. In explicit notation, eqn (4.2.9) is equivalent to

$$\dot{M}_x^r(t) = \gamma[M_y^r B_z^r - M_z^r B_y^r] - M_x^r/T_2,$$
$$\dot{M}_y^r(t) = \gamma[M_z^r B_x^r - M_x^r B_z^r] - M_y^r/T_2,$$
$$\dot{M}_z^r(t) = \gamma[M_x^r B_y^r - M_y^r B_x^r] - (M_z^r - M_0)/T_1. \qquad (4.2.12)$$

Since all computations in the following sections are carried out in the rotating frame, the index r will be omitted.

4.2.2. Ideal pulse experiment

An r.f. pulse of duration τ_p with pulse rotation angle β

$$\beta = -\gamma B_1 \tau_p \qquad (4.2.13)$$

rotates the equilibrium magnetization M_0 about the direction of the applied r.f. field B_1 irrespective of the resonance offset, if we assume a sufficiently strong r.f. field. For an r.f. field applied along the y-axis ($\varphi = \pi/2$ in eqn (4.2.10)), the initial magnetization after the pulse is

$$M_x(0_+) = M_0 \sin \beta,$$
$$M_y(0_+) = 0,$$
$$M_z(0_+) = M_0 \cos \beta. \qquad (4.2.14)$$

The subsequent free induction decay can be described in terms of the two components

$$M_x(t) = M_0 \sin \beta \cos(\Omega t) \exp(-t/T_2),$$
$$M_y(t) = M_0 \sin \beta \sin(\Omega t) \exp(-t/T_2) \qquad (4.2.15)$$

or, in complex notation,

$$M^+(t) = M_x(t) + iM_y(t) = M_0 \sin \beta \exp\{i\Omega t - t/T_2\}. \quad (4.2.16)$$

The imaginary component vanishes for $t = 0$ if the pulse is applied along the y-axis of the rotating frame.

The complex signal $s^+(t)$, obtained by simultaneous observation of both x- and y-components by quadrature detection, is directly proportional to the complex magnetization $M^+(t)$. This signal may be subjected to a complex Fourier transformation

$$S(\omega) = \int_0^\infty s^+(t) \exp\{-i\omega t\} \, dt. \quad (4.2.17)$$

One obtains the complex spectrum

$$S(\omega) = v(\omega) + iu(\omega) \quad (4.2.18)$$

with

$$v(\omega) = M_0 \sin \beta \, a(\Delta\omega),$$
$$u(\omega) = -M_0 \sin \beta \, d(\Delta\omega)$$

with the frequency offset $\Delta\omega = \omega - \Omega$ measured with respect to the centre of the resonance. The functions $a(\Delta\omega)$ and $d(\Delta\omega)$ represent absorption and dispersion signals, respectively,

$$a(\Delta\omega) = \frac{1/T_2}{(1/T_2)^2 + (\Delta\omega)^2}, \quad d(\Delta\omega) = \frac{\Delta\omega}{(1/T_2)^2 + (\Delta\omega)^2}. \quad (4.2.19)$$

Obviously, the maximum signal amplitude is obtained for a pulse rotation angle $\beta = \pi/2$. All resonance lines appear with the same phase, which may be in absorption or dispersion depending on the selection of the component of $S(\omega)$.

4.2.3. Off-resonance effects due to finite pulse amplitude

In the previous section, we assumed that the amplitude of the applied pulse is strong enough to neglect all off-resonance effects, i.e.

$$|\gamma B_1| \gg |\Omega| = |\omega_0 - \omega_{\text{r.f.}}|. \quad (4.2.20)$$

In many experimental situations, the width of the spectra to be investigated is comparable to the maximum available r.f. field strength γB_1, and off-resonance effects cannot be neglected. The magnetization is rotated about a tilted effective field, which depends on the resonance frequency Ω in the rotating frame. Intensity and phase anomalies have to be expected for resonance lines that are at large offsets from the carrier frequency $\omega_{\text{r.f.}}$.

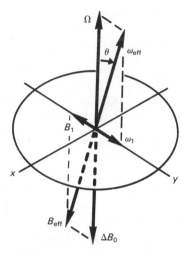

FIG. 4.2.2. Tilted effective field in the rotating frame: the residual z-component of the magnetic field $\Delta \mathbf{B}_0 = \mathbf{B}_0 + \boldsymbol{\omega}_{\mathrm{r.f.}}/\gamma$ and the related free precession frequency vector $\boldsymbol{\Omega} = -\gamma \Delta \mathbf{B}_0 = \boldsymbol{\omega}_0 - \boldsymbol{\omega}_{\mathrm{r.f.}}$ are shown for a case where the carrier frequency is placed above the resonance frequency ($|\omega_{\mathrm{r.f.}}| > |\omega_0|$) for $\gamma > 0$. The effective magnetic field vector $\mathbf{B}_{\mathrm{eff}}$ is indicated for an r.f. field $\mathbf{B}_1$ applied along the negative y-axis. The rotation vector $\boldsymbol{\omega}_1 = -\gamma \mathbf{B}_1$ points along the positive y-axis (for $\gamma > 0$). If the r.f. field is on resonance ($\Delta B_0 = 0$), the magnetization nutates in the xz-plane (positive rotation about the +y-axis: $z \to x \to -z \to -x$). In the absence of r.f. irradiation, free precession in the rotating frame also corresponds to a positive rotation about the +z-axis ($x \to y \to -x \to -y$).

The tilted effective field about which the rotation proceeds can be deduced from Fig. 4.2.2. It is determined by the offset field

$$\Delta B_0 = B_0 + \omega_{\mathrm{r.f.}}/\gamma = -\Omega/\gamma \qquad (4.2.21)$$

along the z-axis and by the r.f. field B_1 in the transverse plane. It has the amplitude

$$B_{\mathrm{eff}} = \{B_1^2 + (\Delta B_0)^2\}^{\frac{1}{2}} \qquad (4.2.22)$$

and is tilted by an angle θ with respect to the z-axis with

$$\tan \theta = B_1/\Delta B_0. \qquad (4.2.23)$$

The effective nutation angle β_{eff} during a pulse of duration τ_p is

$$\beta_{\mathrm{eff}} = -\gamma B_{\mathrm{eff}} \tau_\mathrm{p}. \qquad (4.2.24)$$

It is remarkable that this angle actually increases with increasing offset Ω, as shown in Fig. 4.2.3(b). However, the rotation of a component originally along the z-axis describes a narrower cone.

The components of the magnetization $\mathbf{M}(0_+)$ immediately after the pulse can easily be computed by multiplication of the corresponding

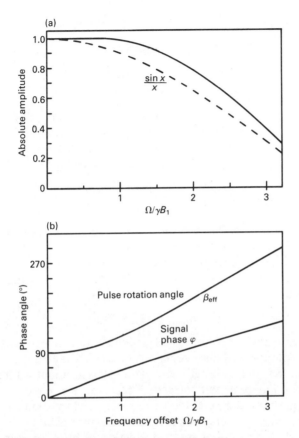

FIG. 4.2.3. Dependence of signal amplitude and signal phase on the frequency offset Ω for a single-pulse Fourier experiment. An on-resonance pulse rotation angle $\beta = 90°$ has been assumed. (a) Absolute value of signal amplitude as a function of $\Omega/\gamma B_1$. For comparison the corresponding $\sin x/x$ dependence of eqn (4.2.31) is included. (b) Signal phase φ and effective pulse rotation angle β_{eff} as functions of $\Omega/\gamma B_1$.

rotation matrices

$$\mathbf{M}(0_+) = \mathbf{R}_x^{-1}(\theta)\mathbf{R}_z(\beta_{\text{eff}})\mathbf{R}_x(\theta)\mathbf{M}(0_-) \qquad (4.2.25)$$

with

$$\mathbf{R}_x(\theta) = \begin{pmatrix} 1 & 0 & 0 \\ 0 & \cos\theta & -\sin\theta \\ 0 & \sin\theta & \cos\theta \end{pmatrix}$$

and

$$\mathbf{R}_z(\beta_{\text{eff}}) = \begin{pmatrix} \cos\beta_{\text{eff}} & -\sin\beta_{\text{eff}} & 0 \\ \sin\beta_{\text{eff}} & \cos\beta_{\text{eff}} & 0 \\ 0 & 0 & 1 \end{pmatrix}.$$

With $\mathbf{M}(0_-) = M_0\mathbf{e}_z$, we find

$$M_x(0_+) = M_0 \sin \beta_{\text{eff}} \sin \theta,$$
$$M_y(0_+) = M_0(1 - \cos \beta_{\text{eff}})\sin \theta \cos \theta,$$
$$M_z(0_+) = M_0[\cos^2 \theta + \cos \beta_{\text{eff}} \sin^2 \theta]. \qquad (4.2.26)$$

The transverse magnetization immediately after the pulse is no longer along the x-axis as in the case of on-resonance irradiation ($\theta = \pi/2$, eqn (4.2.14)), but has a phase shift φ which depends on the resonance offset Ω as well as on the effective pulse rotation angle β_{eff}

$$\tan \varphi = \frac{M_y(0_+)}{M_x(0_+)} = \frac{(1 - \cos \beta_{\text{eff}})\sin \theta}{\sin \beta_{\text{eff}}} \frac{\Omega}{(-\gamma B_1)}. \qquad (4.2.27)$$

Figure 4.2.3(b) shows that the phase shift increases almost linearly with the offset Ω. The resulting phase error in the spectrum can easily be compensated by an appropriate frequency-dependent phase correction.

At the same time, the amplitude response of the transverse magnetization

$$M_{\text{trans}} = (M_x^2 + M_y^2)^{\frac{1}{2}} \qquad (4.2.28)$$

decreases with increasing offset as demonstrated in Fig. 4.2.3(a). Note that the amplitude response remains almost constant up to an offset Ω equal to the r.f. field strength γB_1. The rotation about a tilted effective field has an inherent self-correcting feature: the effect of the tilted rotation is partially compensated by the increased rotation angle β_{eff} (eqn 4.2.24). For larger offsets, however, the response goes to zero, becomes negative and starts to oscillate as a function of offset, as can be seen in Fig. 4.2.4 (4.84). Vanishing transverse magnetization implies that the effective pulse rotation angle β_{eff} about the tilted axis is a multiple of 2π, and the magnetization vector is rotated back along the z-axis at the end of the r.f. pulse.

If the effective r.f. field is tilted ($\theta \neq \pi/2$), the orbit described by the magnetization vector, initially along the z-axis, no longer cuts through the 'south pole' of the unit sphere. This implies that it is not possible to achieve a proper inversion of the magnetization with a single pulse. As discussed in §4.2.7, this problem can be circumvented by using composite r.f. pulses (4.85–4.87). If the tilt of the r.f. field is $\theta < \pi/4$, the orbit of the magnetization no longer intersects the xy-plane, and the excitation of transverse magnetization becomes inefficient.

The vanishing of the transverse magnetization at certain offsets has been exploited to zero the response of strong solvent lines, in particular the water line in aqueous solutions of biomolecules (4.88). To suppress the response of an undesired line, it is sufficient to select the r.f. carrier

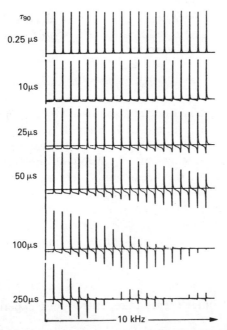

FIG. 4.2.4. Experimental offset dependence of the lineshapes for pulses with six different pulse lengths of 0.25, 10, 25, 50, 100, and 250 μs with on-resonance r.f. rotation angles $\beta = \pi/2$. (Reproduced from Ref. 4.84.)

such that the offset Ω is

$$\Omega = \pm\sqrt{15}\,\gamma B_1 \qquad (4.2.29)$$

if the on-resonance rotation angle is $\beta = \pi/2$. The phase shifts and amplitude distortions for the remaining lines can be corrected (4.89), although the sensitivity will deteriorate. By using composite pulses, it is possible to improve the efficiency of suppression (4.90–4.92).

On superficial grounds, one may be tempted to predict the offset-dependence of the pulse rotation angle by considering the frequency spectrum of a pulse envelope $p(t)$ of finite length,

$$p(t) = \Pi(t/\tau_p) = \begin{cases} 1 & \text{for } |t| < \tau_p/2 \\ 0 & \text{for } |t| > \tau_p/2. \end{cases} \qquad (4.2.30)$$

The Fourier transform is given by

$$P(\Omega) = \frac{\sin(\Omega\tau_p/2)}{\Omega/2} \qquad (4.2.31)$$

which would seem to imply that the response decreases like $\sin x / x$ for increasing offset Ω. The rotation angle β_{eff} should decrease correspondingly and reach a value of zero for

$$\Omega = 2\pi N / \tau_p, \quad N = 1, 2, \ldots . \tag{4.2.32}$$

For comparison, the offset dependence based on this spectral model has been included in Fig. 4.2.3(a). It should be emphasized that the qualitative agreement is largely fictitious. In general, the interpretation of the effects of pulse sequences based entirely on spectral arguments leads to incorrect predictions. A nuclear spin system is strongly non-linear with regard to its response to r.f. perturbations. The flaw in spectral arguments lies in the fact that they rely on linear response theory (§ 4.1). This linear approximation fails as soon as pulse rotation angles with $\beta \gtrsim 10°$ are employed, and the spectral model of r.f. pulses should be used with extreme care.

4.2.4. Longitudinal interference in repetitive pulse experiments

The need for signal averaging and repeated experiments is typical of most Fourier experiments. The measurement of the response of an equilibrium system to a single pulse is rather atypical. It is often not possible to neglect the interference of successive experiments and it is necessary to consider the response to a repetitive pulse sequence where the free induction decays are recorded in the intervals between the pulses.

To maximize the accumulated signal, it is necessary to use a fast repetition rate. We assume in this section that the transverse magnetization decays irreversibly during the recycle time T. However, the recovery of the longitudinal magnetization towards its equilibrium value M_0 in the pulse interval T is normally not complete. After a small number of pulses, a dynamic equilibrium is established which can easily be computed by equating the z-magnetization before a pulse, $M_z(0_-)$ and the z-magnetization at the end of the recycle time, $M_z(T)$. Neglecting off-resonance effects, one obtains

$$M_z(0_+) = M_z(0_-)\cos \beta,$$
$$M_z(T) = M_z(0_+)E_1 + M_0(1 - E_1) \tag{4.2.33}$$

with $E_1 = \exp(-T/T_1)$. From $M_z(T) = M_z(0_-)$, we immediately find

$$M_z(0_-) = M_0 \frac{1 - E_1}{1 - E_1 \cos \beta}. \tag{4.2.34}$$

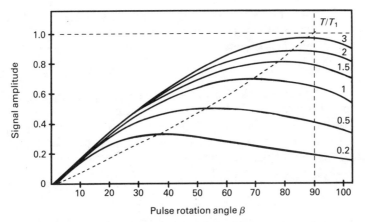

FIG. 4.2.5. Normalized peak amplitude of the absorption-mode signal v_{max}/M_0T_2 in a repetitive Fourier experiment with negligible transverse interference as a function of the pulse rotation angle β for various interpulse spacings T normalized by the longitudinal relaxation time T_1. The broken line connects the maximum amplitudes and indicates the optimum pulse rotation angle.

The initial amplitude of the free induction decay $M_x(0_+)$ is

$$M_x(0_+) = M_0 \frac{1-E_1}{1-E_1 \cos \beta} \sin \beta. \tag{4.2.35}$$

The maximum signal amplitude is no longer obtained for $\beta = \pi/2$ but for β_{opt} determined by the relation

$$\cos \beta_{opt} = E_1 = \exp(-T/T_1). \tag{4.2.36}$$

Figure 4.2.5 shows the signal amplitude as a function of the pulse rotation angle β for various ratios T/T_1. It is apparent that for shorter recycle times, the optimum pulse rotation angle decreases. For $T \geq 3T_1$, there is little interference and more than 95 per cent of the equilibrium magnetization can be brought into the transverse plane.

4.2.5. Transverse interference in repetitive pulse experiments

For long transverse relaxation times T_2 and short recycle times T, some transverse magnetization may remain from the preceding pulse, leading to transverse interference. This situation is quite typical for many pulse experiments with fast recycling, for example in the sensitive-point imaging method (see § 10.2.1). In analogy to longitudinal interference, a dynamic equilibrium is established which depends not only on T_1/T but also on T_2/T and on the resonance offset Ω.

The dynamic equilibrium can be computed in analogy to the previous section. Taking into consideration the effect of tilted r.f. fields, the nutation induced by an r.f. pulse is given by eqn (4.2.26), and the evolution of the magnetization in the pulse interval is described by

$$\mathbf{M}(t) = \mathbf{R}_z(\phi)\exp(-\mathbf{R}t)\mathbf{M}(0_+) + \{1 - \exp(-\mathbf{R}t)\}\mathbf{M}_0 \quad (4.2.37)$$

where $\mathbf{R}_z(\phi)$ expresses the Larmor precession through the angle $\phi = \Omega T$ in the pulse interval. This should not be confused with φ used to denote the phase of the r.f. field and of the magnetization. The term $\exp(-\mathbf{R}t)$ represents transverse and longitudinal relaxation (eqn (4.2.2)). By equating $\mathbf{M}(T) = \mathbf{M}(0_-)$ we find

$$\mathbf{M}(0_+) = \{\mathbf{R}_x^{-1}(\theta)\mathbf{R}_z^{-1}(\beta_{\text{eff}})\mathbf{R}_x(\theta) - \mathbf{R}_z(\phi)\exp(-\mathbf{R}T)\}^{-1}\{1 - \exp(-\mathbf{R}T)\}\mathbf{M}_0.$$
$$(4.2.38)$$

The tilt angle θ is given in eqn (4.2.23) and the effective pulse rotation angle β_{eff} in eqn (4.2.24).

In the steady-state, the three magnetization components immediately after a y-pulse ($\varphi = \pi/2$) are

$$M_x(0_+) = M_0(1 - E_1)\frac{1}{D}[(1 - E_2\cos\phi)\sin\beta_{\text{eff}}\sin\theta$$
$$+ E_2\sin\phi(1 - \cos\beta_{\text{eff}})\sin\theta\cos\theta],$$

$$M_y(0_+) = M_0(1 - E_1)\frac{1}{D}[E_2\sin\beta_{\text{eff}}\sin\phi\sin\theta$$
$$+ (1 - \cos\beta_{\text{eff}})(1 + E_2\cos\phi)\sin\theta\cos\theta],$$

$$M_z(0_+) = M_0(1 - E_1)\frac{1}{D}[-E_2\cos\phi(\sin^2\theta + \cos\beta_{\text{eff}} + \cos^2\theta\cos\beta_{\text{eff}})$$
$$+ E_2^2 + 2E_2\sin\phi\cos\theta\sin\beta_{\text{eff}} + \cos^2\theta + \sin^2\theta\cos\beta_{\text{eff}}] \quad (4.2.39)$$

with
$$D = A + B\cos\theta + C\cos^2\theta + F\cos^3\theta + G\cos^4\theta$$

and
$$A = (\sin^2\theta - E_1\cos\beta_{\text{eff}})(\sin^2\theta - E_2\cos\phi)$$
$$- E_2(E_1 - \sin^2\theta\cos\beta_{\text{eff}})(E_2 - \sin^2\theta\cos\phi),$$
$$B = 2E_2\sin\phi\sin\beta_{\text{eff}}(\sin^2\theta - E_1),$$
$$C = E_2(E_2 - \cos\phi\cos\beta_{\text{eff}}) + (1 - E_2\cos\phi\cos\beta_{\text{eff}})(2\sin^2\theta - E_1),$$
$$F = 2E_2\sin\phi\sin\beta_{\text{eff}},$$
$$G = 1 - E_2\cos\phi\cos\beta_{\text{eff}},$$
$$E_1 = \exp(-T/T_1), \quad E_2 = \exp(-T/T_2).$$

(The corresponding expression in ref 4.93 is partially incorrect.)

In the absence of interference ($E_1 = E_2 = 0$), these equations reduce to eqns (4.2.26).

Consider the effects of transverse interference for ideal pulses, i.e. for $\theta = \pi/2$ and $\beta_{\text{eff}} = \beta$. In this case, eqn (4.2.39) simplifies to

$$M_x(0_+) = M_0(1 - E_1)(1 - E_2 \cos \phi) \sin \beta / D,$$
$$M_y(0_+) = M_0(1 - E_1) E_2 \sin \beta \sin \phi / D,$$
$$M_z(0_+) = M_0(1 - E_1)\{E_2(E_2 - \cos \phi) + (1 - E_2 \cos \phi) \cos \beta\}/D \quad (4.2.40)$$

with
$$D = (1 - E_1 \cos \beta)(1 - E_2 \cos \phi) - (E_1 - \cos \beta)(E_2 - \cos \phi) E_2.$$

The phase and amplitude of the steady-state magnetization depends on the free precession angle $\phi = \Omega T$ in the interval T between successive pulses. Two extreme cases can be distinguished (4.94).

1. If $\phi = n2\pi$ ($n = 0, \pm 1, \ldots$), the effects of successive pulses will be additive, and the saturation effects are most pronounced. (In this case, the resonance frequency Ω coincides with one of the sideband frequencies of the r.f. pulse sequence.)

2. If $\phi = (2n + 1)\pi$ ($n = 0, \pm 1, \ldots$), the precession robs the successive pulses of their cumulative effect, and saturation is minimized. (This condition is fulfilled when the resonance frequency Ω occurs half-way between sidebands of the excitation sequence.)

The optimum pulse rotation angle β_{opt} for maximum signal amplitude depends on the offset and is determined (4.95) by

$$\cos \beta_{\text{opt}} = \frac{E_1 + E_2(\cos \phi - E_2)/(1 - E_2 \cos \phi)}{1 + E_1 E_2 (\cos \phi - E_2)/(1 - E_2 \cos \phi)}. \quad (4.2.41)$$

The optimum angle β_{opt} is plotted in Fig. 4.2.6 for three pulse separations

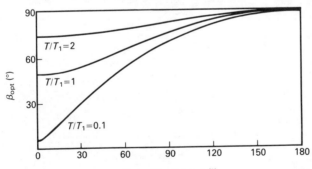

FIG. 4.2.6. Optimum pulse rotation angle β_{opt} as a function of the precession angle $\phi = \Omega T (\text{mod } 2\pi)$ for three pulse separations, assuming $T_1 = T_2$ (eqn (4.2.41)). $\phi = 0$ corresponds to a resonance line coincident with a sideband of the excitation spectrum, while $\phi = 180°$ indicates a line position in the middle between two pulse sidebands.

T, assuming $T_1 = T_2$. The dependence on the line position is particularly pronounced for short pulse separation $T < T_1$. Obviously the pulse rotation angle cannot be optimized simultaneously for lines with different offsets.

In the presence of transverse interference ($E_2 > 0$), *both* initial components after the r.f. pulse, $M_x(0_+)$ and $M_y(0_+)$ in eqn (4.2.40), are different from zero even for ideal pulses with $\theta = \pi/2$ and $\beta_{\text{eff}} = \beta$. The corresponding phase angle φ is

$$\tan \varphi = \frac{M_y(0_+)}{M_x(0_+)} = \frac{E_2 \sin \phi}{1 - E_2 \cos \phi}. \quad (4.2.42)$$

The phases in the resulting spectrum will therefore be dependent on the line position. It is interesting to note that φ is independent of the pulse rotation angle β. The dependence of the phase φ of the magnetization on the free precession angle $\phi = \Omega T$ is illustrated in Fig. 4.2.7. The phase φ varies very strongly for resonance frequencies lying near a pulse modulation sideband ($\phi \simeq 0, 2\pi$) whereas for resonance frequencies centred between two pulse modulation sidebands a weaker, nearly linear dependence on the precession angle ϕ is observed. The resulting frequency-dependent phase shifts in the Fourier spectrum are undesired.

The amplitude of the transverse magnetization is

$$M_{\text{trans}}(0_+) = M_0 \frac{1}{D}(1 - E_1)\sin \beta (1 + E_2^2 - 2E_2 \cos \phi)^{\frac{1}{2}} \quad (4.2.43)$$

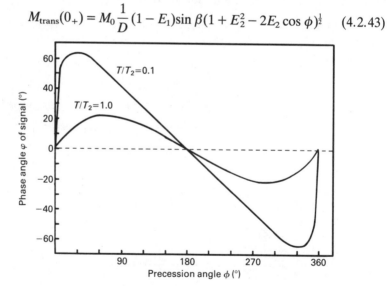

FIG. 4.2.7. Dependence of signal phase φ on the free precession angle $\phi = \Omega T$ (mod 2π) in the pulse interval T for two ratios $T/T_2 = 1$ and 0.1. $\phi = 0$ corresponds to resonance frequencies coinciding with an r.f. pulse modulation sideband; $\phi = 180°$ indicates resonance frequencies centred between two r.f. pulse modulation sidebands.

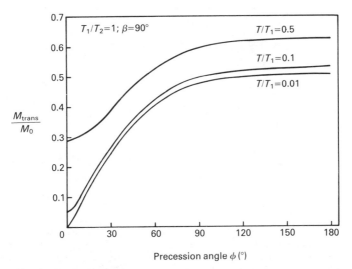

FIG. 4.2.8. Absolute transverse magnetization (eqn (4.2.43)) plotted as a function of the precession angle $\phi = \Omega T$ for $\beta = 90°$ and $T_1/T_2 = 1$. Three pulse intervals have been assumed: $T/T_1 = 0.5$; 0.1; and 0.01.

where D is defined in eqn (4.2.40). The signal intensities derived from this equation are shown in Fig. 4.2.8 for $T_1 = T_2$ and $\beta = 90°$. There is a very strong dependence of the signal amplitude on the precession angle ϕ, i.e. on the line position relative to the pulse sidebands. As pointed out before, maximum saturation occurs for $\phi = 0$, while for $90° < \phi < 270°$, the dependence on ϕ becomes quite weak. For very fast pulsing, the transverse magnetization reaches the value $M_{\text{trans}} = 0.5 M_0$ at $\phi = 180°$. This demonstrates the astonishing fact that at least half of the frequency range hardly shows any saturation even for extremely fast pulsing with $\beta = 90°$. This is actually a well-known phenomenon, called 'steady-state free precession', which was observed in 1951 by Bradford et al. (4.96) and analysed by Carr in 1958 (4.97). It is important for the detection of nuclei with low sensitivity (4.98, 4.99) and for imaging with the sensitive-point method proposed by Hinshaw described in Section 10.2.1 (4.100, 4.101).

In the case of short T_2, the dependence of the signal amplitude on the precession angle ϕ is less pronounced than the case of $T_1 = T_2$ shown in Fig. 4.2.8. The amplitude reached for very fast pulsing at $\phi = 180°$ is given by

$$M_{\text{trans}} = M_0 \frac{T_2}{T_1 + T_2}. \quad (4.2.44)$$

It is convenient to visualize phase and intensity anomalies due to transverse interference by means of curves that depict the locus of the

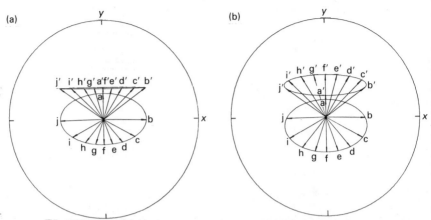

FIG. 4.2.9. Projections on to the xy-plane of the steady-state magnetization vectors in a sequence of equidistant r.f. pulses as a function of the free precession angle $\phi = \Omega T$. Symbols without primes indicate $M_{xy}(0_-)$ (i.e. just before the pulse: a through j correspond to $\phi = n\, 2\pi/10$ with $n = 0, 1, \ldots 9$); primed letters correspond to $M_{xy}(0_+)$ immediately after the pulse (these phases and amplitudes determine the form of the spectrum after Fourier transformation of the free induction decay). In both cases, $T_1 = T_2$ and $T/T_1 = 0.2$. In (a), $\beta = 34°$ corresponds to the value given by eqn (4.2.45) while in (b), $\beta = 52°$. Pulses are applied along the $-x$ axis, in contrast to the y-pulses discussed in the text. (Reproduced from Ref. 4.94.)

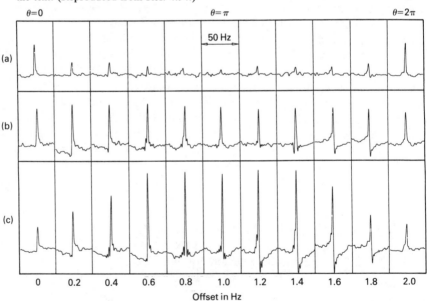

FIG. 4.2.10. Dependence of peak amplitude and phase on the frequency offset relative to the pulse sidebands. For an offset of 0 Hz, the resonance coincides with a pulse sideband, while for an offset of 1 Hz, the resonance is centred between two pulse sidebands. In (b), the rotation angle β_{opt} of the r.f. pulse agrees with eqn (4.2.45). In (a) and (c), the flip angles are $\beta = \beta_{\text{opt}}/6$ and $\beta = 6\beta_{\text{opt}}$, respectively. (Reproduced from Ref. 4.94.)

transverse steady-state magnetization as a function of the free precession angle ϕ (4.94). Two examples are shown in Fig. 4.2.9 for different pulse rotation angles β. The larger the diameter of the locus of $M_{xy}(0_+)$ immediately after the pulse (primed symbols in Fig. 4.2.9), the more pronounced the phase and intensity anomalies. For a pulse rotation angle

$$\cos \beta_{\text{opt}} = \exp(-T/T_1), \qquad (4.2.45)$$

the $M_x(0_+)$ components are independent of the angle $\phi = \omega_0 T$, although the $M_y(0_+)$ components (and therefore the phases and amplitudes in the spectrum) still remain offset-dependent (Fig. 4.2.9(a)). Experimental examples of phase and intensity anomalies for three different flip angles are shown in Fig. 4.2.10.

4.2.6. Remedies for phase and intensity anomalies due to transverse interference

The elimination of phase and intensity anomalies in Fourier spectroscopy is of practical relevance, particularly when quantitative intensities are required. Very often rapid pulsing and the resulting transverse and longitudinal interference cannot be avoided when the sensitivity is to be optimized. Fortunately, a number of techniques has been proposed to eliminate the undesired anomalies.

4.2.6.1. *Quenching of transverse magnetization by magnetic field gradient pulses*

By applying a strong magnetic field gradient pulse before each r.f. pulse, as shown in Fig. 4.2.11, the remaining transverse magnetization can be defocused. However, it should be remembered that defocusing does not imply irreversible destruction of transverse magnetization, since it can be refocused by subsequent gradient pulses. For gradients to be effective in

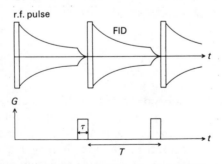

FIG. 4.2.11. Quenching of transverse magnetization by magnetic field gradient pulses applied just before each r.f. pulse to eliminate phase and amplitude distortions associated with transverse interference. (Reproduced from Ref. 4.102.)

eliminating phase and intensity anomalies, it is essential to make the process irreversible (4.102).

An irreversible destruction is caused by translational diffusion during an extended time period, which leads to a loss of the phase memory of the individual spins. It turns out that for efficient quenching the required field gradient G applied for the time τ_g must have at least the value

$$\gamma G = \left(\frac{3}{D\tau_g^2 T}\right)^{\frac{1}{2}} \qquad (4.2.46)$$

where D is the translational diffusion constant of the solution under investigation (4.102). To give an example, for $T = 1$ s and $\tau_g = 0.1$ s, and for a typical diffusion constant $D = 2.5 \times 10^{-5}$ cm^2/s, a field gradient $\gamma G/2\pi$ of at least 550 Hz/cm is required to quench the transverse magnetization. This can readily be achieved with shim coils.

The main drawback of this technique is that the field gradient pulse also affects the field-frequency lock, making it necessary to disable the lock temporarily during the field gradient pulses.

4.2.6.2. Scrambling of transverse interference by random pulse intervals

The angle of free precession $\phi = \Omega T$, which is responsible for the phase and amplitude anomalies illustrated in Figs. 4.2.7–4.2.10, can be varied randomly by using pulse intervals $T = T_0 + \varepsilon_k$ with a large set $\{\varepsilon_k\}$ of randomly generated increments. This technique is most effective if large numbers of free induction decays must be averaged for sensitivity reasons (4.94).

4.2.6.3. Quadriga Fourier spectroscopy

To eliminate phase and intensity anomalies in experiments with fast pulsing rates, Schwenk (4.103) proposed to add signals obtained in four different steady-state Fourier experiments, each with a slightly different carrier frequency

$$\omega'_{\text{r.f.}} = \omega_{\text{r.f.}} + \frac{\pi}{2T}k, \qquad k = 0, 1, 2, 3, \qquad (4.2.47)$$

in such a way that the sideband spectrum of the repetitive pulse sequence is moved by $\frac{1}{4}$ of the sideband spacing from one sequence to the next. One particular resonance line thus appears at four different positions with respect to the sideband spectrum and each time experiences different phase and intensity anomalies. A quadriga spectrum contains less than 1 per cent phase and intensity variations (4.103). The practical realization of this proposal requires four different pulse carrier frequencies.

4.2.6.4. Four-phase Fourier spectroscopy

In a modified form of quadriga spectroscopy, four complementary experiments are performed with the same carrier frequency, but the phase is shifted from pulse to pulse (4.104, 4.105).

Consider four r.f. pulses with phases 0°, 90°, 180°, and 270°, denoted by A, B, C, D respectively. It is possible to perform the sequences

Experiment 1: A A A A A A A...
Experiment 2: A B C D A B C...
Experiment 3: A C A C A C A...
Experiment 4; A D C B A D C...

with a constant pulse spacing T. In the first sequence without phase-shifts, the effective r.f. frequency is $\omega'_{r.f.} = \omega_{r.f.}$. In sequence 2, the effective r.f. frequency is $\omega'_{r.f.} = \omega_{r.f.} + \pi/(2T)$. For experiment 3 we obtain $\omega'_{r.f.} = \omega_{r.f.} + \pi/(T)$, and in experiment 4, $\omega'_{r.f.} = \omega_{r.f.} + 3\pi/(2T)$. These are exactly the frequencies required in quadriga Fourier spectroscopy. The phase-sensitive detector always uses the same phase and the signals can be added directly. The interference terms cancel upon co-addition to the same degree as in the quadriga experiment.

4.2.6.5. Phase-alternated pulse sequences

A simple but nevertheless effective technique to suppress interference effects is based on a periodic 180° phase alternation of the r.f. pulses and alternate addition and subtraction of the resulting signals. This idea relies on the change of sign of the dominant interference effects which are cancelled upon addition of the signals. Pair-wise alternation in the form $(+ + - -)$ proved to be quite satisfactory. Higher-order interferences extending over several pulse periods could be eliminated by more extended phase alternation schemes, for example by eight-pulse sequences of the type $(+ + + - + - - -)$ and $(+ + + - - + - -)$ where the ± indicate the phases of the pulses and the corresponding addition and subtraction of the signals.

4.2.7. Remedies for anomalies due to non-ideal pulses: composite pulses

Since their introduction in 1979 (4.85), composite pulses have become versatile tools for the correction of the effects of non-ideal pulses. In particular, anomalies due to the inhomogeneity of the r.f. field across the sample volume and to off-resonance effects (tilted r.f. fields) can be removed efficiently by composite pulses.

Composite pulses are sequences of closely-spaced pulses designed to be equivalent to a single pulse under ideal conditions, but which have a greater tolerance to imperfection. A wide variety of composite pulses have been proposed, which pursue distinct purposes. They include composite pulses designed to:

(a) Nutate the longitudinal magnetization into the transverse plane in such a way that the residual z-component is minimized in spite of inhomogeneous B_1 fields (4.86, 4.87, 4.106) and tilted effective fields (4.86), as discussed in § 4.2.7.1. This is of particular importance for the measurement of spin–lattice relaxation times by saturation-recovery or by progressive saturation (see § 4.6.1).

(b) Excite transverse magnetization vectors with phases that are 'bunched' together within a narrow range, by minimizing the offset-dependence of the phase of the magnetization, as discussed in § 4.2.7.2. This is relevant for pulse Fourier experiments, for coherence transfer and for spin-locking experiments (4.86).

(c) Achieve accurate inversion ($M_z \rightarrow -M_z$), as discussed in §§ 4.2.7.3 and 4.2.7.4 (4.85–4.87, 4.106, 4.107). This is essential for inversion-recovery relaxation measurements and for many experiments where the spin states of a coupling partner must be inverted (e.g. $I_z \rightarrow -I_z$), as required for heteronuclear decoupling by a single π-pulse during the evolution period in two-dimensional experiments.

(d) Achieve proper refocusing of transverse magnetization in spin-echo experiments (4.108), as described in § 4.2.7.5.

(e) Give accurate rotation angles β in spite of inhomogeneous r.f. fields (4.87), as described in § 4.2.7.4. This is relevant if the β-dependence of coherence transfer amplitudes is exploited (see § 4.5.6 and Chapter 8).

(f) Mimic the effect of phase shifts through arbitrary rotation angles (so-called 'z-pulses'; see § 4.2.7.6).

(g) Generate sequences that are *cyclic* in spite of offset effects and r.f. inhomogeneity to avoid cumulative errors (4.109–4.116) (see § 4.2.7.7). This requirement must be fulfilled to achieve efficient decoupling, as discussed in § 4.7.6.

(h) Induce coherence transfer processes in a uniform manner over a wide range of offsets (4.117).

(i) Inhibit the effects of dipolar or quadrupolar couplings during the pulse to obtain distortion-free solid-state spectra (4.118, 4.119).

Obviously, an exhaustive discussion would be beyond the scope of this chapter. We shall focus on a few of the most promising proposals that deal with cases (a)–(g) mentioned above. These cases can be handled with classical magnetization vectors and Bloch equations, although an equivalent quantum mechanical notation (using $\langle I_x \rangle$, $\langle I_y \rangle$, $\langle I_z \rangle$ instead of M_x, M_y, M_z) is often convenient.

In all applications of composite pulses, it is important to identify the dominant source of non-ideal behaviour, i.e. inhomogeneous or tilted r.f. fields. In many cases, it is not possible to compensate for both types of errors at once.

Off-resonance effects are readily described in terms of an effective r.f. field with a rotation angle $\beta_{\text{eff}} = -\gamma B_{\text{eff}}\tau_p$ defined in eqn (4.2.24) and a tilt θ of the rotation axis defined in eqn (4.2.23). Note that $\theta = \pi/2$ for $\Delta B_0 = 0$ in our notation, in contrast to Ref. 4.86.

In the presence of inhomogeneous r.f. fields, it is convenient to introduce a *nominal* value of the field intensity B_1^0 and a nominal rotation angle $\beta^0 = -\gamma B_1^0 \tau_p$, referring, for example, to the centre of the sample. The r.f. field inhomogeneity may be caused by the geometric arrangement of the r.f. coil or by the dielectric properties of the sample, particularly in the case of high electric conductivity.

In the following sections, we must distinguish the *applied* sequence of pulses with nominal rotation angles β^0, $\beta^{0'}$, $\beta^{0''}$ ... and r.f. phases φ, φ', φ'' ... from the actual rotations experienced by the various magnetization components. The sequence will be denoted

$$P = (\beta)_\varphi (\beta')_{\varphi'} (\beta'')_{\varphi''} \ldots \quad (4.2.48)$$

The actual motion of the magnetization vector $M(t)$ under each constituent pulse

$$\mathbf{M}(0_+) = \mathbf{R}_\varphi(\beta)\mathbf{M}(0_-) \quad (4.2.49a)$$

can be represented by a sequence of rotations

$$\mathbf{R}_\varphi(\beta) = \mathbf{R}_z(\varphi)\mathbf{R}_y(-\pi/2 + \theta)\mathbf{R}_x(\beta_{\text{eff}})\mathbf{R}_y(\pi/2 - \theta)\mathbf{R}_z(-\varphi). \quad (4.2.49b)$$

The overall rotation produced by a sequence of two successive rotations can be described using the quaternion formalism (4.293, 4.294). An equivalent formulation (4.295) avoids the use of matrix multiplications: If a rotation through an angle β_1 about an axis $\mathbf{n}_1$ is followed by a rotation through an angle β_2 about an axis $\mathbf{n}_2$, the overall rotation is through an angle β_{12} about an axis $\mathbf{n}_{12}$ given (4.296) by the equations

$$c_{12} = c_1 c_2 - s_1 s_2 \mathbf{n}_1 \cdot \mathbf{n}_2, \quad (4.2.50a)$$

$$s_{12}\mathbf{n}_{12} = s_1 c_2 \mathbf{n}_1 + c_1 s_2 \mathbf{n}_2 - s_1 s_2 \mathbf{n}_1 \times \mathbf{n}_2 \quad (4.2.50b)$$

where $c_i = \cos \beta_i/2$ and $s_i = \sin \beta_i/2$ and where $\times$ denotes a vector product. These equations may be extended to a sequence of three or more non-commuting rotations (4.295).

In the figures shown in this section, taken from various publications, the sense of rotation is opposite to the convention used throughout this work, i.e. an x-pulse leads to the transformation $M_z \to M_y \to -M_z \to -M_y$. This does not affect the outcome of the experiments.

4.2.7.1. Minimum residual M_z component after a $\pi/2$-pulse

For on-resonance irradiation ($\Delta B_0 = 0$) the residual longitudinal magnetization after a pulse obviously depends on the actual rotation angle β. This is detrimental in experiments such as progressive-saturation and saturation-recovery measurements (§ 4.6.1). For an uncompensated pulse with $0.7\ \pi/2 < \beta < 1.3\ \pi/2$, the residual z-component (4.86) varies in the interval $-0.4 < M_z(0_+)/M_z(0_-) < +0.4$. Provided off-resonance effects can be neglected, the composite pulse (4.86)

$$P = (\beta)_0(\beta)_{\pi/2} \tag{4.2.51}$$

yields a reduced residual component $0 < M_z(0_+)/M_z(0_-) < 0.2$ for the same interval $0.7\ \pi/2 < \beta < 1.3\ \pi/2$, since the z-component that remains after the first pulse is rotated towards the transverse plane while the transverse component remains unaffected.

With on-resonance pulses, better performance can be achieved with the composite sequence (4.106)

$$P = (\beta)_0(2\beta)_{2\pi/3} \tag{4.2.52}$$

with $\beta \simeq \pi/2$. The trajectories of the magnetization vectors are shown in Fig. 4.2.12 for $0.8\ \pi/2 < \beta < \pi/2$. Note that the arc lengths of the two rotations are approximately equal. The vectors are 'bunched' very near to the xy-plane with a phase $\varphi \simeq 5\pi/6$.

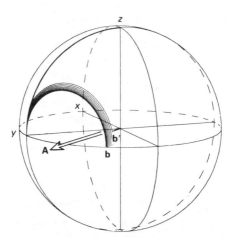

FIG. 4.2.12. Trajectories of a set of magnetization vectors initially aligned along the z-axis under the composite pulse sequence of eqn (4.2.52). The r.f. rotation angle β varies in the interval $0.8\ \pi/2 < \beta < \pi/2$. Note the 'bunching' of the vectors near the equatorial plane, which is achieved because the arc lengths of the two successive trajectories are approximately equal for all β, while the sense of rotation is reversed in the vicinity of the y-axis, where the magnetization follows nearly the same path in both rotations (Reproduced from Ref. 4.106).

The unusual r.f. phase-shift required for the sequence in eqn (4.2.52) can be avoided with a modified sequence (4.106) that is also designed for on-resonance pulses

$$P = (\beta)_{3\pi/2}(2\beta)_0(2\beta)_{\pi/2}(\beta)_0 \quad (4.2.53)$$

with the nominal rotation angle $\beta = \pi/4$. The corresponding trajectories can be found in Ref. (4.106). If offset effects are negligible, it is possible to take the magnetization to the $-z$-axis by linking two composite sequences of this type, the second one with the phases in reversed order (4.106).

If offset effects are important and if an ideal rotation angle $\beta = \pi/2$ can be assumed, the residual longitudinal magnetization after a single uncompensated pulse is fairly small over a wide range of offsets, because of the self-compensating effect of the increased effective rotation angle (see Fig. 4.2.3). If a further reduction of the residual M_z-component is desired in situations where off-resonance effects are important, the 'spin-knotting' sequence is particularly effective (4.86)

$$P = (\beta)_0 - \tau - (\beta')_\pi - \tau' - (\beta'')_0 \quad (4.2.54)$$

with $\beta = 10°$, $\beta' = 60°$, $\beta'' = 140°$, $\tau = 40\ \mu s$, and $\tau' = 11\ \mu s$, optimized for $\gamma B_1/2\pi = 10$ kHz. In an offset range $-0.2 < \Delta B_0/B_1 < 0.2$, the residual component is $-0.01 < M_z(0_+)/M_z(0_-) < 0$, provided r.f. inhomogeneity can be neglected. This composite sequence allows the magnetization to evolve freely in the intervals τ and τ' to make up for the phase errors caused by tilted r.f. fields.

4.2.7.2. Minimum phase dispersion of transverse magnetization after a $\pi/2$-pulse

The phase of the transverse magnetization excited by a single, uncompensated $\pi/2$-pulse has an approximately linear dependence on the offset parameter $\Delta B_0/B_1$ (Fig. 4.2.3). Although this does not pose a problem in ordinary Fourier spectroscopy (where the resulting phase errors of the signals can be corrected mathematically), the dispersion of the phases is undesirable if the initial excitation is followed by another r.f. irradiation sequence, particularly if the magnetization must be spin-locked. The 'spin-knotting' in eqn (4.2.54) has been shown to be effective in this context: over a range $-0.5 < \Delta B_0/B_1 < +0.5$, the phase dispersion of the signal is confined to an interval $-5° < \varphi < +5°$, which may be compared with a range $-30° < \varphi < 30°$ obtained after a single $\pi/2$-pulse.

4.2.7.3. Accurate inversion; sequences optimized by calculation of trajectories

Because the inversion of the magnetization $(M_z \rightarrow -M_z)$ is very sensitive to offset effects and to pulse errors, composite pulses turned out to be

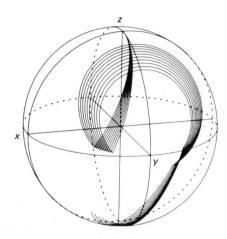

FIG. 4.2.13. Trajectories of a set of magnetization vectors under the composite inversion pulse of eqn (4.2.55), corresponding to offset parameters $0.4 < \Delta B_0/B_1 < 0.6$ (Reproduced from Ref. 4.86).

particularly fruitful in this context. Proper inversion is useful for inversion–recovery relaxation measurements, and for many experiments requiring the inversion of spin states ($I_{kz} \to -I_{kz}$), including a wide variety of two-dimensional techniques described in Chapters 7 and 8.

The mechanism of the compensated inversion pulse sequence first proposed (4.85)

$$P = (\beta)_0 (\beta')_{\pi/2} (\beta)_0 \qquad (4.2.55)$$

with nominal rotation angles $\beta = \pi/2$ and $\beta' = \pi$, can easily be understood for variations of the actual pulse rotation angle β. After the first pulse with $\beta < \pi/2$, the magnetization vector is brought into a location above the xy-plane, is then rotated by the second pulse into an approximately symmetrical position below the xy-plane, from where it is transported by the third pulse into a position close to the 'south pole' of the unit sphere. The compensation for $\beta > \pi/2$ operates in an analogous manner.

To grasp the compensation of off-resonance effects, numerical integrations of the Bloch equations are mandatory. The trajectories described by the magnetization in the sequence of eqn (4.2.55) for $\beta = \pi/2$ and $\beta' = 1.33\pi = 240°$ are shown in Fig. 4.2.13 under off-resonance conditions.

The efficiency of this sequence may be appreciated by considering the percentage of inverted magnetization shown in Fig. 4.2.14. Note that by increasing the rotation angle β' of the central pulse from π to 1.33π (240°), the tolerance with respect to small offsets is greatly increased.

The compensation can be further enhanced with a five-pulse inversion sequence (4.107)

$$P = (\beta)_0(\beta')_{\pi/2}(\beta'')_{3\pi/2}(\beta')_{\pi/2}(\beta)_0 \qquad (4.2.56)$$

where $\beta \simeq \pi/2$, $\beta' \simeq 1.12\pi$, and $\beta'' \simeq 0.44\pi$.

An alternative sequence for inversion was proposed by Baum et al. (4.119)

$$P = (\beta)_\varphi(\beta')_{\varphi'}(\beta'')_{\varphi''}(\beta''')_{\varphi'''}(\beta'''')_0(\beta''')_{\varphi'''}(\beta'')_{\varphi''}(\beta')_{\varphi'}(\beta)_\varphi \qquad (4.2.57)$$

where $\beta \simeq 0.22\pi$, $\beta' = 0.3\pi$, $\beta'' = 0.37\pi$, $\beta''' = 0.47\pi$, $\beta'''' = 1.48\pi$, $\varphi = 2.33\pi$, $\varphi' = 1.16\pi$, $\varphi'' = 0.77\pi$, and $\varphi''' = 0.39\pi$.

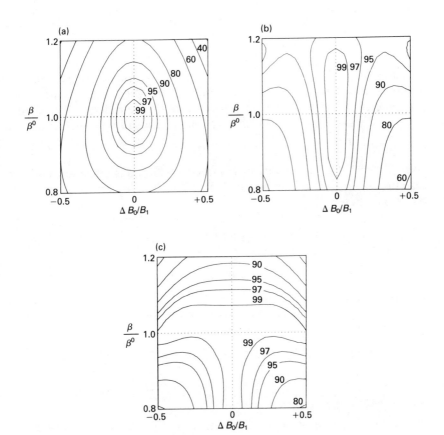

FIG. 4.2.14. Fraction of inverted magnetization $-M_z(0_+)/M_z(0_-)$ (expressed in percentage) as a function of the deviation β/β^0 from the ideal r.f. rotation angle and of the resonance offset parameter $\Delta B_0/B_1$: (a) conventional pulse with $\beta \simeq \pi$; (b) composite pulse $(\beta)_0(\beta')_{\pi/2}(\beta)_0$ with $\beta \simeq \pi/2$ and $\beta' \simeq \pi$; (c) same sequence, but with the central pulse increased to $\beta' \simeq 1.33\pi$ (240°). (Reproduced from Ref. 4.86.)

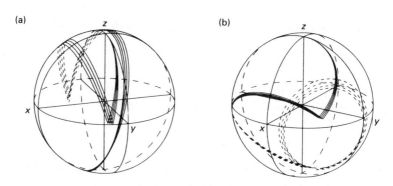

FIG. 4.2.15. Trajectories of magnetization vectors under the 'WALTZ' inversion sequence of eqn (4.2.58). (a) For small offset parameters ($\Delta B_0/B_1 \simeq 0.25$) the compensation is only moderate. (b) For large offsets, i.e. $0.75 < \Delta B_0/B_1 < 0.88$, the first two pulses achieve adequate inversion, while the last pulse merely leads to a rotation through 2π about the tilted effective field (Reproduced from Ref. 4.116).

Sequences which employ r.f. pulses with $\pi/2$ phase increments are often quite sensitive to deviations from ideal $\pi/2$ r.f. phase shifts. This problem can be largely circumvented with the so-called $1\bar{2}3$ sequence ('WALTZ') (4.120, 4.121) which uses only phase shifts through π

$$P = (\beta)_0 (2\beta)_\pi (3\beta)_0 \qquad (4.2.58)$$

with $\beta \simeq \pi/2$. The trajectories described by the magnetization and the resulting inversions are shown in Fig. 4.2.15 for different ranges of the offset parameter.

If offset effects can be neglected, the r.f. inhomogeneity can be compensated effectively with a sequence derived from the composite pulse in eqn (4.2.52)

$$P = (\beta)_0 (4\beta)_{2\pi/3} (\beta)_0 \qquad (4.2.59)$$

with $\beta \simeq \pi/2$ (4.106). A similar degree of compensation can be achieved with a nine-pulse sequence (4.106) which employs the element in eqn (4.2.53) and avoids $2\pi/3$ phase shifts

$$P = (\beta)_{3\pi/2}(2\beta)_0(2\beta)_{\pi/2}(\beta)_0(4\beta)_{\pi/2}(\beta)_0(2\beta)_{3\pi/2}(2\beta)_0(\beta)_{\pi/2} \qquad (4.2.60)$$

with $\beta \simeq \pi/4$.

4.2.7.4. Recursive expansion procedures

A special notation will be used in this section: $P^{(m)}$ represents a composite pulse with mth-order error compensation, which approximates the function of an ideal $\pi/2$-pulse. We shall denote composite pulses which approximate ideal π-pulses (inversion) by the symbol R.

It is possible to achieve higher orders of compensation by recursive expansion of a zero-order pulse (4.87, 4.121). Consider a composite pulse sequence denoted $P^{(m)}$ (the zeroth-order 'sequence' may consist of a single uncompensated pulse, i.e. $P^{(m=0)} = (\pi/2)_\varphi$). To achieve a higher stage of compensation, it is necessary to define *inverse* sequences denoted $(P^{(m)})^{-1}$ such that the corresponding propagators fulfil the conditions

$$(P^{(m)})^{-1}P^{(m)} = P^{(m)}(P^{(m)})^{-1} = \mathbb{1}. \qquad (4.2.61)$$

If resonance offset effects can be neglected, an inverse sequence can be derived simply by reversing the order of all component pulses and shifting the phases through an angle π. Thus for $P^{(m)} = (\beta)_0(\beta)_{\pi/2}$ with $\beta = \pi/2$ (see eqn (4.2.51)), one obtains the inverse

$$(P^{(m)})^{-1} = (\beta)_{3\pi/2}(\beta)_\pi. \qquad (4.2.62)$$

If resonance offset effects cannot be neglected, a good approximation to the inverse sequence $(P^{(m)})^{-1}$ is obtained by taking any cyclic composite pulse (such as a WALTZ-16 sequence, see § 4.2.7.7) which terminates with the element $P^{(m)}$, and then amputating this element from the end of the sequence.

In a first step, the order of compensation is increased for a composite pulse which has again the net effect of a $\pi/2$ pulse. There are four possibilities of linking two mth-order elements to achieve error compensation to order $(m+1)$, which are based on the idea of applying two $\pi/2$-pulses (compensated to mth order) phase-shifted by 90°, in analogy to eqn (4.2.51)

$$P^{(m+1)} = (P^{(m)}_{+\pi/2})^{-1}P^{(m)}, \qquad P^{(m+1)} \xrightarrow{\text{equiv.}} P_z^{-1}P_{\varphi-m\pi/2} \qquad (4.2.63a)$$

$$P^{(m+1)} = (P^{(m)}_{-\pi/2})^{-1}P^{(m)}, \qquad P^{(m+1)} \xrightarrow{\text{equiv.}} P_zP_{\varphi+m\pi/2} \qquad (4.2.63b)$$

$$P^{(m+1)} = P^{(m)}(P^{(m)}_{+\pi/2})^{-1}, \qquad P^{(m+1)} \xrightarrow{\text{equiv.}} P_zP_{\varphi-(m-1)\pi/2} \qquad (4.2.63c)$$

$$P^{(m+1)} = P^{(m)}(P^{(m)}_{-\pi/2})^{-1}, \qquad P^{(m+1)} \xrightarrow{\text{equiv.}} P_z^{-1}P_{\varphi+(m-1)\pi/2}. \qquad (4.2.63d)$$

The order m can be increased recursively, as shown schematically in Fig. 4.2.16.

It should be noted that the resulting pulses are equivalent to a phase-shifted $\pi/2$ pulse, preceded by an additional phase shift $\pm\pi/2$, expressed by the z-pulse $P_z^{\pm 1}$. This phase-shift can be compensated by shifting the entire pulse sequence which precedes the composite pulse by $\pi/2$. The phase $\varphi \mp m\pi/2$ or $\varphi \mp (m-1)\pi/2$ of the resulting effective pulse refers to the phase of the zero-order pulse $P^{(0)}$. Each additional order of compensation shifts the phase by $\mp\pi/2$.

At *any* stage of the recursion procedure, it is possible to derive a composite pulse denoted $R(\beta_{\text{eff}})$, corresponding to an arbitrary rotation

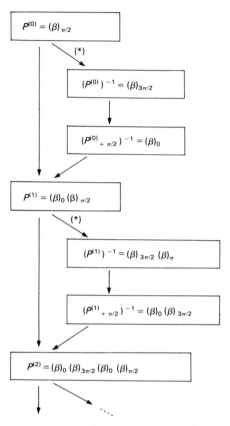

FIG. 4.2.16. Example of the recursive procedure according to eqn (4.2.63a), used to construct composite pulses that approximate the function of an ideal $(\pi/2)$ pulse with mth-order error compensation. The steps marked with an asterisk are valid only for $\Delta B_0 = 0$. Composite inversion pulses can be derived at each stage with the recipe in eqn (4.2.64).

angle β_{eff}, with the prescription

$$R^{(m)}(\beta_{\text{eff}}) = (P^{(m)}_{+\beta_{\text{eff}}})^{-1} P^{(m)}, \qquad R^{(m)}(\beta_{\text{eff}}) \xrightarrow{\text{equiv.}} R_z(-\beta_{\text{eff}}) R_{\varphi - m\pi/2}(\beta_{\text{eff}}).$$

(4.2.64)

The first pulse of the equivalent sequence represents again a phase shift, here $-\beta_{\text{eff}}$, imposed upon the density operator after the actual pulse with flip angle β_{eff} and phase $\varphi - m\pi/2$.

To give an example of this construct, one finds the equivalence

$$(\pi/2)_\beta (\pi/2)_x \xrightarrow{\text{equiv.}} (\beta)_z (\beta)_y. \qquad (4.2.65)$$

4.2 CLASSICAL DESCRIPTION

The product of the two $\pi/2$-pulses with a phase difference $-\beta$ corresponds, except for a preceding phase-shift β, to a rotation by the angle β.

This type of composite pulse is useful when it is important to exploit the β-dependence of coherence transfer amplitudes. One example is the 'distortionless enhancement of polarization transfer' sequence (DEPT) discussed in § 4.5.6; other examples are found in two-dimensional spectroscopy (Chapter 8).

For an r.f.rotation angle $\beta_{\mathrm{eff}} = \pi$, eqn (4.2.64) leads to an inversion sequence compensated to mth order, which will be denoted $R^{(m)}(\beta_{\mathrm{eff}} = \pi)$.

The recursive expansion procedure leads to the following increasingly accurate approximations to a $\pi/2$-pulse (starting with $P^{(0)} = (\beta)_{\pi/2}$, $\beta \simeq \pi/2$, see Fig. 4.2.16), keeping in mind the equivalences given in eqn (4.2.63a),

$$P^{(1)} = (\beta)_0(\beta)_{\pi/2},$$
$$P^{(2)} = (\beta)_0(\beta)_{3\pi/2}(\beta)_0(\beta)_{\pi/2},$$
$$P^{(3)} = (\beta)_0(\beta)_{3\pi/2}(\beta)_\pi(\beta)_{3\pi/2}(\beta)_0(\beta)_{3\pi/2}(\beta)_0(\beta)_{\pi/2}, \quad (4.2.66)$$

which can be used to derive increasingly accurate inversion pulses

$$R^{(0)}(\beta_{\mathrm{eff}} = \pi) = (\beta)_{\pi/2}(\beta)_{\pi/2} = (2\beta)_{\pi/2},$$
$$R^{(1)}(\beta_{\mathrm{eff}} = \pi) = (\beta)_{\pi/2}(\beta)_0(\beta)_0(\beta)_{\pi/2} = (\beta)_{\pi/2}(2\beta)_0(\beta)_{\pi/2},$$
$$R^{(2)}(\beta_{\mathrm{eff}} = \pi) = (\beta)_{\pi/2}(\beta)_0(\beta)_{3\pi/2}(\beta)_0(\beta)_0(\beta)_{3\pi/2}(\beta)_0(\beta)_{\pi/2},$$
$$R^{(3)}(\beta_{\mathrm{eff}} = \pi) = (\beta)_{\pi/2}(\beta)_0(\beta)_{3\pi/2}(\beta)_0(\beta)_{3\pi/2}(\beta)_\pi(\beta)_{3\pi/2}(\beta)_0$$
$$(\beta)_0(\beta)_{3\pi/2}(\beta)_\pi(\beta)_{3\pi/2}(\beta)_0(\beta)_{3\pi/2}(\beta)_0(\beta)_{\pi/2}. \quad (4.2.67)$$

We may note in passing that the first sequence in eqn (4.2.66) is identical to eqn (4.2.51), while the second sequence in eqn (4.2.67) corresponds to eqn (4.2.55). The practical performance of the composite inversion pulses in eqn (4.2.67) is shown in Fig. 4.2.17.

The performance of composite pulses can be improved by employing inverse sequences that fulfil eqn (4.2.62) over a range of offsets. Shaka *et al.* (4.121) proposed a sequence for offset-compensated inversion for $\beta \simeq \pi/2$

$$R(\beta_{\mathrm{eff}} = \pi) = (3\beta)_\pi(4\beta)_0(\beta)_{\pi/2}(3\beta)_{3\pi/2}(4\beta)_{\pi/2}(\beta)_0. \quad (4.2.68)$$

which is equivalent to

$$R(\beta_{\mathrm{eff}} = \pi) = \overline{3X}\,4X\,Y\,\overline{3Y}\,4Y\,X$$

in the shorthand notation of Ref. (4.121).

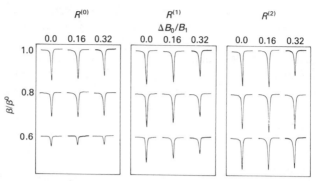

FIG. 4.2.17. Experimental test of the inversion of magnetization achieved with three composite pulse sequences $R^{(0)}$, $R^{(1)}$, and $R^{(2)}$ defined in eqn (4.2.67) as a function of the rotation angle $0.6\, \pi/2 < \beta < \pi/2$ and the offset parameter $0 < \Delta B_0/B_1 < 0.32$. Note that the performance of $R^{(2)}_{\pi/2}$ for large offsets improves if the rotation angle is below its nominal value of $\pi/2$ (Reproduced from Ref. 4.87).

4.2.7.5. Accurate refocusing

In sequences involving multiple refocusing (see § 4.6.2), non-ideal pulses lead to severe cumulative errors. If the echoes are not modulated by homonuclear scalar or dipolar interactions, errors due to imperfect rotation angles $\beta \neq \pi$ can be compensated by shifting the phase of the refocusing pulses relative to the first pulse, as suggested by Meiboom and Gill (4.122). However, echo modulation interferes with the compensation mechanism, and one must resort to composite pulses.

Contrary to earlier apprehensions (4.86), the composite pulses designed for ideal inversion ($M_z \rightarrow -M_z$) discussed in the previous sections can also be used to refocus transverse magnetization (4.108). Any composite pulse that transforms M_z into $-M_z$ can be described by a rotation through an angle π about a vector in the xy-plane. The phase of this effective rotation axis depends on the nature of the imperfections and on the choice of the composite pulse.

For the composite sequence $P = (\beta)_0 (2\beta)_{\pi/2} (\beta)_0$ (see eqn (4.2.55)), it can be shown that for a rotation angle $\beta = \pi/2 + \Delta$ and $\Delta B_0 = 0$, the effective rotation axis lies in the equatorial plane with a phase $\varphi = \pi/2 + \Delta$.

If offset effects must be considered, the composite sequence in eqn (4.2.55) with $\beta = \pi/2$ and $\beta' = 3\pi/2$ can be used, leading to a nearly ideal π-rotation about an axis lying in the xy-plane with a phase $\varphi = \pi/2 + \Delta$ where $\Delta \simeq \pi/4$.

In either case, the phase of the magnetization at the time of the first echo deviates by 2Δ from its ideal position. However, this phase error is compensated at the time of the second echo and generally for all even-numbered echoes (4.108).

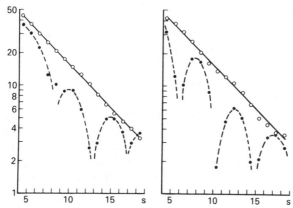

FIG. 4.2.18. Absolute value of the intensities of even-numbered echoes in a Carr–Purcell pulse sequence with multiple refocusing (intervals $2\tau = 0.5$ s). The signals stem from the doublet in the A_2X system of 1,1,2-trichloroethane, which is caused by a homonuclear scalar coupling. (Left: high-field doublet component; right: low-field line.) Open circles: smooth decays proportional to $\exp(-t/T_2)$ obtained with composite pulses as defined in eqn (4.2.55). Black symbols: oscillatory behaviour (distorted by the absolute-value presentation) obtained with conventional pulses. In both cases, the r.f. rotation angles were deliberately misset ($\beta = 0.8\pi$). (Reproduced from Ref. 4.108.)

The efficiency of composite pulses for the compensation of non-ideal rotation angles is illustrated in Fig. 4.2.18.

4.2.7.6. Composite z-pulses

Many sophisticated experiments call for phase-shifts φ of the r.f. carrier through arbitrary angles. An equivalent effect can be achieved with a composite z-pulse (4.123)

$$P = (\pi/2)_{\pi/2}(\beta)_0(\pi/2)_{3\pi/2} \qquad (4.2.69)$$

where β corresponds to the desired phase-shift φ. The net effect is equivalent to what would be obtained if an r.f. pulse with rotation angle β could be applied *along the z-axis* of the rotating frame.

Thus all magnetization vectors experience a forced precession in the xy-plane through an angle β, regardless of their actual offset. For multiple-quantum coherence, the angle of forced precession is proportional to the order p, a property which can be exploited for the separation of different orders (§ 6.3).

One may consider z-pulses as a poor man's phase shifter, which can be used in spectrometers that are not designed to produce phase increments smaller than $\pi/2$. The accuracy of z-pulses is however limited by the inhomogeneity of the r.f. field, which may lead to poor performance in sophisticated experiments.

4.2.7.7. Cyclic composite pulses

Composite pulses are of central importance for efficient heteronuclear decoupling (§ 4.7.6). It has been inferred in §3.3 that it is possible to refocus the effects of heteronuclear IS couplings (at least for weak II interactions) by an inversion pulse applied to the I spins. By means of a repetitive sequence of accurate inversion pulses, virtually continuous refocusing is feasible, leading to spin decoupling. The use of composite pulses can dramatically improve the decoupling efficiency. Ultimate efficiency is obtained by combining inversion pulses to *cyclic* sequences, described by a propagator that approximates the unity operator $\mathbb{1}$ to the highest possible degree over a range of offsets, r.f. rotation angles, and coupling constants (4.112–4.114).

Such cyclic sequences can be constructed from any one of the composite inversion pulses described in §§ 4.2.7.3 and 4.2.7.4. The most effective inversion elements are those of eqn (4.2.55), used in the MLEV decoupling sequences (4.109–4.112) and of eqn (4.2.58), used in WALTZ decoupling sequences (4.115, 4.116). For consistency with the literature, elements that achieve accurate inversion will henceforth be denoted R.

To obtain effective decoupling sequences, the inversion elements R must be combined into a *cycle*

$$C = RR\bar{R}\bar{R} \quad (4.2.70)$$

where $\bar{R}$ is derived from R by inversion of all radio-frequency phases. Such cycles are used in MLEV-4 and WALTZ-4 decoupling sequences. This choice can be rationalized by average Hamiltonian theory (4.112) or by considering the trajectories of the magnetization (4.110).

To improve the cyclicity of the sequence, it is possible to derive a permuted sequence from the cycle in eqn (4.2.70)

$$P = \bar{R}RR\bar{R} \quad (4.2.71)$$

and combine the cycles C and P into a *supercycle*

$$S_1 = CP = RR\bar{R}\bar{R}\ \bar{R}RR\bar{R}, \quad (4.2.72)$$

used for example in the MLEV-8 decoupling sequence.

The next stage of refinement utilizes the phase-inverted sequences $\bar{C}$ and $\bar{P}$, generating the MLEV-16 supercycle

$$S_2 = CP\bar{C}\bar{P} = RR\bar{R}\bar{R}\ \bar{R}RR\bar{R}\ \bar{R}\bar{R}RR\ R\bar{R}\bar{R}R. \quad (4.2.73)$$

The trajectories described by two magnetization vectors (belonging to the I (proton) doublet of a scalar-coupled IS system), considering the magnetization at the end of each cycle in the sequence of eqn (4.2.73),

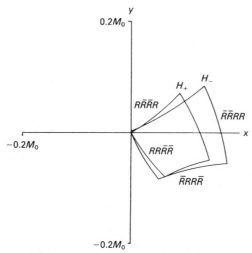

FIG. 4.2.19. Trajectories representative of the four component cycles of the MLEV-16 supercycle (eqn (4.2.73)) projected on to the xy-plane. The excursions are all small and close to the $+z$-axis. There is one trajectory for each proton satellite line (H_+ and H_-). Each arc visualizes the deviation of the MLEV-4 sequence from exact cyclicity, while the very small residual arc required to close the loop represents the deviation of the MLEV-16 supercycle from cyclicity. (Reproduced from Ref. 4.116.)

are shown in Fig. 4.2.19. As shown by Waugh (4.113, 4.114), decoupling is most effective if the difference $(\beta_+ - \beta_-)$ between the overall angles of rotation of the two magnetization vectors belonging to the two doublet components is small. This condition can be fulfilled if β varies only slowly with offset, and if the cyclicity ensures that the overall angles β_+ and β_- are kept small.

The effectiveness of a given cycle is maintained not only if an inversion element R is permuted, as in eqn (4.2.71), but also if a $(\pi/2)$-element is permuted from one end of the cycle to the other, as suggested by Waugh (4.113, 4.114). In this case, the residual angles of rotation (deviations from closed loops, see Fig. 4.2.19) are retained, but the rotation occurs about an axis that lies approximately in the xy-plane. Starting from a cycle C which leads to an overall rotation about an axis close to the $+z$-axis, a new cycle M can be derived by permuting a $(\pi/2)$-element, and the cycle M can be combined with $\bar{M}$. This expansion procedure is employed for the WALTZ decoupling sequences.

The WALTZ-sequences are based on the inversion element of eqn (4.2.58), which has better offset-compensation than the inversion element of eqn (4.2.55) used in the MLEV-sequences, and which has the additional advantage of being less sensitive to misadjustments of the relative r.f. phases. The WALTZ-4 sequence is obtained by combining eqns (4.2.58) and (4.2.70)

$$C = RR\bar{R}\bar{R} = (\beta)_0(2\beta)_\pi(3\beta)_0 \ (\beta)_0(2\beta)_\pi(3\beta)_0$$
$$(\beta)_\pi(2\beta)_0(3\beta)_\pi \ (\beta)_\pi(2\beta)_0(3\beta)_\pi \qquad (4.2.74a)$$

where $\beta \simeq \pi/2$. In shorthand notation, this is equivalent to

$$C = 1\bar{2}3 \ 1\bar{2}3 \ \bar{1}2\bar{3} \ \bar{1}2\bar{3} \qquad (4.2.74b)$$
$$= 1\bar{2}4\bar{2}3 \ \bar{1}2\bar{4}2\bar{3}. \qquad (4.2.74c)$$

In the last expression, contiguous pulses of the same phase have been combined for simplicity.

The next stage of the expansion uses a cyclic permutation of a $(\pi/2)_0$ pulse from the beginning to the end of the sequence of eqn (4.2.74c)

$$C' = \bar{2}4\bar{2}3\bar{1} \ 2\bar{4}2\bar{3}1. \qquad (4.2.75)$$

The sequence C' can be broken down in two subunits, $C' = K\bar{K}$, where the unit
$$K = \bar{2}4\bar{2}3\bar{1} \qquad (4.2.76)$$

is a spin inversion sequence. The sequence $C' = K\bar{K}$ may now be combined with its phase-inverted counterpart to give WALTZ-8

$$K\bar{K}\bar{K}K = \bar{2}4\bar{2}3\bar{1} \ 2\bar{4}2\bar{3}1 \ 2\bar{4}2\bar{3}1 \ \bar{2}4\bar{2}3\bar{1}. \qquad (4.2.77)$$

In the next step, a $(\pi/2)_\pi$ pulse is permuted from the end to the beginning of the sequence of eqn (4.2.77), and a phase-inverted sequence is appended to give WALTZ-16

$$Q\bar{Q}\bar{Q}Q = 3\bar{4}2\bar{3}1\bar{2}4\bar{2}3 \ \bar{3}4\bar{2}3\bar{1}2\bar{4}2\bar{3} \ \bar{3}4\bar{2}3\bar{1}2\bar{4}2\bar{3} \ 3\bar{4}2\bar{3}1\bar{2}4\bar{2}3. \qquad (4.2.78)$$

A careful evaluation of the relative merits of these sequences is given by Shaka *et al.* (4.116), and composite pulse decoupling is briefly compared with conventional modulation techniques in § 4.7.6. Efficient decoupling is particularly useful in carbon-13 spectroscopy, where good line-narrowing is of great practical importance for obtaining high sensitivity.

4.3. Sensitivity of Fourier spectroscopy

In this section we derive analytical expressions for the sensitivity advantage of Fourier spectroscopy in comparison to slow-passage experiments. We have to take into account the non-linear properties of the spin system as well as the discrete sampling of the free induction decay signal.

4.3.1. Signal-to-noise ratio of Fourier spectra

4.3.1.1. The signal

We consider the free induction decay of a single peak at an offset Ω, recorded with a quadrature detection scheme. The complex signal can be

4.3 SENSITIVITY OF FOURIER SPECTROSCOPY

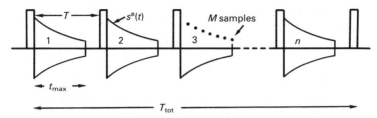

FIG. 4.3.1. A total of n free induction decays are accumulated, consisting of M samples each, excited by r.f. pulses at intervals T. The envelope $s^e(t)$ of the signal is usually truncated at $t_{max} \leq T$.

described by

$$s(t) = s^e(t)\exp\{i\Omega t\} \quad (4.3.1)$$

with the envelope function $s^e(t)$ which is responsible for the peak shape in the resulting spectrum. Usually, the signal function $s(t)$ is multiplied by a suitable weighting function $h(t)$ in order to enhance the sensitivity or resolution of the spectrum. Weighting in time domain is known to be equivalent to convolution filtering in frequency domain (eqns (4.1.27) and (4.1.28)).

We assume that n complex free induction decays are co-added, each represented by M equidistant samples between 0 and t^{max} (Fig. 4.3.1). The peak height in the resulting spectrum is given by the discrete Fourier transform of the weighted envelope of the free induction decay

$$S = \mathcal{F}\{ns^e(t)h(t)\}_{\omega=0} = n\sum_{k=0}^{M-1} s^e\left(k\frac{t^{max}}{M}\right)h\left(k\frac{t^{max}}{M}\right). \quad (4.3.2)$$

It is usually permitted to replace the discrete sum by an integral,

$$S = n\frac{M}{t^{max}}\int_0^{t^{max}} s^e(t)h(t)\,dt. \quad (4.3.3)$$

To simplify the notation, we define an average of the weighted time-domain signal envelope

$$\overline{sh} = \frac{1}{t^{max}}\int_0^{t^{max}} s^e(t)h(t)\,dt \quad (4.3.4)$$

and write the peak height S in the frequency domain in the form

$$S = nM\overline{sh} \quad (4.3.5)$$

which is proportional to the total number of samples that are recorded and to the average height of the weighted envelope of the free induction decay within the limits $0 < t < t^{max}$.

The envelope function $s^e(t)$ to be inserted in eqns (4.3.1–4.3.4) is dependent on experimental circumstances. In the simplest case, the signal decays exponentially,

$$s^e(t) = s^e(0)\exp(-t/T_2). \qquad (4.3.6)$$

4.3.1.2. The noise

The r.m.s. amplitude of the random noise $n(t)$ in time domain depends on the bandwidth of the spectrometer. Assuming white random noise limited by an analogue audio filter with the cut-off frequency f_c, the r.m.s. noise amplitude σ_n is

$$\sigma_n = \langle n(t)^2 \rangle^{\frac{1}{2}} = F^{\frac{1}{2}} \rho_n \qquad (4.3.7)$$

where ρ_n is the square root of the frequency-independent power spectral density. The bandwidth F of the spectrometer is twice the cut-off frequency f_c of the analogue audio filter, because positive and negative frequencies can be distinguished with quadrature phase detection. Weighting with the function $h(t)$ leads to a time-dependent r.m.s. noise amplitude

$$\sigma_n(t) = F^{\frac{1}{2}} \rho_n |h(t)|. \qquad (4.3.8)$$

In the frequency domain, the resulting r.m.s. noise amplitude σ_N is obtained by summing the contributions of all nM time-domain samples (which are assumed to be statistically independent)

$$\sigma_N = (nMF)^{\frac{1}{2}} [\overline{h^2}]^{\frac{1}{2}} \rho_n \qquad (4.3.9)$$

where $[\overline{h^2}]^{\frac{1}{2}}$ is the r.m.s. amplitude of the weighting function

$$[\overline{h^2}]^{\frac{1}{2}} = \left[\frac{1}{t^{\max}} \int_0^{t^{\max}} h(t)^2 \, dt \right]^{\frac{1}{2}}. \qquad (4.3.10)$$

To avoid an increase of the noise amplitude by down-conversion of high-frequency noise, it is advisable to set the cut-off frequency f_c of the audio filter equal to the Nyquist frequency $f_N = \frac{1}{2}\Delta t^{-1}$ of the sampling process with the time increment[1] Δt, leading to

$$F = 2f_c = 1/\Delta t = M/t^{\max}. \qquad (4.3.11)$$

Hence eqn (4.3.9) takes the form

$$\sigma_N = M(n/t^{\max})^{\frac{1}{2}} [\overline{h^2}]^{\frac{1}{2}} \rho_n.$$

For the experimental determination of the r.m.s. amplitude σ_n of the noise, it is normally assumed that the random process is ergodic and that

---

[1] For quadrature detection, Δt is the interval between samples consisting of complex pairs $s_x(t) + is_y(t)$.

the ensemble average in eqn (4.3.7) can be replaced by a time average

$$\sigma_n^2 = \lim_{T\to\infty} \frac{1}{2T} \int_{-T}^{T} n(t)^2 \, dt. \qquad (4.3.12)$$

The r.m.s. amplitude can then either be estimated by a numerical evaluation of eqn (4.3.12) of a sufficiently extensive set of noise samples, or less accurately from the measured peak-to-peak noise amplitude. It has been found (4.124) that for noise with a Gaussian amplitude distribution the expectation value of the peak-to-peak amplitude $n_{max} - n_{min}$ in a set of 100 independent sample values is

$$\langle n_{ptp} \rangle_{100} = 5.0\, \sigma_n. \qquad (4.3.13)$$

The r.m.s. noise amplitude σ_n can be estimated from the peak-to-peak amplitude n_{ptp} of a noise record with 100 zero crossings, which corresponds to approximately 100 independent samples. The same holds for the r.m.s. noise amplitude σ_N in frequency domain.

4.3.1.3. *The sensitivity*

The signal-to-noise ratio is defined as the ratio of the peak height S of a reference signal to the r.m.s. noise amplitude σ_N

$$S/\sigma_N = \frac{\text{signal peak amplitude}}{\text{r.m.s. noise amplitude}}. \qquad (4.3.14)$$

In practical NMR spectroscopy, it has become customary to include an additional factor 2 in the definition of the signal-to-noise ratio

$$S/N = \frac{S}{2\sigma_N} \simeq 2.5 \frac{S}{\langle N_{ptp} \rangle}.$$

This factor 2 is immaterial for comparing relative sensitivities and will not be taken into account here.

Combining eqns (4.3.5), (4.3.9), and (4.3.11) we find the signal-to-noise ratio of the spectrum

$$S/\sigma_N = (nt^{max})^{\frac{1}{2}} \frac{\overline{sh}}{[\overline{h^2}]^{\frac{1}{2}} \rho_n} \frac{1}{}. \qquad (4.3.15)$$

The total time required for accumulating n signals is

$$T_{tot} = nT \qquad (4.3.16)$$

where T is the interval between the pulses of the basic Fourier experiment ($T \geq t^{max}$).

To obtain a general expression for the standard sensitivity of a Fourier experiment, it is convenient to introduce the signal-to-noise ratio per unit time $S/(\sigma_N T_{tot}^{\frac{1}{2}})$, which will simply be called *sensitivity* in the following

$$\frac{S}{\sigma_N T_{tot}^{\frac{1}{2}}} = \frac{\overline{sh}}{[\overline{h^2}]^{\frac{1}{2}}} \left(\frac{t^{max}}{T}\right)^{\frac{1}{2}} \frac{1}{\rho_n}. \quad (4.3.17)$$

Thus the relevant factor that must be optimized is the ratio of the average weighted signal amplitude $\overline{sh}$ and the r.m.s. amplitude of the weighting function $[\overline{h^2}]^{\frac{1}{2}}$. The factor t^{max}/T in eqn (4.3.17) reflects the duty ratio of the receiver in the experiment.

4.3.1.4. Optimization of the weighting function

The optimum weighting function $h(t)$ should maximize the ratio $\overline{sh}/[\overline{h^2}]^{\frac{1}{2}}$ in eqn (4.3.17). With eqns (4.3.4) and (4.3.10) the optimum weighting function can be found by optimizing the expression

$$L = \int_0^{t^{max}} s^e(t) h(t) \, dt + \lambda \int_0^{t^{max}} h(t)^2 \, dt \quad (4.3.18)$$

where λ is a Lagrange multiplier. Applying the procedure of variational calculus (4.125) and taking a functional derivative with respect to $h(t)$ leads to the optimum weighting function $h_m(t)$

$$h_m(t) = s^e(t). \quad (4.3.19)$$

This function is called *matched weighting function* and maximizes the sensitivity. Matched filtering is of central importance in signal processing (4.2, 4.7, 4.8, 4.25).

For matched weighting, we find the sensitivity

$$\left(\frac{S}{\sigma_N T_{tot}^{\frac{1}{2}}}\right)_{matched} = [\overline{s^2}]^{\frac{1}{2}} \left(\frac{t^{max}}{T}\right)^{\frac{1}{2}} \frac{1}{\rho_n} \quad (4.3.20)$$

where (t^{max}/T) is the duty cycle of the receiver, and where $\overline{s^2}$ is the average signal power in the sampling window

$$\overline{s^2} = \frac{1}{t^{max}} \int_0^{t^{max}} s^e(t)^2 \, dt. \quad (4.3.21)$$

Equation (4.3.20) expresses the basic and general result that the maximum achievable sensitivity per unit time is equal to the square root of the ratio of the average signal power $\overline{s^2} t^{max}/T$ and the noise power per unit bandwidth ρ_n^2 (4.2, 4.7, 4.8)

$$\left[\frac{S}{\sigma_N T_{tot}^{\frac{1}{2}}}\right]_{matched} = \left\{\frac{\text{signal power}}{\text{noise power/bandwidth}}\right\}^{\frac{1}{2}}. \quad (4.3.20a)$$

Matched weighting may be understood by noting that, due to the decay of the signal envelope, the 'local sensitivity' varies from sample to sample. With matched weighting, each sample is weighted by its own sensitivity, such that samples with high sensitivity dominate in the weighted average.

4.3.1.5. Optimization of the signal energy

Consider a homogenous line that can be described by classical Bloch equations, with a signal envelope of the simple exponential form of eqn (4.3.6). The average power $\overline{s^2}$ is given by

$$\overline{s^2} = s^e(0)^2 \frac{T_2}{2t^{\max}} [1 - E_2^2] \qquad (4.3.22)$$

where $E_2 = \exp(-t^{\max}/T_2)$. (Note that this definition is different from § 4.2.5, where $E_2 = \exp(-T/T_2)$.)

The initial signal amplitude $s^e(0)$ depends on T_1, on the pulse rotation angle β, and on the recycle time T of the repetitive pulse experiment as discussed in § 4.2.4. In general, the resonance frequency ω_0 is also relevant, as it determines the effect of transverse interference (§ 4.2.5). We shall neglect such transverse interference effects in this section, since the remaining transverse magnetization at the end of the free precession interval T can be destroyed by one of the techniques discussed in § 4.2.6. The initial amplitude $s^e(0)$ is then given by eqn (4.2.35). For the optimum pulse rotation angle

$$\cos \beta_{\text{opt}} = E_1 = \exp(-T/T_1) \qquad (4.3.23)$$

shown in Fig. 4.3.2, the initial amplitude reaches the maximum value

$$s^e(0)_{\max} = M_0 \left[\frac{1 - E_1}{1 + E_1} \right]^{\frac{1}{2}}. \qquad (4.3.24)$$

For fast repetitive pulse experiments, small pulse rotation angles are required to achieve maximum signal amplitude.

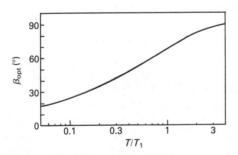

FIG. 4.3.2. Optimum pulse rotation angle β_{opt} as a function of the ratio of the pulse interval T and the spin–lattice relaxation time T_1. (Reproduced from Ref. 4.1.)

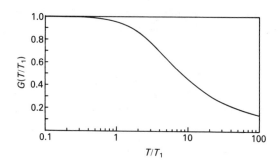

FIG. 4.3.3. Function $G(T/T_1)$ of eqn (4.3.26) which describes the dependence of the sensitivity on the pulse interval T, provided the r.f. rotation angle is optimized according to eqn (4.3.23) and the observation period t^{max} is kept constant. (Reproduced from Ref. 4.1.)

The optimized sensitivity can be derived from eqns (4.3.20), (4.3.22), and (4.3.24)

$$\left(\frac{S}{\sigma_N T_{tot}^{\frac{1}{2}}}\right)_{matched} = M_0^{\frac{1}{2}} \left(\frac{T_2}{T_1}\right)^{\frac{1}{2}} (1 - E_2^2)^{\frac{1}{2}} G(T/T_1) \frac{1}{\rho_n} \quad (4.3.25)$$

with the function

$$G(x) = \left[2\frac{1-e^{-x}}{x(1+e^{-x})}\right]^{\frac{1}{2}} \quad (4.3.26)$$

which describes the influence of the repetition time T, assuming that the observation time t^{max} and hence $E_2^2 = \exp(-2t^{max}/T_2)$ are kept constant. The function $G(T/T_1)$ approaches unity for $T < T_1$ as shown in Fig. 4.3.3. The sensitivity increases for decreasing pulse interval T. The selection of the repetition time T is usually determined by a compromise between resolution and sensitivity. Sensitivity considerations tend towards shorter T, while high resolution requires long observation times t^{max} and hence longer intervals $T \geq t^{max}$.

Equation (4.3.25) shows that the sensitivity is proportional to $(T_2/T_1)^{\frac{1}{2}}$, reflecting the advantage of slow transverse relaxation and fast longitudinal recovery.

Instead of optimizing the pulse rotation angle β for fixed repetition time T according to eqn (4.3.24), it is sometimes more convenient to fix the desired pulse rotation angle β and to optimize the repetition time T for maximum sensitivity (4.126). The optimum repetition times are shown in Table 4.3.1, where the sensitivity is normalized with respect to the maximum sensitivity reached for $T \to 0$, $\beta \to 0$. Note that even with a 90° pulse rotation angle, the loss in sensitivity is insignificant if the repetition time T is optimized.

Table 4.3.1
Sensitivity $(S/N)_T$ for a repetition time T/T_1 optimized for a fixed pulse rotation angle β. The sensitivity is normalized with respect to the sensitivity achievable for $T \to 0$.

$\beta(°)$	$(T/T_1)_{opt}$	$(S/N)_T/(S/N)_0$
10	0.015	1.000
30	0.143	0.999
50	0.421	0.992
70	0.827	0.966
90	1.269	0.902

4.3.2. Signal-to-noise ratio in slow passage spectra

For comparison, we calculate the signal-to-noise ratio for a slow-passage experiment with matched filtering. The absorption mode signal, recorded in time domain with a sweep rate a (in Hz/s) is (4.127)

$$s(t) = M_0 \gamma B_1 \frac{1/T_2}{(1+S)/T_2^2 + (2\pi at)^2} \quad (4.3.27)$$

where S is the saturation parameter $S = (\gamma B_1)^2 T_1 T_2$. The time axis is defined such that resonance is reached at $t = 0$. Matched filtering amounts to a convolution of the time domain signal with the matched filter function (4.2)

$$h_m(t) = s(-t) \quad (4.3.28)$$

leading to the filtered signal

$$s_m(t) = s(t) * s(-t). \quad (4.3.29)$$

The peak amplitude $s_m(0)$ is equal to the signal energy

$$s_m(0) = \int_{-\infty}^{\infty} s(t)^2 \, dt \quad (4.3.30)$$

and the maximum achievable signal-to-noise ratio is

$$\left[\frac{s(0)}{\sigma_n}\right]_{matched} = \left[\int_{-\infty}^{\infty} s(t)^2 \, dt\right]^{\frac{1}{2}} \frac{1}{\rho_n}. \quad (4.3.31)$$

The maximum absorption-mode signal amplitude is obtained for $S = 1$. However, the maximum signal energy is reached for $S = 2$ (4.2, 4.127), in which case the increased line-width outweighs the loss in signal amplitude.

If the sweep rate is expressed by the total spectral range Ω_{tot} covered in the total time T_{tot}, $2\pi a = \Omega_{tot}/T_{tot}$, the optimum signal can be written in the form

$$s(t) = M_0 \left[\frac{T_2}{T_1}\right]^{\frac{1}{2}} \frac{\sqrt{2}}{3 + (\Omega_{tot} T_2 T_{tot}^{-1} t)^2} \qquad (4.3.32)$$

with the signal energy

$$\int_{-\infty}^{\infty} s(t)^2 \, dt = M_0^2 \frac{\pi}{3\sqrt{3}} \frac{T_{tot}}{\Omega_{tot} T_1}. \qquad (4.3.33)$$

This leads to the maximum sensitivity (signal-to-noise ratio per unit time) with matched filtering

$$\left(\frac{s(0)}{\sigma_n T_{tot}^{\frac{1}{2}}}\right)_{matched} = M_0 \left(\frac{\pi}{3\sqrt{3}} \frac{1}{\Omega_{tot} T_1}\right)^{\frac{1}{2}} \frac{1}{\rho_n}. \qquad (4.3.34)$$

It is remarkable that the signal energy and therefore also the achievable sensitivity are independent of the transverse relaxation time T_2 and of the natural line-width $\Delta\omega = 2/T_2$ observed in the absence of saturation.

To improve the baseline stability, it is common practice to employ a modulation technique in slow-passage experiments (4.128, 4.129). The expression in eqn (4.3.34) applies to sideband detection. When centre-band detection is utilized instead, the signal energy is reduced by a factor 0.65.

4.3.3. Sensitivity comparison of Fourier and slow-passage experiments

The ratio of the sensitivities of optimized Fourier and slow-passage experiments, obtained from eqns (4.3.25) and (4.3.34), is

$$\frac{(S/N)_{Fourier}}{(S/N)_{SP}} = \left[\frac{3\sqrt{3}}{2\pi}\right]^{\frac{1}{2}} [1 - \exp(-2t^{max}/T_2)]^{\frac{1}{2}} \left[\frac{\Omega_{tot}}{\Delta\omega}\right]^{\frac{1}{2}} G(T/T_1) \quad (4.3.35)$$

where $\Delta\omega = 2/T_2$ is the full line-width at half-height of the resonance under investigation. This expression assumes a modulation scheme with sideband detection for the slow-passage experiment and quadrature detection for the Fourier experiment. In both cases a matched filter is used.

The ratio $\Omega_{tot}/\Delta\omega$ represents the *number of spectral elements* in the spectrum. The sensitivity advantage of a Fourier experiment is thus proportional to the square root of the number of spectral elements. This can be understood by comparing the Fourier experiment, which excites all resonances simultaneously, with a (hypothetical) multiple-channel slow-passage experiment with $\Omega_{tot}/\Delta\omega$ independent channels. It is clear

4.3 SENSITIVITY OF FOURIER SPECTROSCOPY

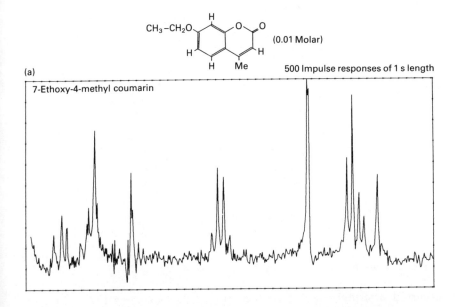

FIG. 4.3.4. 60-MHz proton magnetic resonance spectra of 7-ethoxy-4-methyl coumarin. (a) Fourier transform of 500 free induction signals recorded in 500 s. (b) Single scan recorded in 500 s by slow passage on the same instrument. (Reproduced from Ref. 4.130.)

that the sensitivity improvement will be particularly pronounced for spectra with narrow lines covering a wide spectral range.

An early example of a Fourier spectrum recorded in 1966 (4.130) is shown in Fig. 4.3.4. The experimental sensitivity advantage of the Fourier experiment over the slow-passage experiment, both performed in 500s, is a factor 10. This value compares with a maximum theoretical improvement (eqn 4.3.35) of 12.6 for $\Omega_{tot}/\Delta\omega = 250$ spectral elements, taking into account the centreband detection scheme used in the slow-passage experiment and the single-channel version of the Fourier experiment.

4.3.4. Sensitivity enhancement by recycling of magnetization

In spin systems without couplings, e.g. in proton-decoupled carbon-13 systems, it is possible to enhance the sensitivity by recycling the magnetization. One can exploit the steady-state magnetization that

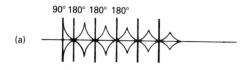

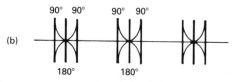

FIG. 4.3.5. Fourier techniques for sensitivity enhancement in inhomogeneous static fields, applicable to systems without homonuclear couplings. (a) Spin–echo Fourier transform (SEFT) method where the signal is observed in a train of Carr–Purcell echoes. (b) Driven equilibrium Fourier transform (DEFT) technique with a $(\pi/2)$-pulse to return the residual transverse magnetization to the z-axis after refocusing.

occurs if a string of $\pi/2$-pulses is applied (4.96, 4.98, 4.137, 4.138), as described in § 4.2.5, or alternatively observe a series of echoes excited by a $\pi/2 - (\tau - \pi - \tau)_n$ Carr–Purcell sequence (4.139) with the so-called Spin Echo Fourier Transform or SEFT method (4.126, 4.140), as shown in Fig. 4.3.5. An alternative approach, known as Driven Equilibrium Fourier Transform (DEFT) (4.126, 4.141), uses a $[\pi/2 - \tau - \pi - \tau - \pi/2]$ sequence, where the signal is observed in the τ-intervals, and where the last pulse restores part of the magnetization along the z-axis (Fig. 4.3.5(b)). In practice, these schemes are not as effective as one might expect from theory (4.126), because translational diffusion through static field gradients and transverse relaxation impede refocusing. The decay of transverse magnetization can also be caused by imperfect decoupling or by couplings to rapidly relaxing protons or quadrupolar nuclei (4.142). Both DEFT and SEFT methods fail in the presence of homonuclear couplings and can only be used for dilute nuclei such as carbon-13.

4.4. Quantum-mechanical description of Fourier spectroscopy

In § 4.2 we gave a classical description of the behaviour of isolated spins in Fourier experiments. In coupled spin systems, additional effects are expected to occur due to more complicated transformation properties under the influence of radio-frequency pulses. It is no longer possible to consider the individual transitions separately, since the coherent excitation affects the entire spin system. Particularly in view of more advanced pulse techniques, a detailed quantum mechanical treatment of Fourier experiments is in order.

We give a density operator formulation of Fourier spectroscopy and address the question of equivalence of slow-passage and Fourier spectra in terms of signal intensities, resonance frequencies, and line-widths.

4.4.1. Density operator formalism applied to Fourier spectroscopy

Consider the basic Fourier pulse experiment shown in Fig. 4.4.1. The density operator of the spin system immediately before applying a non-selective r.f. pulse at time $t = 0$ shall be denoted by $\sigma(0_-)$. Consider an r.f. pulse that causes a positive rotation by the angle β about the y-axis, leading to the density operator

$$\sigma(0_+) = R_y(\beta)\sigma(0_-)R_y^{-1}(\beta) \tag{4.4.1}$$

with the rotation operator

$$R_y(\beta) = \exp\{-i\beta F_y\}. \tag{4.4.2}$$

The subsequent free evolution is governed by the density operator equation (2.1.34)

$$\dot{\sigma}(t) = -i[\mathcal{H}, \sigma] - \hat{\Gamma}\{\sigma(t) - \sigma_0\} \tag{4.4.3}$$

with the Hamiltonian $\mathcal{H}$ and the relaxation superoperator $\hat{\Gamma}$ (cf. § 2.3). Because the equilibrium density operator σ_0 commutes with the unperturbed Hamiltonian $\mathcal{H}$, we can rewrite eqn (4.4.3) in simplified form

$$\dot{\sigma}(t) = \hat{L}\{\sigma(t) - \sigma_0\} \tag{4.4.4}$$

with the Liouville superoperator

$$\hat{L} = -i\hat{\mathcal{H}} - \hat{\Gamma}. \tag{4.4.5}$$

The solution of this equation for the free evolution after pulse excitation is

$$\sigma(t) = \exp(\hat{L}t)\{\sigma(0_+) - \sigma_0\} + \sigma_0. \tag{4.4.6}$$

The complex magnetization $M^+(t)$, which can be measured by quadra-

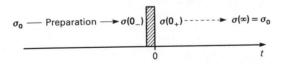

FIG. 4.4.1. Designation of the density operator in the course of a basic Fourier experiment with a single pulse.

ture detection, is proportional to the expectation value of the operator F^+

$$\begin{aligned} M^+(t) &= M_x(t) + iM_y(t) \\ &= N\gamma\hbar\{\langle F_x\rangle(t) + i\langle F_y\rangle(t)\} \\ &= N\gamma\hbar\langle F^+\rangle(t) \\ &= N\gamma\hbar\,\mathrm{tr}\{F^+\sigma(t)\} \end{aligned} \qquad (4.4.7)$$

where N is the number spin systems per unit volume.

In a high magnetic field, the Liouville operator $\hat{L}$ is invariant under rotations about the z-axis and does not mix components of σ belonging to different coherence orders p and the term $\exp\{\hat{L}t\}\sigma_0$ in eqn (4.4.6) does not lead to observable transverse magnetization. We may therefore drop the term σ_0 when combining eqns (4.4.6) and (4.4.7) to obtain a general representation of the complex free induction decay

$$M^+(t) = N\gamma\hbar\,\mathrm{tr}\{F^+ \exp(\hat{L}t)\sigma(0_+)\}. \qquad (4.4.8)$$

This important equation can be written in explicit form by using the Redfield matrix representation of the relaxation superoperator (§ 2.3.2). In the absence of degenerate transitions, each off-diagonal matrix element of $\sigma(t)$ in the eigenbasis of $\mathcal{H}$ evolves independently:

$$M^+(t) = N\gamma\hbar \sum_{rs} F^+_{sr}\sigma_{rs}(0_+)\exp\{(-i\omega_{rs} - \lambda_{rs})t\} \qquad (4.4.9)$$

with the transition frequencies

$$\omega_{rs} = \mathcal{H}_{rs\,rs} = \mathcal{H}_{rr} - \mathcal{H}_{ss} = \langle r|\,\mathcal{H}\,|r\rangle - \langle s|\,\mathcal{H}\,|s\rangle \qquad (4.4.10)$$

and the relaxation rates

$$\lambda_{rs} = -R_{rs\,rs} = 1/T_2^{(rs)}. \qquad (4.4.11)$$

Thus the free induction signal described by eqn (4.4.9) consists of a sum of decaying oscillations. To obtain positive frequencies, which are more convenient in the subsequent calculations, we may substitute $-\omega_{rs} = \omega_{sr}$ in eqn (4.4.9) or, equivalently, drop the minus sign in front of $i\omega_{rs}$ and interchange the indices of the matrix elements of F^+ and $\sigma(0_+)$

$$M^+(t) = N\gamma\hbar \sum_{rs} F^+_{rs}\sigma_{sr}(0_+)\exp\{(i\omega_{rs} - \lambda_{rs})t\}. \qquad (4.4.12)$$

The complex Fourier transformation (eqn (4.2.17)) of the free induction signal yields the complex spectrum

$$S(\omega) = N\gamma\hbar \sum_{rs} F^+_{rs}\sigma_{sr}(0_+)\frac{1}{(i\Delta\omega_{rs} + \lambda_{rs})} \qquad (4.4.13)$$

where the frequency variable

$$\Delta\omega_{rs} = \omega - \omega_{rs} \qquad (4.4.14)$$

is the offset with respect to the centre of the resonance, in analogy to the classical description of eqn (4.2.18).

To describe the intensities and phases of the lines in the spectrum, we define the complex integrated intensitities

$$L^{(rs)} = N\gamma\hbar F_{rs}^+ \sigma_{sr}(0_+) \int_{-\infty}^{+\infty} \frac{1}{(i\Delta\omega_{rs} + \lambda_{rs})} d\omega$$
$$= \pi N\gamma\hbar F_{rs}^+ \sigma_{sr}(0_+). \quad (4.4.15)$$

The real part of $L^{(rs)}$ is associated with the absorption mode, while the imaginary part corresponds to a dispersive contribution to the lineshape. The phase $\varphi^{(rs)}$, defined by

$$\tan \varphi^{(rs)} = \frac{\text{Im}\{\sigma_{sr}(0_+)\}}{\text{Re}\{\sigma_{sr}(0_+)\}}, \quad (4.4.16)$$

gives a measure of the admixture of dispersion mode in the lineshape. The phase may vary from line to line within a spectrum, depending on the elements of the density operator $\sigma(0_+)$ immediately after the pulse.

The absolute intensity

$$|L^{(rs)}| = \pi N\gamma\hbar |F_{rs}^+| |\sigma_{sr}(0_+)| \quad (4.4.17)$$

corresponds to the maximum integrated intensity that can be obtained by proper phase adjustment.

In the simplest application of Fourier spectroscopy, the r.f. pulse is applied to a system in thermodynamic equilibrium ($\sigma(0_-) = \sigma_0$). In the high-temperature approximation ($\hbar|\mathcal{H}| \ll kT$), the equilibrium density operator in eqn (2.1.25) can be expanded

$$\sigma_0 \simeq \frac{1}{\text{tr}\{\mathbb{1}\}} \left\{ 1 - \frac{\mathcal{H}\hbar}{kT} \right\}$$
$$= \frac{1}{\text{tr}\{\mathbb{1}\}} \{1 - \beta_T \mathcal{H}\} \quad (4.4.18)$$

with the inverse spin temperature

$$\beta_T = \frac{\hbar}{kT}. \quad (4.4.19)$$

In the high-field approximation, the laboratory frame Hamiltonian is dominated by the Zeeman interaction

$$\mathcal{H}_z = -\gamma B_0 F_z = \omega_0 F_z \quad (4.4.20)$$

where ω_0 is the Larmor frequency. Hence one obtains

$$\sigma_0 \simeq \frac{1}{\text{tr}\{\mathbb{1}\}} \{1 - \beta_T \omega_0 F_z\}. \quad (4.4.21)$$

The state of the system immediately after an r.f. pulse with rotation angle $\beta = -\gamma B_1 \tau_p$ applied along the y-axis is

$$\sigma(0_+) = \frac{1}{\text{tr}\{\mathbb{1}\}} \{1 - \beta_T \omega_0 (F_z \cos \beta + F_x \sin \beta)\}. \quad (4.4.22)$$

The resulting lines in the frequency domain, eqn (4.4.15), can be adjusted to pure absorption-mode lineshapes ($\varphi^{(rs)} = 0$) with intensities that are proportional to the square of the transition matrix elements F_{rs}^+

$$L^{(rs)} \propto -\frac{\pi N \gamma \hbar \beta_T \omega_0}{2\,\text{tr}\{\mathbb{1}\}} |F_{rs}^+|^2 \sin \beta. \quad (4.4.23)$$

The relative intensitities are obviously independent of the pulse rotation angle β.

4.4.2. Equivalence of slow-passage and Fourier spectroscopy

Fourier spectroscopy is a very universal technique which can be used to investigate arbitrary non-equilibrium states $\sigma(0_-)$. Slow-passage experiments on the other hand can only be used if the system does not change with time. The comparison of slow-passage and Fourier spectroscopy must therefore be restricted to systems in steady-state (4.131). We must exclude coherent non-equilibrium states, where the matrix representation of $\sigma(0_-)$ in the eigenbasis of $\mathcal{H}$ contains off-diagonal elements, which evolve under the Hamiltonian. On the other hand, we may include cases where $\sigma(0_-)$ involves arbitrary populations that deviate from the Boltzmann distribution (so-called 'non-equilibrium states of the first kind' (4.131) or 'incoherent non-equilibrium'). Such states can be caused, for example, by chemically induced dynamic polarization or by nuclear or electronic Overhauser effects. The system may also be subject to chemical exchange processes in dynamic equilibrium.

We shall investigate the equivalence of Fourier and slow-passage spectroscopy for the following conditions:

1. High-temperature approximation;
2. High-field approximation;
3. The system is initially in a steady state $\sigma(0_-) = \sigma^{ss}$ with $\hat{L}\sigma^{ss} = 0$;
4. The Fourier experiment uses only non-selective pulses.

The Liouville superoperator $\hat{L}$ is written in the form (see eqn (2.4.34))

$$\hat{L} = -i\hat{\mathcal{H}} - \hat{\Gamma} + \hat{\Xi}. \quad (4.4.24)$$

4.4 QUANTUM-MECHANICAL DESCRIPTION

The high-field approximation implies rotational invariance of $\hat{L}$. No further restrictions are imposed on the Hamiltonian $\mathcal{H}$. The relaxation superoperator $\hat{\Gamma}$ may comprise, in addition to pure relaxation, terms representing the change of populations due to chemically induced dynamic nuclear polarization and to r.f. irradiation applied to obtain Overhauser effects. The exchange superoperator $\hat{\Xi}$ describes exchange in chemical equilibrium. The superoperator $\hat{L}$ drives the system towards the steady-state σ^{ss} rather than to the equilibrium state σ_0.

4.4.2.1. Fourier spectroscopy

From eqns (4.4.1) and (4.4.8) we obtain the complex magnetization

$$M^+(t) = N\gamma\hbar \operatorname{tr}\{F^+ \exp(\hat{L}t)R(\beta_y)\sigma(0_-)R(\beta_y)^{-1}\}. \tag{4.4.25}$$

The induced signal may formally be Fourier-transformed with respect to t, leading to the complex spectrum[1]

$$S(\omega)^{\mathrm{FT}} = \mathcal{F}\{M^+(t)\} = -N\gamma\hbar \operatorname{tr}\{F^+(\hat{L} - i\omega\mathbb{1})^{-1}R(\beta_y)\sigma(0_-)R(\beta_y)^{-1}\} \tag{4.4.26}$$

where the density operator before the pulse $\sigma(0_-)$ represents the steady-state σ^{ss}.

4.4.2.2. Slow-passage spectroscopy

The interaction with a weak r.f. field $B_1(t)$ is described by the term $\mathcal{H}_1(t)$ in the Hamiltonian or by the commutator superoperator $\hat{\mathcal{H}}_1(t)$ in the density operator equation

$$\dot{\sigma}(t) = \hat{L}\{\sigma(t) - \sigma^{ss}\} - i\hat{\mathcal{H}}_1(t)\sigma(t). \tag{4.4.27}$$

We transform this equation into a frame rotating with the r.f. frequency $\omega_{\mathrm{r.f.}}$ and obtain for the transformed density operator $\sigma^{\mathrm{T}}(t)$ the differential equation

$$\dot{\sigma}^{\mathrm{T}}(t) = (\hat{L} + i\omega_{\mathrm{r.f.}}\hat{F}_z)\{\sigma^{\mathrm{T}}(t) - \sigma^{ss}\} - i\hat{\mathcal{H}}_1\sigma^{\mathrm{T}}(t) \tag{4.4.28}$$

where $\hat{\mathcal{H}}_1$ is now time-independent. To determine the stationary solutions with $\dot{\sigma}^{\mathrm{T}}(t) = 0$, we expand $\sigma(t)$ in powers of the perturbation $\mathcal{H}_1$

$$\sigma^{\mathrm{T}} = \sigma_0^{\mathrm{T}} + \sigma_1^{\mathrm{T}} + \ldots, \tag{4.4.29}$$

and neglect higher-order terms. By inserting eqn (4.4.29) into eqn (4.4.28), we find

$$\sigma_0^{\mathrm{T}} = \sigma^{ss}$$

---

[1] The inversion of the superoperator $(\hat{L} - i\omega\mathbb{1})$, resp. $(\hat{L} + i\omega\hat{F}_z)$, presupposes that the eigenvalue zero has been eliminated by an appropriate reduction of the dimension of the Liouville space. A strict proof of eqn (4.4.26) can be derived by using a projection superoperator projecting on to the subspace of coherences with order $p = -1$.

and
$$\sigma_1^T = (\hat{L} + i\omega_{r.f.}\hat{F}_z)^{-1}i[\mathcal{H}_1, \sigma^{ss}]. \quad (4.4.30)$$

The resulting complex spectrum is (see footnote to eqn (4.4.26))

$$\begin{aligned}S(\omega_{r.f.})^{SP} &= N\gamma\hbar\, \text{tr}\{F^+\sigma_1^T(\omega_{r.f.})\}\\ &= N\gamma\hbar\, \text{tr}\{F^+(\hat{L} + i\omega_{r.f.}\hat{F}_z)^{-1}i[\mathcal{H}_1, \sigma^{ss}]\}.\end{aligned} \quad (4.4.31)$$

In the high-field approximation, $\hat{L}$ commutes with $\hat{F}_z$ and only the coherence components with $p = -1$ of the term $[\mathcal{H}_1, \sigma^{ss}]$ are relevant for the expectation value of F^+. It is therefore allowed to replace $i\omega_{r.f.}\hat{F}_z$ by $-i\omega_{r.f.}\mathbb{1}$ in the above equation, leading to the slow-passage spectrum

$$S(\omega_{r.f.})^{SP} = N\gamma\hbar\, \text{tr}\{F^+(\hat{L} - i\omega_{r.f.}\mathbb{1})^{-1}i[\mathcal{H}_1, \sigma^{ss}]\} \quad (4.4.32)$$

in the linear response approximation.

4.4.2.3. Comparison of Fourier and slow-passage spectra

Comparing eqns (4.4.26) and (4.4.32), it is obvious that the two expressions differ only in the 'initial conditions'

$$\sigma(0_+)^{FT} = R(\beta_y)\sigma^{ss}R(\beta_y)^{-1}, \quad (4.4.33)$$
$$\sigma(0_+)^{SP} = -i[\mathcal{H}_1, \sigma^{ss}]. \quad (4.4.34)$$

For small pulse rotation angle β, the expression for $\sigma(0_+)^{FT}$ may be expanded

$$\sigma(0_+)^{FT} = \sigma^{ss} - i\beta[F_y, \sigma^{ss}] - \frac{\beta^2}{2}[F_y, [F_y, \sigma^{ss}]] + \ldots. \quad (4.4.35)$$

Since we have assumed that σ^{ss} contains only terms with coherence order $p = 0$, only odd powers in β contribute to the observable spectrum $S(\omega)$. For a small pulse rotation angle, the term $-i\beta[F_y, \sigma^{ss}]$ is dominant.

The perturbation $\mathcal{H}_1$ in the slow-passage experiment is

$$\mathcal{H}_1 = -\gamma B_1 F_y \quad (4.4.36)$$

leading to

$$\sigma(0_+)^{SP} = i\gamma B_1[F_y, \sigma^{ss}]. \quad (4.4.37)$$

The comparison of eqns (4.4.35) and (4.4.37) leads to:

Conclusion I. Fourier and slow-passage spectra are identical (apart from a multiplicative factor) for *any* steady-state incoherent non-equilibrium σ^{ss}, provided the pulse rotation angle is small, and provided the r.f. perturbation in the slow-passage experiment is weak.

For a system in Boltzmann equilibrium, described by the density operator σ_0 given in eqn (4.4.21), the equivalence holds for arbitrary pulse rotation angles β

$$S(\omega)^{FT} = \frac{-\beta_T \omega_0 N\gamma\hbar}{\text{tr}\{\mathbb{1}\}} \text{tr}\{F^+(\hat{L} - i\omega\mathbb{1})^{-1}F_x\}\sin\beta, \qquad (4.4.38)$$

$$S(\omega)^{SP} = \frac{\beta_T \omega_0 N\gamma\hbar}{\text{tr}\{\mathbb{1}\}} \text{tr}\{F^+(\hat{L} - i\omega\mathbb{1})^{-1}F_x\}\gamma B_1. \qquad (4.4.39)$$

This leads to:

Conclusion II. The low-power slow-passage spectrum and the corresponding Fourier spectrum are identical, (except for a proportionality constant) for all high-field Hamiltonians, all relaxation mechanisms, and in the presence of arbitrary equilibrium chemical exchange, provided that the system is in thermodynamic Boltzmann equilibrium at high temperature prior to the application of a non-selective r.f. pulse or of a weak continuous-wave r.f. field.

4.4.3. Fourier spectroscopy of non-equilibrium systems

It is essential to distinguish two cases.

1. *Incoherent non-equilibrium state.* Each member of the ensemble of independent spin systems is in an eigenstate of the Hamiltonian or in a superposition state with random phase across the ensemble. The probability distribution of occupation of the various eigenstates does not obey a Boltzmann distribution.

The density operator of the system commutes with the Hamiltonian and does not evolve under the Hamiltonian. There is no coherence. However, the density operator evolves under the relaxation superoperator $\hat{\Gamma}$ towards thermal equilibrium. The matrix representation of the density operator in the eigenbasis of the Hamiltonian is diagonal (see eqn (2.1.10)). This state has been referred to as 'non-equilibrium state of the first kind' (4.131).

2. *Coherent non-equilibrium state.* The system includes a coherent superposition of states, i.e. zero-, single-, or multiple-quantum coherence. The density operator does not commute with the Hamiltonian, and its matrix representation in the eigenbasis contains off-diagonal elements This case has been referred to as 'non-equilibrium state of the second kind' (4.131).

If r.f. pulses are applied to systems in *coherent* non-equilibrium, a variety of phenomena may occur that will be discussed extensively in

Chapter 8 under the heading 'Coherence transfer'. In this section, we shall focus on the implications of *incoherent* non-equilibrium.

An *incoherent* non-equilibrium state $\sigma(0_-)$ with populations P_r can be expressed in terms of the single-element polarization operators $I^{(rr)}$ defined in eqn (2.1.135)

$$\sigma(0_-) = \sum_r P_r I^{(rr)} = \sum_r P_r |r\rangle\langle r|. \qquad (4.4.40)$$

In weakly coupled systems, one may factorize the polarization operators in terms of polarization operators associated with individual spins k, defined in eqn (2.1.114)

$$\sigma(0_-) = \sum_r P_r \prod_k I_k^{\mu_{kr}} \qquad (4.4.41)$$

with μ_{kr} being one of the magnetic quantum numbers M_k, $-I_k \leq M_k \leq I_k$, of spin I_k (for $I_k = \frac{1}{2}$, $\mu_{kr} = \alpha$ or β). To demonstrate the transformation properties of polarization operators, it is convenient to express them by products of Cartesian I_{kz} operators. Thus in a weakly-coupled two-spin $\frac{1}{2}$ system we have with eqn (2.1.114)

$$I^{(1,1)} = I_1^\alpha I_2^\alpha = \tfrac{1}{2}(+I_{1z} + I_{2z} + 2I_{1z}I_{2z} + \tfrac{1}{2}\mathbb{1}),$$
$$I^{(2,2)} = I_1^\alpha I_2^\beta = \tfrac{1}{2}(+I_{1z} - I_{2z} - 2I_{1z}I_{2z} + \tfrac{1}{2}\mathbb{1}),$$
$$I^{(3,3)} = I_1^\beta I_2^\alpha = \tfrac{1}{2}(-I_{1z} + I_{2z} - 2I_{1z}I_{2z} + \tfrac{1}{2}\mathbb{1}),$$
$$I^{(4,4)} = I_1^\beta I_2^\beta = \tfrac{1}{2}(-I_{1z} - I_{2z} + 2I_{1z}I_{2z} + \tfrac{1}{2}\mathbb{1}). \qquad (4.4.42)$$

Thus the density operator of a weakly-coupled two-spin system can be expressed in terms of populations and I_{kz} operators

$$\sigma(0_-) = \tfrac{1}{2}[(P_{\alpha\alpha} + P_{\alpha\beta} - P_{\beta\alpha} - P_{\beta\beta})I_{1z} + (P_{\alpha\alpha} - P_{\alpha\beta} + P_{\beta\alpha} - P_{\beta\beta})I_{2z}$$
$$+ (P_{\alpha\alpha} - P_{\alpha\beta} - P_{\beta\alpha} + P_{\beta\beta})2I_{1z}I_{2z} + (P_{\alpha\alpha} + P_{\alpha\beta} + P_{\beta\alpha} + P_{\beta\beta})\tfrac{1}{2}\mathbb{1}]. \qquad (4.4.43)$$

Extensions to larger spin systems are straightforward (4.132). In a system with N coupled spins with $I = \frac{1}{2}$, the expansion involves 2^N products of the form I_{kz}, $2I_{kz}I_{lz}$, $4I_{kz}I_{lz}I_{mz}$, etc. An operator product of the type $2I_{kz}I_{lz}$ is referred to as 'longitudinal two-spin order' (sometimes called J-order, scalar, or dipolar order), which should not be confused with zero-quantum coherence (see § 4.4.5).

Each of the operator products in eqn (4.4.43) has a characteristic transformation behaviour under the action of a non-selective pulse with rotation angle β. In general, various orders of multiple-quantum coherence are created (e.g. terms with $2I_{kx}I_{lx}$, etc.). However, in a basic Fourier experiment with a single pulse, only observable terms (i.e. products with a single transverse operator) need to be retained. With the arrow notation of eqn (2.1.65), one obtains

$$I_{kz} \xrightarrow{\beta I_{ky}} I_{kx} \sin \beta + \text{non-observable terms}, \quad (4.4.44)$$

$$2I_{kz}I_{lz} \xrightarrow{\beta(I_{ky}+I_{ly})} (2I_{kx}I_{lz} + 2I_{kz}I_{lx})\sin \beta \cos \beta$$
$$+ \text{non-observable terms}, \quad (4.4.45)$$

$$4I_{kz}I_{lz}I_{mz} \xrightarrow{\beta(I_{ky}+I_{ly}+I_{my})} (4I_{kx}I_{lz}I_{mz} + 4I_{kz}I_{lx}I_{mz} + 4I_{kz}I_{lz}I_{mx}) \sin \beta \cos^2\beta$$
$$+ \text{non-observable terms}. \quad (4.4.46)$$

The in-phase coherence terms I_{kx} lead to multiplets with binomial relative intensities, which reach maximum overall intensity for $\beta = \pi/2$. However, antiphase multiplet coherence terms, such as $2I_{kx}I_{lz}$ and $4I_{kx}I_{lz}I_{mz}$, which stem from products of q operators, have amplitudes proportional to $\sin \beta \cos^{q-1}\beta$ and assume maximum intensity for smaller pulse rotation angles

$$\tan \beta_{\text{opt}} = \frac{1}{(q-1)^{\frac{1}{2}}}. \quad (4.4.47)$$

For $q = 2, 3, 4,$ and 5, optimum amplitudes are obtained with $\beta_{\text{opt}} = 45°$, 35.3°, 30°, and 26.6°, respectively. The relative weights of in-phase and antiphase multiplet contributions depend on the pulse rotation angle β.

For $\beta = \pi/2$, only in-phase coherence survives and undistorted multiplets are obtained irrespective of the initial populations. On the other hand, if small rotation angles are used ($\cos \beta \approx 1$), all products of I_{kz} operators produce observable transverse magnetization and, according to Conclusion I of the previous section, the Fourier transform of the free induction signal is equivalent to the slow-passage spectrum.

If the initial density operator contains only terms linear in I_{kz}

$$\sigma(0_-) = \sum_k b_k I_{kz}, \quad (4.4.48)$$

each spin of a weakly coupled system can be considered to have its own spin temperature, that is to say, the same population difference prevails across all transitions belonging to a given spin k. In this case, the relative intensities of the individual lines are independent of the rotation angle β.

To compute the amplitude of individual transitions in a multiplet as a function of the populations before the r.f. pulse, it is possible to express the density operator $\sigma(0_-)$ by eqn (4.4.41) and to apply the transformations of eqns (2.1.118) and (2.2.119) to each constituent operator. Only single-quantum operators with coherence order $p = -1$ must be retained.

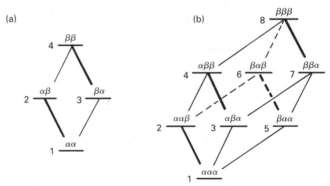

FIG. 4.4.2. Eigenstates in weakly-coupled AX and AMX systems with conventional numbering used in Chapters 2, 4, and 8.

For a weakly coupled two-spin system, for example, one finds

$$\sigma(0_-) \xrightarrow{\beta \Sigma I_{ky}} I_1^\alpha I_2^- L^{(1,2)} + I_1^\beta I_2^- L^{(3,4)}$$
$$+ I_1^- I_2^\alpha L^{(1,3)} + I_1^- I_2^\beta L^{(2,4)} + \text{non-observable terms} \quad (4.4.49)$$

where the states are numbered as shown in Fig. 4.4.2(a), in agreement with eqn (2.1.134). The coefficients $L^{(rs)}$ in eqn (4.4.49) describe the intensities of the corresponding multiplet lines

$$L^{(1,2)} = \tfrac{1}{2} \sin \beta [\cos^2(\beta/2)(P_1 - P_2) + \sin^2(\beta/2)(P_3 - P_4)],$$
$$L^{(3,4)} = \tfrac{1}{2} \sin \beta [\cos^2(\beta/2)(P_3 - P_4) + \sin^2(\beta/2)(P_1 - P_2)],$$
$$L^{(1,3)} = \tfrac{1}{2} \sin \beta [\cos^2(\beta/2)(P_1 - P_3) + \sin^2(\beta/2)(P_2 - P_4)],$$
$$L^{(2,4)} = \tfrac{1}{2} \sin \beta [\cos^2(\beta/2)(P_2 - P_4) + \sin^2(\beta/2)(P_1 - P_3)]. \quad (4.4.50)$$

The multiplicative constants which appear in eqn (4.4.17) have been omitted here. Note that for small rotation angles β, the line intensity is proportional to the population difference across the observed transition. As β approaches $\pi/2$, there is an increasing admixture of population difference across the transition that is parallel to the transition under investigation.

For arbitrary weakly coupled systems with N spins $\tfrac{1}{2}$, the intensity of a line connecting the states $|r\rangle$ and $|s\rangle$ is given by the general expression

$$L^{(rs)} = \tfrac{1}{2} \sin \beta \sum_{(tu)} (\cos \beta/2)^{2(N-1-\Delta_{rs\,tu})} (\sin \beta/2)^{2\Delta_{rs\,tu}} (P_t - P_u) \quad (4.4.51)$$

where the summation extends over all transitions (tu) that are parallel to the transition (rs), i.e. over all transitions within the same multiplet (4.131). The 'spin-flip number' $\Delta_{rs\,tu}$ corresponds to the number of spins that must be reversed ($I_k^\alpha \rightleftarrows I_k^\beta$) in order to bring the transitions (tu) and

(rs) to coincide. Note that the index pairs rs and tu must occur in ordered form.

By way of example, consider the eigenstates of the AMX system in Fig. 4.4.2(b). The line intensity $L^{(1,2)}$ depends on the population differences across the four parallel transitions $(1,2)$, $(3,4)$, $(5,6)$, and $(7,8)$. The spin flip numbers relevant in this example are $\Delta_{12\,12} = 0$, $\Delta_{12\,34} = 1$, $\Delta_{12\,56} = 1$, $\Delta_{12\,78} = 2$; hence eqn (4.4.51) yields for $N = 3$

$$L^{(1,2)} = \tfrac{1}{2}\sin\beta[\cos^6(\beta/2)(P_1 - P_2)$$
$$+ \cos^4\beta/2\,\sin^2(\beta/2)(P_3 - P_4)$$
$$+ \cos^4\beta/2\,\sin^2(\beta/2)(P_5 - P_6)$$
$$+ \cos^2\beta/2\,\sin^4(\beta/2)(P_7 - P_8)]. \quad (4.4.52)$$

If $\beta = \pi/2$, one obtains the usual binomial intensity distribution of the lines within each multiplet, regardless of the population distribution before the pulse, since

$$L^{(rs)}(\beta = \pi/2) = (\tfrac{1}{2})^N \sum_{(tu)} (P_t - P_u); \quad (4.4.53)$$

hence

$$L^{(rs)}(\beta = \pi/2) = L^{(tu)}(\beta = \pi/2) \quad (4.4.54)$$

for all parallel transitions.

The essential feature of eqn (4.4.51) is that in a general case *all* populations P_r contribute to the intensity of each line. For small rotation angles β however, the terms with $\Delta_{rs\,tu} > 0$ may be neglected

$$L^{(rs)}(0 < \beta \ll \pi/2) \simeq \tfrac{1}{2}\sin\beta(P_r - P_s). \quad (4.4.55)$$

Thus for small β, only the populations of the two states $|r\rangle$ and $|s\rangle$ determine the line intensity $L^{(rs)}$, in analogy to slow-passage spectroscopy and in agreement with Conclusion I of the previous section.

The general equation (4.4.51) can easily be derived by simple physical reasoning. The crucial trick is to divide the non-selective pulse into a 'cascade' of semi-selective pulses, each rotating one particular spin through the angle β (4.133). Because the corresponding operators commute, one may factorize the rotation operator

$$\exp\left\{-i\beta \sum_k I_{kv}\right\} = \prod_k \exp\{-i\beta I_{kv}\}. \quad (4.4.56)$$

Although the sequence of pulses is immaterial, it is convenient to apply the last pulse to the spin of which the signal intensitities must be calculated. The preceding pulses merely redistribute populations.

Consider for simplicity an AX system where the eigenstates are numbered as in Fig. 4.4.2(a). If we wish to compute the intensities of the two lines $L^{(1,3)}$ and $L^{(2,4)}$ belonging to spin A, we first apply a semi-selective pulse to the X spin with the transitions (1, 2) and (3, 4), causing a redistribution of populations

$$P_1(0_+) = P_1(0_-)\cos^2\beta/2 + P_2(0_-)\sin^2\beta/2,$$
$$P_2(0_+) = P_2(0_-)\cos^2\beta/2 + P_1(0_-)\sin^2\beta/2,$$
$$P_3(0_+) = P_3(0_-)\cos^2\beta/2 + P_4(0_-)\sin^2\beta/2,$$
$$P_4(0_+) = P_4(0_-)\cos^2\beta/2 + P_3(0_-)\sin^2\beta/2. \quad (4.4.57)$$

At the same time, the semi-selective X pulse creates transverse magnetization, but this can be disregarded for the purpose of our argument. The second semi-selective pulse in the cascade is now applied to the A spin with the transitions (1, 3) and (2, 4), exciting transverse magnetization that leads to the line intensities

$$L^{(1,3)} = \tfrac{1}{2}\sin\beta[P_1(0_+) - P_3(0_+)],$$
$$L^{(2,4)} = \tfrac{1}{2}\sin\beta[P_2(0_+) - P_4(0_+)]. \quad (4.4.58)$$

in agreement with eqn (4.4.50). The extension to larger spin systems allows one to verify eqn (4.4.51).

An example of the β-dependence of Fourier spectra of a system in incoherent non-equilibrium is shown in Fig. 4.4.3. The simulated spectra show the four lines of a weakly coupled AX system, where the populations across two transitions appear inverted after chemically induced nuclear polarization (CIDNP), assuming parameters appropriate for a triplet precursor (4.134). For small β, the line intensities reflect the corresponding population differences, but the intensities within each doublet are equalized for $\beta = \pi/2$.

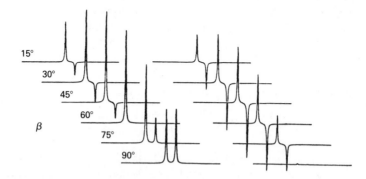

FIG. 4.4.3. Simulated Fourier spectra of an AX system in an incoherent non-equilibrium state (brought about by CIDNP) as a function of the r.f. rotation angle β. Note that the line intensities within each doublet are equalized for $\beta = \pi/2$ and no longer reflect the population differences across the observed transition. (Adapted from Ref. 4.131.)

4.4.4. Selective and semi-selective pulses

With regard to the selectivity of excitation, four types of r.f. pulses can be distinguished. In all cases, the induced transformation of the density operator can be described by

$$\sigma(0_+) = \exp\{-i\beta G_v\}\sigma(0_-)\exp\{i\beta G_v\} \qquad (4.4.59)$$

or, in shorthand notation,

$$\sigma(0_-) \xrightarrow{\beta G_v} \sigma(0_+) \qquad (4.4.60)$$

where β is the rotation angle, $v = x, y$ indicates the r.f. phase, and the form of the operator G characterizes the selectivity.

1. *Non-selective pulses* affect all transitions uniformly

$$G_v = \sum_k I_{kv} \qquad (4.4.61)$$

where the summation extends over all nuclear spins, except in heteronuclear systems where the summation may be restricted to one of the nuclear species (e.g. abundant spins I or rare spins S only).

2. *Semi-selective pulses* affect one particular nucleus k within a weakly coupled system

$$G_v = I_{kv}. \qquad (4.4.62)$$

All transitions within the multiplet belonging to spin k are excited uniformly. Semi-selective pulses are only feasible in weakly coupled spin systems.

3. *Selective pulses* affect individual transitions

$$G_v = I_v^{(rs)}. \qquad (4.4.63)$$

It should be noted that the rotation angle β for non-selective and semi-selective pulses of length τ is given by

$$\beta = -\gamma_k B_1 \tau, \qquad (4.4.64)$$

while the efficiency of selective pulses also depends on the matrix element of the angular momentum operator F^+

$$\beta^{(rs)} = -\gamma B_1 (F^+)_{rs} \tau. \qquad (4.4.65)$$

For selective pulses in weakly coupled systems with spins $I = \frac{1}{2}$, the relevant matrix elements are all unity, but for strong coupling and for systems with $I > \frac{1}{2}$, the effective rotation angle varies from line to line.

For example, in a strongly coupled two-spin system, the rotation angles of selective pulses are

Inner lines: $\beta^{(1,2)} = \beta^{(2,4)} = -\gamma B_1(\cos\theta + \sin\theta)\tau,$

Outer lines: $\beta^{(1,3)} = \beta^{(3,4)} = -\gamma B_1(\cos\theta - \sin\theta)\tau$ (4.4.66)

with $\tan 2\theta = 2\pi J/(\Omega_A - \Omega_B)$. Selective excitation of a single transition in an $I = 1$ one-spin system proceeds with $\beta = -2^{\frac{1}{2}}\gamma B_1\tau$.

Selective pulses in weakly coupled spin systems can be broken down into a sequence of transformations by expressing a single-transition operator $I_\nu^{(rs)}$ in terms of single-spin operators

$$I_\nu^{(rs)} = I_{k\nu} \prod_{l \neq k} I_l^{\mu_l^{(rs)}} \qquad (4.4.67)$$

where $\mu_l^{(rs)} = \alpha$ or β for spins $I_l = \frac{1}{2}$, or $-I_l \leq \mu_l^{(rs)} \leq I_l$ for arbitrary spins I_l. In a system with two spins $\frac{1}{2}$ for example (see Fig. 4.4.2), one finds with eqn (2.1.114)

$$I_x^{(1,3)} = I_{kx}I_l^\alpha = \tfrac{1}{2}(I_{kx} + 2I_{kx}I_{lz}),$$
$$I_x^{(2,4)} = I_{kx}I_l^\beta = \tfrac{1}{2}(I_{kx} - 2I_{kx}I_{lz}). \qquad (4.4.68)$$

The effect of such a pulse on the density operator

$$\sigma(0_-) \xrightarrow{\beta\frac{1}{2}(I_{kx} \pm 2I_{kx}I_{lz})} \sigma(0_+) \qquad (4.4.69)$$

can be expressed as a sequence of rotations (4.132)

$$\sigma(0_-) \xrightarrow{-(\pi/2)I_{ky}} \xrightarrow{(\beta/2)I_{kz}} \xrightarrow{\pm(\beta/2)2I_{kz}I_{lz}} \xrightarrow{(\pi/2)I_{ky}} \sigma(0_+). \qquad (4.4.70)$$

In this notation, it becomes apparent that a selective pulse is equivalent to a free precession period at an appropriate offset bracketed by two semi-selective pulses (4.135).

In systems with more than two weakly-coupled spins, the description of selective pulses involves propagators with products of more than two Cartesian operators. The resulting rotations are analogous to those described in eqn (2.1.93) and illustrated in Fig. 2.1.4. Thus one obtains, for example,

$$I_{ky} \xrightarrow{\beta\frac{1}{4}4I_{kx}I_{lz}I_{mz}} I_{ky}\cos(\beta/4) + 4I_{kz}I_{lz}I_{mz}\sin(\beta/4). \qquad (4.4.71)$$

4. *Simultaneous selective pulses* on connected transitions require special care in their treatment. For example, two selective pulses with equal rotation angle β applied to two progressively connected transitions

$$G_\nu = I_\nu^{(rs)} + I_\nu^{(st)}, \qquad (4.4.72)$$

may be considered as a rotation in a virtual spin 1 subspace spanned by the three energy levels r, s, and t, provided the matrix elements $(F^+)_{rs}$ and $(F^+)_{st}$ are identical (4.136). The effective rotation angle is then

$$\beta = -\gamma B_1 \tfrac{1}{2}(F^+)_{rs}\tau. \tag{4.4.73}$$

4.4.5. Identification of the terms in the density operator

The various terms of the density operator expanded in products of Cartesian operators I_{kx}, I_{ky}, and I_{kz} have a simple physical meaning. In a system with N weakly coupled spins $\tfrac{1}{2}$ (k, l, m, ...), the following classes of product operators can be distinguished (4.132).

1. *One-spin operators*

   I_{kz}: polarization of spin k (z-magnetization),

   I_{kx}: in-phase x-coherence of spin k (x-magnetization),

   I_{ky}: in-phase y-coherence of spin k (y-magnetization).

The operator I_{kz} describes a state with equal population differences across all transitions of spin k. The transverse operators I_{kx} and I_{ky} are representative for the spin-k multiplet with all multiplet components in-phase along the x- or y-axes of the rotating frame.

2. *Two-spin product operators*

   $2I_{kx}I_{lz}$: x-coherence of spin k in antiphase with respect to spin l,

   $2I_{ky}I_{lz}$: y-coherence of spin k in antiphase with respect to spin l,

   $2I_{kx}I_{lx}$, $2I_{ky}I_{ly}$, $2I_{kx}I_{ly}$, and $2I_{ky}I_{lx}$: two-spin coherence of spins k and l,

   $2I_{kz}I_{lz}$: longitudinal two-spin order of spins k and l.

Antiphase coherence represents multiplets where the individual lines have opposite phases, depending on the polarization ($M_l = \pm\tfrac{1}{2}$) of the coupling partner. Two-spin coherence terms consist of superpositions of $p = 0$ and $p = \pm 2$ quantum coherence, as may be appreciated by a transformation into shift operators (see eqn (2.1.107)). Longitudinal two-spin order $2I_{kz}I_{lz}$ refers to a non-equilibrium population distribution without net polarization and without observable magnetization, as shown in Fig. 4.4.4.

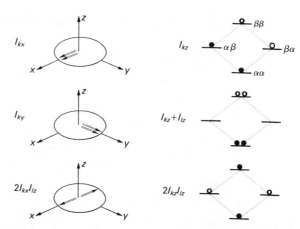

FIG. 4.4.4. Graphical representation of product operators representing single-quantum coherence and longitudinal polarization in a system of two weakly-coupled nuclei with $I = \frac{1}{2}$. The arrows correspond to semiclassical magnetization vectors. Populations are represented by open symbols for states that are depleted, and filled symbols for states that are more populated than in the demagnetized saturated state. (Reproduced from Ref. 4.132.)

3. Three-spin product operators

$4I_{kx}I_{lz}I_{mz}$: x-coherence of spin k, in antiphase with respect to the spins l and m,

$4I_{kx}I_{lx}I_{mz}$, $4I_{ky}I_{ly}I_{mz}$, etc: two-spin coherence of spins k and l, in antiphase with respect to spin m,

$4I_{kx}I_{lx}I_{mx}$, $4I_{ky}I_{ly}I_{my}$, etc: three-spin coherence,

$4I_{kz}I_{lz}I_{mz}$: longitudinal three-spin-order.

Antiphase two-spin coherence $4I_{kx}I_{lx}I_{mz}$ comprises zero- and double-quantum coherence involving two 'active' spins k and l with multiplet components that have opposite phases depending on the polarization of the 'passive' spin m. Three-spin coherence terms consist of a superposition of single-quantum coherence (combination lines) and triple-quantum coherence.

In an analogous manner it is possible to attribute physical significance to products involving shift operators, e.g.:

I_k^+: in-phase +1-quantum coherence (+1QC) of spin k
I_k^-: in-phase −1-quantum coherence (−1QC) of spin k
$I_k^\pm I_{lz}$: ±1-quantum coherence of spin k in antiphase with respect to spin l
$I_k^+ I_l^+$: in-phase +2-quantum coherence (+2QC) of spins k and l
$I_k^- I_l^-$: in-phase −2-quantum coherence (−2QC) of spins k and l

$I_k^+ I_l^-$, $I_k^- I_l^+$: in-phase zero-quantum coherence (ZQC) of spins k and l
$I_k^+ I_l^+ I_{mz}$: +2-quantum coherence (+2QC) of spins k and l in antiphase with respect to spin m

It is often helpful to give a pictoral representation of product operators, in order to emphasize their relationship with semi-classical vector models. Figure 4.4.4 shows how in-phase and antiphase single-quantum coherence (transverse magnetization) and longitudinal terms can be represented graphically.

The use of vector models should preferably be restricted to observable magnetization. To represent multiple-quantum coherence, one may resort to the energy-level diagram (Fig. 4.4.5), and use wavy lines to represent coherence between pairs of eigenstates (dashed wavy lines can be used for imaginary components).

There is a direct relation between product operators involved in the expansion of the density operator and the spectrum obtained after Fourier transformation of the free induction signal. The complex signal observed by quadrature detection is proportional to the complex mag-

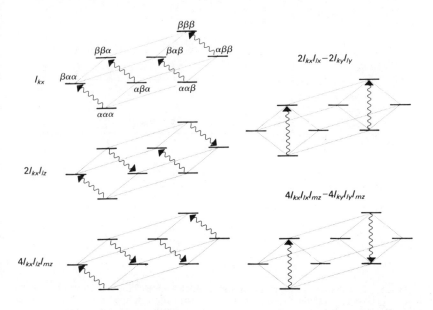

FIG. 4.4.5. Graphical representation of some product operators involving in-phase and antiphase single- and double-quantum coherence in a system of three weakly coupled nuclei with $I = \frac{1}{2}$. The eigenstates (e.g. $|\alpha\beta\alpha\rangle$) have been labelled by the spin states of nuclei k, l, and m in that order. The arrows indicate parallel and antiparallel coherence components. Note that each term represents an entire multiplet (i.e. a set of parallel transitions) rather than an individual transition. (Reproduced from Ref. 4.132.)

netization (eqn (4.4.7))

$$M^+(t) = N\gamma\hbar \, \text{tr}\{F^+\sigma(t)\}$$

$$= N\gamma\hbar \, \text{tr}\left\{\left[\sum_k I_k^+\right]\sigma(t)\right\}. \quad (4.4.74)$$

For each value of k, the trace in eqn (4.4.74) projects only components in the density operator proportional to $I_k^- = I_{kx} - iI_{ky}$. Thus, if we focus attention on systems with spins $I = \frac{1}{2}$, only one-spin operators I_{kx} and I_{ky} give rise to observable magnetization. Products like $2I_{kx}I_{lz}$, representing antiphase coherence, are not observable in a strict sense. In the course of the detection period, however, such antiphase product operators may evolve into observable in-phase coherence

$$2I_{kx}I_{lz} \xrightarrow{\pi J_{kl}\tau 2I_{kz}I_{lz}} [2I_{kx}I_{lz}\cos(\pi J_{kl}\tau) + I_{ky}\sin(\pi J_{kl}\tau)]. \quad (4.4.75)$$

After Fourier transformation, the observable term $I_{ky}\sin(\pi J_{kl}\tau)$ leads to an *antiphase* doublet centred at the chemical shift frequency Ω_k.

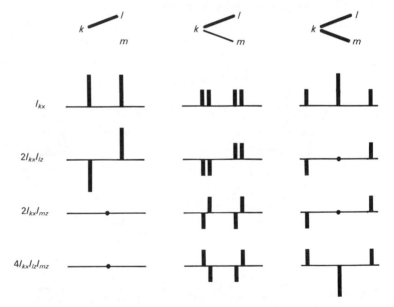

FIG. 4.4.6. Schematic stick spectra of the multiplet of a spin k in a weakly-coupled three-spin system, obtained after Fourier transformation of the free induction signals induced by some typical product operators. If $J_{km} = 0$ (left column), the coherence of spin k in antiphase with respect to spin m is not observable (bottom left). In systems with $J_{kl} = J_{km}$ (right column), the central line vanishes if two components with opposite phases are superimposed. Single-transition operators can be expressed by linear combinations of product operators. Thus, by analogy to eqn (4.4.68), the sum of all four stick spectra in the central column yields a single line on the right (multiplet with intensities 0:0:0:1). The absolute frequency increases from right to left. The multiplet patterns correspond to a positive gyromagnetic ratio. (Reproduced from Ref. 4.132.)

Obviously, such a doublet is only observable when the coupling is resolved, since the two antiphase components cancel if the line-width exceeds the magnitude of the coupling. All product operators which evolve into observable magnetization in the course of the detection period may be referred to as 'observables'. In particular, all product operators containing a single transverse component I_{kx} or I_{ky} and an arbitrary number of longitudinal components, such as $2I_{kx}I_{lz}$, $4I_{kx}I_{lz}I_{mz}\ldots$, become observable in the course of the free induction decay, provided all couplings J_{kl}, $J_{km}\ldots$ are resolved. However, if one of the couplings is insufficiently resolved, observation will be inhibited.

The relative amplitudes and phases of the spectral lines can be derived immediately from the form of the operator products. Consider a system of three weakly coupled spins k, l, and m with $I = \frac{1}{2}$. Figure 4.4.6 illustrates the one-dimensional spectra of spin k obtained after Fourier transformation of a free induction signal induced by a number of typical operator products. For $J_{kl} = J_{km}$, the multiplet of nucleus k collapses to a triplet.

For heteronuclear IS systems with $S = 1$ nuclei, the situation is slightly more complicated. In addition to I_{kx} and I_{ky}, product operators of the type $I_{kx}S_{lz}^2$ and $I_{kx}S_{lx}^2$ or $I_{kx}S_{ly}^2$ are true observables and lead to in-phase magnetization in the form of $1:0:1$ and $\frac{1}{2}:1:\frac{1}{2}$ triplets, respectively.

4.4.6. Composite rotations

Many experiments involve a sequence of pulses with suitable intervals designed, for example, to induce coherence transfer. It is often convenient to treat such sequences as single units which have a simple overall effect. We shall restrict our discussion to a few cases relevant for weakly coupled systems (4.132).

4.4.6.1. Intervals with a central refocusing pulse

If it is required that the coherence evolves in an interval τ under the exclusive influence of coupling terms, the chemical shifts can be refocused in a weakly coupled spin system with the sequence

$$P = \tau/2 - (\pi)_x - \tau/2. \qquad (4.4.76)$$

The effect on the density operator is described as usual by a cascade of unitary transformations

$$\sigma(t) \xrightarrow{\sum \pi J_{kl}\frac{1}{2}\tau 2I_{kz}I_{lz}} \xrightarrow{\sum \Omega_k \frac{1}{2}\tau I_{kz}} \xrightarrow{\pi \sum I_{kx}}$$

$$\times \xrightarrow{\sum \Omega_k \frac{1}{2}\tau I_{kz}} \xrightarrow{\sum \pi J_{kl}\frac{1}{2}\tau 2I_{kz}I_{lz}} \sigma(t+\tau). \qquad (4.4.77)$$

By noting that the shift and coupling terms commute, and that

$$\exp\{i\Omega_k \tfrac{1}{2}\tau I_{kz}\}\exp\{i\pi I_{kx}\} = \exp\{i\pi I_{kx}\}\exp\{-i\Omega_k \tfrac{1}{2}\tau I_{kz}\}, \quad (4.4.78)$$

eqn (4.4.77) may be written in condensed form

$$\sigma(t) \xrightarrow{\pi \sum I_{kx}} \xrightarrow{\sum \pi J_{kl}\tau 2I_{kz}I_{lz}} \sigma(t+\tau). \quad (4.4.79)$$

4.4.6.2. Bilinear rotations with transverse components

A growing class of pulse experiments, ranging from multiple quantum excitation to relayed magnetization transfer and heteronuclear decoupling (see Chapter 8) employ a refocusing sequence bracketed by two $\pi/2$ pulses

$$P = (\pi/2)_x - \tau/2 - (\pi)_x - \tau/2 - (\pi/2)_x. \quad (4.4.80)$$

Using eqn (4.4.79), the effect on the density operator may be written

$$\sigma(t) \xrightarrow{(3\pi/2)\sum I_{kx}} \xrightarrow{\sum \pi J_{kl}\tau 2I_{kz}I_{lz}} \xrightarrow{(\pi/2)\sum I_{kx}} \sigma(t+\tau). \quad (4.4.81)$$

This expression can be simplified to

$$\sigma(t) \xrightarrow{\sum \pi J_{kl}\tau 2I_{ky}I_{ly}} \sigma(t+\tau). \quad (4.4.82)$$

Thus the overall effect of the sequence in eqn (4.4.80) amounts to a bilinear rotation with transverse components $2I_{ky}I_{ly}$, leading to rotations in three-dimensional operator subspaces defined in eqn (2.1.100) and illustrated in Fig. 2.1.5.

4.4.6.3. Sequences without 'sandwich symmetry'

The examples discussed so far involve composite sequences which have the form of symmetrical 'sandwiches' (4.132). Sequences that do not possess suitable symmetry may be modified by inserting 'dummy' pulses that have no effect on the density operator but introduce the symmetry required for simplifying the expressions.

Consider for example the sequence commonly employed for excitation of multiple-quantum coherence of odd orders

$$P = (\pi/2)_x - \tau/2 - (\pi)_y - \tau/2 - (\pi/2)_y. \quad (4.4.83)$$

The sequence can be expanded by inserting two dummy pulses

$$P' = (\pi/2)_x(\pi/2)_{-y}(\pi/2)_y - \tau/2 - (\pi)_y - \tau/2 - (\pi/2)_y. \quad (4.4.84)$$

In this modified form, the sandwich-sequence of eqn (4.4.80) is easily recognized (apart from an overall phase-shift), and the effect of the entire sequence is

$$\sigma(t) \xrightarrow{(\pi/2) \sum I_{kx}} \xrightarrow{-(\pi/2) \sum I_{ky}} \xrightarrow{\sum \pi J_{kl}\tau 2I_{kx}I_{lx}} \sigma(t+\tau). \quad (4.4.85)$$

Since

$$\exp\left\{i\frac{\pi}{2}I_x\right\}\exp\left\{-i\frac{\pi}{2}I_y\right\} = \exp\left\{i\frac{\pi}{2}I_z\right\}\exp\left\{i\frac{\pi}{2}I_x\right\}, \quad (4.4.86)$$

eqn (4.4.85) can be simplified to

$$\sigma(t) \xrightarrow{(\pi/2) \sum I_{kz}} \xrightarrow{(\pi/2) \sum I_{kx}} \xrightarrow{\sum \pi J_{kl}\tau 2I_{kx}I_{lx}} \sigma(t+\tau). \quad (4.4.87)$$

In this form, it is apparent that the first term in eqn (4.4.87) (or, equivalently, the second pulse in eqn (4.4.84)) may be dropped if $\sigma(t)$ contains only longitudinal polarization and zero quantum coherence, which are invariant to z-rotations.

4.4.6.4. Phase cycles

The separation of different orders of single- and multiple-quantum coherence is usually achieved by cycling the r.f. phase of a pulse (or a series of pulses) in subsequent experiments, and by computing suitable linear combinations of the signals (see § 6.2).

A pulse with an arbitrary r.f. phase φ, defined in eqn (4.2.11), can be described by

$$\sigma(t_-) \xrightarrow{\beta(\sum I_{kx}\cos\varphi + \sum I_{ky}\sin\varphi)} \sigma(t_+) \quad (4.4.88)$$

or, in expanded form,

$$\sigma(t_-) \xrightarrow{-\varphi \sum I_{kz}} \xrightarrow{\beta \sum I_{kx}} \xrightarrow{\varphi \sum I_{kz}} \sigma(t_+). \quad (4.4.89)$$

The same notation can be used if entire groups of pulses are shifted in phase. Thus the phase-shifted form of eqn (4.4.80)

$$P = (\pi/2)_\varphi - \tau/2 - (\pi)_\varphi - \tau/2 - (\pi/2)_\varphi \quad (4.4.90)$$

leads to a modification of eqn (4.4.82)

$$\sigma(t) \xrightarrow{-\varphi \sum I_{kz}} \xrightarrow{\sum \pi J_{kl}\tau 2I_{ky}I_{ly}} \xrightarrow{\varphi \sum I_{kz}} \sigma(t+\tau). \quad (4.4.91)$$

If the initial density operator contains only longitudinal polarization and zero-quantum coherence, the first transformation has no effect, and the phase shift boils down to a z-rotation applied *after* the actual pulse sequence.

4.4.6.5. Phase shifts and r.f. rotation angles

Some experiments employ arbitrary r.f. rotation angles $\beta \neq \pi/2$, for example, the DEPT sequence discussed in § 4.5.6 which distinguishes different coupling networks through the β-dependence of coherence transfer. To clarify the mechanism of such experiments, it is useful to expand a single pulse $\beta \sum I_{kx}$ into a sandwich

$$\sigma(t_-) \xrightarrow{-(\pi/2) \sum I_{ky}} \xrightarrow{\beta \sum I_{kz}} \xrightarrow{(\pi/2) \sum I_{ky}} \sigma(t_+). \qquad (4.4.92)$$

Thus a single pulse $(\beta)_x$ has a similar effect to two $(\pi/2)_{\pm y}$ rotations bracketing a $(\beta)_z$ rotation, which in turn is equivalent to a phase shift. This can be exploited in practice to circumvent problems with inaccurate angles β due to inhomogeneous r.f. fields.

4.4.6.6. Heteronuclear systems

Systems containing both abundant I nuclei and dilute S nuclei offer many possibilities for designing sophisticated pulse experiments. To highlight a specific feature of heteronuclear systems, consider the sequence

$$P = (\pi/2)_x^I - \tau/2 - (\pi)_x^{I,S} - \tau/2 - (\pi/2)_x^I \qquad (4.4.93)$$

which involves two π-pulses applied simultaneously to both species to prevent refocusing of the heteronuclear coupling term. This sequence is a constituent part of the so-called INEPT sequence discussed in § 4.5.5. The fate of the density operator is described in analogy to eqn (4.4.82)

$$\sigma(t) \xrightarrow{\pi \sum S_{mx}} \xrightarrow{\sum \pi J_{kl} \tau 2 I_{ky} I_{ly}} \xrightarrow{\sum \pi J_{km} \tau 2 I_{ky} S_{mz}} \sigma(t + \tau). \qquad (4.4.94)$$

Note the appearance of terms $I_{k\nu} S_{m\mu}$ with $\nu \neq \mu$, which occur because the $\pi/2$-pulses act only on the I spins. The resulting rotations take place in three-dimensional operator spaces defined by the commutators

$$[I_{k\lambda}, 2I_{k\mu} I_{l\xi}] = i2I_{k\nu} I_{l\xi} \qquad (4.4.95)$$

with $\lambda, \mu, \nu = x, y, z$, and cyclic permutations, and $\xi = x, y$, or z (4.132). Such rotations can be represented diagrammatically in analogy to Fig. 2.1.4.

4.5. Heteronuclear polarization transfer

Heteronuclear polarization transfer techniques have greatly contributed to the success of NMR of rare and low-γ nuclei which have inherently a very low sensitivity. The basic trick involves the 'borrowing' of polarization from wealthier spins with higher sensitivity. Polarization transfer can be used for three purposes:

1. Enhancement of the initial polarization of low-sensitivity nuclei;
2. Indirect detection of the resonances of low-sensitivity nuclei;

3. 'Editing' of spectra by selecting resonances belonging to specific subunits in a spin system, such as CH, CH_2, and CH_3 groups.

There are many effective ways of enhancing the sensitivity or the information content by polarization transfer. Early examples are found in solid-state NMR, where cross-polarization (4.143, 4.144) and adiabatic demagnetization (4.145, 4.146) offer a unique means of transferring polarization from abundant I to dilute S spins (see §§ 4.5.3 and 4.5.4).

In heteronuclear systems in isotropic solution, polarization can be transferred by the Overhauser effect (4.147) (§ 4.5.2), by selective population inversion (SPI) or selective population transfer (SPT) (4.148, 4.149), and more recently by a sequence for observing Insensitive Nuclei Enhanced by Polarization Transfer (INEPT) (4.150, 4.151). This latter method, which is closely related to heteronuclear two-dimensional spectroscopy (§ 8.5), combines the enhancement of selective population inversion with the advantages of non-selective pulses, and will be discussed in some detail in § 4.5.5.

In isotropic solution, it is of practical importance to identify the number of abundant nuclei that are coupled to each dilute spin. Thus in carbon-13 NMR, assignment is greatly facilitated if CH_n subunits with $n = 0, 1, 2$, and 3 can be separated by editing. This goal can be achieved without involving polarization transfer with the so-called 'Attached Proton Test' (APT) and related methods (4.152–4.159), which are often derived from two-dimensional J-spectroscopy (see § 7.2), or by more advanced methods such as SEMUT (4.160, 4.161). Similar editing effects can be obtained in conjunction with heteronuclear polarization transfer with INEPT, or more accurately with 'Distortionless Enhancement of Polarization Transfer' (DEPT) (4.162–4.164), as discussed in § 4.5.6.

Many polarization transfer techniques originally developed for heteronuclear systems can be adapted to investigate *homonuclear* coupling networks. A variety of editing methods have been developed that rely on 'spin-pattern recognition'. These techniques are sensitive to the topology of the coupling network, and are useful to simplify the analysis of complex overlapping proton spectra. Since many of these techniques are derived from two-dimensional spectroscopy, they will be dealt with more explicitly in Chapter 8. Suffice it to mention here that multiple-quantum filters make it possible to retain selectively signals that stem from coupled subunits which contain at least a specified *minimum* number of coupled nuclei. Thus double-quantum filtration can be used to retain signals stemming from coupled pairs of carbon-13 nuclei (4.165), and from coupled systems with at least two nuclei (4.166–4.170). Higher-order multiple-quantum filtering has been used for extracting signals belonging to more complex spin systems (4.171–4.173). With so-called p-spin filtering methods, it is possible in favourable cases to suppress signals from systems with $N<p$ and $N>p$ nuclei (4.173). Finally, spin-pattern specific sequences (4.174, 4.175) allow one to separate signals associated with coupling networks with different topologies but with the same number of nuclei. For example, four-spin systems of the type A_3X and A_2X_2 may be distinguished. It is to be expected that many additional methods will be forthcoming in the context of signal enhancement and editing, and any attempt to give an exhaustive review is not only beyond the scope of this chapter, but is also bound to become obsolete soon. We shall therefore only discuss a few selected procedures that may help to understand the principles.

4.5.1. Transfer of spin order

Polarization is a particular manifestation of spin order, and polarization transfer involves a conversion between different ordered states. Each type of spin order is associated with a characteristic term in the density operator as discussed in § 4.4.5. Besides Zeeman polarization (I_z, S_z, or F_z), spin order can occur in the form of two-spin order (e.g. I_zS_z), in-phase coherence (e.g. I_x), antiphase coherence (e.g. I_xS_z), and multiple quantum coherence (e.g. I_xS_x), to name just the most common forms. In heteronuclear polarization transfer, the overall transformation $I_z \to S_z$ is usually the relevant process, though it may proceed through intermediate states such as heteronuclear two-spin order (e.g. $I_z \to 2I_zS_z \to S_z$).

The most relevant aspect is the efficiency of the transfer. Spin order can be measured by the spin entropy S. In terms of the density operator σ, the entropy is defined (4.146) by

$$S = -k \, \mathrm{tr}\{\sigma \ln \sigma\}. \quad (4.5.1)$$

In the high-temperature limit, where σ does not deviate appreciably from unity, we may use the approximation

$$\sigma \simeq (\mathbb{1} + B)/\mathrm{tr}\{\mathbb{1}\}$$

where B is a small traceless operator expressing the deviation from the fully saturated state. A brief calculation leads to the entropy

$$S = k \ln(\mathrm{tr}\{\mathbb{1}\}) - k \, \mathrm{tr}\{B^2\}/(2 \, \mathrm{tr}\{\mathbb{1}\}). \quad (4.5.2)$$

The first term represents the entropy of an equally populated state at infinite spin temperature with maximum disorder, while the second term expresses the reduction of entropy due to any form of order within the spin system.

4.5 HETERONUCLEAR POLARIZATION TRANSFER

When the density operator σ is expanded in a set of orthogonal basis operators $\{B_j\}$, eqn (2.1.45),

$$\sigma = \{\mathbb{1} + \sum_j b_j B_j\}/\text{tr}\{\mathbb{1}\} \tag{4.5.3}$$

where each term may represent a particular type of spin order, it is possible to divide S into contributions of the various terms

$$S = k \ln(\text{tr}\{\mathbb{1}\}) + \sum_j S_j$$

with

$$S_j = -kb_j^2 \, \text{tr}\{B_j^+ B_j\}/(2 \, \text{tr}\{\mathbb{1}\}) \tag{4.5.4}$$

In the course of a reversible transfer, the entropy S remains constant. The process is then adiabatic. The ultimate goal in polarization transfer is the transformation of the full polarization from one spin species to another. The optimum which can be achieved is limited by the transfer of the entire entropy contribution S_I^i of the initial spin species I into entropy S_S^f of the final species S

$$S_S^f = S_I^i. \tag{4.5.5}$$

Consider the transfer of spin order expressed by the operator B_k into spin order represented by the operator B_l

$$\sigma^i = b_k^i B_k \rightarrow \sigma^f = b_l^f B_l.$$

Entropy conservation leads to a general relation for optimum transfer: the maximum value of the coefficient b_l^f in the final density operator σ^f is given by

$$b_l^f = b_k^i [\text{tr}\{B_k^2\}/\text{tr}\{B_l^2\}]^{\frac{1}{2}}. \tag{4.5.6a}$$

If the number N_k of spin systems associated with the operator B_k and the number N_l of spin systems associated with the operator B_l are different, the relation is modified

$$b_l^f = b_k^i [N_k \, \text{tr}\{B_k^2\}/N_l \, \text{tr}\{B_l^2\}]^{\frac{1}{2}}. \tag{4.5.6b}$$

To give an example, consider a transfer of polarization in a system with N_I spins I and N_S spins S with arbitrary spin quantum numbers. If we assume an initial equilibrium state for the I spins

$$\sigma^i = \sigma_0 = [1 + \beta_L \gamma_I B_0 I_z]/\text{tr}\{\mathbb{1}\} \tag{4.5.7}$$

with the inverse lattice temperature

$$\beta_L = \hbar/(kT_L), \tag{4.5.8}$$

and if we desire to obtain a final state of the form

$$\sigma^f = [1 + \beta_S \gamma_S B_0 S_z]/\text{tr}\{\mathbb{1}\}, \tag{4.5.9}$$

we can easily compute with eqn (4.5.6b) the resulting inverse spin temperature β_s of the S spins after an adiabatic transfer

$$\beta_S = \beta_L \frac{\gamma_I}{\gamma_S} \left(\frac{N_I I(I+1)}{N_S S(S+1)} \right)^{\frac{1}{2}}. \quad (4.5.10)$$

The maximum achievable polarization enhancement of the S spins, compared with the equilibrium polarization, is therefore given by the factor

$$\eta = \frac{\gamma_I}{\gamma_S} \left(\frac{N_I I(I+1)}{N_S S(S+1)} \right)^{\frac{1}{2}}. \quad (4.5.11)$$

Expressed in terms of the equilibrium magnetization of the I spins $M_{I0} = N_I \gamma_I^2 \hbar^2 I(I+1) B_0/(3kT)$, we find the final S-spin magnetization

$$M_S^f = M_{I0} \frac{\gamma_S}{\gamma_I} \left(\frac{N_S S(S+1)}{N_I I(I+1)} \right)^{\frac{1}{2}}. \quad (4.5.12)$$

This shows that even under idealized conditions it is not possible to transfer the entire I-spin magnetization to the S spins. The enhancement factor η of eqn (4.5.11) is proportional to $(N_I/N_S)^{\frac{1}{2}}$ instead of the expected N_I/N_S proportionality.

By way of example, consider the maximum enhancement of carbon-13 polarization in a $^{13}CH_n$ fragment. From eqn (4.5.11), we obtain

$$\eta = \frac{\gamma_I}{\gamma_S} \sqrt{n}. \quad (4.5.13)$$

The maximum achievable ^{13}C spin polarization is $\eta = 4$ for a CH group, $\eta = 5.66$ for a CH_2 group, and $\eta = 6.93$ for a CH_3 group. If the intermolecular dipolar interactions are effective in transferring spin order between molecules (as is the case in solids), we also have to consider the natural isotopic ratio, $N_I/N_S \simeq 100$, and the maximum enhancement factors increase by another factor 10. This is the basis for transfer of spin order by adiabatic demagnetization and remagnetization in solids (see §4.5.4).

4.5.2. Polarization transfer by nuclear Overhauser effect

By far the simplest procedure for enhancing the polarization of low-γ spins S is the application of a saturating r.f. field to the high-γ spins I. The resulting redistribution of the populations leads to a polarization enhancement of the S spins, provided the relaxation processes are favourable. This transfer of polarization is called nuclear Overhauser enhancement (4.147).

4.5 HETERONUCLEAR POLARIZATION TRANSFER

For an optimum transfer, the S-spin relaxation must be due entirely to dipolar interaction with the I spins, a situation which is often fulfilled for rare and low-γ S spins. The achievable enhancement can be deduced easily from the simple case of an IS spin system with $I = S = \frac{1}{2}$. The relaxation rates caused by the IS dipolar interaction under extreme narrowing conditions (short correlation times and fast rotational diffusion) are determined by the ratios $W_0 : W_1 : W_2 = 1 : 1.5 : 6$. The saturation of the two I-spin transitions leads to a new population distribution and to an enhancement of the S-spin polarization by a factor

$$\eta = 1 + \frac{1}{2}\frac{\gamma_I}{\gamma_S}. \tag{4.5.14}$$

For carbon-13 coupled to protons, the Overhauser enhancement takes the value $\eta = 3$ while for nitrogen-15 coupled to protons one obtains $\eta = -4$.

No advantage can be taken from the fact that the I spins may be in overwhelming majority, since the enhancement factor is independent of the ratio N_I/N_S. The enhancement depends only on the ratio γ_I/γ_S and is much smaller than the optimum predicted by eqn (4.5.11).

Whenever additional mechanisms are involved in the relaxation of the S spins, the enhancement factor η will be reduced because of competitive leakage pathways. In particular, relaxation by chemical shielding anisotropy and spin–rotation interaction may quench the Overhauser enhancement.

The build-up of the enhancement is often quite slow since it proceeds with the longitudinal relaxation rate of the S spins due to the IS interactions. An extended presaturation period must therefore be used to obtain adequate enhancement.

4.5.3. Cross-polarization in the rotating frame

Cross-polarization in the rotating frame has been introduced by Hartmann and Hahn as a means of transferring polarization between different nuclear species in solids (4.143), and has become of central importance for obtaining spectra of rare spins with low gyromagnetic ratios such as ^{13}C, since a significant sensitivity enhancement may be achieved. Cross-polarization can be used either for direct observation of low-sensitivity nuclei (4.143) or for indirect detection of such nuclear species via high-sensitivity nuclei such as protons (4.176, 4.177).

More recently, it has been demonstrated that cross-polarization can also be applied for sensitivity enhancement of scalar-coupled nuclei in isotropic solutions (4.178–4.181), although cross-polarization in liquids has not reached the importance that it holds currently in solid-state spectroscopy.

To understand the basic phenomena of cross-polarization in solids, it is sufficient to consider a thermodynamic treatment, while in liquids a more detailed quantum mechanical treatment is in order. We shall first develop the thermodynamic framework.

It has been shown by Hartmann and Hahn (4.143) that nuclear spin systems consisting of spin species I and S can exchange spin energy when two strong r.f. fields B_{1I} and B_{1S} are applied simultaneously at the I and S Larmor frequencies. The rate of energy exchange strongly depends on the magnitudes of B_{1I} and B_{1S} and reaches a maximum when the so-called Hartmann–Hahn condition is fulfilled

$$\gamma_I B_{1I} = \gamma_S B_{1S}. \qquad (4.5.15)$$

This condition guarantees that the nutation frequencies ω_{1I} and ω_{1S} of the two nuclear species in their respective rotating frames are equal. This leads to a maximum interaction and therefore to a maximum exchange rate $1/T_{IS}$. A detailed treatment of the exchange rate, determined by the dipolar II and IS interactions, can be found elsewhere (4.182, 4.183).

The basic experiment is shown in Fig. 4.5.1(a). The I spins are first allowed to equilibrate, yielding a magnetization

$$M_I^{(0)} = \beta_L C_I B_0 \qquad (4.5.16)$$

where $\beta_L = \hbar/(kT_L)$ is the inverse lattice temperature and C_I is the Curie constant

$$C_I = \tfrac{1}{3}\gamma_I^2 \hbar I(I+1) N_I \qquad (4.5.17)$$

with the number of I spins N_I. By means of a $\pi/2$-pulse followed by the application of an r.f. field B_{1I} phase-shifted by $90°$, the magnetization is spin-locked in the rotating frame.

Since the polarizing r.f. field is relatively small, $B_{1I} \ll B_0$, the locked magnetization $M_I^{(0)}$ appears to correspond to a much higher inverse spin temperature β_0, i.e.

$$M_I^{(0)} = \beta_0 C_I B_{1I} \qquad (4.5.18)$$

with

$$\beta_0 = \beta_L B_0 / B_{1I}. \qquad (4.5.19)$$

The energy of the spin system at this moment is

$$E = -\beta_0 C_I B_{1I}^2. \qquad (4.5.20)$$

After a sufficiently long contact between I and S spins, a thermodynamic equilibrium with equal inverse spin temperatures $\beta_{1I} = \beta_{1S}$ is

4.5 HETERONUCLEAR POLARIZATION TRANSFER

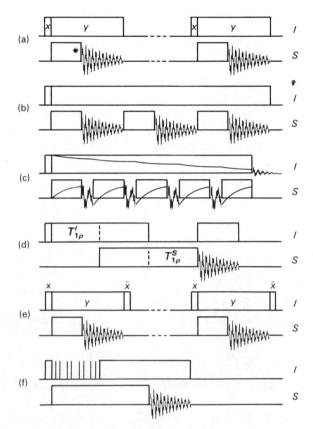

FIG. 4.5.1. (a) Basic scheme for cross-polarization in the rotating frame: after a $(\pi/2)_x$-pulse, the magnetization of the abundant I spins (e.g. protons) is spin-locked with a field B_{1I} along the y-axis, and polarization is transferred to the dilute S spins (e.g. carbon-13, nitrogen-15, etc.) by applying an r.f. field B_{1S} such as to fulfil the Hartmann–Hahn condition, eqn (4.5.15). Decoupling during the observation of the free induction signal of the S spins is optional. (b) Multiple-contact scheme where a series of free induction signals are observed after repeated cross-polarization periods. (c) Indirect detection of S-spin precession by monitoring the loss of I-spin magnetization (4.176). (d) Scheme for the measurement of $T_{1\rho}^I$ and $T_{1\rho}^S$. (e) The residual I magnetization that remains locked after the experiment may be returned to z-axis with a 'flip-back' pulse (4.184). (f) Use of cross-polarization as mixing process in heteronuclear two-dimensional correlation spectroscopy.

reached. Since the energy E is conserved, one obtains (disregarding dipolar energy)

$$E = -\beta_1 C_I B_{1I}^2 - \beta_1 C_S B_{1S}^2 = -\beta_0 C_I B_{1I}^2. \qquad (4.5.21)$$

If the Hartmann–Hahn condition is fulfilled, eqn (4.5.15), we can immediately calculate the resulting inverse spin temperature β_1

$$\beta_1 = \beta_0 (1 + \varepsilon)^{-1} \qquad (4.5.22)$$

with

$$\varepsilon = \frac{\gamma_I^2 C_S}{\gamma_S^2 C_I} = \frac{S(S+1)N_S}{I(I+1)N_I}. \quad (4.5.23)$$

The S-spin magnetization finally becomes

$$M_S^{(1)} = \beta_1 C_S B_{1S} = \frac{\gamma_I}{\gamma_S}(1+\varepsilon)^{-1} \beta_L C_S B_0. \quad (4.5.24)$$

Thus the enhancement with respect to the equilibrium S-spin magnetization is

$$\eta = M_S^{(1)}/M_S^{(0)} = (\gamma_I/\gamma_S)(1+\varepsilon)^{-1}. \quad (4.5.25)$$

For cross-polarization from abundant I spins to dilute S spins, $\varepsilon \ll 1$, hence the sensitivity enhancement is given by the ratio of gyromagnetic ratios. Thus for cross-polarization from protons to carbon-13 or to nitrogen-15, enhancement factors of 4 and 10, respectively, can be obtained. This enhancement is somewhat higher than can be achieved with the Overhauser effect, but does not reach the maximum predicted for an adiabatic transfer. On the other hand, it should be noted that only a small part of the I-spin polarization is actually transferred to the S spins, the remaining I-spin magnetization being

$$M_I^{(1)} = M_I^{(0)}(1+\varepsilon)^{-1}. \quad (4.5.26)$$

This magnetization can be utilized in a multiple-contact experiment where the I-spin magnetization is kept locked over an extended period while the S-magnetization is repeatedly enhanced by cross-polarization, with observation of the S-spin free induction decays in the intervals (Fig. 4.5.1(b)).

Multiple-contact schemes can also be used to detect the S spins indirectly by monitoring the decay of the I-magnetization, as shown in Fig. 4.5.1(c) (4.176). Multiple-contact methods are sensitive to the $T_{1\rho}^I$-decay of the spin-locked I magnetization. This rate constant can be measured with the scheme shown in Fig. 4.5.1(d). If $T_{1\rho}$ processes are too fast for multiple-contact schemes to be effective, it is of advantage to employ a flip-back pulse (4.184) (Fig. 4.5.1(e)) to restore the magnetization along the z-axis between experiments.

In many cases, the sensitivity gain obtained with a single contact is greater than implied in eqn (4.5.25), since the waiting time between experiments can be reduced, because the relevant I-spin magnetization recovers between experiments with T_{1I} which is often much shorter than T_{1S}.

Without going into detailed calculations of the cross-polarization rate constant $1/T_{IS}$, we note that this rate is proportional to the square of the dipolar *IS* interaction (4.182, 4.183). If the *IS* pair interactions are dominant, the rate constant has a typical $[1 - 3\cos^2\theta_{IS}]^2$ dependence on the orientation (polar angle θ_{IS}) of the internuclear vector $\mathbf{r}_{IS}$ with respect to the magnetic field. In a powder spectrum, crystallites with different orientations are associated with different cross-polarization efficiencies. In particular, there is often a 'hole' in the lineshape corresponding to the magic angle orientation of the *IS* vector. In general, intensities and lineshapes in spectra obtained with cross-polarization tend to be unreliable.

To understand cross-polarization in liquids (4.178–4.181) and in solids with a resolved dipolar structure (4.185), a full quantum mechanical treatment of the molecular spin system is in order. We shall restrict the discussion to a two-spin *IS* system which shows many of the essential features.

We start with the Hamiltonian $\mathcal{H}$ in the doubly rotating frame, rotating at the frequencies of the two applied r.f. fields

$$\mathcal{H} = \Omega_I I_z + \Omega_S S_z + 2\pi J_{IS} I_z S_z + \omega_{1I} I_x + \omega_{1S} S_x \quad (4.5.27)$$

with the resonance offsets Ω_I and Ω_S and the two r.f. field strengths ω_{1I} and ω_{1S}. It is convenient to transform this Hamiltonian into a tilted frame in which the two effective fields

$$\Omega_{I\,\text{eff}} = (\Omega_I^2 + \omega_{1I}^2)^{\frac{1}{2}} \quad \text{and} \quad \Omega_{S\,\text{eff}} = (\Omega_S^2 + \omega_{1S}^2)^{\frac{1}{2}} \quad (4.5.28)$$

point along the new z-axes (4.179)

$$\mathcal{H}_{\text{eff}} = \Omega_{I\,\text{eff}} I_z + \Omega_{S\,\text{eff}} S_z + \cos\theta_I \cos\theta_S\, 2\pi J_{IS} I_z S_z \\ + \tfrac{1}{4} \sin\theta_I \sin\theta_S\, 2\pi J_{IS} [I^+ S^- + I^- S^+]. \quad (4.5.29)$$

Spin-locking of the *I*-magnetization brings this magnetization parallel to the effective field $\Omega_{I\,\text{eff}}$. The initial state in the tilted frame is thus proportional to I_z. This term does not commute with the Hamiltonian $\mathcal{H}_{\text{eff}}$ and therefore evolves in time. It is found that the density operator evolves in a space spanned by the four orthogonal single-transition operators

$$I_z^{(1,4)} = \frac{1}{2}(I_z + S_z),$$

$$I_z^{(2,3)} = \frac{1}{2}(I_z - S_z),$$

$$I_x^{(2,3)} = \frac{1}{2}(I^+ S^- + I^- S^+),$$

$$I_y^{(2,3)} = -\frac{i}{2}(I^+ S^- - I^- S^+). \quad (4.5.30)$$

The evolution can be visualized as a rotation of the difference magnetization vector starting with $\frac{1}{2}(I_z - S_z)$ in the three-dimensional subspace of the $(2, 3)$ transition, while the sum magnetization proportional to $\frac{1}{2}(I_z + S_z)$ remains invariant.

The precession frequency ω_p is given by

$$\omega_p = [(\Omega_{I\,\text{eff}} - \Omega_{S\,\text{eff}})^2 + (\pi J_{IS} \sin \theta_I \sin \theta_S)^2]^{\frac{1}{2}} \tag{4.5.31}$$

with the tilt angles of the effective fields θ_I and θ_S given by

$$\theta_I = \tan^{-1} \frac{\omega_{1I}}{\Omega_I}, \qquad \theta_S = \tan^{-1} \frac{\omega_{1S}}{\Omega_S}. \tag{4.5.32}$$

The relevant component is the z-component $\langle I_z^{(2,3)} \rangle$. At the beginning of the cross-polarization period, the state vector is aligned along the positive z-axis. Complete transfer of magnetization to the S spins would correspond to an inversion of the state vector into a position parallel to the $-z$-axis, i.e. leading to $\langle I_z^{(2,3)} \rangle(\tau_m) = -\langle I_z^{(2,3)} \rangle(0)$. Obviously this is only feasible when the rotation axis lies in the xy-plane. Defining an angle ϕ by the relationship

$$\tan \phi = \frac{\pi J_{IS} \sin \theta_I \sin \theta_S}{\Omega_{I\,\text{eff}} - \Omega_{S\,\text{eff}}}, \tag{4.5.33}$$

the coefficients of the two relevant magnetization components become

$$a_{I_z}(\tau_m) = 1 - \sin^2\phi \, \sin^2(\omega_p \tau_m / 2) \tag{4.5.34}$$

and

$$a_{S_z}(\tau_m) = \sin^2\phi \, \sin^2(\omega_p \tau_m / 2). \tag{4.5.35}$$

It is apparent that complete transfer can be achieved with $\phi = \pi/2$. This condition is fulfilled only for exact Hartmann–Hahn matching, i.e. for

$$\Omega_{I\,\text{eff}} = \Omega_{S\,\text{eff}}. \tag{4.5.36}$$

It is obvious that for an isolated two-spin system, cross-polarization is a periodic oscillatory process. In reality, there are damping mechanisms which attenuate the oscillation, and lead to a state

$$a_{S_z}(\infty) = a_{I_z}(\infty) = \tfrac{1}{2} \tag{4.5.37}$$

with equal polarization on both spins. Damping can be caused by inhomogeneous r.f. fields, relaxation, and interactions with further spins.

Equation (4.5.33) demonstrates that the smaller the coupling constant J_{IS}, the more sensitive the amplitude of coherence transfer is on the mismatch $\Omega_{I\,\text{eff}} - \Omega_{S\,\text{eff}}$. This explains why other transfer methods are often preferred for isotropic liquids.

Since cross-polarization with an r.f. phase along the y-axis of the rotating frame only transfers the I_y component of the abundant magnetization, it can be used much like a phase-sensitive detector to monitor the evolution of transverse I-magnetization. This idea can be incorporated into two-dimensional schemes as shown in Fig. 4.5.1(f). In this case, the observable S-magnetization is modulated by I-spin precession, and can be used for the indirect detection of the spectrum of the I spins (4.178, 4.179, 4.186, 4.187). Such schemes will be discussed more extensively in § 8.5.6.

The Hartmann–Hahn experiment is quite demanding from an instrumental point of view. To obtain proper spin-locking, high r.f. fields are required which in solids must exceed the local fields B_{SL} and B_{IL} (given by the respective dipolar fields) and in liquids must exceed the largest offsets (chemical shifts). Fields of 10–20 G are usually necessary, requiring transmitters of 200–1000 W. Cross-polarization times can be of the order of 20 ms, which puts high demands on the probe assembly as well.

For efficient cross-polarization, the deviations from the Hartmann–Hahn condition must be less than the dipolar width of the I and S spin lines. This requirement is not easily met for samples with narrow resonance lines. Double-tuned single-coil arrangements are mandatory to guarantee matching throughout the entire sample volume.

4.5.4. Adiabatic polarization transfer

The transfer processes discussed in the preceeding paragraphs are far from adiabatic and do not allow one to approach the optimum transfer efficiency. We shall briefly discuss some of the procedures that have been devised for an optimum adiabatic and reversible transfer.

In solids, adiabatic polarization transfer between two spin species I and S is feasible via adiabatic demagnetization and remagnetization in the rotating frame (4.145, 4.146). The I-spin magnetization is first locked along the radio-frequency field in analogy to the initial step in Hartmann–Hahn cross-polarization. The amplitude of the r.f. field is then slowly reduced to zero in an adiabatic manner such that the system constantly remains very near to equilibrium. During this process the heat capacity $C_I B_{1I}^2$ of the Zeeman interaction is reduced to zero while the heat capacity $(C_I + C_S) B_L^2$ of the dipolar reservoir remains constant (B_L is the effective local field). The entire spin entropy is thereby transferred into dipolar order. In a final step, the amplitude of an r.f. field applied to the S spins is adiabatically increased from zero to a value that exceeds the local field B_L. This creates magnetization of the S spins. Virtually all spin entropy is concentrated in the S-spin Zeeman interaction, and the optimum transfer expressed by eqns (4.5.10)–(4.5.12) can thus be achieved.

In practice, ideal efficiency of the transfer can often not be attained. The establishment of thermal equilibrium between the Zeeman and dipolar interactions may be exceedingly slow, particularly when B_{1I} or $B_{1S} \gg B_L$, making energy-conserving flip-flop processes unlikely. Adiabatic transfer therefore requires a very slow variation of the r.f. fields, and $T_{1\rho}$ relaxation may lead to an irreversible decay of the magnetization.

In contrast to Hartmann–Hahn cross-polarization, adiabatic transfer does not require matching of r.f. field amplitudes, thus making the experiment less critical. In practice, however, it is not easy to perform a controlled variation of the r.f. amplitude, particularly if non-linear power amplifiers are used. In this case, it is possible to employ a pulsed version of adiabatic demagnetization where the *average* r.f. field strength is varied (4.297). It is also possible to replace the adiabatic demagnetization process by a Jeener–Broekaert pulse sequence (4.298), albeit at the expense of some sensitivity.

In isotropic liquids, it is possible to use scalar spin–spin interactions for adiabatic transfer (4.180, 4.181). After spin-locking of the I spins, the I-spin r.f. field B_{1I} is reduced to zero, while the S-spin r.f. field B_{1S} is simultaneously increased, in such a way that the Hartmann–Hahn condition is crossed in the middle of the process. The method is best explained as a level anti-crossing experiment which leads to an exchange of the populations of the states $|\alpha\beta\rangle$ and $|\beta\alpha\rangle$ in a two-spin IS system. This is equivalent to a full transfer of polarization from spin I to spin S. Critical matching of r.f. fields can again be avoided, and the overall efficiency of the process is good, although it is not possible to achieve full transfer of entropy in systems with equivalent I-spins. Nevertheless, the adiabatic transfer procedure is demanding with regard to r.f. power requirements, a concern which can be fully eliminated by the pulse techniques described in the following two sections.

4.5.5. Polarization transfer by radio-frequency pulses

The procedures discussed so far utilize an extended period of r.f. irradiation during which the scalar or dipolar interaction leads to a transfer of polarization. In the pulse methods to be discussed in this section, precession under the coupling Hamiltonian and external perturbations are separated in time. Free precession under the coupling Hamiltonian is required to create a correlated state of I and S spins in the form of antiphase I spin coherence. This state is then transformed by a pair of r.f. pulses into antiphase S-spin coherence, which can finally be refocused by a further free precession period if desired.

4.5 HETERONUCLEAR POLARIZATION TRANSFER

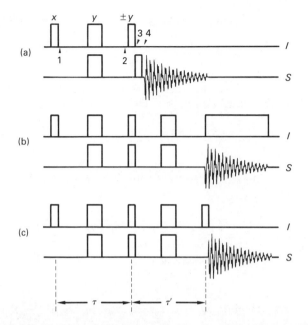

FIG. 4.5.2. (a) Sequence for polarization transfer by r.f. pulses in heteronuclear systems (INEPT), as discussed in the text. Narrow and thick pulses in the diagram represent r.f. rotation angles of $\pi/2$ and π respectively. (b) Analogous sequence with rephasing interval and decoupling in the detection period (refocused INEPT). (c) Sequence with a rephasing interval and with a purging pulse but without proton decoupling, suitable to obtain undistorted multiplets (INEPT$^+$).

In terms of operators, the basic scheme can be described by the sequence of transformations

$$I_{ky} \xrightarrow{(\pi/2)2I_{kz}S_{mz}} -2I_{kx}S_{mz} \xrightarrow{(\pi/2)I_{ky}} 2I_{kz}S_{mz} \xrightarrow{(\pi/2)S_{my}}$$

$$2I_{kz}S_{mx} \xrightarrow{(\pi/2)2I_{kz}S_{mz}} S_{my}. \quad (4.5.38)$$

A full transfer of polarization can thus be achieved for a two-spin system.

We shall discuss in some detail the transfer process involved in the sequence of Fig. 4.5.2, which has become known as INEPT ('Insensitive Nuclei Enhanced by Polarization Transfer') (4.150, 4.151) and which may be derived from heteronuclear two-dimensional correlation spectroscopy (see § 8.5). In a system with a single abundant spin I_k and a dilute spin S_m, the equilibrium density operator may be written (omitting multiplicative constants common to both terms)

$$\sigma^{\mathrm{eq}} = I_{kz} + \frac{\gamma_S}{\gamma_I} S_{mz}. \quad (4.5.39)$$

After the first $(\pi/2)_x^I$-pulse, one obtains

$$\sigma_1 = -I_{ky} + \frac{\gamma_S}{\gamma_I} S_{mz}. \qquad (4.5.40)$$

The net effect of the interval τ in Fig. 4.5.2(a) with the two simultaneous π-pulses can be described by the cascade of transformations of eqn (4.4.79)

$$\sigma_1 \xrightarrow{\pi \Sigma I_{ky}} \xrightarrow{\pi S_{my}} \xrightarrow{\Sigma \pi J_{km}\tau 2I_{kz}S_{mz}} \sigma_2. \qquad (4.5.41)$$

Just before the final pair of pulses in Fig. 4.5.2(a), we obtain in a two-spin system

$$\sigma_2 = +I_{ky} \cos \pi J_{km}\tau - 2I_{kx}S_{mz} \sin \pi J_{km}\tau - \frac{\gamma_S}{\gamma_I} S_{mz}. \qquad (4.5.42)$$

After a pulse $(\pi/2)_\varphi^I$ with phase $\varphi = \pm\pi/2$ (i.e. along the $\pm y$ axes) one has

$$\sigma_3(\varphi = \pm\pi/2) = +I_{ky} \cos \pi J_{km}\tau \pm 2I_{kz}S_{mz} \sin \pi J_{km}\tau - \frac{\gamma_S}{\gamma_I} S_{mz} \qquad (4.5.43)$$

and, after the final $(\pi/2)_x^S$ pulse,

$$\sigma_4(\varphi = \pm\pi/2) = +I_{ky} \cos \pi J_{km}\tau \mp 2I_{kz}S_{my} \sin \pi J_{km}\tau + \frac{\gamma_S}{\gamma_I} S_{my}. \qquad (4.5.44)$$

The last term in this equation stems from 'native' S-polarization, and can be eliminated by subtracting two experiments with alternating phases, leading to the observable terms

$$\sigma^{\text{obs}} = \frac{1}{2}\left[\sigma_4\left(\varphi = +\frac{\pi}{2}\right) - \sigma_4\left(\varphi = -\frac{\pi}{2}\right)\right]$$
$$= -2I_{kz}S_{my} \sin \pi J_{km}\tau. \qquad (4.5.45)$$

This term represents antiphase single-quantum coherence of the dilute spins. The Fourier transform of the induced signal leads to two components at $\Omega_m \pm \pi J_{km}$ with opposite phases and equal amplitudes. If we neglect relaxation and couplings to further nuclei, it is clear that for $\tau = (2J_{km})^{-1}$ the amplitude in eqn (4.5.45) is enhanced by a factor γ_I/γ_S with respect to the native S_{my} signal which appears in eqn (4.5.44). This sensitivity advantage is further enhanced in systems where $T_1^I < T_1^S$, since the experiment can be repeated at intervals of the order of T_1^I. An experimental example is shown in Fig. 4.5.3, where the enhancement achieved with a transfer from ^{1}H to ^{15}N ($\gamma_I/\gamma_S \simeq 10$) is a factor 17.

It can easily be shown for a system with two protons I_k and I_l that the

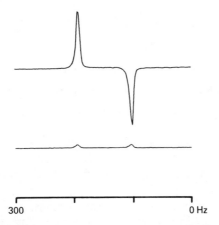

FIG. 4.5.3. Nitrogen-15 spectra of the NH group in N-acetyl valine: Below: normal spectrum obtained with direct ^{15}N observation without proton irradiation. Top: enhanced antiphase doublet obtained with the INEPT sequence of Fig. 4.5.2(a). The signals are enhanced by a factor 17; both experiments were optimized and involve signal-averaging over the same total time.

observable S-magnetization in the difference experiment is

$$\sigma^{obs} = -2I_{kz}S_{my} \sin \pi J_{km}\tau \cos \pi J_{kl}\tau$$
$$-2I_{lz}S_{my} \sin \pi J_{lm}\tau \cos \pi J_{kl}\tau. \quad (4.5.46)$$

Although the total signal is enhanced by transfer from *both* protons, the homonuclear coupling J_{kl} may lead to an attenuation.

If the two nuclei I_k and I_l are magnetically equivalent, the situation may be described by setting $J_{kl} = 0$ and $J_{km} = J_{lm} = J$ in eqn (4.5.46). In this manner, we obtain for an A_2X system

$$\sigma^{obs}(A_2X) = -2(I_{1z} + I_{2z})S_{my} \sin \pi J\tau. \quad (4.5.47)$$

In similar fashion, an A_3X system is treated in effect as an AA'A"X system with vanishing homonuclear couplings, leading to

$$\sigma^{obs}(A_3X) = -2(I_{1z} + I_{2z} + I_{3z})S_{my} \sin \pi J\tau. \quad (4.5.48)$$

The observable antiphase magnetization $-2I_{kz}S_{my}$ in eqns (4.5.45)–(4.5.48) may be partly converted into in-phase S_{mx}-magnetization by delaying the acquisition of the free induction decay by a period τ' as shown in Fig. 4.5.2(b). In this 'refocused INEPT' experiment (4.151), the free induction decay of the dilute S nuclei can be observed while the abundant I nuclei are decoupled. In this case, only in-phase S-magnetization can be observed

$$\sigma_5^{\text{obs}}(\text{AX}) = -S_{mx} \sin \pi J\tau \sin \pi J\tau',$$
$$\sigma_5^{\text{obs}}(\text{A}_2\text{X}) = -2S_{mx} \sin \pi J\tau \sin \pi J\tau' \cos \pi J\tau',$$
$$\sigma_5^{\text{obs}}(\text{A}_3\text{X}) = -3S_{mx} \sin \pi J\tau \sin \pi J\tau' \cos^2 \pi J\tau'. \qquad (4.5.49)$$

The functional dependence of the amplitudes on τ' may be used to identify A_nX groups with $n = 1, 2, 3$, although experiments involving multiple-quantum coherence (see § 4.5.6) appear more suitable for this purpose.

If the *coupled* spectrum of the dilute S spins is to be investigated, additional antiphase terms must be considered. For systems containing doublets, triplets, and quartets, it is not possible to select an interval τ' such that only in-phase S_{mx}-magnetization remains. In the INEPT$^+$ experiment (4.164) shown in Fig. 4.5.2(c), an additional $\pi/2$ 'purging' pulse is applied to the I spins at the beginning of the acquisition period, which converts all antiphase terms of the type $2I_{1z}S_{my}$, $4I_{1z}I_{2z}S_{mx}$, and $8I_{1z}I_{2z}I_{3z}S_{my}$ into non-observable multiple-quantum coherence, leaving only in-phase S_{mx}-magnetization with undistorted amplitude ratios $1:2:1$ and $1:4:4:1$ for triplets and quartets. The success of this procedure is illustrated in Fig. 4.5.4.

Although the more subtle aspects are best described in terms of the operator treatment given above, the basic magnetization transfer step can be readily understood in terms of semi-classical vector models. The state of the system at the end of the τ period, which is formally described by eqn (4.5.42), takes a simple form if $\tau = (2J_{km})^{-1}$ and if population differences across the transitions of the dilute S-spin are neglected

$$\sigma_2 = -2I_{kx}S_{mz}. \qquad (4.5.50)$$

In semi-classical terms, the product $2I_{kx}S_{mz}$ corresponds to two proton magnetization vectors pointing in opposite directions along the $\pm x$-axes of the rotating frame. The $(\pi/2)_y$-pulse rotates the vector which is along the $-x$-axis back towards the z-axis, while the vector along the $+x$-axis is rotated towards the $-z$-axis. Thus the net effect of the last $(\pi/2)^I$-pulse is equivalent to that obtained by a selective inversion of one of the doublet components of the abundant I-spins. The resulting population distribution, formally described by the term $\pm 2I_{kz}S_{mz}$ (longitudinal scalar two-spin order, with a sign depending on the phase $\varphi = \pm \pi/2$ of the last proton pulse) is superimposed on to the 'native' Zeeman polarization across the S-transitions, described by the term $-(\gamma_S/\gamma_I)S_{mz}$ in eqn (4.5.43). For a proton/carbon-13 pair $(\gamma_S/\gamma_I \simeq \frac{1}{4})$, the relative populations

4.5 HETERONUCLEAR POLARIZATION TRANSFER

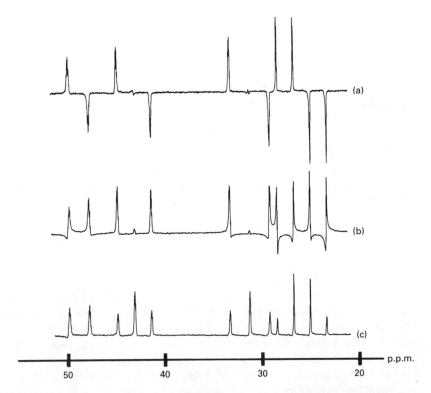

FIG. 4.5.4. Carbon-13 spectra of 1,3-dibromobutane. (a) Antiphase multiplets obtained with the basic INEPT sequence of Fig. 4.5.2(a). (b) Partially rephased multiplets obtained with the refocused INEPT sequence of Fig. 4.5.2(b) without proton decoupling in the detection period. Note the admixture of absorption and dispersion for the quartet centred at 26 p.p.m., and the missing central component of the triplet centred at 43 p.p.m. (c) Pure absorption spectrum with normal binomial multiplet intensities obtained with the INEPT$^+$ sequence of Fig. 4.5.2(c). (Reproduced from Ref. 4.164.)

just before the last $(\pi/2)^S$ pulse in the INEPT experiment are

$$P_{\alpha\alpha} : P_{\alpha\beta} : P_{\beta\alpha} : P_{\beta\beta} = 3 : -3 : -5 : +5 \quad \text{for} \quad \varphi = +\pi/2,$$
$$= -5 : +5 : +3 : -3 \quad \text{for} \quad \varphi = -\pi/2. \quad (4.5.51)$$

The population differences across the carbon-13 transitions are therefore

$$(P_{\alpha\alpha} - P_{\alpha\beta}) = +6 \quad \text{for} \quad \varphi = +\pi/2,$$
$$= -10 \quad \text{for} \quad \varphi = -\pi/2,$$
$$(P_{\beta\alpha} - P_{\beta\beta}) = -10 \quad \text{for} \quad \varphi = +\pi/2,$$
$$= +6 \quad \text{for} \quad \varphi = -\pi/2. \quad (4.5.52)$$

By taking the difference between the two experiments, weighted by $\frac{1}{2}$, one obtains the average population differences across the carbon-13

transitions
$$(P_{\alpha\alpha} - P_{\alpha\beta}) = 8,$$
$$(P_{\beta\alpha} - P_{\beta\beta}) = -8. \qquad (4.5.53)$$

This distribution is quite different from the Boltzmann equilibrium, which may be written

$$P_{\alpha\alpha} : P_{\alpha\beta} : P_{\beta\alpha} : P_{\beta\beta} = +5 : +3 : -3 : -5 \qquad (4.5.54)$$

where the population differences across the carbon transitions are $P_{\alpha\alpha} - P_{\alpha\beta} = P_{\beta\alpha} - P_{\beta\beta} = +2$. Thus the net effect of selective population inversion achieved in the INEPT experiment amounts to a fourfold enhancement of the carbon-13 polarization.

4.5.6. Editing procedures based on polarization transfer

The characteristic τ'-dependence of polarization transfer in CH, CH$_2$, and CH$_3$ fragments of eqn (4.5.49) can be exploited for the identification of AX$_n$ groups. With the selection $\tau' = (2J)^{-1}$, only AX groups produce a non-vanishing response, while for $\tau' = 3(4J)^{-1}$, the signals of AX$_2$ and AX$_3$ systems appear with opposite phases. By suitable linear combinations of experiments with different τ' values, it is possible to separate the three classes of signals.

A closer inspection of eqn (4.5.49) reveals, however, that the separation is feasible only if all AX$_n$ groups have the same coupling constant. In practice, the variation of the magnitude of one-bond J_{CH} constants leads to imperfect separation. An improved procedure has been proposed under the acronym DEPT (Distortionless Enhancement by Polarization Transfer) (4.162, 4.163). This method allows one to achieve a more reliable differentiation of CH, CH$_2$, and CH$_3$ fragments, based on the dependence of the signals on an r.f. rotation angle β rather than the dependence on the duration of an interval of free precession. While the INEPT experiment can be adequately described with semi-classical magnetization vectors, the DEPT method involves multiple-quantum coherence and is therefore better described in terms of density operators.

The basic pulse sequence for DEPT is shown in Fig. 4.5.5(a). In contrast to INEPT, the $(\pi/2)^S$ pulse *precedes* the last $(\beta)^I$ pulse, which, as we shall see, implies that heteronuclear multiple-quantum coherence is involved in the experiment.

Consider the effect of the sequence on the three systems CH, CH$_2$, and CH$_3$. For the sake of clarity, we shall assume that $\tau = (2J_{\text{CH}})^{-1}$, and obtain the density operator terms in Table 4.5.1. Note that all three systems lead to the *same* heteronuclear two-spin coherence $2I_{kx}S_{my}$ after the first $(\pi/2)^S$ pulse. In the two-spin system, this coherence does not

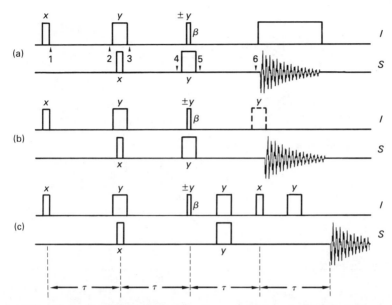

FIG. 4.5.5. (a) Basic sequence for polarization transfer by r.f. pulses in heteronuclear systems (DEPT), as discussed in text. Apart from the pulse with variable r.f. rotation angle β, the $\pi/2$- and π-pulses are represented as in Fig. 4.5.2 by the width of the pulses. (b) Pulse scheme for DEPT$^+$ with a $(\pi)_y$-pulse inserted at the beginning of the detection period on alternate experiments (dashed pulse). (c) Sequence for DEPT^{++} as described in text.

evolve under J_{CH} in the second τ period, while in three- and four-spin systems, the $2I_{kx}S_{my}$ term is transformed into terms that are in antiphase with respect to the additional protons. Since terms with transverse I_{kx} and I_{ky} components are not observable in the S-spectrum, it is clear that the functional dependence of the signal on the r.f. rotation angle β allows one to separate the signals from CH, CH$_2$, and CH$_3$ groups by suitable linear combinations of spectra obtained with different values of β (typically $\beta_1 = \pi/4$, $\beta_2 = \pi/2$, and $\beta_3 = 3\pi/4$).

Table 4.5.1.

Density operator terms arising in the DEPT experiment of Fig. 4.5.5(a) in CH, CH$_2$, and CH$_3$ systems with $\tau = (2\,J_{CH})^{-1}$. The subscripts of σ_i refer to the points on the time axis indicated in the figure.

	σ_2	σ_3	σ_4	$\sigma_5^{(obs)}$
CH	$2I_{kx}S_{mz}$	$2I_{kx}S_{my}$	$2I_{kx}S_{my}$	$\mp 2I_{kz}S_{my}\sin\beta$
CH$_2$	$2I_{kx}S_{mz}$	$2I_{kx}S_{my}$	$-4I_{kx}I_{k'z}S_{mx}$	$\mp 4I_{kz}I_{k'z}S_{mx}\sin\beta\cos\beta$
CH$_3$	$2I_{kx}S_{mz}$	$2I_{kx}S_{my}$	$-8I_{kx}I_{k'z}I_{k''z}S_{my}$	$\pm 8I_{kz}I_{k'z}I_{k''z}S_{my}\sin\beta\cos^2\beta$

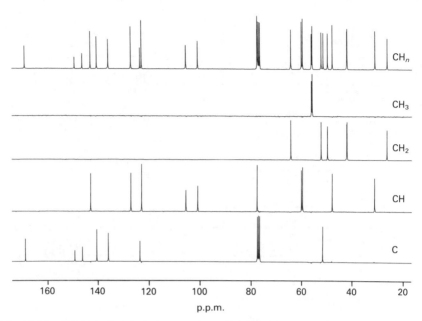

FIG. 4.5.6. Editing of proton-decoupled carbon-13 spectra of a mixture of brucine and 2-bromothiazole, which has a spread of J-couplings from ca. 125 to 192 Hz. Note the clean separation of the signals associated with CH_n groups and the absence of cross-talk. (Reproduced from Ref. 4.161.)

The basic idea of 'editing' carbon-13 spectra by separating signals belonging to CH_n groups with $n = 0$, 1, 2, and 3 has been refined, notably to allow for variations of the magnitude of the J-coupling. Figure 4.5.6 shows an example of editing of a proton-decoupled carbon-13 spectrum with the SEMUT-GL sequence (4.161). This experiment starts with ^{13}C magnetization and allows the observation of *all* carbons including the quarternary ones. The reader is referred to the original literature for a discussion of the relative merits of various sequences (4.161).

With the DEPT sequence, some precautions must be taken if one wishes to obtain *proton-coupled* edited spectra, since the rephasing of the $2I_{kz}S_{my}$ terms in the third τ-interval is dependent on J and on the number of equivalent protons. In the DEPT$^+$ sequence of Fig. 4.5.5(b), a $(\pi)_y^I$ pulse is inserted just before acquisition on every second transient to remove antiphase terms. The DEPT^{++} method of Fig. 4.5.5(c) has the additional advantage that normal binomial multiplet amplitudes are obtained, as shown in Fig. 4.5.7. This technique uses an I-spin refocusing pulse before applying a $(\pi/2)_x$ purging pulse to the I spins. It warrants that all I-spin operators contained in heteronuclear multiple-quantum coherence are in x-phase and are not affected by the purging pulse (4.164).

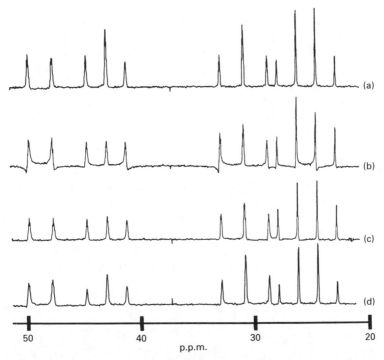

FIG. 4.5.7. Proton-coupled carbon-13 spectra of 1,3-dibromobutane obtained with polarization transfer. (a) Basic DEPT sequence of Fig. 4.5.5(a) with $\tau \simeq (2J)^{-1}$ (matched). (b) DEPT spectrum with $\tau \simeq (4J)^{-1}$ (mismatched). Note the phase and amplitude distortions. (c) DEPT$^+$ sequence of Fig. 4.5.5(b) with $\tau \simeq (4J)^{-1}$ (mismatched). The spectrum appears in pure absorption, but the multiplets do not have a binomial amplitude distribution. (d) DEPT^{++} sequence of Fig. 4.5.5(c) with $\tau \simeq (4J)^{-1}$ (mismatched): the amplitudes appear without distortion. (Reproduced from Ref. 4.164.)

4.6. Investigation of dynamic processes, relaxation, and chemical exchange

Since Fourier spectroscopy is a time-domain method, it provides a direct means of monitoring the recovery towards equilibrium after a perturbation, and is therefore particularly suited for studying dynamic processes such as relaxation and chemical exchange.

In many systems, the recovery is multi-exponential and a detailed analysis is far from trivial for chemical exchange in networks with more than two sites, and for spin–lattice relaxation in multilevel systems. In such cases, two-dimensional exchange studies (Chapter 9) often allow one to determine the rates k_{AB} of chemical exchange processes and the relaxational transition probabilities $W_{\alpha\beta} = R_{\alpha\alpha\beta\beta}$ between all pairs of eigenstates $|\alpha\rangle$ and $|\beta\rangle$ (see § 2.3.2).

Many one-dimensional experiments can be carried out with selective pulses. Provided the spectra lend themselves to selective perturbations, such experiments can yield similar information as provided by the two-dimensional methods described in Chapter 9. Thus one-dimensional selective polarization transfer allows one to study complex exchange networks and to determine the elements of the relaxation matrix. Several reviews (4.188–4.191) about experimental methods for studying dynamic systems give a more detailed treatment than the following brief survey.

4.6.1. Longitudinal relaxation

In a typical T_1 measurement, the recovery of the z-magnetization after an initial perturbation is measured as a function of time. The perturbation may be selective or non-selective. For a full characterization of the relaxation matrix, it is usually necessary to combine numerous experiments with different selective perturbations (4.188). Some of the most common experimental schemes for T_1-measurements are shown in Fig. 4.6.1.

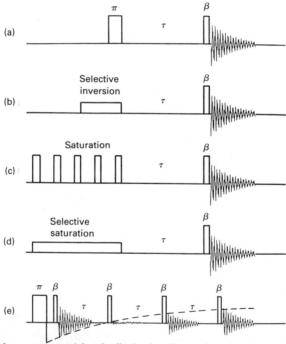

FIG. 4.6.1. Measurements of longitudinal relaxation and exchange processes. (a) Non-selective inversion recovery; (b) selective inversion recovery; (c) non-selective saturation recovery; (d) selective saturation recovery; (e) single-scan saturation recovery. Scheme (b) is closely related to two-dimensional exchange spectroscopy (Chapter 9) and may be used for transient Overhauser measurements and for selective polarization transfer in exchange networks. Scheme (d) may be used for saturation transfer and, if $\tau = 0$, for truncated driven Overhauser studies (TOE).

4.6.1.1. Inversion recovery methods

The maximum excursion from equilibrium is obtained by inverting the z-magnetization with a selective or non-selective π-pulse (4.192–4.194). The recovery towards equilibrium is sampled after a delay τ by a 'monitoring pulse' with rotation angle β, which creates transverse magnetization. For systems with isolated spins, one obtains

$$M_x(\tau) = \{M_z(0_+)\exp(-\tau/T_1) + M_0[1 - \exp(-\tau/T_1)]\}\sin\beta. \quad (4.6.1)$$

Three parameters have to be determined for an isolated spin: the equilibrium magnetization M_0, the initial magnetization after the inversion pulse $M_z(0_+)$, and the longitudinal relaxation time T_1. In the case that an ideal π-pulse has been applied to a system fully in equilibrium, $M_z(0_+) = -M_0$ and eqn (4.6.1) reduces to

$$M_x(\tau) = M_0[1 - 2\exp(-\tau/T_1)]\sin\beta. \quad (4.6.2)$$

The relaxation time T_1 can be determined by a two- or three-parameter fit.

In systems with isolated spins for which the above equations apply, the rotation angle of the monitoring pulse in Fig. 4.6.1(a) can be set to $\beta = \pi/2$, but in coupled systems, the mixing effect of such a pulse entails a loss of information (4.131). As discussed in §4.4.3, a small angle $0 < \beta \ll \pi/2$ leads to line intensities that are proportional to the population differences across the corresponding transitions, and therefore allows one to monitor relaxation processes in multilevel systems.

By using monitoring pulses with small rotation angles β, it is possible to measure the entire T_1-recovery in a single experiment. This can be achieved with the 'single-scan method' (4.195, 4.196) shown in Fig. 4.6.1(e). If $M_z(0_+)$ is the longitudinal magnetization after the π-pulse, one obtains the recovery

$$M_z(n\tau) - M_\infty = (M_z(0_+) - M_0)\exp\{-n\tau/T_1^*\} \quad (4.6.3)$$

where $n\tau$ is the interval between the π-pulse and the $(n+1)$th monitoring pulse, and T_1^* is an *apparent* relaxation time

$$\frac{1}{T_1^*} = \frac{1}{T_1} - \frac{\ln\cos\beta}{\tau}, \quad (4.6.4)$$

while M_∞ is the steady-state magnetization reached after a large number of β-pulses (see eqn (4.2.34))

$$M_\infty = M_0 \frac{1 - \exp(-\tau/T_1)}{1 - \exp(-\tau/T_1)\cos\beta}. \quad (4.6.5)$$

If β approaches $\pi/2$, a steady state is reached too rapidly, while small angles give poor sensitivity. A reasonable compromise is $\beta \simeq 30°$.

The accuracy of the inversion of the longitudinal magnetization can be improved by using composite pulses (§ 4.2.7). The sequence of eqn (4.2.55) has the advantage of compensating both for r.f. inhomogeneity and for resonance offset effects.

In practice, the inversion pulse (whether composite or not) is likely to excite some transverse magnetization, which may interfere with the signal acquired after the monitoring pulse if $\tau < T_2$. In coupled spin systems, the first pulse can also excite some undesirable multiple-quantum coherence, particularly if the system is not allowed to come to complete equilibrium before applying the pulse. The resulting artefacts can be eliminated by phase-cycling (4.197, 4.198). Single-quantum interference, which in the terminology of § 6.3 corresponds to coherence transfer pathways $p = 0 \rightarrow \pm 1 \rightarrow -1$, can be eliminated by alternating the phase of the monitoring pulse together with the receiver phase, while double-quantum interference ($p = 0 \rightarrow \pm 2 \rightarrow -1$) can be suppressed with a cycle of $N \geq 3$ steps. A four-step cycle can eliminate interference associated with $p = \pm 1$, ± 2, and ± 3 in the τ-interval (4.198).

The interference of transverse magnetization and multiple quantum coherence can also be cancelled with field gradient pulses as discussed in § 4.2.6.1. It should be remembered however that the loss of phase coherence produced by an inhomogeneous field is reversible, unless sufficient time is allowed for translational diffusion through field gradients.

Systematic errors may occur if the interval T between subsequent experiments is insufficient for complete recovery (4.199). Provided the transverse magnetization is allowed to decay or eliminated between subsequent experiments, it is possible to obtain accurate T_1-values even if the interval T is reduced well below the condition for negligible interference, $T > 5T_1$. By using accurate $\pi/2$ monitoring pulses it is possible to obtain a perfectly saturated state which recovers in each waiting period T to the same extent. This approach has been called 'fast inversion recovery' (4.200–4.202).

4.6.1.2. Saturation recovery methods

The interference of subsequent scans can be avoided altogether with the saturation recovery method of Fig. 4.6.1(c) (4.203). In the simplest case, the magnetization is brought into the transverse plane by an accurate $\pi/2$-pulse or by a composite pulse of the type described in § 4.2.7.1. The transverse component is destroyed by a field gradient pulse or cancelled by phase alternation. Alternatively (and more reliably), a burst of non-selective pulses can be used to saturate the spins. The recovery is analogous to the inversion recovery of eqn (4.6.2), except for a loss of a factor two in dynamic range

$$M_x(\tau) = M_0[1 - \exp\{-\tau/T_1\}]\sin\beta. \quad (4.6.6)$$

The saturation recovery method is faster, since there is no need for a long delay T between experiments, but inaccuracies may occur due to the difficulty of achieving complete saturation, particularly in coupled spin systems and for wide spectral ranges. Saturation may be ineffective if spin-locking effects occur.

4.6.1.3. Progressive saturation

It is possible to measure the relaxation time T_1 with a simple sequence of equidistant β-pulses, either by studying the saturation as a function of the pulse interval (the so-called 'progressive saturation' method (4.204)) or as a function of β ('variable nutation angle' method (4.205–4.207)). These methods are only reliable if transverse interference is completely suppressed by diffusion through (pulsed) field gradients.

4.6.1.4. Selective perturbations

As illustrated in Fig. 4.6.1, both inversion and saturation recovery experiments may employ selective or semi-selective r.f. pulses, such as to affect single lines or selected multiplets respectively (4.188, 4.208). This type of experiment is useful for measuring both slow chemical exchange and cross-relaxation processes.

In the context of chemically exchanging systems, the sequence of Fig. 4.6.1(b) is often referred to as Hoffman–Forsén method (4.209–4.211). Consider for simplicity a system in chemical and magnetic equilibrium with two sites A and B. By selective inversion of the A spins, one may prepare a non-equilibrium state at the beginning of the recovery period with $M_z^A(\tau = 0) = -M_0^A$ and $M_z^B(\tau = 0) = M_0^B$. The 'labelling' (i.e. inversion) of site A is partly transferred to site B through exchange and cross-relaxation, as illustrated in Fig. 4.6.2 (4.212). This experiment is closely related to two-dimensional exchange spectroscopy, where the labelling is achieved by two non-selective $\pi/2$-pulses separated by an evolution period t_1, instead of a selective π-pulse. Since the dynamics of the exchanging magnetization obey the same principles in both one- and two-dimensional experiments, the reader is referred to § 9.3 for a discussion of the interplay of exchange, cross-relaxation, and spin–lattice relaxation.

In the context of the measurement of nuclear Overhauser effects, the sequence of Fig. 4.6.1(b) is usually known as 'transient NOE method' (4.213, 4.214). Like its two-dimensional counterpart, the NOESY experiment discussed in § 9.3, the transient one-dimensional method is most useful for studying large molecules with long correlation times $\tau_c > \omega_0^{-1}$

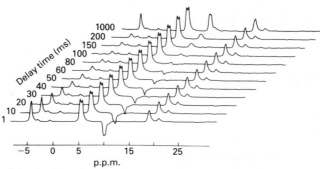

FIG. 4.6.2. Selective inversion recovery (selective polarization transfer) of phosphorus-31 magnetization in the adenylate kinase reaction ATP + AMP ⇌ 2 ADP (adenosine tri-, mono-, and diphosphates) in chemical equilibrium. The inverted signal at 10 p.p.m. corresponds to a superposition of the α-phosphates of ADP and ATP. The reaction leads to a loss of signal amplitude of the phosphate resonance of AMP at −4 p.p.m. for intermediate delay times. (Reproduced from Ref. 4.212.)

where spin–lattice relaxation is slow and does not compete significantly with cross-relaxation.

The sequence of Fig. 4.6.1(d) can be used in two different ways for cross-relaxation measurements: if the recovery is monitored as a function of the length of the selective r.f. pulse (with $\tau = 0$), the experiment is referred to as 'truncated driven NOE method' (TOE) (4.215, 4.216). The selective saturation pulse should be long enough so that transient oscillations are damped by r.f. inhomogeneity, but short on the time-scale of cross-relaxation. This approach is also suitable for studying large molecules.

If the selective irradiation in Fig. 4.6.1(d) extends over a period longer than the inverse relaxation rates, and if $\tau = 0$, the method is referred to as 'steady-state Overhauser measurement' (4.147). This approach is most suitable for studying small molecules in the extreme narrowing limit ($\tau_c \ll \omega_0^{-1}$). In this case, the competition of spin–lattice and cross-relaxation implies that cross-peaks in two-dimensional exchange spectra are extremely weak (see Chapter 9), and the steady-state one-dimensional method remains the most effective approach to Overhauser studies.

4.6.2. Transverse relaxation

In favourable cases, the transverse relaxation time T_2 can be obtained simply by measuring the full line-width at half height in units of Hz

$$\Delta \nu = \frac{\Delta \omega}{2\pi} = \frac{1}{\pi T_2^*} = \frac{1}{\pi} \left[\frac{1}{T_2} + \frac{1}{T_2^+} \right]. \qquad (4.6.7)$$

The effective decay rate $1/T_2^*$ is the sum of the natural relaxation rate

$1/T_2$ and the inhomogeneous broadening contribution $1/T_2^+$. If the latter is negligible or if it can be measured from a reference line that is known to have a negligible homogeneous line-width contribution $1/T_2$, the natural T_2 can be obtained immediately.

The inhomogeneous broadening $1/T_2^+$ arises because the magnetizations associated with different volume elements (so-called 'isochromats') experience different static magnetic fields $B_0(\mathbf{r})$. The resulting dephasing of the magnetization is reversible, and refocusing is possible with suitable pulse sequences (4.139, 4.217).

The spin–echo sequence of Fig. 4.6.3(a) with a single refocusing pulse

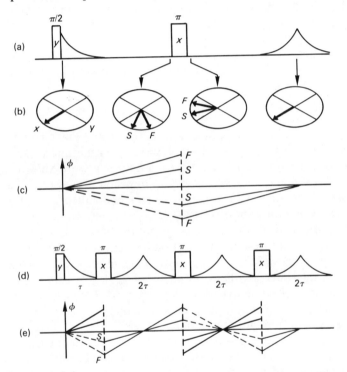

FIG. 4.6.3. (a) Spin–echo sequence with a single refocusing pulse, and with a phase shift between the initial $(\pi/2)_y$-pulse and the $(\pi)_x$-refocusing pulse. (b) Evolution of two magnetization vectors ('isochromats'), labelled **F** and **S** for 'fast' and 'slow' with precession frequencies that are slightly above or slightly below the average resonance offset because of the inhomogeneous static field. Initially both aligned along the x-axis, the two vectors diverge in the defocusing period, are brought by the $(\pi)_x$ pulse into a symmetrical position with respect to the x-axis, and converge in the refocusing period. (c) The phases of the same two isochromats as a function of time (solid lines; ϕ = excursion from the x- towards the y-axis). Note the discontinuity at the time of the π-pulse. In the toggling frame (where the precession frequency appears reversed in the defocusing period, as shown by dashed lines), the trajectories appear continuous. (d) Spin–echo sequence with multiple refocusing. (e) Phases of two isochromats in the course of multiple refocusing (solid lines). In the toggling frame, the precession frequency appears reversed in odd-numbered intervals (dashed lines).

of rotation angle $\beta = \pi$ is often referred to as 'Carr–Purcell method A', or sometimes as 'Hahn echo' (although the latter designation is often reserved for a sequence with $\beta = \pi/2$). Figure 4.6.3(b) shows the evolution of two typical magnetization components or isochromats in the xy-plane (labelled F and S for 'fast' and 'slow').

The diagram of Fig. 4.6.3(c) illustrates how the phases of the magnetization components ($\tan \varphi = M_y/M_x$) diverge in the first period, and converge after a π-rotation about the x-axis ($M_y(\tau^+) = -M_y(\tau^-)$), hence $\varphi(\tau^+) = -\varphi(\tau^-)$. The phenomenon can also be rationalized by assuming that the sense of precession (i.e. the sign of the Hamiltonian) is reversed in the first period. In this description, which corresponds to the 'toggling frame' (4.5, 4.6), the phase of the magnetization evolves as shown by dashed lines in Fig. 4.6.3.

The amplitude of the echoes obtained with a single refocusing pulse ($n = 1$) can be strongly attenuated by translational diffusion through field gradients g (4.139, 4.218)

$$M_x(2n\tau) = M_x(0)\exp(-2n\tau/T_2)\exp(-\tfrac{2}{3}D\gamma^2 g^2 n\tau^3). \qquad (4.6.8)$$

This effect can be largely removed with the 'Carr–Purcell method B' sequence of Fig. 4.6.3(d), which employs n refocusing pulses spaced by intervals 2τ. The phases of two typical isochromats are shown in Fig. 4.6.3(e). This sequence is less sensitive to diffusion processes: for a given total delay $t_{tot} = 2n\tau$ between the nth echo and initial excitation, the relevant factor in eqn (4.6.8) is $n\tau^3 = t_{tot}^3/(8n^2)$. The echo amplitudes are however attenuated if the refocusing pulses have non-ideal rotation angles $\beta \neq \pi$, or if off-resonance effects (tilted effective fields) occur. The $\pi/2$ phase shift between the initial pulse and the refocusing pulses (Fig. 4.6.3(d)) partly alleviates these problems (4.122), but composite pulses (4.108) are more effective, particularly in coupled spin systems where the r.f. phase shift fails to cancel pulse imperfections.

In systems with homonuclear scalar or dipolar couplings, non-selective refocusing pulses with $\beta = \pi$ do not affect the (bilinear) coupling Hamiltonian, and the echoes are therefore modulated (4.139, 4.189). The Fourier transform of the echo envelope (i.e. of the signals $S(2n\tau)$ with $n = 0, 1, 2, \ldots$) leads to a 'spin–echo spectrum' or 'J-spectrum' (4.219, 4.220) which reveals the multiplet structure without chemical shifts and with line-widths determined by $1/T_2$ instead of $1/T_2^*$. If the pulse repetition rate $(2\tau)^{-1}$ is comparable to or larger than the chemical shift difference, the multiplets in the spin–echo spectrum become distorted and, for very fast pulsing, the multiplet splittings disappear completely for scalar J coupling (4.221–4.224) but remain for dipolar coupling.

In experiments with two or more refocusing pulses with $\beta \neq \pi$, additional phenomena can be observed which are known as 'stimulated

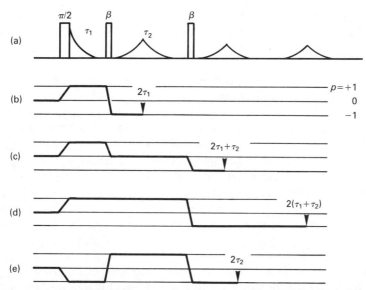

FIG. 4.6.4. Echoes and stimulated echoes. (a) Sequence with two refocusing pulses with $\beta \neq \pi$. (b) Primary echo at $t = 2\tau_1$, analogous to the Carr–Purcell echo of Fig. 4.6.3(b), which can be described by the coherence pathway $p = 0 \to +1 \to -1$ (see § 6.3). (c) Stimulated echo at $t = 2\tau_1 + \tau_2$ where the information is stored in longitudinal magnetization in the τ_2-interval (pathway $p = 0 \to +1 \to 0 \to -1$). (d) Primary echo at $t = 2(\tau_1 + \tau_2)$, arising from magnetization that remains unaffected by the second pulse and is refocused by the third pulse (pathway $p = 0 \to +1 \to +1 \to -1$). (e) Secondary echo at $t = 2\tau_2$, arising from magnetization that experiences both β-pulses as refocusing pulses (pathway $p = 0 \to -1 \to +1 \to -1$).

echoes' (4.217). The coherence transfer pathways in Fig. 4.6.4 provide a simple rationalization: in addition to the conventional echo at $t = 2\tau$, which arises because the β-pulse partly reverses the sense of precession in the transverse plane (pathway $p = 0 \to +1 \to -1$, Fig. 4.6.4(b)), a fraction of the magnetization is converted into longitudinal magnetization ($p = 0$) in the τ_2-interval, with an amplitude that reflects the defocusing in τ_1. The third pulse can convert this into transverse magnetization (overall pathway $p = 0 \to +1 \to 0 \to -1$ in Fig. 4.6.4(c)) which leads to a stimulated echo at $t = 2\tau_1 + \tau_2$. As may be inferred from Fig. 4.6.4, various types of echoes occur when the sum of the intervals in which $p = +1$ quantum coherence evolves is equal to the sum of the $p = -1$ quantum intervals.

4.6.3. Chemical reactions and exchange processes

The NMR lineshape provides a very sensitive measure for slow dynamic processes, and can be analysed both to study chemical equilibria and transient reactions. Equilibrium systems can be studied either by slow-passage or by Fourier spectroscopy techniques, while fast transient reactions can only be investigated by pulse spectroscopy.

1. *Study of chemical equilibrium.* The theory of NMR lineshapes in the presence of equilibrium chemical exchange processes has been worked out in great detail for slow-passage spectroscopy and has been applied to almost all branches of chemistry (4.225–4.231). Most of the studies have been confined to low-power CW NMR techniques, measuring the linear response of the spin system. In addition, it has been shown that further information can be obtained from slow-passage saturation experiments performed on exchanging systems (4.230, 4.232–4.236).

Fortunately, most of the principles of NMR in exchanging systems developed for slow-passage spectroscopy apply equally well to Fourier spectroscopy. In particular, it was shown in § 4.4.2 that Fourier spectra of equilibrium systems are fully equivalent to low-power slow-passage spectra. On first sight, it may be astonishing that this equivalence holds even for systems with slow chemical processes, although the time-scale of the chemical reaction may be comparable to that of the free induction decay. The equivalence arises because the exchange superoperator $\hat{\Xi}$ in eqn (4.4.24) is explicitly time-independent provided the system is in chemical equilibrium.

The literature on NMR in equilibrium exchange systems is quite extensive, and we refer to previous treatises (4.225–4.231) for the analysis of exchange-broadened lineshapes.

2. *Study of transient chemical reactions.* In general, a relatively long observation time is required in NMR, and only slow transient reactions proceeding in a time-scale of milliseconds to minutes can be studied. For very slow reactions in the range of minutes, both slow-passage and Fourier techniques can be applied to sample the state of the system. However, to study faster reactions which evolve in the time range of seconds, the use of fast Fourier experiments is imperative to record snapshots of the momentary state.

If the reaction times are on the order of milliseconds, the reaction will proceed in the course of a single FID, and will affect the lineshape (4.237, 4.238). This occurs frequently in fast stopped-flow experiments as well as in pulsed CIDP studies where fast secondary reactions may take place.

We assume that immediately before the exciting r.f. pulse, the system has been prepared in a chemical non-equilibrium state. The system evolves chemically towards its equilibrium state in the course of the FID. This has the effect that a certain spin, precessing with a reactant

4.6 INVESTIGATION OF DYNAMIC PROCESSES

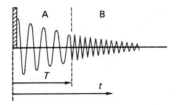

FIG. 4.6.5. Free induction decay of an isolated spin in a molecule undergoing a one-sided reaction $A \rightarrow B$ at time T. In the experiment, an ensemble of molecules with different reaction times T is observed. The finite lifetime of A leads to lifetime broadening while the delayed appearance of B creates phase anomalies. (Reproduced from Ref. 4.237.)

frequency at the beginning of the FID, will suddenly switch its precession to a product frequency (Fig. 4.6.5).

The main complication arises from the fact that the system is in a non-equilibrium state both in terms of chemical concentrations and in terms of the magnetization. Fortunately, chemical processes are independent of the magnetic state, and it is therefore possible to solve the chemical problem first and use this solution as a base to understand the magnetic evolution, as was discussed in § 2.4.

4.6.3.1. One-sided first-order reaction

Consider a one-sided chemical reaction proceeding with rate constant k_1 from state A to state B with negligible backward reaction,

$$A \xrightarrow{k_1} B. \qquad (4.6.9)$$

Equilibrium measurements would not provide any kinetic information in such a system, because the component A completely disappears. However, transient experiments can deliver the desired information.

We assume that the system is initially prepared in state A containing isolated magnetic nuclei with resonance frequency Ω_A. In the course of the FID, the reaction proceeds towards the state B with the resonance frequency Ω_B, as illustrated by Fig. 4.6.5.

The modified Bloch equations for the transverse magnetization components, eqn (2.4.19), take the simple form

$$\dot{M}_A^+ = [i\Omega_A - 1/T_{2A} - k_1]M_A^+,$$
$$\dot{M}_B^+ = [i\Omega_B - 1/T_{2B}]M_B^+ + k_1 M_A^+ \qquad (4.6.10)$$

written in terms of the complex magnetization $M_A^+ = M_{Ax} + iM_{Ay}$ and with the transverse relaxation times T_{2A} and T_{2B}. The magnetization evolves in the course of the free induction decay (4.237)

$$M_A^+(t) = M_A^+(0)\exp\{(i\Omega_A - 1/T_{2A} - k_1)t\},$$

$$M_B^+(t) = M_B^+(0)\exp\{(i\Omega_B - 1/T_{2B})t\} + M_A^+(0)\frac{k_1}{i(\Omega_B - \Omega_A) + \frac{1}{T_{2A}} - \frac{1}{T_{2B}} + k_1}$$

$$\times \left[-\exp\left\{\left(i\Omega_A - \frac{1}{T_{2A}} - k_1\right)t\right\} + \exp\left\{\left(i\Omega_B - \frac{1}{T_{2B}}\right)t\right\}\right]. \quad (4.6.11)$$

The real part of the Fourier spectrum $S(\omega)$ is

$$S(\omega) = \operatorname{Re} \mathscr{F}\{M_A^+(t) + M_B^+(t)\}$$

$$= \frac{B_0(\gamma\hbar)^2 N}{4kT}\Bigg[[A](0)\frac{1/T_{2A} + k_1}{(1/T_{2A} + k_1)^2 + (\omega - \Omega_A)^2}$$

$$+ [B](0)\frac{1/T_{2B}}{(1/T_{2B})^2 + (\omega - \Omega_B)^2}$$

$$+ k_1[A](0)\frac{1/T_{2A} - 1/T_{2B} + k_1}{(1/T_{2A} - 1/T_{2B} + k_1)^2 + (\Omega_A - \Omega_B)^2}$$

$$\times \left\{\frac{1/T_{2B}}{(1/T_{2B})^2 + (\omega - \Omega_B)^2} - \frac{1/T_{2A} + k_1}{(1/T_{2A} + k_1)^2 + (\omega - \Omega_A)^2}\right\}$$

$$+ k_1[A](0)\frac{\Omega_A - \Omega_B}{(1/T_{2A} - 1/T_{2B} + k_1)^2 + (\Omega_A - \Omega_B)^2}$$

$$\times \left\{\frac{\omega - \Omega_B}{(1/T_{2B})^2 + (\omega - \Omega_B)^2} - \frac{\omega - \Omega_A}{(1/T_{2A} + k_1)^2 + (\omega - \Omega_A)^2}\right\}\Bigg]. \quad (4.6.12)$$

The first term of eqn (4.6.12) (see Fig. 4.6.6a) represents the absorption line of the reactant, broadened by the finite lifetime of the molecule A. The second term vanishes for $[B](0) = 0$. The third term (Fig. 4.6.6(b)) describes an absorption contribution to both lines, emissive for the reactant and absorptive for the product. Broadening makes the recognition of the emissive reactant signal difficult. Finally, the fourth term in eqn (4.6.12) (Fig. 4.6.6(c)) is responsible for dispersive contributions of opposite sign centred at the reactant and product frequencies, respectively. Figure 4.6.6(d) represents the actual spectrum which arises from the summation of the three contributions in Fig. 4.6.6(a), (b), and (c).

A set of lineshapes for various reaction rate constants k_1, calculated from eqn (4.6.12), are shown in Fig. 4.6.7. The characteristic features of the lineshapes for one-sided chemical reactions are the broadening and abnormal phase of the reactant line and the pronounced dispersive character of the product line for slow and intermediate reactions. These

4.6 INVESTIGATION OF DYNAMIC PROCESSES

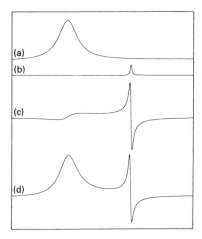

FIG. 4.6.6. Visualization of the three terms of eqn (4.6.12) contributing to the lineshape of a one-spin system involved in a one-sided chemical reaction (for [B](0) = 0). (a) Lifetime-broadened reactant line; (b) broad emissive contribution to reactant line and narrow absorptive contribution to product line with vanishing total integral; (c) dispersive contributions of opposite sign to reactant and product lines; (d) composite signal. (Reproduced from Ref. 4.237.)

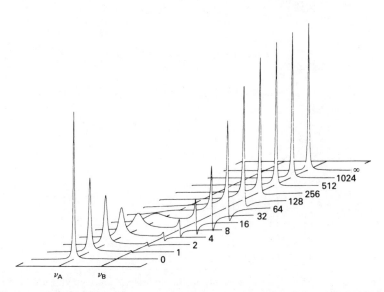

FIG. 4.6.7. Theoretical lineshapes resulting from a one-sided chemical reaction A → B for various reaction rate constants k_1 expressed in s^{-1}. The difference of the resonance frequencies is $|\omega_B - \omega_A|/2\pi = 10$ Hz. (Reproduced from Ref. 4.237.)

dispersive lineshapes have their origin in the delayed creation of the B magnetization in the course of the free induction decay. The line-width of the product line is not affected by the chemical reaction, since the backward reaction has been neglected.

The theoretical lineshapes of Fig. 4.6.7 were verified experimentally by preparing a chemical non-equilibrium state by rapid mixing in a stopped flow probe. In a mixture of methyl formate and sodium hydroxide in D_2O, a hydrolysis reaction occurs

$$HCOOCH_3 + OD^- \xrightarrow{k^{(2)}} HCOO^- + CH_3OD.$$
$$\text{(A)} \qquad \text{(X}_3\text{)} \qquad \text{(A')} \qquad \text{(X}_3'\text{)}$$

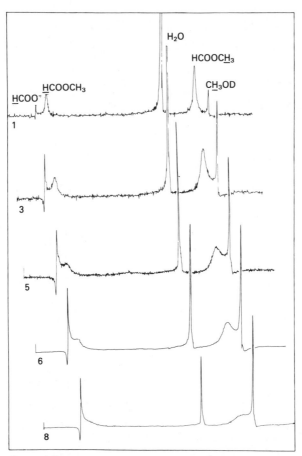

FIG. 4.6.8. Stopped-flow proton spectra of the hydrolysis of methyl formate with deuterated sodium hydroxide in D_2O for various reaction rates increasing from trace 1 to trace 8. Details about these experiments are given in Ref. 4.237.

An excess of NaOD was used such that the reaction is pseudo-first-order. The hydrolysis provides two examples of one-sided first-order reactions

$$A \xrightarrow{k^{(2)}[OD^-]} A' \quad \text{and} \quad X_3 \xrightarrow{k^{(2)}[OD^-]} X_3'.$$

Figure 4.6.8 shows a set of spectra obtained with increasing $[OD^-]$ concentration (and thus with increasing pseudo-first-order rate constant). Both the broadening of the reactant lines and the varying dispersive character of the product lines are clearly visible.

4.6.3.2. Two-sided first-order reaction

Systems with two-sided reactions can, in principle, be investigated in chemical equilibrium by traditional lineshape studies. In some cases however, *transient* two-sided reactions cannot be avoided during the observation of the FID. This situation is frequent in pulsed CIDNP experiments where secondary reactions occur.

In contrast to one-sided reactions, the resonance frequencies observed in the presence of two-sided reactions depend on the reaction rates, and product resonance lines will be broadened because of the backward reaction which limits the lifetime of the products.

To demonstrate these effects, consider a molecule with an isolated spin involved in a first-order chemical reaction

$$A \underset{k_2}{\overset{k_1}{\rightleftarrows}} B$$

in a non-equilibrium situation. The dynamic system of equations is given by eqn (2.4.20)

$$\frac{d}{dt}\begin{bmatrix} M_A^+ \\ M_B^+ \end{bmatrix} = -\begin{bmatrix} -i\Omega_A + 1/T_{2A} + k_1 & -k_2 \\ -k_1 & -i\Omega_B + 1/T_{2B} + k_2 \end{bmatrix} \cdot \begin{bmatrix} M_A^+ \\ M_B^+ \end{bmatrix}. \quad (4.6.13)$$

It is then straightforward to compute the absorption-mode spectrum (4.237)

$$S(\omega) = \mathrm{Re}\, \mathscr{F}\{\mathbf{EU}\, e^{\Lambda t}\mathbf{U}^{-1}\mathbf{M}^+(0)\}, \quad (4.6.14)$$

where $\mathbf{E}$ is a unit vector, $\mathbf{\Lambda}$ is the diagonalized dynamic matrix $\mathbf{L}^+$ with the eigenvalues

$$\Lambda_{11,22} = -\tfrac{1}{2}\{1/T_{2A} + 1/T_{2B} + k_1 + k_2 - i(\Omega_A + \Omega_B)\} \mp \tfrac{1}{2}q,$$
$$q = \{\Delta^2 + 4k_1k_2\}^{\tfrac{1}{2}},$$
$$\Delta = 1/T_{2A} - 1/T_{2B} + k_1 - k_2 - i(\Omega_A - \Omega_B), \quad (4.6.15)$$

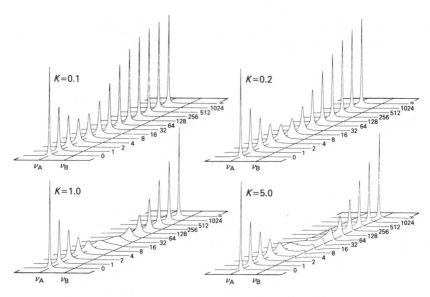

FIG. 4.6.9. Four sets of lineshapes resulting from a two-sided chemical reaction $A \rightleftarrows B$ for four different equilibrium constants $K = 0.1$, $K = 0.2$, $K = 1.0$, and $K = 5.0$ and for various forward rate constants k_1 (in s^{-1}) indicated on the right of each spectrum. The frequency difference $|\Omega_B - \Omega_A|/2\pi$ is 10 Hz. (Reproduced from Ref. 4.237.)

and the diagonalization matrix **U** is

$$\mathbf{U} = \mathbf{U}^{-1} = \{(q + \Delta)^2 + 4k_1 k_2\}^{-\frac{1}{2}} \begin{bmatrix} -(q + \Delta) & 2k_2 \\ 2k_1 & (q + \Delta) \end{bmatrix}. \quad (4.6.16)$$

The initial transverse magnetization components are proportional to the initial concentrations,

$$\mathbf{M}^+(0) = \begin{bmatrix} M_A^+(0) \\ M_B^+(0) \end{bmatrix} \propto \begin{bmatrix} [A](0) \\ [B](0) \end{bmatrix}. \quad (4.6.17)$$

Four sets of spectra calculated with eqn (4.6.14) are shown in Fig. 4.6.9 for different equilibrium constants $K = k_1/k_2$ and various rate constants k_1. The difference of the chemical shift frequencies $(\Omega_A - \Omega_B)/2\pi$ has been assumed to be 10 Hz. In all computations the initial concentration of B has been set equal to zero.

For an equilibrium constant $K = 0.1$, no resonance at ν_B is visible, irrespective of the reaction rate, but at intermediate reaction rates a significant broadening of the resonance line near ν_A occurs. For k_1, $k_2 \to \infty$, the resonance shifts from Ω_A to $(\Omega_A + K\Omega_B)/(1 + K)$. Qualitatively the same behavior is seen for $K = 0.2$. For $K = 1$, a weak dispersive B resonance line can be detected at low reaction rates, and at medium rates an asymmetric lineshape is visible. For $K = 5$, lineshapes qualitatively similar to those for a one-sided reaction (Fig. 4.6.7) are obtained. Figure 4.6.9 demonstrates that, with increasing equilibrium constant K, the reaction rate k_1 for maximum broadening shifts towards higher values. This can also be seen from eqn (4.6.15) as the broadening is, to a first approximation, given by $\frac{1}{2}(k_1 + k_2) = \frac{1}{2}k_1(1 + 1/K)$.

4.6.3.3. Transient chemical reactions with coupled spin systems

To illustrate the effects encountered with coupled spin systems involved in transient chemical reactions, consider a one-sided first-order reaction

$$A \xrightarrow{k_1} B.$$

According to § 2.4.3, it is most convenient to describe the evolution by the concentration-dependent density operators, $\sigma_A^\square = \sigma_A \cdot [A(t)]$. From the general relation eqn (2.4.43) we find for the coherence of order $p = -1$ the equations of motion in terms of the density operators $\sigma_A^{\square(-1)}$, $\sigma_B^{\square(-1)}$ which are responsible for the transverse magnetization

$$\dot\sigma_A^{\square(-1)} = \hat{L}_A^{(-1)}\sigma_A^{\square(-1)}, \qquad (4.6.18)$$

$$\dot\sigma_B^{\square(-1)} = \hat{L}_B^{(-1)}\sigma_B^{\square(-1)} + k_1 R_1 \sigma_A^{\square(-1)} R_1^{-1} \qquad (4.6.19)$$

with the superoperators

$$\hat{L}_A^{(-1)} = -i\hat{\mathcal{H}}_A - \hat{\Gamma}_A - k_1,$$
$$\hat{L}_B^{(-1)} = -i\hat{\mathcal{H}}_B - \hat{\Gamma}_B. \qquad (4.6.20)$$

Equation (4.6.18) can be integrated directly and inserted in eqn (4.6.19), which can be integrated in turn, with the result

$$\sigma_B^{\square(-1)}(t) = \exp\{\hat{L}_B^{(-1)}t\}\sigma_B^{\square(-1)}(0) + k_1 \int_0^t \exp\{\hat{L}_B^{(-1)}(t-t')\}$$
$$\times R_1[\exp\{\hat{L}_A^{(-1)}t'\}\sigma_A^{\square(-1)}(0)]R_1^{-1} \, dt'. \qquad (4.6.21)$$

For the evaluation of eqn (4.6.21), we assume that the relaxation superoperators $\hat{\Gamma}_A$ and $\hat{\Gamma}_B$ commute with the respective Liouville operators $\hat{\mathcal{H}}_A$ and $\hat{\mathcal{H}}_B$. In this case, each transition $|l\rangle \to |k\rangle$ has its own transverse relaxation rate $1/T_{2lk} = -R_{lk\,lk}$ (see eqn (2.3.26a)).

We assume that before application of a 90° pulse the system is in magnetic (but not in chemical) equilibrium. At the beginning of the free induction decay, immediately after the pulse, we have

$$\sigma_A^{\square(-1)}(0) = q_A F^{-(A)}, \qquad \sigma_B^{\square(-1)}(0) = q_B F^{-(B)} \qquad (4.6.22)$$

with

$$q_A = \frac{[A](0)}{\mathrm{tr}\{\mathbb{1}\}} \frac{\hbar \gamma B_0}{2kT}, \qquad q_B = \frac{[B](0)}{\mathrm{tr}\{\mathbb{1}\}} \frac{\hbar \gamma B_0}{2kT}. \qquad (4.6.23)$$

Expressed in the eigenbases of $\mathcal{H}_A$ and $\mathcal{H}_B$, we find the matrix elements of the density operators

$$\sigma_A^{\square(-1)}(t)_{mn} = \exp\{\Lambda_{mn}^{(A)}t\}q_A F_{mn}^{-(A)} \quad (4.6.24)$$

and

$$\sigma_B^{\square(-1)}(t)_{mn} = \exp\{\Lambda_{mn}^{(B)}t\}q_B F_{mn}^{-(B)}$$
$$- q_A k_1 \sum_{l,k} D_{ml} D_{nk}^* \frac{\exp\{\Lambda_{mn}^{(B)}t\} - \exp\{\Lambda_{lk}^{(A)}t\}}{-\Lambda_{mn}^{(B)} + \Lambda_{lk}^{(A)}} F_{lk}^{-(A)} \quad (4.6.25)$$

with the eigenvalues of the dynamic matrices $\mathbf{L}_A^{(-1)}$ and $\mathbf{L}_B^{(-1)}$, respectively,

$$\Lambda_{lk}^{(A)} = -i\omega_{lk}^{(A)} - \frac{1}{T_{2lk}^{(A)}} - k_1, \quad \Lambda_{mn}^{(B)} = -i\omega_{mn}^{(B)} - \frac{1}{T_{2mn}^{(B)}}. \quad (4.6.26)$$

The transformation from the eigenbasis $\psi^{(A)}$ of $\mathcal{H}_A$ to that of $\mathcal{H}_B$ is described by the matrix $\mathbf{D}$

$$(\psi^{(B)}) = \mathbf{D}(\psi^{(A)}). \quad (4.6.27)$$

The rearrangement operator R_1 in eqn (4.6.21) merely effects a permutation of the basis functions of molecule A and has been incorporated in $\mathbf{D}$.

The observable magnetization can be calculated from eqns (4.6.24) and (4.6.25)

$$\langle M^+\rangle(t)/(N_A\gamma\hbar) = \text{tr}\{F^{+(A)}\sigma_A^{\square(-1)}(t) + F^{+(B)}\sigma_B^{\square(-1)}(t)\}. \quad (4.6.28)$$

After Fourier transformation, the spectrum is obtained in the form

$$\langle M^+\rangle(\omega)/(N_A\gamma\hbar) = q_A \sum_{l,k} \frac{(F_{lk}^{-(A)})^2}{i(\omega_{lk}^{(A)} + \omega) + \frac{1}{T_{2lk}^{(A)}} + k_1}$$
$$+ q_B \sum_{m,n} \frac{(F_{mn}^{-(B)})^2}{i(\omega_{mn}^{(B)} + \omega) + \frac{1}{T_{2mn}^{(B)}}}$$
$$- q_A k_1 \sum_{m,n} \sum_{l,k} \frac{F_{mn}^{-(B)} F_{lk}^{-(A)} D_{ml} D_{nk}^*}{i(\omega_{mn}^{(B)} - \omega_{lk}^{(A)}) + \frac{1}{T_{2mn}^{(B)}} - \frac{1}{T_{2lk}^{(A)}} - k_1}$$
$$\times \left[\frac{1}{i(\omega_{mn}^{(B)} + \omega) + \frac{1}{T_{2mn}^{(B)}}} - \frac{1}{i(\omega_{lk}^{(A)} + \omega) + \frac{1}{T_{2lk}^{(A)}} + k_1}\right]. \quad (4.6.29)$$

This equation demonstrates the main features of transient Fourier spectra of coupled systems with one-sided chemical reactions and represents a generalization of eqn (4.6.12) which describes systems without couplings. The spectra involve only the natural transition frequencies $\omega_{lk}^{(A)}$ and $\omega_{mn}^{(B)}$ of molecules A and B. The first term of eqn (4.6.29) describes lifetime-broadened absorption lines of the reactant molecule A. The second term, not affected by exchange, results from product molecules B already present at $t = 0$. It is also absorptive. The third term is proportional to the reaction rate k_1 and describes the coherence transfer from molecule A to molecule B during the FID. It causes mixed phase signals at the A and B resonance frequencies with opposite signs such that the integrated intensity of this term vanishes. For slow reactions, $k_1 \ll |\omega_{mn}^B - \omega_{kl}^A|$, this term gives purely dispersive contributions. At intermediate reaction rates, the phase may vary from line to line. For fast reactions, $k_1 \gg |\omega_{mn}^B - \omega_{lk}^A|$, the phase will turn into pure absorption. The product lines are never broadened in one-sided chemical reactions.

We illustrate these effects for the simplest non-trivial situation, a chemical reaction involving a strongly coupled two-spin system. We assume that during the reaction AB $\longrightarrow$ CD both nuclei change chemical shifts while the coupling constant $J_{AB} = J_{CD}$ remains invariant.

A set of representative spectra for different rate constants k_1 is shown in Fig. 4.6.10, in qualitative agreement with the features which have been found for a one-spin system. Exchange broadening is equal for all reactant lines, while the width of the product lines remains unaffected by the chemical reaction.

Of particular interest are the relative signal intensities and the signal phases. It is immediately obvious in Fig. 4.6.10 that the intensities of the product lines are severely perturbed by the chemical reaction. A detailed analysis of the observed intensity changes has been given in Ref. (4.237).

4.6.3.4. Experimental preparation of chemical non-equilibrium states

A great number of techniques have been proposed and utilized for the creation of the required initial non-equilibrium state of the chemical system. Temperature-jump (4.239) and pressure-jump methods (4.240) are frequently used in connection with optical detection. Flash photolysis (4.241) and pulse radiolysis (4.242, 4.243) are often employed to initiate fast reactions, and stopped-flow techniques (4.244–4.246) have found widespread application. However, only a few of these techniques are useful in the context of NMR because the initiation of the reaction must be performed within the NMR probe assembly and because a significant portion of the material present has to participate in the chemical reaction to obtain sufficient sensitivity. Flow techniques have been found most useful in NMR (4.237, 4.238, 4.247–4.254). Similar lineshape phenomena should also be observable in light-induced chemical reactions, particularly in connection with chemically induced dynamic nuclear polarization

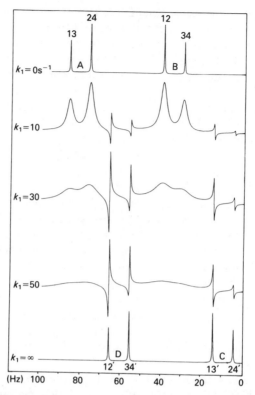

FIG. 4.6.10. Lineshapes of a two-spin system involved in a one-sided chemical reaction AB → CD for five different reaction rate constants k_1. The chosen spectral parameters are $\Omega_A/2\pi = 80$ Hz, $\Omega_B/2\pi = 35$ Hz, $J_{AB} = 10$ Hz, $\Omega_C/2\pi = 10$ Hz, $\Omega_D/2\pi = 60$ Hz, and $J_{CD} = 10$ Hz. Note that the order of the resonance frequencies changes in the course of the reaction. (Reproduced from Ref. 4.237.)

(CIDNP) (4.255, 4.256) when pulsed light excitation is used (4.238, 4.257).

4.7. Fourier double resonance

Double resonance was developed initially in connection with CW techniques, but is equally applicable in one- and two-dimensional Fourier experiments.

Three characteristic double-resonance effects are observable when the strength of the double-resonance r.f. field B_2 is increased.

1. *Weak perturbation* $T_1^{-1} < \gamma B_2 < T_2^{*-1}$. The saturation of selected transitions leads to a non-equilibrium state of the system which manifests itself by perturbed peak intensities, a phenomenon which is well known

as nuclear Overhauser effect (NOE) (4.147, 4.258, 4.259), and which provides a powerful means of studying molecular structure.

2. *Intermediate perturbation strength* $T_2^{*-1} < \gamma B_2 < 2\pi J, D$. For an r.f. field strength which exceeds the line width T_2^{*-1} but is small compared to the scalar interaction J or dipolar interaction D, line splittings are observed which are known as *tickling* effects (4.260, 4.261). The splittings are on the order of γB_2. They carry information on the connectivity of transitions and allow one to trace out the complete energy level diagram. Similar information can be obtained from two-dimensional coherence transfer experiments, as discussed in Chapter 8.

3. *Strong perturbation* $\gamma B_2 > J, D$. Strong perturbations induce drastic changes of the effective Hamiltonian. In particular, certain interactions may become ineffective, leading to *spin decoupling* (4.65, 4.261, 4.262).

Double-resonance experiments may have two diverging goals: Situations 1 and 2 permit one to obtain additional information not available from the single resonance spectrum, while situation 3 leads to a simplification of spectra with a corresponding loss of information.

The three stages of double resonance require different theoretical treatments. For a weak perturbation, it is sufficient to consider the redistribution of populations in terms of a modified master equation, eqn (2.3.3),

$$\frac{d}{dt}\mathbf{P}(t) = \mathbf{W}\{\mathbf{P}(t) - \mathbf{P}_0\} + \mathbf{S}\mathbf{P}(t) \qquad (4.7.1)$$

where $\mathbf{S}$ is the saturation matrix which represents the effects of the r.f. irradiation. Its elements indicate the transition probabilities induced by the irradiation. For example, in a four-level system where transition $|1\rangle \leftrightarrow |3\rangle$ is irradiated, $\mathbf{S}$ is given by

$$\mathbf{S} = |\gamma B_2 (F_x)_{1,3}|^2 \begin{pmatrix} -1 & 0 & 1 & 0 \\ 0 & 0 & 0 & 0 \\ 1 & 0 & -1 & 0 \\ 0 & 0 & 0 & 0 \end{pmatrix}. \qquad (4.7.2)$$

In equilibrium, we obtain the population vector

$$\mathbf{P} = (\mathbf{W} + \mathbf{S})^{-1}\mathbf{W}\mathbf{P}_0. \qquad (4.7.3)$$

In Fourier experiments using non-selective r.f. pulses with small rotation angles β, the signal intensities are proportional to the corresponding population differences $P_j - P_k$, and to the square of the transition matrix elements $(F_x)_{jk}$. However, it should be remembered that for r.f. pulses with large rotation angles β, the signal intensities depend on the populations of *all* energy levels due to the mixing effect of an r.f. pulse (4.131) as described in detail in § 4.4.3.

Spin tickling effects are coherent effects and cannot be described by a master equation in terms of populations alone. A full quantum mechanical treatment is required. In the case of very strong double-resonance irradiation, the application of average Hamiltonian theory provides a convenient theoretical approach (§ 3.2.4).

4.7.1. Theoretical formulation of Fourier double resonance

The complications of double resonance are caused by the presence of two time-dependent perturbations, the double-resonance irradiation of frequency ω_2 and amplitude B_2 (4.131, 4.263–4.266):

$$\mathcal{H}_2(t) = -\tfrac{1}{2}\gamma B_2[F^+ \exp(-i\omega_2 t) + F^- \exp(i\omega_2 t)] \qquad (4.7.4)$$

with $F^+ = \sum_k I_k^+$, and the pulsed r.f. field of frequency ω_1 used to excite coherence for detection. The same frequency ω_1 is also used as reference for the phase-sensitive detector (Fig. 4.7.1). The two frequencies define two independently rotating frames $F^{(1)}$ and $F^{(2)}$ which are assumed to have a phase difference φ_0 at $t = 0$. The phase difference between the two frames increases at time t to

$$\varphi = \varphi_0 + (\omega_1 - \omega_2)t. \qquad (4.7.5)$$

Excitation and detection occur in the frame $F^{(1)}$ while the evolution is best visualized in frame $F^{(2)}$

$$\begin{array}{ccc} \text{Excitation} & \text{Evolution} & \text{Detection} \\ \downarrow & \downarrow & \downarrow \\ F^{(1)} \longrightarrow & F^{(2)} \longrightarrow F^{(2)} \longrightarrow & F^{(1)}. \end{array}$$

The Hamiltonian describing the evolution is time-independent in the frame $F^{(2)}$

$$\mathcal{H}^{(2)} = \mathcal{H}_0 - \omega_2 F_z - \gamma B_2 F_x. \qquad (4.7.6)$$

In the course of the calculation, we have to change frames whenever required.

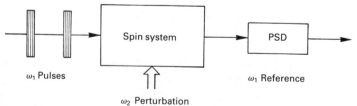

FIG. 4.7.1. Block diagram of a Fourier double-resonance experiment. (Reproduced from Ref. 4.269.)

4.7.1.1. Double-resonance irradiation during detection

We first consider an experiment where the initial $(\pi/2)_y$-pulse of frequency ω_1 is applied to a system in thermal equilibrium. The double-resonance r.f. field of frequency ω_2 is switched on immediately after the r.f. pulse.

The density operator after the r.f. pulse in the frame $F^{(1)}$ is

$$\sigma^{(1)}(0) = F_x \tag{4.7.7}$$

which, transformed into the frame $F^{(2)}$, becomes

$$\sigma^{(2)}(0) = \tfrac{1}{2}\{F^+ e^{-i\varphi_0} + F^- e^{i\varphi_0}\}. \tag{4.7.8}$$

At time t, we have

$$\sigma^{(2)}(t) = \exp\{-i\mathcal{H}^{(2)}t\}\tfrac{1}{2}\{F^+ e^{-i\varphi_0} + F^- e^{i\varphi_0}\}\exp\{i\mathcal{H}^{(2)}t\}. \tag{4.7.9}$$

To compute the signal $s(t)$, we may transform $\sigma^{(2)}(t)$ back to the frame F_1. It turns out to be easier to transform the observable operator $Q^{(1)}$

$$Q^{(1)} = F^+ \tag{4.7.10}$$

from frame $F^{(1)}$ to frame $F^{(2)}$

$$Q^{(2)} = F^+ \exp\{-i(\omega_1 - \omega_2)t - i\varphi_0\} \tag{4.7.11}$$

and to evaluate its expectation value in frame $F^{(2)}$

$$\begin{aligned}
s^+(t) &= \text{tr}\{Q^{(2)}\sigma^{(2)}(t)\} \\
&= \tfrac{1}{2}\exp\{-i(\omega_1 - \omega_2)t\}\text{tr}\{F^+ \exp(-i\mathcal{H}^{(2)}t)F^- \exp(i\mathcal{H}^{(2)}t)\} \\
&\quad + \tfrac{1}{2}\exp\{-i(\omega_1 - \omega_2)t\}\text{tr}\{F^+ \exp(-i\mathcal{H}^{(2)}t)F^+ \exp(i\mathcal{H}^{(2)}t)\} \\
&\quad \times \exp(-2i\varphi_0).
\end{aligned} \tag{4.7.12}$$

Obviously, the signal consists of two parts, a term which is independent of the phase difference φ_0 and thus independent of the instant at which the r.f. pulse is applied, and a term which depends on φ_0. The phase of this second term is determined by the relative phase of the two frequencies at the time of the r.f. pulse.

Two different situations can be conceived in a signal-averaging sequence.

1. If no special precautions are taken, the phase φ_0 will vary randomly from experiment to experiment and the second term in eqn (4.7.12) will average to zero in the course of signal-averaging.

2. If the r.f. pulse of frequency ω_1 is triggered at a particular value of φ_0, the second term of eqn (4.7.12) will also give a contribution. The triggering can be achieved in practice by mixing the two frequencies and releasing the r.f. pulse at a selected phase angle of the difference frequency.

The spectra obtained with the two methods will differ in the intensities and phases of the various peaks. The phase-dependent term in eqn (4.7.12) disappears if the irradiation is weak or far from the observation frequency ω_1, because this term involves two F^+ operators and the trace vanishes unless $\mathcal{H}^{(2)}$ mixes F^+ and F^-. This term is irrelevant in heteronuclear double-resonance experiments and becomes effective only for double-resonance irradiation near the observed transitions. It follows from eqn (4.7.12) that CW and Fourier double-resonance experiments yield identical results if the phase-dependent term can be neglected.

For the explicit evaluation of eqn (4.7.12), it is best to diagonalize $\mathcal{H}^{(2)}$. We denote its diagonal form by $\mathcal{H}^D$

$$\mathcal{H}^{(2)} = D^{-1} \mathcal{H}^D D \tag{4.7.13}$$

and find

$$\text{tr}\{F^+ \exp(-i\mathcal{H}^{(2)}t) F^- \exp(i\mathcal{H}^{(2)}t)\}$$
$$= \text{tr}\{F^+ D^{-1} \exp(-i\mathcal{H}^D t) D F^- D^{-1} \exp(i\mathcal{H}^D t) D\}. \tag{4.7.14}$$

The observed signal can be represented as a sum

$$s^+(t) = \sum_{kl} G_{kl} \exp\{i(\omega_2 - \omega_{kl} - \omega_1)t\} \tag{4.7.15}$$

with the eigenfrequencies $\omega_{kl} = \varepsilon_k - \varepsilon_l$ of the Hamiltonian $\mathcal{H}^{(2)}$ and with the complex intensities

$$G_{kl} = \frac{1}{2}\left\{\sum_{pq\,rs} F^+_{qp} D^{-1}_{pk} D_{kr} F^-_{rs} D^{-1}_{sl} D_{lq}\right.$$
$$\left. + \sum_{pq\,rs} F^+_{qp} D^{-1}_{pk} D_{kr} F^+_{rs} D^{-1}_{sl} D_{lq} \exp(-2i\varphi_0)\right\}. \tag{4.7.16}$$

It is important to note that the sum over the index pairs (k, l) in eqn (4.7.15) extends over all possible combinations, including pairs (l, k) which are associated with frequencies that have opposite signs. This implies that the spectrum of a system with n energy levels consists of $n(n-1)$ transitions symmetrically disposed about the frequency ω_2 (or about $\omega_2 - \omega_1$ in the observation frame $F^{(1)}$). However, the intensity factors G_{kl} and G_{lk} are normally different and the transitions at $\omega_2 \pm \omega_{kl}$ have unequal intensities. Examples are shown in Fig. 4.7.2.

Without going into more detail, we can summarize the general features of Fourier spectra with double-resonance irradiation during detection only.

1. Double resonance leads to a spectrum that is symmetrical in frequency but not in amplitude with respect to ω_2.

4.7 FOURIER DOUBLE RESONANCE

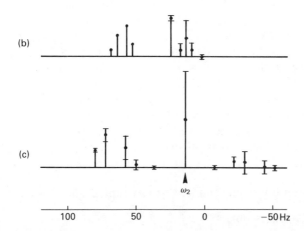

FIG. 4.7.2. Schematic spin-tickling Fourier spectra of a two-spin system ($\Omega_A/2\pi = 60$ Hz, $\Omega_B/2\pi = 20$ Hz, $J = 10$ Hz). (a) Unperturbed single-resonance spectrum; (b) and (c), spectra obtained with a double-resonance field applied during the detection period only. Before the pulse, the system is assumed to be in thermal equilibrium. The strength of the double-resonance field is $\gamma B_2/2\pi = 5$ and 40 Hz respectively. The intensities depend on the relative phases of the pulsed r.f. field and the double-resonance r.f. field at the time of the pulse. The range of variation as a function of the phase difference is indicated by horizontal bars. The solid dot in the centre of this range marks the phase-independent part of the intensity, which can be observed if several free-induction decays are co-added without phase synchronization.

2. The intensities of Fourier double-resonance spectra depend on the phase difference φ_0 of the two r.f. frequencies ω_1 and ω_2 at $t = 0$.

3. In general, CW and Fourier double-resonance spectra exhibit different intensities. To obtain intensities proportional to those of CW experiments, it is necessary to average numerous free induction decays without phase-synchronization of the two r.f. frequencies at $t = 0$.

4.7.1.2. Continuous Fourier double resonance

Instead of applying the double resonance irradiation *after* the initial r.f. pulse, one may allow the system to reach a steady state under the influence of continuous double-resonance irradiation prior to the application of an r.f. pulse. In this case, the state before the pulse $\sigma^{(2)}(0_-)$ contains coherent components as well as perturbed populations which will lead to further phase and intensity anomalies that depend on the phase difference φ_0 at the time of the r.f. pulse. The steady state before the pulse can be computed, at least in principle, from the equation

$$\sigma^{(2)}(0_-) = \{i\hat{\mathscr{H}}^{(2)} + \hat{\Gamma}\}^{-1}\hat{\Gamma}\sigma_0$$

where $\hat{\mathscr{H}}^{(2)}$ represents the commutator with the Hamiltonian expressed in the rotating frame $F^{(2)}$, and $\hat{\Gamma}$ is the relaxation superoperator.

If signal-averaging is used *without* synchronization of the r.f. pulse, the coherent components of $\sigma^{(2)}(0_-)$ are cancelled. However, the perturbed populations still lead to intensity variations due to Overhauser effects. Furthermore, a non-selective r.f. pulse acting on a non-equilibrium system not only creates off-diagonal elements $\sigma(0)_{rs}$ with $\Delta M_{rs} = M_r - M_s = \pm 1$, (observable magnetization), but also zero- and multiple-quantum coherence ($\Delta M_{rs} = 0, \pm 2, \pm 3, \ldots$).

If the double-resonance irradiation is removed immediately before applying the r.f. pulse, it is possible to observe pure nuclear Overhauser effects unperturbed by tickling or decoupling effects (4.267).

4.7.2. Fourier double resonance of two coupled spins $I = 1/2$

4.7.2.1. Strong coupling

The characteristic features of Fourier double resonance become apparent for strongly-coupled systems, or in weakly-coupled systems when the irradiated multiplet is observed. To illustrate the dependence of the intensities on the phase difference φ and the occurrence of frequency pairs $\{(\omega_2 - \omega_{kl}), (\omega_2 - \omega_{lk})\}$, double-resonance spectra have been calculated for a system of two spins with $I = \frac{1}{2}$ with moderately weak coupling ($\Delta\Omega/2\pi J = 4$). In Fig. 4.7.2, the range of the intensity variations with angle φ has been indicated by horizontal bars, with the dot in the centre corresponding to the phase-independent part.

For a moderate r.f. field strength ($\gamma B_2/(2\pi) = J/2$), the transitions (2, 4) and (1, 3) that are connected to the irradiated transition (1, 2) are split into doublets. Two pairs of signals symmetric with respect to ω_2 are visible. Furthermore, it is evident that only lines in the vicinity of ω_2 have a phase-dependent intensity. If the r.f. field strength is increased to $\gamma B_2/(2\pi) = 4J$, five pairs of signals appear with strongly phase-dependent intensities. The intensities of the sixth pair, at frequencies 116 Hz and −87 Hz, are unobservably small. Obviously, for strong r.f. perturbations the variations with φ can become quite significant.

It is important to note that the intensities obtained by co-adding FIDs with non-synchronized r.f. pulses are equivalent to those obtained in a CW double-resonance experiment, disregarding possible Overhauser effects occurring in the course of slow-passage observation.

4.7.2.2. Weak coupling

When considering a weakly-coupled two-spin $\frac{1}{2}$ system, it is sufficient to use the simple arguments of Bloom and Shoolery (4.268) to compute the double-resonance spectrum. The physical origin of the mirror-image frequency pairs, $(\omega_2 - \omega_{kl})$, $(\omega_2 - \omega_{lk})$ predicted in eqn (4.7.16) will become evident.

Let us assume that the double-resonance r.f. field is applied in the vicinity of the Larmor frequency of the B spin, neglecting its direct influence on the A spin. The Hamiltonian in the frame $F^{(2)}$ can then be simplified to

$$\mathcal{H}^{(2)} = (\Omega_A - \omega_2)I_{Az} + (\Omega_B - \omega_2 + 2\pi J I_{Az})I_{Bz} - \gamma_B B_2 I_{Bx}. \quad (4.7.17)$$

We define for each magnetic quantum number M_A of spin A an effective field $\mathbf{B}_{\text{eff}}(M_A)$ acting on the B spin, as shown in Fig. 4.7.3. This field is tilted at an angle

$$\theta(M_A) = \tan^{-1}\{-\gamma_B B_2/(\Omega_B - \omega_2 + 2\pi J M_A)\} \quad (4.7.18)$$

with respect to the z-axis, and has the magnitude

$$\gamma_B B_{\text{eff}}(M_A) = \{(\Omega_B - \omega_2 + 2\pi J M_A)^2 + (\gamma_B B_2)^2\}^{\frac{1}{2}}. \quad (4.7.19)$$

The crucial point is that $\mathbf{B}_{\text{eff}}$ depends on the polarization of the A nucleus, i.e. there are two effective fields for $M_A = +\frac{1}{2}$ and $-\frac{1}{2}$, differing both in magnitude and direction.

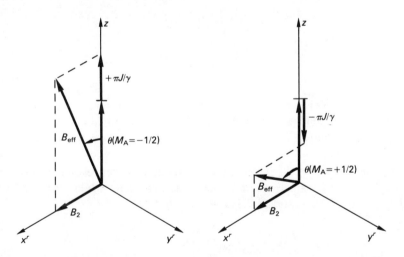

FIG. 4.7.3. Double resonance in a weakly coupled two-spin system. In a frame (x^r, y^r, z^r), rotating at the double-resonance frequency ω_2 with respect to the laboratory frame, the irradiated spin B is quantized along an effective field $B_{\text{eff}}(M_A)$ which depends, both in magnitude and direction, on the spin quantum number $M_A = \pm\frac{1}{2}$ of nucleus A.

With the B spin quantized along one of these effective fields, we obtain the following eigenstates $\psi(M_A, M_B)$ and energies $E(M_A, M_B)$ in the rotating frame $F^{(2)}$

$$\psi_1 = \psi(\tfrac{1}{2}, \tfrac{1}{2}) = \alpha_A \alpha_B \cos(\tfrac{1}{2}\theta(\tfrac{1}{2})) + \alpha_A \beta_B \sin(\tfrac{1}{2}\theta(\tfrac{1}{2})),$$
$$E(\tfrac{1}{2}, \tfrac{1}{2}) = \tfrac{1}{2}(\Omega_A - \omega_2) - \tfrac{1}{2}\gamma_B B_{\text{eff}}(\tfrac{1}{2}),$$
$$\psi_2 = \psi(\tfrac{1}{2}, -\tfrac{1}{2}) = -\alpha_A \alpha_B \sin(\tfrac{1}{2}\theta(\tfrac{1}{2})) + \alpha_A \beta_B \cos(\tfrac{1}{2}\theta(\tfrac{1}{2})),$$
$$E(\tfrac{1}{2}, -\tfrac{1}{2}) = \tfrac{1}{2}(\Omega_A - \omega_2) + \tfrac{1}{2}\gamma_B B_{\text{eff}}(\tfrac{1}{2}),$$
$$\psi_3 = \psi(-\tfrac{1}{2}, \tfrac{1}{2}) = \beta_A \alpha_B \cos(\tfrac{1}{2}\theta(-\tfrac{1}{2})) + \beta_A \beta_B \sin(\tfrac{1}{2}\theta(-\tfrac{1}{2})),$$
$$E(-\tfrac{1}{2}, \tfrac{1}{2}) = -\tfrac{1}{2}(\Omega_A - \omega_2) - \tfrac{1}{2}\gamma_B B_{\text{eff}}(-\tfrac{1}{2}),$$
$$\psi_4 = \psi(-\tfrac{1}{2}, -\tfrac{1}{2}) = -\beta_A \alpha_B \sin(\tfrac{1}{2}\theta(-\tfrac{1}{2})) + \beta_A \beta_B \cos(\tfrac{1}{2}\theta(-\tfrac{1}{2})),$$
$$E(-\tfrac{1}{2}, -\tfrac{1}{2}) = -\tfrac{1}{2}(\Omega_A - \omega_2) + \tfrac{1}{2}\gamma_B B_{\text{eff}}(-\tfrac{1}{2}). \quad (4.7.20)$$

The A spin signal intensities are proportional to the squares of the matrix elements $\langle \psi_i | I_A^+ | \psi_j \rangle$. Because of the mixing of the product functions in ψ_k, all six possible transitions become allowed. Four of these are A spin transitions, with the frequencies and intensities

$$\omega_{13}^{(2)} = \Omega_A - \omega_2 - \tfrac{1}{2}\gamma_B \{B_{\text{eff}}(\tfrac{1}{2}) - B_{\text{eff}}(-\tfrac{1}{2})\},$$
$$I_{13} \propto \cos^2 \xi,$$
$$\omega_{24}^{(2)} = \Omega_A - \omega_2 + \tfrac{1}{2}\gamma_B \{B_{\text{eff}}(\tfrac{1}{2}) - B_{\text{eff}}(-\tfrac{1}{2})\},$$
$$I_{24} \propto \cos^2 \xi,$$
$$\omega_{14}^{(2)} = \Omega_A - \omega_2 - \tfrac{1}{2}\gamma_B \{B_{\text{eff}}(\tfrac{1}{2}) + B_{\text{eff}}(-\tfrac{1}{2})\},$$
$$I_{14} \propto \sin^2 \xi,$$
$$\omega_{23}^{(2)} = \Omega_A - \omega_2 + \tfrac{1}{2}\gamma_B \{B_{\text{eff}}(\tfrac{1}{2}) + B_{\text{eff}}(-\tfrac{1}{2})\},$$
$$I_{23} \propto \sin^2 \xi \quad (4.7.21)$$

where $\xi = \tfrac{1}{2}\{\theta(\tfrac{1}{2}) - \theta(-\tfrac{1}{2})\}$. In the course of an A spin transition, the B spin quantization axis changes its direction and the effective field $B_{\text{eff}}(M_A)$ its value. The frequencies in the laboratory frame are $\omega_{ij}^L = \omega_{ij}^{(2)} + \omega_2$.

For spin tickling, the irradiation frequency ω_2 is on-resonance with one of the B spin transitions, e.g. $\Omega_B + \pi J - \omega_2 = 0$, while $|\gamma B_2| \ll |2\pi J|$. From eqns (4.7.18) and (4.7.19) we see that

$$\theta(\tfrac{1}{2}) = \frac{\pi}{2}, \quad \theta(-\tfrac{1}{2}) \approx 0, \quad B_{\text{eff}}(\tfrac{1}{2}) = B_2, \quad B_{\text{eff}}(-\tfrac{1}{2}) = 2\pi J/\gamma_B.$$

In this case all intensities are equal to $\tfrac{1}{2}$, and the four frequencies are

$$\omega^L = \Omega_A \pm \pi J \pm \tfrac{1}{2}\gamma B_2. \quad (4.7.22)$$

4.7 FOURIER DOUBLE RESONANCE

Thus each of the A spin lines is split into a doublet with splitting parameter γB_2, as will be discussed in more general terms in the following section.

To predict the frequencies of the irradiated B spin, it is sufficient to consider the classical motion of a magnetization vector $\mathbf{M}_B = (M_{Bx}, M_{By}, M_{Bz})$ about the effective fields $B_{eff}(M_A = \pm\frac{1}{2})$, shown in Fig. 4.7.3. When the motion is transformed back into the laboratory frame, one obtains for the y-component of the B spin magnetization

$$M_{By}^L(t) = \tfrac{1}{2}[1 + \cos\theta(M_A)]\{\sin(\alpha + \gamma)[\cos\theta(M_A)M_{Bx}(0) - \sin\theta(M_A)M_{Bz}(0)] + \cos(\alpha + \gamma)M_{By}(0)\}$$
$$+ \tfrac{1}{2}[1 - \cos\theta(M_A)]\{\sin(\alpha - \gamma)[-\cos\theta(M_A)M_{Bx}(0) + \sin\theta(M_A)M_{Bz}(0)] + \cos(\alpha - \gamma)M_{By}(0)\}$$
$$+ \sin\theta(M_A)\sin\alpha[\sin\theta(M_A)M_{Bx}(0) + \cos\theta(M_A)M_{Bz}(0)] \quad (4.7.23)$$

with $\alpha = \omega_2 t$ and $\gamma = -\gamma_B B_{eff}(M_A)t$. One immediately sees that there are three precession frequencies

$$\omega_B = \omega_2$$

and

$$\omega_B(M_A) = \omega_2 \mp \gamma_B B_{eff}(M_A). \quad (4.7.24)$$

Taking into account the two possible values of M_A, a total of five lines is observed in the B part of the spectrum (Fig. 4.7.2(b)).

The occurrence of five lines in the B spin spectrum can be rationalized by considering the motion of the magnetization $\mathbf{M}_B$ about one of the effective fields $\mathbf{B}_{eff}(M_A = \pm\frac{1}{2})$ in Fig. 4.7.3. The projection of the trajectory of $\mathbf{M}_B(t)$ on to the $x^r y^r$ plane gives an ellipse which is displaced from the origin. In the laboratory frame, the component of $\mathbf{M}_B(t)$ corresponding to the offset of the ellipse rotates with ω_2 while the elliptic motion can be decomposed into counter-rotating components with frequencies $\omega_2 \pm \gamma B_{eff}(M_A = \frac{1}{2})$ and $\omega_2 \pm \gamma B_{eff}(M_A = -\frac{1}{2})$.

This simple picture can also be used to predict the relative intensities within the two pairs of satellites. When the phase φ_0 between r.f. pulse and double-resonance field at $t = 0$ is set to $\varphi_0 = 0$, eqn (4.7.23) yields the intensity ratio

$$r(M_A) = \frac{I[\omega_2 - \gamma_B B_{eff}(M_A)]}{I[\omega_2 + \gamma_B B_{eff}(M_A)]} = \frac{1 + \cos\theta(M_A)}{1 - \cos\theta(M_A)}, \quad (4.7.25)$$

in agreement with both theoretical and experimental results (4.263–4.265). On the other hand, when a sufficient number of experiments

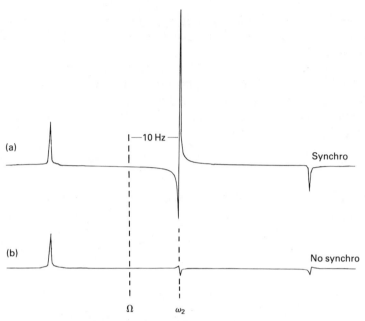

FIG. 4.7.4. Fourier double-resonance experiments on a one-spin system. An r.f. field ω_2 with amplitude $\gamma B_2/(2\pi) = 23$ Hz was applied 10 Hz off-resonance with respect to the Larmor frequency Ω. In (a), the r.f. pulse was phase-synchronized to the double-resonance frequency, whereas in (b) no synchronization was employed. 64 scans were co-added. (Adapted from Ref. 4.264.)

without phase synchronization is co-added, the ratio is given by

$$r(M_A) = \frac{[1 + \cos\theta(M_A)]^2}{[1 - \cos\theta(M_A)]^2}, \quad (4.7.26)$$

again in agreement with the experiment (4.264). The latter intensity ratio is also obtained when observing the slow-passage CW spectrum of the perturbed system (4.264). Experimental Fourier double-resonance spectra corresponding to the two cases mentioned are shown in Fig. 4.7.4. For the one-spin system shown, $M_A = 0$, and the tilt angle θ becomes $\theta = \tan^{-1}\{-\gamma_B B_2/(\Omega_B - \omega_2)\}$. Equations (4.7.25) and (4.7.26) apply also to this case and explain the observed relative intensities of the two satellites.

4.7.3. Spin tickling

Spin tickling experiments employ selective perturbations of individual resonance lines (4.260, 4.261). The r.f. field must be weak compared to the multiplet splittings but strong compared to the line-width in order to obtain observable splittings of the lines. Such experiments can be used to elucidate the connectivity of transitions in coupled spin systems.

Identical tickling spectra are obtained in Fourier and CW modes, except for possible Overhauser effects and for sidebands of the irradiated transition. Synchronization of the two r.f. frequencies is of no importance. We can therefore apply the CW tickling theory to Fourier experiments. A comprehensive explanation of spin tickling effects has been given by Freeman and Anderson (4.260). We reformulate their treatment in a single-transition operator form which leads to a particularly concise notation (4.269).

Let us assume that the double-resonance field acts selectively on the transition (ab). Instead of transforming into a frame $F^{(2)}$ rotating at ω_2 for all spins, it is sufficient to transform into a frame by means of the single-transition operator $T = \exp\{-i\omega_2 I_z^{(ab)}t\}$. Instead of eqn (4.7.6), one obtains

$$\mathcal{H}^T = \mathcal{H}_0 - \omega_2 I_z^{(ab)} + h_{ab} I_x^{(ab)}, \qquad (4.7.27)$$

with the r.f. field

$$h_{ab} = -2B_2 \left(\sum_{k=1}^{N} \gamma_k I_{kx} \right)_{ab}. \qquad (4.7.28)$$

Note that h_{ab} is expressed in frequency units and depends on the transition matrix element.

Represented in the eigenbasis of $\mathcal{H}_0$, $\mathcal{H}^T$ consists of a 2×2 block $\mathcal{H}^{(ab)}$, containing the off-diagonal elements of $I_x^{(ab)}$, and a diagonal matrix $\mathcal{H}_0^{(-)}$ involving all other eigenstates

$$\mathcal{H}^T = \mathcal{H}^{(ab)} \oplus \mathcal{H}_0^{(-)}, \qquad (4.7.29)$$

$$\mathcal{H}^{(ab)} = \mathcal{H}_0^{(ab)} - \omega_2 I_z^{(ab)} + h_{ab} I_x^{(ab)}. \qquad (4.7.30)$$

By diagonalizing $\mathcal{H}^{(ab)}$, the eigenvalues of $\mathcal{H}^T$ are obtained in the form

$$\varepsilon_\kappa = \mathcal{H}_{0\,kk}, \quad k \neq a, b$$

$$\varepsilon_{\alpha,\beta} = \tfrac{1}{2}(\mathcal{H}_{0\,aa} + \mathcal{H}_{0\,bb}) \mp \tfrac{1}{2}q. \qquad (4.7.31)$$

The parameter

$$q = \{(\omega_2 - \omega_{ab})^2 + h_{ab}^2\}^{\frac{1}{2}} \qquad (4.7.32)$$

measures the strength of the effective field acting on the transition (ab). Greek indices in eqn (4.7.31) refer to the eigenbasis of $\mathcal{H}^T$, while Roman indices pertain to the eigenbasis of $\mathcal{H}_0$.

The tickling spectrum can readily be predicted by computing differences of the eigenvalues in eqn (4.7.31), and then transforming back into the laboratory frame. The principal result is that each transition connected to the irradiated transition, i.e. involving either one of the two levels $|a\rangle$ or $|b\rangle$, is split into a tickling doublet,

$$\omega_{ac}^{\pm} = \mathcal{H}_{0\,aa} - \mathcal{H}_{0\,cc} + \tfrac{1}{2}(\omega_2 - \omega_{ab} \pm q),$$
$$\omega_{bd}^{\pm} = \mathcal{H}_{0\,bb} - \mathcal{H}_{0\,dd} - \tfrac{1}{2}(\omega_2 - \omega_{ab} \pm q). \tag{4.7.33}$$

The relative intensities of the two lines are given by

$$I_{ac}^{+}/I_{ac}^{-} = (1 - \cos\theta)/(1 + \cos\theta), \tag{4.7.34}$$

with the angle θ between effective field and z-axis,

$$\theta = \tan^{-1}[h_{ab}/(\omega_2 - \omega_{ab})]. \tag{4.7.35}$$

Transitions that are *not* connected to (ab) remain unaffected by the irradiation, as can be deduced from eqn (4.7.31). The irradiated transition itself splits into a triplet of lines at frequencies

$$\omega_{ab} = \omega_2, \qquad \omega_2 \pm q. \tag{4.7.36}$$

Besides the fact that the splitting of a line in a tickling spectrum indicates that it is connected to the irradiated one, the *line-widths* of the tickling doublets (4.260) allow one to determine whether the connectivity is progressive or regressive. For progressive connections, e.g. $(ca) - (ab)$ with $M_c - M_b = 2$, the line-widths in the tickling doublets are larger than for an unconnected single-quantum transition, while for regressive connections, e.g. for $(ab) - (db)$ with $M_d = M_a$, the lines in the doublet are narrower as compared to unsplit transitions.

4.7.4. Spin decoupling treated by average Hamiltonian theory

The motion of a system in a double-resonance experiment cannot always be described by a modified Hamiltonian in the laboratory frame. For spin decoupling with a strong r.f. field, it is however possible to formulate an effective Hamiltonian, and it is feasible to apply average Hamiltonian theory.

We can start with the Hamiltonian in the rotating frame $F^{(2)}$, eqn (4.7.6). We identify the perturbation $\mathcal{H}_1(t)$ with the double-resonance interaction $-\gamma B_2 F_x$ and $\mathcal{H}_0$ with $\mathcal{H}_0^T = \mathcal{H}_0 - \omega_2 F_z$. We obtain with eqn (3.2.28) the zero-order average Hamiltonian

$$\bar{\mathcal{H}}_0^{(0)} = \frac{1}{t_c} \int_0^{t_c} dt_1 \, \mathcal{H}_0^T(t_1) \tag{4.7.37}$$

with

$$\mathcal{H}_0^T(t) = U_1^{-1}(t)\{\mathcal{H}_0 - \omega_2 F_z\} U_1(t) \tag{4.7.38}$$

and

$$U_1(t) = \exp\{+i\gamma B_2 F_x t\}. \tag{4.7.39}$$

The cyclic condition, which is necessary for eqn (4.7.37) to hold, can be fulfilled by selecting the 'cycle time' t_c

$$t_c = 2\pi/(\gamma B_2). \quad (4.7.40)$$

Stroboscopic sampling, however, is not necessary in most practical cases of spin decoupling. In the following, we shall apply eqn (4.7.37) to heteronuclear spin decoupling and to off-resonance decoupling.

4.7.4.1. Heteronuclear spin decoupling

Consider a system containing two spin species I and S with scalar couplings $2\pi J_{kl}$ and dipolar couplings a_{kl} with the Hamiltonian

$$\mathcal{H}_0 = \mathcal{H}_I + \mathcal{H}_{IS} + \mathcal{H}_S \quad (4.7.41)$$

with

$$\mathcal{H}_I = \sum_k \omega_{Ik} I_{kz} + \sum_{k<k'} \{2\pi J_{kk'} \mathbf{I}_k \mathbf{I}_{k'}$$
$$+ a_{kk'}[3I_{kz}I_{k'z} - \mathbf{I}_k \mathbf{I}_{k'}]\}, \quad (4.7.42)$$

$$\mathcal{H}_{IS} = \sum_{k,l} (2\pi J_{kl} + 2a_{kl}) I_{kz} S_{lz}, \quad (4.7.43)$$

$$\mathcal{H}_S = \sum_l \omega_{Sl} S_{lz} + \sum_{l<l'} \{2\pi J_{ll'} \mathbf{S}_l \mathbf{S}_{l'}$$
$$+ a_{ll'}[3S_{lz}S_{l'z} - \mathbf{S}_l \mathbf{S}_{l'}]\} \quad (4.7.44)$$

with $a_{kl} = b_{kl} \frac{1}{2}(1 - 3\cos^2\theta_{kl})$ (see eqns (2.2.18) and (2.2.19)). We assume that the I spins are irradiated while the S spin resonances are observed. In this case, it is sufficient to transform into a frame rotating with respect to the I spins alone, i.e.

$$\mathcal{H}_0^T = \mathcal{H}_0 - \omega_2 F_z \quad \text{with} \quad F_z = \sum_k I_{kz}. \quad (4.7.45)$$

Equation (4.7.37) immediately leads to the zero-order average Hamiltonian

$$\mathcal{H}_0^{(0)} = \mathcal{H}_S + \sum_{k<k'} \{2\pi J_{kk'} \mathbf{I}_k \mathbf{I}_{k'} - \tfrac{1}{2} a_{kk'}[3I_{kx}I_{k'x} - \mathbf{I}_k \mathbf{I}_{k'}]\}. \quad (4.7.46)$$

While the scalar coupling terms among the I spins remain invariant, only those parts of the I spin dipolar interactions which are secular with respect to the strong r.f. field applied along the x-axis are retained. The dipolar coupling coefficients $a_{kk'}$ are therefore multiplied by a factor of $(-\tfrac{1}{2})$. The heteronuclear couplings and the perturbing r.f. field no longer appear in the average Hamiltonian (see also § 4.7.7).

By considering the higher-order terms, it is possible to check the convergence of the Magnus expansion. Assuming weak coupling both among the I spins and among the S spins and disregarding for the present discussion dipolar interactions, we obtain the first- and second-order terms

$$\bar{\mathcal{H}}_0^{(1)} = \frac{1}{\gamma B_2}\left\{-\tfrac{1}{2}\sum_k Q_k^2 I_{kx} + \sum_{k\neq k'} 2\pi J_{kk'} Q_k I_{kx} I_{k'z}\right\}, \quad (4.7.47)$$

$$\bar{\mathcal{H}}_0^{(2)} = \frac{1}{2(\gamma B_2)^2}\left\{\sum_k Q_k^3 I_{kz} + \tfrac{3}{2}\sum_{k\neq k'} 2\pi J_{kk'}[Q_k Q_{k'} I_{kx} I_{k'x} - Q_k^2 I_{kz} I_{k'z}]\right\} \quad (4.7.48)$$

with

$$Q_k = (\omega_{Ik} - \omega_2) + \sum_l 2\pi J_{kl} S_{lz}. \quad (4.7.49)$$

$\bar{\mathcal{H}}_0^{(1)}$ and $\bar{\mathcal{H}}_0^{(2)}$ contain operators of the form $S_{lz}I_{kx}$ and $S_{lz}S_{l'z}I_{kx}I_{k'x}$, which cause a transformation of transverse S spin coherence S^+ into heteronuclear multiple-quantum coherence. The first- and second-order terms $\bar{\mathcal{H}}_0^{(1)}$ and $\bar{\mathcal{H}}_0^{(2)}$ are smaller than the zero-order term $\bar{\mathcal{H}}_0^{(0)}$ by factors of $2\pi J_{kl}/(\gamma B_2)$ and $[2\pi J_{kl}/(\gamma B_2)]^2$, respectively. Thus the condition for a rapid convergence of the Magnus expansion is simply that the decoupling r.f. field should be much larger than the heteronuclear spin–spin couplings, $\gamma B_2 \gg 2\pi J_{kl}$, $2a_{kl}$ and $\gamma B_2 \gg |\omega_{Ik} - \omega_2|$. If this is fulfilled, it is legitimate to neglect higher-order terms and to restrict oneself to the Hamiltonian $\bar{\mathcal{H}}_0^{(0)}$ given by eqn (4.7.46).

4.7.4.2. Off-resonance decoupling

Off-resonance decoupling can be used for assignments in heteronuclear spin systems by scaling the spin–spin interactions (4.270–4.272). A strong r.f. field is applied outside the relevant I-spin spectrum, and partial decoupling leads to scaling of the S-spin multiplets dependent on the I-spin resonance offset. To apply average Hamiltonian theory, it is necessary to redefine the Hamiltonian. We incorporate the resonance offset term of the I spins in the perturbation operator $\mathcal{H}_1$

$$\mathcal{H}_1 = \sum_k (\omega_{Ik} - \omega_2)I_{kz} - \gamma B_2 F_x. \quad (4.7.50)$$

The 'unperturbed' Hamiltonian $\mathcal{H}_0$ is then

$$\mathcal{H}_0 = \mathcal{H}_S + \mathcal{H}_{IS} + \mathcal{H}_{II}. \quad (4.7.51)$$

The spins I_k are quantized along different effective fields $\mathbf{B}_{\text{eff},k}$, as shown in Fig. 4.7.5. Expressing I_{kz} in terms of the operators $I_{kz'}$, $I_{kx'}$ appropriate

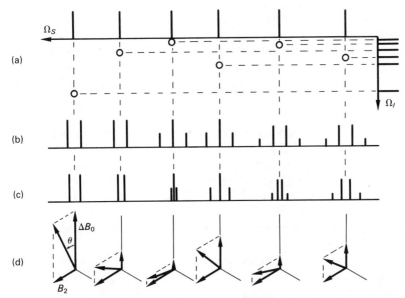

FIG. 4.7.5. Quantization in off-resonance decoupling. (a) Schematic heteronuclear two-dimensional correlation spectrum (see § 8.5.3) showing the carbon-13 chemical shifts Ω_{Sk} and corresponding proton shifts Ω_{Ik} of six arbitrary CH_n groups in a hypothetical molecule. (b) Proton-coupled carbon-13 spectrum with two doublets, two triplets, and two quartets, all assumed to have the same coupling strength J_{IS}. (c) Carbon-13 spectrum in the presence of off-resonance proton decoupling, showing different scaling factors that depend on the offset Ω_{Ik} of the attached protons with respect to the decoupler frequency. (d) The irradiated spins I_k (protons) are quantized along different effective fields $\mathbf{B}_{\text{eff},k}$ which are the resultant of the r.f. field $\mathbf{B}_2$ and the offset field $\Delta\mathbf{B}_{0,k} = (\omega_2 - \omega_{Ik})/\gamma$ with the tilt angle θ_k of eqn (4.7.53).

to the new quantization axis, we obtain

$$\tilde{\mathcal{H}}_0^T(t) = \mathcal{H}_S + \sum_{k,l} 2\pi J_{kl} S_{lz}$$
$$\times [I_{kz'} \cos\theta_k - I_{kx'} \sin\theta_k \cos(\gamma_k B_{\text{eff},k} t)$$
$$- I_{ky} \sin\theta_k \sin(\gamma_k B_{\text{eff},k} t)] + \mathcal{H}_{II} \quad (4.7.52)$$

with

$$\theta_k = \tan^{-1}[\gamma B_2/(\omega_2 - \omega_{Ik})]. \quad (4.7.53)$$

There is no single time τ_c for which the cyclic condition can be fulfilled for all spins. However, for a sufficiently strong r.f. field and hence short cycle times, the time-dependent terms in eqn (4.7.52) do not contribute to the average Hamiltonian

$$\tilde{\mathcal{H}}_0^{(0)} = \mathcal{H}_S + \sum_{k,l} 2\pi J_{kl} \cos\theta_k I_{kz'} S_{lz} + \tilde{\mathcal{H}}_{II}^{(0)} \quad (4.7.54)$$

The term $\tilde{\mathcal{H}}_{II}^{(0)}$ describes the modified II interactions.

Each heteronuclear spin–spin coupling J_{kl} is scaled by a factor $\cos\theta_k$, which may be different for each spin I_k. For strong r.f. fields, $|\gamma B_2| \gg |(\omega_2 - \omega_{Ik})| = |\Omega_{Ik}|$, the residual coupling constants J'_{kl} can be approximated by

$$J'_{kl} = J_{kl} \left| \frac{\Omega_{Ik}}{\gamma B_{\text{eff},k}} \right| \approx J_{kl} \left| \frac{\Omega_{Ik}}{\gamma B_2} \right|. \qquad (4.7.55)$$

Off-resonance decoupling has been used extensively to assign complicated ^{13}C spectra (4.273). By varying the position of the irradiation frequency ω_2 relative to the I spin spectrum, and observing the resulting changes in the scaling factors $\cos\theta_k$, correlations between the I and S spectra can be established (4.272, 4.274). The same information can however be obtained in a more systematic and unambiguous way by heteronuclear two-dimensional spectroscopy, as will be discussed in § 8.5.

The offset-dependence inherent to off-resonance decoupling can be removed by using multiple-pulse sequences (4.275, 4.276) to achieve a uniform scaling of all heteronuclear couplings.

4.7.5. Time-shared decoupling

The strong radio-frequency fields necessary for decoupling can cause severe leakage problems between transmitter and receiver in homonuclear experiments. To prevent leakage into the receiver, it has become customary to employ time-shared decoupling (4.277). As indicated in Fig. 4.7.6, the decoupler field is applied in the form of short r.f. pulses of duration τ_p between the sampling intervals. The reduction in sensitivity caused by the restricted sampling time, given by the factor

$$r = \sqrt{(\tau_s/\tau)} \qquad (4.7.56)$$

can be made negligible by applying sufficiently short r.f. pulses.

To estimate the efficiency of a decoupling sequence with small flip

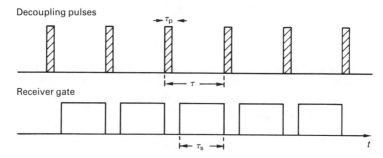

FIG. 4.7.6. Sequence for time-shared decoupling where the transmitter and receiver are operating in alternation.

angles and high repetition rates, it is allowed to consider the frequency spectrum of the pulse sequence. The amplitudes a_n of the pulse modulation sidebands at the frequencies $\omega_n = \omega_2 + 2\pi n/\tau$, $n = 0, \pm 1, \pm 2, \ldots$ are given by

$$a_n = B_p \frac{\sin(\pi n \tau_p/\tau)}{\pi n}, \qquad (4.7.57)$$

where B_p represents the peak value of the r.f. field. In particular, the amplitude a_0 of the centreband at the r.f. frequency ω_2 is just determined by the duty cycle τ_p/τ,

$$a_0 = B_p \frac{\tau_p}{\tau}, \qquad (4.7.58)$$

which represents the time average of the r.f. field. For a sufficiently high sampling rate $1/\tau$, only one of the pulse modulation sidebands lies within the spectrum of interest and all other sidebands may be neglected to a good approximation.

The influence of the additional sidebands has been discussed in connection with several related experiments. Pulsed r.f. fields have been used for homonuclear spin locking (4.278, 4.279) and ADRF experiments (4.280, 4.281), as well as for heteronuclear cross-polarization, using Hartmann–Hahn matching or ADRF (4.282, 4.283). Experimental evidence suggests that it is legitimate to picture the pulse train as a superposition of CW frequencies, and to retain only the frequencies that fall into the spectral range of interest, provided the repetition rate $1/\tau$ is sufficiently high, and the rotation angle $\theta = -\gamma B_p \tau_p$ of the individual pulses is small. Deviations from the simple picture arise when θ approaches an integer multiple of π; in this case, the time development of the spin system under the influence of the pulse train must be calculated explicitly (4.280, 4.281). The simple Fourier picture is inadequate to describe multiple-pulse sequences designed for the elimination of homonuclear dipolar couplings and for scaling of heteronuclear interactions.

4.7.6. Broadband decoupling and scaling of heteronuclear interactions

Spin decoupling with a monochromatic r.f. field is very sensitive to the resonance offset of the r.f. irradiation, and leads to inefficient decoupling when the I-spin resonance frequencies cover a wide range. The introduction of broadband decoupling techniques was of central importance for the development of carbon-13 resonance as a practical tool in organic chemistry. Techniques that are optimized for minimum r.f. power allow one to avoid intolerable heating of the sample.

The development of broadband decoupling techniques started with the introduction of noise decoupling (4.65), where a random or pseudo-random sequence of closely spaced r.f. pulses is applied, in order to chase the I-spin magnetization vectors randomly about the unit sphere. The effect is rather similar to that of rapid relaxation or chemical exchange of the I spins. Indeed, exchange broadening and exchange narrowing are phenomena that are also encountered in noise decoupling (4.65). In genuine exchange or relaxation processes, however, the random motion of each spin is independent, leading to a perfect ensemble average for the observable quantities. In the case of stochastic excitation, by contrast, all spins in the sample move coherently, and it is the average *in time* which is responsible for the decoupling effect (see § 4.1.6).

Originally, a rapid pseudo-random phase inversion was used to generate a stochastic irradiation (4.65). Numerous other modulation procedures have been proposed since, including square-wave and chirp modulation of the radio-frequency (4.262, 4.284, 4.285), however without significant improvement over noise decoupling.

These developments were often guided by the assumption that it would be sufficient to generate a sufficiently wide frequency spectrum to cover the relevant resonance frequencies. Such arguments implicitly assume a linear response of the spin system. Non-linear effects can be partly taken into account (4.65). However, the true transformation properties of the spins, such as the invariance under 2π-pulses, cannot be exploited with such a treatment. It is therefore not astonishing that none of these modulation techniques approaches ideal decoupling efficiency.

4.7.6.1. Multiple-pulse decoupling techniques

A novel approach, which no longer relies on spectral arguments, finally led to techniques with much improved decoupling efficiency (4.109–4.116). By applying sequences of composite pulses (see § 4.2.7) it is possible to achieve homogeneous decoupling over a wide frequency range with limited r.f. power. Of particular importance in the design of such sequences are the cyclic properties and the expansion procedures discussed in § 4.2.7.7. The MLEV-16 sequence of eqn (4.2.73) and the WALTZ-16 sequence of eqn (4.2.78) achieve high-order compensation of non-ideal behaviour. The WALTZ sequences have the practical advantage that the performance is not very sensitive to the adjustment of the 180° phase shift, while the MLEV methods rely on accurate 90° phase shifts. A comparison of the efficiency of the MLEV-64 sequence and of various modulation schemes is shown in Fig. 4.7.7.

4.7.6.2. Scaling of heteronuclear couplings

In some situations, complete decoupling may not be desirable, since all information about the magnitude of the heteronuclear couplings and the

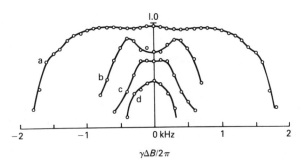

FIG. 4.7.7. Comparison of the performance of various decoupling schemes as a function of the offset of the decoupled nuclei. The experimental points indicate the height of the proton-decoupled carbon-13 signal of formic acid, which gives a measure of the line-narrowing efficiency. The amplitude of the decoupling field is $\gamma B_2/(2\pi) = 1.5$ kHz throughout. (a) MLEV-64 sequence with the composite inversion sequence $R = (\pi/2)_0(\pi)_{\pi/2}(\pi/2)_0$ of eqn (4.2.55). (b) Frequency-swept square-wave phase modulation. (c) 100-Hz square-wave phase modulation. (d) Noise decoupling with 1-kHz clock rate. (Reproduced from Ref. 4.112.)

multiplicities is lost. If the multiplet splittings are *scaled* by a uniform factor, this loss of information is avoided, while overlapping multiplets can be unravelled. The application of scaling techniques may be considered as an alternative to off-resonance decoupling (§ 4.7.4.2). The same information can be obtained with two-dimensional separation techniques (§ 7.2.2).

Uniform scaling of heteronuclear multiplet splittings can be achieved with multiple-pulse techniques (4.275, 4.288). These methods are related to the sequences suggested by Ellett and Waugh (4.286) for scaling of chemical shifts in liquids ('chemical shift concertina') and to the sequences recently introduced for chemical shift scaling in the context of magic angle spinning (4.287). Two-pulse cycles have been optimized for arbitrary scaling factors, leading to 'joint' and 'separated' pulse sequences (4.275).

More recently, it has been shown (4.276) that composite pulses can greatly improve the performance of scaling experiments with regard to the uniformity of the scaling factor with respect to offset, and with regard to the sensitivity to inhomogeneous r.f. fields.

4.7.7. Illusions of decoupling

Under ideal conditions, double resonance can lead to complete decoupling of I and S spins, and the resulting S spin spectrum does not reflect the presence of I spins or of the applied r.f. field. It is therefore tempting to simplify the Hamiltonian

$$\mathcal{H}(t) = \mathcal{H}_{ZI} + \mathcal{H}_{II} + \mathcal{H}_{IS} + \mathcal{H}_{ZS} + \mathcal{H}_{SS} + \mathcal{H}_I^{\text{r.f.}}(t) \qquad (4.7.61)$$

by neglecting all terms involving I spins, in order to obtain an effective Hamiltonian

$$\mathcal{H}_S = \mathcal{H}_{ZS} + \mathcal{H}_{SS}. \qquad (4.7.62)$$

As we shall demonstrate below, this simplification is not always justified (4.289, 4.290).

Consider for simplicity decoupling by continuous irradiation. With average Hamiltonian theory or perturbation theory, one can derive the following average Hamiltonian

$$\bar{\mathcal{H}} = \mathcal{H}_S + \bar{\mathcal{H}}_I \qquad (4.7.63)$$

with

$$\bar{\mathcal{H}}_I = \mathcal{H}_{II} + \mathcal{H}_I^{\text{r.f.}} \qquad (4.7.64)$$

where the I spin terms are expressed in a frame rotating with the r.f. frequency. In contrast to eqn (4.7.46), the r.f. perturbation term $\mathcal{H}_I^{\text{r.f.}}$ must in general be retained, since it cannot be assumed that the perturbation is cyclic and that the sampling is stroboscopic, as in eqn (4.7.40). Only terms that are 'orthogonal' to the perturbation, like $\mathcal{H}_{ZI}$ and $\mathcal{H}_{IS}$, are suppressed.

The question arises: under what circumstances can $\bar{\mathcal{H}}_I$ be disregarded? Let us consider the general situation of Fig. 4.7.8. The spin system is prepared in an initial state $\sigma(0)$ by some particular excitation sequence of duration τ_p before applying I-spin decoupling. Before the detection with the observable Q, another delay τ_d is inserted. Instead of transforming the density operator to account for the transformation in the τ_d-interval, we can use a transformed effective observable Q_{eff}

$$Q_{\text{eff}} = P^{-1}QP. \qquad (4.7.65)$$

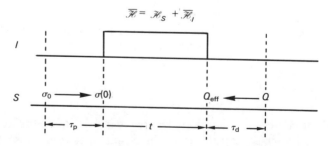

FIG. 4.7.8. Heteronuclear experiment which may be subject to illusions of decoupling. The decoupling period t is preceded and followed by periods τ_p and τ_d without decoupling. The initial density operator is denoted $\sigma(0)$ and the effective observable is Q_{eff}.

with the propagator P acting during the time τ_d. The expectation value of Q is

$$\langle Q \rangle(t) = \text{tr}\{Q_{\text{eff}} \exp(-i\mathcal{H}_S t)\exp(-i\bar{\mathcal{H}}_I t)\sigma(0)$$
$$\times \exp(i\bar{\mathcal{H}}_I t)\exp(i\mathcal{H}_S t)\}. \quad (4.7.66)$$

By using the invariance of the trace to cyclic permutations, we immediately find from eqn (4.7.66) that $\bar{\mathcal{H}}_I$ must affect the time evolution of the expectation value $\langle Q \rangle(t)$ if two conditions are simultaneously fulfilled (4.289, 4.290)

$$(1) \ [\bar{\mathcal{H}}_I, \sigma(0)] \neq 0,$$
$$(2) \ [\bar{\mathcal{H}}_I, Q_{\text{eff}}] \neq 0. \quad (4.7.67)$$

Thus it is only legitimate to drop the term $\bar{\mathcal{H}}_I$ and to use the simplified Hamiltonian $\bar{\mathcal{H}} = \mathcal{H}_S$ of eqn (4.7.62) if either the initial density operator $\sigma(0)$ or the effective observable Q_{eff} commutes with $\bar{\mathcal{H}}_I$. Because $\bar{\mathcal{H}}_I$ is a pure I spin Hamiltonian, both $\sigma(0)$ and Q_{eff} must contain I spin operators to make $\bar{\mathcal{H}}_I$ effective.

The most likely situation which leads to illusions of decoupling occurs for an initial state $\sigma(0)$ with antiphase coherence,

$$\sigma(0) = 2S_x I_z \quad (4.7.68)$$

which can be prepared by free precession under the heteronuclear J-coupling for a time $\tau_p = 1/(2J_{SI})$. In the course of spin decoupling, this coherence periodically oscillates between antiphase S spin coherence $2S_x I_z$ and heteronuclear two-spin coherence $2S_x I_y$

$$\sigma(t_d) = 2S_x I_z \cos \omega_{1I} t - 2S_x I_y \sin \omega_{1I} t \quad (4.7.69)$$

with $\omega_{1I} = -\gamma_I B_{1I}$. The oscillation frequency is proportional to the applied r.f. field strength. An inhomogeneous r.f. field will lead to a rapid decay of antiphase multiplet coherence. It should be noted that the term $2S_x I_z$ in eqn (4.7.69) can only be observed in the absence of decoupling. Two-dimensional experiments of this type can easily be conceived (4.289–4.291). In some cases, it is also possible to use this effect to suppress undesired components, for example in experiments using spherical or planar randomization (4.290, 4.292). Additional complications may occur in the presence of II interactions.

5
MULTIPLE QUANTUM TRANSITIONS

In the past, multiple-quantum transitions have often been neglected in favour of the more easily accessible single-quantum transitions. Many traditional texts on the practical aspects of magnetic resonance hardly mention the existence of the numerous stepbrothers of the one-quantum transitions. Although it is possible to observe multiple-quantum transitions by conventional slow-passage techniques, it is a thorny path and few spectroscopists have ventured to animate the sleeping beauty. Some of the brave scientists were Yatsiv (5.1) and Freeman, Anderson, and Reilly (5.2).

The situation did not much improve with the introduction of Fourier spectroscopy. On the contrary, one of the basic laws of Fourier spectroscopy states that only single-quantum coherence can be observed directly. However, it was soon recognized that indirect detection schemes, which may be considered as special forms of two-dimensional spectroscopy, permit elegant and convenient observation of all orders of multiple-quantum coherence. Two-dimensional spectroscopy initiated a real renaissance of multiple-quantum spectroscopy. Many contributions were made in the past few years by the research groups of Hashi (5.3–5.6), Pines (5.7–5.21), Ernst (5.22–5.41), Vold (5.42–5.47), Freeman (5.48–5.56), and Vega (5.57–5.59), while Bodenhausen (5.60) and Weitekamp (5.61) have written review articles on multiple-quantum NMR.

The discussion of the technical aspects of two-dimensional multiple-quantum NMR will be deferred to § 8.4. In this chapter, we shall discuss basic properties of multiple-quantum transitions which can make their observation desirable. Some of these interesting features are summarized below.

1. *Systems with $I = \frac{1}{2}$ spins in isotropic phase.* In scalar-coupled spin-$\frac{1}{2}$ systems in liquids, multiple-quantum frequencies contain additional information on the topology of the energy-level scheme. The information content is similar to that obtainable from double-resonance experiments. It allows, for example, the determination of the relative signs of coupling constants, the identification of magnetically equivalent spins, and the characterization of linear coupling networks (5.37) of the type AMX with $J_{AX} = 0$.

2. *Systems with $I = \frac{1}{2}$ spins in oriented phase.* The single-quantum spectra of dipolar-coupled spin-$\frac{1}{2}$ systems (solutes in liquid crystals, etc.)

are in most cases excessively complex and highly redundant. Because multiple-quantum spectra of high orders contain far fewer transitions, it is possible to match the number of observable lines to the number of unknown parameters (internuclear distances, bond angles, and order parameters) (5.12).

3. *Systems with spins $I > \frac{1}{2}$ in oriented phase.* Under favourable conditions, certain multiple-quantum transitions are insensitive to the nuclear quadrupole interaction. These transitions permit the observation of spectral features which are normally masked in the single-quantum spectrum by the much stronger quadrupolar interaction. This property has been exploited in deuterium double-quantum spectroscopy (5.7, 5.8) and in nitrogen-14 double-quantum spectroscopy (5.27, 5.30).

4. *Relaxation.* The transverse relaxation rates of multiple-quantum coherences (in particular the adiabatic contributions to the line-widths) contain relevant information about the power spectral densities that characterize the relaxation processes. Multiple-quantum transitions can reveal the correlation of relaxation mechanisms acting on different nuclei, which can often not be determined from single-quantum spectra (5.13, 5.25, 5.42–5.45).

5. *Lack of inhomogeneous broadening.* An additional feature of interest is the insensitivity of zero-quantum coherence to magnetic field inhomogeneity, which can be exploited to record high-resolution spectra in inhomogenous magnetic fields (5.22, 5.23).

6. *Decoupling.* Double-quantum transitions are important for decoupling quadrupolar nuclei like deuterium or nitrogen-14. Careful adjustment of the decoupler frequency to double-quantum transitions allows efficient decoupling with limited r.f. power (5.62–5.64).

Today, multiple-quantum spectroscopy has lost much of its mystical character and has become an integral part of pulse methodology. For this reason, we have integrated the majority of the subject matter into the following chapters 6–9. Coherence transfer through multiple-quantum order is discussed in § 6.3, and is exploited in multiple-quantum filtering (§ 8.3.3). Two-dimensional multiple-quantum spectroscopy is treated in detail in § 8.4. Multiple-quantum coherence can lead to artefacts in 2D exchange-type experiments as will be explained in § 9.4. On the other hand, it is also possible to exploit multiple-quantum order for the study of exchange networks (§ 9.4.3).

In this chapter we shall present some more basic aspects of multiple-quantum spectroscopy. We shall briefly discuss the number of transitions expected for various systems (§ 5.1), and review the traditional continuous-wave (CW) approach to multiple-quantum NMR in § 5.2. The limitations of this method can be largely overcome with two-dimensional indirect detection schemes. Various approaches for the excitation of multiple-quantum coherence are discussed in § 5.3, while transverse relaxation of multiple-quantum coherence is discussed in § 5.4.

5.1. Number of transitions

Consider a general system of N spins having a total of Λ energy levels. Except for studies in low magnetic fields and for very strong quadrupolar interaction, the Zeeman interaction gives the dominant contribution to the energy, and each eigenstate $|a\rangle$ may be characterized by its total magnetic quantum number

$$M_a = \langle a| F_z |a\rangle, \qquad (5.1.1)$$

defined as the expectation value of the z-component of the total spin operator

$$\mathbf{F} = \sum_{i=1}^{N} \mathbf{I}_i. \qquad (5.1.2)$$

M_a may take one of the values $-L, -L+1, \ldots, L-1, L$, where the total spin quantum number L is the sum of the spin quantum numbers of the component nuclei,

$$L = \sum_i I_i. \qquad (5.1.3)$$

A total of $\binom{\Lambda}{2}$ transitions is possible between the Λ levels. These transitions may be classified according to the magnetic quantum number change $p_{ab} = M_a - M_b$ which we shall call the *order* of the transition. Transitions with $|p| = 1$ are known as single-quantum transitions, abbreviated as 1QTs. Correspondingly, double-quantum, triple-quantum, and higher-order transitions will be referred to as 2QTs, 3QTs, and pQTs, respectively. There is a special class of transitions for which the two levels connected by the transition lie within a group of levels with the same magnetic quantum number M. These zero-quantum transitions are abbreviated as ZQTs.

In weakly-coupled spin systems, the quantum number M_k of the operator I_{kz} of spin k is a good quantum number, and a transition between eigenstates $|a\rangle$ and $|b\rangle$ can be characterized by a vector $\{\Delta M_1, \Delta M_2 \ldots \Delta M_k \ldots \Delta M_N\}$. In this case, we may distinguish combination lines, which are referred to as q-spin–pQTs, where q is the number of actively involved spins with $\Delta M_k \neq 0$ and the overall change of magnetic quantum number is $p = \sum \Delta M_k$. For spin-$\frac{1}{2}$ systems, the relation $q = p, p+2, p+4, \ldots$ holds.

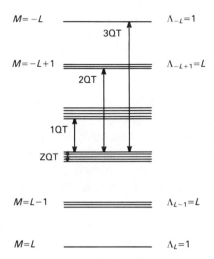

FIG. 5.1.1. Energy-level scheme of a multi-spin system with a total of Λ energy levels arranged in groups of Λ_M levels with magnetic quantum number M. Some examples of zero quantum transitions (ZQTs), single-quantum transitions (1QTs), and multiple-quantum transitions (2QTs, 3QTs) are indicated.

We shall see in § 6.3 that it is often necessary to distinguish orders p with different sign, particularly in view of experimental procedures for the separation of various orders. In this case, one must distinguish for each pair of eigenstates $|a\rangle$ and $|b\rangle$ two coherences (matrix elements σ_{ab} and σ_{ba} of the density operator), which are associated with $p_{ab} = M_a - M_b$ and $p_{ba} = M_b - M_a = -p_{ab}$, respectively. In this section, however, we focus attention on positive orders p, and count the number of transitions accordingly.

The energy level scheme of a system with Λ levels is shown schematically in Fig. 5.1.1. Note that the density of states, and hence the number of transitions, is greatest in the centre. Let us denote by the symbol Λ_M the number of energy levels with a certain value of M. The number Z_p of transitions of a given order p is

$$Z_p = \sum_{M=-L}^{L-p} \Lambda_M \Lambda_{M+p}, \qquad p = 1, 2, \ldots 2L, \qquad (5.1.4)$$

$$Z_0 = \sum_{M=-L+1}^{L-1} \binom{\Lambda_M}{2}. \qquad (5.1.5)$$

The class of transitions with the highest possible quantum number change $|p| = 2L$ contains only one single transition, $Z_{2L} = 1$.

For a system consisting of N non-equivalent spins-$\frac{1}{2}$, we obtain

$$Z_p = \binom{2N}{N-p}, \qquad p = 1, 2, \ldots, N, \tag{5.1.6}$$

$$Z_0 = \frac{1}{2}\left\{\binom{2N}{N} - 2^N\right\}. \tag{5.1.7}$$

The number of transitions for systems containing up to six non-equivalent spins is shown in Table 5.1.1. Provided the orders p can be observed separately (which is not possible in CW multiple-quantum NMR, as discussed in §5.2, but can be achieved by indirect two-dimensional detection as shown in §6.3), it is clear from this table that the number of

Table 5.1.1

Number of transitions in systems with N coupled $I = \frac{1}{2}$ spins. (Adapted from Ref. 5.25)

Spin system	Number of spins	Number of transitions						
		ZQT	1QT	2QT	3QT	4QT	5QT	6QT
AB	2	1	4	1				
ABC	3	6	15	6	1			
A_2B	3	2	9	4	1			
ABCD	4	27	56	28	8	1		
A_2BC	4	13	34	18	6	1		
A_2B_2	4	5	20	12	4	1		
A_3B	4	4	16	9	4	1		
ABCDE	5	110	210	120	45	10	1	
A_2BCD	5	60	128	76	31	8	1	
A_2B_2C	5	30	78	48	21	6	1	
A_3BC	5	24	61	38	18	6	1	
A_3B_2	5	10	37	24	12	4	1	
ABCDEF	6	430	792	495	220	66	12	1
A_2B_2CD	6	138	296	195	96	34	8	1
A_3BCD	6	109	232	155	80	31	8	1
$A_2B_2C_2$	6	72	180	123	62	24	6	1
A_3B_2C	6	57	142	97	52	21	6	1
A_3B_3	6	21	68	48	28	12	4	1

resonance lines can be reduced at will. Thus the complexity of the spectrum can be adapted to the number of parameters that have to be determined. In an isotropic scalar-coupled five-spin system for example, the 45 lines of the triple-quantum spectrum in principle suffice to determine all five shifts and 10 coupling constants. The single-quantum spectrum of the same system is highly redundant.

In systems containing magnetically equivalent nuclei, the number of transitions is reduced, as shown in Table 5.1.1. These numbers can be derived by noting that the eigenfunctions may be classified in terms of *group spin* quantum numbers (5.65, 5.66). Since all external magnetic or electric fields used for creating coherence and for observing resonance are homogeneous over molecular dimensions, the corresponding perturbation operator is always totally symmetric with respect to permutation of equivalent nuclei. As a consequence, it is not possible to induce coherence between states with different group spin quantum numbers, and the spectrum can be represented as a superposition of subspectra (5.65, 5.66). Equations (5.1.4) and (5.1.5) are valid for each subspectrum, and the results can be summed over all possible combinations of the group spin quantum numbers, leading to the numbers of transitions in Table 5.1.1.

5.2. Detection of multiple-quantum transitions by continuous-wave NMR

The well-known selection rule of magnetic resonance at high static magnetic fields

$$\Delta M = \pm 1 \tag{5.2.1}$$

originates from first-order time-dependent perturbation theory. When sufficiently strong r.f. fields B_1 are applied, it is necessary to consider the higher-order terms in the perturbation expansion. These terms will violate the first-order selection rule and allow one to observe multiple-quantum transitions (MQTs). The general rule is that MQTs can be observed whenever several connected single-quantum transitions are simultaneously excited by the applied strong r.f. field.

We do not intend to give a detailed account of CW detection of MQTs here, but will focus on those aspects which are relevant for a comparison with indirect detection schemes described in §§ 8.4 and 8.5. The basic theory of CW MQT detection was developed by Yatsiv (5.1). Bucci *et al.* (5.67) introduced a notation which is particularly suitable for practical calculations. The application of CW multiple-quantum NMR to high-resolution liquid-state spectroscopy was promoted by Anderson *et al.* (5.2), Kaplan and Meiboom (5.68), Cohen and Whiffen (5.69), Musher (5.70), and Martin *et al.* (5.71).

Transitions in CW spectroscopy can in general be explained by the crossing of two energy levels in the frame rotating with the applied r.f. frequency $\omega_{\text{r.f.}}$. The crossing (or in effect the avoided crossing in the presence of a perturbation) of two energy levels occurs when a multiple of the applied frequency $\omega_{\text{r.f.}}$ matches the corresponding transition frequency. Let us denote by $\{E_a\}$ the set of energy eigenvalues in the laboratory frame (measured in frequency units). In the rotating frame we obtain

$$E_a^{\text{r}} = E_a - M_a \omega_{\text{r.f.}} \tag{5.2.2}$$

where M_a is the magnetic quantum number of state $|a\rangle$. The crossing of the two levels $|a\rangle$ and $|b\rangle$ implies

$$E_a^{\text{r}} - E_b^{\text{r}} = E_a - E_b - (M_a - M_b)\omega_{\text{r.f.}} = 0 \tag{5.2.3}$$

or

$$\omega_{\text{r.f.}} = \omega_{ab}^{\text{CW}} = (E_a - E_b)/p_{ab} \tag{5.2.4}$$

with

$$p_{ab} = M_a - M_b.$$

Thus the transition frequency ω_{ab}^{CW} between the levels $|a\rangle$ and $|b\rangle$ is equal to the energy difference $E_a - E_b$ in the laboratory frame divided by the difference of magnetic quantum numbers $M_a - M_b$. In CW spectroscopy, the MQTs appear therefore within the same frequency range as the 1QTs. This is illustrated in Fig. 5.2.1 for a two-spin $\frac{1}{2}$ system. This figure

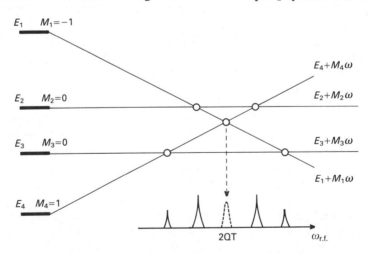

FIG. 5.2.1. Energy levels of a two-spin $\frac{1}{2}$ system in the rotating frame as a function of the radio-frequency $\omega_{\text{r.f.}}$. The crossing of two levels indicates a possible transition. A total of five transitions is possible including one double-quantum transition.

also demonstrates the general fact that zero-quantum transitions cannot be observed by CW techniques, as no level crossing is possible for states with $M_a - M_b = 0$.

It is well known that non-overlapping single-quantum transitions can be described in terms of 'simple lines' (5.72). Such transitions behave according to the Bloch equations for arbitrary r.f. field strength, provided the applied r.f. field only excites the considered transition. The motion involves an effective r.f. field and longitudinal and transverse relaxation times $T_1^{(ab)}$ and $T_2^{(ab)}$. This behaviour is not restricted to 1QTs, and equally applies to non-overlapping MQTs. They appear, therefore, as Lorentzian lines.

Yatsiv (5.1) derived a general expression for the shapes of MQTs observed by CW NMR. Following loosely the notation of Bucci et al. (5.67), we obtain for the lineshape of transition (ab)

$$v^{(ab)}(\omega) = N\gamma\hbar^2\omega^2(kT)^{-1}p_{ab}\alpha_{ab}^2(\gamma B_1)^{2p_{ab}-1} \frac{1/T_2^{\prime(ab)}}{(\omega - \omega_{ab}^{\prime CW})^2 + \left(\frac{1}{T_2^{\prime(ab)}}\right)^2(1+S)}. \tag{5.2.5}$$

The modified resonance frequency $\omega_{ab}^{\prime CW}$ of the p-quantum transition includes a level shift d_{ab} due to the effect of the strong r.f. field on the various energy levels

$$\omega_{ab}^{\prime CW} = \omega_{ab}^{CW} + d_{ab} = \frac{E_a - E_b}{p_{ab}} + d_{ab}. \tag{5.2.6}$$

The multiple-quantum line-width $1/T_2^{\prime(ab)}$ is scaled by the order of the transition

$$1/T_2^{\prime(ab)} = (1/T_2^{(ab)})/p_{ab} = R_{ab\,ab}/p_{ab} \tag{5.2.7}$$

where $R_{ab\,ab}$ is the corresponding matrix element of the Redfield matrix (see § 2.3.2). The spin–lattice relaxation rate $1/T_1^{\prime(ab)}$, on the other hand, occurs with reversed scaling

$$1/T_1^{\prime(ab)} = p_{ab}/T_1^{(ab)} \tag{5.2.8}$$

where $1/T_1^{(ab)}$ is the relaxation rate computed from the network of transitions in complete analogy to Abragam's treatment of the simple line (5.72). The rate $1/T_1^{\prime(ab)}$ appears in the saturation factor S of eqn (5.2.5), to be discussed in § 5.2.2. The transition matrix element α_{ab} is given in the next section. We now discuss separately the intensity, saturation, level shift, and line-width of multiple-quantum transitions.

5.2.1. Intensity of multiple-quantum transitions

The essential factors which determine the intensity of an MQT of order p_{ab} for low r.f. power are contained in the expression

$$I^{(ab)} \propto \alpha_{ab}^2 (\gamma B_1)^{2p_{ab}-1}. \tag{5.2.9}$$

The signal intensity increases proportionally to $B_1^{2p_{ab}-1}$. For higher transitions, therefore, higher r.f. fields are necessary to achieve an intensity comparable to that of single-quantum transitions.

The factor α_{ab}^2 represents the transition matrix element which is obtained by nth-order perturbation theory. It depends on the frequency differences involved and is given by

$$\alpha_{ab} = \left| \sum_{i,j,\ldots k} \frac{(I_{ai}I_{ij}\ldots I_{kb})^{\frac{1}{2}}}{p_{ib}(\omega_{ab}^{CW} - \omega_{ib}^{CW})p_{jb}(\omega_{ab}^{CW} - \omega_{jb}^{CW})\ldots p_{kb}(\omega_{ab}^{CW} - \omega_{kb}^{CW})} \right|$$

$$= \left| \frac{1}{(p_{ab}-1)!} \sum_{i,j,\ldots k} \frac{(I_{ai}I_{ij}\ldots I_{kb})^{\frac{1}{2}}}{(\omega_{ab}^{CW} - \omega_{ib}^{CW})(\omega_{ab}^{CW} - \omega_{jb}^{CW})\ldots (\omega_{ab}^{CW} - \omega_{kb}^{CW})} \right| \tag{5.2.10}$$

where I_{ij} are single-quantum transition intensities

$$I_{ij} = |\langle i| F^+ |j\rangle|^2 \tag{5.2.11}$$

and the summation extends over all possible 'ladders' of progressively connected single-quantum transitions leading from level $|a\rangle$ to level $|b\rangle$, as shown in Fig. 5.2.2. Note that eqn (5.2.10) only involves frequencies which are normalized according to eqn (5.2.4). These are MQT frequencies which would be observed in CW detection. The factor α_{ab}

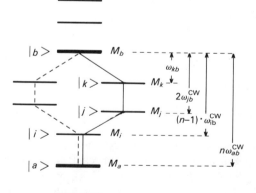

FIG. 5.2.2. Two progressive sequences of 1QTs leading from level $|a\rangle$ to level $|b\rangle$. All possible progressive sequences of this kind enter into the summation of eqn (5.2.10).

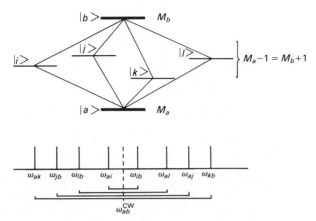

FIG. 5.2.3. The excitation of a double-quantum transition in a multi-level system requires the simultaneous perturbation of pairs of connected transitions lying symmetrically with respect to the frequency $\omega_{ab}^{\rm CW}$. The contributions are given by eqn (5.2.12). In a system with two spins $I = \frac{1}{2}$, there are only two intermediate states $|j\rangle$ and $|k\rangle$.

becomes large only when all transition frequencies ω_{ib}, ω_{jb}, ..., ω_{kb} of one branch of the ladder $a, i, j, k, \ldots, b$ lie near to the observed MQT frequency $\omega_{ab}^{\rm CW}$. The wider the spread of these frequencies, the weaker the MQT intensity. In principle, more extensive pathways from level $|a\rangle$ to level $|b\rangle$, which may include regressive connectivities, can also contribute to the MQT intensity. However, these terms appear with higher powers in γB_1 and can be neglected in most cases.

Consider for illustration a double-quantum transition in an arbitrary multilevel system. In this case, pairs of consecutive transitions between $|a\rangle$ and $|b\rangle$ contribute as indicated in Fig. 5.2.3, and we obtain the simple expression

$$\alpha_{ab} = 2\left|\left(\frac{(I_{ai}I_{ib})^{\frac{1}{2}}}{\omega_{ai} - \omega_{ib}} + \frac{(I_{aj}I_{jb})^{\frac{1}{2}}}{\omega_{aj} - \omega_{jb}} + \frac{(I_{ak}I_{kb})^{\frac{1}{2}}}{\omega_{ak} - \omega_{kb}} + \ldots\right)\right|. \quad (5.2.12)$$

The double-quantum transition appears in the centre between pairs of single-quantum transitions $(\omega_{ai}, \omega_{ib})$, $(\omega_{aj}, \omega_{jb})$, ... with $\omega_{ai} = \omega_{ai}^{\rm CW}$. The pair with the smallest frequency difference usually contributes most to the 2QT intensity. To excite a double-quantum transition, it is necessary to perturb simultaneously two connected single-quantum transitions. The nearer their resonance frequencies, the more effective is the perturbation.

5.2.2. Saturation of multiple-quantum transitions

The saturation behaviour differs from that of single-quantum transitions. For a single-quantum transition, the saturation parameter S in eqn (5.2.5)

is given by
$$S = T_1^{(ab)} T_2^{(ab)} (\gamma B_1)^2. \tag{5.2.13}$$

For a p-quantum transition, one obtains[1]
$$S = T_1'^{(ab)} T_2'^{(ab)} (\gamma B_1)^{2p_{ab}} \alpha_{ab}^2. \tag{5.2.14}$$

The optimum r.f. field strength for maximum signal intensity is
$$(\gamma B_1)_{\text{opt}} = [(2p_{ab} - 1)/(T_1'^{(ab)} T_2'^{(ab)} \alpha_{ab}^2)]^{1/(2p_{ab})}. \tag{5.2.15}$$

Again, for higher-order transitions, higher r.f. fields are necessary to reach the maximum signal amplitude. To obtain a rough estimate of the required r.f. field strength, we assume that the involved frequency differences $\omega_{ab}^{\text{CW}} - \omega_{ib}^{\text{CW}}$ are all of the same order of magnitude
$$\omega_{ab}^{\text{CW}} - \omega_{ib}^{\text{CW}} \approx \Delta, \tag{5.2.16}$$

and that only a single term contributes to the sum in eqn (5.2.10). One obtains then the approximate expression
$$(\gamma B_1)_{\text{opt}} = \left[(2p_{ab} - 1)\{(p_{ab} - 1)!\}^2 \frac{\Delta^{2p_{ab}-2}}{T_1^{(ab)} T_2^{(ab)}} \right]^{1/(2p_{ab})}. \tag{5.2.17}$$

For single-quantum transitions, this is consistent with the solution of the Bloch equations
$$(\gamma B_1)_{\text{opt}} = [T_1^{(ab)} T_2^{(ab)}]^{-\frac{1}{2}}. \tag{5.2.18}$$

For double-quantum transitions, we obtain from eqn (5.2.17)
$$(\gamma B_1)_{\text{opt}} = [3\Delta^2/(T_1^{(ab)} T_2^{(ab)})]^{\frac{1}{4}}. \tag{5.2.19}$$

The optimum r.f. field is proportional to the geometric mean of the frequency difference Δ and the mean relaxation rate $[T_1^{(ab)} T_2^{(ab)}]^{-\frac{1}{2}}$.

5.2.3. Level shift of multiple-quantum transitions

The strong r.f. field necessary to excite MQTs causes Bloch–Siegert-type shifts of neighbouring single-quantum resonance lines. Whenever these transitions involve one of the two levels $|a\rangle$ and $|b\rangle$ of the observed MQT, the latter will experience a shift d_{ab}. For single-quantum transitions and usually also for 2QTs, the level shift lies well within the line-width. For higher transitions, however, an observable shift may be produced.

From eqn (2.33) of Yatsiv (5.1) one can deduce the following

---

[1] Note that $T_1'^{(ab)} T_2'^{(ab)} = T_1^{(ab)} T_2^{(ab)}$ (see eqns (5.2.7) and (5.2.8)).

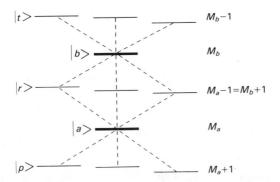

FIG. 5.2.4. Energy levels $|p\rangle$, $|r\rangle$, and $|t\rangle$ contributing to the level shift of the 2QT between $|a\rangle$ and $|b\rangle$ according to eqn (5.2.21).

expression[1] for the shift of an MQT of order p_{ab}

$$d_{ab} = \frac{1}{p_{ab}}(\gamma B_1)^2 \sum_k{}' \frac{I_{ak} - I_{kb}}{p_{kb}(\omega_{kb}^{CW} - \omega_{ab}^{CW})}. \tag{5.2.20}$$

Only those energy levels $|k\rangle$ that are connected by single-quantum transitions to the levels $|a\rangle$ and/or $|b\rangle$ contribute to the shift. The reason for this restriction lies in the fact that eqn (5.2.20) has been derived by second-order perturbation theory. The higher-order terms are usually negligible.

Anderson et al. (5.2) simplified eqn (5.2.20) for 2QTs. They distinguish three classes of energy levels

r-levels with $M_r = M_a - 1 = M_b + 1$

t-levels with $M_t = M_b - 1$

p-levels with $M_p = M_a + 1$.

This classification, which is illustrated in Fig. 5.2.4, leads to the expression

$$d_{ab} = -(\gamma B_1)^2 \left\{ \sum_r \frac{I_{ar} - I_{rb}}{\omega_{ar}^{CW} - \omega_{rb}^{CW}} + \sum_t \frac{I_{tb}}{2(\omega_{ab}^{CW} - \omega_{bt}^{CW})} + \sum_p \frac{I_{ap}}{2(\omega_{ab}^{CW} - \omega_{pa}^{CW})} \right\}. \tag{5.2.21}$$

The first term of this expression shows that an r-level contributes to the level shift only if transitions (ar) and (rb) have different intensities. In a two-spin $\frac{1}{2}$ system, this term is cancelled because the intensities are equal. In this case, the 2QT will not experience a shift, irrespective of the

---

[1] It is essential to recognize that $\omega_{kb}^{CW} = \omega_{bk}^{CW}$ according to the definition, eqn (5.2.4), and that all ω_{ik}^{CW} have the same sign, provided only gyromagnetic ratios of equal sign are involved.

applied r.f. field strength. However, d_{ab} can usually not be neglected in a three-spin system.

It should be noted that level shifts are rather undesirable and make precise measurements with CW methods difficult. This problem is eliminated in the indirect detection schemes that are described in §§ 5.3 and 8.4.

5.2.4. Line-widths of multiple-quantum transitions

An intriguing fact with regard to line-widths is the appearance of the factor $1/p_{ab}$ in eqn (5.2.7), which indicates that for a homogeneous magnetic field multiple-quantum transitions of order p are narrower by a factor p than the corresponding single-quantum transitions, bearing in mind that the matrix elements $R_{ab\,ab}$ are usually of the same order of magnitude for all transitions.

The factor $1/p$ has its origin in the fact that one observes only one of the p quanta that make up the transition. In this respect, CW techniques seem to have advantages over the indirect detection schemes discussed in §§ 5.3 and 8.4, where this line-narrowing factor $1/p$ is lost. However, in two-dimensional multiple-quantum NMR, the full precession of all p quanta determines the observed time-dependence, while in CW detection the precession frequencies are scaled also by the factor p. The effective resolution is therefore the same in either mode of observation.

In experimental CW spectra, one normally observes nearly equal widths for resonance lines of all orders (5.2), because the dominant line-broadening mechanism is the magnetic field inhomogeneity.

5.2.5. Applications of CW multiple-quantum NMR

For the sake of illustration, we consider the application of CW multiple-quantum NMR to the analysis of scalar-coupled systems in isotropic phase. In complicated spectra, multiple-quantum transitions make it possible to identify pairs of progressively connected single-quantum transitions and hence to correlate the conventional spectrum with the energy-level scheme.

The conventional 1QT proton spectrum (5.2) of the $(ABC)_3X$ system of trivinyl phosphine is shown in the top trace of Fig. 5.2.5. The assignment to the energy level scheme can be made by increasing the r.f. field strength. The double-quantum transitions (marked by vertical lines) appear at frequencies that correspond to the average of two progressively connected single-quantum transitions, in accordance with Fig. 5.2.3. The connected single-quantum frequencies are indicated by the tips of the horizontal bars. At still higher r.f. field amplitudes, two triple-quantum

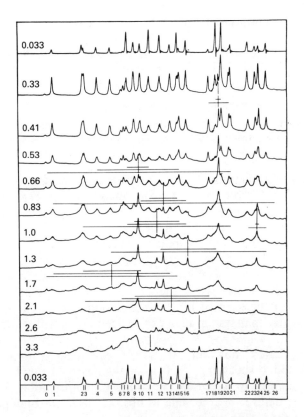

FIG. 5.2.5. Proton spectra of the $(ABC)_3X$ system of trivinyl-phosphine recorded at 60 MHz. The r.f. field value $\gamma B_1/2\pi$ in units of Hz is given at the left of each trace. The double-quantum transitions are marked by vertical lines with crossbars to identify the associated progressive pairs of transitions, and triple-quantum transitions by vertical lines without crossbars. (Reproduced from Ref. 5.2.)

transitions appear (vertical lines without crossbars) at the arithmetic mean of three progressively connected single-quantum transitions.

Although the information contained in Fig. 5.2.5 suffices for an unambiguous assignment, the CW approach suffers from a number of drawbacks: the required r.f. field amplitudes are not known *a priori*, and a single setting is usually not sufficient to observe all multiple-quantum transitions. The different orders ($p = 1, 2, 3$) cannot be separated, and the narrow multiple-quantum lines overlap with saturation-broadened single-quantum lines. Most of these practical problems can be circumvented with the indirect two-dimensional detection schemes (§§ 5.3 and 8.4).

5.3. Time-domain multiple-quantum spectroscopy

In time-domain multiple-quantum spectroscopy, multiple-quantum transitions are detected indirectly in a three-step process, in contrast to CW detection.

1. Excitation of multiple-quantum coherence;
2. Free evolution, normally in the absence of r.f. perturbations;
3. Conversion into observable single-quantum coherence which is subsequently detected.

This naturally leads to the basic form of two-dimensional spectroscopy to be discussed in Chapter 6. Step 1 corresponds to the preparation period, step 2 to the evolution period, and step 3 comprises the mixing and detection periods (see e.g. Fig. 6.1.2).

A complete time-domain multiple-quantum experiment produces a two-dimensional spectrum where the multiple-quantum frequencies appear along the ω_1-axis while the ω_2-domain contains the single-quantum frequencies. In many cases, the correlation of multiple- and single-quantum frequencies in such a two-dimensional representation provides insight into the spin system under investigation, as will be demonstrated in § 8.4.

In some cases, however, it is not essential to display the full two-dimensional spectrum, and a one-dimensional multiple-quantum spectrum suffices to represent the relevant information, such as the multiplet structure or the line-widths of selected multiple-quantum transitions. One-dimensional multiple-quantum spectra can be obtained by projecting a two-dimensional spectrum on to the ω_1-axis, or, more easily, by monitoring the amplitude of the single-quantum magnetization at a fixed point in time after the reconversion step as a function of the evolution period t_1, and converting the signal $s(t_1)$ into a spectrum $S(\omega_1)$ by a one-dimensional Fourier transformation (see § 6.5.5).

In this section, we discuss the various 'building blocks' consisting of pulse sequences used for the creation of multiple-quantum coherence and for its reconversion into observable single-quantum coherence.

In contrast to CW detection, time-domain spectroscopy allows one to separate signals associated with different orders of multiple-quantum transitions by shifting the signals along the frequency axis, either by exploiting the offset-dependence (§ 5.3.2), or by time-proportional phase incrementation (§ 6.6.3). The spectra can be simplified by selecting signals from a unique order and eliminating all other signals by a phase cycle, as described in § 6.3. These procedures allow one to draw advantage of the statistics of the number of transitions in Table 5.1.1.

Basic properties of multiple-quantum coherence, such as the characteristic multiplet structure, and the relaxation of multiple-quantum

5.3 TIME-DOMAIN MULTIPLE-QUANTUM SPECTROSCOPY

coherence will be discussed in §§ 5.4 and 5.5, respectively, while the use of multiple-quantum spectroscopy for the analysis of complex spin systems will be deferred to § 8.4.

5.3.1. Excitation and detection of multiple-quantum coherence

The basic scheme used for two-dimensional multiple-quantum NMR of homonuclear systems is shown in Fig. 5.3.1(a). Acting on a system in

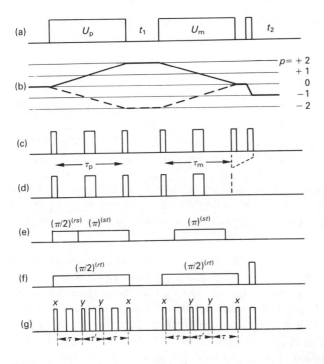

FIG. 5.3.1. (a) Basic scheme for homonuclear multiple-quantum NMR. The preparation and mixing propagators U_p and U_m, respectively, transform Zeeman polarization into multiple-quantum coherence and vice versa; the resulting t_1-modulated longitudinal polarization is converted into observable, transverse magnetization by a 'read' pulse. (b) Coherence transfer pathways for double-quantum NMR. Phase-cycling schemes allow either selective detection of a unique $p = 0 \to +2 \to -1$ pathway, or else retain a superposition of two pathways $p = 0 \to \pm 2 \to -1$, as described in § 6.3. (c) In systems where non-selective pulses can be applied, the propagators U_p and U_m can consist of two identical sequences $[(\pi/2) - \tau/2 - (\pi) - \tau/2 - (\pi/2)]$. (d) In practical realizations, the last pulse of the reconversion sequence in (c) and the read pulse can be dropped. The reconversion boils down to a single $\pi/2$-pulse followed by a $[\tau/2 - (\pi) - \tau/2]$-sequence which allows antiphase coherence to be converted into in-phase coherence. The pulse phases are defined in the text. (e) Cascade of selective $\pi/2$- and π-pulses applied to connected single-quantum transitions (§ 5.3.1.2). (f) Selective pulses applied at CW multiple-quantum frequencies (§ 5.3.1.3). (g) Spin-connectivity selective sequence: this example is effective in A_nX-type coupling networks (§ 5.3.1.4).

thermal equilibrium, the preparation propagator U_p excites the desired multiple-quantum coherences, which freely precess in the evolution period t_1 and are transferred back into (t_1-modulated) longitudinal polarization ($p = 0$) by a mixing propagator U_m. This can be converted into observable magnetization ($p = -1$) by a 'read' pulse, which in practice can be merged with the sequence U_m. Phase-cycling is used either to select a unique pathway $0 \rightarrow +p \rightarrow -1$, or else to allow two mirror-image pathways simultaneously ($0 \rightarrow \pm p \rightarrow -1$), as discussed in § 6.3 and illustrated in Fig. 5.3.1(b) for the coherence transfer process $p = 0 \rightarrow \pm 2 \rightarrow -1$.

The experimental realization of the propagators U_p and U_m depends on the system under investigation. In homonuclear spin systems, at least six strategies may be employed.

1. A sequence of two or more non-selective pulses separated by free evolution intervals (Fig. 5.3.1(c), (d)).

2. A cascade of two or more selective pulses applied to connected single-quantum transitions (Fig. 5.3.1(e)).

3. A selective multiple-quantum pulse applied at the frequency $\omega_{r.f.}^{CW} = (E_r - E_t)/p_{rt}$ where the multiple-quantum transition appears in CW spectroscopy (Fig. 5.3.1(f)).

4. A sequence of non-selective pulses designed to be effective in coupled systems with a particular coupling topology (Fig. 5.3.1(g)).

5. A sequence of non-selective pulse sandwiches for selective excitation of a specified order p (Fig. 5.3.2).

Additional excitation schemes can be designed for heteronuclear systems, as will be discussed in § 8.5.5. All methods listed above can be adapted for studies in isotropic or oriented media. Selective pulse methods obviously require some knowledge of the frequencies and connectivities of the transitions. This requirement is often fulfilled in oriented systems with quadrupolar spins (^{2}D, ^{14}N, ^{27}Al, etc.), but rarely in dipolar- or scalar-coupled $I = \frac{1}{2}$ systems. Multiple-quantum spectra of unknown systems are best obtained with schemes that employ non-selective pulses.

5.3.1.1. Non-selective pulses

The pulse sequence

$$P = [(\pi/2)_x - \tau/2 - (\pi)_x - \tau/2 - (\pi/2)_x] \qquad (5.3.1)$$

employed in Fig. 5.3.1(c) can be used as excitation sequence in a wide range of systems involving scalar, dipolar, or quadrupolar couplings. The central π-pulse removes, for weak coupling, the effect of chemical shifts, without affecting the couplings. These lead to a transformation of

5.3 TIME-DOMAIN MULTIPLE-QUANTUM SPECTROSCOPY

in-phase single-quantum coherence (created by the first $\pi/2$-pulse) into antiphase coherence which is converted into multiple-quantum coherence by the third pulse at the end of the τ-interval. In a system with two weakly scalar-coupled spins $I_k = I_l = \frac{1}{2}$, the excitation sequence is described by the following cascade of transformations:

$$\sigma(0_-) = I_{kz} + I_{lz}$$
$$\downarrow (\pi/2)_x$$
$$\sigma(0_+) = -I_{ky} - I_{ly}$$
$$\downarrow [\tau/2 - (\pi)_x - \tau/2] \text{ with } \tau = \tau_{opt} = (2J_{kl})^{-1}$$
$$\sigma(\tau_-) = -2I_{kx}I_{lz} - 2I_{kz}I_{lx}$$
$$\downarrow (\pi/2)_x$$
$$\sigma(\tau_+) = 2I_{kx}I_{ly} + 2I_{ky}I_{lx}$$
$$= \frac{1}{i}(I_k^+ I_l^+ - I_k^- I_l^-). \quad (5.3.2)$$

Thus one obtains a superposition of $p = +2$ and $p = -2$ quantum coherences.

In general, a more concise description of the excitation can be obtained if the overall effect of the pulse sandwich $[(\pi/2)_x - \tau/2 - (\pi)_x - \tau/2 - (\pi/2)_x]$ is formulated in terms of a bilinear rotation (see eqn (2.1.99) and Fig. 2.1.5)

$$\sigma(0_-) \xrightarrow{\Sigma \pi J_{kl} \tau 2 I_{ky} I_{ly}} \sigma(\tau_+). \quad (5.3.3)$$

For a two-spin system initially in thermal equilibrium one obtains after the pulse sandwich

$$\sigma(\tau_+) = (I_{kz} + I_{lz})\cos \pi J_{kl} \tau$$
$$+ (2I_{kx}I_{ly} + 2I_{ky}I_{lx})\sin \pi J_{kl} \tau. \quad (5.3.4)$$

Clearly, the polarization is entirely converted into double-quantum coherence if $\tau = \tau_{opt} = (2J_{kl})^{-1}$. In this case, which is sometimes referred to as 'matched preparation delay', one may speak of a yield of conversion of 100 per cent.

In practice however, one is confronted with an unknown distribution of scalar, dipolar, or quadrupolar coupling constants, and the sinusoidal dependence in eqn (5.3.4) is a hindrance to uniform excitation. This problem can be alleviated by variation of the interval τ in concert with the evolution time t_1 of the two-dimensional experiment (5.37). Uniform excitation can also be achieved by co-addition of a set of experiments where the interval τ is varied simultaneously in *both* excitation and reconversion sequences (5.14, 5.19, 5.35, 5.73).

In systems with weak couplings, the sandwich $[(\pi/2)_x - \tau/2 - (\pi)_x - \tau/2 - (\pi/2)_x]$ can only excite coherences of *even* orders. To excite coherences of odd orders (e.g. triple-quantum coherence), one may use the phase-shifted sequence

$$P = [(\pi/2)_x - \tau/2 - (\pi)_y - \tau/2 - (\pi/2)_y]. \quad (5.3.5)$$

This sequence can be expanded formally by inserting dummy pulses (5.38)

$$P = [(\pi/2)_x (\pi/2)_{-y} (\pi/2)_y - \tau/2 - (\pi)_y - \tau/2 - (\pi/2)_y]. \quad (5.3.6)$$

In this modified form, it is obvious that the last three pulses form again a bilinear rotation, and the entire transformation is described by

$$\sigma(0) \xrightarrow{(\pi/2)\Sigma I_{kx}} \xrightarrow{-(\pi/2)\Sigma I_{ky}} \xrightarrow{\Sigma \pi J_{kl}\tau 2I_{kx}I_{lx}} \sigma(\tau). \quad (5.3.7)$$

If we start from thermal equilibrium, the first pulse generates in-phase $-I_{ky}$ coherence which is not affected by the second term in the cascade. In fact, these two pulses may be considered as a composite $\pi/2$ pulse, eqn (4.2.51). For the sake of illustration, consider a three-spin system with $\sigma(0) = I_{kz} + I_{lz} + I_{mz}$. Assuming for simplicity that $\pi J_{kl}\tau = \pi J_{km}\tau = \pi J_{lm}\tau = \pi/2$, one obtains after the sandwich of eqn (5.3.5) a superposition of triple-quantum coherence and three-spin–single-quantum coherence (leading to combination lines)

$$\begin{aligned}\sigma(\tau) &= 4I_{ky}I_{lx}I_{mx} + 4I_{kx}I_{ly}I_{mx} + 4I_{kx}I_{lx}I_{my} \\ &= \frac{3}{2i}(I_k^+ I_l^+ I_m^+ - I_k^- I_l^- I_m^-) \\ &\quad + \frac{1}{2i}[(I_k^+ I_l^+ I_m^- - I_k^- I_l^- I_m^+) \\ &\quad + (I_k^+ I_l^- I_m^+ - I_k^- I_l^+ I_m^-) \\ &\quad + (I_k^- I_l^+ I_m^+ - I_k^+ I_l^- I_m^-)]. \end{aligned} \quad (5.3.8)$$

The appearance of three-spin–single-quantum combination coherences highlights the difficulties that are inherent to the excitation of pure higher-order coherences.

The sandwich of non-selective pulses of eqn (5.3.1) is applicable for the excitation of multiple-quantum coherence of quadrupolar spins in oriented phase, provided the amplitude of the r.f. pulses exceeds the quadrupolar splittings. Thus in oriented spins with $S = 1$ with an axially symmetric quadrupole tensor ($\eta = 0$) (eqn (2.2.24)), the pulse sequence

5.3 TIME-DOMAIN MULTIPLE-QUANTUM SPECTROSCOPY

$[(\pi/2)_x - \tau/2 - (\pi)_x - \tau/2 - (\pi/2)_x]$ leads to the transformation (5.38)

$$\sigma(0) \xrightarrow{\omega_Q \tau S_y^2} \sigma(\tau). \qquad (5.3.9)$$

Starting from thermal equilibrium ($\sigma(0) = S_z$), one obtains

$$\sigma(\tau) = S_z \cos \omega_Q \tau + \{S_x S_y + S_y S_x\} \sin \omega_Q \tau \qquad (5.3.10)$$

where the splitting of the two single-quantum transitions is $2\omega_Q$, and the term in curly brackets corresponds to pure double-quantum coherence (5.38). If relaxation is neglected, the conversion yield is 100 per cent for $\tau = \pi/(2\omega_Q)$.

5.3.1.2. Selective single-quantum pulses

If the transitions are well-resolved, and if their assignment in terms of the energy-level scheme is known, multiple-quantum coherence can be excited efficiently with a cascade of selective pulses applied in succession to allowed transitions that share common energy levels. Consider for simplicity a three-level system with the eigenstates $|r\rangle$, $|s\rangle$, and $|t\rangle$ with two allowed single-quantum transitions at ω_{rs} and ω_{st}. Two strategies can be employed for the excitation of the forbidden (double- or zero-quantum) coherence $|r\rangle\langle t|$.

1. A selective $(\pi/2)_y^{(rs)}$-pulse creates single-quantum coherence $|r\rangle\langle s|$, which is converted by a selective $(\pi)_y^{(st)}$ pulse into the desired coherence $|r\rangle\langle t|$, as may be derived from eqn (2.1.129) (Fig. 5.3.1(e)). This procedure can be extended to larger spin systems by applying a cascade of selective π-pulses to a 'ladder' of connected transitions. This method has been used in ^{27}Al ($I = \frac{5}{2}$) in single crystals, where the quadrupole interaction leads to well-separated single-quantum transitions (5.3).

2. An alternative approach uses first a $(\pi)^{(rs)}$-pulse to invert $I_z^{(rs)}$ selectively, in order to create a non-equilibrium population distribution, which is then converted into multiple quantum coherence by a *non-selective* $\pi/2$-pulse (5.22). In weakly-coupled systems, all multiple-quantum coherences in which the spin that has an inverted transition is actively involved are excited uniformly in this manner (5.74).

5.3.1.3. Selective multiple-quantum pulses

It was shown in § 5.2 that multiple-quantum transitions may be excited by CW spectroscopy. If we consider a p-quantum transition with $p = p_{ab} = M_a - M_b$ between two states $|a\rangle$ and $|b\rangle$, it is possible to excite the coherence $|a\rangle\langle b|$ by a selective pulse at the frequency $\omega_{ab}^{CW} = (E_a - E_b)/p_{ab}$ (eqn (5.2.4)). In the context of two-dimensional spectroscopy, such pulses can be used to convert polarization into coherence and vice versa (Fig. 5.3.1(f)) (5.8).

Basically, the external r.f. field can interact only with single-quantum transitions. By the concerted and coherent action of the pulse on 'ladders' of allowed, connected transitions, the coherence is transferred to the multiple-quantum transition. Thus in a three-level system $|a\rangle$, $|i\rangle$, and $|b\rangle$, the double-quantum coherence $|a\rangle\langle b|$ is excited by the concerted effect of the r.f. field on the two single-quantum transitions (a, i) and (i, b). To derive the effective rotation angle associated with a selective p-quantum pulse by perturbation theory of higher order, it is convenient to use the formalism of Yatsiv (5.1), which has been reviewed in §5.2. In the rotating frame, one obtains for a density matrix element σ_{ab} associated with a p-quantum transition between the states $|a\rangle$ and $|b\rangle$

$$\dot{\sigma}_{ab}(t) = -[i(\omega_{ab}^{\text{CW}} + d_{ab} - \omega) - R_{ab\,ab}]\sigma_{ab}(t)$$
$$+ \frac{(\gamma B_1)^{p_{ab}} \alpha_{ab}}{p_{ab}} [\sigma_{aa}^0 - \sigma_{bb}^0] \quad (5.3.11)$$

where d_{ab} is the level shift defined in eqn (5.2.20), $R_{ab\,ab}$ describes relaxation, and the factor α_{ab} is given by eqn (5.2.10). The effective r.f. field describing the excitation of p-quantum coherence is determined by

$$\gamma B_{\text{eff}} = \frac{(\gamma B_1)^{p_{ab}} \alpha_{ab}}{p_{ab}}$$
$$\propto \gamma B_1 \left[\frac{\gamma B_1}{\Delta}\right]^{p_{ab}-1} \quad (5.3.12)$$

with the frequency difference $\Delta \approx (\omega_{ab}^{\text{CW}} - \omega_{ib}^{\text{CW}})$ defined in eqn (5.2.16).

In the case of a spin $I = 1$ oriented in an anisotropic environment, assuming that the r.f. carrier is positioned in the middle between the two allowed transitions that are separated by $2\omega_Q$, and neglecting second-order quadrupole effects, the excitation of double-quantum coherence by a selective pulse of duration τ involves the effective r.f. rotation angle (5.24, 5.57)

$$\beta_{\text{eff}} = -\gamma B_{\text{eff}}\tau = \frac{\omega_1}{\omega_Q} \omega_1 \tau. \quad (5.3.13)$$

For triple-quantum coherence in an oriented $S = \frac{3}{2}$ spin, one obtains an effective rotation angle
$$\beta_{\text{eff}} = \frac{3}{8} \frac{\omega_1^2}{\omega_Q^2} \omega_1 \tau. \quad (5.3.14)$$

In the case of a pair of scalar-coupled $I = \frac{1}{2}$ nuclei with a chemical shift difference $\Delta\Omega = \Omega_A - \Omega_B$, one finds for the double-quantum transition

$$\beta_{\text{eff}} = \frac{2\pi J}{\Delta\Omega/2} \frac{\omega_1}{\Delta\Omega/2} \omega_1 \tau. \quad (5.3.15)$$

These three cases, which can be calculated in closed analytical form, are consistent with the general behaviour described by eqn (5.3.12). Note that in eqn (5.3.15), two factors contribute to reduce the efficiency of the r.f. field: a factor $\omega_1/(\Delta\Omega/2)$ that corresponds to the factor ω_1/ω_Q in eqn (5.3.13), and an additional factor $2\pi J/(\Delta\Omega/2)$ which expresses the fact that selective excitation of double-quantum coherence is only possible if the two spins are sufficiently strongly coupled.

5.3.1.4. Spin-connectivity selective excitation

The excitation of multiple-quantum coherence is strongly dependent on the structure of the spin system. This feature can be exploited for distinguishing and separating various subsystems in a complex spectrum. It will be shown in § 8.3.3 that filtering procedures of this type can be applied to one- and two-dimensional spectra. In a p-quantum filter, for example, p-quantum coherence is excited, rejecting all those spin systems which are not capable of carrying p-quantum coherence, in particular systems with less than p coupled $I = \frac{1}{2}$ spins. Phase cycles can be used to select the desired coherence transfer pathways (§ 6.3).

In isotropic phase, it is often the *topology* of the network of scalar couplings, rather than the total number of coupled spins, that is characteristic of the molecular species. Thus four-spin systems may occur as AMKX, A_2MX, A_2X_2, or A_3X systems. The coupling network of the A_3X system may be represented as a three-pointed star, with the X proton interacting equally with all three A protons, which have apparently no interactions among each other. The A_2X_2 network, on the other hand, may be thought of as a square with the equivalent nuclei at opposite vertices.

It is possible to design pulse sequences which efficiently excite multiple-quantum coherence in coupling networks of specified topology. If the system departs from this topology, the creation of coherence is inhibited. It is therefore possible to distinguish molecular fragments having the same number of spins but different coupling patterns (5.39, 5.75).

Consider by way of example the pulse sequence shown in Fig. 5.3.1(g), which is 'tailored' for excitation of multiple-quantum coherence in star-like networks with an odd number of apices. To excite $n+1$-quantum coherence in systems of the A_nX type with n odd, it is necessary to choose $\tau = (2J)^{-1}$ and $\tau' = \tau/n$. The central part of the excitation sequence in Fig. 5.3.1(g), $[(\pi/2)_y - \tau'/2 - (\pi)_y - \tau'/2 - (\pi/2)_y]$, is described for a weakly-coupled spin system by the propagator

$$U' = \exp\left\{-i\sum_{kl} \pi J_{kl} 2I_{kx}I_{lx}\tau'\right\} \quad (5.3.16)$$

with $\tau' = \tau_{\text{opt}}/n = (2nJ)^{-1}$ which is transformed by the embracing free precession periods with $\tau = \tau_{opt} = (2J)^{-1}$ and finally by the $(\pi/2)_x$ pulses. For an A_3X system, the complete propagator U contains only terms in which all four spins are involved. If the spins are numbered I_1, I_2, I_3, and I_4 for spins X, A, A', and A'', respectively, one obtains with $\tau' = (6J)^{-1}$ and $\tau = (2J)^{-1}$

$$U = \exp\left\{-i\frac{\pi}{2}\tfrac{1}{2}(I_1^+I_2^+I_3^+I_4^+ + I_1^-I_2^-I_3^-I_4^-)\right\}$$

$$\times \exp\left\{-i\frac{\pi}{2}\tfrac{1}{2}(I_1^-I_2^+I_3^+I_4^+ + I_1^+I_2^-I_3^-I_4^-)\right\}$$

$$\times \exp\left\{+i\frac{\pi}{6}[\tfrac{1}{2}(I_1^+I_2^-I_3^+I_4^+ + I_1^-I_2^+I_3^-I_4^-)\right.$$

$$+ \tfrac{1}{2}(I_1^+I_2^+I_3^-I_4^+ + I_1^-I_2^-I_3^+I_4^-)$$

$$\left. + \tfrac{1}{2}(I_1^+I_2^+I_3^+I_4^- + I_1^-I_2^-I_3^-I_4^+)]\right\}$$

$$\times \exp\left\{+i\frac{\pi}{6}[\tfrac{1}{2}(I_1^+I_2^+I_3^-I_4^- + I_1^-I_2^-I_3^+I_4^+)\right.$$

$$+ \tfrac{1}{2}(I_1^+I_2^-I_3^+I_4^- + I_1^-I_2^+I_3^-I_4^+)$$

$$\left. + \tfrac{1}{2}(I_1^+I_2^-I_3^-I_4^+ + I_1^-I_2^+I_3^+I_4^-)]\right\}. \quad (5.3.17)$$

It should be noted that all operator product pairs in eqn (5.3.17) commute with each other, since they represent single transition operators acting on non-connected transitions. Only the first term, which can be written in the form

$$U^{(\alpha\alpha\alpha\alpha,\beta\beta\beta\beta)} = \exp\left\{-i\frac{\pi}{2}I_x^{(\alpha\alpha\alpha\alpha,\beta\beta\beta\beta)}\right\} \quad (5.3.18)$$

is responsible for the excitation of four-quantum coherence. This term converts the population difference between the two levels of maximum and minimum magnetic quantum number into pure four-quantum coherence. The remaining terms in the propagator of eqn (5.3.17) generate in addition four-spin–two-quantum coherence and, if the initial state is not in equilibrium, four-spin–zero-quantum coherence.

The excitation will be equally effective in other star-like networks where X interacts through roughly uniform couplings with n peripheral spins. However, the pulse sequence tailored for A_3X systems fails to excite four-quantum coherence in an A_2X_2 system. On the other hand, if all pulses in the sequence of Fig. 5.3.1(g) were applied with the *same* phase, the situation would be reversed, with A_2X_2 resonances excited and A_3X signals rejected (5.39).

5.3 TIME-DOMAIN MULTIPLE-QUANTUM SPECTROSCOPY

Experimental evidence for the selectivity of these procedures is shown in the filtered two-dimensional correlation spectrum of the protein basic pancreatic trypsin inhibitor (BPTI) in Fig. 8.3.8. Only the A_3X systems of the alanine residues give rise to strong cross- and diagonal peaks in the filtered COSY-spectrum, while threonine and leucine produce only weak responses, and none of the other amino acids give a significant contribution.

5.3.1.5. Selective excitation of specific orders

In contrast to isotropic scalar coupled systems, dipolar coupled spins in liquid crystalline phase feature well-resolved couplings between all spins. Furthermore, it is possible to reverse the sign of the effective (dipolar) Hamiltonian experimentally, in such a way that true time reversal can be achieved (5.76, 5.77). In these circumstances, it is possible to design schemes for the selective excitation of coherences of a given order p (5.11, 5.14–5.16, 5.19, 5.61). The basic building block of the pulse sequences, shown in Fig. 5.3.2(b), is a brief interval of free precession $\Delta\tau_p$ bracketed by two propagators U and $(U')^{-1}$. In the simplest case, the average Hamiltonians $\bar{\mathcal{H}}_p$ and $-\bar{\mathcal{H}}'_p$ prevailing in these intervals may be related by $\bar{\mathcal{H}}_p = \frac{1}{2}\bar{\mathcal{H}}'_p$, applied for durations T and $T' = T/2$. Such a sandwich excites multiple-quantum coherence of all orders. By repeating the cycle N times in immediate succession, with all pulses in the basic building block shifted through $\phi = k2\pi/N$, $k = 0, 1, \ldots, N-1$, only coherences of order $p = 0$, $\pm N$, $\pm 2N \ldots$ will remain. The entire procedure is repeated to ensure effective averaging of the undesired orders. An experimental example is shown in Fig. 5.3.2(d) where the orders $p = 0$ and 4 in oriented benzene have been excited selectively. Further details of this elegant but demanding procedure are described in a review by Weitekamp (5.61).

5.3.2. Offset-dependence of multiple-quantum frequencies and separation of orders

There is a remarkable difference between CW detection and indirect two-dimensional detection of multiple-quantum transitions: in the former case, one observes reduced transition frequencies $\omega_{ab}^{CW} = (E_a - E_b)/p_{ab}$, while in the latter case one observes free precession that involves the true multiple-quantum frequencies $\omega_{ab} = (E_a - E_b)$.

The detection is normally performed in a frame rotating with the carrier frequency $\omega_{r.f.}$ of the pulses that are used to excite and reconvert the multiple-quantum coherence. This frequency $\omega_{r.f.}$ is also used as reference frequency of the phase-sensitive detector. In this rotating

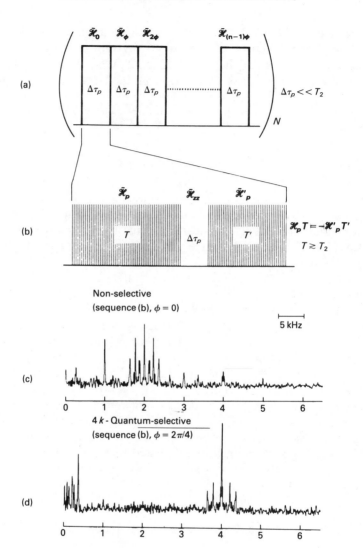

FIG. 5.3.2. (a) Schematic sequence for selective excitation of multiple-quantum coherence of specified order p: a basic unit or building block is repeated N times in immediate succession, the phases of all pulses within the basic unit being advanced in steps of $\Delta\phi = 2\pi/N$. (b) Building block employed in (a), consisting of a brief period of free precession $\Delta\tau_p$ bracketed by two intervals of duration T and T' with average Hamiltonians $\bar{\mathcal{H}}_p$ and $\bar{\mathcal{H}}'_p$ such that $\bar{\mathcal{H}}_p T = -\bar{\mathcal{H}}'_p T'$. (c) Multiple-quantum spectrum of benzene in liquid crystalline solution, obtained with non-selective excitation. All orders $p = 0, 1, \ldots, 6$ appear in the spectrum, which corresponds to the projection of a two-dimensional spectrum on to the ω_1-axis. The different orders p are separated by time-proportional phase increments. (d) Same as (c), but with selective excitation of the orders $p = 0$ and 4, achieved with sequence (a) with $\Delta\phi = 2\pi/4$. (Reproduced from Ref. 5.11.)

5.3 TIME-DOMAIN MULTIPLE-QUANTUM SPECTROSCOPY

frame, the Hamiltonian is

$$\mathcal{H}^r = \mathcal{H} - \omega_{r.f.} F_z. \qquad (5.3.19)$$

A coherence σ_{ab}^r between two arbitrary states $|a\rangle$ and $|b\rangle$ evolves in the rotating frame according to

$$\sigma_{ab}^r(t) = \sigma_{ab}(0) \exp\{-i(E_a - E_b - p_{ab}\omega_{r.f.})t\}\exp(-t/T_2^{(ab)}); \qquad (5.3.20)$$

hence the apparent multiple-quantum transition frequency ω_{ab} observed by indirect detection is

$$\omega_{ab} = (E_a - E_b) - p_{ab}\omega_{r.f.} \qquad (5.3.21)$$

where $p_{ab} = M_a - M_b$ is the order of the transition. This formula implies a very peculiar dependence of the apparent frequencies on the carrier frequency $\omega_{r.f.}$. Whenever the carrier frequency is shifted by $\Delta\omega_{r.f.}$, the multiple-quantum transition frequencies shift by $p\Delta\omega_{r.f.}$ relative to the frequency origin in the rotating frame. This property is illustrated in Fig. 5.3.3. Note that the zero-quantum transitions occur always at the same frequencies irrespective of the choice of the carrier frequency $\omega_{r.f.}$.

The characteristic offset dependence can be used to separate multiple-quantum signals of different orders, simply by setting the r.f. carrier well outside the single-quantum spectrum. If the chemical shifts Ω_k are contained in the interval $\Omega_{min} < \Omega_k < \Omega_{max}$ in the single-quantum spectrum, the transition frequencies $\omega^{(p)}$ in the p-quantum spectrum are contained in the interval $p\Omega_{min} < \omega^{(p)} < p\Omega_{max}$. However, in many cases,

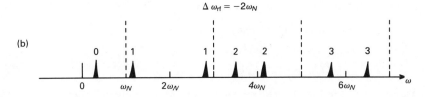

FIG. 5.3.3. Frequency shift of multiple-quantum transitions of various orders, observed upon shifting the r.f. carrier frequency in indirect two-dimensional detection. (a) Carrier within the single-quantum spectrum; (b) carrier shifted by $2\omega_N$ to the left with respect to (a). The axis shown corresponds to the ω_1-domain of a (projected) two-dimensional spectrum.

the p-quantum spectrum is not much wider than the single-quantum spectrum. Note in particular that there is only one MQT of the highest order, and that the width of the $(N-1)$-quantum spectrum equals that of the 1QT spectrum.

It is also possible to separate different orders (even if the carrier is positioned within the single-quantum spectrum) with time-proportional phase increments (5.10, 5.44), as discussed in § 6.6.3.

Finally, phase-cycling procedures can be employed to eliminate all signals except for those associated with a particular order p. These procedures exploit the characteristic behaviour of p-quantum coherence under rotations about the z-axis, which allows one to select a particular coherence transfer pathway by taking linear combinations of signals obtained from experiments with phase-shifted preparation and mixing propagators (see eqns (5.3.24)–(5.3.26) and § 6.3).

The experimental procedures for the separation of different orders are reflected in the separation of the density operator into contributions belonging to different orders p

$$\sigma = \sum_{p=-2L}^{+2L} \sigma^p \tag{5.3.22}$$

where the maximum total spin quantum number L is given in eqn (5.1.3). The term σ^p in eqn (5.3.22) can be written in the form

$$\sigma^p = \sum_{a,b}{}' \sigma_{ab} |a\rangle\langle b| \tag{5.3.23}$$

where the restricted sum extends over the pairs (a, b) with $p_{ab} = p$.

The decomposition of the density matrix is shown schematically in Fig. 5.3.4 for a system with three spins $I = \frac{1}{2}$. The reduction of the number of matrix elements with increasing order p is clearly apparent, in agreement with Table 5.1.1. The submatrices corresponding to σ^0 comprise diagonal elements (populations of the eigenstates) and zero-quantum coherences.

The representation of σ^p in eqn (5.3.23) allows one to deduce the characteristic transformation properties of a p-quantum coherence $|a\rangle\langle b|$ under r.f. phase shifts, i.e. under rotations about the z-axis

$$\exp\{-i\varphi F_z\} |a\rangle\langle b| \exp\{i\varphi F_z\}$$
$$= \exp\{-i\varphi M_a\} |a\rangle\langle b| \exp\{i\varphi M_b\}$$
$$= |a\rangle\langle b| \exp\{-ip_{ab}\varphi\}. \tag{5.3.24}$$

The same holds for all coherences belonging to a given order p

$$\exp\{-i\varphi F_z\}\sigma^p \exp\{i\varphi F_z\} = \sigma^p \exp\{-ip\varphi\} \tag{5.3.25}$$

or, in shorthand notation,

$$\sigma^p \xrightarrow{\varphi F_z} \sigma^p \exp\{-ip\varphi\}. \tag{5.3.26}$$

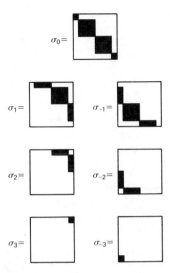

FIG. 5.3.4. Matrix representations of the density operator components σ^p classified according to order p for a three-spin $\frac{1}{2}$ system. The order of basis functions is assumed to be $\alpha\alpha\alpha$, $\alpha\alpha\beta$, $\beta\alpha\alpha$, $\alpha\beta\beta$, $\beta\alpha\beta$, $\beta\beta\alpha$, and $\beta\beta\beta$.

This property is of fundamental importance for the experimental separation of coherences of various orders p, as will be discussed in § 6.3.

The transformation properties of multiple-quantum coherence can also be put into evidence by expressing coherence in terms of irreducible tensor operators (5.11–5.21, 5.78–5.86) (see § 2.1.10)

$$\sigma = \sum_{k,l,m} b_{lm}^{(k)} T_{lm}^{(k)}. \qquad (5.3.27)$$

In this case, the quantum number $m = -l, -l+1, \ldots +l$ can be identified with the coherence order p. The rank of the tensors extends over $l = 0, 1, \ldots, 2L$ where L is given in eqn (5.1.3). The index k distinguishes different operators with the same transformation properties. Under rotations about the z-axis one obtains in analogy to eqn (5.3.26)

$$T_{lm} \xrightarrow{\varphi F_z} T_{lm} \exp\{-im\varphi\}. \qquad (5.3.28)$$

The effect of a non-selective r.f. pulse is equivalent to a rotation in three-dimensional space and transforms the order $p = m$ without affecting the rank l

$$T_{lm}^{(k)} \xrightarrow{\Sigma_j \beta I_{jy}} \sum_{m'} d_{m'm}^l(\beta) T_{lm'}^{(k)} \qquad (5.3.29)$$

with the elements $d_{m'm}^l(\beta)$ of the Wigner rotation matrix.

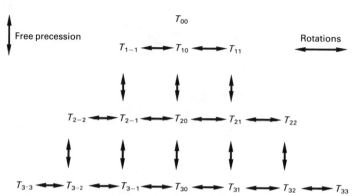

FIG. 5.3.5. Representation of coherence by irreducible tensor operators: transformations along horizontal lines (one or more steps) may be induced by r.f. pulses; transformations along vertical lines (one or more steps, depending on the coupling network) occur in free precession intervals. In thermal equilibrium, the density operator contains only T_{00} and T_{10} terms. Terms with T_{lm} correspond to m-quantum coherence. To excite $m = +3$ quantum coherence, for example, one can follow the path $T_{10} \rightarrow T_{11} \rightarrow T_{31} \rightarrow T_{33}$ with a pulse sequence of the type $(\pi/2) - \tau - (\pi/2)$, or alternatively the path $T_{10} \rightarrow T_{11} \rightarrow T_{21} \rightarrow T_{22} \rightarrow T_{32} \rightarrow T_{33}$ with a sequence of three pulses and two intervals.

On the other hand, free precession in the absence of r.f. perturbations may change the rank l without affecting the order

$$T_{lm}^{(k)} \xrightarrow{\mathcal{H}t} \sum_{k'l'} b_{l'm}^{(k')} T_{l'm}^{(k')}. \tag{5.3.30}$$

These properties become apparent in experiments consisting of r.f. pulses separated by free precession intervals, as illustrated in Fig. 5.3.5.

Irreducible tensor operators have been used extensively by Sanctuary (5.78–5.82), by Bain (5.83–5.86), and by Pines and co-workers (5.11–5.21, 5.61), with emphasis on quadrupolar, scalar, and dipolar couplings, respectively.

5.3.3. Structure of multiple-quantum spectra

Multiple-quantum spectra have a similar, though normally simpler structure than the corresponding single-quantum spectra. In solids and liquid crystals, the spectra exhibit dipolar or quadrupolar splittings or broadening, while in isotropic liquids one observes multiplet structures due to scalar couplings.

In this section, we concentrate the discussion on scalar-coupled spin systems with $I = \frac{1}{2}$ in isotropic solution. We may address either individual coherences corresponding to single lines in the multiple-quantum spectrum, or else consider entire multiplets originating from a set of coherences.

5.3 TIME-DOMAIN MULTIPLE-QUANTUM SPECTROSCOPY

To determine the precession frequency ω_{ab} of a coherence $|a\rangle\langle b|$ in a weakly coupled system, we may represent the coherence by a product of single-element operators as in eqn (2.1.116)

$$|a\rangle\langle b| = \prod_{k=1}^{N} I_k^{\mu_k} \qquad (5.3.31)$$

with $\mu_k = +$ or $-$ for the q 'active spins' with the changes in magnetic quantum number $\Delta M_k = \pm 1$ and $\mu_k = \alpha, \beta$ for the $N - q$ 'passive spins' with the magnetic quantum numbers $M_k = \pm\frac{1}{2}$.

By way of example, a two-spin $I = \frac{1}{2}$ system may have $p = +2$ quantum coherence $I_k^+ I_l^+$, $p = -2$ quantum coherence $I_k^- I_l^-$, and two zero-quantum coherences $I_k^+ I_l^-$ and $I_k^- I_l^+$. In larger spin systems, we may encounter p-spin–p-quantum coherences, e.g. $I_k^+ I_l^+ I_m^\alpha I_n^\beta$, as well as q-spin–p-quantum coherences, with $p = q - 2, q - 4, \ldots$, such as the 4-spin–2-quantum term $I_k^+ I_l^+ I_m^+ I_n^-$.

The coherence $|a\rangle\langle b|$ has the precession frequency

$$\omega_{ab} = \underbrace{\sum_k \Omega_k \Delta M_k}_{\text{active}} + \underbrace{\sum_k \sum_l 2\pi J_{kl} \Delta M_k M_l}_{\text{active passive}} \qquad (5.3.32)$$

where the summations extend over all active spins k and over all passive spins l, respectively. Couplings between active spins or between passive spins do not affect the precession frequency.

Of particular interest are coherences where all N spins actively participate (all $\mu_k = \pm 1$ in eqn (5.3.31), $q = N$, and $p = N, N - 2, N - 4, \ldots$). These coherences are associated with a pair of eigenstates that are related through the inversion of all spins, and have been called 'spin-inversion coherences' (5.61) or 'class 1 coherences' (5.44). The precession frequencies of these spin-inversion coherences, according to eqn (5.3.32), are determined only by the chemical shifts and are not affected by the coupling terms, provided the coupling is weak. The 'total spin coherences' (5.17, 5.61) with $q = p = N$ (all $\mu_k = +$ or all $\mu_k = -$ in eqn (5.3.31)) are a special case of spin-inversion coherences.

Coherences which involve a number of active spins $q < N$ belong to multiplets. It is often convenient to collect such terms in operators which describe multiplets rather than individual transitions. For example, in a three-spin system there are two $p = +2$ coherences where k and l are active and m passive, described by $I_k^+ I_l^+ I_m^\alpha$ and $I_k^+ I_l^+ I_m^\beta$. The corresponding multiplet operators representing in-phase or antiphase two-quantum doublets are in this case

$$I_k^+ I_l^+ (I_m^\alpha + I_m^\beta) = I_k^+ I_l^+,$$
$$I_k^+ I_l^+ (I_m^\alpha - I_m^\beta) = 2I_k^+ I_l^+ I_{mz} \qquad (5.3.33)$$

with possible extensions for larger spin systems.

The multiplet structure of multiple-quantum spectra is analogous to single-quantum spectra. The effective multiple-quantum chemical shift

$$\Omega_{\text{eff}} = \sum_{\substack{k \\ \text{active}}} \Delta M_k \Omega_k \qquad (5.3.34)$$

corresponds to a linear combination of the shifts of all active nuclei, while the effective coupling constant to a passive spin m

$$J_{\text{eff}} = \sum_{\substack{k \\ \text{active}}} \Delta M_k J_{km} \qquad (5.3.35)$$

represents a linear combination of the couplings between the passive spin m and each of the active spins k. Each passive spin leads to an effective coupling constant. It is apparent that the effective coupling constants J_{eff} depend on the relative signs of the couplings J_{km}, which can thus be determined from a multiple-quantum spectrum.

An example of a zero-quantum spectrum in a weakly coupled three-spin system k, l, m is shown in Fig. 5.3.6. A comparison with the

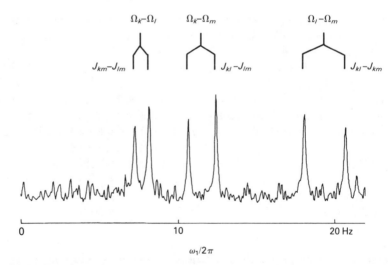

FIG. 5.3.6. Experimental zero-quantum spectrum of the weakly scalar-coupled three-spin system of 2-furancarboxylic acid methyl ester in isotropic phase. The spectrum was obtained by indirect detection, and corresponds to a projection of an absolute-value two-dimensional spectrum on to the ω_1-axis. The order $p = 0$ was selected by allowing all other orders of single- and multiple-quantum coherence to decay in an inhomogeneous static field. The signals stem from six zero-quantum coherences that can be described by single-element operators (from left to right) $I_k^+ I_l^- I_m^\beta$, $I_k^+ I_l^- I_m^\alpha$, $I_k^\beta I_l^+ I_m^-$, $I_k^\alpha I_l^+ I_m^-$, $I_k^+ I_l^\beta I_m^-$, and $I_k^+ I_l^\alpha I_m^-$ where the spins k, l, and m are assigned in the order of increasing Larmor frequency. Aliasing about the Nyquist frequency has been utilized to enhance digital resolution. (Adapted from Ref. 5.23.)

5.3 TIME-DOMAIN MULTIPLE-QUANTUM SPECTROSCOPY

single-quantum spectrum shows immediately that all three coupling constants have equal sign.

It is often convenient to convert multiplet coherences of the type of eqn (5.3.33) into Cartesian operator products, in order to simplify the description of coherence transfer processes. We represent double-quantum coherence (more accurately, in-phase multiplet coherence of order $p = \pm 2$) by the two Cartesian components (5.38)

$$\{2QT\}_x = \tfrac{1}{2}(I_k^+ I_l^+ + I_k^- I_l^-) = \tfrac{1}{2}(2I_{kx}I_{lx} - 2I_{ky}I_{ly}),$$

$$\{2QT\}_y = \frac{1}{2i}(I_k^+ I_l^+ - I_k^- I_l^-) = \tfrac{1}{2}(2I_{kx}I_{ly} + 2I_{ky}I_{lx}). \quad (5.3.36)$$

The corresponding zero-quantum in-phase multiplet coherence is represented by

$$\{ZQT\}_x = \tfrac{1}{2}(I_k^+ I_l^- + I_k^- I_l^+) = \tfrac{1}{2}(2I_{kx}I_{lx} + 2I_{ky}I_{ly})$$

$$\{ZQT\}_y = \frac{1}{2i}(I_k^+ I_l^- - I_k^- I_l^+) = \tfrac{1}{2}(2I_{ky}I_{lx} - 2I_{kx}I_{ly}). \quad (5.3.37)$$

Arbitrary q-spin–p-quantum coherences can be defined by analogy.

The precession under chemical shifts proceeds in analogy to the rotations in the single-spin subspaces (I_{kx}, I_{ky}, I_{kz}) of eqn (2.1.92)

$$\{q\text{-spin-}pQT\}_x \xrightarrow{\sum \Omega_k \tau I_{kz}} \{q\text{-spin-}pQT\}_x \cos \Omega_{\text{eff}} \tau$$

$$+ \{q\text{-spin-}pQT\}_y \sin \Omega_{\text{eff}} \tau \quad (5.3.38)$$

with the effective chemical shift defined in eqn (5.3.34).

Couplings to passive nuclei lead to a conversion of in-phase into antiphase multiplet coherence in analogy to eqn (2.1.94)

$$\{q\text{-spin-}pQT\}_x \xrightarrow{\sum_k \pi J_{km} \tau 2 I_{kz} I_{mz}} \{q\text{-spin-}pQT\}_x \cos \pi J_{\text{eff}} \tau$$

$$+ 2I_{mz}\{q\text{-spin-}pQT\}_y \sin \pi J_{\text{eff}} \tau \quad (5.3.39)$$

with J_{eff} defined in eqn (5.3.35).

5.3.4. Multiple-quantum double resonance

Double-resonance effects similar to those explained for single-quantum spectroscopy in § 4.7 also appear in multiple-quantum spectroscopy when a strong r.f. field is applied continuously to a resonance line in the spectrum. In the tickling regime, the multiple-quantum lines exhibit a splitting behaviour analogous to single-quantum transitions. Additional features, which have been termed 'images' and 'satellites', appear in the spectrum. For a detailed discussion, the reader is referred to Ref. 5.26.

5.4. Relaxation of multiple-quantum coherence

Multiple-quantum spectroscopy opens new perspectives for relaxation studies, since the information contained in the relaxation rates of multiple- and single-quantum coherences is often complementary. Provided none of the precession frequencies are degenerate, a coherence between an arbitrary pair of states $|a\rangle$ and $|b\rangle$ decays exponentially

$$\sigma_{ab}(t) = \sigma_{ab}(0)\exp\{-i\omega_{ab}t - t/T_2^{(ab)}\}. \quad (5.4.1)$$

The decay rate $1/T_2^{(ab)}$ is composed of adiabatic and non-adiabatic contributions (eqns (2.3.27)–(2.3.29))

$$1/T_2^{(ab)} = \frac{1}{2}\left[\sum_{c\neq a} W_{ac} + \sum_{c\neq b} W_{bc}\right]$$
$$+ \frac{1}{2}\int_{-\infty}^{+\infty} \overline{\{\langle a|\mathcal{H}_1(t)|a\rangle - \langle b|\mathcal{H}_1(t)|b\rangle\}}$$
$$\times \{\langle a|\mathcal{H}_1(t+\tau)|a\rangle - \langle b|\mathcal{H}_1(t+\tau)|b\rangle\}\,d\tau. \quad (5.4.2)$$

The adiabatic contribution (last term) is caused by fluctuations of the energy difference between the states $|a\rangle$ and $|b\rangle$. It is clear that only *non-concerted* fluctuations, which shift the energy of the two states by unequal amounts, contribute to the adiabatic relaxation. Because different combinations of energy levels $|a\rangle$ and $|b\rangle$ are involved in single- and in multiple-quantum transitions, the adiabatic term is sensitive to different types of interactions, and multiple-quantum relaxation may provide additional information that cannot be obtained from conventional single-quantum T_2 measurements.

5.4.1. Correlated external random fields

The unique properties of multiple-quantum relaxation can be exploited to study the correlation of external random fields that relax spin $I = \frac{1}{2}$ nuclei, for example in the case of relaxation induced by paramagnetic agents (5.25).

Consider a system of two weakly coupled spins A and B, exposed to external fluctuating fields $\mathbf{B}_A(t)$ and $\mathbf{B}_B(t)$ with the Hamiltonian

$$\mathcal{H}_1(t) = -\gamma_A \mathbf{I}_A \mathbf{B}_A(t) - \gamma_B \mathbf{I}_B \mathbf{B}_B(t). \quad (5.4.3)$$

To explore the information content of single-quantum relaxation, we compute the relaxation rate $1/T_2^{(1,2)}$ of the B-spin transition between the states $|1\rangle = |\alpha\alpha\rangle$ and $|2\rangle = |\alpha\beta\rangle$ and find

$$1/T_2^{(1,2)} = W_{12} + \tfrac{1}{2}W_{13} + \tfrac{1}{2}W_{24} + \frac{\gamma^2}{2}\int_{-\infty}^{\infty} \overline{B_{Bz}(t)B_{Bz}(t+\tau)}\,d\tau. \quad (5.4.4)$$

5.4 RELAXATION OF MULTIPLE-QUANTUM COHERENCE

The transition probabilities W_{kl} correspond to single-quantum transitions and are determined by the power spectral densities of the individual random fields (see eqns (2.3.4) and (2.3.5)). The last (adiabatic) term in eqn (5.4.4) depends only on the field $\mathbf{B}_B(t)$. No information on the correlation of the two random fields $\mathbf{B}_A(t)$ and $\mathbf{B}_B(t)$ can be obtained from the single-quantum relaxation rate $1/T_2^{(1,2)}$. Similarly, the longitudinal relaxation rates are not informative in this respect.

However, the double-quantum relaxation rate of the coherence between the states $|1\rangle = |\alpha\alpha\rangle$ and $|4\rangle = |\beta\beta\rangle$ is

$$1/T_2^{(1,4)} = \tfrac{1}{2}(W_{12} + W_{13} + W_{24} + W_{34})$$
$$+ \frac{\gamma^2}{2} \int_{-\infty}^{+\infty} \overline{\{B_{Az}(t) + B_{Bz}(t)\}\{B_{Az}(t+\tau) + B_{Bz}(t+\tau)\}}\, d\tau. \quad (5.4.5)$$

With the assumption of an exponential correlation function we obtain the relevant averages

$$\overline{B_{Bz}(t)B_{Bz}(t+\tau)} = \overline{B_{Bz}^2} \exp\{-|\tau|/\tau_c\} \quad (5.4.6)$$

and

$$\overline{B_{Az}(t)B_{Bz}(t)} = C_{AB}(\overline{B_{Az}^2}\,\overline{B_{Bz}^2})^{\frac{1}{2}} \quad (5.4.7)$$

leading to the double-quantum relaxation rate

$$1/T_2^{(1,4)} = \tfrac{1}{2}(W_{12} + W_{13} + W_{24} + W_{34}) + \gamma^2 \tau_c [\overline{B_{Az}^2} + \overline{B_{Bz}^2} + 2C_{AB}(\overline{B_{Az}^2}\,\overline{B_{Bz}^2})^{\frac{1}{2}}] \quad (5.4.8)$$

which obviously contains the relevant correlation coefficient C_{AB} of the two random fields. The full expressions for all transverse relaxation rates in a weakly coupled two-spin system in the extreme narrowing regime are collected in Table 5.4.1. Corresponding expressions for pure dipolar relaxation in a two-spin system can be found in Chapter 9 (eqn (9.4.7)).

Table 5.4.1
Relaxation of two non-equivalent protons by partially correlated random fields in the extreme narrowing approximation (5.25)

Single-quantum relaxation

$$1/T_2^{(1,2)} = 1/T_2^{(3,4)} = \gamma^2 \tau_c [\overline{B_{Ax}^2} + \overline{B_{Bx}^2} + \overline{B_{Bz}^2}]$$
$$1/T_2^{(1,3)} = 1/T_2^{(2,4)} = \gamma^2 \tau_c [\overline{B_{Ax}^2} + \overline{B_{Bx}^2} + \overline{B_{Az}^2}]$$

Double-quantum relaxation

$$1/T_2^{(1,4)} = \gamma^2 \tau_c [\overline{B_{Ax}^2} + \overline{B_{Bx}^2} + \overline{B_{Az}^2} + \overline{B_{Bz}^2} + 2C_{AB}(\overline{B_{Az}^2}\,\overline{B_{Bz}^2})^{\frac{1}{2}}]$$

Zero-quantum relaxation

$$1/T_2^{(2,3)} = \gamma^2 \tau_c [\overline{B_{Ax}^2} + \overline{B_{Bx}^2} + \overline{B_{Az}^2} + \overline{B_{Bz}^2} - 2C_{AB}(\overline{B_{Az}^2}\,\overline{B_{Bz}^2})^{\frac{1}{2}}]$$

An example is provided by the relaxation of the two protons in 2,3-dibromothiophene induced by paramagnetic molecules in solution (5.25). Although the relaxation is actually dipolar in origin, an adequate description of the intermolecular relaxation is obtained by considering the fluctuating magnetic fields induced at the sites of the protons A and B. By selecting paramagnetic molecules of different sizes, such as oxygen and 1,1-diphenyl-2-picrylhydrazyl (DPPH), different degrees of correlation of the random fields are expected. The bulky DPPH molecule, with the unpaired electron spin density delocalized over three aromatic ring systems, is expected to relax both protons simultaneously, while the small oxygen molecule should be capable of approaching and relaxing each proton separately. The effects on zero-, single- and double-quantum line-widths of the two protons in 2,3-dibromothiophene is shown in Fig. 5.4.1. The analysis shows that the correlation coefficients C_{AB} are 0.89 and 0.79 for DPPH and oxygen respectively.

Tang and Pines (5.13) studied the effect of paramagnetic relaxation on the protons of a rapidly rotating methyl group in solutes dissolved in liquid crystals. If the relaxation-effective process breaks the C_3 symmetry, relaxation pathways which connect irreducible representations of the molecular symmetry group become allowed. A correlation parameter ξ is introduced which is equal to unity when the relaxation conserves the C_3 symmetry, and drops to zero when each proton is relaxed independently. The multiple-quantum line-widths of CH_3CN in a nematic solvent were measured in the presence of di-t-butylnitroxide radicals (5.13). The concentration dependence indicates that the correlation parameter ξ is equal to unity, which is consistent with a picture where the time-scale of the methyl group rotation is much shorter than the lifetime of the interaction between the protons and the paramagnetic molecule.

5.4.2. Quadrupolar relaxation

Deuterium NMR is an attractive probe for motional studies, because the relaxation is dominated by the quadrupolar interaction, which is entirely intramolecular in origin (5.87).

In systems containing a single oriented deuteron, the transverse relaxation rate $1/T_2$ of the double-quantum coherence spanning the states $|1\rangle = |M=+1\rangle$ and $|3\rangle = |M=-1\rangle$ has a different dependence on the spectral densities than the spin–lattice relaxation rate $1/T_1$ (5.44)

$$1/T_2^{(1,3)} = \frac{3\pi^2}{2}\left(\frac{e^2qQ}{h}\right)^2 \{J_1(\omega_0) + 2J_2(2\omega_0)\}, \tag{5.4.9}$$

$$1/T_1 = \frac{3\pi^2}{2}\left(\frac{e^2qQ}{h}\right)^2 \{J_1(\omega_0) + 4J_2(2\omega_0)\}. \tag{5.4.10}$$

5.4 RELAXATION OF MULTIPLE-QUANTUM COHERENCE

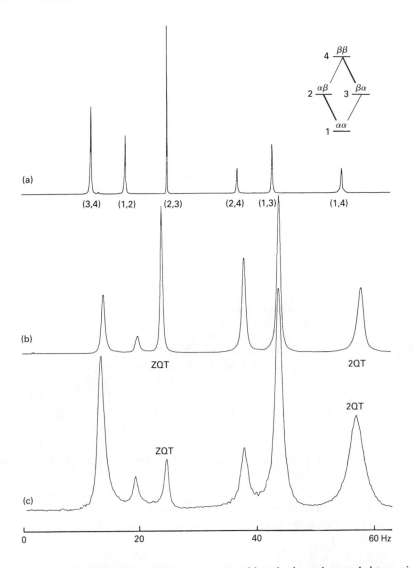

FIG. 5.4.1. Zero-, single-, and double-quantum transitions in the scalar-coupled two-spin system of 2,3-dibromothiophene, obtained by projecting a two-dimensional spectrum on to the ω_1-axis. The line-widths of the zero- and double-quantum transitions $|2\rangle \leftrightarrow |3\rangle$ and $|1\rangle \leftrightarrow |4\rangle$ provide a measure of the correlation of random-field fluctuations induced by oxygen (middle) and 1,1-diphenyl-2-picryl-hydrazyl (bottom). Inhomogeneous broadening is accounted for by subtracting the line-widths of the degassed sample (top). (Reproduced from Ref. 5.25.)

These two equations can readily be solved to determine $J_1(\omega_0)$ and $J_2(2\omega_0)$ (5.43). It is worth noting that the transverse decay of the double-quantum coherence is slower than the longitudinal relaxation, and is quite distinct from single-quantum decay rates, which also depend on $J_0(0)$. The separation of the spectral densities allows one to determine the diffusion constants of anisotropic molecular rotation.

In saturated hydrocarbon chains, the selective deuteration of one methylene group is a powerful technique for motional studies (5.87, 5.88). The relaxation of oriented CD_2 and CD_3 groups have been described in detail by Vold and co-workers (5.42, 5.45, 5.47).

To describe the motion of a CD_2 group, it is necessary to define two spectral density functions, $J_q^A(\omega)$ for the auto-correlation of the reorientation of a C–D bond, and $J_q^C(\omega)$ for the cross-correlation of the motion of the two deuterons (5.42)

$$J_q^A(\omega) = \int_0^\infty \exp(-i\omega\tau)(F_q(t) - \bar{F}_q)(F_q^*(t+\tau) - \bar{F}_q^*) \, d\tau, \quad (5.4.11)$$

$$J_q^C(\omega) = \int_0^\infty \exp(-i\omega\tau)(F_q(t) - \bar{F}_q)(E_q^*(t+\tau) - \bar{E}_q^*) \, d\tau \quad (5.4.12)$$

where $F_q(t)$ and $E_q(t)$ are spherical harmonics of rank q describing the orientation of the first and the second C–D bond, assumed to be parallel to the symmetry axes of the field gradient tensors. It turns out that the complete set of six independent values of the spectral density functions can be determined with T_2-measurements of single- and multiple-quantum coherences combined with selective inversion-recovery experiments (5.45).

An example of a two-dimensional double-quantum spin–echo spectrum of CD_2Cl_2 in a nematic solvent is shown in Fig. 5.4.2. The line-widths in the ω_1-domain yield the relaxation rates $1/T_2^{(ab)}$ of the four double-quantum coherences σ_{14}, σ_{46}, σ_{25}, and σ_{79}, which together with spin–lattice relaxation rates allow one to determine all spectral densities of the molecular motion.

5.4.3. Measurement of multiple-quantum relaxation rates and effects of magnetic field inhomogeneity

Although CW multiple-quantum NMR can, at least in principle, provide relaxation information (5.89), indirect detection of free multiple-quantum precession greatly simplifies the task of measuring transverse multiple-quantum relaxation rates. The technical aspects of two-dimensional spectroscopy will be deferred to Chapter 8. This section briefly reviews features that are specific to relaxation studies.

5.4 RELAXATION OF MULTIPLE-QUANTUM COHERENCE

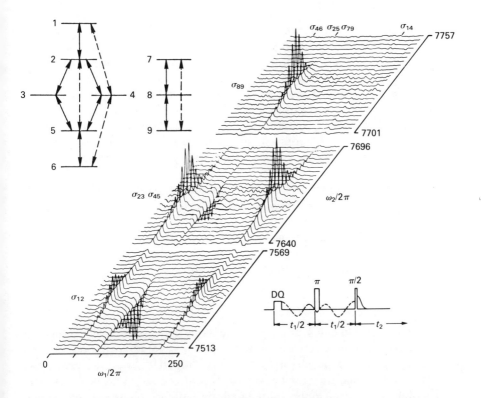

FIG. 5.4.2. Energy-level diagram of a system of two coupled spins $I = 1$ (e.g. CD_2 system), and phase-sensitive two-dimensional double-quantum spectrum. Four of the eight allowed transitions appear in the $F_2 = \omega_2/2\pi$ domain, while the four double-quantum (DQ) transitions that are not affected by first-order quadrupole splitting (dotted lines in the energy-level diagram) appear in the $F_1 = \omega_1/2\pi$ domain. (Reproduced from Ref. 5.45.)

In order to measure the natural line-widths of multiple-quantum transitions by two-dimensional spectroscopy, it is important to consider the effects of inhomogeneous static fields. Equation (5.3.25) shows that a p-quantum coherence experiences a p-fold dependence on the static field inhomogeneity $\Delta B_0(\mathbf{r})$

$$\sigma^p \xrightarrow{-\gamma \Delta B_0 \tau F_z} \sigma^p \exp\{ip\gamma \Delta B_0 \tau\}. \qquad (5.4.13)$$

This implies that higher-order multiple-quantum transitions are often significantly broadened by magnetic field inhomogeneity and that the increased spread obtained by the additive chemical shifts is counterbalanced by the increased line-width.

Zero-quantum coherences ($p=0$), on the other hand, are completely insensitive to field inhomogeneity and offer in principle an attractive possibility of recording high-resolution spectra in inhomogeneous fields. An example of a zero-quantum spectrum of an AMX spin system is shown in Fig. 5.3.6. It must however be remembered that the detection pulse converts zero-quantum coherence into *antiphase* single-quantum magnetization, which cannot be observed in a very inhomogeneous magnetic field. This requirement limits somewhat the prospects of obtaining high-resolution spectra in arbitrarily inhomogeneous magnetic fields.

For the measurement of multiple-quantum relaxation rates for $p \geq 2$, the use of refocusing techniques is often essential, except for rare cases where the inhomogeneous broadening is insignificant compared to the natural multiple-quantum line-width (5.25, 5.43).

If the inhomogeneous contribution $1/T_2^+$ can be measured separately, it can be subtracted from the observed width to obtain the true relaxation rate

$$1/T_2^{(ab)} = 1/T_2^{(ab)*} - p/T_2^+. \tag{5.4.14}$$

In less favourable cases, the inhomogeneous dephasing can be refocused by inserting non-selective π-pulses in the evolution period of the two-dimensional experiment (5.25, 5.44). Imperfect π-pulses lead however to coherence transfer between different orders p and $p' \neq -p$. The resulting artefacts can be eliminated by suitable phase-cycling, which relies on the fact that the desired inversion ($p \rightarrow p' = -p$) is associated with a phase shift $2p\varphi_k$ if the r.f. phase of the π-pulse is shifted through φ_k (5.44). The procedures are easily derived from the prescriptions in § 6.3 for the selection of coherence transfer pathways.

In strongly coupled systems (e.g. dipolar-coupled protons in anisotropic phase), even ideal refocusing pulses lead to undesired coherence transfer phenomena within the same order of coherence, similar to those discussed in § 7.2.3 for single-quantum refocusing. In this case, the ω_1-domain reveals linear combinations of eigenfrequencies of the free precession Hamiltonian.

It must be noted that, in spite of refocusing, the observable decay rates remain susceptible to irreversible dephasing associated with molecular diffusion in the inhomogeneous magnetic field.

Figure 5.4.3 shows a typical double-quantum decay of a single oriented deuteron. In this case, the decay has been measured over an extended period $t_1^{max} \simeq 5T_2$, and the Fourier transformation yields an essentially undistorted Lorentzian line. In cases with a single ω_1-frequency component, the decay rate can be obtained by a weighted least-squares fit of the time-domain envelope. The Fourier transformation with respect to t_1

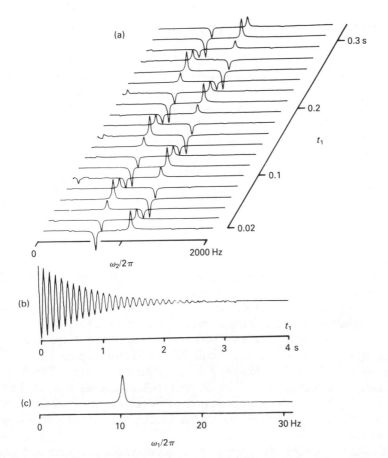

FIG. 5.4.3. Decay of double-quantum coherence in isolated, oriented deuterium nuclei, recorded with a sequence $(\pi/2) - \tau - (\pi/2) - \frac{1}{2}t_1 - (\pi) - \frac{1}{2}t_1 - (\pi/2) - t_2$. Inhomogeneous dephasing is refocused by the π-pulse in the middle of the evolution period (if diffusion can be neglected). The t_1-modulation is introduced artificially with time-proportional phase increments (see § 6.5.2). (a) shows the time–frequency domain $s(t_1, \omega_2)$, (b) a section parallel to t_1 at a peak position in ω_2, while (c) presents its Fourier transform. (Reproduced from Ref. 5.44.)

is, however, useful if a manifold of signals must be separated, as in Fig. 5.4.2. If the signals are truncated in t_1, care must be taken to deconvolute $\sin \omega_1/\omega_1$ distortions.

In systems with spectra that are too wide to be covered by non-selective refocusing pulses, it is possible to use a selective multiple-quantum refocusing pulse (5.4) at $\omega_{r.f.} = (E_a - E_b)/p_{ab}$. Selective r.f. fields also allow one to achieve spin-locking of multiple-quantum coherence and to measure a relaxation time $T_{1\rho}^{(ab)}$ in the rotating frame (5.6, 5.90). In rotating frame experiments, inhomogeneous dephasing can be compensated by suitable phase shifts which lead to rotary echoes (5.5, 5.90).

It is well known that refocusing can be initiated not only by pulses with flip angles $\beta = \pi$, but also by pulses with arbitrary values β. If, however, the pulse induces coherence transfer at the same time, new effects, so-called 'coherence transfer echoes', can appear (5.3, 5.91). Defocusing and refocusing proceed on different coherences which may exhibit unequal sensitivity to magnetic field inhomogeneity such that the required refocusing time $\tau^{(r)}$ can differ from the defocusing time $\tau^{(d)}$. Their relation takes the general form (5.91)

$$\frac{\tau^{(r)}}{\tau^{(d)}} = -\frac{\sum_k \gamma_k p_k^{(d)}}{\sum_l \gamma_l p_l^{(r)}} \qquad (5.4.15)$$

where $p_k^{(d,r)}$ is the order of coherence of nuclear species k with gyromagnetic ratio γ_k involved in the coherence during defocusing and refocusing, respectively (see also § 6.5.2, eqn (6.5.12)). This equation is also applicable to heteronuclear multiple-quantum coherence.

Coherence transfer echoes have been observed for the transfer of multiple-quantum coherence into single-quantum coherence (5.91), where $\tau^{(r)} = 2\tau^{(d)}$ for double-quantum coherence, and $\tau^{(r)} = 3\tau^{(d)}$ for triple-quantum coherence. The different echoes appear separated on the time axis. By Fourier transforming the nth echo, one obtains a spectrum with contributions originating only from molecules with at least n coupled spins.

Similarly, echoes can appear after heteronuclear coherence transfer. For example, after transferring coherence from ^{1}H to ^{13}C, an echo is observed at $\tau^{(r)} = (\gamma_H/\gamma_{^{13}C})\tau^{(d)} \simeq 4\tau^{(d)}$ (5.91).

6

TWO-DIMENSIONAL FOURIER SPECTROSCOPY

The extension of spectroscopy to two and more dimensions stems from the recognition that the properties of a molecular system cannot be fully characterized by a conventional one-dimensional (1D) spectrum. Even an ensemble of one-spin systems is a non-linear system, as mentioned in Chapter 4, and is only incompletely characterized by its impulse response or transfer function. This holds even more so for coupled spin systems. Many ambiguities remain in 1D spectra, such as the assignment of multiplets and the identification of connected transitions. Two-dimensional (2D) spectroscopy is a general concept which makes it possible to acquire more detailed information about the system under investigation. Since the first proposal in 1971 (6.1) and the first experimental realization in 1974 (6.2–6.5) an astonishing number of powerful 2D techniques have been invented, tested, and applied to problems in physics, chemistry, biology, and medicine.

Before presenting in Chapters 7–9 a survey of the various 2D techniques, the present chapter describes basic aspects of 2D spectroscopy that are common to various realizations, such as coherence transfer pathways, two-dimensional Fourier transformations, peak shapes, various means for manipulating 2D spectra, and the sensitivity of 2D spectroscopy.

6.1. Basic principles

A two-dimensional (2D) spectrum represents, in our terminology, a signal function $S(\omega_1, \omega_2)$ of two independent frequency variables. This definition does not include stacked plots $S(\tau, \omega)$ of 1D spectra, which are often used in the context of relaxation studies to represent the time evolution of a spectrum after a perturbation.

Several strategies can be employed for obtaining 2D spectra with two frequency variables, as shown schematically in Fig. 6.1.1.

1. *Frequency-domain experiments.* The direct characterization of a system as a function of two frequency variables is the purpose of conventional double-resonance experiments. By sweeping the observation frequency ω_1 one measures the response (slow-passage spectrum) of a system which is perturbed by a second frequency ω_2. Systematic variation of the double-resonance frequency ω_2 leads directly to a 2D spectrum $S(\omega_1, \omega_2)$.

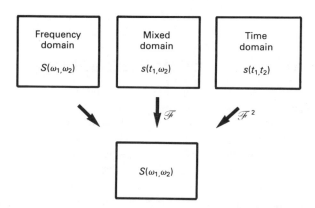

FIG. 6.1.1. Three strategies for obtaining 2D spectra with two independent frequency variables.

2. *Mixed time-domain/frequency-domain experiments.* A spin system modified by a double-resonance irradiation with a frequency ω_2 can also be tested by the impulse response $s(t_1, \omega_2)$ as a function of the time variable t_1. A 2D spectrum is then obtained by Fourier transformation with respect to t_1, as discussed in § 4.7.

3. *2D time-domain experiments.* A signal $s(t_1, t_2)$ is measured as a function of two independent time variables defined by a suitable segmentation of the time axis, and is converted by a 2D Fourier transformation into a 2D frequency-domain spectrum $S(\omega_1, \omega_2)$. In most experiments discussed in this monograph, the signal $s(t_1, t_2)$ is obtained by incrementing the interval t_1 parametrically from experiment to experiment, and recording the free induction signal as a function of t_2.

4. *Stochastic excitation.* An alternative to the segmentation of the time axis is used in stochastic 2D spectroscopy. The system is perturbed by a stationary random input process $x(t)$, and the response $y(t)$ is correlated with the product of two delayed functions $g_1(x(t-t_1))$ and $g_2(x(t-t_2))$ derived from the input process $x(t)$. For a time-invariant system, the correlation function

$$c(t_1, t_2) = \langle y(t)g_1(x(t-t_1))g_2(x(t-t_2))\rangle \qquad (6.1.1)$$

depends on two independent time variables t_1 and t_2. In principle, this approach can be extended to arbitrary dimensions, and can deliver most of the information obtainable with pulsed 2D spectroscopy, although in a less direct way. For a detailed discussion of stochastic multidimensional spectroscopy, the reader is referred to Refs. 6.6 and 6.7.

In this chapter, we shall be concerned primarily with 2D time-domain experiments where the signal $s(t_1, t_2)$ is obtained by segmentation of the

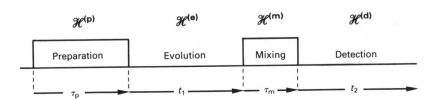

FIG. 6.1.2. Basic scheme for 2D time-domain spectroscopy, with four distinct intervals leading to a time-domain signal $s(\tau_p, t_1, \tau_m, t_2)$. Suitable manipulations make it possible to have different effective Hamiltonians in each interval.

time axis. In a general case, as shown in Fig. 6.1.2, we distinguish four intervals: the preparation period τ_p, the evolution time t_1, the mixing period τ_m, and the detection period t_2; hence the time-domain signal is more accurately described as $s(\tau_p, t_1, \tau_m, t_2)$. For special purposes, it may be necessary to introduce even more time parameters.

The four basic intervals will be discussed in detail in § 6.2. Let us briefly sum up their qualitative features:

1. *Preparation period.* In the course of the preparation period τ_p, the spin system is prepared in a coherent non-equilibrium state, which will evolve in the subsequent periods. In the simplest experiments, the preparation period consists of a single pulse. However, the preparation may also involve more sophisticated pulse sequences, including polarization transfer, excitation of multiple-quantum coherence, and enhancement of the polarization of nuclei with low sensitivity. The preparation period is normally of fixed length τ_p for all experiments in a 2D sequence.

2. *Evolution period.* In the course of the evolution period t_1, the spin system freely evolves under the influence of the Hamiltonian $\mathcal{H}^{(e)}$, which may be modified by decoupling, sample spinning, or periodic pulse sequences. In addition, aperiodic perturbations, e.g. refocusing by a π-pulse, may also be used in the evolution period. The evolution during t_1 determines the frequencies in the ω_1-domain. To sample the t_1-evolution, a series of experiments must be carried out with systematic incrementation of t_1.

3. *Mixing period.* In all experiments based on coherence or polarization transfer, the design of the mixing period is the key to enhancing the information content of the spectra. The mixing period may consist of one or more pulses, separated by intervals, usually with fixed duration, although some experiments involve a variable mixing time τ_m. The mixing process transforms single-, multiple-, or zero-quantum coherence into observable transverse magnetization, often through intermediate stages involving longitudinal polarization or multiple-quantum coherence.

The transfer of coherence or polarization induced by the mixing process is characteristic of the system under investigation. In many cases, the 2D spectrum may be regarded as a visual representation of the pathways of the mixing process.

4. *Detection period.* In the course of the detection period, the transverse magnetization is measured as a function of t_2. The system evolves under the Hamiltonian $\mathcal{H}^{(d)}$, which may be modified by means similar to those employed for $\mathcal{H}^{(e)}$.

It is easy to conceive cases where additional time intervals are needed. Depending on the effects one wishes to observe, some of the intervals may be constant throughout the experiment. In other situations three or more parameters must be varied independently and Fourier transformation leads to a three- or higher-dimensional spectrum. In practice, simplified forms of such spectra can be obtained by simultaneous variation of two or more time parameters. For example, if $\tau_m = \chi t_1$, the frequency domain $\omega_m = \omega_1/\chi$ contains information on the evolution of the system in *two* distinct intervals t_1 and τ_m. From the point of view of signal processing, such 'projected' 3D spectra are not distinct from 2D spectra, although their information content can be enhanced (see § 9.6).

It is a characteristic feature of 2D spectroscopy that the signals are recorded only in the detection period. The behaviour in the evolution period is monitored indirectly through the phase or amplitude modulation imposed upon the initial condition at $t_2 = 0$ by the systematic variation of t_1 from experiment to experiment.

In Chapters 7–9, we have adopted a classification of 2D experiments into three groups.

1. Experiments designed to *separate* different interactions (e.g. shifts and couplings) in orthogonal frequency dimensions, with the purpose of resolving 1D spectra by spreading overlapping resonances in a second dimension. These experiments require a means of modifying the effective Hamiltonian, in such a way that the spectra in the evolution and detection periods (determined by $\mathcal{H}^{(e)}$ and $\mathcal{H}^{(d)}$, respectively) contain different information (Chapter 7).

2. Experiments designed to *correlate* transitions of coupled spins by transferring transverse magnetization or multiple-quantum coherence from one transition to another in the course of a suitably designed mixing process (Chapter 8).

3. A third class of 2D time-domain experiments is concerned with studying *dynamic processes* such as chemical exchange, cross-relaxation, transient Overhauser effects, and spin-diffusion in solids (Chapter 9).

6.2. Formal theory of two-dimensional spectroscopy

Before analysing in more detail the fate of the coherence in 2D experiments in terms of coherence transfer pathways, we present a formal treatment based on density operator theory (6.5). A 2D experiment generally consists of a series of radio-frequency pulses $P_1 \ldots P_n$, separating n free precession periods of length τ_k, each with a characteristic Hamiltonian $\mathcal{H}^{(k)}$. One or possibly several of these periods will be identified with the evolution period with the time variable t_1. The final period after the last r.f. pulse is the detection period with the time variable t_2. This is schematically indicated in Fig. 6.2.1. The Hamiltonians $\mathcal{H}^{(k)}$ may either represent the natural Hamiltonian, or else a Hamiltonian that is modified by applying a CW r.f. field (e.g. decoupling) or a pulse sequence leading to an average Hamiltonian which in some cases requires stroboscopic sampling or incrementation of τ_k.

For the sake of economy of notation, we shall make use in the present section of the superoperator notation discussed in § 2.1.4. The evolution in the interval τ_k between pulses is described by the density operator equation, eqn (2.1.34),

$$\dot{\sigma}(t) = -i[\mathcal{H}^{(k)}, \sigma(t)] - \hat{\Gamma}^{(k)}\{\sigma(t) - \sigma_0^{(k)}\}$$
$$= -\{i\hat{\mathcal{H}}^{(k)} + \hat{\Gamma}^{(k)}\}\sigma(t) + \hat{\Gamma}^{(k)}\sigma_0^{(k)}. \quad (6.2.1)$$

The superoperator $\hat{\Gamma}^{(k)}$ includes the effects of relaxation and, if applicable, of chemical exchange. In some cases, $\hat{\Gamma}^{(k)}$ may depend on externally applied conditions, hence the index (k). The solution of eqn (6.2.1) takes the form

$$\sigma(\tau_k) = \exp\{-(i\hat{\mathcal{H}}^{(k)} + \hat{\Gamma}^{(k)})\tau_k\}[\sigma(\tau_k = 0) - \sigma_0^{(k)}] + \sigma_0^{(k)}. \quad (6.2.2)$$

The r.f. pulse P_k causes a transformation which may be represented by the superoperator $\hat{P}_k$

$$\sigma(\tau_k = 0) = \hat{P}_k \sigma(\tau_{k-1}) = P_k \sigma(\tau_{k-1}) P_k^{-1}. \quad (6.2.3)$$

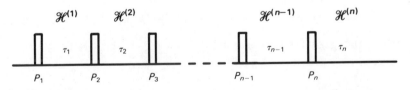

FIG. 6.2.1. Segmentation of the time axis in a general pulse experiment with two or more dimensions. Each free precession interval τ_k between two r.f. pulses P_k and P_{k+1} is associated with a Hamiltonian $\mathcal{H}^{(k)}$.

The combination of eqns (6.2.2) and (6.2.3) describes the entire time evolution in the course of a 2D experiment.

It should be noted that the terms $\sigma_0^{(k)}$ in eqn (6.2.2) arise because of relaxation towards equilibrium in the course of free precession. These terms lead to contributions in $\sigma(\tau_k)$ that are independent of the evolution in previous precession periods. Such contributions are, in essence, characteristic of a shorter experiment starting with pulse P_{k+1}. In the final spectrum, these terms normally lead to undesirable features which do not reflect the entire pulse sequence. They often appear in the form of so-called axial peaks, which appear on the ω_2-axis at $\omega_1 = 0$. In a well-designed experiment, these terms are suppressed by suitable phase cycles (see § 6.3). We shall therefore systematically drop the terms $\sigma_0^{(k)}$ (except, of course, for the very first term σ_0 which describes the initial state before the pulse P_1). We must however keep in mind that their experimental elimination has to be carefully considered when designing the pulse sequence.

This simplification leads to a concise equation that represents the evolution in an arbitrary 2D experiment

$$\sigma(\tau_1, \ldots, \tau_n) = \prod_{k=1}^{n} \exp\{-(i\hat{\mathcal{H}}^{(k)} + \hat{\Gamma}^{(k)})\tau_k\}\hat{P}_k\sigma_0 \qquad (6.2.4)$$

where the product is time-ordered in a chronological sequence from right to left.

Consider for clarity the standard form of many simple 2D experiments, shown in Fig. 6.1.2. The sequence starts either with a single preparation pulse or a sequence of pulses, represented by the superoperator $\hat{P}$. The free evolution in t_1 is governed by the Hamiltonian $\mathcal{H}^{(e)}$. The mixing process (if present) may be brought about by a single mixing pulse or by an extended mixing period sandwiched between two r.f. pulses; the overall effect being represented by the mixing or coherence transfer superoperator $\hat{R}$. In the detection period, the system evolves freely under the Hamiltonian $\mathcal{H}^{(d)}$. This leads to the general expression

$$\begin{aligned}\sigma(t_1, t_2) = {}&\exp\{-(i\hat{\mathcal{H}}^{(d)} + \hat{\Gamma}^{(d)})t_2\}\\&\times \hat{R}\exp\{-(i\hat{\mathcal{H}}^{(e)} + \hat{\Gamma}^{(e)})t_1\}\hat{P}\sigma_0.\end{aligned} \qquad (6.2.5)$$

The complex signal is proportional to

$$s^+(t_1, t_2) = \operatorname{tr}\{F^+\sigma(t_1, t_2)\}. \qquad (6.2.6)$$

Equation (6.2.6) can be evaluated by an explicit matrix representation or equivalently by expanding the density operator in single-transition operators. To obtain easily understandable results, it may be necessary to make simplifying assumptions: we often neglect chemical exchange

6.2 FORMAL THEORY OF TWO-DIMENSIONAL SPECTROSCOPY

processes and cross-relaxation in t_1 and t_2 (but not during a possibly extended mixing period) and assume a spectrum with non-overlapping lines. In this case, the superoperators $\hat{\mathcal{H}}$ and $\hat{\Gamma}$ commute and can be separated.

6.2.1. Explicit matrix representation

To obtain a physically meaningful matrix representation, it is convenient to express operators and superoperators in the eigenbasis of the Hamiltonian $\mathcal{H}^{(k)}$. The eigenvalues of $\hat{\mathcal{H}}^{(k)}$ are differences of eigenvalues of $\mathcal{H}^{(k)}$ (see eqn (2.1.85))

$$(\hat{\mathcal{H}}^{(k)})_{rs\,rs} = \omega^{(k)}_{rs} = \mathcal{H}^{(k)}_{rr} - \mathcal{H}^{(k)}_{ss} = \langle r| \mathcal{H}^{(k)} |r\rangle - \langle s| \mathcal{H}^{(k)} |s\rangle. \quad (6.2.7)$$

These correspond to the transition frequencies appearing in the single- or multiple-quantum spectrum. The eigenvalues of the superoperator $\hat{\Gamma}^{(k)}$ relevant for the evolution of coherent components are the transverse relaxation rates (see § 2.3.2)

$$(\hat{\Gamma}^{(k)})_{rs\,rs} = \lambda^{(k)}_{rs} = 1/T^{(k)}_{2rs}. \quad (6.2.8)$$

With these expressions, the complex signal may be written, in analogy to eqn (4.4.9)

$$s^+(t_1, t_2) = \sum_{rs}\sum_{tu} F^+_{sr}\exp\{(-i\omega^{(d)}_{rs} - \lambda^{(d)}_{rs})t_2\}$$

$$\times R_{rs\,tu}\exp\{(-i\omega^{(e)}_{tu} - \lambda^{(e)}_{tu})t_1\}(\hat{P}\sigma_0)_{tu} \quad (6.2.9)$$

or, in a more convenient form,

$$s^+(t_1, t_2) = \sum_{rs}\sum_{tu} \exp\{(-i\omega^{(d)}_{rs} - \lambda^{(d)}_{rs})t_2\}Z_{rs\,tu}\exp\{(-i\omega^{(e)}_{tu} - \lambda^{(e)}_{tu})t_1\}$$

$$(6.2.10a)$$

with the complex amplitude

$$Z_{rs\,tu} = F^+_{sr}R_{rs\,tu}(\hat{P}\sigma_0)_{tu}. \quad (6.2.11)$$

To obtain consistent expressions with positive frequencies after 2D Fourier transformation (see § 6.5), one may substitute $-\omega^{(d)}_{rs} = \omega^{(d)}_{sr}$, and $-\omega^{(e)}_{tu} = \omega^{(e)}_{ut}$,

$$s^+(t_1, t_2) = \sum_{rs}\sum_{tu} \exp\{(i\omega^{(d)}_{sr} - \lambda^{(d)}_{rs})t_2\}Z_{rs\,tu}\exp\{(i\omega^{(e)}_{ut} - \lambda^{(e)}_{tu})t_1\}$$

$$(6.2.10b)$$

Equation (6.2.10a) can be considered as a product of a row vector containing the frequencies and line-widths that are relevant in the detection period, a supermatrix **Z**, and a column vector containing the frequencies and line-widths characteristic of the evolution period

$$s^+(t_1, t_2) = \boxed{\omega_{rs}^{(d)}, \lambda_{rs}^{(d)}} \cdot \boxed{\mathbf{Z}} \cdot \boxed{\begin{array}{c} \omega_{tu}^{(e)} \\ \lambda_{tu}^{(e)} \end{array}}. \quad (6.2.10c)$$

The spectrum $S(\omega_1, \omega_2)$, obtained by a 2D Fourier transformation of the signal $s^+(t_1, t_2)$ (see below), contains peaks centred at the frequencies $\omega_1 = -\omega_{tu}^{(e)}$ and $\omega_2 = -\omega_{rs}^{(d)}$ with line-widths $\lambda_{tu}^{(e)}$ and $\lambda_{rs}^{(d)}$. The complex element $Z_{rs\,tu}$ of the supermatrix **Z** determines the intensity and phase of a peak with frequency coordinates $(\omega_1, \omega_2) = (-\omega_{tu}^{(e)}, -\omega_{rs}^{(d)})$. It is therefore appropriate to refer to the supermatrix **Z** as the *complex intensity matrix* and the 2D spectrum can be regarded as a visual representation of the matrix **Z**.

The complex intensities $Z_{rs\,tu}$ defined in eqn (6.2.11) contain three factors. The matrix element $(\hat{P}\sigma_0)_{tu}$ represents the initial amplitude and phase of a coherence component associated with a transition $|t\rangle \leftrightarrow |u\rangle$ at the beginning of the evolution period. The selection rules of observation, imposed by the observable operator, are represented by the matrix element F_{sr}^+. The most important factor in many experiments is the complex *coherence transfer amplitude* $R_{rs\,tu}$ which describes the transfer of coherence from transition $|t\rangle \leftrightarrow |u\rangle$ to the transition $|r\rangle \leftrightarrow |s\rangle$ during the mixing process. The mixing or coherence transfer superoperator $\hat{R}$ and its matrix elements $R_{rs\,tu}$ are characteristic for various forms of 2D spectroscopy.

In *2D separation* experiments (Chapter 7), the mixing period is normally skipped; hence

$$\hat{R} = \mathbb{1} \quad \text{with} \quad R_{rs\,tu} = \delta_{r,t}\delta_{s,u}. \quad (6.2.12)$$

In *2D correlation* experiments (Chapter 8), the mixing process is often induced by a single pulse or by a pulse sequence during which relaxation can be disregarded. The coherence transfer is then equivalent to a unitary transformation

$$\sigma(t_1, t_2 = 0) = R\sigma(t_1)R^{-1} \quad (6.2.13)$$

and the coherence transfer matrix element is simply equal to the product of two rotation matrix elements,

$$R_{rs\,tu} = R_{rt}R_{us}^{-1} = R_{rt}R_{su}^*. \quad (6.2.14)$$

In the case of *2D exchange* experiments (Chapter 9), $\hat{R}$ expresses the sequence of two r.f. pulses separated by an extended mixing period which can often not be represented by a unitary transformation.

6.2 FORMAL THEORY OF TWO-DIMENSIONAL SPECTROSCOPY

In very rare situations, the Hamiltonians $\mathcal{H}^{(e)}$ and $\mathcal{H}^{(d)}$ do not commute, and the mixing process is accompanied by a change of eigenbasis. This can be taken into account by a modified definition (6.5) of the elements $R_{rs\,tu}$.

6.2.2. Expansion of the density operator in single-transition operators

Instead of using explicit matrix elements, it is fully equivalent to represent coherence by single-transition operators. For example, the coherence between eigenstates $|t\rangle$ and $|u\rangle$ of the Hamiltonian can be represented by the operator $|t\rangle\langle u|$. Depending on the magnetic quantum numbers M_t and M_u, this operator can be identified with the single-transition operators $I^{+(tu)}$ or $I^{-(ut)}$ defined in eqn (2.1.131)

$$|t\rangle\langle u| = I^{+(tu)} \quad \text{for} \quad M_t > M_u,$$
$$|t\rangle\langle u| = I^{-(ut)} \quad \text{for} \quad M_t < M_u. \qquad (6.2.15)$$

(For zero-quantum transitions, the identification is arbitrary.) For $t = u$, the operator $|t\rangle\langle t|$ represents the polarization operator of eigenstate $|t\rangle$ defined in eqn (2.1.115).

The density operator $\sigma(t)$ can be expanded in terms of the $|t\rangle\langle u|$ operators

$$\sigma(t) = \sum_{tu} \sigma_{tu} |t\rangle\langle u| \qquad (6.2.16)$$

where the coefficients σ_{tu} are the matrix elements of σ in the basis $\{|t\rangle, |u\rangle, \ldots\}$. We may now follow the individual coherence components through the 2D experiment.

The coherence $|t\rangle\langle u|$, excited in the preparation period with a complex amplitude $\sigma_{tu}(t_1 = 0)$, evolves in the course of the evolution period t_1 under the Hamiltonian $\mathcal{H}^{(e)}$. The coefficient of $|t\rangle\langle u|$ is

$$\sigma_{tu}(t_1) = \sigma_{tu}(t_1 = 0)\exp\{(-i\omega_{tu}^{(e)} - \lambda_{tu}^{(e)})t_1\} \qquad (6.2.17)$$

with the frequency $\omega_{tu}^{(e)}$ given by eqn (6.2.7) and with the relaxation rate $\lambda_{tu}^{(e)}$ of eqn (6.2.8). The mixing process partially transforms the coherence $|t\rangle\langle u|$ into the coherence $|r\rangle\langle s|$ with a coefficient $b_{rs\,tu}$

$$b_{rs\,tu}(t_1, t_2 = 0) = R_{rs\,tu}\sigma_{tu}(t_1). \qquad (6.2.18)$$

In the detection period t_2, the coherence $|r\rangle\langle s|$ continues to evolve under the Hamiltonian $\mathcal{H}^{(d)}$. The coefficient of $|r\rangle\langle s|$ is

$$b_{rs\,tu}(t_1, t_2) = \exp\{(-i\omega_{rs}^{(d)} - \lambda_{rs}^{(d)})t_2\}R_{rs\,tu}$$
$$\times \exp\{(-i\omega_{tu}^{(e)} - \lambda_{tu}^{(e)})t_1\}\sigma_{tu}(t_1 = 0). \qquad (6.2.19)$$

Thus the relevant part of the density operator is

$$\sigma(t_1, t_2) = \sum_{rs} \sum_{tu} b_{rs\,tu}(t_1, t_2) |r\rangle\langle s| \qquad (6.2.20)$$

and the complex observable signal is given by

$$s^+(t_1, t_2) = \mathrm{tr}\{F^+ \sigma(t_1, t_2)\}. \qquad (6.2.21)$$

With the expansion

$$F^+ = \sum_{sr} F^+_{sr} |s\rangle\langle r|, \qquad (6.2.22)$$

we finally obtain the complex signal

$$\begin{aligned} s^+(t_1, t_2) &= \sum_{rs} \sum_{tu} F^+_{sr} b_{rs\,tu}(t_1, t_2) \\ &= \sum_{rs} \sum_{tu} \exp\{(-i\omega_{rs}^{(d)} - \lambda_{rs}^{(d)})t_2\} Z_{rs\,tu} \\ &\quad \exp\{(-i\omega_{tu}^{(e)} - \lambda_{tu}^{(e)})t_1\} \end{aligned} \qquad (6.2.23)$$

with the complex amplitude

$$Z_{rs\,tu} = F^+_{sr} R_{rs\,tu} \sigma_{tu}(t_1 = 0) \qquad (6.2.24)$$

in agreement with eqns (6.2.10a) and (6.2.11a).

In applications to weakly coupled spin systems, it is possible to move between various sets of basis operators discussed in §§ 2.1.5–2.1.9. Thus in a system with two spins $I_1 = I_2 = \tfrac{1}{2}$, where the eigenstates are numbered as in Fig. 4.4.2, one obtains with eqns (2.1.134)

$$\begin{aligned} I_1^+ &= |1\rangle\langle 3| + |2\rangle\langle 4|, \\ I_1^- &= |3\rangle\langle 1| + |4\rangle\langle 2| \end{aligned} \qquad (6.2.25)$$

which represent 'multiplet coherences' involving both doublet components associated with spin I_1. The related Cartesian operators correspond to superpositions,

$$\begin{aligned} I_{1x} &= \tfrac{1}{2}\{|1\rangle\langle 3| + |2\rangle\langle 4| + |3\rangle\langle 1| + |4\rangle\langle 2|\}, \\ I_{1y} &= \frac{1}{2i}\{|1\rangle\langle 3| + |2\rangle\langle 4| - |3\rangle\langle 1| - |4\rangle\langle 2|\}. \end{aligned} \qquad (6.2.26)$$

It is often useful to employ different sets of basis operators in order to analyse free precession, coherence transfer, and phase cycles.

6.3. Coherence transfer pathways

To obtain a better understanding of the fate of the coherence in the course of a pulse experiment, it is useful to describe its path through various orders of coherence (6.8, 6.9). This approach allows a unified treatment of multiple- and single-quantum experiments.

In high-field NMR, each eigenstate $|t\rangle$ of the Hamiltonian is characterized by a magnetic quantum number M_t, and each coherence $|t\rangle\langle u|$ is associated with a 'coherence order' $p_{tu} = M_t - M_u$. The quantities M_t and p_{tu} are both 'good' quantum numbers under free precession, because the high-field Hamiltonian has rotational symmetry, and the eigenstate $|t\rangle$ transforms according to the irreducible representation M_t of the one-dimensional rotation group. Hence a coherence $|t\rangle\langle u|$ transforms according to the representation $p_{tu} = M_t - M_u$.

Each transition between eigenstates is associated with two coherences $|t\rangle\langle u|$ and $|u\rangle\langle t|$ which have coherence orders p_{tu} and p_{ut} of opposite sign. At first sight, it may appear artificial to distinguish these terms. We shall see below that it is in fact meaningful to design experiments which single out a pathway involving a coherence order p_{tu} with a definite sign in a given interval while discarding signal contributions that stem from a 'mirror image' pathway through the order $-p_{tu}$.

In the course of free precession, the quantum number p_{tu} is conserved, while r.f. pulses may induce a change in coherence order $p_{tu} \to p_{rs}$ by transferring coherence from $|t\rangle\langle u|$ to $|r\rangle\langle s|$.

To describe pulse experiments in terms of a pathway through various orders of coherence, we classify the density operator terms according to the order p (see eqn (5.3.22))

$$\sigma(t) = \sum_p \sigma^p(t). \qquad (6.3.1)$$

The summation extends over $-2L \leq p \leq 2L$, where $L = \sum I_k$ is the sum of the quantum numbers of all spins. For a system of K spins $\frac{1}{2}$, p extends from $-K$ to $+K$. This classification can be carried out explicitly if the density operator is expressed in single-transition shift operators (e.g. $I^{+(tu)}$), in products of shift operators (e.g. $I_j^+ I_k^+$) or in irreducible tensor operators (e.g. $T_p^{(jk)}$).

It is revealing to represent the sequence of events in various experiments by 'coherence transfer pathways' such as shown in Fig. 6.3.1. Free precession proceeds within the levels of this diagram, while pulses may induce 'transitions' between different coherence orders. All coherence transfer pathways start at thermal equilibrium with $p = 0$ and must end with single-quantum coherence ($p = \pm 1$) to be detectable. We choose to observe a complex signal by quadrature detection,

$$\begin{aligned} s^+(t) &= s_x(t) + \mathrm{i} s_y(t) \\ &= \mathrm{tr}\{\sigma(t) F_x\} + \mathrm{i}\,\mathrm{tr}\{\sigma(t) F_y\} \\ &= \mathrm{tr}\{\sigma(t) F^+\}. \end{aligned} \qquad (6.3.2)$$

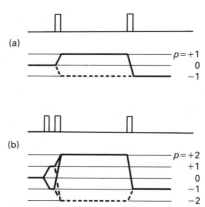

FIG. 6.3.1. Coherence transfer pathways in some typical 2D experiments. (a) Homonuclear 2D correlation spectroscopy (see § 8.2): the preparation period (typically a single $\pi/2$-pulse) excites single-quantum coherence of orders $p = \pm 1$; these coherences are transferred into observable coherence ($p = -1$) by a mixing propagator which normally consists of a single r.f. pulse with rotation angle β. (b) In double-quantum spectroscopy, the preparation propagator (typically a series of pulses interleaved with free precession periods) excites coherence of order $p = \pm 2$, which is reconverted into observable $p = -1$ coherence by a suitable mixing propagator.

Only density operator components proportional to $I^{-(rs)}$ with $p = -1$ can contribute to the signal, and all pathways that do not lead to $p = -1$ can be disregarded.

6.3.1. Selection of pathways

Coherence can be transferred between different orders by a propagator U_i, which may represent a single pulse or a sequence of pulses, such as the composite sequence $[(\pi/2) - \tau/2 - (\pi) - \tau/2 - (\pi/2)]$ commonly used for multiple-quantum excitation. Free precession intervals can also be represented by a propagator U_i.

A complete pulse experiment may consist of a series of such processes

$$\sigma_0 \xrightarrow{U_1} \xrightarrow{U_2} \cdots \xrightarrow{U_n} \sigma(t). \qquad (6.3.3)$$

Each propagator U_i may cause a transfer of a particular order of coherence $\sigma^p(t_i^-)$ into numerous different orders $\sigma^{p'}(t_i^+)$, leading to a 'branching' or 'fanning out' of the coherence transfer pathways

$$U_i \sigma^p(t_i^-) U_i^{-1} = \sum_{p'} \sigma^{p'}(t_i^+) \qquad (6.3.4)$$

where the arguments t_i^- and t_i^+ refer to the state just before and immediately after the transformation by U_i.

A p-quantum coherence has a characteristic transformation behaviour under rotations about the z-axis (see eqn (5.3.25))

$$\exp\{-i\varphi F_z\}\sigma^p \exp\{i\varphi F_z\} = \sigma^p \exp\{-ip\varphi\} \quad (6.3.5a)$$

or, in symbolic notation,

$$\sigma^p \xrightarrow{\varphi F_z} \sigma^p \exp\{-ip\varphi\}. \quad (6.3.5b)$$

This property makes it possible to separate different orders (6.10).

The key to the separation of coherence transfer pathways is the use of propagators $U_i(\varphi_i)$ that are shifted in phase

$$U_i(\varphi_i) = \exp\{-i\varphi_i F_z\} U_i(0) \exp\{i\varphi_i F_z\}. \quad (6.3.6)$$

If a propagator is made up of a sequence of pulses, each constituent pulse must be shifted in phase. For example, the sequence for double-quantum excitation takes the phase shifted form $[(\pi/2)_\varphi - \tau/2 - (\pi)_\varphi - \tau/2 - (\pi/2)_\varphi]$.

Under a phase-shifted propagator, the transformation in eqn (6.3.4) is modified

$$U_i(\varphi_i)\sigma^p(t_i^-)U_i(\varphi_i)^{-1} = \sum_{p'} \sigma^{p'}(t_i^+)\exp\{-i\Delta p_i \varphi_i\} \quad (6.3.7)$$

where

$$\Delta p_i = p'(t_i^+) - p(t_i^-) \quad (6.3.8)$$

corresponds to the change in coherence order under the propagator U_i. Thus the phase shift of a coherence component which has undergone a change of order $p \to p'$ under a propagator $U_i(\varphi_i)$ is given by

$$\sigma(\Delta p_i, \varphi_i) = \sigma(\Delta p_i, \varphi_i = 0)\exp\{-i\Delta p_i \varphi_i\}. \quad (6.3.9)$$

This phase-shift will be carried over to the $p = -1$ coherence that is observed in the detection period. The corresponding contribution to the time-domain signal therefore reflects the change Δp_i under the phase-shifted propagator $U_i(\varphi_i)$

$$s(\Delta p_i, \varphi_i, t) = s(\Delta p_i, \varphi_i = 0, t)\exp\{-i\Delta p_i \varphi_i\}. \quad (6.3.10)$$

The total signal is made up of contributions from all pathways

$$s(\varphi_i, t) = \sum_{\Delta p_i} s(\Delta p_i, \varphi_i, t)$$

$$= \sum_{\Delta p_i} s(\Delta p_i, \varphi_i = 0, t)\exp\{-i\Delta p_i \varphi_i\}. \quad (6.3.11)$$

To restrict the coherence transfer under U_i to a particular change Δp_i in coherence order, we may perform N_i experiments with systematic increments of the r.f. phase φ_i of the propagator

$$\varphi_i = k_i 2\pi/N_i, \qquad k_i = 0, 1, \ldots, N_i - 1. \tag{6.3.12}$$

Each of the N_i signals $s(\varphi_i, t)$ observed in the detection period consists of a superposition, as expressed in eqn (6.3.11). This superposition can be unravelled by a discrete Fourier analysis (6.10) with respect to the phase φ_i

$$s(\Delta p_i, t) = \frac{1}{N_i} \sum_{k_i=0}^{N_i-1} s(\varphi_i, t) \exp\{i\Delta p_i \varphi_i\}. \tag{6.3.13}$$

This weighted linear combination of the signals retains only contributions associated with coherence that has undergone a change Δp_i under the propagator U_i. However, the procedure does not automatically reject all other pathways: by carrying out a series of N_i experiments, one selects a series of values

$$\Delta p_i^{(\text{selected})} = \Delta p_i^{(\text{desired})} \pm nN_i, \qquad n = 0, 1, 2, \ldots. \tag{6.3.14}$$

This situation is reminiscent of aliasing in frequency-domain Fourier analysis, and is a consequence of the sampling theorem, eqn (4.1.40). Clearly, if a *unique* Δp_i value must be selected from a range of possible values, N_i must be sufficiently large, and the phase increment $\Delta \varphi = 2\pi/N_i$ correspondingly small.

In many applications of 2D spectroscopy, it turns out that it is essential to select two Δp_i-values simultaneously. We shall see in § 6.5.3 that this is a prerequisite for obtaining pure 2D absorption spectra. In this context, it is possible to take advantage of the aliasing in Δp-space (6.9).

To obtain a clear picture of this aliasing behaviour, it is useful to list all possible changes in coherence order, emphasizing the *desired* Δp_i-value by bold typeface, and indicating the Δp_i-values that must be blocked by setting them in brackets. Consider for example the 'Δp-list' appropriate to select a pathway $p = 0 \rightarrow +1 \rightarrow -1$ in homonuclear 2D correlation spectroscopy (see Fig. 6.3.1a). It is sufficient to cycle the phase of the mixing pulse to select

$$\Delta p_2 = \cdots -3, \mathbf{-2}, (-1), (0), 1, 2, 3, \ldots. \tag{6.3.15}$$

This notation implies that the values $\Delta p_2 = -3, +1, +2, +3 \ldots$ are immaterial in this case: although such coherence transfer processes may take place, they do not affect the outcome of the experiment. The requirements summed up in eqn (6.3.15) can be fulfilled by a three-step phase cycle with the phases $\varphi_2 = 0°, 120°, 240°$ ($N_2 = 3$, $\varphi_2 = k_2 2\pi/3$), which leads to the selection

$$\Delta p_2 = \cdots (-3), \mathbf{-2}, (-1), (0), 1, (2), (3), \ldots. \tag{6.3.16a}$$

6.3 COHERENCE TRANSFER PATHWAYS

If 120° phase shifts are not available, the same selection requirement can be fulfilled with $N_2 = 4$, i.e. $\varphi_2 = 0°, 90°, 180°, 270°$.

$$\Delta p_2 = \cdots (-3), -2, (-1), (0), (1), 2, (3), \ldots \quad (6.3.16\text{b})$$

In this case, one may speak of an 'overkill', since the suppression of $\Delta p = +1$ is not essential.

The weighting of the signals in eqn (6.3.13), which is the key to the selection procedure, can be achieved by one of three strategies.

1. Multiplication of the complex recorded signals with complex phase factors $\exp\{i\Delta p_i \varphi_i\}$. This corresponds to a software phase correction.
2. Phase-shifting of *all* pulses in the sequence by an additional phase increment $\varphi = \Delta p_i \varphi_i$, and addition of the signals without weighting.
3. Shifting of the phase of the receiver reference channel through

$$\varphi^{\text{ref}} = -\Delta p_i \varphi_i. \quad (6.3.17)$$

(If the observable operator in eqn (6.3.2) is defined as F^- instead of F^+, the opposite shift must be applied, and all pathways terminate at the level $p = +1$.)

For the sake of clarity, consider again the selection of a pathway $p = 0 \rightarrow +1 \rightarrow -1$ in homonuclear 2D correlation spectroscopy (see Fig. 6.3.1(a)). The selectivity requirement, summed up in eqn (6.3.15), shall be fulfilled by cycling the mixing pulse through $N_2 = 4$ steps with $\varphi_2 = k_2 \pi/2$, $k_2 = 0, 1, 2, 3$. At the same time, we shift the phase of the receiver reference $\varphi^{\text{ref}} = -\Delta p_2 \varphi_2 = +2\varphi_2 = +k_2 \pi$, which amounts to alternating addition and subtraction, as shown in Table 6.3.1.

This cycle is equivalent to 'Exorcycle' (6.11) and has been used in spin–echo correlation spectroscopy (6.12) and in heteronuclear correlation spectroscopy (6.13, 6.14). Because it does not retain the mirror-image pathway shown in Fig. 6.3.1(a), it is not suitable for obtaining pure 2D absorption peakshapes (see § 6.5.3).

Table 6.3.1
Phase cycle for the selection of the pathway $p = 0 \rightarrow +1 \rightarrow -1$ (Exorcycle)

$(\pi/2)_{\varphi_1} - t_1 - (\beta)_{\varphi_2} - t_2$		
$\varphi_1 = 0$	$\varphi_2 = 0$	$\varphi^{\text{ref}} = 0$
0	$\pi/2$	π
0	π	0
0	$3\pi/2$	π
$\Delta p_1 = $ free	$\Delta p_2 = -2$	

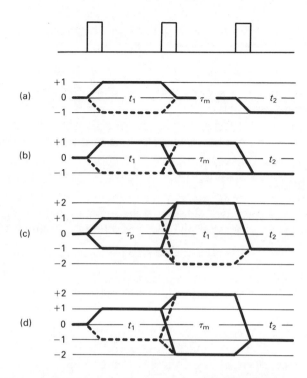

FIG. 6.3.2. Coherence transfer pathways for various experiments involving three consecutive r.f. pulses. (a) 2D exchange spectroscopy (Chapter 9); (b) relayed 2D correlation spectroscopy (§ 8.3.4); (c) 2D double-quantum spectroscopy (§ 8.4); and (d) 2D correlation spectroscopy with double-quantum filter (§ 8.3.3). The distinction between these experiments lies in the incrementation of the intervals and in the selection of coherence transfer pathways. If composite lineshapes or absolute-value plots are acceptable (see § 6.5), it is sufficient to select the pathways shown by solid lines. For pure phase spectra (i.e. pure 2D absorption lineshapes), it is essential to retain 'mirror image' pathways as well, which are indicated by dashed lines (§ 6.5.3). (Reproduced from Ref. 6.9.)

6.3.2. Multiple transfer

Many experiments involve several consecutive coherence transfer steps, as shown in Fig. 6.3.2. Clearly, the pulse sequence alone does not completely characterize the experiment: it is important to specify which time variables are to be incremented from experiment to experiment, and which pathways are to be selected.

If the experiment consists of a series of n propagators $U_1, U_2, \ldots, U_n$, describing n pulses or composite rotations, each coherence transfer pathway is unambiguously specified by a vector

$$\Delta \mathbf{p} - \{\Delta p_1, \Delta p_2, \ldots, \Delta p_n\}. \qquad (6.3.18)$$

Since all pathways must begin with $p = 0$ and are assumed to end with $p = -1$ to be observable (see eqn 6.3.2), the sum of the components of the vector $\mathbf{\Delta p}$ is fixed,

$$\sum_{i=1}^{n} \Delta p_i = -1. \tag{6.3.19}$$

Thus if $(n - 1)$ values of Δp_i are specified by $(n - 1)$ independent phase cycles, the entire vector $\mathbf{\Delta p}$ and hence the complete pathway are defined unambiguously. Because the r.f. phase shifts required for the pathway selection are often subject to systematic errors, it may however be advisable in practice to employ independent phase cycles to select the desired Δp_i values for *all* n coherence transfer steps.

The r.f. phases of the n propagators can also be written in vector form in analogy to eqn (6.3.18)

$$\mathbf{\varphi} = \{\varphi_1, \varphi_2, \ldots, \varphi_n\}. \tag{6.3.20}$$

The accumulated phase of a coherence component which has proceeded along a pathway specified by $\mathbf{\Delta p}$ is obtained by generalizing eqn (6.3.9)

$$\sigma(\mathbf{\Delta p}, \mathbf{\varphi}) = \sigma(\mathbf{\Delta p}, \mathbf{\varphi} = \mathbf{0})\exp\{-i\mathbf{\Delta p} \cdot \mathbf{\varphi}\}$$
$$= \sigma(\mathbf{\Delta p}, \mathbf{\varphi} = \mathbf{0})\exp\{-i(\Delta p_1 \varphi_1 + \Delta p_2 \varphi_2 + \cdots + \Delta p_n \varphi_n)\}. \tag{6.3.21}$$

This phase shift is carried over to the $p = -1$ coherence that is observed in the detection period. The corresponding contribution to the time-domain signal therefore reflects the pathway $\mathbf{\Delta p}$ and the phase vector $\mathbf{\varphi}$

$$s(\mathbf{\Delta p}, \mathbf{\varphi}, t) = s(\mathbf{\Delta p}, \mathbf{\varphi} = \mathbf{0}, t)\exp\{-i\mathbf{\Delta p} \cdot \mathbf{\varphi}\}. \tag{6.3.22}$$

The total signal is again made up of contributions from all possible pathways

$$s(\mathbf{\varphi}, t) = \sum_{\mathbf{\Delta p}} s(\mathbf{\Delta p}, \mathbf{\varphi}, t). \tag{6.3.23}$$

To separate these pathways, it is necessary to cycle the phases of the propagators independently

$$\varphi_1 = k_1 2\pi/N_1, \ldots, \varphi_n = k_n 2\pi/N_n \tag{6.3.24}$$

for $k_1 = 0, 1, \ldots, N_1 - 1; \ldots; k_n = 0, 1, \ldots, N_n - 1$. The total number of experiments to be performed is $N = N_1 \cdot N_2 \cdot \ldots \cdot N_n$. Because of the sum rule in eqn (6.3.19), it is possible to skip the phase cycling for one of the propagators as mentioned before.

To select the desired pathway characterized by the vector $\mathbf{\Delta p}$, the signals must be combined according to a discrete n-dimensional Fourier

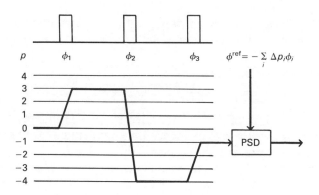

FIG. 6.3.3. Schematic representation of the parameters that are relevant in pathway selection procedures: the r.f. phases $\varphi_1, \varphi_2, \ldots, \varphi_n$ of the propagators $U_1, U_2, \ldots, U_n$, and the receiver reference phase $\varphi^{\text{ref}} = -\Delta\mathbf{p} \cdot \boldsymbol{\varphi} = -\sum_i \Delta p_i \varphi_i$. (Reproduced from Ref. 6.9.)

analysis which is a direct extension of eqn (6.3.13)

$$s(\Delta\mathbf{p}, t) = \frac{1}{N} \sum_{k_1=0}^{N_1-1} \sum_{k_2=0}^{N_2-1} \cdots \sum_{k_n=0}^{N_n-1} s(\boldsymbol{\varphi}, t) \exp\{i\Delta\mathbf{p} \cdot \boldsymbol{\varphi}\}. \quad (6.3.25)$$

The weighted linear combination can again be achieved by one of the three strategies outlined in § 6.3.1. A convenient approach relies on shifting the receiver reference channel by

$$\varphi^{\text{ref}} = -\Delta\mathbf{p} \cdot \boldsymbol{\varphi} = -\sum_i \Delta p_i \varphi_i. \quad (6.3.26)$$

The parameters involved in this procedure are shown schematically in Fig. 6.3.3 for a hypothetical experiment involving a pathway $p = 0 \to +3 \to -4 \to -1$.

Since the selectivity under each propagator U_i is determined by the number N_i of phase values, there is a manifold of pathways that survive the selection process, with

$$\Delta\mathbf{p} = \{\Delta p_1 \pm n_1 N_1, \Delta p_2 \pm n_2 N_2, \ldots, \Delta p_n \pm n_n N_n\} \quad (6.3.27)$$

with $n_1 = 0, 1, 2, \ldots, n_2 = 0, 1, 2, \ldots, \ldots, n_n = 0, 1, 2, \ldots$. The resulting branching of the selected pathways is illustrated in Fig. 6.3.4. Since the maximum order of coherence is limited (in a system with K spins $\frac{1}{2}$, $|p_{\max}| \leq K$), and since the efficiency of coherence transfer into high orders is often very small, it is usually possible to select a unique pathway by using a relatively small number N_i of phases in the phase cycle.

6.4 TWO-DIMENSIONAL FOURIER TRANSFORMATION

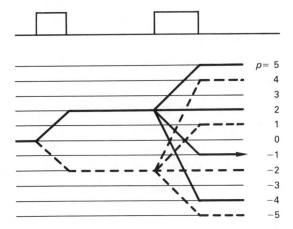

FIG. 6.3.4. Branching of pathways that are retained in a hypothetical experiment with two propagators U_1 and U_2 in a system with $|p_{\max}| = 5$. The phases are cycled with $N_1 = 4$ ($\varphi_1 = k_1 \pi/2$) to select $\Delta p_1 = +2$, and $N_2 = 3$ ($\varphi_2 = k_2 2\pi/3$) for the selection of $\Delta p_2 = -3$. Note that only one pathway contributes to observable $p = -1$ quantum coherence.

6.4. Two-dimensional Fourier transformation

The 2D Fourier transformation, which plays a central role in 2D spectroscopy, is a straightforward generalization of the one-dimensional Fourier transformation discussed in § 4.1.2 (6.15, 6.16).

The complex 2D Fourier transformation of the time-domain signal $s(t_1, t_2)$ is defined by

$$S(\omega_1, \omega_2) = \mathcal{F}\{s(t_1, t_2)\} = \mathcal{F}^{(1)}\mathcal{F}^{(2)}\{s(t_1, t_2)\}$$
$$= \int_{-\infty}^{\infty} dt_1 \exp(-i\omega_1 t_1) \int_{-\infty}^{\infty} dt_2 \exp(-i\omega_2 t_2) s(t_1, t_2) \quad (6.4.1)$$

with the inversion relation

$$s(t_1, t_2) = \mathcal{F}^{-1}\{S(\omega_1, \omega_2)\} = \mathcal{F}^{(1)-1}\mathcal{F}^{(2)-1}\{S(\omega_1, \omega_2)\}$$
$$= \frac{1}{4\pi^2} \int_{-\infty}^{\infty} d\omega_1 \exp(i\omega_1 t_1) \int_{-\infty}^{\infty} d\omega_2 \exp(i\omega_2 t_2) S(\omega_1, \omega_2). \quad (6.4.2)$$

A 2D Fourier transformation can therefore be considered as a succession of two 1D Fourier transformations.

The time signal $s(t_1, t_2)$ can be a complex function

$$s(t_1, t_2) = \text{Re}\{s(t_1, t_2)\} + i\,\text{Im}\{s(t_1, t_2)\}$$
$$= s_r(t_1, t_2) + i s_i(t_1, t_2) \quad (6.4.3)$$

recorded by quadrature detection in the t_2-period. The 2D spectrum $S(\omega_1, \omega_2)$ is a complex function even if the time-domain signal is real

$$S(\omega_1, \omega_2) = \text{Re}\{S(\omega_1, \omega_2)\} + i\,\text{Im}\{S(\omega_1, \omega_2)\}$$
$$= S_r(\omega_1, \omega_2) + i\,S_i(\omega_1, \omega_2). \quad (6.4.4)$$

In explicit calculations, it is often convenient to express the complex Fourier integrals by real transforms performed on the real component functions

$$S(\omega_1, \omega_2) = (\mathcal{F}_c^{(1)} - i\mathcal{F}_s^{(1)})(\mathcal{F}_c^{(2)} - i\mathcal{F}_s^{(2)})\{s_r(t_1, t_2) + is_i(t_1, t_2)\}. \quad (6.4.5)$$

The real and imaginary parts of the spectrum are

$$S_r(\omega_1, \omega_2) = \mathcal{F}^{cc}\{s_r\} - \mathcal{F}^{ss}\{s_r\} + \mathcal{F}^{cs}\{s_i\} + \mathcal{F}^{sc}\{s_i\}, \quad (6.4.6)$$
$$S_i(\omega_1, \omega_2) = -\mathcal{F}^{cs}\{s_r\} - \mathcal{F}^{sc}\{s_r\} + \mathcal{F}^{cc}\{s_i\} - \mathcal{F}^{ss}\{s_i\} \quad (6.4.7)$$

with the following definitions

$$\mathcal{F}^{cc}\{s_r\} = \mathcal{F}_c^{(1)}\mathcal{F}_c^{(2)}\{s_r(t_1, t_2)\}$$
$$= \int_{-\infty}^{\infty} dt_1 \cos \omega_1 t_1 \int_{-\infty}^{\infty} dt_2 \cos \omega_2 t_2 \, s_r(t_1, t_2), \quad (6.4.8)$$

$$\mathcal{F}^{cs}\{s_r\} = \mathcal{F}_c^{(1)}\mathcal{F}_s^{(2)}\{s_r(t_1, t_2)\}$$
$$= \int_{-\infty}^{\infty} dt_1 \cos \omega_1 t_1 \int_{-\infty}^{\infty} dt_2 \sin \omega_2 t_2 \, s_r(t_1, t_2), \quad (6.4.9)$$

$$\mathcal{F}^{sc}\{s_r\} = \mathcal{F}_s^{(1)}\mathcal{F}_c^{(2)}\{s_r(t_1, t_2)\}$$
$$= \int_{-\infty}^{\infty} dt_1 \sin \omega_1 t_1 \int_{-\infty}^{\infty} dt_2 \cos \omega_2 t_2 \, s_r(t_1, t_2), \quad (6.4.10)$$

$$\mathcal{F}^{ss}\{s_r\} = \mathcal{F}_s^{(1)}\mathcal{F}_s^{(2)}\{s_r(t_1, t_2)\}$$
$$= \int_{-\infty}^{\infty} dt_1 \sin \omega_1 t_1 \int_{-\infty}^{\infty} dt_2 \sin \omega_2 t_2 \, s_r(t_1, t_2). \quad (6.4.11)$$

6.4.1. Properties of the complex 2D Fourier transformation

6.4.1.1. Vector notation

By using time- and frequency-domain vectors, with a tilde to represent row vectors,

$$\mathbf{t} = \begin{bmatrix} t_1 \\ t_2 \end{bmatrix}, \quad \tilde{\boldsymbol{\omega}} = (\omega_1, \omega_2), \quad (6.4.12)$$

a compact notation of 2D Fourier transformations can be obtained

$$S(\tilde{\boldsymbol{\omega}}) = \int_{-\infty}^{\infty} d\mathbf{t} \, \exp(-i\tilde{\boldsymbol{\omega}}\mathbf{t}) \, s(\mathbf{t}) \quad (6.4.13)$$

and
$$s(t) = \frac{1}{4\pi^2} \int_{-\infty}^{\infty} d\omega \exp(i\vec{\omega}t) \, S(\vec{\omega}). \tag{6.4.14}$$

The equivalent expressions in terms of a frequency vector $\tilde{\mathbf{f}} = \vec{\omega}/(2\pi)$ are

$$S(\tilde{\mathbf{f}}) = \int_{-\infty}^{\infty} dt \exp(-i2\pi \tilde{\mathbf{f}} \mathbf{t}) \, s(\mathbf{t}) \tag{6.4.15}$$

and

$$s(\mathbf{t}) = \int_{-\infty}^{\infty} d\mathbf{f} \exp(i2\pi \tilde{\mathbf{f}} \mathbf{t}) \, S(\tilde{\mathbf{f}}). \tag{6.4.16}$$

Equations (6.4.15) and (6.4.16) apply to Fourier transformations in arbitrary dimensions.

6.4.1.2. Similarity theorem

A transformation of the variables in one domain is reflected by a corresponding transformation in the other domain. If the two functions $s(\mathbf{t})$ and $S(\vec{\omega})$ form a Fourier transform pair (indicated by $\leftrightarrow$), and the matrix $\mathbf{A}$ represents a transformation of the time variables, we obtain the new Fourier transform pair

$$s'(\mathbf{t}) = s(\mathbf{A}\mathbf{t}) \leftrightarrow S'(\vec{\omega}) = \frac{1}{|A|} S(\vec{\omega}\mathbf{A}^{-1}). \tag{6.4.17}$$

where $|A|$ is the determinant of $\mathbf{A}$. This theorem holds for arbitrary linear transformations in arbitrary dimensions.

Consider the special case of an orthogonal transformation $\mathbf{A}$ in time domain,

$$\mathbf{t}' = \mathbf{A}\mathbf{t} \quad \text{with} \quad |A| = 1 \quad \text{and} \quad \mathbf{A}^{-1} = \tilde{\mathbf{A}}. \tag{6.4.18}$$

In this case, we find immediately

$$\vec{\omega}' = \vec{\omega}\mathbf{A}^{-1} \tag{6.4.19}$$

or
$$\omega' = \mathbf{A}\omega. \tag{6.4.20}$$

Thus an orthogonal transformation in one domain induces the *same* orthogonal transformation in the other domain, as shown in Fig. 6.4.1. This property provides a very convenient means of computing Fourier transforms of functions in rotated coordinate systems. It also allows one to compute the effects of a shearing operation in one domain, which is often used in 2D spectroscopy (see § 6.6.1).

6.4.1.3. Convolution theorem

The one-dimensional convolution theorem can easily be generalized to two dimensions

$$s_1(\mathbf{t}) \ast\ast s_2(\mathbf{t}) \leftrightarrow S_1(\vec{\omega}) \cdot S_2(\vec{\omega}) \tag{6.4.21}$$

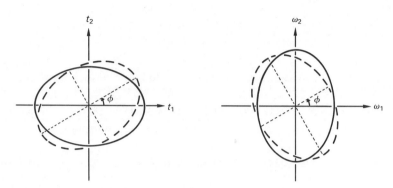

FIG. 6.4.1. The similarity theorem of 2D complex Fourier transformation. A rotation of a function in one domain induces the same rotation of its Fourier transform in the other domain.

and

$$s_1(\mathbf{t}) \cdot s_2(\mathbf{t}) \leftrightarrow \frac{1}{4\pi^2} S_1(\tilde{\omega}) ** S_2(\tilde{\omega}) \qquad (6.4.22)$$

where the 2D convolution integral is defined by

$$s_1(\mathbf{t}) ** s_2(\mathbf{t}) = \int_{-\infty}^{\infty} d\tau_1 \int_{-\infty}^{\infty} d\tau_2 s_1(\tau_1, \tau_2) s_2(t_1 - \tau_1, t_2 - \tau_2). \qquad (6.4.23)$$

If eqn (6.4.22) is written in terms of the frequency vector $\tilde{\mathbf{f}}$, the factor $4\pi^2$ must be dropped.

Equation (6.4.22) implies that convolution filtering in the frequency domain, which is of central importance for sensitivity and resolution enhancement (§ 6.8), can be achieved by straightforward multiplication of the time-domain signal.

6.4.1.4. Power theorem

The two-dimensional integrals of the absolute squares of the time- and frequency-domain functions are related

$$\int_{-\infty}^{\infty} |s(\mathbf{t})|^2 \, d\mathbf{t} = \frac{1}{4\pi^2} \int_{-\infty}^{\infty} |S(\tilde{\omega})|^2 \, d\omega. \qquad (6.4.24)$$

This theorem is essential for computing signal and noise energy in the context of sensitivity considerations.

6.4.1.5. Projection cross-section theorem

In some applications of 2D spectroscopy (see § 6.5.5), it is necessary to calculate *skew* projections of the spectrum $S(\omega_1, \omega_2)$ on to an axis which subtends an angle ϕ with respect to the ω_1-axis, as illustrated in Fig.

6.4 TWO-DIMENSIONAL FOURIER TRANSFORMATION

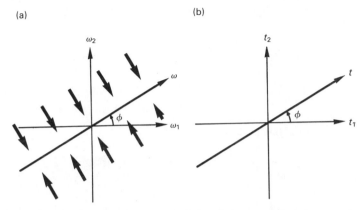

FIG. 6.4.2. (a) A skew projection $P(\omega, \phi)$ of the 2D frequency domain on to an axis that subtends an angle ϕ with respect to the ω_1-axis (eqn (6.4.25)) corresponds to the Fourier transformation of a skew cross-section taken through the origin of the 2D time-domain (b) that subtends the same angle ϕ with respect to the t_1-axis (eqn (6.4.26)). Note that the limit $\phi = \pi/2$ (projection on to the ω_2-axis) corresponds to eqn (6.4.30), while $\phi = 0$ (projection on to the ω_1-axis) is consistent with eqn (6.4.31).

6.4.2(a). Such a skew projection, denoted $P(\omega, \phi)$, is a function of a single frequency variable ω and is given by

$$P(\omega, \phi) = \int_{-\infty}^{\infty} d\omega' S(\omega \cos \phi - \omega' \sin \phi, \omega \sin \phi + \omega' \cos \phi). \quad (6.4.25)$$

The projection cross-section theorem (6.17, 6.19, 6.20) states that $P(\omega, \phi)$ forms a Fourier transform pair with a cross-section through the time-domain signal $s(t_1, t_2)$ passing through the origin $t_1 = t_2 = 0$, and subtending an angle ϕ with respect to the t_1-axis (see Fig. 6.4.2(b))

$$c(t, \phi) = s(t \cos \phi, t \sin \phi) \quad (6.4.26)$$

with the Fourier transform relations

$$P(\omega, \phi) = \mathcal{F}\{c(t, \phi)\}, \quad (6.4.27)$$
$$c(t, \phi) = \mathcal{F}^{-1}\{P(\omega, \phi)\}. \quad (6.4.28)$$

In 2D spectroscopy, causality (6.18) implies that the time-domain signal vanishes for negative times

$$s(t_1, t_2) = 0, \quad \text{for} \quad t_1 < 0 \quad \text{and} \quad t_2 < 0. \quad (6.4.29)$$

Hence the time-domain cross-section in eqn (6.4.26) vanishes if $\pi/2 < \phi < \pi$, except for the point $s(0, 0)$. The skew frequency-domain projections in eqn (6.4.25) therefore also vanish for $\pi/2 < \phi < \pi$. This has important practical implications in 2D spectroscopy which will be discussed in § 6.5.5.

For $\phi = \pi/2$ one obtains the projection of the spectrum $S(\omega_1, \omega_2)$ on to the ω_2-axis

$$P(\omega, \phi = \pi/2) = \int_{-\infty}^{+\infty} S(\omega_1, \omega_2) \, d\omega_1$$
$$= \mathscr{F}\{s(t_1 = 0, t_2)\}$$
$$= S(t_1 = 0, \omega_2). \qquad (6.4.30)$$

The projection on the ω_1-axis corresponds to $\phi = 0$

$$P(\omega, \phi = 0) = \int_{-\infty}^{+\infty} S(\omega_1, \omega_2) \, d\omega_2$$
$$= \mathscr{F}\{s(t_1, t_2 = 0).$$
$$= S(\omega_1, t_2 = 0). \qquad (6.4.31)$$

Note that the projection of eqn (6.4.30) is equivalent to the 1D Fourier transform with respect to t_2 of a single free induction signal obtained with $t_1 = 0$ (first row of the matrix $S(t_1, \omega_2)$). On the other hand, to obtain an ω_1-spectrum by the projection in eqn (6.4.31), a complete series of t_1-increments must be measured, but the signal acquisition may be reduced to a single point in the t_2-domain.

Projections of the time-domain signal can be related in a similar manner to cross-sections of the frequency-domain signal.

6.4.1.6. Kramers–Kronig relations in two dimensions

For a causal 1D impulse response $s(t)$ with

$$s(t) = 0 \quad \text{for} \quad t < 0, \qquad (6.4.32)$$

the real and imaginary parts of the Fourier transform $S(\omega)$ form a Hilbert transform pair

$$S_i(\omega) = \mathscr{H}\{S_r(\omega)\}, \qquad (6.4.33)$$
$$S_r(\omega) = -\mathscr{H}\{S_i(\omega)\}. \qquad (6.4.34)$$

with the Hilbert transformation $\mathscr{H}$ defined in eqn (4.1.18). These relationships are known as Kramers–Kronig relations. The two components $S_r(\omega)$ and $S_i(\omega)$ are two spectra that differ in phase by an angle $\phi = -\pi/2$. A Hilbert transformation can therefore be considered as inducing a phase shift of $-\pi/2$. Linear combinations of one component and its Hilbert transform allow one to achieve a phase shift through an arbitrary angle (6.18).

The 2D time-domain function $s(t_1, t_2)$ is usually causal in both time variables and vanishes for $t_1 < 0$ and $t_2 < 0$ (eqn (6.4.29)), except if the time-domain is extended to encompas $t_2 < 0$. The Kramers–Kronig relations are normally applicable to both time variables

$$\mathcal{H}^{(1)}\{S_r(\omega_1, \omega_2)\} = S_i\{\omega_1, \omega_2\}, \quad (6.4.35)$$

$$\mathcal{H}^{(2)}\{S_r(\omega_1, \omega_2)\} = S_i\{\omega_1, \omega_2\} \quad (6.4.36)$$

where the Hilbert transformations $\mathcal{H}^{(1)}$ and $\mathcal{H}^{(2)}$ refer to the variables ω_1 and ω_2, respectively. It is important to note that both transformations have the same effect and that the application of both transformations in sequence leads, except for a sign change, back to the original spectrum. There are only two orthogonal components $S_r(\omega_1, \omega_2)$ and $S_i(\omega_1, \omega_2)$ and only one phase variable. The lack of a second independent phase variable is also the reason that a full separation of absorption and dispersion mode signals is often impossible in 2D spectroscopy (see § 6.5).

6.4.2. Hypercomplex two-dimensional Fourier transformation

The complex 2D Fourier transformation discussed above has shortcomings insofar as only two independent components $S_r(\omega_1, \omega_2)$ and $S_i(\omega_1, \omega_2)$ can be distinguished. In order to allow for an independent phase adjustment with respect to the two frequency variables, it appears necessary to distinguish four signal components in such a way that real and imaginary parts can be combined independently in both dimensions. This requirement can be fulfilled by introducing a hypercomplex 2D Fourier transformation (6.64) which is briefly discussed in this section but will not be applied explicitly in the remainder of this volume.

The hypercomplex Fourier transformation converts a four-component time-domain signal $^4s(t_1, t_2)$ into a four-component spectrum $^4S(\omega_1, \omega_2)$. Defining two independent imaginary units i and j which refer to the two orthogonal axes,

$$i^2 = j^2 = -1, \quad (6.4.37)$$

the hypercomplex time-domain function may be written

$$^4s(t_1, t_2) = s_{rr}(t_1, t_2) + is_{ri}(t_1, t_2) + js_{jr}(t_1, t_2) + ijs_{ji}(t_1, t_2). \quad (6.4.38)$$

The real and imaginary parts in t_2 (second index of the signal functions) are measured by quadrature detection in t_2, while the distinction of the real and imaginary parts in t_1 is achieved with two independent experiments which differ in the r.f. phase of one or more pulses (typically, the phase φ^{prep} of the entire preparation sequence may be shifted by $\Delta\varphi = \pi/(2\,|p|)$, where p is the order of coherence evolving in t_1).

Thus one obtains the four signal components from two complementary experiments

Experiment 1 $\quad$ $\begin{cases} \text{quadrature component } s_x\colon s_{rr}(t_1, t_2) \\ \text{quadrature component } s_y\colon s_{ri}(t_1, t_2) \end{cases}$
$(\varphi^{\text{prep}} = 0)$

Experiment 2 $\quad$ $\begin{cases} \text{quadrature component } s_x\colon s_{jr}(t_1, t_2) \\ \text{quadrature component } s_y\colon s_{ji}(t_1, t_2) \end{cases}$
$\left(\varphi^{\text{prep}} = \dfrac{\pi}{2|p|}\right)$

The hypercomplex Fourier transformation is then defined by

$$^4S(\omega_1, \omega_2) = \mathscr{F}\{^4s(t_1, t_2)\} = \mathscr{F}_j^{(1)}\mathscr{F}_i^{(2)}\{^4s(t_1, t_2)\}$$
$$= \int_{-\infty}^{\infty} dt_1 \exp(-j\omega_1 t_1) \int_{-\infty}^{\infty} dt_2 \exp(-i\omega_2 t_2)\, ^4s(t_1, t_2) \quad (6.4.39)$$

together with the corresponding inversion relation. The frequency-domain signal $^4S(\omega_1, \omega_2)$ is also a hypercomplex function with four components

$$^4S(\omega_1, \omega_2) = S_{rr}(\omega_1, \omega_2) + iS_{ri}(\omega_1, \omega_2) + jS_{jr}(\omega_1, \omega_2)$$
$$jiS_{ji}(\omega_1, \omega_2). \quad (6.4.40)$$

In analogy to eqns (6.4.6)–(6.4.11), the hypercomplex Fourier transformation of eqn (6.4.39) may be expressed by real cosine- and sine-Fourier transforms of the real time components s_{rr}, s_{ri}, s_{jr}, and s_{ji}

$$S_{rr}(\omega_1, \omega_2) = +\mathscr{F}^{cc}\{s_{rr}\} + \mathscr{F}^{cs}\{s_{ri}\} + \mathscr{F}^{sc}\{s_{jr}\} + \mathscr{F}^{ss}\{s_{ji}\},$$
$$S_{ri}(\omega_1, \omega_2) = -\mathscr{F}^{cs}\{s_{rr}\} + \mathscr{F}^{cc}\{s_{ri}\} - \mathscr{F}^{ss}\{s_{jr}\} + \mathscr{F}^{sc}\{s_{ji}\},$$
$$S_{jr}(\omega_1, \omega_2) = -\mathscr{F}^{sc}\{s_{rr}\} - \mathscr{F}^{ss}\{s_{ri}\} + \mathscr{F}^{cc}\{s_{jr}\} + \mathscr{F}^{cs}\{s_{ji}\},$$
$$S_{ji}(\omega_1, \omega_2) = +\mathscr{F}^{ss}\{s_{rr}\} - \mathscr{F}^{sc}\{s_{ri}\} - \mathscr{F}^{cs}\{s_{jr}\} + \mathscr{F}^{cc}\{s_{ji}\}. \quad (6.4.41)$$

We now have at our disposal two independent phase variables associated with the imaginary units j and i, and it is possible to adjust the phase independently along both frequency axes by linear combinations of the four components of $^4S(\omega_1, \omega_2)$.

The hypercomplex Fourier transformation is a concise mathematical concept which takes into account that the two Fourier transformations in t_1 and t_2 are independent. It avoids the superposition of real–real and imaginary–imaginary parts inherent in the complex 2D Fourier transformation.

It is however possible to avoid the use of hypercomplex transformations if the phase-shifted experiments required to obtain the four components in eqn (6.4.38) are considered as part of a phase cycle, and if coherence transfer pathways are separated by a discrete Fourier analysis with respect to phase as described by eqns (6.3.13) and (6.3.25).

6.5. Peak shapes in two-dimensional spectra

The basic complex time-domain signal associated with a single coherence transfer pathway is described by (see eqn 6.2.10b)

$$s(t_1, t_2) = s_{rs\,tu}(0, 0)\exp\{(-i\omega_{tu}^{(e)} - \lambda_{tu}^{(e)})t_1\}\exp\{(-i\omega_{rs}^{(d)} - \lambda_{rs}^{(d)})t_2\} \quad (6.5.1)$$

with $t_1, t_2 \geq 0$ and the complex amplitude $s_{rs\,tu}(0, 0) = Z_{rs\,tu}$ defined in eqn (6.2.11).

This time-domain signal is converted into a complex 2D Lorentzian spectrum by the complex 2D Fourier transformation of eqn (6.4.1)

$$S(\omega_1, \omega_2) = s_{rs\,tu}(0, 0) \frac{1}{i\Delta\omega_{tu}^{(e)} + \lambda_{tu}^{(e)}} \cdot \frac{1}{i\Delta\omega_{rs}^{(d)} + \lambda_{rs}^{(d)}} \quad (6.5.2)$$

with the frequency variables (using $\omega_{ut}^{(e)} = -\omega_{tu}^{(e)}$, $\omega_{sr}^{(d)} = -\omega_{rs}^{(d)}$)

$$\Delta\omega_{tu}^{(e)} = \omega_1 - \omega_{ut}^{(e)} \quad \text{and} \quad \Delta\omega_{rs}^{(d)} = \omega_2 - \omega_{sr}^{(d)} \quad (6.5.3)$$

which correspond to the offsets with respect to the centre of the resonance, in analogy with the treatment of 1D spectra in eqn (4.2.19).

By separating the absorptive and dispersive contributions, eqn (6.5.2) can be written

$$S(\omega_1, \omega_2) = s_{rs\,tu}(0, 0)[a_{tu}(\omega_1) - id_{tu}(\omega_1)][a_{rs}(\omega_2) - id_{rs}(\omega_2)]$$
$$= s_{rs\,tu}(0, 0)\{[a_{tu}(\omega_1)a_{rs}(\omega_2) - d_{tu}(\omega_1)d_{rs}(\omega_2)] \quad (6.5.4a)$$
$$+ i[-a_{tu}(\omega_1)d_{rs}(\omega_2) - d_{tu}(\omega_1)a_{rs}(\omega_2)]\} \quad (6.5.4b)$$

with the Lorentzian absorption and dispersion components

$$a_{tu}(\omega_1) = \frac{\lambda_{tu}^{(e)}}{(\Delta\omega_{tu}^{(e)})^2 + (\lambda_{tu}^{(e)})^2}, \quad a_{rs}(\omega_2) = \frac{\lambda_{rs}^{(d)}}{(\Delta\omega_{rs}^{(d)})^2 + (\lambda_{rs}^{(d)})^2},$$
$$d_{tu}(\omega_1) = \frac{\Delta\omega_{tu}^{(e)}}{(\Delta\omega_{tu}^{(e)})^2 + (\lambda_{tu}^{(e)})^2}, \quad d_{rs}(\omega_2) = \frac{\Delta\omega_{rs}^{(d)}}{(\Delta\omega_{rs}^{(d)})^2 + (\lambda_{rs}^{(d)})^2}. \quad (6.5.5)$$

By adjusting the phase-sensitive detector or by calculating a suitable phase correction, it is possible to select either real or imaginary part of the signal in eqn (6.4.4). Two special cases can be distinguished: if the amplitude $s_{rs\,tu}(0, 0)$ is real, the real part of the spectrum is

$$\text{Re}\{S(\omega_1, \omega_2)\} = s_{rs\,tu}(0, 0)[a_{tu}(\omega_1)a_{rs}(\omega_2) - d_{tu}(\omega_1)d_{rs}(\omega_2)]. \quad (6.5.6)$$

This peak shape is said to have a 'mixed phase', since it consists of a superposition of a pure 2D absorption peak and a pure 2D dispersion peak.

If the amplitude $s_{rs\,tu}(0, 0)$ is imaginary, the real part of the signal is

$$\text{Re}\{S(\omega_1, \omega_2)\} = -is_{rs\,tu}(0, 0)[a_{tu}(\omega_1)d_{rs}(\omega_2) + d_{tu}(\omega_1)a_{rs}(\omega_2)] \quad (6.5.7)$$

which corresponds to a superposition of two mixed absorptive–dispersive peak shapes.

It is a characteristic feature of 2D spectroscopy that a single coherence transfer pathway invariably leads to a mixed phase peak. It will be shown in §6.5.3 how the absorptive and dispersive contributions can be separated to obtain pure phase peaks.

6.5.1. Basic peakshapes

The pure 2D absorption peakshape

$$S(\omega_1, \omega_2) = a_{tu}(\omega_1)a_{rs}(\omega_2), \tag{6.5.8}$$

which is illustrated by Fig. 6.5.1(a), has the property that all sections taken parallel to either axis yield pure 1D absorption Lorentzian lineshapes. The amplitude drops to half-height for $\Delta\omega_{tu}^{(e)} = 0$, $\Delta\omega_{rs}^{(d)} = \lambda_{rs}^{(d)}$ and $\Delta\omega_{tu}^{(e)} = \lambda_{tu}^{(e)}$, $\Delta\omega_{rs}^{(d)} = 0$. The drop in amplitude is steeper in the bisecting planes: if the line-widths are equal in both domains, the signal drops to half-height for $\Delta\omega_{tu}^{(e)} = \pm\Delta\omega_{rs}^{(d)} = \pm((2)^{\frac{1}{2}} - 1)^{\frac{1}{2}}\lambda \simeq 0.64\lambda$, i.e. at a radial distance of 0.91λ from the centre of the peak. The asymptotic decay is proportional to $(\Delta\omega_{tu}^{(e)})^{-2}$ and $(\Delta\omega_{rs}^{(d)})^{-2}$ on sections parallel to one of the frequency axes, while it is proportional to the inverse fourth

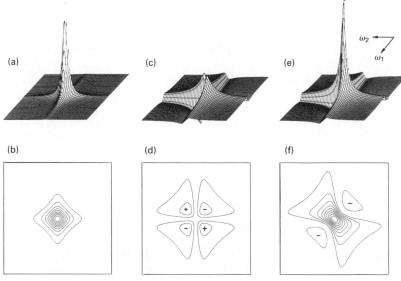

FIG. 6.5.1. Peakshapes in 2D spectra: (a) and (b) pure 2D absorption $a_{tu}(\omega_1)a_{rs}(\omega_2)$; (c) and (d) pure negative 2D dispersion $-d_{tu}(\omega_1)d_{rs}(\omega_2)$; (e) and (f) mixed phase peakshape, also known as 'phase-twisted' peakshape, consisting of a superposition $a_{tu}(\omega_1)a_{rs}(\omega_2) - d_{tu}(\omega_1)d_{rs}(\omega_2)$.

power in the bisecting planes. This lack of cylindrical or elliptical symmetry has been called 'star effect' (6.22), and may be appreciated in Fig. 6.5.1(b). This effect can be removed by a 2D Lorentz–Gauss transformation (§ 6.5.6.2).

The pure (negative) 2D dispersion peakshape

$$S(\omega_1, \omega_2) = -d_{tu}(\omega_1)d_{rs}(\omega_2) \qquad (6.5.9)$$

is shown in Fig. 6.5.1(c) and (d). It may be appreciated in the contour plot of Fig. 6.5.1(d) that the dispersion peak has vanishing signal contribution for $\Delta\omega_{tu}^{(e)} = 0$ or $\Delta\omega_{rs}^{(d)} = 0$. In the four quadrants there are positive and negative lobes. These lobes are negative for $\Delta\omega_{tu}^{(e)} \cdot \Delta\omega_{rs}^{(d)} > 0$ (e.g. on the diagonal line bisecting the ω_1- and ω_2-axes), whereas positive lobes occur for $\Delta\omega_{tu}^{(e)} \cdot \Delta\omega_{rs}^{(d)} < 0$. It is apparent that the decay of the dispersive peaks is much slower than that of the absorptive peaks, in analogy to 1D Lorentzian lineshapes. The suppression of the dispersive component is therefore desirable for improving resolution.

The mixed peakshape of eqn (6.5.6)

$$S(\omega_1, \omega_2) = a_{tu}(\omega_1)a_{rs}(\omega_2) - d_{tu}(\omega_1)d_{rs}(\omega_2) \qquad (6.5.10)$$

is quite common in 2D spectroscopy. This peakshape, which results from the superposition of pure absorption and negative dispersion peak contributions, is illustrated in Fig. 6.5.1(e) and (f) and is often referred to as 'phase-twist' (6.22) since sections taken parallel to the ω_2-axis below resonance ($\omega_1 < \omega_{ut}^{(e)}$) show a positive dispersion, which acquires an increasing admixture of absorption mode as one approaches resonance, and turns into negative dispersion above resonance ($\omega_1 > \omega_{ut}^{(e)}$).

We may note in passing that the lineshapes in Fig. 6.5.1 are shown in a coordinate system where the ω_2-frequency runs horizontally from right to left, and the ω_1-frequency increases from top to bottom. The frequency origin may be at the top right or in the centre. This frame has been chosen to fulfil two criteria.

1. The conventional spectrum, which appears in the ω_2-domain, should be plotted in accordance with universal practice in high-resolution NMR with frequencies (in Hz or p.p.m.) increasing from right to left.

2. The positive diagonal ($\Delta\omega_1/\Delta\omega_2 = +1$) should run from top right to bottom left.

In some cases, it may be desirable to interchange the ω_1- and ω_2-axes if the lineshape or multiplet structure in the ω_1-domain must be made apparent in stacked plots. Criterium 1 remains fulfilled if the spectrum is viewed from the right. Various authors have used diverging conventions, which may have their own justifications.

6.5.2. Inhomogeneous broadening and interference of neighbouring peaks with mixed phases

Peakshapes with mixed phases are not only a hindrance to resolution, since the wings of the dispersion components fall off with $\Delta\omega^{-1}$ rather than with $\Delta\omega^{-2}$, but also give rise to complicated interference effects.

Consider a superposition of two signals in a 2D spectrum that are separated in their coordinates by $\Delta\Omega_1$ and $\Delta\Omega_2$ (see Fig. 6.5.2). If the separation between the peaks is comparable to the widths $\lambda^{(e)}$ and $\lambda^{(d)}$, the dispersion lobes will interfere. The form of this interference depends on the relative position of the two peaks determined by the ratio $\Delta\Omega_1/\Delta\Omega_2$. If this ratio is positive, the negative lobes of the two signals partly cancel the absorption components, leading to a loss in amplitude (destructive interference). If, on the other hand, the ratio $\Delta\Omega_1/\Delta\Omega_2$ is

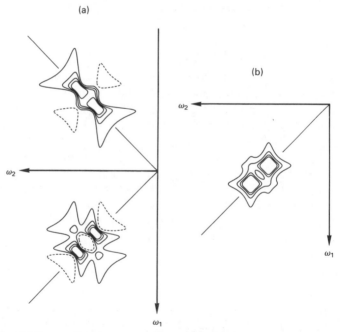

FIG. 6.5.2. (a) Superposition of a pair of lines with mixed phases: in the lower quadrant, the resonance positions are displaced along the positive diagonal ($\Delta\Omega_1 = \Delta\Omega_2 = 4\lambda_1 = 4\lambda_2$, i.e. displaced by twice the full line-width at half-height), leading to destructive interference due to the overlap of negative dispersion lobes with positive absorption components. In the upper quadrant, the resonance positions are displaced along the negative diagonal ($\Delta\Omega_1 = -\Delta\Omega_2 = 4\lambda_1 = 4\lambda_2$), leading to constructive interference. (b) Superposition of two pure 2D absorption lines, displaced by $\Delta\Omega_1 = \Delta\Omega_2 = 4\lambda_1 = 4\lambda_2$. The contour levels shown correspond to 22, 16, 10, 4, and −4 per cent of the maximum height of an isolated peak. Negative contours are drawn with dashed lines.

negative, the positive lobes of the dispersion components interfere constructively with the absorption signals (Fig. 6.5.2(a)). It is apparent that in this case a rather elongated ridge is obtained, while if $\Delta\Omega_1/\Delta\Omega_2$ is positive, the extension is similar in all directions.

It should be emphasized that destructive interference can be removed if pure 2D absorption lineshapes are obtained by one of the techniques described in § 6.5.3. On the other hand, absolute value displays do not provide safeguards against cancellation effects.

Interference phenomena may occur in any crowded 2D spectrum. They are particularly severe when inhomogeneous broadening leads to a continuous distribution of signals along a 'ridge' in 2D frequency domain. This situation is encountered in solid-state spectra of powders and in inhomogeneous static magnetic fields. In homonuclear J-spectroscopy (see § 7.2), a similar situation arises when ill-resolved long-range couplings lead to line broadening with $\Delta\Omega_1/\Delta\Omega_2 = +1$, resulting in abnormal amplitudes losses.

The parameter $\Delta\Omega_1/\Delta\Omega_2$, relevant for the peak shape in the presence of inhomogeneous broadening, depends on the order of coherence in the evolution and detection periods (6.23). It is convenient to define the ratio

$$\kappa = \frac{\gamma^{(1)}p^{(1)}}{\gamma^{(2)}p^{(2)}} \quad (6.5.11)$$

where $p^{(1)}$ is the (signed) order of coherence in the t_1-interval (i.e. plus or minus p-quantum, including plus or minus single quantum), $p^{(2)} = -1$ represents the single-quantum coherence observed in the detection period (see eqn 6.3.2), while $\gamma^{(1)}$ and $\gamma^{(2)}$ are the gyromagnetic ratios of the nuclei that carry the coherence in the evolution and detection periods, respectively.

Some examples of inhomogeneously broadened 2D spectra are shown in Fig. 6.5.3. It is clear that destructive interference of mixed-phase lines will occur if $\kappa > 0$, while cancellation effects are avoided when $\kappa < 0$, i.e. if $\gamma^{(1)}p^{(1)}$ and $\gamma^{(2)}p^{(2)}$ are of opposite sign.

This is illustrated in Fig. 6.5.4. In a heteronuclear 2D correlation spectrum obtained by transferring coherence from ^{1}H to ^{31}P (see § 8.3), there are two types of signal with $\kappa = \pm\gamma(^1\text{H})/\gamma(^{31}\text{P}) \simeq \pm 2.5$. In the absolute-value 2D spectrum of Fig. 6.5.4, there are two patterns with four signals, each spanning a square in frequency domain of side length $2\pi J_{\text{PH}}$. The signals with $\kappa = -2.5$ (upper half) show constructive interference, while the other set ($\kappa = +2.5$, lower half) suffers from cancellation effects.

The improved resolution for $\kappa < 0$ can be explained by the reversal of the sense of Zeeman precession: the inhomogeneous dephasing in the

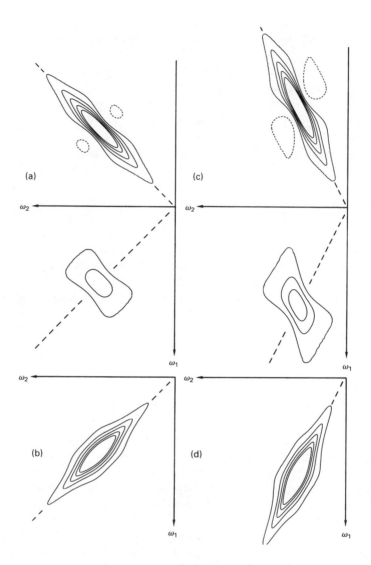

FIG. 6.5.3. Inhomogeneous broadening in 2D spectra. (a) Superposition of mixed-phase peakshapes with a Lorentzian envelope, and with slopes $\Delta\Omega_1/\Delta\Omega_2 = \kappa = \pm 1$, appropriate for single-quantum 2D correlation spectroscopy. The inhomogeneous width of the envelope projected onto the ω_2-axis is 10 times the full width at half-height of the homogeneous peak. Note the constructive interference in the upper quadrant ('N peak' or 'echo signal'), and the destructive interference in the lower quadrant ('P peak' or 'antiecho signal'). (b) Superposition of pure 2D absorption peakshapes with the same parameters as in (a). The contours in (a) and (b) are drawn at 38, 29, 21, 14, 7, and −7 per cent of the maximum height of the peak in (b). Negative contours are drawn with dashed lines. (c) Superposition of mixed phase peakshapes with slopes $\Delta\Omega_1/\Delta\Omega_2 = \kappa = \pm 2$, appropriate for double-quantum spectroscopy. The width of the envelope projected on to the ω_2-axis is the same as in (a). (d) Superposition of pure 2D absorption peakshapes with $\kappa = \pm 2$ as in (c). The contours in (c) and (d) are drawn at 29, 22, 16, 10, 5, and −5 per cent of the maximum height of the peak in (d). The maximum in the upper quadrant of (c) is 3.7 times higher than the maximum in the lower quadrant of (c).

6.5 PEAK SHAPES IN TWO-DIMENSIONAL SPECTRA

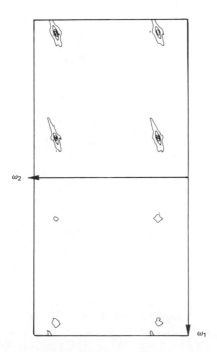

FIG. 6.5.4. Inhomogeneous broadening in heteronuclear 2D correlation spectra obtained by transferring coherence from ^{1}H to ^{31}P (pathways $p_I = \pm 1$, $p_S = 0 \to p_I = 0$, $p_S = -1$) in the AX system in phosphorous acid $D_2(HPO_3)$ ($J_{HP} \simeq 600$ Hz). (Adapted from Ref. 6.14.)

evolution period is refocused to form a coherence transfer echo (6.24) in the course of the detection period.

Coherence transfer echo effects can be understood by considering the signal envelope in 2D time domain. Since the inhomogeneous dephasing in the evolution period is refocused in the course of the detection period at a point in time (see also eqn (5.4.15))

$$t_2 = -\frac{\gamma^{(1)}p^{(1)}}{\gamma^{(2)}p^{(2)}}t_1 = -\kappa t_1, \tag{6.5.12}$$

the signal envelope in time domain features a 'ridge' of maximum amplitude with a slope

$$\frac{\Delta t_2}{\Delta t_1} = -\kappa \tag{6.5.13}$$

which is orthogonal to the slope of the elongated lineshape in frequency domain. The ridge in the 2D time-domain ($t_1, t_2 > 0$) is only evident if κ is negative, i.e. if $\gamma^{(1)}p^{(1)}$ and $\gamma^{(2)}p^{(2)}$ are of opposite sign. In this case,

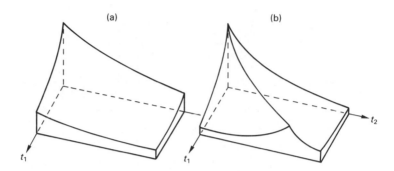

FIG. 6.5.5. Time-domain signal envelopes typical for inhomogeneous static fields. (a) Signal components without refocusing ($\kappa > 0$), with envelopes that decay exponentially in both time domains (eqn (6.5.15)). (b) Signal components with $\kappa = -2$ (appropriate for homonuclear double-quantum spectroscopy with a coherence transfer pathway $p = 0 \rightarrow +2 \rightarrow -1$) which lead to coherence transfer echoes and hence to a ridge in time domain with a slope $\Delta t_2 / \Delta t_1 = 2$ (eqn (6.5.14)).

the signal envelope function takes the form ($\kappa < 0$)

$$s^e(t_1, t_2) = s^e(0, 0)\exp(-\lambda^{(e)}t_1)\exp(-\lambda^{(d)}t_2)\exp\{-\lambda^+ |t_2 + \kappa t_1|\} \quad (6.5.14)$$

where the inhomogeneous distribution of frequency components is described as an additional Lorentzian broadening $\lambda^+ = 1/T_2^+$. Components with positive κ do not lead to refocusing in the time domain, hence their envelope function takes a simple biexponential form ($\kappa > 0$)

$$s^e(t_1, t_2) = s^e(0, 0)\exp\{-(\lambda^{(e)} + \kappa\lambda^+)t_1\}\exp\{-(\lambda^{(d)} + \lambda^+)t_2\}. \quad (6.5.15)$$

The signal envelope functions typical of inhomogeneous static fields are illustrated in Fig. 6.5.5. They are relevant for the sensitivity that can be achieved in 2D spectra (see § 6.8).

In the context of homonuclear single-quantum 2D correlation spectroscopy, components with $\kappa > 0$, which correspond to a coherence transfer pathway $p = 0 \rightarrow -1 \rightarrow -1$, have been called 'antiechoes' (6.25) or 'P signals' for positive κ (6.12). Components with $\kappa < 0$, corresponding to $p = 0 \rightarrow +1 \rightarrow -1$, are sometimes referred to as 'echo signals' or 'N signals' for negative κ. Phase-cycling procedures can be used to select either family of signals. To obtain pure 2D absorption lineshapes, it is essential to retain both pathways as discussed in § 6.5.3.

In 2D spectra of polycrystalline powders or amorphous materials (see § 7.3), inhomogeneous line-widths arising from anisotropic interactions also lead to interference phenomena, which can be avoided by obtaining pure absorption lineshapes.

6.5.3. Techniques for obtaining pure two-dimensional absorption peaks

In order to improve resolution and to avoid cancellation effects, it is desirable to suppress the dispersive components in peaks with mixed phases. In this section, we discuss three distinct strategies for obtaining pure phase spectra (i.e. either pure 2D absorption or pure 2D dispersion peaks):

1. Real (cosine) Fourier transformation with respect to t_1;
2. Time reversal in a complementary experiment;
3. Combination of two experiments in quadrature.

The three strategies are applicable to different classes of 2D experiments. It should be noted that in some techniques, such as in homonuclear 2D J-spectroscopy (§ 7.2), and for diagonal peaks in correlation spectroscopy with r.f. pulses with $\beta \neq \pi/2$, none of these techniques are applicable. In such circumstances, it is possible to take recourse to signal-processing techniques:

4. Pseudo-echo filtration (§ 6.5.6.3);
5. Subtraction of the dispersive component after locating the peaks by computer procedures (6.32).

6.5.3.1. Real Fourier transformation in t_1

To define the conditions that must be fulfilled to obtain pure phase spectra, it is useful to distinguish *amplitude modulation* and *phase modulation* (6.22). These expressions refer to the t_1-dependence of the signal $s(t_1, t_2)$ which becomes particularly obvious after 1D Fourier transformation with respect to t_2: depending on the type of 2D experiment the spectrum $s(t_1, \omega_2)$ may appear modulated either in amplitude or in phase as a function of t_1. We shall see that pure spectra can only be obtained for amplitude modulation (6.21, 6.22).

A single coherence transfer pathway invariably leads to a signal of the form of eqn (6.5.1). Omitting the factor $s_{rs\,tu}(0, 0)$, one obtains

$$s(t_1, t_2) = \exp\{-i(\omega_{tu}^{(e)}t_1 + \omega_{rs}^{(d)}t_2)\}\exp\{-\lambda_{tu}^{(e)}t_1 - \lambda_{rs}^{(d)}t_2\}. \quad (6.5.16)$$

Clearly, the evolution in t_1 leads to a phase modulation of the signal with a phase $-\omega_{tu}^{(e)}t_1$, but if we disregard the attenuation by transverse relaxation, the amplitude of the signal is not affected by the t_1-evolution. This type of signal is often encountered in 2D separation experiments, as discussed in Chapter 7.

In other types of 2D experiments, the signal is modulated in amplitude, i.e. proportional to the cosine or sine of the precession angle,

$$s(t_1, t_2) = \cos(\omega_{tu}^{(e)}t_1)\exp\{-i\omega_{rs}^{(d)}t_2\}\exp\{-\lambda_{tu}^{(e)}t_1 - \lambda_{rs}^{(d)}t_2\} \quad (6.5.17)$$

or
$$s(t_1, t_2) = \sin(\omega_{tu}^{(e)}t_1)\exp\{-i\omega_{rs}^{(d)}t_2\}\exp\{-\lambda_{tu}^{(e)}t_1 - \lambda_{rs}^{(d)}t_2\}. \quad (6.5.18)$$

Such signals are obtained for example in 2D correlation experiments (Chapter 8) based on the pulse sequence $(\pi/2) - t_1 - (\pi/2)_x - t_2$. Since the y-component of the magnetization of an isolated spin at the end of t_1 is converted into longitudinal magnetization, the only remaining observable component is the x-component, which is amplitude-modulated by the t_1 evolution. A similar behaviour occurs in 2D exchange experiments (Chapter 9).

Amplitude modulation can be understood as a superposition of two equally weighted coherence transfer pathways with the frequencies $\omega_{tu}^{(e)}$ and $-\omega_{tu}^{(e)}$ in the evolution period, since the sum of two exponential functions of the form of eqn (6.5.16) is proportional to the cosine term in eqn (6.5.17). Indeed, a complex Fourier transformation of eqn (6.5.17) would produce a symmetrical pair of two peaks with mixed phases with frequency coordinates $(\omega_1, \omega_2) = (\omega_{tu}^{(e)}, \omega_{sr}^{(d)})$ and $(-\omega_{tu}^{(e)}, \omega_{sr}^{(d)})$.

The amplitude-modulated time-domain signal may be converted into a pure phase 2D spectrum by a 2D Fourier transformation that is real with respect to t_1 and complex with respect to t_2

$$S(\omega_1, \omega_2) = \int_0^\infty \int_0^\infty s(t_1, t_2)\cos(\omega_1 t_1)\exp(-i\omega_2 t_2)\, dt_1\, dt_2 \quad (6.5.19)$$

or

$$S(\omega_1, \omega_2) = \int_0^\infty \int_0^\infty s(t_1, t_2)\sin(\omega_1 t_1)\exp(-i\omega_2 t_2)\, dt_1\, dt_2. \quad (6.5.20)$$

The cosine transform, eqn (6.5.19) of the cosine amplitude-modulated signal in eqn (6.5.17) leads to a spectrum

$$S(\omega_1, \omega_2) = \tfrac{1}{2}\{a_{tu}(\omega_1) + a_{tu}(-\omega_1)\}\{a_{rs}(\omega_2) - id_{rs}(\omega_2)\}. \quad (6.5.21)$$

The 2D spectrum is fully symmetric with respect to $\omega_1 = 0$ and we focus our attention in the following on positive ω_1 values. If the real part of the signal is selected, one obtains a pure 2D absorption peak. The elimination of the dispersive component of the mixed phase peak shape can be understood in terms of a superposition of two signals, as explained in Fig. 6.5.6.

In many types of 2D spectra, signals occur in pairs that are symmetrical in frequency but unequal in amplitude. This situation is typical for diagonal multiplets in homonuclear correlation spectra obtained with a mixing pulse $\beta \neq \pi/2$, and for many signals in multiple-quantum spectra obtained with a mixing pulse $\beta \neq \pi/2$. In this case, the peakshape obtained after a real cosine transformation still consists of a mixture of 2D absorption and 2D dispersion, and may be represented (6.5) by an

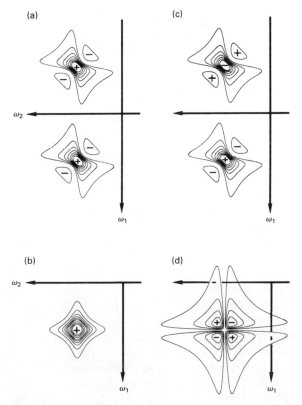

FIG. 6.5.6. Peakshapes obtained by calculating complex or real Fourier transformations with respect to t_1 of amplitude-modulated signals. (a) A complex transform of a cosine-modulated signal, eqn (6.5.17), leads to a pair of peaks with mixed phases with the same signs, symmetrically disposed at $\omega_1 = \pm\omega_{tu}^{(c)}$. (b) A cosine transform, eqn (6.5.19), of the same signal leads to pure 2D absorption, since the dispersion lobes (which are antisymmetrical with respect to $\omega_1 = 0$) cancel because the $\omega_1 < 0$ quadrants are 'folded' on to the $\omega_1 > 0$ quadrants. (c) A complex transform of a sine-modulated signal, eqn (6.5.18), leads to a pair of peaks with mixed phases with opposite signs. (d) A cosine transformation of a sine-modulated signal leads to pure 2D dispersion (a sine transform would yield pure 2D absorption).

expression of the form

$$S(\omega_1, \omega_2) = +A\, a_{tu}(\omega_1)a_{rs}(\omega_2)$$
$$- B\, d_{tu}(\omega_1)d_{rs}(\omega_2)$$
$$+ C\, a_{tu}(\omega_1)d_{rs}(\omega_2)$$
$$+ D\, d_{tu}(\omega_1)a_{rs}(\omega_2) \quad (6.5.22)$$

where the coefficients reflect a superposition of two coherence transfer pathways, one involving a transfer of coherence $|t\rangle\langle u| \rightarrow |r\rangle\langle s|$, the other a transfer $|m\rangle\langle n| \to |r\rangle\langle s|$. The coherences[1] $|m\rangle\langle n|$ and $|t\rangle\langle u|$ are associated with opposite frequencies $\omega_{mn}^{(e)} = -\omega_{tu}^{(e)}$. These coefficients depend on the complex amplitudes of the two pathways

$$A = \tfrac{1}{2}\operatorname{Re}\{Z_{rs\,tu} + Z_{rs\,mn}\}$$
$$B = \tfrac{1}{2}\operatorname{Re}\{Z_{rs\,tu} - Z_{rs\,mn}\}$$
$$C = \tfrac{1}{2}\operatorname{Im}\{Z_{rs\,tu} + Z_{rs\,mn}\}$$
$$D = \tfrac{1}{2}\operatorname{Im}\{Z_{rs\,tu} - Z_{rs\,mn}\} \quad (6.5.23)$$

with the complex amplitudes defined in eqn (6.2.11)

$$Z_{rs\,tu} = F_{sr}^{+} R_{rs\,tu} (\hat{P}\sigma_0)_{tu}$$

where $(\hat{P}\sigma_0)_{tu}$ represents the phase and amplitude of the coherence $|t\rangle\langle u|$ at the beginning of the evolution period, $R_{rs\,tu}$ describes the complex amplitude of coherence transfer, and F_{sr}^{+} is the matrix element of the observable operator.

It can be shown (6.26) that the coherence transfer factor $R_{rs\,tu}$ is real for a transfer of multiple-quantum coherence $|t\rangle\langle u|$ of odd orders p (including single-quantum coherence) into observable magnetization, while $R_{rs\,tu}$ is imaginary for a transfer from even orders p. The term $(\hat{P}\sigma_0)_{tu}$, which expresses the initial phase, can be made real or imaginary by a suitable choice of the r.f. phase of the preparation propagator. Thus, if we exclude multiple-quantum spectra with signals stemming from both even and odd orders, we can arrange all factors $Z_{rs\,tu}$ to be real (hence $C = D = 0$). Pure 2D absorption peaks can then be obtained by a cosine transformation if the two transfer factors have equal amplitudes and equal signs ($B = 0$), as illustrated in Fig. 6.5.6(b).

We shall encounter examples in Chapter 8 where mixed peakshapes cannot be avoided, such as in 2D correlation spectra of strongly coupled systems (see § 8.2.4). Mixed peakshapes are also obtained for diagonal peaks in 2D correlation spectra of weakly-coupled systems obtained with a mixing pulse $\beta \neq \pi/2$. Consider the two processes illustrated in Fig. 6.5.7. If the coefficients in eqn (6.5.23) are expressed in terms of matrix elements of the mixing operator defined in eqn (6.2.14) we obtain

$$A = \tfrac{1}{2} F_{sr}^{+} (\hat{P}\sigma_0)_{tu} \{R_{rt}R_{us}^{-1} + R_{ru}R_{ts}^{-1}\},$$
$$B = \tfrac{1}{2} F_{sr}^{+} (\hat{P}\sigma_0)_{tu} \{R_{rt}R_{us}^{-1} - R_{ru}R_{ts}^{-1}\}. \quad (6.5.24)$$

---

[1] In some applications of 2D spectroscopy, such as in heteronuclear J-spectroscopy with decoupling in the detection period, the two signals with opposite frequencies arise from coherences between different pairs of states $|t\rangle\langle u|$ and $|m\rangle\langle n|$, whereas in 2D correlation and multiple-quantum spectroscopy, the symmetrical signals are associated with the coherences $|t\rangle\langle u|$ and $|u\rangle\langle t|$ in the evolution period.

6.5 PEAK SHAPES IN TWO-DIMENSIONAL SPECTRA

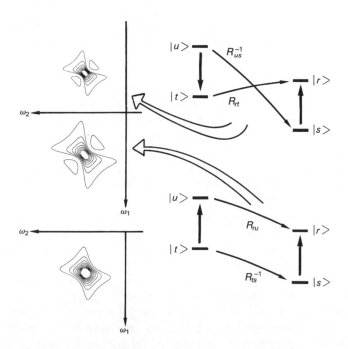

FIG. 6.5.7. Origin of signals that appear symmetrically at $\omega_1 = \omega_{tu}^{(e)}$ and $\omega_1 = \omega_{ut}^{(e)} = -\omega_{tu}^{(e)}$ in 2D spectra obtained with complex Fourier transformation with respect to t_1 (top left). In 2D correlation and multiple-quantum spectra, the mixed phase peak in the upper quadrant results from coherence transfer $|t\rangle\langle u| \to |r\rangle\langle s|$ with an intensity proportional to $R_{rt}R_{us}^{-1}$ while the mixed phase peak in the lower quadrant arises from the transfer $|u\rangle\langle t| \to |r\rangle\langle s|$ with an intensity proportional to $R_{ru}R_{ts}^{-1}$ (eqn (6.5.24)). In a 2D spectrum obtained with a real (cosine) transformation (bottom left), only one peak appears, generally with an admixture of 2D absorption and 2D dispersion peakshapes given by eqns (6.5.24) and (6.5.22). The arrows point from the bra $\langle u|$ to the ket $|t\rangle$ of the coherence $|t\rangle\langle u|$.

The factors involved in this equation are shown schematically in Fig. 6.5.7.

The mixed peak shapes obtained can be described in terms of a phase ϕ defined by

$$\tan \phi = \frac{B}{A}. \tag{6.5.25}$$

The following special cases are often encountered.

1. $\phi = 0$: pure positive 2D absorption;
2. $\phi = \pi/2$: pure negative 2D dispersion;
3. $\phi = \pi/4$: mixed peakshape described by eqn (6.5.10), also known as 'phase-twisted peak';
4. $\phi = 3\pi/4$: mixed peakshape with opposite 'twist'.

In Fig. 6.5.7, it is clear that the order of coherence is preserved for the pathway $|u\rangle\langle t| \to |r\rangle\langle s|$, since $p_{ut} = p_{rs} = -1$, while the mirror-image pathway involves a change in coherence order from $p_{tu} = +1$ to $p_{rs} = -1$ (crossed arrows). This aspect must be carefully considered in the design of phase-cycles for the selection of coherence transfer pathways. To obtain pure phase peakshapes, it is essential to select two 'mirror-image' pathways involving orders p and $p' = -p$ in the evolution period (see Fig. 6.3.2). In practice, this can be achieved by cycling the phase of the mixing propagator in increments $\Delta\varphi = 2\pi/N$ with $N = 2p$. In the case of conventional (single-quantum) correlation spectroscopy, this boils down to a simple alternation of the phase of the mixing pulse ($\varphi_m = 0, \pi$) with addition of the signals (6.9).

In 2D correlation spectroscopy of weakly coupled spin systems, it turns out that the amplitudes of the mirror-image pathways are equal for all cross-peaks (i.e. if the coherences $|t\rangle\langle u|$ and $|r\rangle\langle s|$ belong to different spins) regardless of the mixing pulse angle β, while for diagonal multiplets (i.e. if the coherences $|t\rangle\langle u|$ and $|r\rangle\langle s|$ are associated with parallel transitions), the amplitudes are equal only for $\beta = \pi/2$ (6.5).

In the case of multiple-quantum spectroscopy, the amplitudes for mirror-image coherence transfer pathways are equal for so-called 'remote connectivity' signals (i.e. if $|t\rangle\langle u|$ and $|r\rangle\langle s|$ involve different active spins) for all β, while 'direct connectivity' signals (i.e. if the multiple- and single-quantum coherences involve a common active spin), the signals are equal only for $\beta = \pi/2$ (6.26).

The calculation of a real Fourier transformation with respect to t_1 leads to undesirable folding of signals if the carrier is positioned within the spectrum. We shall see in § 6.6.3 that this problem can be circumvented, provided the signals are shifted in ω_1 with *time-proportional phase increments* (TPPI), i.e. if the r.f. phase of the preparation propagator is shifted in subsequent t_1-increments

$$\varphi^{(\text{prep})} = \frac{\pi}{2|p|} \frac{t_1}{\Delta t_1} \qquad (6.5.26)$$

where p is the order of the coherence evolving in the t_1-period. The net effect of this procedure is that, while the observed single-quantum spectrum may contain signals on either side of $\omega_2 = 0$, the p-quantum peaks will all appear on one side of $\omega_1 = 0$, and a real Fourier transformation may be calculated with respect to t_1 without entailing confusing overlaps. It is interesting to compare this approach with the use of two complementary experiments discussed in § 6.5.3.3.

6.5.3.2. Time-reversal in a complementary experiment

In some 2D experiments, it is possible to record two distinct spectra $S^A(\omega_1, \omega_2)$ and $S^B(\omega_1, \omega_2)$ in such a way that to each resonance frequency $\omega_1^A = \omega_{tu}^{(e)}$ corresponds a resonance frequency $\omega_1^B = -\omega_{tu}^{(e)}$. The signals obtained from the two experiments differ in the phase-modulated frequency factors $\exp\{-i(\omega_{tu}^{(e)}t_1 + \omega_{rs}^{(d)}t_2)\}$ and $\exp\{-i(-\omega_{tu}^{(e)}t_1 + \omega_{rs}^{(d)}t_2)\}$ which produce, when added, the amplitude-modulated signal $2\cos(\omega_{tu}^{(e)}t_1)\exp(-i\omega_{rs}^{(d)}t_2)$ and effectively lead from eqn (6.5.16) to eqn (6.5.17).

This can be achieved experimentally by reversing the sign of the effective Hamiltonian $\mathcal{H}^{(e)}$ that governs the evolution in experiment B. An example is found in heteronuclear J-spectroscopy (§ 7.2), where the decoupler can be switched off either in the defocusing or in the refocusing period of the experiment (6.27). An alternative procedure, applicable to systems without homonuclear couplings, consists in applying a π-pulse just before the detection period, which has the effect of reversing the apparent sense of precession in t_1 (6.21).

It is important to note that these procedures are also applicable to spectra that do *not* possess inherent symmetry about $\omega_1 = 0$, such as heteronuclear J-spectra of strongly coupled spin systems, or 2D spectra of polycrystalline powders designed to separate $\mathcal{H}_{IS}$ and $\mathcal{H}_{ZS}$ interactions (see §§ 7.2.3 and 7.3.4).

The lineshape contributions from the two complementary experiments are shown schematically in Fig. 6.5.8. The spectrum resulting from experiment B is mirror-imaged with respect to $\omega_1 = 0$ to obtain a matrix $S^{B'}(\omega_1, \omega_2)$, and the linear combinations

$$S^C(\omega_1, \omega_2) = S^A(\omega_1, \omega_2) \pm S^{B'}(\omega_1, \omega_2) \qquad (6.5.27)$$

yield pure phase peakshapes, i.e. pure 2D absorption or pure 2D dispersion depending on the the sign of the linear combination.

In 2D experiments involving coherence transfer, the spectra $S^A(\omega_1, \omega_2)$ and $S^B(\omega_1, \omega_2)$ can be obtained by separating two mirror image pathways with the orders p and $p' = -p$ in the evolution period. Such spectra can be derived from a single set of experiments where the phase of the mixing propagator is cycled in increments $\varphi = 2\pi/N$ with $N \geq 2p + 1$ (6.9).

6.5.3.3. Combination of two experiments in quadrature

To obtain pure phase spectra and at the same time distinguish positive and negative ω_1 frequencies, it is also possible to record two complementary spectra which are distinct in the phase of the coherence

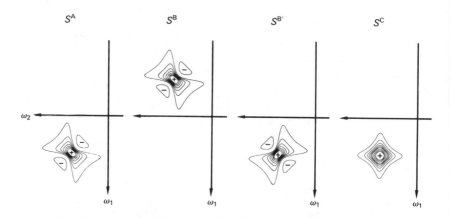

FIG. 6.5.8. Pure phase peakshapes obtained by time-reversal and subsequent linear combination of complementary 2D spectra. The mixed phase peak in $S^A(\omega_1, \omega_2)$ is obtained with the basic pulse sequence. The complementary spectrum $S^B(\omega_1, \omega_2)$, where the resonance appears at a mirror-image position with respect to $\omega_1 = 0$, is obtained either by reversing the sign of the effective Hamiltonian or by inserting a π-pulse at the beginning of the detection period. Note the signs of the dispersion lobes which have translational symmetry with respect to $S^A(\omega_1, \omega_2)$. The spectrum $S^{B'}(\omega_1, \omega_2)$ is the mirror-image of the spectrum $S^B(\omega_1, \omega_2)$ with respect to $\omega_1 = 0$. The dispersion contributions are cancelled in the sum $S^C(\omega_1, \omega_2) = S^A(\omega_1, \omega_2) + S^{B'}(\omega_1, \omega_2)$.

excited at the beginning of the evolution period. This can be achieved by shifting the phase of the preparation sequence

$$\varphi_A^{(\text{prep})} = 0 \quad \text{and} \quad \varphi_B^{(\text{prep})} = \frac{\pi}{2|p|} \tag{6.5.28}$$

where p is the order of coherence in the evolution period. There is an obvious analogy with the time-proportional phase increments, eqn (6.5.26), used in conjunction with real Fourier transformations. Like in the latter approach, the quadrature method requires that the two transfer processes $|t\rangle\langle u| \to |r\rangle\langle s|$ and $|u\rangle\langle t| \to |r\rangle\langle s|$ are associated with equal coherence transfer amplitudes $R_{rs\,tu} = R_{rs\,ut}$, and the method presupposes that these two coherence transfer pathways are both retained by the phase-cycling procedure.

The data derived from the two complementary experiments can be treated according to Bachmann *et al.* (6.21), or by the 'extraction procedure' described by States *et al.* (6.28), as described in the following. Note that both approaches are fully equivalent.

The two complementary experiments differ in the phase of the coherence at the beginning of the evolution period. Since multiple or single-quantum coherence of (signed) order $p_{tu} = M_t - M_u$ experiences a phase shift equal to $-p_{tu}\varphi^{(\text{prep})}$, the initial phases of the coherences in the two experiments with r.f. phases defined in eqn (6.5.28) are related according to

$$(\hat{P}_B \sigma_0)_{tu} = (\hat{P}_A \sigma_0)_{tu} \exp\{-ip_{tu}(\varphi_B^{(\text{prep})} - \varphi_A^{(\text{prep})})\}$$

$$= -i \frac{p_{tu}}{|p_{tu}|} (\hat{P}_A \sigma_0)_{tu} \tag{6.5.29}$$

where the superoperators $\hat{P}_A$ and $\hat{P}_B$ represent the phase-shifted preparation propagators of the two experiments. Thus the complex signal amplitudes defined in eqn (6.2.11b) are also related

$$Z_{rs\,tu}^{(B)} = -i \frac{p_{tu}}{|p_{tu}|} Z_{rs\,tu}^{(A)}. \tag{6.5.30}$$

The superposition of the two pathways leads to a sum of two signals of the form of eqn (6.2.10b) with $\omega_{ut}^{(e)} = -\omega_{tu}^{(e)}$ and $p_{ut} = -p_{tu}$. Provided that $R_{rs\,ut} = R_{rs\,tu}$, one obtains

$$S_A(t_1, t_2) = 2\cos(\omega_{tu}^{(e)} t_1)\exp(-\lambda_{tu}^{(e)} t_1)\exp\{(-i\omega_{rs}^{(d)} - \lambda_{rs}^{(d)})t_2\} Z_{rs\,tu}^{(A)},$$

$$S_B(t_1, t_2) = 2\sin(\omega_{tu}^{(e)} t_1)\exp(-\lambda_{tu}^{(e)} t_1)\exp\{(-i\omega_{rs}^{(d)} - \lambda_{rs}^{(d)})t_2\} Z_{rs\,tu}^{(A)}. \tag{6.5.31}$$

After a complex Fourier transformation with respect to t_2 and appropriated phase correction with respect to ω_2 we obtain (eqn 6.5.5)

$$S_A(t_1, \omega_2) = 2[a_{rs}(\omega_2) - id_{rs}(\omega_2)]\cos(\omega_{tu}^{(e)} t_1)\exp(-\lambda_{tu}^{(e)} t_1) Z_{rs\,tu}^{(A)},$$

$$S_B(t_1, \omega_2) = 2[a_{rs}(\omega_2) - id_{rs}(\omega_2)]\sin(\omega_{tu}^{(e)} t_1)\exp(-\lambda_{tu}^{(e)} t_1) Z_{rs\,tu}^{(A)}. \tag{6.5.32}$$

We can drop the dispersion terms $d_{rs}(\omega_2)$ by extracting the real parts of the two signals. A new complex signal is defined by combining the two real parts

$$S_C(t_1, \omega_2) = \frac{1}{2}\text{Re}\{S_A(t_1, \omega_2)\} + \frac{i}{2}\text{Re}\{S_B(t_1, \omega_2)\}$$

$$= a_{rs}(\omega_2)\exp\{(-i\omega_{tu}^{(e)} - \lambda_{tu}^{(e)})t_1\} Z_{rs\,tu}^{(A)}. \tag{6.5.33}$$

A complex transform with respect to t_1 and subsequent phase correction with respect to ω_1 leads to the spectrum

$$S_C(\omega_1, \omega_2) = a_{rs}(\omega_2)[a_{tu}(\omega_1) - id_{tu}(\omega_1)] Z_{rs\,tu}^{(A)}. \tag{6.5.34}$$

The real component of this spectrum has the desired pure 2D absorption peak shape. It must be emphasized that all four quadrants of the 2D frequency domain may contain different information. The approach outlined here has been applied to 2D NOE spectroscopy (6.28) as shown in Fig. 6.5.9, and can be extended to arbitrary coherence transfer

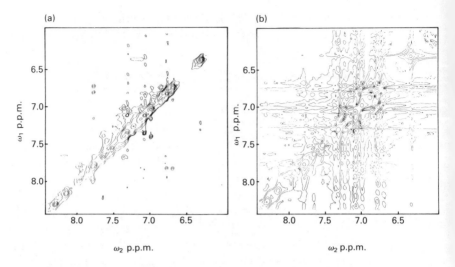

FIG. 6.5.9. Comparison of 2D spectra presented in pure 2D absorption mode (a) and in absolute-value mode (b), illustrating the advantage of pure phase spectra for enhancing resolution. These 2D NOE spectra from the protein basic pancreatic trypsin inhibitor show only the aromatic region. Both spectra were computed from the same data, with the same Gaussian filtration, and the contour levels are drawn at 0.15, 0.3, 0.6, 1, 2.5, 5, and 10 per cent of the maximum peak. (Reproduced from Ref. 6.28.)

experiments, provided that the transfer processes are symmetrical ($R_{rs\,ut} = R_{rs\,tu}$).

6.5.4. Absolute-value spectra

To circumvent the need for phase corrections, it is possible to display the absolute amplitude of the 2D spectra (square root of the power spectrum)

$$|S|(\omega_1, \omega_2) = \{[\text{Re}\{S(\omega_1, \omega_2)\}]^2 + [\text{Im}\{S(\omega_1, \omega_2)\}]^2\}^{\frac{1}{2}}. \quad (6.5.35)$$

Substituting eqns (6.5.4) and (6.5.5) one obtains the peakshape function

$$|S|(\omega_1, \omega_2) = |s_{rs\,tu}(0, 0)|$$

$$\times \frac{1}{[(\Delta\omega_{tu}^{(e)})^2 + (\lambda_{tu}^{(e)})^2]^{\frac{1}{2}} \cdot [(\Delta\omega_{rs}^{(d)})^2 + (\lambda_{rs}^{(d)})^2]^{\frac{1}{2}}}. \quad (6.5.36)$$

Absolute-value displays have much lower resolution than pure 2D absorption lineshapes, as illustrated in Fig. 6.5.9. Along the ridges, the signal drops to half-height for $\Delta\omega_{rs}^{(d)} = 3^{\frac{1}{2}}\lambda_{rs}^{(d)}$ or $\Delta\omega_{tu}^{(e)} = 3^{\frac{1}{2}}\lambda_{tu}^{(e)}$. If the

6.5 PEAK SHAPES IN TWO-DIMENSIONAL SPECTRA

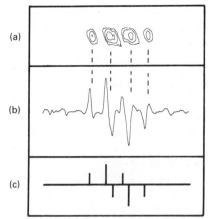

FIG. 6.5.10. Comparison of an absolute-value display (a) and absorption-mode sections (b) taken from a heteronuclear zero-quantum 2D correlation spectrum of phosphothreonine, showing typical distortions of the ill-resolved multiplet structure in the absolute-value mode. (Adapted from Ref. 6.29.)

line-widths in both domains are equal, the half-height signal occurs on the bisecting axes at $\Delta\omega_{rs}^{(d)} = \pm\Delta\omega_{tu}^{(e)} = \pm\lambda$, i.e. at a radial distance $2^{\frac{1}{2}}\lambda$ from the centre of the peak. These values should be compared with those for pure 2D absorption signals, given in § 6.5.1. The amplitude of $|S|(\omega_1, \omega_2)$ drops most rapidly along skew sections, leaving dominant ridges parallel to either frequency axis. This is illustrated in Fig. 6.5.14 in § 6.5.6.2. This 'star effect' (6.22) may be perturbing when two closely-spaced resonances interfere, since the superposition of two orthogonal ridges may easily be misinterpreted as a significant cross-peak.

Since the absolute-value representation is inherently non-linear, care should be taken when considering interference of closely-spaced resonances. In particular, absolute-value representations can be misleading if the spectra contain closely-spaced peaks with opposite phases, as frequently occurs in coherence transfer experiments. An example is shown in Fig. 6.5.10: the sections taken through the phase-sensitive spectrum show six lines (an antiphase doublet of in-phase triplets), with absorption-mode Lorentzian lineshapes. The inner four lines form closely-spaced 'up–down' patterns that resemble dispersion lines. In the absolute-value spectrum (Fig. 6.5.10(a)), the inner four lines are represented by only two peaks (6.29).

6.5.5. Projections of 2D spectra

One-dimensional spectra obtained by projecting 2D spectra along a suitable direction often contain information that cannot be obtained from a conventional 1D spectrum. Such projections are defined in eqns (6.4.25), (6.4.30), and (6.4.31). Typical examples are encountered in the following applications.

1. *Multiple-quantum spectroscopy.* If the emphasis is put on the determination of multiple-quantum precession frequencies and lineshapes, without regard to the correlation with the single-quantum spectrum that appears in the ω_2-domain, a 1D multiple-quantum spectrum can be obtained by a projection of a 2D spectrum on to the ω_1-axis. This projection corresponds to $\phi = 0$ as in eqn (6.4.31). Applications are described in § 8.4.

2. *Filtered single-quantum spectra.* A (weighted) projection of a 2D spectrum on to the ω_2-axis ($\phi = \pi/2$ as in eqn (6.4.30)) may lead to a filtered single-quantum 1D spectrum. This situation is encountered in 1D coherence transfer experiments (§ 4.5) and in 1D multiple-quantum filtering experiments such as in the INADEQUATE method (§ 8.4.2).

3. *Homonuclear J-spectroscopy.* By means of a skew projection ($\phi = 3\pi/4$ in eqn (6.4.25)), it is possible to obtain a 1D spectrum where homonuclear interactions are fully decoupled. By projecting along other directions, spectra with scaled couplings can be constructed (§ 7.2.1).

Special attention has to be paid to possible cancellation effects of positive and negative intensities. Contributions to the projection of opposite sign may be due to differential coherence transfer. Thus in the case of multiple-quantum spectroscopy, the total integrated signal intensity of the 2D spectrum is normally zero. Hence a projection on to the ω_1-axis leads to zero intensity, unless an absolute-value spectrum is projected or some means of refocusing is used in the experimental sequence, in order to convert antiphase multiplets in the 2D spectrum into in-phase signals prior to projection.

Vanishing projections can also be due to the mixed peak shape of eqn (6.5.10), which is obtained by a complex Fourier transformation. This can easily be understood with the projection cross-section theorem of § 6.4.1.5. The skew projection that would be desirable in homonuclear 2D J-spectroscopy (see § 7.2.1) corresponds to an angle $\phi = 3\pi/4$ (135°) between the ω_1-axis and the direction on to which the spectrum must be projected, as illustrated in Fig. 6.5.11. According to eqn (6.4.27), this projection is equivalent to the Fourier transformation of a skew section taken through 2D time domain along an axis which also subtends an angle of $3\pi/4$ with respect to the t_1-axis. This skew section vanishes except for the point $s(t_1 = 0, t_2 = 0)$, since the signal is zero for $t_1 < 0$ and $t_2 < 0$. Viewed in frequency domain, the integral vanishes because the negative dispersion lobes precisely cancel the positive absorption component, as may be appreciated in Fig. 6.5.11.

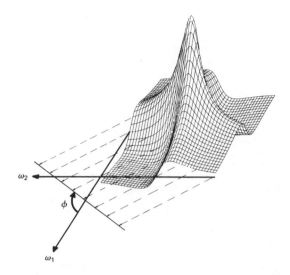

FIG. 6.5.11. Skew projection of a mixed phase ('phase-twisted') peak, eqn (6.5.10), by integration parallel to the positive diagonal ($\phi = 3\pi/4$ in eqn (6.4.25)). The integral of the negative dispersion lobes exactly cancels the integral of the absorption component, as implied by the projection cross-section theorem. (Adapted from Ref. 6.17.)

The following procedures have been proposed to prevent cancellation of skew projections.

1. Projection of absolute-value spectra (6.17, 6.46).
2. Elimination of the dispersive component of the mixed-phase peak-shape by pseudo-echo filtering (6.30, 6.31), as described in § 6.5.6.3.
3. Subtraction of dispersive component after locating the peaks with iterative computer procedures (6.32).
4. 'Skyline projection' (see Fig. 6.5.12). Such a projection corresponds to the shadow that would be cast if a 3D model of the spectrum were illuminated by a distant light source (6.33). The skyline projection suffers from the disadvantage that the intensities are no longer related to the number of spins involved. In particular, multiplet splittings and line broadening in the direction of projection will reduce the intensity of the projection.

The signal-to-noise ratio of a projection can be improved by suitable weighting of the 2D spectrum before projecting.

6.5.6. Two-dimensional filtering

The convolution theorem, eqn (6.4.22), implies that 2D frequency-domain filtering by convolution with a suitable filter function $H(\omega_1, \omega_2)$

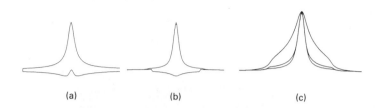

FIG. 6.5.12. Skyline projections (shadows) of a 2D peak with mixed phase (eqn (6.5.10)): (a) projection of minima and maxima on to the ω_1-axis ($\phi = 0$ in Fig. 6.5.11); (b) projection of minima and maxima with $\phi = 3\pi/4$ (which is desirable in homonuclear 2D J-spectroscopy); (c) comparison of lineshapes of different projections with $\phi = 3\pi/4$: the outer curve shows the integral projection of the absolute-value spectrum $|S|(\omega_1, \omega_2)$, the middle curve the skyline projection of $|S|(\omega_1, \omega_2)$, and the inner curve the skyline projection of the (mixed phase) absolute value peak shape $|\text{Re}\{S(\omega_1, \omega_2)\}|$. (Reproduced from Ref. 6.33.)

is equivalent to a multiplication of the time-domain signal $s(t_1, t_2)$ with a weighting function $h(t_1, t_2)$.

The application of resolution enhancement has become standard practice in 2D spectroscopy (6.43). Resolution enhancement is of particular importance in 2D correlation spectroscopy where cross-peak multiplets appear in antiphase in both dimensions and vanish if the resolution in either dimension is insufficient. The corresponding time-domain signal components are proportional to $\sin \pi J_{kl} t_1 \sin \pi J_{kl} t_2$ (see § 8.2.1). For small t_1- and t_2-values, the contribution to the cross-peak amplitude is therefore insignificant. For this reason, it is advisable to weight the time-domain signal $s(t_1, t_2)$ by a weighting function $h(t_1, t_2)$ which de-emphasizes the signal obtained for small t_1- and t_2-values.

Any of the resolution enhancement functions described in § 4.1.3.2 can be applied in the two domains. The most common resolution enhancement functions are the sine-bell functions (6.40–6.42) and the Lorentz–Gauss transformation (6.34, 6.38). In some cases, it may be desirable to treat the two time domains with different weighting functions. It is also conceivable that the weighting function $h(t_1, t_2)$ is not separable into a product $h(t_1)h(t_2)$.

In the present section, we shall focus attention on three further important aspects of 2D filtering.

1. *Sensitivity enhancement.* The signal-to-noise ratio is optimized by matched filtration (6.34–6.38).

2. *Removal of the star effect.* A 2D absorption mode peakshape with cylindrical or elliptical symmetry can be obtained by a Lorentz–Gauss transformation (6.22, 6.34–6.37).

3. *Cancellation of dispersion components.* Pure 2D absorption peak-shapes can be obtained by pseudo-echo filtration (6.30, 6.31).

6.5.6.1. Matched filter

In complete analogy to sensitivity optimization in 1D spectroscopy (§ 4.3.1.4), the signal-to-noise ratio can be maximized by multiplying the time-domain signal with a matched weighting function $h(t_1, t_2)$ that is proportional to the signal envelope $s^e(t_1, t_2)$ (see also § 6.8). If coherence transfer echoes can be neglected, the signal envelope may often be assumed to have a biexponential form

$$s^e(t_1, t_2) = s^e(0, 0)\exp(-\lambda^{(e)}t_1)\exp(-\lambda^{(d)}t_2); \quad (6.5.37)$$

hence the matched weighting function can be factorized

$$\begin{aligned}h(t_1, t_2) &= \exp(-\lambda^{(e)}t_1)\exp(-\lambda^{(d)}t_2)\\ &= h_1(t_1)h_2(t_2).\end{aligned} \quad (6.5.38)$$

Thus the weighting can be applied in two consecutive stages, once before the first, and once before the second Fourier transformation

$$\begin{aligned}s'(t_1, t_2) &= s(t_1, t_2)h_2(t_2),\\ s'(t_1, \omega_2) &= \mathscr{F}^{(2)}\{s'(t_1, t_2)\},\\ s''(t_1, \omega_2) &= s'(t_1, \omega_2)h_1(t_1),\\ S(\omega_1, \omega_2) &= \mathscr{F}^{(1)}\{s''(t_1, \omega_2)\}.\end{aligned} \quad (6.5.39)$$

In many cases, however, the signal envelope and the weighting function cannot be factorized, and the 2D time domain must be multiplied by a 2D weighting function prior to 2D Fourier transformation. This situation arises for example if coherence transfer echoes lead to a 'ridge' in the signal envelope, eqn (6.5.14), as illustrated in Fig. 6.5.5(b).

In all cases, matched filtration broadens the line-widths in both frequency dimensions. For a Lorentzian lineshape, the broadening is by a factor 2, and for a Gaussian lineshape a factor $\sqrt{2}$ (6.34).

6.5.6.2. Lorentz–Gauss transformation

A time-domain envelope that decays exponentially in both dimensions, eqn (6.5.37), lacks cylindrical symmetry about the origin $t_1 = t_2 = 0$, as may be appreciated in Fig. 6.5.5(a). The corresponding 2D absorption

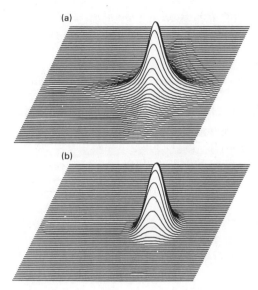

FIG. 6.5.13. Computer simulations of a phase-sensitive pure 2D absorption peak derived from (a) an exponentially decaying envelope, eqn (6.5.37); and (b) a Gaussian envelope, eqn (6.5.41). Note the star shape of the Lorentzian signal and the cylindrical symmetry of the Gaussian signal. (Reproduced from Ref. 6.22.)

Lorentzian peak shape also lacks cylindrical symmetry in the frequency domain, and leads to the 'star effect' illustrated in Fig. 6.5.1(b). This feature, which is sometimes undesirable, can be removed by converting the exponential decay into a Gaussian envelope (6.22, 6.34) by means of the weighting function

$$h(t_1, t_2) = \exp(+\lambda^{(e)}t_1)\exp(+\lambda^{(d)}t_2)\exp(-\sigma_1^2 t_1^2/2)\exp(-\sigma_2^2 t_2^2/2). \quad (6.5.40)$$

With the envelope of eqn (6.5.37), this leads to

$$s^{e'}(t_1, t_2) = s^e(0, 0)\exp(-\sigma_1^2 t_1^2/2)\exp(-\sigma_2^2 t_2^2/2). \quad (6.5.41)$$

After 2D Fourier transformation, one obtains a Gaussian lineshape

$$s'(\omega_1, \omega_2) = s^e(0, 0)\left(\frac{2\pi}{\sigma_1 \sigma_2}\right)\exp\left(\frac{-\Delta\omega_1^2}{2\sigma_1^2}\right)\exp\left(\frac{-\Delta\omega_2^2}{2\sigma_2^2}\right). \quad (6.5.42)$$

The contours are circular for $\sigma_1 = \sigma_2$ and elliptical for unequal widths. The effect of the transformation may be appreciated in Fig. 6.5.13, where a Lorentzian 2D absorption signal is compared with the corresponding Gaussian function. This procedure is useful only if the dispersion contributions are eliminated by one of the strategies described in § 6.5.3.

It should be mentioned that the Lorentz–Gauss transformation is at the same time one of the best resolution enhancement procedures

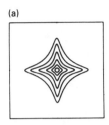 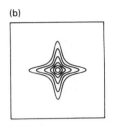 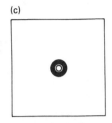

FIG. 6.5.14. Absolute-value mode lineshapes calculated according to eqn (6.5.35): (a) signal envelope with exponential decay in both dimensions, described by eqn (6.5.36); (b) signal envelope with Gaussian decay in both dimensions (note the lack of cylindrical symmetry, in contrast to the corresponding phase-sensitive display in Fig. 6.5.13(b); (c) signal envelope with Gaussian pseudo-echo characteristics, described in § 6.5.6.3. (Reproduced from Ref. 6.39.)

(6.34–6.38). By selecting σ_1 and σ_2 in eqn (6.5.40) sufficiently small, it is possible to narrow the 2D peak in one or both dimensions at will, though it must be remembered that the resolution can only be enhanced at the expense of sensitivity (6.34).

It must be emphasized that the elliptical symmetry of Gaussian signals is obtained only in phase-sensitive displays. If the absolute amplitude of a Gaussian signal is calculated according to eqn (6.5.35), a peakshape is obtained which features again a star effect, as shown in Fig. 6.5.14(b).

6.5.6.3. Pseudo-echo transformation

The dispersion-mode contributions in a mixed-phase peak can be eliminated by rendering the signal envelope symmetrical about the centre of the sampled interval in time domain, i.e. symmetrical with respect to $t_1 = \frac{1}{2}t_1^{\max}$ and $t_2 = \frac{1}{2}t_2^{\max}$

$$s_1^e(t_1) = s_1^e(t_1^{\max} - t_1),$$
$$s_2^e(t_2) = s_2^e(t_2^{\max} - t_2). \qquad (6.5.43)$$

To understand the effect, consider a 1D signal which is defined for both negative and positive times and has a symmetrical envelope

$$s^e(t) = s^e(-t). \qquad (6.5.44)$$

Such a signal can be obtained in principle by recording the rising and decaying part of an echo, assuming that the natural decay $1/T_2$ can be neglected.

If the signal envelope is exponential

$$s^e(t) = s^e(0) \exp\{-\lambda |t|\}, \qquad (6.5.45)$$

the complex Fourier transformation yields a pure absorption-mode

Lorentzian lineshape

$$S(\omega) = \int_{-\infty}^{+\infty} s^e(t) \exp\{-i\omega t\}\, dt$$
$$= 2s^e(0) \frac{\lambda}{\lambda^2 + \omega^2}. \qquad (6.5.46)$$

The imaginary (dispersion) component vanishes, since the sine transformation of a symmetrical function is zero. This also applies to non-exponential envelopes, and remains valid if the signal is truncated symmetrically in the time domain, with restriction of the Fourier integral to the interval $-\frac{1}{2}t^{max} < t < \frac{1}{2}t^{max}$.

In addition to the envelope, the actual signal is characterized by a phase φ and a precession frequency ω_0

$$s(t) = s^e(0) \exp\{-\lambda |t|\} \exp\{i\omega_0 t\} \exp\{i\varphi\}. \qquad (6.5.47)$$

The complex Fourier transformation yields the spectrum

$$S(\omega) = 2s^e(0) \frac{\lambda}{\lambda^2 + (\Delta\omega)^2} \exp\{i\varphi\}$$
$$= 2s^e(0)\, a(\omega) \exp\{i\varphi\} \qquad (6.5.48)$$

where $a(\omega)$ is used to denote the absorption component, eqn (4.2.19), and $\Delta\omega = \omega - \omega_0$ is the offset with respect to the centre of the resonance.

Thus in the particular case of a symmetrical envelope, we obtain

$$\text{Re}\{S(\omega)\} = 2s^e(0) \cos\varphi\, a(\omega),$$
$$\text{Im}\{S(\omega)\} = 2s^e(0) \sin\varphi\, a(\omega). \qquad (6.5.49)$$

Hence, barring overlapping lines, a pure absorption Lorentzian signal is obtained by calculating the square root of the power spectrum (absolute value)

$$|S|(\omega) = \{[\text{Re}\{S(\omega)\}]^2 + [\text{Im}\{S(\omega)\}]^2\}^{\frac{1}{2}}$$
$$= 2s^e(0)\, a(\omega). \qquad (6.5.50)$$

In most applications, the signal is not defined for $t < 0$, since we are not usually dealing with echoes. However, a signal envelope defined only for $t > 0$ can be transformed into a so-called 'pseudo-echo' envelope. If the original envelope of a signal acquired in the interval $0 < t < t^{max}$ is assumed to decay exponentially

$$s^e(t) = s^e(0)\exp\{-\lambda t\}, \qquad (6.5.51)$$

a Lorentzian pseudo-echo envelope with a line-width $2\lambda_f$ is obtained by multiplying the signal with a weighting factor

$$h(t) = \exp\{+\lambda t\}\exp\{-|t - \tfrac{1}{2}t^{\max}|\,\lambda_f\} \tag{6.5.52}$$

leading to a modified envelope

$$s^{e'}(t) = s^e(0)\exp\{-|t - \tfrac{1}{2}t^{\max}|\,\lambda_f\}. \tag{6.5.53}$$

Taking into account the initial phase and the precession frequency, the filtered signal is

$$s(t) = s^e(0)\exp\{-|t - \tfrac{1}{2}t^{\max}|\,\lambda_f\}\exp\{i\omega_0 t\}\exp\{i\varphi\}. \tag{6.5.54}$$

The origin of the time-domain can be shifted by substituting $t = t' + \tfrac{1}{2}t^{\max}$

$$s(t' + \tfrac{1}{2}t^{\max}) = s^e(0)\exp\{-|t'|\,\lambda_f\}\exp\{i\omega_0 t'\}\exp\{i(\varphi + \tfrac{1}{2}\omega_0 t^{\max})\} \tag{6.5.55}$$

which is identical to eqn (6.5.47) apart from the frequency-dependent phase factor. A complex Fourier transformation in the interval $-\tfrac{1}{2}t^{\max} < t' < \tfrac{1}{2}t^{\max}$ (equivalent to $0 < t < t^{\max}$) yields the spectrum

$$S(\omega) = 2s^e(0)\,a(\omega)\exp\{i(\varphi + \tfrac{1}{2}\omega_0 t^{\max})\}. \tag{6.5.56}$$

The phase factor can be removed by calculating the absolute value according to eqn (6.5.50). Note that for non-overlapping lines this does *not* lead to a degradation of the resolution, unlike the absolute-value spectra discussed in § 6.5.4. Since all signals appear in positive absorption, the information about the *sign* of the signals (which is important in 2D coherence transfer experiments) is however lost.

In the context of 2D spectroscopy, pseudo-echo filtration allows one to obtain pure 2D absorption peaks even if none of the methods described in § 6.5.3 are applicable. Thus for an exponentially decaying signal, eqn (6.5.37), a 2D Lorentzian pseudo-echo filter is obtained by extending eqn (6.5.52) to two dimensions

$$h(t_1, t_2) = h_1(t_1) \cdot h_2(t_2) \tag{6.5.57}$$

with

$$h_1(t_1) = \exp\{+\lambda^{(e)} t_1\}\exp\{-|t_1 - \tfrac{1}{2}t_1^{\max}|\,\lambda_f^{(e)}\},$$
$$h_2(t_2) = \exp\{+\lambda^{(d)} t_2\}\exp\{-|t_2 - \tfrac{1}{2}t_2^{\max}|\,\lambda_f^{(d)}\}.$$

The 2D absorption peaks obtained in this manner suffer again from a lack of cylindrical symmetry (Fig. 6.5.13(a)). This can be avoided by using a Gaussian pseudo-echo filter

$$h_1(t_1) = \exp\{+\lambda^{(e)} t_1\}\exp\{-(t_1 - \tfrac{1}{2}t_1^{\max})^2(\sigma^{(e)})^2/2\},$$
$$h_2(t_2) = \exp\{+\lambda^{(d)} t_1\}\exp\{-(t_2 - \tfrac{1}{2}t_2^{\max})^2(\sigma^{(d)})^2/2\}, \tag{6.5.58}$$

which leads to peak shapes with cylindrical or elliptical symmetry, as illustrated in Fig. 6.5.13(b).

Since pseudo-echo filters strongly alternate the signal for small t_1 and t_2 values, the sensitivity is significantly reduced. It is therefore recommended to use one of the procedures discussed in § 6.5.3 for obtaining pure 2D absorption whenever possible.

6.6. Manipulations of two-dimensional spectra

To simplify the analysis of 2D spectra, it is sometimes desirable to modify the presentation of the spectra, either by manipulating the data matrix in computer memory, or else by changing the data acquisition procedures. Some of the manipulations described in this section take advantage of the similarity theorem of Fourier transformations, eqn (6.4.17), which implies that a transformation of frequency variables is reflected by a corresponding transformation of time variables. Other methods exploit the inherent symmetry properties of 2D spectra. More advanced techniques attempt to identify characteristic peak patterns, with the ultimate goal of a fully automated interpretation of 2D spectra.

6.6.1. Shearing transformations

In some types of 2D experiments, the peaks do not occupy the entire frequency domain. This is illustrated in Fig. 6.6.1: while the frequencies $\omega_{tu}^{(e)}$ and $\omega_{rs}^{(d)}$ may take arbitrary values, the frequency difference $\omega_{tu}^{(e)} - \omega_{rs}^{(d)}$ is limited, and all signals fall within a band near to the diagonal. This situation is typical for homonuclear correlation spectra (§ 8.3.1) in the absence of couplings between nuclei with large shift differences. The band structure also occurs in 2D J-spectra obtained with a sequence $\pi/2 - t_1 - \pi - t_2$ (i.e. without refocusing period (6.43)), and

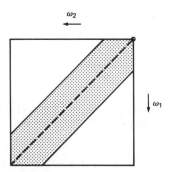

FIG. 6.6.1. Band structure that occurs in certain types of 2D spectra: all signals fall within the shaded area, because the maximum difference in resonance frequencies $\omega_{tu}^{(e)} - \omega_{rs}^{(d)}$ that are connected by cross-peaks is limited.

6.6 MANIPULATIONS OF TWO-DIMENSIONAL SPECTRA

in double-quantum spectra of two-spin systems (6.9, 6.44). In such cases, it is possible to reduce the spectral width in ω_1 without causing a loss of information by folding.

Folding in the ω_1 domain of the spectrum occurs if the sampling theorem is violated with respect to t_1, i.e. if a resonance frequency $\omega_{tu}^{(e)}$ exceeds the Nyquist frequency $\omega_1^N = \pi/\Delta t_1$, where Δt_1 is the t_1 increment. If a complex Fourier transformation is calculated with respect to t_1, folding leads to the apparent frequency

$$\omega_{tu}^{(e)\text{apparent}} = (\omega_{tu}^{(e)} + \omega_1^N)_{\text{mod}(2\omega_1^N)} - \omega_1^N. \quad (6.6.1)$$

Thus one must subtract $2\omega_1^N = 2\pi/\Delta t_1$ an integer number of times from $\omega_{tu}^{(e)}$ until the apparent frequency appears in the interval $-\omega_1^N < \omega_{tu}^{(e)\text{apparent}} < \omega_1^N$. An example of folding is found in Fig. 6.5.4: part of the elongated inhomogeneously broadened line is cut off at the Nyquist frequency $\omega_1 = -\omega_1^N$ and reappears at the bottom of the spectrum.

Provided the spectral width in ω_1 is sufficient to cover the width of the frequency band shown in Fig. 6.6.1, it is possible to untangle the folded spectrum by foldover correction (6.12). This can be achieved by shearing of the experimental matrix $S(\omega_1, \omega_2)$ stored in the computer memory into a foldover-corrected spectrum $S'(\omega_1', \omega_2)$, in such a way that the diagonal is tilted to appear horizontal. The corrected spectrum is related to the original spectrum through the relation

$$S'(\omega_1', \omega_2) = S(\omega_1, \omega_2)$$

where

$$\omega_1' = \frac{1}{1+|\chi|}[(\omega_1 + \chi\omega_2 + \omega_1^N)_{\text{mod}(2\omega_1^N)} - \omega_1^N] \quad (6.6.2)$$

The proper shearing parameter is $\chi = -\kappa$, eqn (6.5.11). For homonuclear single quantum spectroscopy, one chooses $\chi = \pm 1$ depending on the selection of N or P peaks. An example of this transformation is shown in Fig. 6.6.2. If a homonuclear J-spectrum (§ 7.2.1) is obtained with the sequence $\pi/2 - t_1 - \pi - t_2$ (i.e. without refocusing period), the signals are contained in a narrow band along the negative diagonal (Fig. 6.6.2(a)). By shifting the signals according to eqn (6.6.2), the usual representation of a 2D J-spectrum can be obtained, with the signals lying in a band on either side of $\omega_1 = 0$. The procedure is applicable provided $2\omega_1^N$ is greater than the widest multiplet (6.43).

Depending on the digitization of the data matrix, the transformation in eqn (6.6.2) may require interpolation. This problem can be circumvented by transforming the time-frequency domain $s(t_1, \omega_2)$ into a modified matrix $s'(t_1, \omega_2)$ before calculating the Fourier transformation with respect to t_1 (6.45). To obtain a frequency shift in the ω_1-domain that is

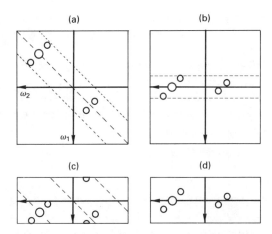

FIG. 6.6.2. Schematic 2D spectra before and after foldover correction. (a) Homonuclear 2D correlation spectrum $S(\omega_1, \omega_2)$ obtained with the pulse sequence $\pi/2 - t_1 - \pi - t_2$, equivalent to the sequence appropriate for J-spectroscopy without refocusing delay. The signals correspond to an AX_2 system. (b) Spectrum $S'(\omega_1', \omega_2)$ derived from (a) by shifting the signals according to eqn (6.6.2) with $\chi = 1$. (c) Same as (a), but with reduced Nyquist frequency ω_1^N in such a way that the signals are folded. (d) Foldover-corrected spectrum derived from (c) with eqn (6.6.2).

proportional to ω_2, each row of the matrix $s(t_1, \omega_2)$ is subjected to a linear phase correction with

$$\varphi^{\text{corr}} = -\chi \omega_2 t_1. \tag{6.6.3}$$

In general, phase-manipulations in the mixed time-frequency domain represent a convenient alternative to shifting the signals in 2D frequency domain.

6.6.2. Delayed acquisition

It is sometimes possible to achieve a 'shearing' of the 2D matrix similar to the procedures described in § 6.6.1 by delaying the acquisition.

Thus the $(\pi/2) - t_1 - (\pi) - t_2$ experiment mentioned above is converted into the well-known 2D J-spectroscopy sequence $(\pi/2) - t_1/2 - (\pi) - t_1/2 - t_2$ (6.46). Homonuclear correlation spectroscopy (COSY) is modified into spin–echo correlation spectroscopy (SECSY) (6.12) with the sequence $(\pi/2) - t_1/2 - (\beta) - t_1/2 - t_2$. Delayed acquisition can also be used in multiple-quantum experiments (6.9) and in heteronuclear correlation spectroscopy (6.14).

In general, the evolution period can be redefined to include part of the interval after the mixing period, while the beginning of the detection period is delayed by χt_1 ($\chi > 0$)

$$t_1' = (1 + \chi)t_1. \tag{6.6.4}$$

In the resulting spectrum, the peaks then appear at frequencies ω_1'

$$\omega_1' = \frac{1}{1+\chi}(\omega_1 + \chi\omega_2). \tag{6.6.5}$$

There is an important distinction between foldover correction and delayed acquisition, as may be appreciated in Fig. 6.6.3. In conventional correlation experiments without delayed acquisition, the pathways $p = 0 \to -1 \to -1$ ('P peaks', with a parameter $\kappa > 0$ in eqn (6.5.11)) and $p = 0 \to +1 \to -1$ ('N peaks', $\kappa < 0$) lead to signals that appear symmetrically with respect to $\omega_1 = 0$. A real cosine Fourier transformation therefore yields pure phase peakshapes (§ 6.5.3.1), and subsequent foldover correction leads to (sheared) pure phase peakshapes.

By contrast, the pathways with $\kappa = \pm 1$ do not lead to symmetrical signals in experiments with delayed acquisition (Fig. 6.6.3(b)). For this reason, it is impossible to cancel the dispersion lobes by calculating a

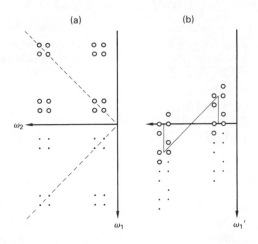

FIG. 6.6.3. The effect of delayed acquisition in homonuclear correlation spectroscopy. (a) With the sequence $\pi/2 - t_1 - \beta - t_2$ ('COSY'), the P and N peaks associated with the coherence transfer pathways $p = 0 \to -1 \to -1$ and $p = 0 \to +1 \to -1$ in a weakly-coupled two-spin system are indicated by dots and open circles respectively. Pure phase spectra can be obtained with a cosine transformation with respect to t_1. (b) With the sequence $\pi/2 - t_1/2 - \beta - t_1/2 - t_2$ ('SECSY'), all signals are shifted to $\omega_1' = \frac{1}{2}(\omega_1 + \omega_2)$, and no longer appear in symmetrical pairs. The signals indicated by dots are normally suppressed by phase-cycling, leaving only the signals represented by open circles. Note that the chemical shift differences are scaled by a factor $\frac{1}{2}$ in ω_1', while the multiplet splittings remain invariant.

cosine Fourier transformation with respect to t_1. The undesired $\kappa = +1$ pathways ('P peaks') are normally suppressed by phase-cycling (6.9, 6.12). Most experiments with delayed acquisition do *not* allow one to obtain pure phase unless additional manipulations are carried out, such as z-filtration (6.62).

6.6.3. Time-proportional phase increments

In many 2D experiments, it is useful to shift the resonance frequencies in the ω_1-domain in a manner that distinguishes the (signed) order p of the coherence evolving in the t_1-interval.

We first describe a scheme which can be used in one-dimensional Fourier spectroscopy to distinguish positive and negative precession frequencies without resorting to a complex data acquisition (6.47, 6.48). The procedure allows one to set the transmitter frequency in the centre of the spectrum and to shift the apparent precession frequencies to one side of the carrier by sampling the x-, y-, $-x$-, and $-y$-components of the signal sequentially. With the sampling interval Δt, one obtains a single real free induction decay of the form

$$s(t) = \{s_x(0), s_y(\Delta t), -s_x(2\Delta t), -s_y(3\Delta t), \ldots\}. \qquad (6.6.6)$$

This amounts to a phase increment of $\pi/2$ between subsequent sampling points or to a frequency shift of $\Delta\omega = \pi/(2\Delta t)$, which is equivalent to a shift by half the Nyquist frequency of the sampling process.

The same procedure can of course be applied to the ω_2-domain of 2D spectra. The extension of this idea to the ω_1-domain naturally leads to time-proportional phase increments (TPPI) (6.49, 6.50). In this method, the phase of the first pulse (or, more generally, of the preparation propagator) is incremented in proportion to t_1

$$\varphi^{(\text{prep})} = \frac{\pi}{2N} \frac{t_1}{\Delta t_1} \qquad (6.6.7)$$

where the parameter N determines the fractional frequency shift. This procedure must not be confused with the phase-cycling of the pulses required for the selection of coherence transfer pathways (§ 6.3).

Because of the phase-shift in eqn (6.6.7), the phase of a coherence $|t\rangle\langle u|$ with the coefficient σ_{tu} is modulated as a function of t_1

$$\sigma_{tu}(t_1, \varphi^{(\text{prep})}) = \sigma_{tu}(t_1 = 0)\exp\{-i\omega_{tu}^{(e)}t_1 - \lambda_{tu}^{(e)}t_1\}\exp\{-ip_{tu}\varphi^{(\text{prep})}\} \qquad (6.6.8)$$

Defining

$$\omega^{(\text{TPPI})} = \varphi^{(\text{prep})} \frac{1}{t_1} = \frac{\pi}{2N\Delta t_1}, \qquad (6.6.9)$$

the coefficient of the coherence $|t\rangle\langle u|$ at the end of the evolution period may be written

$$\sigma_{tu}(t_1) = \sigma_{tu}(t_1 = 0)\exp\{-i(\omega_{tu}^{(e)} + p_{tu}\omega^{(\text{TPPI})})t_1 - \lambda_{tu}^{(e)}t_1\}. \quad (6.6.10)$$

Thus the signal appears shifted in frequency

$$\omega_1 = \omega_{tu}^{(e)} + p_{tu}\omega^{(\text{TPPI})}. \quad (6.6.11)$$

Note that signals stemming from coherences with opposite orders p and $p' = -p$ are shifted in opposite directions.

This idea has found widespread use in multiple-quantum NMR (6.49–6.51), where a suitable choice of N and Δt_1 makes it possible to separate different orders $p = 0, \pm 1, \pm 2, \ldots$ in the ω_1-domain, even in situations where all orders have vanishing effective chemical shifts because of refocusing at the mid-point of the evolution period. The sampling rate $1/\Delta t_1$ must be increased by a factor $2|p_{\max}|$, such that the spectral range is increased to accommodate all orders $-|p_{\max}|\ldots|p_{\max}|$.

The same procedure can be used to obtain pure phase lineshapes by calculating a real cosine transformation with respect to t_1 (§ 6.5.3.1). In this case, one selects $N = |p|$, where p is the order that one wishes to observe (for $p = \pm 1$ coherences, $\varphi^{(\text{prep})} = \frac{1}{2}\pi(t_1/\Delta t_1)$ causes the signals to shift by half the Nyquist frequency). This shift of the signals is useful when the carrier frequency is set within the spectrum, as shown in Fig. 6.6.4. P peak and N peak signals shift in opposite directions, since they are associated with orders p of opposite sign. Clearly, the sampling rate and the Nyquist frequency ω_1^N must be increased by a factor two to avoid folding in Fig. 6.6.4(b). Pure phase peakshapes (i.e. pure 2D absorption or 2D dispersion) are obtained by calculating a real Fourier transformation with respect to t_1, provided the symmetrical signals have equal amplitudes.

6.6.4. Symmetrization

The amplitudes and frequency coordinates of the signals in 2D spectra obey characteristic symmetry rules. In particular, homonuclear 2D correlation spectra (§ 8.2) and 2D exchange spectra (§§ 9.3 and 9.7) are normally symmetrical about the diagonal, i.e. $S(\omega_1, \omega_2) = S(\omega_2, \omega_1)$, thus reflecting the symmetry of coherence transfer in the course of the mixing process.

It can generally be proved that 2D correlation spectra are symmetric if a single non-selective preparation pulse is used and if the mixing sequence is associated with a propagator R with a symmetric matrix representation $R_{rt} = R_{tr}$ (eqn (6.2.14)). This leads to the conclusion that a

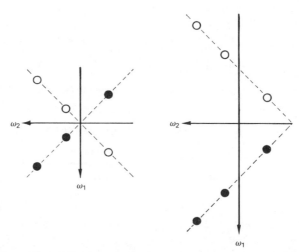

FIG. 6.6.4. Separation of signals stemming from mirror-image coherence transfer pathways $0 \to \pm p \to -1$ by time-proportional phase increments. The symbols represent diagonal multiplets in 2D correlation spectra of a three-spin system, obtained with complex Fourier transformation in both dimensions, with the carrier frequency set within the spectrum. Open symbols correspond to $0 \to +1 \to -1$ pathways ('N peaks'), filled symbols are associated with $0 \to -1 \to -1$ pathways ('P peaks'). The two classes normally overlap (left) but can be separated by incrementing the phase of the preparation pulse according to eqn (6.6.7), i.e. $\varphi^{(\text{prep})} = \pi t_1/(2\Delta t_1)$ (right). If the amplitudes are symmetrical, pure-phase peak shapes are obtained by calculating a real Fourier transformation with respect to t_1 (§ 6.5.3.1).

single (possibly composite) mixing pulse of arbitrary amplitude and duration invariably produces symmetric 2D spectra (unpublished work by C. Griesinger, C. Gemperle, O. W. Sørensen, and R. R. Ernst).

The symmetry can be broken by asymmetric mixing sequences involving free precession periods with several consecutive transfers, such as in relay experiments (§8.3.4). The symmetry is also violated if the preparation is frequency-dependent by inclusion of selective pulses or free precession intervals. Asymmetries may result from a delay between successive experiments too short for full recovery, leading to initial polarizations which depend on the T_1 relaxation times of the different spins. Naturally, processes like constant time spectroscopy with ω_1 decoupling (§8.3.2) also cause asymmetric spectra.

In the case of 2D exchange and 2D NOE spectra (§§ 9.3 and 9.7), the cross-peaks are again symmetrical about the diagonal,

$$I_{AB}(\tau_m) = I_{BA}(\tau_m), \qquad (6.6.12)$$

provided the system is initially in dynamic chemical equilibrium, and provided the longitudinal magnetization is in thermal equilibrium before the first non-selective r.f. pulse of the experiment. Equation (6.6.12) is a

6.6 MANIPULATIONS OF TWO-DIMENSIONAL SPECTRA

consequence of the principle of microscopic reversibility, and implies that all significant cross-peaks should appear in symmetrical pairs.

In 2D multiple-quantum spectra, there is usually no simple symmetry about the (skew) diagonal (see § 8.4), but the signals often have translational symmetry if shifted parallel to the ω_2-axis.

Various *symmetrization procedures* have been designed to exploit the symmetry of the signals in such a way that spurious noise peaks are rejected if they fail to obey the expected symmetry (6.52, 6.53).

1. In 'triangular multiplication', a pair of points in the experimental spectrum that lie symmetrically on either side of the diagonal are replaced by their geometric average

$$S'(\omega_1, \omega_2) = [S(\omega_1, \omega_2) \cdot S(\omega_2, \omega_1)]^{\frac{1}{2}}. \qquad (6.6.13)$$

This scheme is applicable to absolute-value spectra, or to 2D absorption mode spectra if all signals are known to be positive (e.g. pure chemical exchange or cross-relaxation in the slow-motion limit).

2. In phase-sensitive spectra with positive and negative signals, an effective procedure consists in retaining only the signal component with the smallest absolute amplitude

$$\begin{aligned} S'(\omega_1, \omega_2) &= S(\omega_1, \omega_2) \quad \text{if} \quad |S(\omega_1, \omega_2)| \leq |S(\omega_2, \omega_1)| \\ &= S(\omega_2, \omega_1) \quad \text{if} \quad |S(\omega_1, \omega_2)| > |S(\omega_2, \omega_1)|. \end{aligned} \qquad (6.6.14)$$

This procedure, which is also applicable to absolute-value spectra, is particularly efficient in suppressing so-called 't_1-noise'.

It should be recognized that these procedures are non-linear and lead to a correlation of signal and noise. They should not be used in attempts to recover small signals buried in noise, but are useful to suppress spurious noise peaks which might be misinterpreted as signals. An example of the symmetrization of a 2D NOE spectrum is shown in Fig. 6.6.5.

6.6.5. Pattern recognition

A more sophisticated approach to processing 2D spectra in order to simplify the interpretation relies on the identification of multiplets by computer *pattern recognition* procedures. These may be considered as a first step towards the automated analysis of 2D spectra. At the same time, pattern recognition may lead to an improved discrimination of noise and artefacts in 2D spectra. In contrast to 1D spectra, where it is difficult to distinguish a multiplet from an accidental juxtaposition of chemical shifts, 2D spectra yield multiplets that have a sufficiently high information content to avoid pitfalls in pattern recognition procedures.

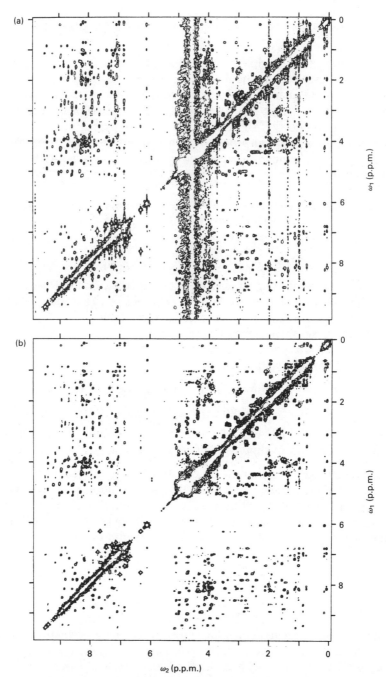

FIG. 6.6.5. The effect of symmetrization, eqn (6.6.14), in an absolute-value 2D NOE spectrum of the protein seminal inhibitor II A (molecular mass $\simeq 6500$) in H_2O, obtained with the sequence $\pi/2 - t_1 - \pi/2 - \tau_m - \pi/2 - t_2$ with a mixing time $\tau_m = 200$ ms. (a) Original spectrum; (b) symmetrized spectrum. Note the prominent t_1-noise ridges in (a), particularly the artefacts in the vicinity of $\omega_2 = 4.6$ p.p.m. which stem from the H_2O resonance. (Reproduced from Ref. 6.43.)

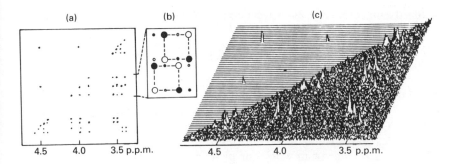

FIG. 6.6.6. Pattern recognition in homonuclear 2D correlation spectra. (a) Lower triangle: spectrum of the three-spin AMX proton system of 2,3-dibromopropanoic acid recorded with a pulse sequence with double-quantum filtration $\pi/2 - t_1 - \pi/4 - \pi/4 - t_2$ (§ 8.3.3). Upper triangle: reduced spectrum obtained by pattern recognition. (b) Schematic cross-peak pattern characteristic of three-spin systems. Filled and open symbols represent positive and negative signals respectively; large symbols correspond to predominant peaks for small r.f. rotation angles. (c) Pattern recognition in a noisy spectrum, obtained by adding computer-generated Gaussian noise to the experimental spectrum in (a). All significant patterns are correctly identified, and in this case no accidental patterns have been identified in the noise. (Reproduced from Ref. 6.54.)

Thus in homonuclear 2D correlation spectra (§ 8.2), the two cross-peaks associated with a two-spin system consist of four signals each, which can be described by 16 frequency coordinates. The redundancy of the multiplet patterns increases in more complex spin systems: for N inequivalent $I = \frac{1}{2}$ spins with non-vanishing couplings, only $N(N + 1)/2$ parameters (shifts and couplings) need to be determined, while a correlation spectrum recorded with a mixing pulse $\beta = \pi/2$ contains $N(N - 1)2^{2(N-1)}$ cross-peak components, each with two coordinates. The number of dominant cross-peaks is reduced to $N(N - 1)2^N$ if small r.f. rotation angles are used (e.g. $\beta = \pi/4$). Under these circumstances, pattern recognition procedures appear promising.

A simple procedure relies on the identification of the proper *sign alternation* of cross-peak multiplet patterns (6.54). In systems with non-degenerate couplings, the cross-peaks are composed of antiphase square patterns (Fig. 6.6.6(b)), while rectangular patterns may occur in systems with equivalent nuclei. The 2D matrix can be searched systematically for the occurrence of suitable patterns, while the test values of the J-couplings are varied in a realistic range. If the proper alternation of signs is found, the amplitude is stored in a reduced 2D matrix at the centre of gravity of the cross-peak multiplet, while the relevant J-couplings are stored in a separate table. An example of pattern recognition in a 2D correlation spectrum of a three-spin system is shown in Fig. 6.6.6. The search procedure is also applicable to 2D spectra with rather low signal-to-noise ratios. These methods can be refined by considering different 2D spectra with complementary information contents (6.54).

6.6.6. Single-channel detection

So far, it has been assumed throughout Chapter 6 that a complex signal $s^+(t_1, t_2) = s_x(t_1, t_2) + is_y(t_1, t_2)$ is recorded by quadrature phase detection in the t_2-period. This allows one to distinguish the sign of a resonance frequency $\omega_{rs}^{(d)}$. A complex Fourier transformation in both dimensions leads to a 2D spectrum with four distinct quadrants.

In some circumstances, one may be restricted to recording a real signal $s_x(t_1, t_2)$ with a single-channel phase detector. In this case, the carrier frequency is normally adjusted such that all resonances $\omega_{rs}^{(d)}$ have the same sign, and a real transformation is calculated with respect to t_2. Single-channel detection does *not* preclude from determining the relative signs of $\omega_{tu}^{(e)}$ and $\omega_{rs}^{(d)}$ by a complex Fourier transformation with respect to t_1 (e.g. in 2D J-spectroscopy), or from obtaining pure phase spectra by a real Fourier transformation with respect to t_1 (e.g. in 2D correlation spectroscopy). Figure 6.6.7 illustrates the implications for a schematic 2D correlation spectrum.

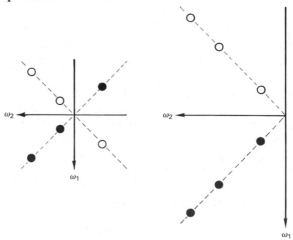

FIG. 6.6.7. Quadrature and single-channel detection in the t_2-period. The symbols represent diagonal multiplets in a 2D correlation spectrum of a three-spin system, in analogy to Fig. 6.6.4. The P and N signals associated with mirror-image pathways are represented by filled and open symbols respectively. On the left, the carrier is positioned within the spectrum, a complex signal is recorded by quadrature detection, and a 2D complex Fourier transform leads to four separate quadrants. The two classes of signals can be separated by time-proportional phase increments as described in § 6.6.3. On the right, a real signal is recorded by single-channel detection, with the carrier set on one side of the spectrum in ω_2, and a real Fourier transformation is calculated with respect to t_2, followed by a complex transform with respect to t_1.

6.7. Operator terms and multiplet structures in two-dimensional spectra

In the context of 1D spectroscopy, the relationship between multiplet structures and Cartesian operator products contained in the expansion of the density operator were discussed in § 4.4.5 (see Fig. 4.4.6).

In 2D spectroscopy, the situation can be described in analogous fashion: the complex signal

$$s^+(t_1, t_2) = \mathrm{tr}\left\{\sigma(t_1, t_2) \sum I_k^+\right\} \qquad (6.7.1)$$

is induced only by $p = -1$ quantum coherence that is proportional to $I_k^- = I_{kx} - iI_{kx}$. If $\sigma(t_1, t_2 = 0)$ contains a product operator like $2I_{kx}I_{lz}$, which is not observable in a strict sense, this term may still evolve into observable in-phase coherence in the course of the detection period t_2, eqn (4.4.75), leading to an antiphase doublet in the ω_2-domain.

The multiplet structure in the ω_1-domain is determined by the Hamiltonian $\mathcal{H}^{(e)}$ and the type of preparation. In general, it is straightforward to translate Cartesian operator products with their associated t_1-dependence into 2D signal patterns.

Consider, by way of example, a density operator term $\sigma(t_1, t_2 = 0) = -2I_{ky}I_{lz} \cos \Omega_k t_1 \sin \pi J_{kl} t_1$. If we choose to detect the complex signal $s^+(t_1, t_2)$ according to eqn (6.7.1), there are eight terms that give rise to observable magnetization in the detection period

$$\begin{aligned}\sigma^{\mathrm{obs}}(t_1, t_2) &= I_{kx} \exp\{+i\Omega_k t_2\}\sin(\pi J_{kl} t_2)\cos(\Omega_k t_1)\sin(\pi J_{kl} t_1) \\
&= -\tfrac{1}{8}I_{kx} \exp\{+i\Omega_k t_2\}[\exp(+i\pi J_{kl} t_2) - \exp(-i\pi J_{kl} t_2)] \\
&\quad \times [\exp(+i\Omega_k t_1) + \exp(-i\Omega_k t_1)] \\
&\quad \times [\exp(+i\pi J_{kl} t_1) - \exp(-i\pi J_{kl} t_1)]. \end{aligned} \qquad (6.7.2)$$

After 2D Fourier transformation, one obtains eight peaks at $(\omega_1, \omega_2) = (\pm\Omega_k \pm \pi J, \Omega_k \pm \pi J)$ with amplitudes of alternating signs, all of which have mixed-phase 2D peakshapes as described by eqn (6.5.10).

In a similar manner, any operator product contained in the density operator $\sigma(t_1, t_2 = 0)$ can be related to a 2D signal pattern. For future reference, the signals associated with some products that may occur in a weakly-coupled two-spin system are shown schematically in Fig. 6.7.1. In deriving these patterns, it has been assumed that a phase correction has been applied such that after 1D Fourier transformation the term I_{ky} leads to a pure absorption 1D Lorentzian lineshape. We employ vector diagrams to represent the 2D signal phase as explained in Fig. 6.7.1.

Note that the operators in $\sigma(t_1, t_2 = 0)$ (I_{kx}, $2I_{ky}I_{lz}$, etc.) are responsible for the phases in the ω_2-domain, while the trigonometric functions

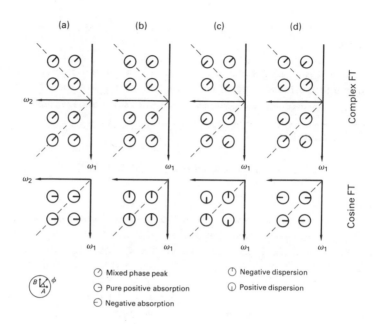

FIG. 6.7.1. 2D signal patterns associated with typical Cartesian operators in the density operator $\sigma(t_1, t_2 = 0)$. (a) $I_{ky} \cos \Omega_k t_1 \cos \pi J_{kl} t_1$; (b) $I_{kx} \sin \Omega_k t_1 \cos \pi J_{kl} t_1$; (c) $2I_{kx}I_{lz} \cos \Omega_k t_1 \sin \pi J_{kl} t_1$; (d) $2I_{ky}I_{lz} \sin \Omega_k t_1 \sin \pi J_{kl} t_1$.

If a complex Fourier transformation (FT) is calculated with respect to t_1 (upper part), all peaks have composite peakshapes $\pm[a(\omega_1)a(\omega_2) - d(\omega_1)d(\omega_2)]$, eqn (6.5.10). A cosine Fourier transformation (lower part) leads to pure 2D absorption if the signals in the upper part are symmetrical with respect to $\omega_1 = 0$ (same symbols for positive and negative ω_1), and to pure 2D dispersion if the signals are antisymmetric with respect to $\omega_1 = 0$.

determine the phases in the ω_1-direction. In the ω_2-domain, one obtains

I_{kx}: dispersive in-phase doublet,
I_{ky}: absorptive in-phase doublet,
$I_{kx}I_{lz}$: dispersive antiphase doublet,
$I_{ky}I_{lz}$: absorptive antiphase doublet.

In the ω_1-domain one obtains

$\cos \Omega_k t_1 \cos \pi J_{kl} t_1$: absorptive in-phase doublet,
$\sin \Omega_k t_1 \cos \pi J_{kl} t_1$: dispersive in-phase doublet,
$\cos \Omega_k t_1 \sin \pi J_{kl} t_1$: dispersive antiphase doublet,
$\sin \Omega_k t_1 \sin \pi J_{kl} t_1$: absorptive antiphase doublet.

In the ω_1-domain the multiplets are centred on the chemical shift which appears in the argument of the t_1-dependent trigonometric function, and in the ω_2-domain on the chemical shift of the spin which appears with a transverse term in the operator product.

6.8. Sensitivity of two-dimensional spectra

Two-dimensional experiments are inherently time-consuming and sensitivity considerations are therefore important. The purpose of this section is twofold. It presents a comparison of the sensitivities of 1D and 2D spectroscopy, and allows the parameters of 2D experiments to be optimized for maximum sensitivity. The computations of the signal envelope and of the random noise proceed in close analogy to the discussion of 1D spectroscopy in § 4.3.

6.8.1. The signal envelope

Consider a 2D peak centred at the coordinates $(\omega_1, \omega_2) = (\omega_{tu}^{(e)}, \omega_{rs}^{(d)})$. The corresponding time-domain signal has the form

$$s(t_1, t_2) = s^e(t_1, t_2)\exp\{-i\omega_{tu}^{(e)}t_1\}\exp\{-i\omega_{rs}^{(d)}t_2\} \qquad (6.8.1)$$

where the envelope function $s^e(t_1, t_2)$ determines the peak shape in the 2D spectrum. In the simplest case, the signal decays exponentially in the two dimensions with the rates $\lambda^{(e)} = 1/T_2^{(e)}$ and $\lambda^{(d)} = 1/T_2^{(d)}$ (eqn (6.5.16)), as illustrated in Fig. 6.5.5(a).

In the presence of inhomogeneous broadening by the external magnetic field or by a continuous distribution of chemical shifts, the time-domain envelope of the signal features a 'ridge' described by eqn (6.5.14), and illustrated in Fig. 6.5.5(b). In the context of this section, it is important to note that this type of envelope cannot be factorized into a product $s^e(t_1) \cdot s^e(t_2)$.

In many 2D experiments, cross-peaks appear as antiphase multiplets in both dimensions, because only antiphase coherence is transferred by r.f. pulses (6.5, 6.57). If the separation $2\pi J$ between these signals exceeds the line-width, each multiplet component can be treated independently. However, if the splitting is barely resolved, the antiphase signals partly cancel each other. In this case, it is of advantage to consider the signal envelope $s^e(t_1, t_2)$ of the entire multiplet. In the simple case of an antiphase doublet with $\omega_1 = \Omega_k \pm \pi J$ and $\omega_2 = \Omega_l \pm \pi J$, the envelope has the form

$$s^e(t_1, t_2) = s^e(0, 0)\sin \pi J t_1 \sin \pi J t_2 \exp\{-t_1/T_2^{(e)}\}\exp\{-t_2/T_2^{(d)}\}, \qquad (6.8.2)$$

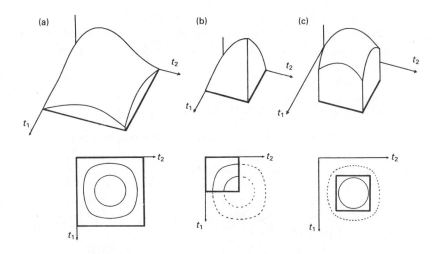

FIG. 6.8.1. (a) A signal envelope function $s^e(t_1, t_2)$ as described by eqn (6.8.2), typical for ill-resolved antiphase doublets encountered in 2D correlation spectroscopy. The envelope is zero along the axes $t_1 = 0$ and $t_2 = 0$. If t_1^{max} and t_2^{max} are restricted (b), the average signal envelope height may be improved by introducing delays before $t_1 = 0$ and $t_2 = 0$ such that the 'captive volume' is roughly symmetric around the time-domain maximum in (c). For clarity, schematic contour plots are shown beneath the perspective views. (Reproduced from Ref. 6.23.)

with vanishing intensity along the t_1- and t_2-axes, as shown in Fig. 6.8.1. This figure also illustrates that the selection of a sampling window allows one to 'capture' different parts of the signal envelope.

We assume from the outset that the signal function $s(t_1, t_2)$ is multiplied by a suitable weighting function $h(t_1, t_2)$ in order to enhance the resolution or sensitivity of the 2D spectrum. If the absorption-mode peak is symmetrical, its height S in the 2D frequency-domain spectrum $S(\omega_1, \omega_2)$ is given by the volume of the weighted envelope function. We assume that the time-domain signal is sampled by taking M_1 increments in t_1 between 0 and t_1^{max}, for each of which n free induction decays are recorded, consisting of M_2 sampling points with t_2 between 0 and t_2^{max}. In this case, the signal height S may be expressed by the sum of the values of the weighted time-domain envelope at all sampling points

$$S = n \sum_{k=0}^{M_1-1} \sum_{l=0}^{M_2-1} s^e\left(k\frac{t_1^{max}}{M_1}, l\frac{t_2^{max}}{M_2}\right) h\left(k\frac{t_1^{max}}{M_1}, l\frac{t_2^{max}}{M_2}\right). \quad (6.8.3)$$

In most practical situations, it is permissible to replace the discrete sums by integrals, giving

$$S = n \frac{M_1}{t_1^{max}} \frac{M_2}{t_2^{max}} \int_0^{t_1^{max}} dt_1 \int_0^{t_2^{max}} dt_2 \, s^e(t_1, t_2) h(t_1, t_2). \quad (6.8.4)$$

6.8 SENSITIVITY OF TWO-DIMENSIONAL SPECTRA

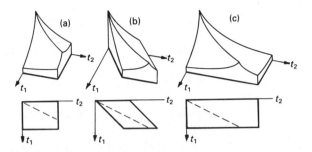

FIG. 6.8.2. Optimization of the captive volume, eqn (6.8.5), for a signal envelope $s^e(t_1, t_2)$ with a ridge as described by eqn (6.5.14) which is typical for signal components that are subject to coherence transfer echo effects with $\kappa < 0$ in eqn (6.5.11). All three diagrams show the same envelope with $\kappa = -2$, corresponding to a homonuclear double-quantum experiment or a hypothetical heteronuclear shift correlation experiment with $\gamma^{(1)} = 2\gamma^{(2)}$. In the projections, the coherence transfer echo ridge is indicated as a dotted line. (a) Captive volume obtained if t_1^{max} and t_2^{max} are restricted because of constraints on data space. (b) By introducing a delay χt_1 before $t_2 = 0$, the captive volume can be increased if the coherence transfer echo ridge is pronounced. (c) If data space is unrestricted, the captive volume can be increased by extending t_2^{max}. (Reproduced from Ref. 6.23.)

The most significant factor influencing the sensitivity of a 2D spectrum is the average time-domain signal amplitude $\overline{sh}$ defined in analogy to eqn (4.3.4)

$$\overline{sh} = \frac{1}{t_1^{max}} \frac{1}{t_2^{max}} \int_0^{t_1^{max}} dt_1 \int_0^{t_2^{max}} dt_2\, s^e(t_1, t_2) h(t_1, t_2). \tag{6.8.5}$$

The peak height S in the frequency domain is then given by

$$S = nM_1 M_2 \overline{sh}, \tag{6.8.6}$$

which is the analogue of eqn (4.3.5) for the 1D experiment. The signal S is proportional to the total number of samples that are recorded and to the average height of the weighted 'captive volume' of the time-domain signal envelope within the limits $0 < t_1 < t_1^{max}$ and $0 < t_2 < t_2^{max}$. Figures 6.8.1 and 6.8.2 illustrate how it is possible to define different captive volumes for a given envelope function by proper selection of the boundaries in the time domain.

6.8.2. Thermal noise and t_1-noise

The random noise in 2D spectra arises primarily from thermal noise sources in the receiver circuits as well as from instrumental instabilities that lead to so-called 't_1-noise'.

The t_1-noise is due to variations of the experimental conditions from experiment to experiment in a 2D sequence, leading to irregular fluctuations of the signal as a function of t_1. The resulting noise bands appear parallel to the ω_1-axis in the 2D spectrum and are highly dependent on technical factors of the instrument. A typical 2D spectrum with pronounced t_1-noise is shown in Fig. 6.6.5. In many cases, t_1-noise is correlated for different resonances in the ω_2-domain. It can be partially cancelled in certain experiments by combining cross-sections parallel to the ω_1-axis in the 2D spectrum (6.58, 6.59). In some cases, t_1-noise can also be reduced by allowing undesirable signal components to defocus in magnetic field gradients and by selecting the relevant coherence transfer pathways by means of static or r.f. field gradient pulses (6.51, 6.60, 6.63) rather than by phase-cycling (6.9, 6.10).

In the following, we concentrate on thermal random noise. In the time domain, the r.m.s. amplitude of the thermal noise is given by eqn (4.3.7). The weighting function $h(t_1, t_2)$ causes the r.m.s. noise amplitude σ_n to become time-dependent, in analogy to eqn (4.3.8),

$$\sigma_n(t_1, t_2) = F^{\frac{1}{2}} \rho_n |h(t_1, t_2)| \tag{6.8.7}$$

where F is again the full bandwidth of the quadrature receiver, equal to twice the cut-off frequency f_c of the audio filter.

In the frequency domain, the r.m.s. noise amplitude σ_N is the result of summing contributions from all nM_1M_2 time-domain points acquired in the experiment. Taking into account the weighting by $h(t_1, t_2)$ this leads to

$$\sigma_N = (nM_1M_2F)^{\frac{1}{2}} [\overline{h^2}]^{\frac{1}{2}} \rho_n. \tag{6.8.8}$$

Here $[\overline{h^2}]^{\frac{1}{2}}$ is the r.m.s. amplitude of the weighting function, given by

$$\overline{h^2} = \frac{1}{t_1^{\max}} \frac{1}{t_2^{\max}} \int_0^{t_1^{\max}} dt_1 \int_0^{t_2^{\max}} dt_2 \, h(t_1, t_2)^2. \tag{6.8.9}$$

To avoid an increase of the noise amplitude by down-conversion of high-frequency noise, it is advisable to set the cut-off frequency f_c of the audio filter equal to the Nyquist frequency of the sampling process such that

$$F = 2f_c = 1/\Delta t_2 = M_2/t_2^{\max}, \tag{6.8.10}$$

as in 1D spectroscopy (eqn (4.3.11).

6.8.3. Sensitivity

Combining eqns (6.8.6), (6.8.8), and (6.8.10), we find for the signal-to-noise ratio of the 2D spectrum

$$S/\sigma_N = (nM_1 t_2^{\max})^{\frac{1}{2}} \frac{\overline{sh}}{[\overline{h^2}]^{\frac{1}{2}}} \frac{1}{\rho_n}. \tag{6.8.11}$$

6.8 SENSITIVITY OF TWO-DIMENSIONAL SPECTRA

Obviously, the signal-to-noise ratio increases with the number of experimental scans nM_1. To obtain an expression for a standard sensitivity, we consider the signal-to-noise ratio per unit time, in analogy to eqn (4.3.17). The total time T_{tot} required for accumulating the 2D data matrix is

$$T_{tot} = nM_1 T \qquad (6.8.12)$$

where T is the average time of each experimental scan, including the waiting time between experiments (see Fig. 4.3.1). This leads to the signal-to-noise ratio per unit time (called 'sensitivity' in the following)

$$\frac{S}{\sigma_N T_{tot}^{\frac{1}{2}}} = \frac{\overline{sh}}{[\overline{h^2}]^{\frac{1}{2}}} \left[\frac{t_2^{max}}{T}\right]^{\frac{1}{2}} \frac{1}{\rho_n} \qquad (6.8.13)$$

where t_2^{max}/T corresponds to the duty ratio of the receiver. In the case that no weighting is applied, eqn (6.8.13) reduces to the simple expression

$$\frac{S}{\sigma_N T_{tot}^{\frac{1}{2}}} = \left[\frac{t_2^{max}}{T}\right]^{\frac{1}{2}} \frac{\bar{s}}{\rho_n}. \qquad (6.8.14)$$

As usual, optimum sensitivity is attained by using a matched weighting function $h(t_1, t_2) \propto s^e(t_1, t_2)$ (6.34, 6.35). In this case,

$$\left[\frac{S}{\sigma_N T_{tot}^{\frac{1}{2}}}\right]_{matched} = \left[\frac{t_2^{max}}{T}\right]^{\frac{1}{2}} \frac{[\overline{s^2}]^{\frac{1}{2}}}{\rho_n} \qquad (6.8.15)$$

with

$$\overline{s^2} = \frac{1}{t_1^{max}} \frac{1}{t_2^{max}} \int_0^{t_1^{max}} dt_1 \int_0^{t_2^{max}} dt_2 \, s^e(t_1, t_2)^2. \qquad (6.8.16)$$

6.8.4. Comparison of the sensitivities of one- and two-dimensional experiments

To compare the optimized sensitivities of 1D experiments, eqn (4.3.20), and of 2D experiments, eqn (6.8.15), we can set $t_1^{max} = t_2^{max}$ if we assume that the same resolution must be achieved in the ω_2-domain as in the 1D experiment. For identical total performance times, we find the sensitivity ratio

$$\frac{\text{2D sensitivity}}{\text{1D sensitivity}} = \frac{[\overline{s^2}(2D)]^{\frac{1}{2}}}{[\overline{s^2}(1D)]^{\frac{1}{2}}}. \qquad (6.8.17)$$

Thus the sensitivity ratio is only determined by average signal power, given by the r.m.s. signal envelope in the two experiments, eqns (4.3.21) and (6.8.16).

The comparison is straightforward when it is possible to separate the envelope function in the 2D experiment

$$s^e(t_1, t_2) = s_1^e(t_1) \cdot s_2^e(t_2) \qquad (6.8.18)$$

where $s_2^e(t_2)$ is normally identical to the envelope $s^e(t)$ in a corresponding 1D experiment. This assumption leads to the sensitivity ratio

$$\frac{\text{2D sensitivity}}{\text{1D sensitivity}} = \left[\frac{1}{t_1^{\max}} \int_0^{t_1^{\max}} dt_1\, s_1^e(t_1)^2 \right]^{\frac{1}{2}}. \qquad (6.8.19)$$

This expression indicates that the same sensitivity can be achieved in the two experiments (6.35), provided

1. $s_1^e(t_1)$ is equal to unity, i.e. transverse relaxation and inhomogeneous decay in the evolution period t_1 can be neglected;

2. The instrumental stability is sufficient to allow one to neglect t_1-noise.

An additional condition, which is not obvious from eqn (6.8.19), requires that

3. The number of peaks in the two experiments is the same, i.e. the intensity of a line in the 1D spectrum is not distributed among several peaks in the 2D spectrum;

Criterion 1 implies low resolution in t_1. In the case that high resolution in t_1 must be achieved, a small sensitivity loss in 2D spectroscopy must be taken into account. While criterion 3 is fulfilled in 2D J-spectra of weakly coupled systems, it is not satisfied in 2D correlation, multiple-quantum, and 2D exchange experiments, and a loss in sensitivity must be accepted in these cases. However, this type of comparison fails to take into account the greatly enhanced information content of 2D spectra. In many cases, extensive series of 1D experiments are necessary to obtain similar information, and the information obtained per unit time is superior in 2D experiments, even if the sensitivity is inferior.

6.8.5. Optimization of two-dimensional experiments

Equations (6.8.11)–(6.8.16) do not explicitly state that the initial amplitudes of the signal $s^e(0, 0)$ and hence $\bar{s}$ are functions of the recovery delay between subsequent experiments. This delay is required to obtain a reproducible initial state at the beginning of each experiment, and should be of the order $3T_1$ to avoid interference between subsequent experiments, which would give rise to t_1-noise in the frequency domain. We shall regard the average time per experiment T as a constant for a given experiment, and concentrate on the optimization of the experimental parameters M_1, M_2, $t_1^{\max}$, and $t_2^{\max}$.

Two simple cases may be distinguished depending on the type of 2D method. The resolution that is required in the ω_1-domain may in some cases be comparable to the natural line-width, while in many cases a much reduced resolution suffices.

6.8.5.1. Low resolution in the ω_1-domain

If poor resolution is acceptable in the ω_1-domain, for example in heteronuclear shift correlation spectroscopy (§ 8.5), in experiments to establish the carbon–carbon connectivities (§ 8.4.2), and in 2D exchange experiments (Chapter 9), the signal envelope decays only minimally in the interval $0 < t_1 < t_1^{max}$ and we can assume a signal envelope

$$s^e(t_1, t_2) = s^e(0, 0)\exp\{-t_2/T_2^{(d)}\}. \qquad (6.8.20)$$

Matched filtering is required in the t_2-domain while in the t_1-domain it is sufficient to apply an apodization window function (see § 4.1.3.1). The resulting sensitivity is

$$\left(\frac{S}{\sigma_N T_{tot}^{\frac{1}{2}}}\right)_{matched} = \frac{\overline{h_1}}{[\overline{h_1^2}]^{\frac{1}{2}}} \frac{s(0,0)}{\rho_n} \left[\frac{T_2^{(d)}}{2T}\right]^{\frac{1}{2}} [1 - \exp\{-2t_2^{max}/T_2^{(d)}\}]^{\frac{1}{2}} \qquad (6.8.21)$$

where

$$\overline{h_1} = \frac{1}{t_1^{max}} \int_0^{t_1^{max}} h_1(t_1)\, dt_1 \qquad (6.8.22)$$

and

$$\overline{h_1^2} = \frac{1}{t_1^{max}} \int_0^{t_1^{max}} h_1(t_1)^2\, dt_1. \qquad (6.8.23)$$

If data storage space is limited, it is advisable to minimize both t_1^{max} and the sampling rate (taking into account the frequency range to be covered) to give the smallest number of rows M_1 in the data matrix. The data space so released can then be used for extending the sampling in the t_2-domain (6.55, 6.56).

Equation (6.8.21) implies that apodization in the evolution period leads to a slight sensitivity loss determined by the factor $\overline{h_1}/[\overline{h_1^2}]^{\frac{1}{2}}$. Suitable apodization functions have been described in § 4.1.3.1. Cosine apodization, for example, leads to a reduction of the sensitivity by 10 per cent and reduces the ripple to 6.7 per cent of the peak amplitude (see Fig. 4.1.5). Gaussian apodization with $h_1(x) = \exp\{-3x^2\}$, on the other hand, reduces the ripple to less than 1 per cent but leads to a sensitivity loss of 15 per cent (6.61).

6.8.5.2. High resolution in the ω_1-domain

In some applications of 2D spectroscopy, it is desirable to digitize the spectrum in ω_1 to an extent comparable with or exceeding the inherent line-width, for example to measure accurate coupling constants, chemical shifts, or line-widths.

For a signal envelope that decays exponentially in the two time-domains with time constants $T_2^{(e)}$ and $T_2^{(d)}$, the sensitivity without filtration is given by

$$\frac{S}{\sigma_N T_{tot}^{\frac{1}{2}}} = \frac{s(0,0)}{\rho_n T^{\frac{1}{2}}} \frac{T_2^{(e)}}{t_1^{max}} \frac{T_2^{(d)}}{(t_2^{max})^{\frac{1}{2}}}$$

$$\times \left[1 - \exp\left(-\frac{t_1^{max}}{T_2^{(e)}}\right)\right]\left[1 - \exp\left(-\frac{t_2^{max}}{T_2^{(d)}}\right)\right]. \quad (6.8.24)$$

The sensitivity without filtering is maximum for $t_2^{max} = 1.3 T_2^{(d)}$ and $t_1^{max} \to 0$.

With matched filtration in both frequency domains, the sensitivity is

$$\left(\frac{S}{\sigma_N T_{tot}^{\frac{1}{2}}}\right)_{matched} = \frac{s(0,0)}{2\rho_n T^{\frac{1}{2}}}$$

$$\times \left\{\frac{T_2^{(e)} T_2^{(d)}}{t_1^{max}}\left[1 - \exp\left(-\frac{2t_1^{max}}{T_2^{(e)}}\right)\right]\left[1 - \exp\left(-\frac{2t_2^{max}}{T_2^{(d)}}\right)\right]\right\}^{\frac{1}{2}}. \quad (6.8.25)$$

Here, the sensitivity can be increased by extending t_2^{max} as much as possible, but maximum sensitivity still requires $t_1^{max} \to 0$. It is clear that an increase in resolution in the ω_1-dimension brought about by lengthening t_1^{max} must be paid for by a decrease in sensitivity per unit time.

If the signal envelope does not decay monotonically in the two time-domains, one must take care that the volume of signal captured within $0 < t_1 < t_1^{max}$ and $0 < t_2 < t_2^{max}$ is maximized, if necessary by inserting delays in the experimental pulse sequence. In the case of a coherence transfer echo (Fig. 6.8.2), it may be advisable to delay the start of acquisition in the t_2-dimension by an amount χt_1 such that the coherence transfer echo is 'captured'. If sensitivity is the ultimate goal, it is *not* recommended to start the acquisition at the *top* of the ridge, since this entails a loss of captured volume in the rising part of the envelope. Whether delayed acquisition schemes are worthwhile depends on the 'sharpness' of the coherence transfer echo, and on practical restrictions on t_2^{max} set by the capacity of the data storage device.

In 2D correlation spectroscopy (Chapter 8), high resolution is essential in both dimensions, since cross-peaks appear in antiphase by the nature of the coherence transfer process, and are associated with time-domain envelopes of the form of eqn (6.8.2), shown in Fig. 6.8.1. For optimum sensitivity, it is necessary to capture the maxima of these envelopes, which occur near $t_1 = t_2 \simeq (2J)^{-1}$ and the acquisition window should be placed as shown in Fig. 6.8.1(c).

6.8.5.3. Practical recommendations

To optimize the sensitivity, only the spin–lattice relaxation time T_1, the form of the signal envelope $s^e(t_1, t_2)$, the weighting function $h(t_1, t_2)$, and the required resolution in the two dimensions as determined by the maximum values t_1^{max} and t_2^{max} are relevant. The following recommendations allow one to optimize the sensitivity of a 2D experiment (6.23).

1. Estimate the expected spin–lattice relaxation times and signal envelope functions $s^e(t_1, t_2)$.

2. Choose an appropriate delay time between experiments. Normally this must exceed the spin lattice relaxation time T_1 to avoid problems of t_1-noise due to signal saturation. This defines the mean time per experiment T.

3. Always use the minimum possible resolution in the ω_1-dimension compatible with experimental requirements. This defines the value of t_1^{max}.

4. If data space is a problem, it is important to minimize the number of rows M_1 accumulated in the t_1-dimension by using the smallest possible sampling rate in the t_1-domain. However, if data space is unrestricted, this sampling rate is irrelevant for sensitivity per unit time.

5. The number M_2 of acquired samples and t_2^{max} should be as large as possible within the constraint $t_2^{max} < T$. The sampling rate in this dimension does not affect sensitivity as long as proper analogue filtration of the signal before analogue-to-digital conversion is employed to avoid down-conversion of high-frequency noise.

6. In those cases where the envelope of the interesting signals does not decay monotonically, it may be desirable to incorporate delays into the pulse sequence so as to capture the maximum signal intensity within the intervals $0 < t_1 < t_1^{max}$, $0 < t_2 < t_2^{max}$.

7. Optimum sensitivity is achieved by multiplication of the time-domain data by a weighting function $h(t_1, t_2)$, shaped to match the 2D time-domain envelope $s^e(t_1, t_2)$.

7
TWO-DIMENSIONAL SEPARATION OF INTERACTIONS

7.1. Basic principles

The interpretation of NMR spectra in liquid or solid phases is often impeded by the complexity of overlapping resonances. When the Hamiltonian is composed of terms of different physical origin, such as chemical shifts or dipolar or scalar couplings, it is often possible to render spectra more intelligible by separating various interactions in orthogonal frequency domains. Two-dimensional separation is preferable to decoupling techniques, since there is no loss of information. In the course of 2D separation, the number of peaks is conserved. The primary advantage lies in the fact that overlapping resonances are unravelled.

The separation of interactions by 2D spectroscopy can be compared with two-dimensional chromatography. In a one-dimensional thin layer or paper chromatogram, the separation of the constituents by elution with a given solvent is often incomplete. Elution with a second solvent in a perpendicular direction may then achieve full separation.

In NMR spectroscopy, the choice of two solvents is replaced by the choice of two suitable (effective) Hamiltonians $\mathcal{H}^{(e)}$ and $\mathcal{H}^{(d)}$ for the evolution and detection periods which allow unique characterization of each line.

A particularly simple example is shown in Fig. 7.1.1. The overlapping multiplets in the normal proton-coupled carbon-13 spectrum shown on top along ω_2 can be unravelled by separating the signals according to their chemical shifts in the ω_1-domain. In this case, $\mathcal{H}^{(e)} = \sum_k \Omega_k S_{kz}$ and $\mathcal{H}^{(d)} = \sum_k \Omega_k S_{kz} + \sum_{kl} 2\pi J_{kl} S_{kz} I_{lz}$ + proton terms. If the Hamiltonians were interchanged, the information contents of the ω_1- and ω_2-domains would simply be interchanged (7.1).

A typical example of a homonuclear 2D J-spectrum is shown in Fig. 7.1.2. In applications to weakly coupled systems, the ω_1-domain in a 2D J-spectrum carries only coupling information, $\mathcal{H}^{(e)} = \sum 2\pi J_{kl} I_{kz} I_{lz}$, while the ω_2-domain shows both couplings and chemical shifts, $\mathcal{H}^{(d)} = \sum \Omega_k I_{kz} + \sum 2\pi J_{kl} I_{kz} I_{lz}$, as discussed in detail in § 7.2.1.

The 2D separation experiment follows the general pattern discussed in §§ 6.1 and 6.2. However, no mixing pulses are required for 2D separation. If the Hamiltonians $\mathcal{H}^{(e)}$ and $\mathcal{H}^{(d)}$ commute, the switching of the Hamiltonian does not induce coherence transfer, and the same coherence $|t\rangle\langle u|$ precessing with the frequency $-\omega_{tu}^{(e)}$ in the evolution

7.1 BASIC PRINCIPLES

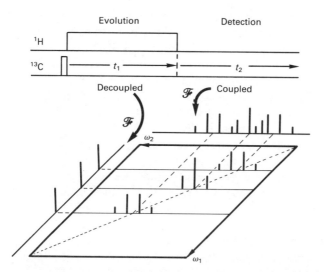

FIG. 7.1.1. A simple example of 2D separation: a proton-coupled carbon-13 spectrum with overlapping multiplets can be unravelled in a 2D spectrum by retaining only the chemical shifts in the ω_1-domain, while both shifts and couplings are allowed to act in the ω_2-domain. This can be achieved by applying broadband decoupling in the evolution period t_1.

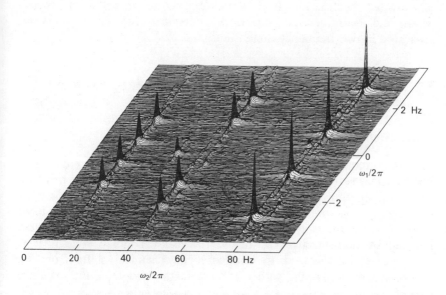

FIG. 7.1.2. Two-dimensional J-spectrum of 2-furanoic acid methyl ester, obtained with the scheme of Fig. 7.2.1(a). The projection on to the ω_2-axis corresponds to the normal 1D spectrum with the chemical shifts of the three ring protons and their couplings, while the ω_1-domain reveals only scalar couplings (a doublet of doublets for each site).

period continues to precess with a modified frequency $-\omega_{tu}^{(d)}$ in the detection period. In this case, the coherence transfer amplitude $R_{rs\,tu}$ in eqns (6.2.11) and (6.2.12) can be set equal to $R_{rs\,tu} = \delta_{rt} \cdot \delta_{su}$. If the two Hamiltonians do not commute, however, coherence transfer processes $|t\rangle\langle u| \to |r\rangle\langle s|$ may be induced without changing the order of coherence ($\Delta p = 0$). In order to obtain pure 2D absorption lineshapes, some experimental schemes require a phase separation pulse between evolution and detection periods, as described in § 6.5.3.

It is also possible to combine 2D separation with coherence transfer. Thus it may be useful to detect multiplets arising from proton–proton interactions indirectly by transferring the proton magnetization to carbon-13 nuclei in order to utilize the greater dispersion of chemical shifts in carbon-13 spectra. Such techniques will be deferred to Chapter 8.

In the following sections, we shall give an account of the separation of shifts and scalar couplings in isotropic phase, which is straightforward in weakly coupled homo- and heteronuclear systems but requires a more careful analysis in the presence of strong coupling. In oriented phase, particularly in static powders or in conjunction with magic-angle sample spinning, the separation of dipolar couplings and (anisotropic) chemical shifts yields structural information that cannot be easily extracted from 1D powder spectra. Finally, it is possible to obtain solid-state spectra where the isotropic and anisotropic chemical shift components are separated in two frequency dimensions.

7.2. Separation of chemical shifts and scalar couplings in isotropic phase

7.2.1. Homonuclear systems

The separation of chemical shifts and scalar spin–spin couplings in homonuclear systems, such as coupled proton spectra, seems at first sight to be hampered by the lack of practical means for homonuclear decoupling. It is not possible to reduce the Hamiltonian to the shift terms alone, because the isotropic coupling terms remain invariant to rotations induced by non-selective r.f. fields. In weakly coupled systems it is, however, straightforward to eliminate the chemical shifts from the ω_1-domain by refocusing with a π-pulse in the middle of the evolution period, as discussed in § 3.3.2. Complicating effects due to strong coupling will be deferred to § 7.2.3. In the present section, the discussion will be restricted to weakly coupled systems.

With the basic pulse sequence for homonuclear 2D separation (7.2),

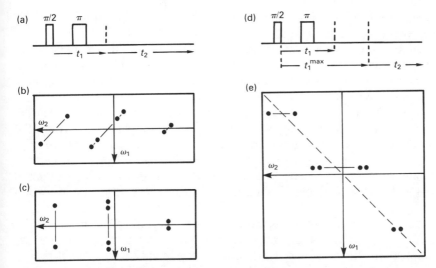

FIG. 7.2.1. (a) Basic scheme for the separation of couplings and shifts in homonuclear spin systems, also referred to as '2D J-spectroscopy', '2D J-resolved spectroscopy', or 'spin–echo spectroscopy'. (b) Schematic 2D J-spectrum of a weakly coupled linear AMX system (with $J_{AX} = 0$). All signals have mixed phases described by eqn (6.5.10). (c) 'Sheared' 2D J-spectrum obtained by aligning the signals by rearrangement of the data matrix. The projection on to the ω_2-axis corresponds to the broadband decoupled spectrum, provided care is taken to avoid cancellation of positive and negative intensities upon integration, as discussed in § 6.5.5. (d) Pulse scheme for 'constant-time' 2D J-spectroscopy. (e) Schematic 2D spectrum obtained in a constant-time experiment with t_1^{max} = constant. The projection on to the ω_1-axis corresponds to the decoupled spectrum.

shown in Fig. 7.2.1(a), one obtains the Hamiltonians (see eqn (3.3.18))

$$\mathcal{H}^{(e)} = \mathcal{H}_{II} = \sum_{k<l} 2\pi J_{kl} I_{kz} I_{lz}, \qquad (7.2.1)$$

$$\mathcal{H}^{(d)} = \mathcal{H}_{ZI} + \mathcal{H}_{II} = \sum_k \Omega_k I_{kz} + \sum_{k<l} 2\pi J_{kl} I_{kz} I_{lz}. \qquad (7.2.2)$$

In the weak-coupling approximation, we have

$$[\mathcal{H}^{(e)}, \mathcal{H}^{(d)}] = 0. \qquad (7.2.3)$$

No coherence transfer occurs and the number of peaks in the 2D separated spectrum is identical to that in the 1D spectrum.

In the 2D spectrum $S(\omega_1, \omega_2)$, the multiplets appear aligned parallel to the positive diagonal, as shown schematically in Fig. 7.2.1(b). All peaks have the same mixed phase or 'phase-twisted' lineshape, described by eqn (6.5.10) with equal 2D absorptive and dispersive contributions. It is clear that the multiplets are separated in the 2D spectrum provided the chemical shifts Ω_k are resolved. This compares favourably with the ordinary 1D spectrum, which corresponds to the projection of the 2D spectrum on to the ω_2-axis.

In the presence of inhomogeneous broadening, the resonances are spread such as to give line-widths in ω_2 determined by T_2^*, but the refocusing leads to natural line-widths in ω_1, provided diffusion effects in t_1 can be neglected.

The 2D data matrix $S(\omega_1, \omega_2)$ may be transformed into a 'sheared' matrix $S(\omega_1, \omega_2')$ where $\omega_2' = \omega_2 - \omega_1$ (7.3). In the sheared spectrum, the multiplets appear aligned parallel to the ω_1-axis, as shown in Fig. 7.2.1(c). The shearing transformation must be carried out by computational means and cannot be achieved by a modified experimental sequence in time domain. Shearing usually calls for data interpolation since the digitization of the two domains may be different (see § 6.6.1).

Sections parallel to the ω_1-axis through a sheared spectrum reveal the multiplet structure with enhanced resolution. Some experimental examples are shown in Fig. 7.2.2, which correspond to the methyl proton resonances of basic pancreatic trypsin inhibitor (BPTI) (7.3).

The projection of a sheared spectrum (Fig. 7.2.1(c)) on to the ω_2-axis yields a 1D spectrum where the splittings are eliminated completely and only chemical shifts remain (7.2). This corresponds to a 1D spectrum that would be obtained, at least in principle, by CW observation with simultaneous broadband decoupling of all nuclei except for the nucleus that is observed.

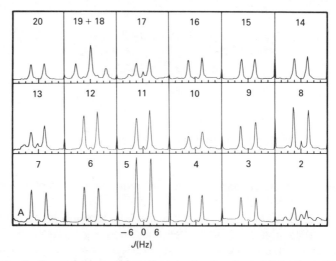

FIG. 7.2.2. Sections taken parallel to the ω_1-axis of a sheared 2D spin–echo spectrum (compare with Fig. 7.2.1(c)) of basic pancreatic trypsin inhibitor (BPTI), representing the multiplets of 19 of the 20 methyl groups in the protein. The signals are shown in absolute value representation (eqn (6.5.35)) (Reproduced from Ref. 7.3.)

7.2 CHEMICAL SHIFTS AND SCALAR COUPLINGS

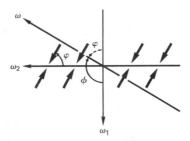

FIG. 7.2.3. The relationship between the angle ϕ that is defined in the projection cross-section theorem of § 6.4.1.5, and the angle φ considered in projections of 2D J-spectra. The angle φ defines the *direction* of the *skew* projection on to the ω_2-axis, while the projection cross-section theorem considers an *orthogonal* projection on to an axis that subtends an angle ϕ with the ω_1-axis (compare with Fig. 6.4.2).

Clearly, the orthogonal projection on to the ω_2-axis of a sheared spectrum is equivalent to a skew projection parallel to the diagonal of the original 2D spectrum. As discussed in § 6.5.5, such skew projections of mixed phase lines vanish identically. A non-vanishing projection can be calculated from the absolute-value spectrum. This procedure leads to a significant loss in resolution, unless heavy filtering is used before projecting.

By skew projection of the original 2D spectrum on to the ω_2-axis at different angles, it is possible to scale the multiplet splittings by arbitrary factors. Figure 7.2.3 shows the relationship between the angle ϕ defined in the context of the projection cross-section theorem, and the angle φ subtended between the *direction* of the projection and the ω_2-axis, which is usually referred to in the context of 2D J-spectra. The scaling factor of the multiplet splittings obtained with skew projections is given by

$$f = 1 + \cot \phi = 1 - \cot \varphi. \tag{7.2.4}$$

This makes it possible to obtain 1D spectra that retain multiplet information without suffering from extensive overlaps, as shown in Fig. 7.2.4. Scaling factors of 0, 0.25, 0.5, and 1 are obtained with $\varphi = 45°$, 53.1°, 63.4°, and 90°, respectively. This type of scaling provides a unique means of circumventing the invariance of the homonuclear scalar coupling under non-selective pulses. In systems with heteronuclear couplings, or in experiments where selective pulses can be applied to homonuclear systems, more direct scaling methods are possible, and it is not necessary to take recourse to 2D spectroscopy (7.4).

An alternative to projection methods to obtain a spectrum where all homonuclear multiplet splittings are eliminated is the 'constant-time experiment' (7.5) shown schematically in Fig. 7.2.1(d). In contrast to the usual scheme of 2D J-spectroscopy, the signal acquisition is started at a

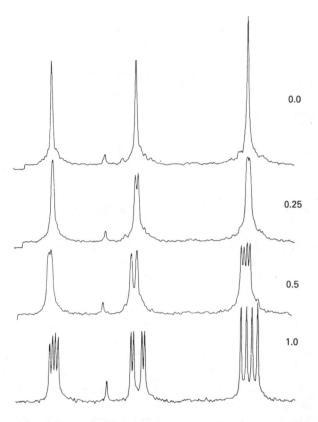

FIG. 7.2.4. Projections of the absolute-value 2D J-spectrum of the AMX system of 2-furanoic acid methyl ester shown in Fig. 7.1.2. The direction of the projection with respect to the ω_1-axis (see Fig. 7.2.3) is given by $\varphi = 45°$, 53.1°, 63.4°, and 90°, leading to scaling factors $f = 0$, 0.25, 0.5, and 1.0, respectively. The spectrum contains a weak solvent peak also visible in Fig. 7.1.2 (Reproduced from Ref. 7.3.)

constant time t_1^{max} after the initial pulse, and a π-pulse is shifted systematically within this interval from experiment to experiment. Since homonuclear couplings are not affected by the position of the π-pulse, the t_1-modulation is determined only by chemical shifts. The resulting 2D constant time spectrum of a typical three-spin system is shown in Fig. 7.2.1(e). The decoupled spectrum can be obtained from a projection on to the ω_1-axis, and can also be obtained directly by Fourier transformation of a one-dimensional signal $s(t_1, t_2 = 0)$ measured by recording a single sample at $t_2 = 0$ for all t_1-increments. It should however be noted that the signal amplitudes in the decoupled spectrum obtained in this way depend on $\cos \pi J_{kl} t_1^{max}$. The selection of the maximum duration of the evolution time t_1^{max} is therefore critical (7.5). Clearly, the transverse relaxation in the t_1^{max} interval leads to a loss in sensitivity. The line-widths in the ω_1-domain, however, can be very narrow.

The basic scheme for 2D J-spectroscopy of Fig. 7.2.1(a) can be modified by using a series of refocusing pulses in analogy to the sequence of Carr and Purcell (7.6)

$$(\pi/2) - [\tfrac{1}{2}\Delta t_1 - \pi - \tfrac{1}{2}\Delta t_1]_n - t_2. \qquad (7.2.5)$$

The effects of translational diffusion through static field gradients on the ω_1-line-width can then be largely suppressed. The repetition rate must be small compared to the differences in chemical shifts of coupled nuclei in order to avoid distortions due to apparent strong coupling under the effective Hamiltonian (7.7).

It is possible to record the entire 2D time-domain matrix in one single experiment (7.8), since each echo in a Carr–Purcell sequence may be considered as an experiment with a different t_1-value. Since the detection interval t_2^{max} is limited by the spacing between the π-pulses, the resolution that can be achieved in the ω_2-domain is severely curtailed. A modified scheme has been described to collect a complete data matrix $S(t_1, t_2)$ with a set of M complementary experiments, each having a pulse spacing $M\Delta t_1$ between π-pulses of the Carr–Purcell sequence to allow for more extended sampling of the echoes that occur between the refocusing pulses (7.8).

It should be emphasized that spin–echo spectra do not contain information that allows one to identify coupling partners, unless the values of the scalar couplings can be utilized for identification. Although the 2D correlation methods described in Chapter 8 are more powerful, coupling partners can also be identified by combining spin–echo spectroscopy with double resonance (7.9). A selective decoupling field is applied in the t_1- and/or t_2-periods. As shown schematically in Fig. 7.2.5(a), irradiation in t_2 leads to multiplets with a slope that depends on the offset of the CW decoupling field from the chemical shifts of the coupling partners. If the decoupling field is applied throughout the experiment, the normal slope of the multiplets parallel to the positive diagonal is retained, but the J-splittings are scaled in analogy to off-resonance decoupling (see § 4.7.4.2), as shown in Fig. 7.2.5(b). It is interesting to note that the introduction of a selective r.f. field amounts to an extension from two to three frequency variables. We shall see in § 9.6 how 3D experiments can be carried out entirely in time domain.

Homonuclear 2D separation methods are subject to artefacts if the refocusing pulse deviates from its nominal rotation angle $\beta = \pi$ (7.10). For sufficiently large deviations, the 2D spectrum corresponds to a correlation spectrum with delayed acquisition (so-called spin–echo correlation spectrum), as discussed in § 8.3.1. Provided the errors are small, it is sufficient to cancel the undesirable coherence transfer pathways with

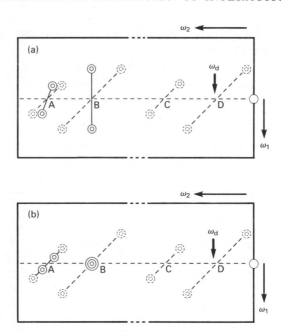

FIG. 7.2.5. Spin decoupling in spin–echo spectra of a mixture of two weakly coupled two-spin systems AC and BD. The double resonance irradiation ω_d is applied at the chemical shift of nucleus D. In (a), decoupling is applied only in the detection period t_2, causing the multiplets of A and B to be tilted. In (b), the decoupling field is applied throughout evolution and detection periods, causing the splittings of the A and B multiplets to be scaled without tilting.

the so-called Exorcycle procedure (7.11) or by a shorter three-step phase cycle (7.12). Complicating effects due to strong coupling will be described in § 7.2.3.

7.2.2. Two-dimensional separation in heteronuclear systems

Heteronuclear systems consisting of rare spins S such as carbon-13, nitrogen-15, phosphorus-31, etc., coupled to abundant spins I such as protons, offer numerous possibilities for 2D separation. Such systems can be described by the Hamiltonian

$$\mathcal{H} = \mathcal{H}_{ZS} + \mathcal{H}_{IS} + \mathcal{H}_{ZI} + \mathcal{H}_{II} \tag{7.2.7}$$

where $\mathcal{H}_{ZS}$ and $\mathcal{H}_{ZI}$ are Zeeman terms,

$$\mathcal{H}_{ZS} = \sum_k \Omega_k S_{kz},$$

$$\mathcal{H}_{ZI} = \sum_l \Omega_l I_{lz}, \tag{7.2.8}$$

and $\mathcal{H}_{IS}$ and $\mathcal{H}_{II}$ are scalar coupling terms

$$\mathcal{H}_{IS} = \sum_{kl} 2\pi J_{kl} S_{kz} I_{lz},$$

$$\mathcal{H}_{II} = \sum_{l<l'} 2\pi J_{ll'} \mathbf{I}_l \cdot \mathbf{I}_{l'}. \quad (7.2.9)$$

If the I spins are weakly coupled among each other, one obtains

$$\mathcal{H}_{II} = \sum_{l<l'} 2\pi J_{ll'} I_{lz} I_{l'z}, \quad (7.2.10)$$

and all terms of $\mathcal{H}$ commute.

In contrast to homonuclear couplings, the effects of heteronuclear couplings can be removed by broadband decoupling of the I spins, and it is possible to achieve direct separation of the chemical shifts of the S spins and the IS multiplet splittings.

7.2.2.1. Separation of S-spin multiplets and S-spin chemical shifts

Multiplets due to heteronuclear IS-couplings that are displaced by S-spin chemical shifts can be separated with the simple experiment of Fig. 7.2.6 (7.1). In the evolution period t_1, the full Hamiltonian of eqn (7.2.7) is allowed to act, while heteronuclear broadband decoupling suppresses the IS interactions in the detection period, leading to the effective Hamiltonian $\mathcal{H}^{(d)} = \mathcal{H}_{ZS}$ (the reader is referred to § 4.7.7 for the limitations of such a description). This leads to a 2D spectrum of the type shown in Fig. 7.2.7, where the chemical shifts cause a separation of the various multiplets in the ω_2-direction.

In some situations, it may be of advantage to decouple in the evolution period rather than in the detection period (Fig. 7.1.1), since it may be sufficient to achieve a rather low resolution in the shift dimension. In this case, a restricted number of t_1-values would be sufficient, while the multiplets can be recorded with maximum resolution in the ω_2-domain.

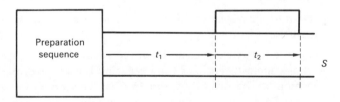

FIG. 7.2.6. Simple scheme for the separation of $\mathcal{H}_{ZS} + \mathcal{H}_{IS}$ (shifts and couplings in ω_1) vs. $\mathcal{H}_{ZS}$ (only chemical shifts in ω_2). The preparation may consist either of a single pulse with rotation angle β applied to the S nuclei (optimized as described in eqn (4.3.23)), or alternatively of a heteronuclear polarization transfer scheme to enhance the S spin polarization (see § 4.5).

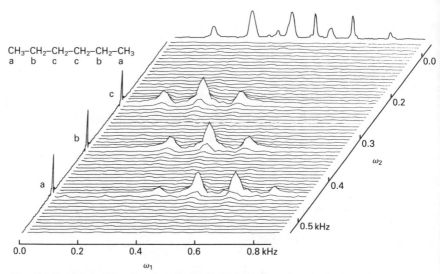

FIG. 7.2.7. Separation of carbon-13 chemical shifts (in ω_2) and shifts plus carbon-proton couplings (in ω_1) for n-hexane. The decoupled spectrum is shown on the left parallel to the ω_2-axis, and the conventional proton-coupled spectrum on top. This is one of the earliest published examples of 2D spectra (Reproduced from Ref. 7.1.)

Decoupling of the heteronuclear interaction in the evolution period can also be achieved by applying a π-pulse to the I spins at the midpoint of the t_1-interval, provided strong coupling effects can be disregarded.

The preparation sequence in Fig. 7.2.6 may consist of a simple r.f. pulse with rotation angle β, which may be optimized as a function of the spin–lattice relaxation rate and of the interval between subsequent scans, as described in § 4.3.1.5. Alternatively, the sensitivity may be enhanced by using polarization transfer methods such as cross-polarization, INEPT, or DEPT (§ 4.5). In contrast to schemes that involve refocusing pulses during evolution, the method of Fig. 7.2.6 is directly applicable to systems with strong $\mathcal{H}_{II}$ interactions without additional complications.

7.2.2.2. Removal of the chemical shifts from the ω_1-domain by shearing of the two-dimensional spectrum

The chemical shifts may be removed from the ω_1-domain of a spectrum obtained by the scheme in Fig. 7.2.6 by shearing the 2D data matrix in the computer memory according to eqn (6.6.2). This operation can be carried out even if the spectral width in ω_1 is deliberately chosen just wide enough to cover the range of heteronuclear couplings $\mathcal{H}_{IS}$, in such a manner that the signals displaced by $\mathcal{H}_{ZS}$ are allowed to fold repeatedly in the ω_1-domain. In this case, the shearing transformation is referred to as 'foldover correction' (7.13, 7.14). It should be noted that the number of required t_1-increments is determined only by the number of data points that is necessary to represent the multiplets with sufficient resolution in the ω_1-frequency domain.

7.2.2.3. *Separation of S-spin multiplets vs. S-spin chemical shifts by refocusing and gated decoupling*

Alternative schemes for separating S-spin multiplets and S-spin chemical shifts that do not require a shearing operation are shown in Fig. 7.2.8 (7.15–7.18). The common feature of these schemes is the removal of the S-spin chemical shifts from the ω_1-domain by a refocusing pulse. To avoid a simultaneous refocusing of the IS couplings, proton decoupling is applied in one-half of the evolution period, as shown in the schemes A and B of Fig. 7.2.8. These sequences lead to the effective Hamiltonians

$$\mathcal{H}_A^{(e)} = \tfrac{1}{2}(\mathcal{H}_{IS} + \mathcal{H}_{ZI} + \mathcal{H}_{II}),$$
$$\mathcal{H}_B^{(e)} = -\mathcal{H}_A^{(e)}. \tag{7.2.11}$$

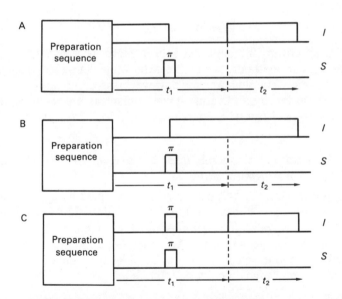

FIG. 7.2.8. Schemes for the separation of $\mathcal{H}_{IS}$ vs. $\mathcal{H}_{ZS}$. (A) A pulse applied at the midpoint of the evolution period refocuses the chemical shifts of the S spins, while decoupling in the defocusing period prevents refocusing of the IS coupling ('gated decoupler' method). (B) Related scheme, where decoupling is applied in the refocusing period. Both schemes A and B yield multiplets that agree with the conventional coupled spectrum regardless of the strength of the coupling. (C) Scheme with refocusing of the S spins and simultaneous inversion of the I spins ('proton flip' method). In strongly coupled systems, this sequence leads to symmetrical multiplets which contain more resonances than the conventional coupled spectrum.

The combination of both experiments makes it possible to obtain pure 2D absorption peakshapes in systems with strong coupling (7.19). In weakly coupled systems, the part of the evolution Hamiltonian relevant for the S spins is

$$\mathcal{H}_A^{(e)} = -\mathcal{H}_B^{(e)} = \tfrac{1}{2}\mathcal{H}_{IS}. \qquad (7.2.12)$$

These methods yield multiplets that are identical to those of the conventional coupled spectrum except for a scaling factor $\tfrac{1}{2}$ in the coupling constants (7.17, 7.18). The resolution can nevertheless exceed the resolution of conventional 1D spectra if $T_2 > 2T_2^*$ although diffusion through static field gradients makes it difficult to approach the natural line-width.

7.2.2.4. Separation of S-spin multiplets vs. S-spin chemical shifts by refocusing and inversion of I nuclei

In weakly coupled systems, the scheme of Fig. 7.2.8(C), which is often referred to as the 'proton flip' method (7.16, 7.17), leads to an effective Hamiltonian

$$\mathcal{H}_C^{(e)} = \mathcal{H}_{IS}. \qquad (7.2.13)$$

In the 2D spectrum, the multiplet splittings are twice as large as for the gated decoupler methods, while the achievable line-width remains the same. In strongly coupled systems however, it is not possible to describe the time evolution by an average Hamiltonian, as the π-pulses in effect induce coherence transfer phenomena (7.17, 7.20), as discussed in § 7.2.3.

7.2.2.5. Separation of S-spin multiplets due to coupling to a selected I-spin vs. S-spin chemical shifts

The method in Fig. 7.2.8(C) can be readily modified by applying a weak, selective π-pulse to a chosen proton with chemical shift Ω_l. In this case, the ω_1-domain carries information about the couplings J_{kl} between the chosen spin I_l and all spins S_k. All other heteronuclear couplings are refocused by the non-selective $(\pi)_S$-pulse. This is an efficient method for investigating long-range couplings (7.21).

7.2.2.6. Separation of long-range IS-couplings vs. S-spin chemical shifts

With the scheme of Fig. 7.2.9, the splittings due to strong one-bond IS couplings can be suppressed in order to focus attention on weak long-range couplings. This scheme can be derived from the sequence of Fig. 7.2.8(C) by replacing the pair of simultaneous π-pulses by a 'bilinear rotation' sandwich (7.22–7.25). For $\varphi = -x$, the net effect for pairs of

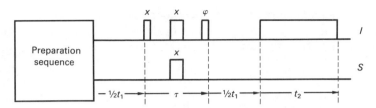

FIG. 7.2.9. Schemes employing bilinear rotation sandwiches (7.22–7.24) to obtain 2D spectra which in the ω_1-domain contain only long-range $\mathcal{H}_{IS}$ interactions (if $\varphi = -x$) or short-range couplings (if $\varphi = x$).

nuclei I_m and S_k with a small coupling $J_{km} \ll 1/\tau$ is exactly the same as in the sequence of Fig. 7.2.8(C), leading to splittings $2\pi J_{km}$ in the ω_1-domain. For pairs of nuclei I_l and S_k with large couplings that are matched to the condition $J_{kl} = 1/\tau$, however, the net effect of the bilinear rotation sandwich is to refocus both S-spin chemical shifts *and* heteronuclear couplings.

The mechanism of this sequence can be readily understood by considering the effective transformation under a bilinear rotation. We shall make use of the following general relations.

1. Two successive rotations by π about orthogonal axes are equivalent to a single π-rotation about an axis that is perpendicular to the first two axes of rotation:

$$\xrightarrow{\pi I_{l\lambda}} \xrightarrow{\pi I_{l\mu}} = \xrightarrow{\pi I_{l\nu}} \qquad (7.2.14)$$

where λ, μ, ν stand for an arbitrary permutation of x, y, z, and where the spin I_l may have an arbitrary spin quantum number. For spins $I_l = \frac{1}{2}$, such a relation holds not only for the unitary transformation but also for the individual operators, provided λ, μ, ν is a *cyclic* permutation of x, y, z

$$\exp\{-i\pi I_{l\lambda}\}\exp\{-i\pi I_{l\mu}\} = \exp\{-i\pi I_{l\nu}\}. \qquad (7.2.15)$$

2. The bilinear transformation $\exp\{-i\pi 2S_{k\varepsilon}I_{l\xi}\}\sigma\exp\{i\pi 2S_{k\varepsilon}I_{l\xi}\}$ can be decomposed into a product of two π-rotations applied to the two spins, provided $S_k = I_l = \frac{1}{2}$

$$\xrightarrow{\pi 2 S_{k\varepsilon} I_{l\xi}} = \xrightarrow{\pi S_{k\varepsilon}} \xrightarrow{\pi I_{l\xi}} \qquad (7.2.16)$$

where ε and ξ represent arbitrary combinations of x, y, and z. Note that this relation applies only to the unitary transformation and *not* to the individual operators, in contrast to eqn (7.2.15).

We assume that the n spins I_l are direct neighbours of the spin S_k, while the remaining spins I_m have only long-range couplings to S_k such that $J_{kl} \gg J_{km}, J_{lm}$. If $\varphi = -x$, the effect of the bilinear rotation sandwich can be described by the transformation (7.26)

$$\sigma(\tfrac{1}{2}t_1^-) \xrightarrow{\pi(S_{kx}+\Sigma I_{lx}+\Sigma I_{mx})} \xrightarrow{\Sigma \pi J_{kl}\tau 2S_{kz}I_{ly}} \xrightarrow{\Sigma \pi J_{km}\tau 2S_{kz}I_{my}}$$

$$\xrightarrow{\Sigma \pi J_{lm}\tau 2I_{ly}I_{my}} \xrightarrow{\Sigma \pi J_{ll'}\tau 2I_{ly}I_{l'y}} \xrightarrow{\Sigma \pi J_{mm'}\tau 2I_{my}I_{m'y}} \sigma(\tfrac{1}{2}t_1^+). \quad (7.2.17)$$

The delay τ is chosen such that $\pi J_{kl}\tau = \pi$, i.e. $\tau = (J_{kl})^{-1}$, while $\pi J_{km}\tau \simeq 0$ and $\pi J_{lm}\tau \simeq 0$. Under these assumptions, the transformation can be simplified using the above relations

$$\sigma(\tfrac{1}{2}t_1^-) \xrightarrow{\pi S_{kv}} \xrightarrow{\pi \Sigma I_{lz}} \xrightarrow{\pi \Sigma I_{mx}} \sigma(\tfrac{1}{2}t_1^+) \quad (7.2.18)$$

where $v = x$ or y for even and odd numbers n_l of direct neighbour spins, respectively (e.g. for CH_2 and CH_3 systems).

If the transformation of eqn (7.2.18) is applied to the Hamiltonian of the weakly coupled system, one obtains immediately

$$\mathcal{H}_k + \mathcal{H}_l + \mathcal{H}_m + \mathcal{H}_{kl} + \mathcal{H}_{km} + \mathcal{H}_{lm} + \mathcal{H}_{ll'} + \mathcal{H}_{mm'} \xrightarrow{\pi S_{kv}} \xrightarrow{\pi \Sigma I_{lz}} \xrightarrow{\pi \Sigma I_{mx}}$$

$$-\mathcal{H}_k + \mathcal{H}_l - \mathcal{H}_m - \mathcal{H}_{kl} + \mathcal{H}_{km} - \mathcal{H}_{lm} + \mathcal{H}_{ll'} + \mathcal{H}_{mm'} \quad (7.2.19)$$

leading to the average Hamiltonian for the full t_1-period

$$\bar{\mathcal{H}} = \mathcal{H}_l + \mathcal{H}_{km} + \mathcal{H}_{ll'} + \mathcal{H}_{mm'}. \quad (7.2.20)$$

In other words, the bilinear rotation leads to refocusing of the chemical shift term $\mathcal{H}_k$ and of the one-bond coupling term $\mathcal{H}_{kl}$ while the long-range coupling term $\mathcal{H}_{km}$ and some I spin coupling terms remain unaffected. Thus only the long-range couplings contribute to the modulation of the signal at the end of the evolution period t_1 (Fig. 7.2.10(d)).

If the τ-delay is not properly matched to the inverse of the dominant coupling, artefacts will appear, which can be removed with a 'compensated bilinear rotation sequence' (7.22) built on similar principles as composite pulses designed for cancelling errors in the r.f. rotation angle (§ 4.2.7). Such procedures are limited by the fact that the pulse sandwich must remain short on the time scale of the homonuclear and long-range couplings. Bilinear rotations will also fail in strongly coupled systems.

7.2.2.7. Separation of one-bond IS-couplings vs. S-spin chemical shifts

If the scheme in Fig. 7.2.9 is used with $\varphi = +x$ (i.e. if the phase of the last I-pulse is reversed with respect to the sequence used in the previous paragraph), the resulting transformation can again be simplified with

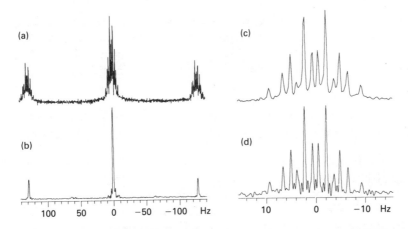

FIG. 7.2.10. (a) Conventional proton-coupled carbon-13 multiplet of $ClCH_2{}^{13}CH_2CH_3$. (b) Simplified triplet that appears in the ω_1-domain of a 2D spin–echo spectrum obtained with the sequence of Fig. 7.2.9 with $\varphi = x$, in such a way that the echo is only modulated by the one-bond couplings to the neighbouring protons. (c) Expansion of the central multiplet of the conventional spectrum in (a). (d) Multiplet in the ω_1-domain of a 2D spectrum obtained with the sequence of Fig. 7.2.9 with $\varphi = -x$, where the carbon-13 echoes are modulated only by the couplings to remote protons. (Reproduced from Ref. 7.26.)

eqns (7.2.14) and (7.2.16)

$$\sigma(\tfrac{1}{2}t_1^-) \xrightarrow{\pi S_{k\nu}} \xrightarrow{\pi \Sigma I_{ly}} \sigma(\tfrac{1}{2}t_1^+) \qquad (7.2.21)$$

with $\nu = x$ or y for even and odd numbers n of direct neighbour spins I_l. As a result, the average Hamiltonian takes the form

$$\bar{\mathcal{H}} = \mathcal{H}_m + \mathcal{H}_{kl} + \mathcal{H}_{mm'}. \qquad (7.2.22)$$

Hence the t_1-evolution is determined only by the one-bond couplings J_{kl} and not by the long-range couplings J_{km}. The 'remote' nuclei I_m in effect experience a pulse with a rotation angle $\beta = 2\pi$. In the ω_1-domain of the 2D spectrum, the multiplet structure due to long-range IS interactions is removed, which is important in large molecules where a manifold of ill-resolved couplings often results in severe line-broadening. By eliminating these couplings, it is possible to measure short-range couplings very accurately and with greatly improved sensitivity. The efficiency of this approach may be appreciated in Fig. 7.2.10.

7.2.2.8. Artefacts

Heteronuclear 2D separated spectra may suffer from imperfect pulses. Artefacts arising from S-spin refocusing pulses that deviate from $\beta = \pi$ can be cancelled by phase cycling (7.11). Additional signals arising from imperfect I-inversion pulses are more pernicious (7.27) but can be largely suppressed by using composite pulses to mimic ideal conditions (7.28). Under special circumstances, further artefacts may arise which are related to the so-called illusions of decoupling (§ 4.7.7).

7.2.2.9. Pure two-dimensional absorption

In contrast to homonuclear spin–echo spectra, heteronuclear separation methods lend themselves to modifications that yield pure 2D absorption lineshapes.

In weakly-coupled systems, the multiplets of the S-spins are necessarily symmetrical about $\omega_1 = 0$. The procedure described in § 6.5.3.1a is therefore applicable to schemes where the chemical shifts are eliminated from the ω_1-domain: it is sufficient to calculate a real (cosine) Fourier transformation with respect to t_1 to obtain pure absorption.

In strongly-coupled systems, the symmetry of the ω_1-domain is broken (see § 7.2.3), and it is necessary to perform two complementary experiments. The time-reversal technique (§ 6.5.3.2) can be implemented in two distinct ways. An elegant approach simply uses two versions of the gated decoupler experiment (A and B in Fig. 7.2.8) with addition of the spectrum $S_A(\omega_1, \omega_2)$ with the reversed spectrum $S_B(-\omega_1, \omega_2)$ to eliminate dispersion components (7.19). Alternatively, two complementary experiments can be carried out, the first spectrum $S_A(\omega_1, \omega_2)$ being obtained with any of the schemes in Figs. 7.2.6, 7.2.8, or 7.2.9. The second spectrum $S_B(\omega_1, \omega_2)$ is obtained with the *same* sequence except that a π-pulse is inserted at the beginning of the detection period (7.29). This procedure leads to a reversal of the apparent sense of precession in the evolution period. Again, pure absorption mode is obtained by adding the spectra $S_A(\omega_1, \omega_2) + S_B(-\omega_1, \omega_2)$. Finally, it is also possible to obtain pure phase lineshapes by inserting a $\pi/2$-pulse at the beginning of the detection period in each experiment and by cycling its phase by 90° between successive experiments (7.29). This approach is more symmetrical than the reversed precession method and less susceptible to artefacts, but the sensitivity is reduced by a factor $\sqrt{2}$.

7.2.3. Strong coupling effects in refocusing experiments

In strongly coupled systems, it is not possible to eliminate chemical shifts by refocusing, nor is it possible to describe the evolution in a $[t_1/2 - \pi - t_1/2]$ period in terms of an effective Hamiltonian (see § 3.3.2). The π-pulse leads to coherence transfer between various transitions, and a multitude of new effective precession frequencies may appear in the ω_1-domain. These new frequencies correspond to the mean between two resonance frequencies of the natural spectrum.

7.2 CHEMICAL SHIFTS AND SCALAR COUPLINGS

The effect of a π-pulse on a strongly coupled spin system can be understood as follows. In terms of product functions such as $|\phi_p\rangle = |\alpha\beta\beta\rangle$, the inversion operator R_{π_x} is in essence a permutation operator[1] which permutes pairs of product functions $|\phi_p\rangle$ and $|\phi_{p'}\rangle$

$$|\phi_p\rangle = |\alpha\beta\beta\rangle \xrightarrow{\pi_x} -i|\phi_{p'}\rangle = -i|\beta\alpha\alpha\rangle \qquad (7.2.23)$$

In weakly coupled systems, the product functions are eigenfunctions of the Hamiltonian, and the inversion operator R_{π_x} also leads to a pair-wise permutation of coherences, e.g.

$$|\phi_p\rangle\langle\phi_q| = |\alpha\beta\beta\rangle\langle\alpha\beta\alpha| \xrightarrow{\pi_x} -|\beta\alpha\alpha\rangle\langle\beta\alpha\beta| = -|\phi_{p'}\rangle\langle\phi_{q'}|. \qquad (7.2.24)$$

In strongly-coupled systems, on the other hand, the eigenfunctions $|\psi_r\rangle$ are linear combinations of product functions $|\phi_p\rangle$

$$|\psi_r\rangle = \sum_p U_{pr}|\phi_p\rangle \qquad (7.2.25)$$

(see eqn (2.1.139)), and permutation normally does not reproduce eigenfunctions. In other words,

$$R_\pi|\psi_r\rangle = \sum_p U_{pr}R_\pi|\phi_p\rangle \qquad (7.2.26)$$

is no longer an eigenfunction but represents a linear combination of eigenfunctions. A π-pulse therefore leads to coherence transfer which can be expressed in the form

$$|t\rangle\langle u| \xrightarrow{R_\pi} \sum_{r,s} R_{rs\,tu}|r\rangle\langle s| \qquad (7.2.27)$$

with the transfer coefficients

$$R_{rs\,tu} = (R_\pi^{(e)})_{rt}(R_\pi^{(e)})_{su}^*. \qquad (7.2.28)$$

The elements of the rotation operators must be evaluated in the eigenbasis, i.e.

$$\mathbf{R}_\pi^{(r)} = \mathbf{U}^{-1}\mathbf{R}_\pi^{(p)}\mathbf{U} \qquad (7.2.29)$$

where $\mathbf{R}_\pi^{(p)}$ is the matrix of pair-wise permutations in the product basis, and the matrix $\mathbf{U}$ expresses the transformation from the product basis $\{|\phi_p\rangle\}$ to the eigenbasis $\{|\psi_r\rangle\}$, eqn (7.2.25).

Consider a strongly coupled homonuclear system with two spins k and l ('AB' system). The transformation matrix $\mathbf{U}$, given in eqn (2.1.143), only

---

[1] Note that $|\alpha\rangle \xrightarrow{\pi_x} i|\beta\rangle$, $|\beta\rangle \xrightarrow{\pi_x} i|\alpha\rangle$, $|\alpha\rangle \xrightarrow{\pi_y} |\beta\rangle$, and $|\beta\rangle \xrightarrow{\pi_y} -|\alpha\rangle$. It is therefore important to take account of the number of spins in the product functions and to distinguish between x- and y-pulses.

mixes the functions $\phi_2 = |\alpha\beta\rangle$ and $\phi_3 = |\beta\alpha\rangle$, leading to the following transformations under a π pulse

$$|1\rangle\langle 2| \xrightarrow{\pi_x} |4\rangle\langle 3| \cos 2\theta + |4\rangle\langle 2| \sin 2\theta,$$

$$|3\rangle\langle 4| \xrightarrow{\pi_x} |2\rangle\langle 1| \cos 2\theta - |3\rangle\langle 1| \sin 2\theta,$$

$$|1\rangle\langle 3| \xrightarrow{\pi_x} |4\rangle\langle 2| \cos 2\theta - |4\rangle\langle 3| \sin 2\theta,$$

$$|2\rangle\langle 4| \xrightarrow{\pi_x} |3\rangle\langle 1| \cos 2\theta + |2\rangle\langle 1| \sin 2\theta, \qquad (7.2.30)$$

where $\tan 2\theta = 2\pi J_{kl}/(\Omega_k - \Omega_l)$.

Thus the coherence is not simply transferred into parallel transitions as in the weak coupling limit ($\theta = 0$), but also into progressively connected transitions. The form of these equations immediately shows that a 2D correlation spectrum of an AB system obtained with the sequence $(\pi/2) - t_1 - (\pi) - t_2$ (see § 8.2) features eight signals as opposed to four in the weakly coupled limit, as shown schematically in Fig. 7.2.11(a). The signal amplitudes are given by the factors $Z_{rs\,tu}$ defined in eqn (6.2.11).

By delaying the acquisition, i.e. with the sequence $(\pi/2) - t_1/2 - (\pi) - t_1/2 - t_2$, one obtains a 2D J-spectrum. The signals are shifted as described by eqn (6.6.5), to yield the customary representation of spin–echo spectra shown in Fig. 7.2.11(b). In the weak-coupling limit ($\theta = 0$), this spectrum takes the simple form shown in Fig. 7.2.1(b). The additional signals associated with strong coupling appear at ω_1-frequencies that correspond to chemical shift differences, as in the delayed acquisition form of correlation spectroscopy ('SECSY', see § 8.3.1). If the 2D data matrix is subjected to a shearing transformation as shown in Fig. 7.2.11(c), the projection features a signal half-way between the two chemical shifts that is characteristic of strongly coupled systems. Since projections are normally calculated from absolute-value spectra, all four signals appearing at $\omega_2 = \frac{1}{2}(\Omega_k + \Omega_l)$ are added together regardless of their alternating signs. Although the occurrence of additional signals might at first sight appear detrimental, it has turned out in practice that the assignment is in fact facilitated, since the additional signals provide direct evidence of the existence of coupling partners.

If the complexity of the strongly coupled spin system is increased from AB to ABX, additional features appear, particularly in the X sub-spectrum (7.30–7.34). These can be encountered in both homo- and heteronuclear systems (in the latter case, A and B are usually protons and X a carbon-13 nucleus).

7.2 CHEMICAL SHIFTS AND SCALAR COUPLINGS

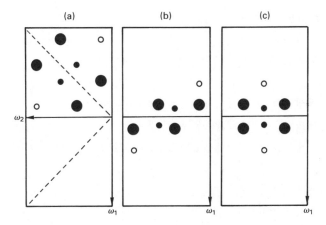

FIG. 7.2.11. Strong coupling effects in 2D correlation and 2D J-spectra. (a) 2D correlation spectrum of a strongly-coupled AB system, obtained with the sequence $(\pi/2) - t_1 - (\pi) - t_2$ (COSY with $\beta = \pi$, as described in § 8.2). Large filled symbols represent amplitudes $Z = (1 - 2\sin^2\theta)\cos 2\theta$; small filled symbols correspond to $Z = (1 + \sin 2\theta)\sin 2\theta$, and small open symbols represent negative signals with $Z = -(1 - \sin 2\theta)\sin 2\theta$. The 2D spectrum is drawn for $2\pi J/(\Omega_k - \Omega_l) = 0.75$. (b) 2D J-spectrum of the same AB system obtained with the sequence $(\pi/2) - t_1/2 - (\pi) - t_1/2 - t_2$, equivalent to the COSY sequence used in (a) combined with delayed acquisition ('SECSY', see § 8.3.1). The signals are shifted in the ω_1-domain according to eqn (6.6.5), but the amplitudes are the same as in (a). In both cases (a) and (b), all signals have mixed phases described by eqn (6.5.10). (c) Sheared 2D spectrum obtained by rearranging the data matrix in computer memory. In the projection on to the ω_2-axis, a signal that is characteristic of strong coupling appears half-way between the apparent chemical shifts of the coupling partners.

In the limit of weak coupling, the eigenstates of the ABX system are numbered in agreement with Fig. 4.4.2(b). It is well known that the two possible polarizations of the X spin lead to two subsets of spin functions for the AB part which mix under strong coupling

$$(\psi_3, \psi_4) = (\phi_3, \phi_4) \begin{pmatrix} c^+ & -s^+ \\ s^+ & c^+ \end{pmatrix},$$

$$(\psi_5, \psi_6) = (\phi_5, \phi_6) \begin{pmatrix} c^- & -s^- \\ s^- & c^- \end{pmatrix} \quad (7.2.3)$$

where $c^{\pm} = \cos\theta^{\pm}$, $s^{\pm} = \sin\theta^{\pm}$, $\tan 2\theta^{\pm} = 2\pi J_{kl}/D^{\pm}$, and $D^{\pm} = \Omega_k - \Omega_l \pm \pi(J_{km} - J_{lm})$, the spins A, B, and X being denoted by I_k, I_l, and S_m, respectively. The product states ϕ_1, ϕ_2, ϕ_7, and ϕ_8 remain invariant.

If we express the six coherence components of the X spin in terms of shift operators $I^{\pm}$ and polarization operators I^{α}, I^{β}, defined in eqn (2.1.114), we find six single-element operators $|r\rangle\langle s|$ that correspond to the six transitions in the X region.

$$|1\rangle\langle 2| = I_k^\alpha I_l^\alpha S_m^+,$$
$$|7\rangle\langle 8| = I_k^\beta I_l^\beta S_m^+,$$
$$|3\rangle\langle 5| = [c^+c^- I_k^\alpha I_l^\beta + s^+s^- I_k^\beta I_l^\alpha + c^+s^- I_k^+ I_l^- + s^+c^- I_k^- I_l^+]S_m^+,$$
$$|4\rangle\langle 6| = [s^+s^- I_k^\alpha I_l^\beta + c^+c^- I_k^\beta I_l^\alpha - s^+c^- I_k^+ I_l^- - c^+s^- I_k^- I_l^+]S_m^+,$$
$$|3\rangle\langle 6| = [-c^+s^- I_k^\alpha I_l^\beta + s^+c^- I_k^\beta I_l^\alpha + c^+c^- I_k^+ I_l^- - s^+s^- I_k^- I_l^+]S_m^+,$$
$$|4\rangle\langle 5| = [-s^+c^- I_k^\alpha I_l^\beta + c^+s^- I_k^\beta I_l^\alpha - s^+s^- I_k^+ I_l^- + c^+c^- I_k^- I_l^+]S_m^+.$$
(7.2.32)

The effect of a non-selective π-pulse amounts to interchanging the indices $(+, -)$ and (α, β) in all operators and, for a π_y pulse, to changing the sign:

$$I^{\alpha,\beta} \xrightarrow{\pi_x} I^{\beta,\alpha}, \qquad I^{\alpha,\beta} \xrightarrow{\pi_y} -I^{\beta,\alpha},$$
$$I^\pm \xrightarrow{\pi_x} I^\mp, \qquad I^\pm \xrightarrow{\pi_y} -I^\mp. \qquad (7.2.33)$$

In the case

$$|1\rangle\langle 2| = I_k^\alpha I_l^\alpha S_m^+ \xrightarrow{\pi_x} I_k^\beta I_l^\beta S_m^- = |8\rangle\langle 7|, \qquad (7.2.34a)$$

the coherence is entirely transferred from one transition to another. In contrast, one obtains

$$|3\rangle\langle 5| \xrightarrow{\pi_x} |6\rangle\langle 4| \cos^2(\theta^+ + \theta^-)$$
$$+ |5\rangle\langle 3| \sin^2(\theta^+ + \theta^-)$$
$$+ \{|5\rangle\langle 4| + |6\rangle\langle 3|\} \cos(\theta^+ + \theta^-)\sin(\theta^+ + \theta^-). \qquad (7.2.34b)$$

In this case, the coherence $|3\rangle\langle 5|$ is transformed into a superposition of four coherences, leading to additional peaks in the 2D spin–echo spectrum.

The schematic spectrum in Fig. 7.2.12 shows a case where one of the AB subspectra is nearly degenerate. Since the couplings J_{AX} and J_{BX} are not refocused, the AB subspectra appear displaced away from the $\omega_1 = 0$ axis. In the X region one observes, in addition to the four peaks expected in a weakly coupled three-spin system, 14 peaks that arise from strong coupling, four of which have negative amplitudes. Decoupling in the detection period leads to a collapse of all ω_2-frequencies on to $\omega_2 = \Omega_m$; in this case the number of additional signals due to strong coupling is reduced to seven.

7.2 CHEMICAL SHIFTS AND SCALAR COUPLINGS

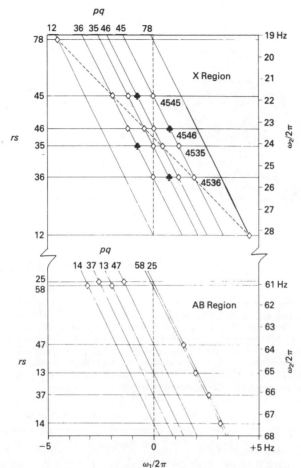

FIG. 7.2.12. Schematic 2D spin–echo spectrum of a homonuclear ABX system. Signals with amplitudes $Z_{rs\,pq}$ are indicated by the intersection of sloping lines with indices pq and horizontal lines with indices rs. Diamonds and clubs represent positive and negative signals respectively. The conventional numbering of the eigenstates is given in Fig. 4.4.2(b) and analytical expressions for the intensities can be found in Ref. 7.30. Because one of the AB subspectra is almost degenerate, the transitions $|2\rangle \leftrightarrow |6\rangle$ and $|6\rangle \leftrightarrow |8\rangle$ have almost vanishing intensity. This is reflected in correspondingly low intensities for eight responses in the AB region of the 2D spectrum, which have been omitted in this figure. Note that the definition of the frequency axes is at variance with the convention used in other figures. (Reproduced from Ref. 7.30.)

In the experimental example of a 2D J-spectrum presented in Fig. 7.2.13, the characteristic features associated with strong coupling are clearly visible in the centre of the AB region of the ABX spectrum. These features also appear in the projection (Fig. 7.2.13(b)). A band of weak peaks half-way between the AB and X regions indicates that the weak coupling approximation is not truly applicable to the X nucleus.

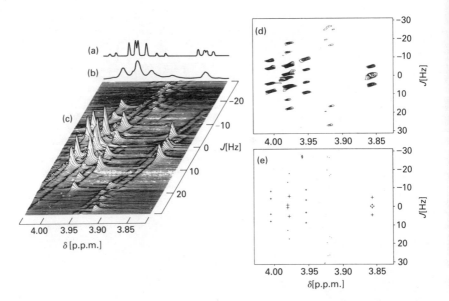

FIG. 7.2.13. Two-dimensional absolute value J-spectra of the strongly-coupled three-spin system of L-serine ($ND_2CHCOODCH_2OD$), obtained with an 0.1 M solution in D_2O at 360 MHz, $pD = 6.5$, $T = 25°C$. (a) Normal 1D spectrum; (b) projection of the sheared 2D spectrum on to the chemical shift axis; (c) stacked plot and (d) contour plot of the sheared spectrum; (e) simulation of the sheared spectrum calculated with the 'Son of Laocoon' program (7.30). Note the weak peaks that appear half-way between the chemical shifts (compared with Fig. 7.2.11(c)). The size of the symbols in (e) indicates the amplitude of the signals. (Reproduced from Ref. 7.31.)

The simulation of spectra of this complexity requires numerical procedures. Examples calculated by the 'Son of Laocoon' program (7.30) are given in Figs. 7.2.13(e) and 7.2.14. In Fig. 7.2.14, experimental and computed spectra of a strongly coupled ABC system are compared.

In 2D J-spectra of heteronuclear systems, additional cross-peaks due to strong coupling occur only for some types of pulse sequences. In the simple scheme of Fig. 7.2.6 and in both gated decoupler methods in Fig. 7.2.8(A) and (B), the ω_1-domain reflects the natural spectrum of the dilute S spins. However, in experiments involving simultaneous π-pulses applied to both I and S spins, strong coupling leads to complications.

In heteronuclear systems with three or more strongly coupled I spins, the natural spectrum of the S nucleus is in general asymmetric with respect to the chemical shift Ω_S. It can be shown (7.20) that the 2D spin–echo spectrum obtained with simultaneous π-pulses applied to both I and S nuclei (Fig. 7.2.8(C)) and with decoupling in the detection period invariably leads to multiplets that are symmetrical about $\omega_1 = 0$.

Figure 7.2.15(a) shows a section through a 2D J-spectrum corresponding to the C_2-resonance of pyridine (a system with five strongly coupled

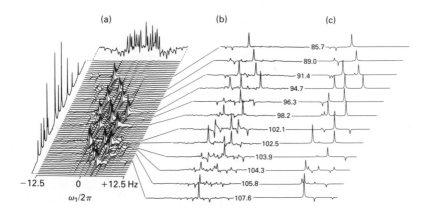

FIG. 7.2.14. Two-dimensional J-spectrum of the strongly coupled ABC system of 2-chlorothiophene. (a) Phase-sensitive experimental spectrum obtained at 80 MHz, with the phase-sensitive projection on to the ω_1-axis (top) and the conventional spectrum (left). (b) Experimental sections taken from the phase-sensitive 2D spectrum. (c) Simulations of phase-sensitive sections. The dispersion peaks in (b) should be disregarded in comparing (b) and (c), since they result from interference of resonances centred on neighbouring sections. All signals have mixed phases described by eqn (6.5.10). (Reproduced from Ref. 7.30.)

protons). The simulated spectra shown in Fig. 7.2.15(b), (c) indicate the presence of a large number of weak lines, some of which are negative. All lines occur at differences of resonance frequencies in the natural proton-coupled carbon spectrum (7.20).

Since the spectra obtained in this manner fulfil the symmetry conditions outlined in § 6.5.3.1, pure 2D absorption signals can be obtained by calculating a real (cosine) Fourier transformation with respect to t_1.

7.2.4. Echo modulation by non-resonant nuclei

In heteronuclear systems with strong coupling between the abundant I-spins, the spin–echoes of the S-spin are modulated even if the π-pulse is applied only to the dilute S-spin, i.e. with the sequence of Fig. 7.2.8(C) without the $(\pi)^I$-pulse. This effect, known as 'echo modulation by non-resonant spins' (7.35–7.37, 7.20), can in principle be used to obtain information on the I-spin spectrum. However, except in favourable cases (7.35), the amplitude of the modulation tends to be small. Figure 7.2.16 shows sections taken from 2D spin–echo spectra of the C_2 resonance in pyridine, where a large number of weak lines associated with strong coupling between the protons leads to a pair of broad satellites. A theoretical treatment of this effect is given in § 3.3.2.

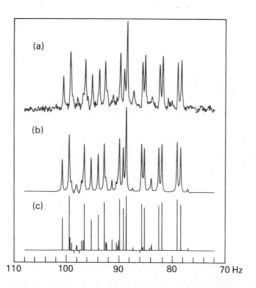

FIG. 7.2.15. Two-dimensional spin–echo spectra in a heteronuclear system with strong coupling: (a) Section parallel to the ω_1-axis taken from a phase-sensitive 2D spin–echo spectrum of carbon C_2 in pyridine (ABCDEX system), obtained with the sequence in Fig. 7.2.8(C) with simultaneous π-pulses applied to both I and S nuclei. Only one-half of the symmetrical spectrum is shown. (b) Simulation with a line-width of 0.18 Hz. (c) Corresponding stick spectrum, which reveals a number of weak lines, some of which have negative amplitudes. (Reproduced from Ref. 7.20.)

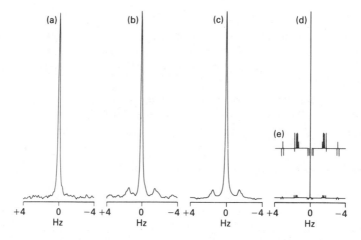

FIG. 7.2.16. Echo modulation by strongly coupled non-resonant spins. All spectra represent sections taken parallel to the ω_1-axis from phase-sensitive 2D spin–echo spectra of the C_2 resonance in pyridine (ABCDEX system). (a) For comparison, a π refocusing pulse was applied to the S spins with proton decoupling throughout the experiment. Note the absence of sidebands. (b) With S-spin refocusing, but without proton irradiation in the evolution period (scheme of Fig. 7.2.8(C), but without $(\pi)^I$-pulse). Coherence transfer induced by the $(\pi)^S$-pulse in the system with strong I–I coupling leads to satellites on either side of the dominant unmodulated peak. (c) Simulation corresponding to (b). (d) Simulated stick spectrum. (e) Inset with fivefold amplification. (Reproduced from Ref. 7.20.)

7.3. Separation of chemical shifts and dipolar couplings in oriented phase

In solids and liquid crystalline phases, 2D spectroscopy may be used to separate the (anisotropic) chemical shifts and dipolar couplings in analogy to 2D separation in liquids. The resulting spectra with dipolar multiplets separated by chemical shifts are known as 'separated local field spectra' (7.38–7.40) and can be used either to obtain structural information or to measure the relative orientation of chemical shielding and dipolar interaction tensors.

7.3.1. Homonuclear separated local field spectra

The Hamiltonian governing an ensemble of spins with $I=\frac{1}{2}$ in a solid or in an anisotropic solution is described by eqns (2.2.2) and (2.2.18)

$$\mathcal{H} = \mathcal{H}_{ZI} + \mathcal{H}_{II} \tag{7.3.1a}$$

with

$$\mathcal{H}_{ZI} = \sum_k \Omega_k I_{kz}, \tag{7.3.1b}$$

where $\Omega_k(\theta, \phi)$ is orientation-dependent (see eqn (2.2.2)), and

$$\mathcal{H}_{II} = \sum_{k<l} b_{kl}(1 - 3\cos^2\theta_{kl})\{I_{kz}I_{lz} - \tfrac{1}{2}(I_{kx}I_{lx} + I_{ky}I_{ly})\} \tag{7.3.1c}$$

with

$$b_{kl} = \frac{\mu_0 \gamma_k \gamma_l \hbar}{4\pi r_{kl}^3}.$$

In contrast to scalar coupled systems, it is possible to design techniques which suppress the dipolar interaction by utilizing the transformation properties of second-rank tensors. The most important tools for manipulating dipolar interactions are multiple-pulse sequences of which numerous designs have been proposed (§ 3.2.3). At this point, it is sufficient to note that the dipolar interaction can be eliminated to an arbitrary degree by suitable pulse sequences, such as WHH-4, MREV-8, and BR-24, (7.41, 7.42), leading to the average Hamiltonian

$$\bar{\mathcal{H}} = \sum_k \kappa \Omega_k I_{kz} \tag{7.3.2}$$

where κ is a scaling factor that depends on the pulse sequence.

In the scheme shown in Fig. 7.3.1(a), the full Hamiltonian without r.f. irradiation is allowed to act in the evolution period, while multiple-pulse

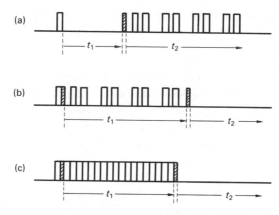

FIG. 7.3.1. Three schemes for obtaining separated local field spectra in solids with homonuclear dipolar couplings: (a) The natural spectrum (anisotropic shifts and dipole couplings) appears in the ω_1-domain, while the scaled chemical shifts determine the ω_2-resonances. (b) Reversed situation. (c) Same as (b), but with a windowless multiple-pulse sequence for dipolar decoupling in the evolution period.

dipolar decoupling is applied in the detection period (7.43). The sampling process must be synchronized with the cyclic pulse sequence. To avoid saturation of the receiver, it is necessary to apply a sequence with windows during which the signal may be observed. The alternative scheme of Fig. 7.3.1(b) employs dipolar decoupling in the evolution period; in this case, windowless sequences (7.44) can be employed, as shown in Fig. 7.3.1(c).

Since the effective axis of precession under multiple-pulse conditions is usually tilted with respect to the z-axis, 'tilt pulses' (7.45, 7.46) should be inserted to maximize the signal amplitude by setting the initial magnetization perpendicular to this axis (shaded pulses in Fig. 7.3.1).

The scheme of Fig. 7.3.1(a) has been applied to a single crystal of malonic acid (7.43). Figure 7.3.2 shows sections taken parallel to the ω_1-axis at four distinct chemical shifts, together with computer simulations. These sections show the dipolar splittings of nuclei with the chemical shifts $\omega_2 = \kappa \Omega_k$. The analysis of the 2D spectrum allows one to assign the resonances to specific lattice sites.

7.3.2. Heteronuclear separated local field spectra

In heteronuclear systems containing dilute S-spins (such as carbon-13, phosphorus-31, etc.) and abundant I spins (protons or fluorine-19), the natural spectrum of the S spins is determined by the Hamiltonian

$$\mathcal{H} = \mathcal{H}_{ZS} + \mathcal{H}_{IS} + \mathcal{H}_{ZI} + \mathcal{H}_{II}. \tag{7.3.3a}$$

7.3 CHEMICAL SHIFTS AND DIPOLAR COUPLINGS

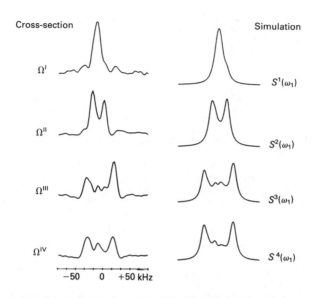

FIG. 7.3.2. Separated local-field spectra of the protons in a single crystal of malonic acid ($CH_2(COOH)_2$). Left: sections parallel to the ω_1-axis taken from an experimental spectrum obtained with the sequence in Fig. 7.3.1(a), at the ω_2-frequencies of the four resolved chemical shifts, which correspond to four magnetically inequivalent proton sites in the crystal. Right: simulated sections obtained by including dipolar interactions between each of the four inequivalent protons and three neighbouring protons in the lattice. The agreement allows one to conclude that the signals with $\omega_2 = \Omega^I$ and Ω^{II} belong to the carboxylic protons, while Ω^{III} and Ω^{IV} are associated with the methylene protons. (Reproduced from Ref. 7.43.)

where

$$\mathcal{H}_{ZS} = \sum_l \Omega_l S_{lz}, \tag{7.3.3b}$$

$$\mathcal{H}_{IS} = \sum_{k,l} b_{kl}(1 - 3\cos^2\theta_{kl}) I_{kz} S_{lz}, \tag{7.3.3c}$$

and b_{kl}, $\mathcal{H}_{ZI}$, and $\mathcal{H}_{II}$ are given in eqns (7.3.1b) and (7.3.1c). Heteronuclear separated local-field spectra allow one to measure the heteronuclear dipolar couplings.

The separation can be achieved with the scheme of Fig. 7.3.3. The multiple-pulse sequence applied to the I-spins in the t_1-period eliminates $\mathcal{H}_{II}$, and leads to a scaling of $\mathcal{H}_{IS}$ by a factor κ characteristic of the particular pulse sequence. Heteronuclear decoupling during detection reduces the Hamiltonian to the S-spin Zeeman interaction. The Hamiltonians that are relevant for the S-spin precession are therefore

$$\mathcal{H}^{(e)} = \mathcal{H}_{ZS} + \kappa \mathcal{H}_{IS}$$

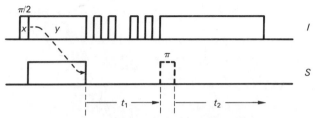

FIG. 7.3.3. Simple scheme for separated local-field spectroscopy in heteronuclear systems, leading to a separation of $\mathcal{H}_{ZS} + \kappa\mathcal{H}_{IS}$ vs. $\mathcal{H}_{ZS}$. The scaling factor κ is determined by the type of multiple-pulse sequence used in the evolution period to suppress the $\mathcal{H}_{II}$ interactions. The S magnetization is prepared by cross-polarization, and observed as usual in the presence of on-resonance I-decoupling in the t_2-period. The $(\pi)^S$-pulse indicated by dashed lines may be inserted on alternate scans to reverse the apparent sense of precession in the t_1-period, in order to obtain pure 2D absorption lineshapes as described in §§ 6.5.3.2 and 7.2.2.9.

and

$$\mathcal{H}^{(d)} = \mathcal{H}_{ZS}. \tag{7.3.4}$$

The effect of the Zeeman term $\mathcal{H}_{ZS}$ on the ω_1-domain can be removed either by refocusing the precession due to the chemical shifts with a $(\pi)^S$-pulse (7.40, 7.47), or simply by 'shearing' the 2D data matrix in complete analogy to the procedure described in § 7.2.2.2.

Pure 2D absorption lineshapes can be obtained by inserting a $(\pi)^S$-pulse just before the detection period to reverse the apparent sense of precession in alternate scans, as described in §§ 6.5.3.2 and 7.2.2.9.

It should be pointed out that multiple-pulse dipolar decoupling is less critical in these experiments than in the homonuclear experiments of Fig. 7.3.1. In the S-spin spectrum, the homonuclear couplings $\mathcal{H}_{II}$ produce only a second-order effect. It is therefore often sufficient to use continuous off-resonance I-spin decoupling in the evolution period with an effective field tilted with respect to the z-axis by the magic angle 54.7°, which eliminates $\mathcal{H}_{II}$ and scales $\mathcal{H}_{IS}$ by a factor $\kappa = 3^{-\frac{1}{2}}$ (7.48, 7.49).

In many cases it is possible to leave out I-spin decoupling in the evolution period altogether. This simplified experiment is applicable to systems where proton–proton distances are large, or where $\mathcal{H}_{II}$ is reduced by motional averaging. In the case of freely rotating methyl groups, for example, it can be shown (7.39) that

$$[\mathcal{H}_{II}^{\text{intra}}, \mathcal{H}_{ZS} + \mathcal{H}_{IS}] = 0; \tag{7.3.5}$$

hence the S-spectrum is not affected by the intramethyl group proton–proton coupling Hamiltonian $\mathcal{H}_{II}^{\text{intra}}$ and yields a quartet with binomial intensities. In a rigid CH_2 group however, the diagonalization of the full Hamiltonian (i.e. in the absence of multiple-pulse decoupling) shows that there are five allowed S-spin transitions (7.39). An example of a separated local-field spectrum of a single crystal of ammonium hydrogen malonate is shown in Fig. 7.3.4. Note that the CH_2 group does not give rise to a doublet of doublets, but to a multiplet with five lines, the central line being very weak in this crystal orientation.

7.3.3. Correlation of chemical shielding and dipolar coupling tensors in static powders

Two-dimensional separated local-field spectra of microcrystalline powders present an attractive alternative to single-crystal studies, since the need for large single crystals is circumvented. In favourable cases, it is possible to determine the orientation of the chemical shielding tensor with respect to the molecular frame without requiring laborious crystal rotations. Since, in the absence of motional averaging, the dipolar coupling tensors are necessarily coaxial with the corresponding internuclear vectors, it is sufficient to determine the relative orientation of the shielding and dipolar tensors to relate the shielding tensor to the molecular frame. In principle, the relative orientation of two tensor interactions can be derived from a 1D spectrum, but the accuracy is greatly improved if the two interactions are separated by 2D spectroscopy (7.49).

If the experimental scheme of Fig. 7.3.3 is applied to a powder sample, one obtains a 2D spectrum where the chemical shielding $\mathcal{H}_{ZS}$ determines the amplitude distribution in the ω_2-domain, while the heteronuclear dipolar interaction $\mathcal{H}_{IS}$ primarily determines the amplitude distribution in the ω_1-domain, since the $\mathcal{H}_{ZS}$ interaction is much weaker. The contribution of the shielding tensor to the ω_1-frequencies can be removed by refocusing or by shearing the 2D matrix in computer memory, in which case one is left with a pure dipolar powder pattern in the ω_1-domain.

The characteristic shape of such two-dimensional powder patterns for a $^{13}C-^{1}H$ two-spin system can be appreciated in Fig. 7.3.5. It is apparent that the spectra strongly depend on the relative orientation of the $^{13}C-^{1}H$ dipolar coupling tensor and the ^{13}C chemical shielding tensor. An asymmetric chemical shielding tensor has been assumed in the simulations (see caption to Fig. 7.3.5), corresponding to the carboxyl carbon of methyl formate ($H^{13}COOCH_3$).

The interpretation of the spectra in Fig. 7.3.5 is straightforward. The dipolar splitting in the (vertical) ω_1-direction leads to a maximum splitting $\Delta\omega_1 = b_{kl}$ if the internuclear vector $\mathbf{r}_{kl}$ is parallel to $\mathbf{B}_0$, and to pronounced ridges for the splitting $\Delta\omega_1 = -b_{kl}/2$, i.e. if the internuclear vector $\mathbf{r}_{kl}$ is perpendicular to $\mathbf{B}_0$. The three principal values of the chemical shielding tensor can easily be recognized by the three pronounced peaks along the horizontal ω_2-direction, which are particularly

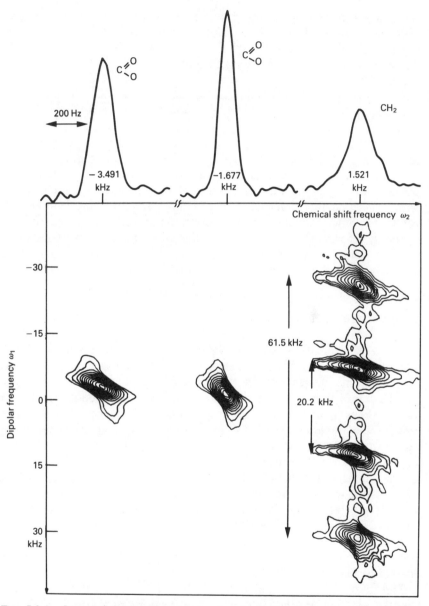

FIG. 7.3.4. Separated local-field spectrum of a single crystal of ammonium hydrogen malonate NH₄OOC CH₂ COOH, obtained with the sequence in Fig. 7.3.3 but without multiple-pulse decoupling in the evolution period. The CH₂ group gives rise to a multiplet of five lines with a weak central component. The two r_{IS} vectors in the CH₂ group subtend angles of 78° and 164° with respect to the static field. (Reproduced from Ref. 7.39.)

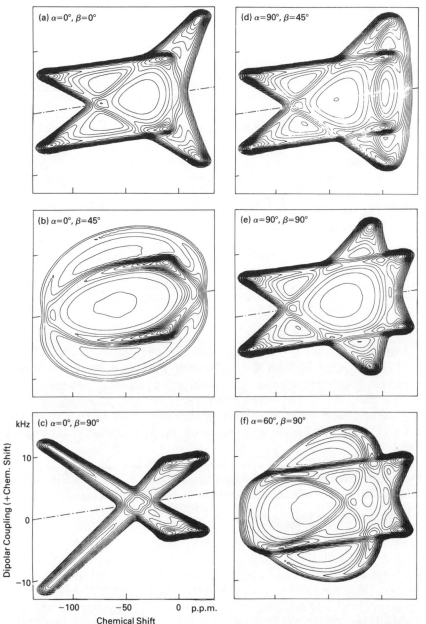

FIG. 7.3.5. Computer simulations of 2D separated local field spectra of a polycrystalline powder containing isolated *IS* pairs, with $\mathcal{H}^{(e)} = \mathcal{H}_{ZS} + \kappa\mathcal{H}_{IS}$ and $\mathcal{H}^{(d)} = \mathcal{H}_{ZS}$. The shielding tensor used in the calculations corresponds to the carboxyl carbon-13 in methyl formate ($\sigma_{11} = -126.6$ p.p.m., $\sigma_{22} = -7.0$ p.p.m., $\sigma_3 = 25.4$ p.p.m.) with $\kappa = 3^{-\frac{1}{2}}$ for off-resonance magic angle decoupling. The maximum dipolar splitting is $D_{\|} = 22.63$ kHz. Gaussian lines were assumed with full width at half-height $\Delta\omega_1 = 1000$ Hz, $\Delta\omega_2 = 250$ Hz. The powder spectra shown correspond to six different relative orientations of the two tensor interactions, described by the polar angle β between $D_{\|}$ and σ_{33} and the azimuthal angle α that is subtended by the projection of $D_{\|}$ on to the (σ_{11}, σ_{22})-plane and the σ_{11}-component. (Reproduced from Ref. 7.49.)

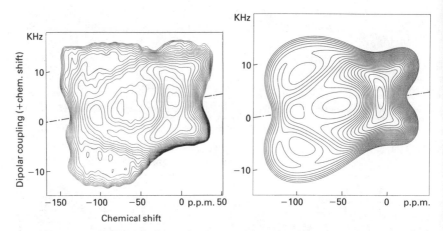

FIG. 7.3.6. Two-dimensional separated local-field spectra of polycrystalline powders. Comparison of an experimental spectrum (left) of ^{13}C enriched methyl formate (H^{13}COOCH$_3$) with a computer simulation (right), assuming a relative orientation of the shielding and dipole tensors defined by $\alpha = 33°$ and $\beta = 85°$ (see Fig. 7.3.5). (Reproduced from Ref. 7.49.)

obvious for $\alpha = 0°$, $\beta = 0°$, for $\alpha = 90°$, $\beta = 90°$, and for $\alpha = 0°$, $\beta = 90°$.

In the case of H^{13}COOCH$_3$, shown in Fig. 7.3.6, a good agreement between the experimental and simulated spectra is obtained for an orientation of the CH vector with respect to the principal axis of the shielding tensor given by $\beta = 85°$ (polar angle with respect to σ_{33}) and $\alpha = 33°$ (azimuthal angle with respect to σ_{11} in the plane spanned by σ_{11} and σ_{22}).

For the analysis of such experimental 2D powder spectra, a comparison with computer simulations such as those shown in Fig. 7.3.5 is indispensable. Two approaches are feasible for a computer calculation of powder plots.

7.3.3.1. Full simulation of powder spectra

Let us start with the computation of a 1D powder spectrum. Assuming a general Hamiltonian $\mathcal{H}(\theta, \phi)$ which depends on the two polar angles θ, ϕ describing the orientation of the magnetic field B_0 with respect to the axis system of one crystallite, the 1D powder spectrum can be expressed in the form

$$S(\omega) = \sum_i \int a_i(\theta, \phi) g_i[\omega - \omega_i(\theta, \phi)] \, d\Omega,$$

$$\text{with} \quad d\Omega = (4\pi)^{-1} \sin\theta \, d\theta \, d\phi. \quad (7.3.6)$$

The summation extends over all transition frequencies $\omega_i(\theta, \phi)$ of the Hamiltonian $\mathcal{H}(\theta, \phi)$. The coefficients $a_i(\theta, \phi)$ express the inherent signal strengths of these transitions. The lineshape functions $g_i(\omega)$ are often identical for all transitions and all orientations. The integration extends over all orientations Ω of the field relative to the crystallite.

The generalization to 2D powder spectra is straightforward

$$S(\omega_1, \omega_2) = \sum_i \sum_j \int a_{ij}(\theta, \phi) g_i^{(e)}[\omega_1 - \omega_i^{(e)}(\theta, \phi)]$$
$$\times g_j^{(d)}[\omega_2 - \omega_j^{(d)}(\theta, \phi)] \, d\Omega \quad (7.3.7)$$

where $\omega_i^{(e)}(\theta, \phi)$ and $\omega_j^{(d)}(\theta, \phi)$ are the transition frequencies of the effective Hamiltonians $\mathcal{H}^{(e)}$ and $\mathcal{H}^{(d)}$ in the evolution and detection periods, respectively. The powder plots of Fig. 7.3.5 have been computed with eqn (7.3.7).

7.3.3.2. Ridge plots

Figure 7.3.5 demonstrates that the 2D powder plots are dominated by a set of pronounced ridges. If the lineshape function becomes infinitely narrow, the ridges grow into singularities of infinite intensity. The ridges can easily be calculated and assembled into a 'ridge plot' as shown in Fig. 7.3.7.

The intensity $S(\omega_1, \omega_2)$ in the 2D spectrum is determined by the area of the surface elements on the unit sphere that contribute to the particular frequency pair (ω_1, ω_2). We can express the intensity in the form

$$S(\omega_1, \omega_2) = \sum_k |\mathbf{grad}[\omega_1(\theta_k, \phi_k)] \times \mathbf{grad}[\omega_2(\theta_k, \phi_k)]|^{-1} \quad (7.3.8)$$

where the sum runs over all those points on the unit sphere that contribute to the frequency pair ω_1 and ω_2. The principal value of the vector product of the gradients is a measure for the area of the contributing surface element. The intensity $S(\omega_1, \omega_2)$ can tend to infinity in two cases: (i) when one of the two gradients is zero; and (ii) when the two gradient vectors are parallel. These two conditions lead to ridges of infinite amplitude and allow a simple construction of the ridge plots. In favourable cases, the relative orientation of the two tensors can be determined by comparing an experimental 2D powder spectrum directly with ridge plots.

It should be noted that 2D powder spectra must be presented in pure 2D absorption mode to avoid severe distortions due to interference of mixed phase lineshapes (see § 6.5.2). The dispersion part of the mixed phase peak shapes can be eliminated by inserting a $(\pi)^S$-pulse on alternate scans as shown in Fig. 7.3.3, with subsequent combination of the spectra as described in § 6.5.3.2. Further intensity variations may occur if the S-magnetization is excited by cross-polarization, since the

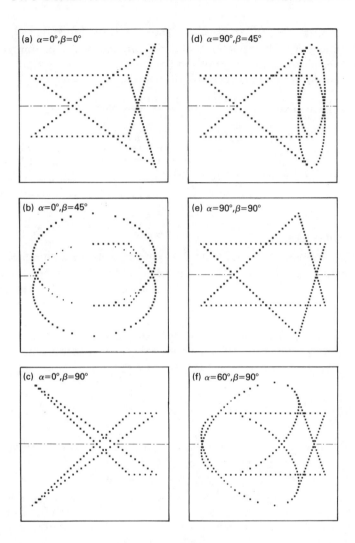

FIG. 7.3.7. Ridge plots of simulated 2D separated local-field spectra corresponding to the parameters in Fig. 7.3.5, except that the chemical shielding contribution was removed from the vertical ω_1-domain. (Reproduced from Ref. 7.49.)

efficiency of the transfer depends on the orientation of the r_{IS} vector with respect to the static field. Preparation by a $(\pi/2)^S$-pulse is to be preferred in spite of the resulting loss in sensitivity, since a uniform excitation of all microcrystallites is essential for obtaining characteristic 2D powder spectra.

7.3.4. Separation of $\mathcal{H}_{IS}$ and $\mathcal{H}_{ZS}$ under magic-angle spinning conditions

Inhomogeneous anisotropic interactions can be averaged by macroscopic rotation of the sample about an axis tilted by the magic angle $\theta = \arccos(3^{-\frac{1}{2}}) = 54.7°$ with respect to the static field. If the magnitude of the interactions exceeds the spinning frequency ω_r, the time evolution of the magnetization is characterized by the appearance of rotational echoes, which lead to spinning sidebands in the frequency domain.

In the context of 2D separation of $\mathcal{H}_{IS}$ and $\mathcal{H}_{ZS}$, four situations can be distinguished.

1. $\omega_r > \|\mathcal{H}_{ZS}\|, \|\mathcal{H}_{IS}\|$. All information about anisotropy is lost, unless the effect of sample rotation is defeated by appropriate r.f. pulses (see § 7.4.1).
2. $\|\mathcal{H}_{IS}\| > \omega_r > \|\mathcal{H}_{ZS}\|$. The chemical shielding anisotropy is averaged out completely, but the magnitude of the dipolar coupling can be determined from an analysis of the sidebands.
3. $\|\mathcal{H}_{IS}\|, \|\mathcal{H}_{ZS}\| > \omega_r$. Both interactions lead to rotational echoes and sidebands. Not only the principal values of both tensors, but also their relative orientation can be determined from an appropriate 2D experiment (7.50–7.53).
4. $\|\mathcal{H}_{IS}\|, \|\mathcal{H}_{ZS}^{\text{aniso}}\| > \omega_r > \|\mathcal{H}_{ZS}^{\text{iso}}\|$. Again, both anisotropic interactions lead to sidebands, but synchronous sampling in the t_2-domain can be used to suppress the chemical shielding sidebands, with the result that the 2D spectrum has the appearance of case 2. The condition $\omega_r > \|\mathcal{H}_{ZS}^{\text{iso}}\|$, which must be fulfilled to avoid aliasing of the isotropic shifts, can be obtained by scaling the $\mathcal{H}_{ZS}^{\text{iso}}$ interaction by appropriate multiple-pulse techniques (7.54, 7.55).

The separation of $\mathcal{H}_{IS}$ vs. $\mathcal{H}_{ZS}$ can be achieved with the sequences shown in Fig. 7.3.8, which are closely related to the scheme of Fig. 7.3.3 used for static samples. The S-magnetization is initially prepared by cross-polarization. As in the gated decoupler method for isotropic systems (§ 7.2.2.3), the $\mathcal{H}_{IS}$ coupling is allowed to act in t_1, but multiple-pulse sequences (not shown) must be used to remove $\mathcal{H}_{II}$ interactions. Unlike most other 2D experiments, the acquisition is started at a fixed point in time $t = 2N\tau_r = 2N/\nu_r$ after initial excitation, which must correspond to an even-numbered rotational echo, i.e. at a point in time where the shift anisotropy is refocused. The evolution period t_1 is defined as the interval where heteronuclear decoupling is interrupted. The isotropic S-spin shifts are removed in the evolution period by a $(\pi)^S$-pulse which must coincide with a rotational echo.

The sequences in Fig. 7.3.8(a) and (b) are complementary, since the apparent sense of precession in ω_1 is opposite, thus making it possible to

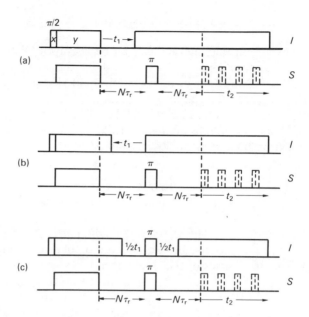

FIG. 7.3.8. Separated local-field spectroscopy under magic angle spinning conditions. (a) Simple scheme with a $(\pi)^S$-pulse synchronized with a rotational echo, where the signal acquisition is started at the time of an even-numbered echo ($t = 2N\tau_r = 2N/\nu_r$, $N = 1, 2, 3, \ldots$). The multiple-pulse sequence required for decoupling the $\mathcal{H}_{II}$ interactions in the t_1-period is not shown for clarity. The dashed pulse pairs in the detection period allow one to scale the isotropic chemical shifts in ω_2 in order to fulfil the condition $\omega_r > \|\bar{\mathcal{H}}_{ZS}^{iso}\|$ if desired. (b) Time-reversed analogue of the experiment, which can be used in combination with (a) to obtain pure 2D absorption peakshapes. (c) 'Double-sided evolution' which leads to symmetrical dipolar sideband patterns with signals spaced at $\Delta\omega_1 = \omega_r/2$ (7.53).

obtain pure 2D absorption lineshapes by a linear combination according to the procedure described in § 6.5.3.2. The scheme of Fig. 7.3.8(c) yields resonance frequencies symmetrical with respect to $\omega_1 = 0$ and thus lends itself to the method of § 6.5.3.1 to obtain pure 2D absorption.

In all three schemes of Fig. 7.3.8, one may exploit the fact that the Hamiltonian is periodic in $1/\nu_r$ if relaxation is disregarded. It is therefore sufficient to measure a limited number n of t_1-increments and begin the acquisition of the signal at the second rotational echo at $t = 2\tau_r = 2/\nu_r$ (7.56).

An experimental example, corresponding to case 3 with $\|\mathcal{H}_{IS}\|$, $\|\mathcal{H}_{ZS}\| > \omega_r$, is shown in Fig. 7.3.9. Selected sections from this pure absorption spectrum are reproduced in Fig. 7.3.10 showing that the spectra are not symmetrical about $\omega_1 = 0$ and that negative lines may appear. As in the case of static powder spectra, the relative orientation of

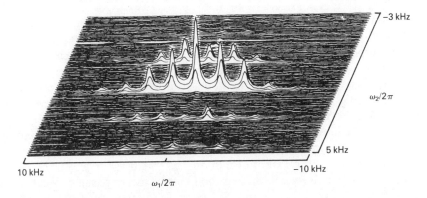

FIG. 7.3.9. Separated local-field spectrum of the amide ^{15}N resonance in diglycine-hydrochloride, obtained by combining the methods in Fig. 7.3.8(a) and (b) to achieve pure 2D absorption peakshapes. The ω_2-domain shows sidebands due to the anisotropy of the nitrogen-15 shielding tensor, while the ω_1-domain shows sidebands associated with the dipolar ^{15}N–^{1}H coupling. (Reproduced from Ref. 7.54.)

the two tensors can be determined by comparison with simulations. The magnitude of the (scaled) dipolar interaction can be determined from the projection onto the ω_1-axis, and the principal values of the shielding tensors can be obtained by analysing the sideband amplitudes in the conventional decoupled 1D spectrum (7.57, 7.58), leaving only the

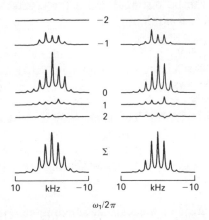

FIG. 7.3.10. Sections taken parallel to the ω_1-axis of the 2D spectrum in Fig. 7.3.9 at ω_2-frequencies corresponding to the centreband and sidebands of the ^{15}N shielding tensor, numbered 0, ±1, and ±2. The experimental sections on the left agree with the simulations on the right which are based on the assumption that the NH bond subtends an angle of 25° with the principal axis of the nearly axial ^{15}N shielding tensor. The projections shown below indicate a dipole coupling with $D_\parallel = 4.9$ kHz. (Reproduced from Ref. 7.54.)

question of relative orientation to be determined from the full 2D sideband pattern.

The dipolar and chemical shielding sideband patterns can provide information on motional processes which partially average the anisotropy. It is well known that motional processes may lead to non-axially symmetric dipole coupling tensors (7.41, 7.68), which are reflected in the sideband patterns. The comparison of the measured dipole couplings in systems with internal motion, such as polymers, with the corresponding dipole couplings in a rigid lattice (e.g. crystalline monomers) yields direct information about the extent of motional averaging (7.56).

In systems with weak dipolar couplings, the separated local field experiment in Fig. 7.3.8(a) makes it possible to measure *scalar* couplings in solids. In favourable cases, it is therefore possible to determine the number of neighbouring protons attached to each carbon-13 by observing doublets, triplets, or quartets in the ω_1-domain (7.59–7.61).

7.4. Separation of isotropic and anisotropic chemical shifts

To determine the anisotropy of chemical shielding tensors from static powder spectra, it is normally necessary to separate overlapping powder patterns. Magic-angle sample spinning allows the separation of different isotropic chemical shifts, while the anisotropy of the shift tensors can be retrieved if the effect of spinning can be inhibited in some way.

Two situations can be distinguished if the spectra are recorded under magic-angle spinning conditions.

1. $\omega_r > \|\mathcal{H}_{ZS}^{aniso}\|, \|\mathcal{H}_{ZS}^{iso}\|$. In the conventional magic-angle spectrum, all information is lost about the anisotropy, but it may be retrieved by applying r.f. pulses that partially inhibit the averaging effect of sample rotation (7.62–7.64).

2. $\|\mathcal{H}_{ZS}^{aniso}\| > \omega_r > \|\bar{\mathcal{H}}_{ZS}^{iso}\|$. The anisotropy leads to a sideband pattern, which can be removed in one dimension of the 2D spectrum by synchronous sampling. Scaling of the shifts may be required to fulfil the condition $\omega_r > \|\bar{\mathcal{H}}_{ZS}^{iso}\|$ to prevent aliasing of the isotropic shifts (7.55, 7.65).

The averaging effect of mechanical spinning can also be inhibited by 'flipping' the axis of rotation away from the magic angle in one interval (say, in the evolution period) of a 2D experiment (7.66).

Finally, it is possible to achieve a separation of isotropic and anisotropic shielding terms without continuous spinning of the sample, by using discrete rotations which mimic the effect of spinning and lead to a removal of the anisotropic interaction (7.67).

7.4.1. Rotation-synchronized pulses

Ideally, the anisotropy could be determined during t_1 while the isotropic shifts could be observed in t_2 if the spinner could be started instantaneously at the end of the evolution period. In practice, a similar effect can be achieved with continuous spinning by periodically inverting the magnetization in spin space in the evolution period to inhibit the averaging effect of macroscopic rotation.

It is possible to reintroduce anisotropic interactions by applying a π-pulse after every half-revolution (7.62), or alternatively by applying a sequence of six π-pulses during each spinner revolution (7.63). In the limit $\omega_r > \|\mathcal{H}_{ZS}^{aniso}\|$, the ω_2-domain carries information only about the isotropic shifts, while the ω_1-domain provides a measure of the anisotropy. Unfortunately, the powder patterns obtained do not correspond to static powder patterns, and the principal values of the shielding tensors can only be obtained by comparison with computer simulations.

This problem can be largely circumvented with the scheme shown in Fig. 7.4.1. In this case, the sense of precession of the transverse S-magnetization is reversed by two π-pulses in a brief interval $\tau \ll 1/\nu_r$. Because the pulses are synchronized with sample rotation, these intervals coincide with a particular position of the rotor. The experiment amounts to observing the evolution of the magnetization for a brief interval in which the rotor can be assumed to be static. In principle, the ω_1-dimension shows the same powder patterns as for a truly static sample, while the ω_2-domain shows a high-resolution isotropic spectrum (7.64).

7.4.2. Synchronous sampling with scaling

If the spinning frequency is low, i.e. if $\omega_r < \|\mathcal{H}_{ZS}^{aniso}\|$, one normally obtains sidebands that are characteristic of the shielding anisotropy in both frequency domains. To simplify the spectra, the sidebands should be suppressed in one domain in order to retain only the isotropic chemical

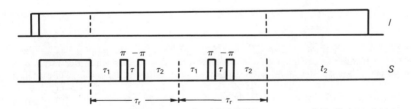

FIG. 7.4.1. Separation of $\mathcal{H}_{ZS}^{aniso}$ vs. $\mathcal{H}_{ZS}^{iso}$ in powders under magic-angle spinning conditions: scheme for retrieving the anisotropy in the ω_1-period by applying rotation-synchronized pulse pairs $(\pi)_x - \tau - (\pi)_{-x}$ (7.64). The duration of the evolution period t_1 is determined by the number of τ-intervals ($t_1 = n\tau$, $n = 2$ in this figure).

shifts. This can be achieved by stroboscopic detection: the signal is sampled in t_2 in synchronism with the rotational echoes. As a result, the sidebands are folded and coincide with the centrebands. Unfortunately, the spectra become unintelligible if the range of isotropic shifts exceeds the spinning frequency. By appropriate scaling of the chemical shifts, it is possible to ensure that $\omega_r > \|\mathcal{H}_{ZS}^{iso}\|$. In practice, this can be achieved by a series of $(\beta)_x(\beta)_{-x}$ pulses in the detection period (7.55, 7.65) in analogy to scaling in liquid phase (7.4).

7.4.3. Magic-angle flipping

Conceptually (though perhaps not technically), a straightforward approach to separating $\mathcal{H}_{ZS}^{iso}$ and $\mathcal{H}_{ZS}^{aniso}$ involves moving the axis of rotation away from the magic angle in the evolution period of the experiment (7.66). If the sample is spun perpendicular to the static field in t_1, the

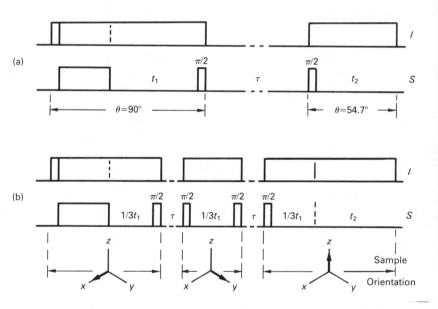

FIG. 7.4.2. Separation of $\mathcal{H}_{ZS}^{aniso}$ vs. $\mathcal{H}_{ZS}^{iso}$. (a) Magic-angle 'flipping' experiment: the magnetization is excited by cross-polarization and precesses while the rotation axis is perpendicular to the static field, stored as longitudinal polarization while the axis of rotation is flipped in the τ-interval along the magic angle, and transferred back into transverse magnetization at the beginning of the detection period. (b) 'Magic-angle hopping' experiment: a static powder is rotated along three orthogonal axes with respect to the laboratory frame, in such a way that the S-magnetization precesses for $\frac{1}{3}t_1$ in each direction. As a result, only $\mathcal{H}_{ZS}^{iso}$ is observed in ω_1, while the ω_2-domain shows the natural static powder spectrum. The magnetization is temporarily stored as longitudinal polarization in the intervals τ_1 and τ_2 while the sample is rotated.

7.4 ISOTROPIC AND ANISOTROPIC CHEMICAL SHIFTS

powder pattern is collapsed to half the width of the pattern obtained from a static sample, while the sense of the frequencies is also reversed (7.42). In the scheme of Fig. 7.4.2(a), the transverse S-magnetization is allowed to evolve while the rotation axis is tilted at $\theta = 90°$ (i.e. perpendicular to the static field). The S_x-component is then stored as longitudinal polarization by a $\pi/2$-pulse, the axis is flipped from $\theta = 90°$ to $\theta = 54.7°$, (typically in ca. 1 s), and the magnetization is brought back into the transverse plane to observe the high-resolution free induction signal at the magic angle.

7.4.4. Magic-angle hopping

The elimination of $\mathcal{H}_{ZS}^{aniso}$ can also be achieved by tipping a static sample sequentially into three orthogonal orientations x, y, and z with respect to the static field. Since the movements can only be done slowly, the transverse magnetization is temporarily stored in longitudinal Zeeman polarization. The reorientation of the sample must be carried out in a time short compared to T_1. The experimental scheme shown in Fig. 7.4.2(b) leads to a 2D spectrum which carries the isotropic shift information in ω_1, and the normal static powder pattern in ω_2 (7.67).

8
TWO-DIMENSIONAL CORRELATION METHODS BASED ON COHERENCE TRANSFER

The interpretation of NMR spectra of networks of coupled spins requires some way of establishing the connectivity of nuclear spins. Coherence transfer provides a powerful and attractive means for investigating nuclear spin systems, and represents an alternative to more conventional schemes of double resonance in a wide variety of applications such as:

1. Correlation of chemical shifts of nuclei that are directly coupled via scalar or dipolar interactions;
2. Identification of the topology of homo- and heteronuclear coupling networks;
3. Elucidation of the connectivity of transitions in the energy-level diagram;
4. Determination of the magnitudes and relative signs of scalar and dipolar couplings.

In double resonance, this type of information is obtained by selectively perturbing the Hamiltonian to modify the spectrum in a characteristic manner, or by selective saturation to modify the signal intensities in a manner which reflects the connectivity of perturbed and observed transitions. On the other hand, 2D correlation spectroscopy is based on the transfer of coherence from one transition to another and allows one to visualize the structure of the spin system in a most direct and informative manner.

The simplest experiment, originally proposed by Jeener (8.1) and first put into practice and analysed in detail by Aue *et al.* (8.2), is based on the sequence $(\pi/2) - t_1 - (\beta) - t_2$. We shall refer to this experiment as 'homonuclear 2D correlation spectroscopy', or COSY, although it is also known under the acronym HOMCOR, or as 'Jeener experiment'. In the basic experiment, the transfer of coherence between single-quantum transitions is induced by a single non-selective r.f. pulse with rotation angle β. In weakly coupled systems, the existence of a non-vanishing coupling between two spins is a necessary condition for coherence transfer to occur. Thus the appearance of cross-peaks in 2D correlation spectra constitutes a proof of the existence of resolved scalar or dipolar couplings, and allows one to identify (or 'correlate') the chemical shifts of the coupling partners. The 2D spectrum may be interpreted as a 'map' which traces out coherence transfer in the mixing process.

Although the basic experiment of 2D correlation spectroscopy has proven to be quite powerful for the interpretation of complex spectra (8.3–8.5), some ambiguities in the assignment may remain. These can often be resolved by using several consecutive coherence transfer processes, as in relayed magnetization transfer (8.6, 8.7) or in multiple-quantum NMR (8.2, 8.8). Single- and multiple-quantum correlation techniques can also be used effectively in heteronuclear systems with abundant I spins, such as protons, and dilute S spins, e.g. carbon-13, nitrogen-15, etc. (8.9–8.11). In heteronuclear systems, coherence transfer methods allow one to enhance the sensitivity by means of indirect detection (8.9, 8.12, 8.13).

Since many recent experiments involve both single- and multiple-quantum precession periods, and since there is no fundamental difference between single- and multiple-quantum coherence, we present 2D correlation spectroscopy in a unified manner irrespective of the order of coherence involved in a particular experiment.

Many important features of 2D correlation experiments can be represented by coherence transfer diagrams which describe the pathways through various orders of multiple-quantum coherence, as discussed in § 6.3. This description provides a unifying picture of a wide variety of experiments, and also leads to practical recipes for the selection of desired orders by phase-cycling procedures.

Coherence transfer pathways do not, however, provide insight into the factors that determine the amplitudes of coherence transfer. It has been shown in § 6.2.1 how explicit representations of the density operator can be used to predict coherence transfer amplitudes (eqn (6.2.11)). This theme will be expanded in § 8.1.

In many systems, the coupling network may be incomplete, with vanishing couplings between certain spins. As a result, some coherence transfer processes have vanishing amplitudes and are said to be forbidden. This phenomenon can be explained in terms of the coherence transfer selection rules discussed in § 8.1.

The basic form of homonuclear correlation spectroscopy is discussed in § 8.2. Various modifications and refinements, such as delayed-acquisition methods, constant-time experiments, multiple-quantum filters, relayed magnetization transfer, and total correlation spectroscopy, are treated in § 8.3. § 8.4 deals with practical aspects of multiple-quantum spectroscopy of homonuclear systems, and is complementary to the principles given in Chapter 5. Heteronuclear systems containing abundant and dilute spin species I and S lend themselves to a variety of manipulations which are the subject of § 8.5.

8.1. Coherence transfer in two-dimensional correlation spectroscopy: amplitudes and selection rules

Coherence transfer, the basic phenomenon on which 2D correlation spectroscopy relies, was discussed in § 2.1.11. Cross-peaks in a 2D correlation spectrum are caused by the transfer of coherence from a transition between two states $|t\rangle$ and $|u\rangle$ to a transition between two other states $|r\rangle$ and $|s\rangle$,

$$|t\rangle\langle u| \rightarrow |r\rangle\langle s|.$$

The peak intensities are determined by the amplitudes of coherence transfer, taking into account the limitations imposed by coherence transfer selection rules.

Coherence transfer can be induced by a single r.f. pulse or by a sequence of pulses. The overall effect may be described by a unitary transformation R (neglecting relaxation effects) as discussed in § 6.2. The amplitude of a coherence transfer $|t\rangle\langle u| \rightarrow |r\rangle\langle s|$ is given by the complex factor $Z_{rs\,tu}$ of eqn (6.2.11), which can be expressed in terms of the matrix elements of **R** with eqn (6.2.14). In some cases, it is more convenient to compute the amplitudes of coherence transfer in terms of single-transition operators. Thus in a two-spin system, the coherence transfer

$$|\beta\beta\rangle\langle\alpha\beta| = I_k^- I_l^\beta \rightarrow |\alpha\beta\rangle\langle\alpha\alpha| = I_k^\alpha I_l^-$$

involves the conversion of operators $I_k^- \rightarrow I_k^\alpha$ and $I_l^\beta \rightarrow I_l^-$. The amplitude can be derived from eqns (2.1.109), (2.1.118), and (2.1.119).

To understand the relevant features of coherence transfer, it is often useful to consider the fate of a group of coherences associated with an entire spin multiplet rather than treating individual transitions. A spin multiplet pattern can be represented by a Cartesian operator product, as shown in §§ 4.4.5 and 6.7. This procedure will be used extensively in the following.

Coherence transfer by a non-selective pulse is intimately related to the existence of antiphase coherence before and after the pulse. In a two-spin system for example, the relevant transformation by a $(\pi/2)_x$-pulse is

$$2I_{ky}I_{lz} \xrightarrow{(\pi/2)(I_{kx}+I_{lx})} -2I_{kz}I_{ly}$$

where antiphase coherence of spin I_k is transformed into antiphase coherence of spin I_l. This is also the basic process involved in heteronuclear polarization transfer techniques like INEPT which can be considered as offsprings of 2D spectroscopy techniques (see § 4.5.5). It may be convenient to consider such transformations as arising from a 'pulse cascade' (8.14) where each pulse acts on one particular spin, a process which is particularly easy to visualize.

It is often possible to represent coherence transfer under an extended mixing sequence by a single propagator. Thus the composite rotation obtained with the 'sandwich' sequence $[(\pi/2)_x - \tau/2 - (\pi)_x - \tau/2 - (\pi/2)_x]$ (§ 4.4.6), which is extensively used for relayed coherence transfer (§ 8.3.4) and for multiple-quantum excitation and detection (§ 8.4.1), can be represented in weakly coupled systems by a single transformation under the propagator $\exp\{-i\sum \pi J_{kl}\tau 2I_{ky}I_{ly}\}$, leading to rotations in operator subspaces given in eqn (2.1.100) and illustrated in Fig. 2.1.5.

We shall now turn our attention to the selection rules that impose restrictions on coherence transfer.

1. A coherence transfer $|t\rangle\langle u| \to |r\rangle\langle s|$ is forbidden regardless of the strength of the coupling unless all four states belong to the same irreducible representation of the symmetry group of the nuclear spin Hamiltonian.

2. Coherence transfer is possible between *any* pair of transitions of a spin system belonging to the same irreducible representation. Indeed, for weak coupling, all coherence transfer amplitudes have the same absolute amplitude under a single $\pi/2$-mixing pulse, as will be explained below.

When two or more transitions are degenerate, i.e. when some couplings are equal or vanishingly small, cancellations of antiphase intensities may occur that lead to apparent restrictions of the allowed coherence transfer processes.

3. A non-selective pulse applied to a weakly coupled spin system can only transfer single-quantum coherence of spin k to observable single quantum coherence of spin l if $J_{kl} \neq 0$.

In systems with strong coupling, the rules are less restrictive.

4. In strongly coupled systems, coherence transfer induced by a single non-selective pulse may occur if the spins involved in the two transitions all belong to a connected network of spin couplings.

For the excitation and detection of multiple-quantum coherence, the following selection rules apply.

5. By applying a non-selective pulse to a weakly coupled spin system, a q-spin–p-quantum coherence involving a set of q active spins $k, l, m, \ldots$ can only be excited from or transferred to observable single-quantum coherence of one of these spins (e.g. spin k) if all $(q-1)$ couplings $J_{kl}, J_{km}, \ldots$ are different from zero.

6. By a non-selective pulse applied to a weakly coupled spin system, a q-spin–p-quantum coherence involving a set of q active spins $k, l, m, \ldots$ can only be excited from or transferred to observable single-quantum coherence of a *passive* spin n, not involved in the q-spin–p-quantum coherence, if all q couplings $J_{kn}, J_{ln}, J_{mn}, \ldots$ are different from zero.

In weakly-coupled systems with magnetically equivalent nuclei, coherence transfer phenomena are often adequately described in terms of product functions of the individual spins rather than in terms of symmetrized basis functions (8.15). The symmetry is taken into account by noting that a coupling $J_{kk'}$ between two equivalent nuclei is ineffective in isotropic solution. The selection rules can therefore be applied by setting $J_{kk'} = 0$ for all pairs of equivalent nuclei. It follows then from rule 5 that multiple-quantum coherence involving two or more equivalent nuclei cannot be transferred to observable single quantum coherence of one of these equivalent spins with a single non-selective pulse. This conclusion may be invalidated in case of multi-exponential relaxation in the system of equivalent spins, calling for a description in terms of symmetry-adapted basis functions.

The vanishing of cross-peaks in 2D spectra for very small couplings J_{kl} can be understood by noting that the rate of creation of antiphase coherence in the course of the evolution period t_1, starting from in-phase coherence $\sigma(t_1 = 0) = I_{kx}$, is proportional to $\sin \pi J_{kl} t_1$. At the same time, transverse relaxation leads to a decay of coherence

$$\sigma(t_1) = [I_{kx} \cos \pi J_{kl} t_1 + 2I_{ky} I_{lz} \sin \pi J_{kl} t_1] \exp(-t_1/T_2). \qquad (8.1.1)$$

If $J_{kl} \ll 1/T_2$, the amplitude of the antiphase coherence $2I_{ky}I_{lz}$ will remain very small and the resulting cross-peaks are attenuated accordingly. It should however be pointed out that even for small coupling constants $J_{kl} \gtrsim (5T_2)^{-1}$, where the multiplet structure is not apparent in a 1D spectrum, it is nevertheless possible to detect small cross-peaks in a 2D correlation spectrum. The determining factor is the natural line-width T_2^{-1}, rather than the inhomogeneous T_2^*-decay, if diffusion through static field gradients may be neglected. This may be appreciated by considering the refocusing of inhomogeneous broadening by the mixing pulse, as discussed in § 6.5.2.

The fact that *all* coherence transfer processes between non-degenerate transitions in weakly coupled systems are feasible under a single pulse with rotation angle β at $t = t_1$ can easily be inferred from the transformation

$$\sigma(t_1^+) = \exp(-i\beta F_x) \sigma(t_1^-) \exp(i\beta F_x) \qquad (8.1.2)$$

For weakly coupled spins $I = \frac{1}{2}$, the propagator can be evaluated explicitly (8.16)

$$\exp(-i\beta F_x) = \prod_{k=1}^{N} \exp(-i\beta I_{kx})$$

$$= \prod_{k=1}^{N} \{\mathbb{1}_k \cos \beta/2 - 2iI_{kx} \sin \beta/2\} \qquad (8.1.3)$$

where $\mathbb{1}_k$ is the unity operator of spin I_k.

The matrix representation of the rotation operator for a single spin $I_k = \frac{1}{2}$ is

$$(\exp(-i\beta I_{kx})) = \begin{pmatrix} \cos \beta/2 & -i \sin \beta/2 \\ -i \sin \beta/2 & \cos \beta/2 \end{pmatrix}, \qquad (8.1.4)$$

and the matrix representation of the rotation operator of a system of N spins with $I_k = \frac{1}{2}$ is obtained from the direct product

$$\left(\prod_{k=1}^{N} \exp(-i\beta I_{kx}) \right) = (\exp(-i\beta I_{1x})) \otimes (\exp(-i\beta I_{2x})) \otimes \ldots \qquad (8.1.5)$$

For $\beta = \pi/2$, all elements in eqn (8.1.4) have the same absolute value $(2)^{-\frac{1}{2}}$, hence all coherence transfer processes have the same amplitudes. For higher spins $I_k \geq 1$, the transfer amplitudes are no longer uniform, but all coherence transfer processes remain feasible, except for special values of β where some matrix elements may disappear.

This shows that the coherence transfer selection rules are not due to vanishing matrix elements, but rather to the coincidence of peaks with opposite amplitudes which occurs when the couplings are not sufficiently resolved. To calculate the general appearance of a 2D spectrum of a weakly coupled system, it is convenient to adopt the following strategy.

1. One assumes at first that all couplings are resolved and that all transitions are non-degenerate. Each coherence transfer process leads to a resolved peak in the 2D spectrum.

2. Vanishing or degenerate couplings are then taken into account by shifting the positions of the peaks in the 2D spectrum accordingly. The resulting overall peak amplitudes are obtained by summation over the amplitudes of the transfer processes that lead to coinciding signals.

This approach is useful for predicting the effect of degenerate coupling constants, particularly the amplitudes in triplets, quartets, etc. Thus the cancellation effects illustrated for 1D spectra in Fig. 4.4.6 can easily be extended to two dimensions.

The antiphase character of cross-peaks is a general feature as long as a single mixing pulse is used. By employing extended mixing periods, which may contain several mixing pulses, it is however possible to induce coherence transfer processes that violate the selection rules which are applicable to a single pulse. Typical examples are relayed magnetization transfer (§ 8.3.4) and total correlation spectroscopy (§ 8.3.5).

8.2. Homonuclear two-dimensional correlation spectroscopy

The basic form of 2D correlation spectroscopy employs a single mixing pulse with rotation angle β to induce coherence transfer (Fig. 8.2.1). This

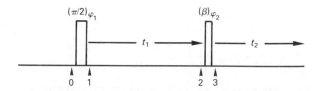

FIG. 8.2.1. Pulse sequence for homonuclear 2D correlation spectroscopy (COSY) with a $\pi/2$ preparation pulse and a mixing pulse with an r.f. rotation angle β. The r.f. phases φ_1 and φ_2 can be cycled to select coherence transfer pathways. The density operators σ_i in eqns (8.2.1)–(8.2.3) correspond to the points labelled by $i = 0, 1, 2, 3$ on the time axis.

simple experiment has proven to be a powerful tool for the analysis of crowded homonuclear spectra, and is particularly useful for the elucidation of proton spectra of macromolecules. The same experiment can also be applied to analyse spectra of other nuclear species with homonuclear spin couplings, such as phosphorus-31, boron-11, tungsten-183, etc.

8.2.1. Weakly coupled two-spin systems

Many features of 2D correlation spectroscopy become apparent by considering a system of two weakly coupled spins with $I_k = I_l = \frac{1}{2}$ and $|\Omega_k - \Omega_l| \gg |2\pi J_{kl}|$. Under the pulse sequence of Fig. 8.2.1 with $\varphi_1 = \varphi_2 = 0$ (x-pulses), the sequence of events can be described by a cascade of transformations

$$\sigma_0 \xrightarrow{(\pi/2)(I_{kx}+I_{lx})} \sigma_1 \xrightarrow{\Omega_k t_1 I_{kz}} \xrightarrow{\Omega_l t_1 I_{lz}} \xrightarrow{\pi J_{kl} t_1 2 I_{kz} I_{lz}} \sigma_2 \xrightarrow{\beta(I_{kx}+I_{lx})} \sigma_3. \quad (8.2.1)$$

The indices of the density operator σ refer to the numbers on the time axis in Fig. 8.2.1. The equilibrium state $\sigma_0 = I_{kz} + I_{lz}$ (dropping unimportant factors as usual) is first transferred into transverse coherence by the preparation pulse, $\sigma_1 = -I_{ky} - I_{ly}$. The precession in the evolution time t_1 generates coherence components that are in-phase (I_{kx}, I_{ky}, etc.) or in antiphase ($2I_{kx}I_{lz}$, etc.)

$$\begin{aligned}\sigma_2 = &-[I_{ky}\cos\Omega_k t_1 + I_{ly}\cos\Omega_l t_1]\cos\pi J_{kl}t_1 \\ &+[I_{kx}\sin\Omega_k t_1 + I_{lx}\sin\Omega_l t_1]\cos\pi J_{kl}t_1 \\ &+[2I_{kx}I_{lz}\cos\Omega_k t_1 + 2I_{kz}I_{lx}\cos\Omega_l t_1]\sin\pi J_{kl}t_1 \\ &+[2I_{ky}I_{lz}\sin\Omega_k t_1 + 2I_{kz}I_{ly}\sin\Omega_l t_1]\sin\pi J_{kl}t_1. \end{aligned} \quad (8.2.2)$$

The transformation under the mixing pulse β leads to the following state

8.2 HOMONUCLEAR 2D CORRELATION SPECTROSCOPY

at the beginning of the detection period

$$\sigma_3 = \sigma(t_1, t_2 = 0)$$

$$= -[\overset{①}{I_{kz} \cos \Omega_k t_1} + \overset{②}{I_{lz} \cos \Omega_l t_1}]\cos \pi J_{kl} t_1 \sin \beta$$

$$+ [\overset{③}{I_{kx} \sin \Omega_k t_1} + \overset{④}{I_{lx} \sin \Omega_l t_1}]\cos \pi J_{kl} t_1$$

$$- [\overset{⑤}{I_{ky} \cos \Omega_k t_1} + \overset{⑥}{I_{ly} \cos \Omega_l t_1}]\cos \pi J_{kl} t_1 \cos \beta$$

$$+ [\overset{⑦}{2I_{kz}I_{lz} \sin \Omega_k t_1} + \overset{⑧}{2I_{kz}I_{lz} \sin \Omega_l t_1}]\sin \pi J_{kl} t_1 \sin \beta \cos \beta$$

$$+ [\overset{⑨}{2I_{kx}I_{lz} \cos \Omega_k t_1} + \overset{⑩}{2I_{kz}I_{lx} \cos \Omega_l t_1}]\sin \pi J_{kl} t_1 \cos \beta$$

$$+ [\overset{⑪}{2I_{ky}I_{lz} \sin \Omega_k t_1} + \overset{⑫}{2I_{kz}I_{ly} \sin \Omega_l t_1}]\sin \pi J_{kl} t_1 \cos^2\beta$$

$$- [\overset{⑬}{2I_{kz}I_{ly} \sin \Omega_k t_1} + \overset{⑭}{2I_{ky}I_{lz} \sin \Omega_l t_1}]\sin \pi J_{kl} t_1 \sin^2\beta$$

$$- [\overset{⑮}{2I_{kx}I_{ly} \cos \Omega_k t_1} + \overset{⑯}{2I_{ky}I_{lx} \cos \Omega_l t_1}]\sin \pi J_{kl} t_1 \sin \beta$$

$$- [\overset{⑰}{2I_{ky}I_{ly} \sin \Omega_k t_1} + \overset{⑱}{2I_{ky}I_{ly} \sin \Omega_l t_1}]\sin \pi J_{kl} t_1 \sin \beta \cos \beta. \quad (8.2.3)$$

The odd-numbered terms lead to peaks with $\omega_1 = \pm\Omega_k \pm \pi J_{kl}$, while the even-numbered terms correspond to peaks with $\omega_1 = \pm\Omega_l \pm \pi J_{kl}$.

It is instructive to consider more closely the transformation from eqn (8.2.2) to eqn (8.2.3) induced by the pulse β_x.

1. In-phase terms that are orthogonal to the phase of the mixing pulse, such as I_{ky} and I_{ly}, are partially transformed into unobservable z-magnetization (terms ① and ② in eqn (8.2.3)). The remaining terms ⑤ and ⑥ contribute to the diagonal multiplets, i.e. precess with the same Zeeman frequency during evolution and detection.

2. In-phase terms that are parallel to the phase of the mixing pulse, such as I_{kx} and I_{lx}, remain invariant (terms ③ and ④) and contribute to the diagonal multiplets.

3. Antiphase terms that are parallel to the phase of the mixing pulse ($2I_{kx}I_{lz}$, $2I_{kz}I_{lx}$) are partially transformed into unobservable zero- and double-quantum coherence (terms ⑮ and ⑯). The remaining terms ⑨ and ⑩ again contribute to the diagonal multiplets.

4. Antiphase terms that are orthogonal to the phase of the mixing

pulse ($2I_{ky}I_{lz}$, $2I_{kz}I_{ly}$) in part remain invariant (terms ⑪ and ⑫, which contribute to the diagonal multiplets); the same terms are in part transformed into unobservable longitudinal two-spin order (terms ⑦ and ⑧) and into unobservable zero-and double-quantum coherence (terms ⑰ and ⑱). Most important of all, the terms ⑬ and ⑭ originate from coherence transfer between the two spins ($2I_{ky}I_{lz} \rightarrow 2I_{kz}I_{ly}$) and are responsible for the cross-peak multiplets. Clearly, cross-peaks occur only when antiphase coherence is formed during the evolution period under the influence of a resolved spin–spin coupling.

The effect of the r.f. rotation angle β on the 2D correlation spectrum of a two-spin system is shown schematically in Fig. 8.2.2(a), assuming a complex Fourier transformation with respect to t_1. The multiplet patterns corresponding to the various operator terms in eqn (8.2.3) can be found in Fig. 6.7.1.

For $\beta = 0$, only the terms ③, ④, ⑤, ⑥, ⑪, and ⑫ in eqn (8.2.3) survive, all of which contribute to the diagonal multiplets. Their superposition leaves only multiplet components which coincide with the diagonal $\omega_1 = \omega_2$, as shown in the left column of Fig. 8.2.2(a).

For $\beta = \pi$, the same terms appear as for $\beta = 0$, however with different coefficients. The superposition of the resulting multiplets only leaves off-diagonal peaks belonging to multiplets centred on the diagonal with $\omega_1 = -\omega_2$, as shown in the right column of Fig. 8.2.2(a). These signals are closely related to those obtained in 2D spin–echo spectra, as discussed in § 7.2.1.

For $\beta = \pi/2$, only the terms ③, ④, ⑬, and ⑭ in eqn (8.2.3) contribute to the 2D spectrum. The terms ③ and ④ are responsible for in-phase multiplets on the diagonals $\omega_1 = \pm \omega_2$, while the terms ⑬ and ⑭ lead to antiphase cross-peak multiplets, as shown in the central column of Fig. 8.2.2(a).

If a complex Fourier transformation is calculated with respect to t_1, as assumed in Fig. 8.2.2(a), all signals have mixed phases $\pm[a(\omega_1)a(\omega_2) - d(\omega_1)d(\omega_2)]$, irrespective of the pulse rotation angle β. The signs of the signals in the multiplets of the cross- and diagonal peaks have different symmetry relations. Consider the signs of pairs of multiplet components within the cross- and diagonal peaks that appear symmetrically with respect to $\omega_1 = 0$. In cross-peak multiplets, such pairs have the same sign, while in diagonal peak multiplets the signs are opposite.

The 2D spectra in Fig. 8.2.2(b), which can be obtained with a *real* cosine Fourier transformation with respect to t_1, can easily be derived from the complex Fourier-transformed spectra of Fig. 8.2.2(a) by adding the contributions from positive and negative ω_1 frequencies. The phases obtained with a real cosine Fourier transformation can also be obtained

8.2 HOMONUCLEAR 2D CORRELATION SPECTROSCOPY

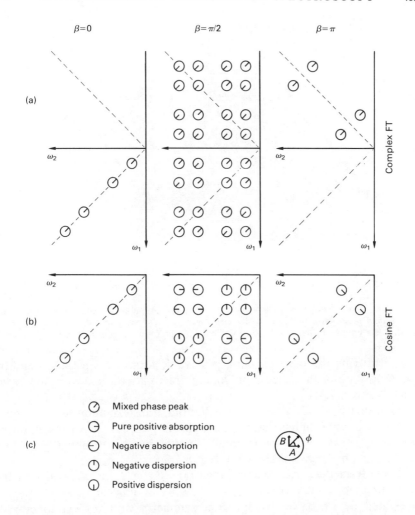

FIG. 8.2.2. (a) Schematic 2D correlation spectra of a weakly-coupled two-spin system for $\beta = 0$, $\pi/2$, and π, assuming a *complex* Fourier transformation with respect to t_1. Since the carrier is placed outside the spectrum, all signals in the lower quadrants are associated with coherence transfer pathways $p = 0 \to -1 \to -1$ ($\kappa = +1$ in eqn (6.5.11), so-called 'P' peaks), while signals in the upper quadrants stem from pathways $p = 0 \to +1 \to -1$ ($\kappa = -1$, so-called 'N' peaks). (b) Schematic 2D spectra obtained after a *real* cosine Fourier transformation with respect to t_1. (c) The mixed peakshapes $S(\omega_1, \omega_2) = Aa(\omega_1)a(\omega_2) - Bd(\omega_1)d(\omega_2)$ are indicated by polar diagrams representing the coefficients A and B which lead to resultant vectors characterized by the phase angle $\phi = \tan^{-1} B/A$. For example for $A = 0$ and $B = 1$, one obtains pure negative 2D dispersion signals, illustrated in Fig. 6.5.1(e) and (f).

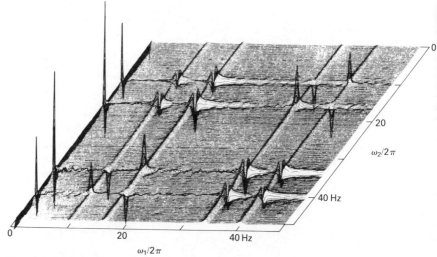

FIG. 8.2.3. Experimental 2D correlation spectrum of the proton AB system in 2,3-dibromothiophene at 60 MHz, obtained with $\beta = \pi/2$ with a cosine transformation with respect to t_1. The cross-peaks appear essentially in pure 2D absorption, while the diagonal peaks have nearly pure 2D dispersion peak shapes (slight deviations are due to strong coupling effects, as discussed in § 8.2.4). (Reproduced from Ref. 8.123.)

by identifying the non-vanishing observable terms in eqn (8.2.3) and by superposition of the corresponding multiplet patterns shown in the lower half of Fig. 6.7.1.

Provided $\beta = \pi/2$, a real cosine Fourier transformation leads to pure 2D absorption antiphase cross-peaks and to pure 2D dispersion in-phase diagonal peaks. An experimental 2D spectrum corresponding to Fig. 8.2.2(b) with $\beta = \pi/2$ is shown in Fig. 8.2.3.

For intermediate r.f. rotation angles $0 < \beta < \pi/2$ and $\pi/2 < \beta < \pi$, the contributions to the cross-peaks (terms ⑬ and ⑭ in eqn (8.2.3)) are weighted by $\sin^2\beta$, but retain their symmetry with respect to $\omega_1 = 0$. Consequently, cross-peaks always appear in pure 2D absorption if a cosine transform is calculated with respect to t_1, provided the spins are weakly coupled. The diagonal peaks on the other hand appear with an admixture of 2D dispersion and 2D absorption, as may be inferred from Fig. 8.2.2(b).

The intensities and phases of individual peaks may also be derived by expanding the density operator in terms of shift and polarization operators. This approach makes it possible to distinguish transfer processes that originate in opposite orders of coherence p and $p' = -p$, i.e. between coherence transfer pathways that lead to signals in the upper and lower quadrants of Fig. 8.2.2(a). It is instructive in this context to consider the amplitudes of coherence transfer between parallel transi-

8.2 HOMONUCLEAR 2D CORRELATION SPECTROSCOPY

tions. The transfer from $|\beta\beta\rangle\langle\beta\alpha| = I_k^\beta I_l^-$ to $|\alpha\beta\rangle\langle\alpha\alpha| = I_k^\alpha I_l^-$, for example, is described with eqns (2.1.109) and (2.1.119) by

$$I_k^\beta I_l^- \xrightarrow{\beta(I_{kx}+I_{lx})} \sin^2(\beta/2)\cos^2(\beta/2)\, I_k^\alpha I_l^- + \text{other terms}. \quad (8.2.4)$$

In this transfer, the order of coherence is conserved (pathway $p = 0 \to -1 \to -1$; $\kappa = +1$ in eqn (6.5.11)). The corresponding signal appears near the diagonal $\omega_1 = \omega_2$ (lower half in Fig. 8.2.2(a)), and may be referred to as a 'P' peak or 'anti-echo' signal. On the other hand, the transfer from $|\beta\alpha\rangle\langle\beta\beta| = I_k^\beta I_l^+$ to $|\alpha\beta\rangle\langle\alpha\alpha| = I_k^\alpha I_l^-$ is determined by

$$I_k^\beta I_l^+ \xrightarrow{\beta(I_{kx}+I_{lx})} \sin^4(\beta/2)\, I_k^\alpha I_l^- + \text{other terms}. \quad (8.2.5)$$

The resulting peak appears close to the diagonal $\omega_1 = -\omega_2$ (upper half of Fig. 8.2.2(a); pathway $p = 0 \to +1 \to -1$, $\kappa = -1$, so-called 'N' peak or 'echo' signal). The two signals associated with eqns (8.2.4) and (8.2.5) appear symmetrically disposed with respect to $\omega_1 = 0$. As described in eqns (6.5.22) and (6.5.23), a real Fourier transformation with respect to t_1 leads to a peak shape with an admixture of pure 2D absorption (coefficient $A = \frac{1}{2}(\sin^2(\beta/2)\cos^2(\beta/2) + \sin^4(\beta/2)) = \frac{1}{2}\sin^2(\beta/2)$) and pure 2D dispersion (coefficient $B = \frac{1}{2}(\sin^2(\beta/2)\cos^2(\beta/2) - \sin^4(\beta/2)) = \frac{1}{2}\sin^2(\beta/2)\cos\beta$). In general, the phases of peak shapes can be readily calculated in this way.

8.2.2. Applications to complicated spectra

The detailed discussion of a weakly-coupled two-spin system might distract from the practical significance of 2D correlation spectroscopy for the elucidation of complicated spectra. The experimental example in Fig. 8.2.4 illustrates an application to a macromolecule. The 2D correlation spectrum of basic pancreatic trypsin inhibitor (BPTI) reveals a multitude of cross-peaks indicating scalar couplings between pairs of protons within each of the 58 amino-acid residues (8.17). In this example, the digital resolution is insufficient to resolve the multiplet structure of the cross-peaks, and the absolute-value representation leads to featureless resonances centred at the coordinates of the chemical shifts of the coupling partners $(\omega_1, \omega_2) = (\Omega_k, \Omega_l)$. In spite of limited resolution, the information content of such 2D correlation spectra is extremely useful for spectral assignment. The partial contour plot in Fig. 8.2.5 shows a region of strategic importance, which reveals the scalar couplings between NH protons and $C^\alpha H$ protons for almost all amino-acid residues of BPTI. The assignment is based on a comparison of 2D correlation spectra and 2D exchange spectra (NOESY), as described in Chapter 9.

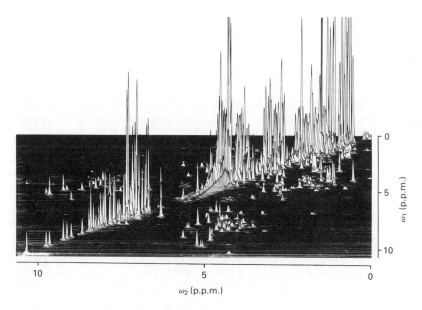

FIG. 8.2.4. Homonuclear 2D correlation spectrum of basic pancreatic trypsin inhibitor (BPTI, with 58 amino-acid residues), obtained at 500 MHz with a 0.02 M solution in 90 per cent H_2O + 10 per cent D_2O at 80°C. The spectrum, which has been symmetrized (§ 6.6.4), is shown in absolute-value mode (§ 6.5.4). The dominant diagonal ridge corresponds to the conventional 1D spectrum, while cross-peaks indicate pairs of scalar coupled protons. Note that the multiplet structure of the cross-peaks is not resolved in this case (digital resolution 5.3 Hz/point). (Reproduced from Ref. 8.17.)

The information content of 2D correlation spectra can be greatly enhanced by plotting phase-sensitive spectra in order to resolve the multiplet structure of the cross-peaks. This makes it possible to observe couplings between the 'active' nuclei and 'passive' spins which are not involved in the coherence transfer. A detailed map of a cross-peak multiplet belonging to cysteine C57 in bull seminal inhibitor (BUSI IIA) is shown in Fig. 8.2.6 (8.18). This cross-peak multiplet arises from coherence transfer from I_l (the $C^\alpha H$ proton) to I_k (the NH proton) and therefore has a signal pattern that is in antiphase with respect to J_{kl} in both frequency dimensions, but is split in addition by the coupling J_{lm} between nucleus l and a 'passive' nucleus m (β proton; the coupling to the β' proton is not resolved). It should be emphasized that detailed coupling information cannot be extracted from the conventional 1D spectrum because of extensive overlap in both regions around Ω_l and Ω_k.

Homonuclear 2D correlation spectroscopy is of course not restricted to proton NMR. Some interesting nuclei with $I = \frac{1}{2}$ that spring to one's mind are phosphorus-31 (for example in biological di- and triphosphates), silicon-29, cadmium-113, mercury-199, and tungsten-183. In the latter

8.2 HOMONUCLEAR 2D CORRELATION SPECTROSCOPY

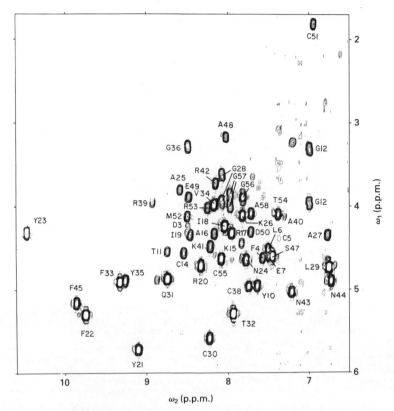

FIG. 8.2.5. Detail of a 2D correlation spectrum of basic pancreatic trypsin inhibitor (BPTI), obtained at 500 MHz in H_2O at 68°C, showing cross-peaks that reveal scalar couplings between NH protons (chemical shifts between 6.6 and 10.6 p.p.m. on the horizontal ω_2-axis) and $C^\alpha H$ protons (between 1.7 and 6.0 p.p.m. on the vertical ω_1-axis). The assignments are given according to IUPAC–IUB conventions. At this digital resolution (5.3 Hz/point), the multiplet structures of the cross-peaks cannot be resolved. (Reproduced from Ref. 8.17.)

case, 2D correlation spectra have proven useful in establishing the structure of clusters (8.19).

Correlation spectroscopy is also not limited to the study of nuclei with $I = \frac{1}{2}$. An instructive example involving homonuclear couplings between ^{11}B nuclei ($I = \frac{3}{2}$) is shown in Fig. 8.2.7, where all scalar couplings in the carborane 1,2-$C_2B_{10}H_{12}$ are revealed (8.20). This spectrum illustrates that coherence transfer can occur quite effectively even if the scalar coupling is not resolved in the 1D spectrum (shown at the bottom of the figure). The transfer of coherence between $I = \frac{3}{2}$ nuclei can be described in terms of the operator set in eqn (2.1.103).

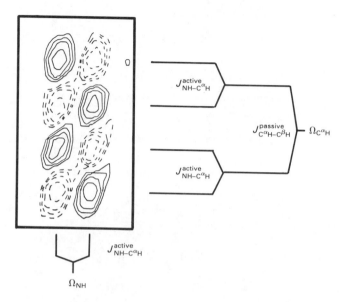

FIG. 8.2.6. Detail of a high-resolution 2D correlation spectrum of the protein BUSI IIA, obtained at 360 MHz with 1024 × 4096 data points with zero-filling to 2048 × 16 384 points, showing the NH–C$^\alpha$H cross-peak of cysteine-57 (HSC$^\beta$HH′C$^\alpha$H(NH$_2$)COOH) in phase-sensitive pure 2D absorption (negative peaks are represented by dashed contours). The coupling between the C$^\alpha$H proton and the (passive) C$^\beta$H proton leads to a duplication of the basic square of four signals with alternating signs. The coupling between C$^\alpha$H and C$^\beta$H′ is not resolved. (Adapted from Ref. 8.18.)

8.2.3. Connectivity and multiplet effects in weakly coupled systems

The peak amplitudes and their signs in 2D correlation spectroscopy are determined by the topology of the energy level diagram. Progressive and regressive connectivities lead to positive and negative cross-peaks, respectively. The signal amplitudes can be readily predicted, provided the concept of connectivity is extended to encompass *remotely connected* transitions (8.2).

1. Two transitions are *parallel* to order q if they belong to the same active spin and differ in the polarization of q passive spins.

2. Two transitions of two different spins I_k and I_l, $|r\rangle\langle s| = I_k^+ I_l^{\gamma'} I_m^{\gamma''} \ldots$ and $|t\rangle\langle u| = I_k^{\gamma''} I_l^+ I_m^{\gamma'''} \ldots$ with γ, γ', γ'', and $\gamma''' = \alpha$ or β, are *regressive* if the polarizations $I_l^{\gamma'}$ and $I_k^{\gamma''}$ are the same, and *progressive* if the polarizations $I_l^{\gamma'}$ and $I_k^{\gamma''}$ are opposite. The connectivity is to order q if the two transitions differ in the polarizations $I_m^{\gamma''}$, $I_n^{\gamma'''}$, ... of q further spins that are passive in both transitions.

8.2 HOMONUCLEAR 2D CORRELATION SPECTROSCOPY

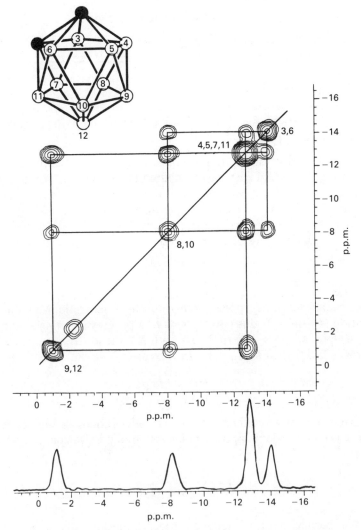

FIG. 8.2.7. Homonuclear 2D correlation spectrum of the ^{11}B resonances ($I = \frac{3}{2}$, 80 per cent natural abundance) in o-carborane 1,2-$C_2B_{10}H_{12}$ in isotropic solution. The 1D spectrum shown below exhibits four signals with relative amplitudes 2:2:4:2, where only the largest peak can be assigned unambiguously on symmetry grounds to the positions 4, 5, 7, and 11. Note that the couplings, which should lead to 1:1:1:1 quartets, are not resolved in the 1D spectrum because the quadrupolar relaxation leads to short T_2. In the 2D spectrum, it is apparent that only the boron atoms 8 and 10 are coupled to all other sites. The assignment can be completed by assuming that the sites 3 and 6 are more shielded than sites 9 and 12, because the former are adjacent to carbon atoms. (Reproduced from Ref. 8.20.)

These definitions can also be reformulated as follows.

1. Two transitions belonging to the same spin k are *parallel to order q* if they can be brought to coincide by inverting the polarization of q passive nuclei $l, m, \ldots$.

2. Two transitions belonging to nuclei k and l are *remotely progressive* (or *regressive*) to order q if they can be made to share a common eigenstate in progressive (or regressive) arrangement by inverting the polarizations of q passive nuclei $m, n, \ldots$.

Examples of direct and remote connectivities in a three-spin system are given in Fig. 8.2.8. An easy way to figure out the connectivity of a pair of transitions is to write down the corresponding single-transition operators in terms of shift and polarization operators. In the case of Fig. 8.2.8(b) for example, we have two transitions belonging to spin 2 and spin 3 respectively, which may be represented by the operator products

$$\begin{pmatrix} I_1^\alpha \cdot I_2^+ \cdot I_3^\beta \\ I_1^\beta \cdot I_2^\alpha \cdot I_3^+ \end{pmatrix}.$$

Since the active spins appear with opposite polarizations in the other transition (I_2^α and I_3^β), this corresponds to a progressive connectivity. The passive spin appears inverted (I_1^α and I_1^β); hence the connectivity is progressive to first order ($q = 1$). In the case of Fig. 8.2.8(d)

$$\begin{pmatrix} I_1^\alpha \cdot I_2^+ \cdot I_3^\beta \\ I_1^\beta \cdot I_2^\beta \cdot I_3^+ \end{pmatrix},$$

we have a regressive connectivity (equal polarizations of the active spins I_2 and I_3) of first order (opposite polarization of passive spin I_1), while the case of Fig. 8.2.8(f)

$$\begin{pmatrix} I_1^\alpha \cdot I_2^+ \cdot I_3^\beta \\ I_1^\beta \cdot I_2^+ \cdot I_3^\alpha \end{pmatrix}$$

shows a parallel pair of second order (opposite polarization of two passive spins I_1 and I_3).

The coherence transfer amplitudes $R_{rs\,tu}$ for a given connectivity can easily be derived from eqn (6.2.14) by representing the rotation operator R by a cascade of rotations $R_1 R_2 R_3 \ldots R_N$ acting on the individual spins and using the relations eqns (2.1.109), (2.1.118), and (2.1.119). If a real cosine transformation is calculated with respect to t_1, one obtains peak shapes with phases described by eqn (6.5.22)

$$S(\omega_1, \omega_2) = A\, a(\omega_1)a(\omega_2) - B\, d(\omega_1)d(\omega_2).$$

With eqn (6.5.23), one finds the following coefficients (8.2).

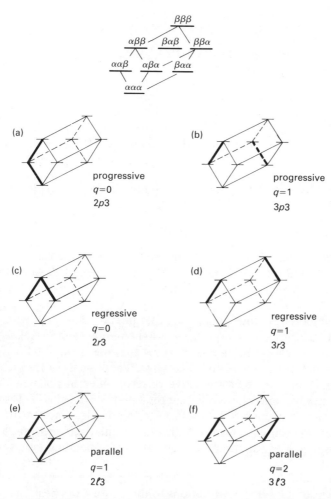

FIG. 8.2.8. Direct and remote connectivities in a three-spin system. Progressive and regressive connectivities can be classified according to the number $q = 0, 1, \ldots, (N-2)$ of passive spins that must be inverted to obtain a configuration with one common eigenstate. Parallel transitions occur with $q = 1, 2, \ldots, (N-1)$. For reference, the designation of the connectivities according to Ref. 8.2 has been included: the symbols p, r, and l, which stand for progressive, regressive, and parallel, respectively, are followed by the total number of spins $N = 3$ and preceded by the number of spins that are either active or change polarization in the course of coherence transfer. (a) Direct progressive connectivity, $q = 0$ ($2p\,3$); (b) remote progressive connectivity of first order, $q = 1$ ($3p\,3$); (c) direct regressive connectivity, $q = 0$ ($2r\,3$); (d) remote regressive connectivity of first order, $q = 1$ ($3r\,3$); (e) parallel connectivity of first order, $q = 1$ ($2l\,3$); (f) Parallel connectivity of second order, $q = 2$ ($3l\,3$).

Progressive and regressive connectivity to order q

$$A = \pm\tfrac{1}{16}\sin^2\beta(\sin\beta/2)^{2q}(\cos\beta/2)^{2(N-2-q)},$$
$$B = 0 \tag{8.2.6}$$

where the positive sign refers to progressive transitions.

Parallel connectivity to order q

$$A = \tfrac{1}{8}\cos\beta(\sin\beta/2)^{2q}(\cos\beta/2)^{2(N-1-q)},$$
$$B = -\tfrac{1}{8}(\sin\beta/2)^{2q}\cos(\beta/2)^{2(N-1-q)}. \tag{8.2.7}$$

Thus the phase ϕ of progressive and regressive connectivities, defined in eqn (6.5.25)

$$\tan\phi = B/A,$$

corresponds to pure positive or negative 2D absorption ($\phi = 0, \pi$) for progressive and regressive connectivities, regardless of β, while the phase of the peaks associated with parallel transitions depends on the rotation angle β.

The amplitudes and phases of the peaks associated with various connectivities in a system with $N = 4$ non-equivalent, weakly coupled $I = \tfrac{1}{2}$ spins are shown in Fig. 8.2.9, using the shorthand notation l, p, and r for parallel, progressive, and regressive. These symbols are preceded by the number of spins that are active or have different polarization in the two transitions, and followed by the total number of spins N. It should be pointed out that eqns (8.2.6) and (8.2.7) refer to a system of N spins where all couplings are resolved. If some of the couplings vanish, two or more lines will be superimposed, and the corresponding intensities must be added.

Instead of focusing on individual multiplet lines, it is often more convenient to consider contributions to the entire cross-peak multiplet, as shown in § 8.2.1 for the two-spin system (see Fig. 8.2.2). Let us consider a system with three weakly coupled nuclei k, l, and m and focus attention on a 2D cross-peak multiplet centred at $(\omega_1, \omega_2) = (\Omega_k, \Omega_l)$. Such a multiplet arises from coherence transfer from k to l, the third nucleus m playing the role of a passive coupling partner. In a 2D correlation experiment, only two terms in $\sigma(t_1, t_2 = 0)$ contribute to this multiplet

$$\begin{aligned}\sigma(t_1, t_2 = 0) = &-2I_{kz}I_{ly}\sin\Omega_k t_1 \sin\pi J_{kl}t_1 \cos\pi J_{km}t_1 \sin^2\beta \\ &- 4I_{kz}I_{ly}I_{mz}\cos\Omega_k t_1 \sin\pi J_{kl}t_1 \sin\pi J_{km}t_1 \cos\beta\sin^2\beta \\ &+ \text{other terms}.\end{aligned} \tag{8.2.8}$$

The signal contributions associated with these two terms are visualized in the schematic 2D spectra of Fig. 8.2.10(a) and (b). The actual cross-peak

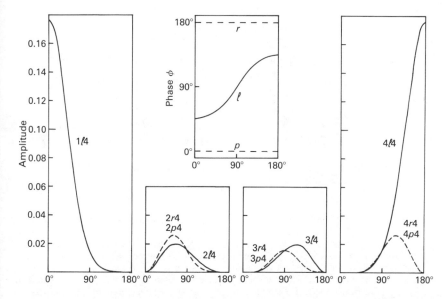

FIG. 8.2.9. Amplitudes and phases of cross- and diagonal peaks as a function of the mixing pulse angle β in a 2D correlation spectrum of a weakly-coupled four-spin system, assuming a cosine transformation with respect to t_1 (compare Fig. 8.2.2(b). Connectivities that are labelled with $1l4$, $2l4$, $3l4$, and $4l4$ correspond to parallel transitions to order $q = 0, 1, 2,$ and 3, respectively ($1l4$ corresponds to the failure of coherence transfer). The symbols $2p4$, $3p4$, and $4p4$ represent progressive connectivities of order $q = 0, 1,$ and 2, while $2r4$, $3r4$, and $4r4$ stand for regressive connectivities of order $q = 0, 1,$ and 2. The phase ϕ of the signals, defined by eqn (6.5.25), i.e., $\tan\phi = B/A$, corresponds to pure positive absorption ($\phi = 0°$) and pure negative absorption ($\phi = 180°$) for progressive and regressive connectivities, regardless of β. Parallel peaks, which appear near to the diagonal of the 2D spectrum, show an equal admixture of absorption and dispersion for $\beta = 0$ ($\phi = 45°$) and $\beta = 180°$ ($\phi = 135°$) and appear in pure dispersion for $\beta = 90°$ ($\phi = 90°$). (Reproduced from Ref. (8.2).)

multiplet centred at $(\omega_1, \omega_2) = (\Omega_k, \Omega_l)$ arises from a weighted superposition. For $0 < \beta < \pi/2$, the dominant symbols in Fig. 8.2.10(c) correspond to the largest amplitudes, which are associated with directly connected progressive and regressive transitions ($q = 0$). The weaker signals arise from remotely connected transitions with $q = 1$.

In non-degenerate systems, the multiplet effects shown schematically in Fig. 8.2.10(c) can be rationalized in terms of *subspectral analysis*. The dominant signals for $0 < \beta < \pi/2$ appear to span two interleaved squares in frequency space. Each square corresponds to one of the polarizations $M_m = \pm\frac{1}{2}$ of the passive nucleus, and represents a *subspectrum* in a two-dimensional sense. For small r.f. rotation angles β, only a small fraction $\sin^2\beta/2 \ll 1$ of the I_{mz} polarization is reversed, hence the transfer of coherence between k and l is essentially confined within the subspectra

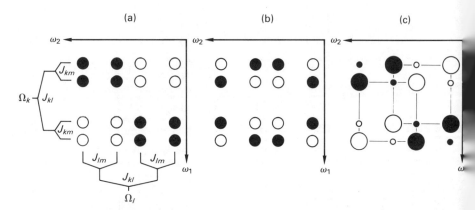

FIG. 8.2.10. The effect of the r.f. rotation angle β on a cross-peak multiplet centred at $(\omega_1, \omega_2) = (\Omega_k, \Omega_l)$ in a system of three weakly coupled spins, assuming a real cosine transformation with respect to t_1 in analogy to Fig. 8.2.2(b). All peaks appear in pure 2D absorption regardless of β; positive and negative peaks are represented by filled and open symbols, respectively. The signals shown correspond to the terms in eqn (8.2.8), drawn according to Fig. 6.7.1: (a) term $-2I_{kz}I_{ly} \sin \Omega_k t_1 \sin \pi J_{kl}t_1 \cos \pi J_{km}t_1$ (weighted by $\sin^2 \beta$); (b) term $-4I_{kz}I_{ly}I_{mz} \cos \Omega_k t_1 \sin \pi J_{kl}t_1 \sin \pi J_{km}t_1$ (weighted by $\cos \beta \sin^2\beta$); (c) weighted superposition corresponding to eqn (8.2.8): large symbols indicate amplitudes $\pm 2\sin^2\beta \cos^2\beta/2 = \pm\sin^2\beta(1 + \cos \beta)$, small symbols correspond to $\pm 2\sin^2\beta \sin^2\beta/2 = \pm \sin^2\beta(1 - \cos \beta)$. Note that for small β, coherence transfer is confined within subspectra corresponding to quantum numbers $M_m = \pm\frac{1}{2}$ of the passive spin m. Hence the cross-peak consists of eight dominant signals spanning two squares determined by the active coupling $2\pi J_{kl}$, displaced by the couplings to the passive spin $2\pi J_{km}$ in ω_1 and $2\pi J_{lm}$ in ω_2.

with $M_m = +\frac{1}{2}$ or $M_m = -\frac{1}{2}$. Note that for $\pi/2 < \beta < \pi$ the situation is reversed, since coherence transfer between k and l is more likely to be accompanied by an inversion of the passive coupling partner m.

If one of the passive couplings J_{km} or J_{lm} vanishes, only the first term in eqn (8.2.8) contributes to the cross-peak centred at $(\omega_1, \omega_2) = (\Omega_k, \Omega_l)$. In this case, all intensities in the multiplet have equal absolute values irrespective of β, and the multiplet structure can be derived from Fig. 8.2.10(a) by bringing rows or columns to superposition to account for the vanishing couplings.

In systems with N weakly coupled nuclei, the cross-peaks connecting two active nuclei k and l are split by couplings to $N - 2$ passive spins. For $0 < \beta \ll \pi/2$, the cross-peak multiplet therefore consists of a superposition of $N - 2$ 'subspectral squares' which arise from coherence transfer between directly progressively and regressively connected transitions ($q = 0$). As one approaches $\beta = \pi/2$, remote connectivities ($1 \leq q \leq N - 2$) lead to additional cross-peaks.

The schematic spectra of Fig. 8.2.10 refer to a situation where the couplings to the passive spin J_{km} and J_{lm} have the same sign. If these couplings have opposite signs, the positions of the high and low intensity

8.2 HOMONUCLEAR 2D CORRELATION SPECTROSCOPY

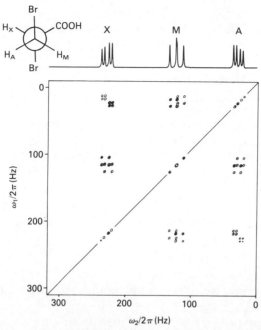

FIG. 8.2.11. Determination of the relative signs of passive couplings from the multiplet structure in a 2D correlation spectrum obtained with $\beta = \pi/4$. The spectrum of 2,3 dibromopropionic acid shows that the geminal coupling J_{AM} has a sign that is opposite to the signs of the vicinal couplings J_{AX} and J_{MX}. Thus the cross-peak centred at $(\omega_1, \omega_2) = (\Omega_A, \Omega_M)$ (middle of top row) consists of two subspectral squares displaced by two passive couplings J_{AX} and J_{MX} of the same sign ($0 < |\phi| < \pi/4$), as in Fig. 8.2.10(c). The cross-peak at $(\omega_1, \omega_2) = (\Omega_A, \Omega_X)$ on the other hand is split by two passive couplings J_{AM} and J_{AX} of opposite sign, leading to a reversal of the relative positions of the two subspectral squares ($\pi/4 < |\phi| < 3\pi/4$). The same observation applies to the cross-peak centred at (Ω_M, Ω_X). (Reproduced from Ref. 8.5.)

squares obtained with $0 < \beta < \pi/2$ are interchanged. For $J_{km} \cdot J_{lm} > 0$, the line connecting the centres of gravity of the two squares subtends an angle $0 < |\phi| < \pi/4$ with respect to the diagonal. For $J_{km} \cdot J_{lm} < 0$, this angle is $\pi/4 < |\phi| < 3\pi/4$. These slopes can be readily identified, even if the algebraic signs are masked in absolute value displays (8.5). Thus all relative signs in a complex coupling network may be ascertained by mere inspection of a 2D correlation spectrum obtained with $\beta \neq \pi/2$. The experimental example in Fig. 8.2.11 indicates that the geminal coupling in dibromopropionic acid has an algebraic sign opposite to that of the vicinal couplings.

Although the simplification of 2D correlation spectra with $0 < \beta < \pi/2$ is often useful in assignment, there are situations where the β-dependence of the intensities is undesirable. It has been shown that proper sums and differences of selected sections taken from 2D spectra can be calculated to obtain equal intensities for all multiplet components irrespective of β (8.21). This can be readily understood by considering sums and differences of the signal components shown schematically in Fig. 8.2.10.

8.2.4. Strong coupling in two-dimensional correlation spectroscopy

Although many qualitative features of weakly coupled systems remain, strong coupling introduces complications in the sense that the selection rules of § 8.1 break down. In addition, the amplitudes and phases no longer obey the simple relations of eqns (8.2.6) and (8.2.7).

The calculation of the coherence transfer amplitudes for strong coupling according to eqn (6.2.14) requires the evaluation of the matrix elements of the rotation operator $\exp\{-i\beta F_x\}$ in the eigenbasis $\{\psi_k\}$ of the Hamiltonian

$$R_{rt} = \langle \psi_r | \exp\{-i\beta F_x\} | \psi_t \rangle. \tag{8.2.9}$$

It is also possible to adopt an alternative approach, and convert the coherences $|\psi_t\rangle\langle\psi_u|$ that are found at the end of the evolution period from the eigenbasis $\{\psi\}$ to the product basis $\{\phi\}$ according to eqns (2.1.141) and (2.1.142)

$$|\psi_t\rangle\langle\psi_u| = \sum_{tu} U_{t'u'\,tu} |\phi_{t'}\rangle\langle\phi_{u'}|$$

$$= \sum_{tu} U_{t't} U^*_{u'u} |\phi_{t'}\rangle\langle\phi_{u'}|.$$

The matrix elements $U_{t't}$ are defined by $|\psi_t\rangle = \sum_{t'} |\phi_{t'}\rangle U_{t't}$. The coherences $|\phi_{t'}\rangle\langle\phi_{u'}|$ in the product basis can be expressed by products of the type $I_k^\pm I_l^\alpha I_m^\beta \ldots$ which transform under r.f. pulses according to eqns (2.1.111), (2.1.118), and (2.1.119). The resulting coherences $|\phi_{r'}\rangle\langle\phi_{s'}|$ must then be reconverted into the eigenbasis coherences $|\psi_r\rangle\langle\psi_s|$.

The characteristic features with regard to amplitudes and phases can be illustrated for the case of a strongly coupled two-spin system. With the eigenbasis defined in eqn (2.1.143), one obtains the matrix representation of the pulse operator $R = \exp\{-i\beta F_x\}$ given in eqn (60) of Ref. 8.2. For $\beta = \pi/2$ one obtains

$$R(\beta = \pi/2) = \frac{1}{2}\begin{pmatrix} 1 & -iu & -iv & -1 \\ -iu & v^2 & -uv & -iu \\ -iv & -uv & u^2 & -iv \\ -1 & -iu & -iv & 1 \end{pmatrix} \tag{8.2.10}$$

where

$$u = \cos\theta + \sin\theta, \quad v = \cos\theta - \sin\theta,$$
$$uv = \cos 2\theta, \quad u^2 = 1 + \sin 2\theta, \quad v^2 = 1 - \sin 2\theta,$$
$$\tan 2\theta = 2\pi J/(\Omega_A - \Omega_B).$$

The complex amplitudes of coherence transfer are derived from eqns (6.2.11) and (6.2.14)

$$Z_{rs\,tu} = F^+_{sr} R_{rs\,tu} \sigma(t_1 = 0)_{tu}$$
$$= -F^+_{sr} R_{rt} R^*_{su} F_{ytu} \tag{8.2.11}$$

with the matrix elements F^+_{sr} of the operator F^+

$$F^+ = \begin{pmatrix} 0 & u & v & 0 \\ 0 & 0 & 0 & u \\ 0 & 0 & 0 & v \\ 0 & 0 & 0 & 0 \end{pmatrix}. \tag{8.2.12}$$

If a complex Fourier transformation is calculated with respect to t_1, 32 peaks are obtained which map the 32 relevant elements of **Z**. In a 2D correlation spectrum of a two-spin system with $\beta = \pi/2$, only four different real amplitudes are obtained, as represented schematically in Fig. 8.2.12. By comparison with the conventional 1D spectrum, where the amplitudes of the inner and outer lines are $u^2 = (1 + \sin 2\theta)$ and $v^2 = (1 - \sin 2\theta)$, respectively, the 2D spectrum in Fig. 8.2.12 features three classes of signals with amplitudes u^4, u^2v^2, and v^4.

As in the weakly coupled limit, progressive and regressive connectivities lead to positive and negative peaks, respectively. All signals in the lower (positive) quadrant correspond to coherence transfer pathways $p = 0 \to -1 \to -1$ ($\kappa = +1$ in eqn (6.5.11), so-called 'P' peaks) while those in the upper (negative) quadrant arise from $p = 0 \to +1 \to -1$ ($\kappa = -1$, or 'N' peaks). Regressive and parallel peaks (marked with 'r' and 'l' respectively) have amplitudes $\pm u^2v^2$ in both quadrants. The amplitudes of progressive and diagonal peaks (labelled with 'p' and 'd' respectively) appear with amplitudes that are *not* symmetrical with respect to $\omega_1 = 0$.

In weakly coupled systems, a real cosine Fourier transformation leads to pure phase signals, i.e. pure 2D absorption or 2D dispersion for cross- and diagonal peaks respectively, as shown in Fig. 8.2.2(b). In strongly coupled systems, mixed phases are obtained even if $\beta = \pi/2$, since the amplitudes $Z_{rs\,tu}$ and $Z_{rs\,ut}$ represented in Fig. 8.2.12 are not necessarily symmetrical. One obtains the peak shapes from eqns (6.5.22) and

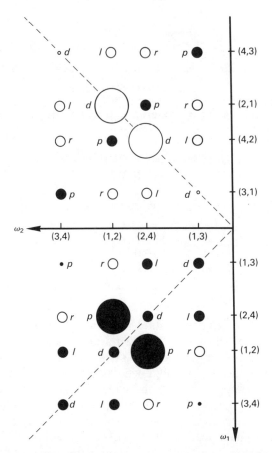

FIG. 8.2.12. Schematic 2D correlation spectrum of a strongly coupled two-spin system for $\beta = \pi/2$, assuming a complex Fourier transformation such that the individual peaks have amplitudes described by the corresponding factors $Z_{rs\,tu}$. Filled and open symbols correspond to positive and negative mixed peak shapes $\pm[a(\omega_1)a(\omega_2) - d(\omega_1)d(\omega_2)]$, i.e. the phase ϕ, defined by $\tan \phi = B/A$, is $\pi/4$ as in Fig. 8.2.2(a). Large, medium, and small symbols represent amplitudes $Z_{rs\,tu}$ proportional to $u^4 = (1 + \sin 2\theta)^2$, $u^2v^2 = (1 - \sin^2 2\theta)$, and $v^4 = (1 - \sin 2\theta)^2$, respectively. The frequency coordinates are drawn for $2\pi J/(\Omega_A - \Omega_B) = 0.75$. The relative amplitudes in the conventional 1D spectrum are $1:4:1$ in this case; in the 2D spectrum, the relative amplitudes of the three classes of peaks are $16:4:1$. The signs of the peaks agree with the patterns obtained in the weakly coupled limit, shown in Fig. 8.2.2(a).

(6.5.23)
$$S(\omega_1, \omega_2) = A\, a(\omega_1)a(\omega_2) - B\, d(\omega_1)d(\omega_2)$$

with the coefficients
$$A = \tfrac{1}{2} \mathrm{Re}\{Z_{rs\,tu} + Z_{rs\,ut}\},$$
$$B = \tfrac{1}{2} \mathrm{Re}\{Z_{rs\,tu} - Z_{rs\,ut}\},$$

8.2 HOMONUCLEAR 2D CORRELATION SPECTROSCOPY

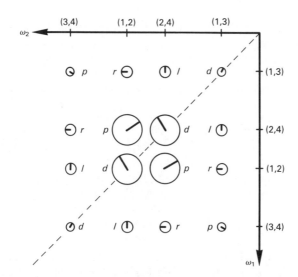

FIG. 8.2.13. Peak shapes in a 2D correlation spectrum of a strongly coupled two-spin system obtained with $\beta = \pi/2$, drawn for $2\pi J/(\Omega_A - \Omega_B) = 0.75$, like in Fig. 8.2.12, except that a real cosine Fourier transformation has been calculated in the t_1-domain. The coefficients A and B of the 2D absorption and 2D dispersion contributions are represented by polar diagrams with a vector with phase $\phi = \tan^{-1} B/A$, as shown in Fig. 8.2.2(c). Note that regressive and parallel peaks (labelled r and l) appear in pure negative absorption and pure negative dispersion respectively regardless of the strength of the coupling.

and the Lorentzian 2D absorption and 2D dispersion functions illustrated in Fig. 6.5.1. The phases obtained in a strongly coupled two-spin system are shown schematically in Fig. 8.2.13. The following properties are typical for strong coupling.

1. Cross-peaks associated with regressive connectivity appear in pure negative absorption regardless of the strength of the coupling;
2. Progressive connectivity peaks are in positive absorption with an admixture of positive dispersion;
3. Parallel connectivity peaks appear in pure dispersion;
4. The signals on the diagonal all have mixed peak shapes.

If a complex Fourier transformation is calculated with respect to t_1, as shown schematically in Fig. 8.2.12, the amplitudes are proportional to the squares of the corresponding intensities in 1D spectra, and the weakest signals might easily be overlooked. By calculating a real cosine Fourier transformation as in Fig. 8.2.13, the dynamic range of the signal amplitudes is reduced. This is another reason to prefer a real Fourier transformation.

In larger spin systems with strongly coupled subunits, cross-peaks can occur which violate the coherence transfer selection rules described in § 8.1. Thus in an ABX system, coherence transfer can occur between A and X transitions even if $J_{AX} = 0$. Two factors contribute to the break-down of the selection rules in strongly coupled systems.

1. Signals with opposite phases that appear at the same frequency coordinates no longer cancel each other because the amplitudes of coherence transfer are no longer exactly opposite, in contrast to the situation that arises in weakly coupled systems.

2. Due to so-called 'virtual couplings', the degeneracy of pairs of transitions is lifted. Thus in an ABX system with $J_{AX} = 0$, two A transitions corresponding to $M_X = \pm\frac{1}{2}$ are *not* degenerate, since the two AB subspectra are characterized by different parameters.

In general, the appearance of cross-peaks at $(\omega_1, \omega_2) = (\Omega_A, \Omega_X)$ in 2D correlation spectra of strongly coupled spin systems cannot be taken as a proof of the existence of a non-vanishing coupling J_{AX}.

8.2.5. Magnetic equivalence

It is known that in general the introduction of symmetry-adapted basis functions can significantly ease quantum mechanical calculations. In the context of 2D spectroscopy, it turns out that symmetry-adapted functions are useful in the case of strong coupling, where the required computer calculations can be significantly reduced. In weakly coupled spin systems, the use of symmetry-adapted wavefunctions and group spins provides little advantage in the discussion of coherence transfer as long as effects of multi-exponential relaxation among the equivalent spins can be excluded. Hence an A_nX system may be treated as an $AA' \ldots A^{n-1}X$ system where all couplings $J_{AX} = J_{A'X} = \ldots = J_{A^{n-1}X}$ are equal and where all $J_{AA'} = \ldots = J_{AA^{n-1}}$ may be neglected. The multiplets obtained in 2D correlation spectra are easily predicted.

1. First one assumes that all couplings are non-degenerate, leading to multiplets such as shown in Fig. 8.2.10.

2. Vanishing and degenerate couplings are then taken into account by shifting the positions of the peaks accordingly.

Schematic 2D correlation spectra of A_2X and A_3X systems are shown in Fig. 8.2.14.

It has recently been found mandatory to employ symmetry-adapted basis functions for understanding the violations of the coherence transfer selection rules, stated in § 8.1, observed for magnetically equivalent nuclei in macromolecules (N. Müller, G. Bodenhausen, K. Wüthrich, and R. R. Ernst, *J. Magn. Reson.* **65**, 531 (1985)). These violations are

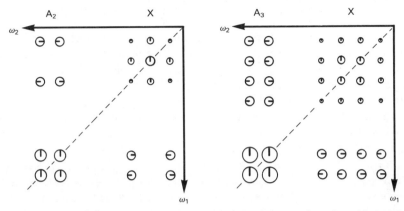

FIG. 8.2.14. Cross- and diagonal multiplets in 2D correlation spectra of weakly coupled systems with magnetic equivalence. The phases of the signals are indicated as in Fig. 8.2.2(c), assuming a real cosine Fourier transformation with respect to t_1 and a mixing pulse with $\beta = \pi/2$. Cross-peaks have pure 2D absorption peakshapes with alternating signs, while the multiplets centred on the diagonal appear in pure negative dispersion (see symbols in Fig. 8.2.2). The amplitudes represented by different diameters of the symbols are 1:2:4:8 for the A_2X system and 1:3:9:12:48 for the A_3X system.

manifested by the appearance of forbidden cross-peaks in multiple-quantum filtered 2D correlation spectra (§ 8.3.3.1) and are due to non-exponential relaxation.

8.3. Modified two-dimensional correlation experiments

The utility of the basic two-pulse correlation experiment can be enhanced by a number of modified schemes. The modifications reviewed in this section aim at the following improvements.

1. Reduction of performance time and data-storage requirements by reducing the spectral width in ω_1 (delayed acquisition, 2D spin–echo correlated spectroscopy, § 8.3.1).

2. Simplification of 2D correlation spectra by spin decoupling in the ω_1-domain via constant-time experiments (§ 8.3.2).

3. Editing of 2D correlation spectra by multiple-quantum filtering (§ 8.3.3).

4. Enhancement of information content by inducing relayed coherence transfer between remotely coupled spins (§ 8.3.4 and § 8.3.5).

While the first two modifications utilize a single coherence transfer pulse combined with a different segmentation of the time axis, the latter two modifications employ modified mixing operators which either restrict coherence transfer to a limited subset of pathways or allow additional coherence transfer pathways.

8.3.1. Delayed acquisition: spin–echo correlation spectroscopy

In many systems of practical importance, pairs of nuclei with large chemical shift differences (e.g. aromatic and aliphatic protons) are not coupled with each other. In such cases, the conventional correlation sequence leads to a 2D spectrum where the cross-peaks lie within a narrow band along the diagonal, as shown in Fig. 6.6.1, and it is possible to reduce the size of the data matrix without loss of information. As described in § 6.6, this can be achieved either by reducing the sampling rate and thus the spectral width in the ω_1-dimension of a conventional COSY experiment with subsequent foldover correction ('FOCSY') (8.22), or by delaying the acquisition as shown in Fig. 8.3.1(a). Because the observation is started at the top of the coherence transfer echo, this experiment has become known as 'spin–echo correlation spectroscopy' ('SECSY') (8.23). In the frequency domain, the effect of delayed

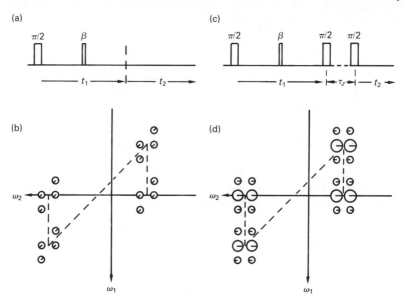

FIG. 8.3.1. (a) Sequence for spin–echo correlation spectroscopy (SECSY), equivalent to that of conventional correlation spectroscopy (COSY) with delayed acquisition ($\chi = 1$ in eqn (6.6.4)). (b) Schematic spectrum of a weakly coupled two-spin system obtained with the sequence in (a). Note that all signals have mixed phases: the cross-peaks have alternating signs ($A = B = \pm 1$), while the multiplets centred on the ω_2-axis are all in-phase ($A = B = -1$). These phases are shown with the polar diagrams defined in Fig. 8.2.2. (c) Modified SECSY-sequence with a z-filter inserted between the evolution and detection periods, where one of the in-phase magnetization components is temporarily stored as longitudinal I_{kz} polarization, while antiphase terms are eliminated by phase-cycling and variation of τ_z. (d) Schematic spectrum of a two-spin system obtained with sequence (c). All signals have pure absorption peakshapes, but the complexity of the multiplets is increased.

acquisition leads to an ω_2-dependent shift of the signals parallel to the ω_1-axis, described by eqn (6.6.5) with $\chi = 1$,

$$\omega_1' = \tfrac{1}{2}(\omega_1 + \omega_2). \tag{8.3.1}$$

The separation between cross- and diagonal peaks in the ω_1'-dimension is halved compared to a conventional COSY spectrum, but the J-splittings, and hence the separation between the antiphase peaks within a multiplet, are not scaled. The homogeneous line-width $1/T_2$ is retained in the ω_1'-domain, while the inhomogeneous contribution $1/T_2^+$ is eliminated if translational diffusion can be neglected.

A phase-cycle (8.23, 8.24) is employed to select 'N' peaks that correspond to the coherence transfer pathway $p = 0 \to +1 \to -1$ (§ 6.3). The 'P' peaks would defeat the purpose of narrowing the ω_1' spectral range and must be eliminated. The peak shapes obtained in the basic SECSY experiment therefore consist of a superposition of 2D absorption and 2D dispersion components with equal weights, eqn (6.5.10). As in spin–echo spectroscopy (§ 7.2), there is no pulse at the end of the evolution period which would allow the selection of a particular phase.

This problem can be overcome by including a 'z-filter' (8.25) in the sequence, as shown in Fig. 8.3.1(c). The z-filter consists of two $\pi/2$-pulses separated by a variable interval τ_z. The first $\pi/2$-pulse converts one of the two in-phase coherence components, say I_{ky}, into polarization I_{kz}. All coherent components remaining in the τ_z interval are eliminated by a suitable phase cycle and by variation of τ_z (8.25). The preserved I_{kz} term is then reconverted into $-I_{ky}$ by the last pulse in the sequence. All coherences are now in-phase and a pure absorption mode spectrum can be obtained. The price that must be paid is a loss in sensitivity by a factor $\sqrt{2}$, a longer phase cycle, and a more complicated multiplet structure due to the elimination of antiphase components by the z-filter, as shown in Fig. 8.3.1(d). These disadvantages are partly compensated by the fact that one obtains in-phase pure 2D absorption peaks.

8.3.2. Constant-time correlation spectroscopy: ω_1-decoupling

It is sometimes desirable to collapse the multiplet structure in the ω_1-domain to avoid overlapping cross-peak multiplets. This can be achieved with the scheme of Fig. 8.3.2(a), known as 'constant-time' or 'ω_1-decoupled' correlation spectroscopy (8.26, 8.27). A fixed interval $\tau_e \geq t_1^{\max}$ separates the preparation and mixing pulses. This interval is chosen to be of the order of $\tau_e \simeq (2J)^{-1}$ (or an odd multiple thereof), where J is the estimated active coupling between the nuclei that should be correlated. A refocusing pulse is applied in the τ_e interval as shown in Fig. 8.3.2(a), in such a way that the precession under chemical shifts is

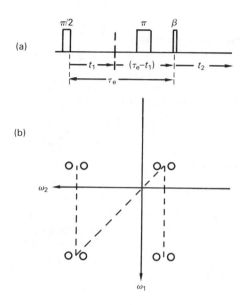

FIG. 8.3.2. (a) Pulse sequence for constant-time correlation spectroscopy, where the initial pulse and the mixing pulse β are separated by a constant interval $\tau_e \geq t_1^{max}$, while the t_1-modulation by chemical shifts is obtained by shifting the position of the π-pulse. (b) Schematic spectrum of a weakly coupled two-spin system obtained with the sequence in (a). Note the absence of scalar splittings in the ω_1-domain, which explains the designation 'ω_1-decoupled correlation spectroscopy'.

effective only in the t_1-period, while refocusing cancels this precession in the two intervals $(\tau_e - t_1)/2$. Since the precession under scalar couplings is not affected by the π-pulse, there is no t_1-dependent J-modulation, and the ω_1-domain does not contain any multiplet structure.

The essential features of the experiment can be illustrated by considering a weakly coupled system with three spins k, l, and m, where we focus attention on the coherence transfer $I_k \to I_l$ which leads to a cross-peak multiplet centred at $(\omega_1, \omega_2) = (\Omega_k, \Omega_l)$. Only two terms at the beginning of the detection period contribute to this cross-peak

$$\sigma(t_1, \tau_e, t_2 = 0) = -2I_{kz}I_{ly} \sin \Omega_k t_1 \sin \pi J_{kl}\tau_e \cos \pi J_{km}\tau_e \sin^2\beta$$
$$- 8I_{kz}I_{ly}I_{mz} \cos \Omega_k t_1 \sin \pi J_{kl}\tau_e \sin \pi J_{km}\tau_e$$
$$\times \sin^2\beta \cos \beta$$
$$+ \text{other terms}. \qquad (8.3.2)$$

Note that only the chemical shift Ω_k causes a t_1-modulation. The dependence on $J_{kl}\tau_e$ and $J_{km}\tau_e$ does not lead to a multiplet structure in

ω_1, but rather to an attenuation of the cross-peak amplitudes that depends on the choice of τ_e.

It can be shown (8.27) that the multiplet centred on the diagonal at $(\omega_1, \omega_2) = (\Omega_k, \Omega_k)$ is determined in general by eight different operator terms. For $\beta = \pi/2$ however, the only contribution to this diagonal peak is

$$\sigma(t_1, \tau_e, t_2 = 0, \beta = \pi/2) = I_{kx} \sin \Omega_k t_1 \cos \pi J_{kl} \tau_e \cos \pi J_{km} \tau_e$$
$$+ \text{other terms.} \quad (8.3.3)$$

To select the optimum rotation angle β of the mixing pulse, the following properties must be considered.

1. For $\beta = \pi/2$, only the first term in eqn (8.3.2) survives, and the amplitude of a cross-peak centred at $(\omega_1, \omega_2) = (\Omega_k, \Omega_l)$ is maximized for $\tau_e = (2J_{kl})^{-1}$ if couplings J_{km} to passive spins can be neglected. However, if J_{km} also fulfils the condition $\tau_e \simeq (2J_{km})^{-1}$, the amplitude is strongly attenuated. Pure 2D absorption cross-peaks can be obtained, and the diagonal peaks have minimum amplitude because only one out of eight terms contributes.

2. For a rotation angle $\beta \simeq \pi/3$, the amplitude of the cross-peak between k and l is less dependent on J_{km}, since both terms in eqn (8.3.2) contribute. All signals have mixed phase peak shapes, and the diagonal peaks will be stronger than for $\beta = \pi/2$.

The advantage of constant-time ω_1-decoupled correlation spectra is illustrated in Fig. 8.3.3, which shows small sections of 2D spectra of the protein basic pancreatic trypsin inhibitor (BPTI). The collapse of the multiplet structure in the ω_1-domain in Fig. 8.3.3(b), obtained with the sequence in Fig. 8.3.2(a) with $\beta = \pi/2$ and $\tau_e = 92$ ms, makes it possible to separate overlapping cross-peaks.

8.3.3. Filtering and editing

In many circumstances, 2D correlation spectra are too complicated for a straightforward analysis and call for filtering procedures to reduce the number of peaks. The basic ideas of filtering are quite general and can be applied to any kind of 1D or 2D spectra as illustrated in Fig. 8.3.4. Multiple-quantum filters of order p (8.28–8.30) eliminate the responses of all spin systems with $N < p$ coupled nuclei and can be considered as a high-pass filter in the spin number domain (Fig. 8.3.4(a)). In favourable cases, it is possible to restrict the response to spin systems with $N = p$, in analogy to a narrow bandpass filter (8.32) (Fig. 8.3.4(b)). Finally, it is

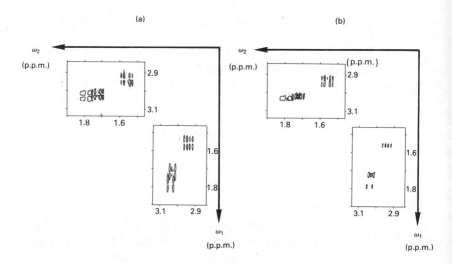

FIG. 8.3.3. (a) Excerpts from a conventional 2D correlation spectrum of basic pancreatic trypsin inhibitor (BPTI), shown in phase-sensitive mode with pure 2D absorption peakshapes (both positive and negative contour levels are drawn). The regions shown are symmetrically located with respect to the diagonal, and contain cross-peaks that originate from the δ–ε couplings of three lysine residues (δ-resonances 1.6–1.8 p.p.m., ε-resonances 2.9–3.1 p.p.m.). (b) Corresponding excerpts from an ω_1- decoupled correlation spectrum obtained with the constant-time experiment of Fig. 8.3.2(a) with $\tau_e = 92$ ms. Note the collapse of the doublet structure in the ω_1-domain, which allows one to unravel overlapping cross-peaks, particularly in the lowest part of the figure. The ω_1-decoupling fails in the upper cross-peak, where strong coupling between two non-equivalent ε protons inhibits ω_1-decoupling (the other ε- and δ-proton pairs seem to be equivalent within experimental resolution). (Adapted from Ref. 8.27.)

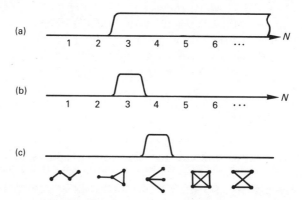

FIG. 8.3.4. Filtering and editing. (a) In systems with N spins, p-quantum filters act as high-pass filters in the spin number domain in the sense that only signals of systems with $N \geq p$ are retained. (b) Bandpass filters can be designed to select only the response of systems with $N = p$ spins. (c) Spin-topology selective propagators U and V allow one to select the response from, say, A_3X systems while suppressing signals from other systems with the same number of spins.

8.3 MODIFIED 2D CORRELATION EXPERIMENTS

also possible to construct filters which are only responsive to certain topologies of the coupling network (8.36, 8.37) (Fig. 8.3.4(c)).

Filtering procedures often involve three steps: (i) conversion into a suitable form of multiple-quantum coherence by means of a single pulse or a pulse sequence; (ii) selection of a particular order of multiple-quantum coherence by phase-cycling or by exploiting the effects of an inhomogeneous static or radio-frequency magnetic field; and (iii) reconversion into the desired form of coherence (usually single-quantum coherence) by another pulse or pulse sequence. Instead of a temporary transfer into p-quantum coherence (8.28–8.30, 8.36, 8.37), some procedures use a transfer into z-magnetization, for example in the so-called z-filter (8.25).

Spin filtering 'building blocks' may be inserted in various ways in 1D and 2D pulse sequences, as shown in Fig. 8.3.5. In 1D experiments, the spin filter replaces the r.f. excitation pulse (Fig. 8.3.5(a)). In 2D experiments, a filter may be inserted in the preparation or mixing period.

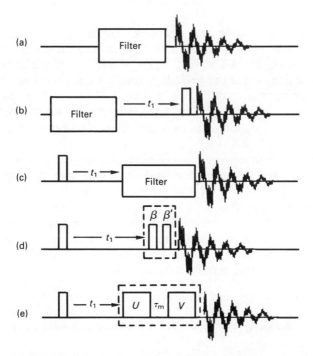

FIG. 8.3.5. Insertion of 'building blocks' to achieve filtering: (a) in 1D spectroscopy; (b) in the preparation period of 2D sequences; and (c) in the mixing period of 2D methods. (d) Sequence for multiple-quantum filtered 2D correlation spectroscopy. (e) Schematic form for spin-topology selective multiple-quantum filters involving two pulse sequences U and V tailored to transfer coherence in spin systems of specified topology.

A typical realization of this idea is shown in Fig. 8.3.5(d): in multiple-quantum filtered correlation spectroscopy, the mixing period consists of a pair of pulses $[(\beta)_\varphi (\beta')_{\varphi'}]$ where the phases must be cycled to select p-quantum coherence in the short interval between the two pulses. In more sophisticated experiments, the mixing period may involve two sequences U and V, normally containing two or more pulses, possibly separated by free precession periods (Fig. 8.3.5(e)).

8.3.3.1. Multiple-quantum filtering

The mixing process in multiple-quantum filtering is most simply effected by a pair of closely spaced r.f. pulses that transfer coherence in two steps, first from single- into p-quantum coherence, and then back into observable magnetization (Fig. 8.3.5(d)).

There are three distinct advantages to this procedure.

1. With a p-quantum filter it is possible to suppress resonances of spin systems with less than p coupled spins, particularly those stemming from solvents, since their magnetization cannot give rise to p-quantum coherence.

2. Because coherence transfer is governed by selection rules (see § 8.1), it is possible to focus on specific coupling patterns: the appearance of a cross-peak centred at $(\omega_1, \omega_2) = (\Omega_k, \Omega_l)$ in a p-quantum filtered spectrum indicates that the two active nuclei k and l must be coupled to a *common* set of at least $p - 2$ passive nuclei $m, n \ldots$. The appearance of a diagonal signal in a p-quantum filtered spectrum indicates the existence of at least $p - 1$ coupling partners. Examples illustrating these rules will be shown in Fig. 8.3.6.

3. In contrast to conventional 2D correlation spectroscopy, all diagonal and cross-peaks obtained with multiple-quantum filtering consist of antiphase multiplets with almost pure 2D absorption peak shapes. Thus partial cancellation due to broad lines reduces the amplitudes of all multiplets to an equal extent. This largely alleviates the problem that conventional correlation spectra are often obscured by dominant diagonal peaks. Some residual dispersive character of the diagonal peaks can originate from transfer processes of the type

$$I_{kx}I_{lz}I_{mz} \xrightarrow{(\pi/2)_y} I_{kz}I_l^+ I_m^+ \xrightarrow{(\pi/2)_y} I_{kx}I_{lz}I_{mz}$$

while the dominant contributions to the diagonal peaks originate from processes

$$I_{ky}I_{lz} \xrightarrow{(\pi/2)_y} I_k^+ I_l^+ \xrightarrow{(\pi/2)_y} I_{ky}I_{lz}$$

which result in absorptive peakshapes. Note that in the former case, the double-quantum coherence does not involve spins that are active in the evolution and detection periods.

Double-quantum filtering has the property that it does not affect the intensities of cross-peaks for a $\pi/2$-mixing pulse except for an overall scaling by a factor $\frac{1}{2}$. This can be shown as follows: in a conventional COSY experiment with a single $(\pi/2)_x$-mixing pulse, the only term at the end of the evolution period that can lead to a cross-peak between spins k and l is $2I_{ky}I_{lz}$ (fourth term in eqn (8.2.2)), which is converted into $-2I_{kz}I_{ly}$ by the mixing pulse, without loss in amplitude. To make a comparison with double-quantum filtered correlation spectroscopy, it is convenient to assume that the first of the two $(\pi/2)$-mixing pulses is applied along the y-axis. The $2I_{ky}I_{lz}$ term is converted into a superposition of zero- and $p = \pm 2$-quantum coherence (8.15)

$$2I_{ky}I_{lz} \xrightarrow{\pi/2(I_{ky}+I_{ly})} 2I_{ky}I_{lx} = \frac{1}{2i}(I_k^+I_l^+ - I_k^-I_l^- + I_k^+I_l^- - I_k^-I_l^+). \quad (8.3.4)$$

The zero-quantum terms are eliminated by phase-cycling, leaving only $p = \pm 2$ terms

$$\frac{1}{2i}(I_k^+I_l^+ - I_k^-I_l^-) = \tfrac{1}{2}(2I_{kx}I_{ly} + 2I_{ky}I_{lx}). \quad (8.3.5)$$

The second $(\pi/2)_y$-mixing pulse converts this into $-\tfrac{1}{2}(2I_{kz}I_{ly} + 2I_{ky}I_{lz})$. The first of these terms is responsible for the cross-peak between k and l, while the second term contributes to the diagonal peak. Note that the amplitude is halved in comparison with the conventional COSY sequence, due to the elimination of the zero-quantum coherences. In practice, this loss is compensated by the elimination of the dispersion components of the diagonal peaks (8.30), and by a partial suppression of t_1-noise.

If the rotation angles of both pulses in the mixing pair differ from $\pi/2$, one obtains cross-peaks in systems with $N > 2$ spins that are analogous to those shown in Fig. 8.2.10. Thus for $\beta \ll \pi/2$, the coherence transfer between two spins k and l is confined within subspectra which correspond to the polarizations $M_m = \pm\tfrac{1}{2}$ of a passive spin m (8.31). In contrast to conventional COSY spectra obtained with $\beta \neq \pi/2$ however, it it is not possible to obtain pure 2D absorption cross-peak shapes in double-quantum filtered COSY spectra with $\beta \neq \pi/2$.

Experimental examples are shown in Fig. 8.3.6. The conventional correlation spectrum of 1,3-dibromobutane shows strong diagonal singlet peaks from DMSO (2.5 p.p.m.) and dioxane (3.5 p.p.m.) (Fig. 8.3.6(a)). In the double-quantum filtered spectrum (Fig. 8.3.6(b)), the solvent peaks have been eliminated without significantly affecting the intensity ratios of the cross-peaks. In the triple-quantum filtered spectrum (Fig. 8.3.6(c)), the coherence transfer selection rules lead to a reduction of the number of signals. Note in particular the disappearance of both cross- and diagonal peaks associated with the CH_3 group.

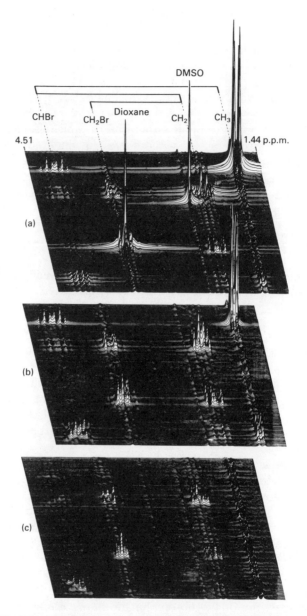

FIG. 8.3.6. Simplification of 2D correlation spectra by multiple-quantum filtration. (a) Conventional COSY spectrum of 1,3-dibromobutane with singlet signals from DMSO and dioxane. (b) Double-quantum filtered COSY spectrum, obtained with the sequence $(\pi/2) - t_1 - (\pi/2)(\pi/2) - t_2$, which shows that the structure of the cross-peaks is retained, while the diagonal peaks are partly eliminated. (c) Triple-quantum filtered correlation spectrum, obtained with the same pulse sequence as (b) but with a phase-cycle to retain the pathways $p = 0 \to \pm 1 \to \pm 3 \to -1$ (§ 6.3). Note that signals involving the methyl group are absent in (c), in accordance with the coherence transfer selection rules. All spectra are displayed in absolute-value mode. (Reproduced from Ref. 8.28.)

8.3 MODIFIED 2D CORRELATION EXPERIMENTS

The simplification that occurs in triple-quantum filtered correlation spectra can be understood in terms of the coherence transfer selection rules of §8.1. In order to transfer coherence from CH to CH_3 in 1,3-dibromobutane, it would be necessary to pass through a triple-quantum coherence involving either all three CH_3 protons or the CH proton plus two of the CH_3 protons. Both coherences can indeed be excited starting from CH coherence. However, a conversion into CH_3 single-quantum coherence is impossible, since the couplings among the CH_3 protons are ineffective because of their magnetic equivalence. For the same reason it is impossible to excite triple-quantum coherence starting from CH_3 single-quantum coherence. This explains the disappearance of all signals involving the methyl group in the triple-quantum-filtered 2D spectrum of Fig. 8.3.6(c). These selection rules involving equivalent spins tend to break down if transverse relaxation is multi-exponential, as in macromolecules in the slow motion limit.

The 2D correlation spectrum obtained in this manner is not merely simplified, but 'edited', in the sense that the elimination of selected signals allows one to draw conclusions about the topology of the coupling network.

The advantage of double-quantum filtered correlation spectroscopy for the resolution of cross- and diagonal peaks can be appreciated in phase-sensitive presentations. Figure 8.3.7 shows two excerpts of spectra

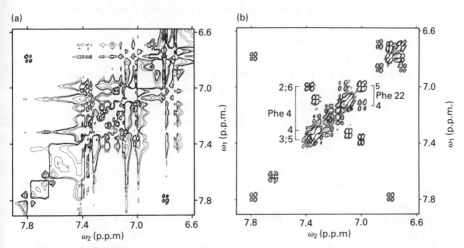

FIG. 8.3.7. Comparison of the aromatic region of phase-sensitive correlation spectra of basic pancreatic trypsin inhibitor: (a) conventional COSY experiment; (b) double-quantum filtered COSY spectrum. Both spectra were processed with identical resolution enhancement functions. Note the dispersion components in (a) which are largely eliminated in (b), thus allowing one to identify unambiguously a number of cross-peaks in the vicinity of the diagonal. (Adapted from Ref. 8.30.)

of basic pancreatic trypsin inhibitor, where it is apparent in the conventional COSY spectrum that the 2D dispersion peak shapes of the diagonal peaks strongly overlap with the 2D absorption cross-peaks. The almost entire elimination of dispersion components in the filtered spectrum allows one to identify cross-peaks that lie close to the diagonal.

8.3.3.2. p-Spin filters

It is often desirable to simplify 2D correlation spectra by suppressing all responses except for those stemming from systems with exactly p coupled spins. This corresponds to the 'bandpass filter' shown schematically in Fig. 8.3.4(b). One possible approach (8.32) towards the design of such a filter exploits the fact that a p-quantum coherence in a system with p spins, which may be referred to as 'total spin coherence' (8.33–8.35), evolves under the sum of all chemical shifts and is *not* affected by couplings, whereas a p-quantum coherence in a system with $N > p$ spins is modulated by couplings to $(N - p)$ passive spins. In the scheme of Fig. 8.3.5(e), it is possible to vary the interval τ_m between the pulses (or sequences of pulses) U and V, inserting a π-pulse at $\frac{1}{2}\tau_m$ to refocus the shifts. Averaging of the signal $s(t_1, \tau_m, t_2)$ over a series of τ_m values leads, in favourable cases, to a destructive interference of J-modulated p-quantum coherence in systems with $N > p$ spins.

This approach does not guarantee an ideal bandpass filter. For once, all q-spin–p-quantum coherences also pass the phase cycle designed to select p-quantum coherence. If $q = N$, these coherences (known as 'spin inversion coherences' (8.35)) are not modulated by couplings, and therefore survive the averaging over τ_m. Furthermore, the multiplet structure of the p-quantum coherence (8.8) may have a component that is not modulated, either because of symmetry or because of accidental degeneracies of J-couplings. In spite of these shortcomings, p-spin filters tend to perform reasonably well, because contributions from large spin systems are generally rather weak (8.32).

8.3.3.3. Filtering according to coupling network connectivity

As described in more detail in § 5.3.1.4, it is possible to design a pulse sequence U which efficiently excites multiple-quantum coherence in a network of specified coupling connectivity (8.36, 8.37). Thus it is not only possible to select the response of systems with exactly p coupled spins, but one can distinguish between various systems which have the same number of spins, and differ only in the type of coupling network. Thus various four-spin systems, such as AMKX, AMX_2, A_2X_2, A_3X systems, can be distinguished.

If connectivity-selective sequences U and V are inserted in the sequence for filtered 2D correlation spectra, only signals from the desired

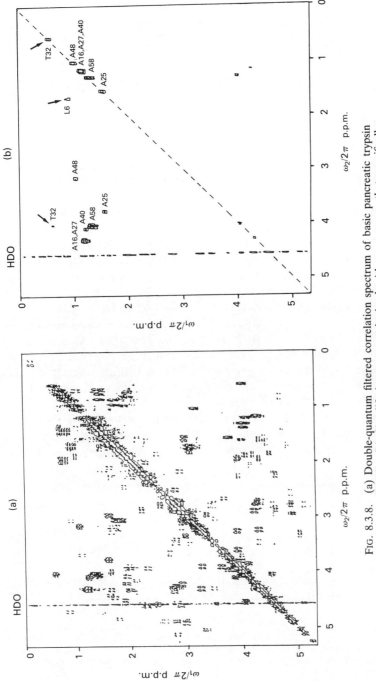

FIG. 8.3.8. (a) Double-quantum filtered correlation spectrum of basic pancreatic trypsin inhibitor (BPTI). (b) Simplified spectrum obtained with a sequence that specifically excites 4-quantum coherence in A_3X systems, and hence eliminates almost all signals except those stemming from the six alanine residues. Some weak responses from threonine and leucine are also visible (marked by arrows). The asymmetry of the spectrum originates from the connectivity-selective preparation sequence used in combination with constant time evolution. (Reproduced from ref. 8.37.)

coupling networks are retained. Figure 8.3.8 shows how the conventional COSY spectrum of basic pancreatic tryspsin inhibitor can be simplified by selecting the responses of the alanine residues, which are distinct from other amino-acids in that they contain spin systems of the A_3X type (8.37). The weak spurious responses of threonine and lysine can be rationalized in terms of the similarity of their coupling networks.

8.3.4. Relayed coherence transfer

Conventional 2D correlation spectroscopy, which is based on a single coherence transfer, can only provide evidence for the existence of direct coupling partners. In larger spin systems, this information is not always sufficient for unambiguous assignment.

Consider a 2D correlation spectrum with three chemical shifts Ω_k, Ω_l, and Ω_m and two pairs of cross-peaks as shown in Fig. 8.3.9(a). The information in this spectrum is insufficient to decide whether all three resonances belong to the same spin system with a linear coupling network $I_k - I_l - I_m$. Indeed, the spectrum could be rationalized by assuming a superposition of two distinct two-spin systems $I_k - I_l$ and $I_{l'} - I_m$ with an accidental degeneracy of the chemical shifts Ω_l and $\Omega_{l'}$.

This ambiguity can be removed if the coherence is transferred in two consecutive steps from k to l and l to m. This type of experiment is known as relayed magnetization transfer (8.6, 8.7). Relayed transfer can be achieved with the pulse sequence in Fig. 8.3.10(a), where the mixing pulse of the conventional 2D correlation experiment is replaced by a $[\pi/2 - \tau/2 - \pi - \tau/2 - \pi/2]$ sequence.

Relayed coherence transfer can easily be understood by considering a linear system k, l, m with $J_{km} = 0$, where we focus attention on coherence which is carried from spin k in t_1 to spin m in t_2. The net

(a)   (b)

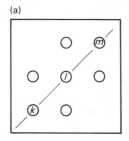 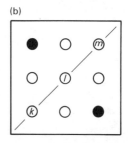

FIG. 8.3.9. (a) Conventional correlation spectrum of a linear three-spin system $I_k - I_l - I_m$ with $J_{km} = 0$ and hence with vanishing cross-peaks between Ω_k and Ω_m. (b) Schematic relayed correlation spectrum of the same system, with cross-peaks centred at $(\omega_1, \omega_2) = (\Omega_k, \Omega_m)$ and (Ω_m, Ω_k) which prove that I_k and I_m possess a common coupling partner and hence belong to the same coupling network.

8.3 MODIFIED 2D CORRELATION EXPERIMENTS

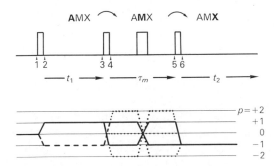

FIG. 8.3.10. (a) Sequence for relayed 2D correlation spectroscopy, where the mixing pulse of the basic COSY experiment is replaced by a $[(\pi/2)_x - \tau_m/2 - (\pi)_x - \tau_m/2 - (\pi/2)_x]$ sequence. In an AMX system with $J_{AX} = 0$, single-quantum coherence (indicated by bold lettering) is transferred first from A to M and subsequently from M to X. (b) Coherence transfer pathways in relayed transfer: only $p = \pm 1$ quantum coherences are relevant, while contributions from longitudinal magnetization ($p = 0$) in τ_m must be eliminated by phase-cycling.

effect of the mixing sequence $[(\pi/2)_x - \tau_m/2 - (\pi)_x - \tau_m/2 - (\pi/2)_x]$ is described by eqn (4.4.82). Only antiphase coherence of spin k at the end of the evolution period can give rise to relayed coherence transfer from spin k to spin m

$$\sigma_3 = I_{ky}I_{lz} \sin(\Omega_k t_1)\sin(\pi J_{kl} t_1). \tag{8.3.5}$$

With the transformations illustrated in Fig. 2.1.5, one obtains

$$2I_{ky}I_{lz} \xrightarrow{\pi J_{kl}\tau_m 2I_{ky}I_{ly}} \xrightarrow{\pi J_{lm}\tau_m 2I_{ly}I_{my}}$$
$$- I_{lz}I_{my} \sin(\pi J_{kl}\tau_m)\sin(\pi J_{lm}\tau_m)$$
$$+ \text{other terms.} \tag{8.3.6}$$

Hence the relevant term at the beginning of the detection period is

$$\sigma_6 = -I_{lz}I_{my} \sin(\Omega_k t_1)\sin(\pi J_{kl} t_1)\sin(\pi J_{kl}\tau_m)\sin(\pi J_{lm}\tau_m). \tag{8.3.7}$$

None of the numerous other terms contribute to a cross-peak between k and m. Equation (8.3.7) describes a cross-peak with antiphase doublet structure in both dimensions. If relaxation effects are included, the amplitude of this cross-peak is determined by the transfer function

$$f(\tau_m) = -\sin(\pi J_{kl}\tau_m)\sin(\pi J_{lm}\tau_m)\exp\{-\tau_m/T_2\}. \tag{8.3.8}$$

In more extensive networks, additional couplings of spins l and m to passive nuclei l' and m', respectively, reduce the transfer according to the transfer function

$$f(\tau_m) = -\sin(\pi J_{kl}\tau_m)\sin(\pi J_{lm}\tau_m)\prod_{l'}\cos(\pi J_{ll'}\tau_m)$$

$$\times \prod_{m'}\cos(\pi J_{mm'}\tau_m)\exp\{-\tau_m/T_2\}. \quad (8.3.9)$$

The coherence transfer pathways that are relevant in relayed transfer (§ 6.3) are shown in Fig. 8.3.10(b). To suppress signal contributions associated with longitudinal magnetization in τ_m (which may migrate because of cross-relaxation, see Chapter 9), the pathways with $p = 0$ in τ_m must be suppressed (8.38). In practical applications, τ_m is normally kept constant and optimized for maximum amplitude of the relayed cross-peaks. In this case, one may select the mirror-image pathways $p = 0 \to +1 \to \pm 1 \to -1$ and $p = 0 \to -1 \to \pm 1 \to -1$.

An experimental example of a relayed correlation spectrum is shown in Fig. 8.3.11. Four sections of a 2D spectrum of the protein basic pancreatic trypsin inhibitor (BPTI) are shown: the diagonal peaks associated with the NH protons, the region of the cross-peaks stemming from a one-step $C^\alpha H \to NH$ transfer, and the region of the relayed signals arising from double transfer $C^\beta H \to C^\alpha H \to NH$. These additional signals are of practical relevance for the assignment of amino acids with nearly degenerate $C^\alpha H$ resonances, and more generally for sequential resonance assignments whenever two linear spin systems $k-l-m$ and $k'-l'-m'$ have degenerate shifts $\Omega_l \simeq \Omega_{l'}$ (8.38–8.40).

As may be appreciated in the phase-sensitive plot of Fig. 8.3.11, the relayed peaks appear in pure 2D absorption, while the cross-peaks associated with a single coherence transfer step and the diagonal peaks appear with mixed phases.

The basic idea of relayed transfer can be extended by including a series of n consecutive $\pi/2$-pulses separated by suitable intervals, in order to transfer coherence across n couplings in a linear coupling network. The intervals should be approximately tailored to the relevant J-couplings. However, the efficiency of multiple-step transfer suffers from numerous competitive transfers and from relaxation in the extended mixing period.

A three-step relayed transfer experiment has been applied to BPTI to transfer magnetization $C^\gamma H \to C^\beta H \to C^\alpha H \to NH$ (8.38) in order to lift ambiguities in spectral assignment.

If the duration of the mixing interval τ_m is constant, the efficiency of relayed transfer is obviously dependent on the magnitude of the coupling constants. This problem may be alleviated by removing the π-pulse from the τ_m interval and by systematic variation of τ_m in concert with t_1, i.e. $\tau_m = \chi t_1$ (8.7). In this case, it is important to restrict the number of coherence transfer pathways. If we focus attention on $p = 0 \to +1 \to$

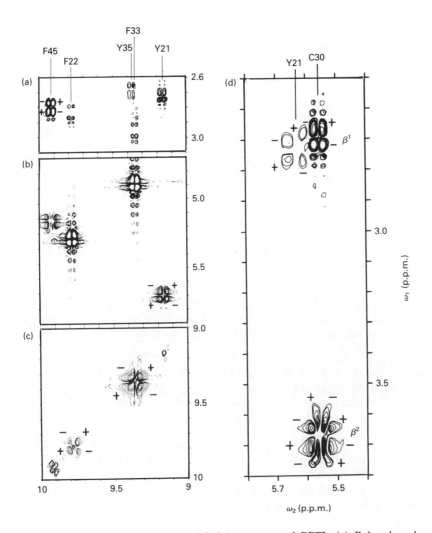

FIG. 8.3.11. Excerpts of a relayed correlation spectrum of BPTI. (a) Relayed peaks associated with the transfer $C^\beta H \to C^\alpha H \to NH$ in three phenylalanine and two tyrosine residues (note the antiphase pure 2D absorption peaks). (b) 'Direct connectivity' signals stemming from a single coherence transfer step $C^\alpha H \to NH$ in the same experiment. (c) Diagonal peaks of the NH region. The peaks in (b) and (c) appear with mixed phases. Note that the spin systems of F33 and Y35 have strongly overlapping chemical shifts of the NH and $C^\alpha H$ protons and can only be distinguished by the shifts of the $C^\beta H$ protons. (d) Direct $\alpha-\beta$ connectivities for Y21 and C30 and relayed connectivity between the β^1 and α protons of C30 via β^2. (Reproduced from Ref. 8.38.)

$+1 \to -1$ and $p = 0 \to -1 \to -1 \to -1$, the effective ω_1-frequency associated with a transfer $k \to l \to m$ is

$$\omega_1^{\text{eff}} = \pm(\Omega_k + \chi\Omega_l). \qquad (8.3.10)$$

If, on the other hand, we select the two pathways $p = 0 \to +1 \to -1 \to -1$ and $p = 0 \to -1 \to +1 \to -1$, the effective precession frequency is

$$\omega_1^{\text{eff}} = \pm(\Omega_k - \chi\Omega_l). \qquad (8.3.11)$$

For $\chi = 1$ (i.e. $\tau_m = t_1$), one obtains sums or differences of chemical shifts, and one may speak of 'pseudo double- or zero-quantum spectra' (8.41). Analogous heteronuclear relayed transfer experiments (e.g. $^1\text{H} \to {}^1\text{H} \to {}^{13}\text{C}$) can be valuable for assignment in heteronuclear spin systems. Independent manipulation of abundant and rare spins allows one to build special features into the experiment, such as the suppression of signals that fail to participate in relayed transfer (8.42) (see § 8.5.4).

It is interesting to compare relayed coherence transfer with triple resonance experiments (8.43, 8.44). Consider a linear system $I_k - I_l - I_m$ with $J_{km} = 0$. In order to prove by conventional multiple resonance that the spins k and m belong to one and the same coupling network, one could envisage the following Gedankenexperiment.

1. An r.f. field B_l with a strength comparable to the coupling J_{kl} is applied at the shift Ω_l of the 'central' nucleus l in order to mix the states.

2. A strong r.f. field B_k is swept through the resonance region of spin k.

3. At the same time, the response of spin m is observed by a CW or Fourier experiment.

In the case that the three spins belong to the same coupling network, the m-spin response depends (in a complicated manner) on the position of the r.f. field B_k. This shows that multiple coherence transfer is closely related to multiple resonance, and the basic relayed transfer experiment may be thought of as a pseudo-3D experiment.

8.3.5. Coherence transfer by an average Hamiltonian in total correlation spectroscopy

In the methods discussed so far, the transfer of coherence between two transitions is brought about by one or more mixing pulses. The preparation of antiphase coherence is a prerequisite for coherence transfer under a pulse.

An alternative approach uses a mixing period of extended duration τ_m with a suitable average mixing Hamiltonian $\bar{\mathcal{H}}^{(m)}$. To obtain a transfer of coherence, $\bar{\mathcal{H}}^{(m)}$ should be tailored to 'mix' the various coherences. By proper selection of $\bar{\mathcal{H}}^{(m)}$ it is possible to transfer all components, including in-phase coherence as well as longitudinal polarization.

If it were possible to eliminate all chemical shifts from $\bar{\mathcal{H}}^{(m)}$ without removing the mutual spin–spin couplings, the coherences would be able to migrate in an oscillatory manner through the entire spin system. In 2D correlation spectra, this implies that cross-peaks would be obtained between all transitions belonging to the same coupled spin system. This may be considered as the ultimate extension of relayed correlation spectroscopy, leading to 'total correlation spectroscopy' (TOCSY) (8.45).

The desired mixing Hamiltonian has the simple form

$$\bar{\mathcal{H}}^{(m)} = \sum_{k<l} 2\pi J_{kl} \mathbf{I}_k \mathbf{I}_l \qquad (8.3.12)$$

and may be called 'isotropic mixing' Hamiltonian. The eigenfunctions of $\bar{\mathcal{H}}^{(m)}$ involve linear combinations of the spin functions of all spins. The evolution in the course of the mixing process can be rationalized in terms of so-called 'collective spin modes' (8.45). These collective modes are responsible for the transfer of coherence. This is in contrast to the high-field Hamiltonian, where the individual spins move independently in 'single-spin modes' provided the couplings are weak.

Various pulse sequences have been proposed (8.45) to obtain the isotropic mixing Hamiltonian of eqn (8.3.12). The simplest approach employs a string of π-pulses with a repetition rate that is fast compared to the largest chemical shift difference (8.45). Additional sequences for efficient coverage of a wide frequency range have been proposed by Bax (8.46).

The features of coherence transfer through isotropic mixing can be appreciated most easily by considering a two-spin-$\frac{1}{2}$ system with $\bar{\mathcal{H}}^{(m)} = 2\pi J_{kl} \mathbf{I}_k \mathbf{I}_l$. The collective spin modes in this simple case correspond to the sum and the difference of single spin operators and to the sum and difference of product spin operators

$$\Sigma_\alpha = \tfrac{1}{2}\{I_{k\alpha} + I_{l\alpha}\},$$
$$\Delta_\alpha = \tfrac{1}{2}\{I_{k\alpha} - I_{l\alpha}\},$$
$$\Sigma_{\alpha\beta} = \{I_{k\alpha} I_{l\beta} + I_{k\beta} I_{l\alpha}\},$$
$$\Delta_{\alpha\beta} = \{I_{k\alpha} I_{l\beta} - I_{k\beta} I_{l\alpha}\}, \qquad \alpha, \beta = x, y, z \qquad (8.3.13)$$

with the commutation relations

$$[\bar{\mathcal{H}}^{(m)}, \Sigma_\alpha] = 0,$$
$$[\bar{\mathcal{H}}^{(m)}, \Sigma_{\alpha\beta}] = 0,$$
$$[\bar{\mathcal{H}}^{(m)}, \Delta_\alpha] = i\Delta_{\beta\gamma},$$
$$[\bar{\mathcal{H}}^{(m)}, \Delta_{\beta\gamma}] = -i\Delta_\alpha \qquad (8.3.14)$$

where (α, β, γ) is a cyclic permutation of (x, y, z). This leads immediately to the following time evolution of the difference terms

$$\Delta_\alpha \xrightarrow{\mathcal{H}^{(m)}\tau_m} \Delta_\alpha \cos(2\pi J_{kl}\tau_m) + \Delta_{\beta\gamma} \sin(2\pi J_{kl}\tau_m)$$

and

$$\Delta_{\beta\gamma} \xrightarrow{\mathcal{H}^{(m)}\tau_m} \Delta_{\beta\gamma} \cos(2\pi J_{kl}\tau_m) - \Delta_\alpha \sin(2\pi J_{kl}\tau_m), \quad (8.3.15)$$

while the sum terms remain invariant.

Because the Hamiltonian $\mathcal{H}^{(m)}$ is isotropic, all three components x, y, z behave identically. For the evolution of the x-component of spin k, we find explicitly

$$I_{kx} \xrightarrow{\mathcal{H}^{(m)}\tau_m} \tfrac{1}{2}I_{kx}\{1 + \cos(2\pi J_{kl}\tau_m)\} + \tfrac{1}{2}I_{lx}\{1 - \cos(2\pi J_{kl}\tau_m)\}$$

$$+ (I_{ky}I_{lz} - I_{kz}I_{ly})\sin(2\pi J_{kl}\tau_m). \quad (8.3.15a)$$

Similar expressions apply also to the other components of in-phase (I_{ky}) and antiphase coherences ($2I_{kx}I_{lz}$, $2I_{ky}I_{lz}$) as well as to the longitudinal polarization (I_{kz}). All of these components are periodically exchanged between spins k and l under the influence of $\mathcal{H}^{(m)}$.

The most important feature in eqn (8.3.15a) is the transfer of *in-phase* I_{kx} into *in-phase* I_{lx} coherence, which implies that the corresponding contributions to the cross-peak multiplets in the 2D spectrum are in-phase. The coherence oscillates back and forth: for a suitable choice of τ_m, only cross-peaks remain while the diagonal peaks vanish altogether, as illustrated by the experimental examples in Fig. 8.3.12.

The transfer of *all* components, achieved with sequences that compensate for r.f. field inhomogeneity (e.g. a string of π-pulses with alternating phases along the $+x$-axis and $-x$-axis), actually precludes one from obtaining pure phase spectra. On the other hand, with sequences that are not compensated for r.f. field inhomogeneity, such as a sequence of in-phase $(\pi)_x$ pulses (Fig. 8.3.12(a)), the terms that do not commute with $I_{kx} + I_{lx}$ decay rapidly because of r.f. inhomogeneity and only the terms I_{kx}, I_{lx}, $I_{kx}I_{lx}$, and $I_{ky}I_{lz} - I_{kz}I_{ly}$ can survive. The antiphase components associated with the $(2I_{ky}I_{lz} - 2I_{kz}I_{ly})$ term can be cancelled by co-addition of signals obtained with different mixing times τ_m. In this case, pure 2D absorption peakshapes can be obtained, as shown in Fig. 8.3.12(b).

Two-dimensional correlation spectra of larger molecules with extensive coupling networks obtained with total correlation spectroscopy can show cross-peaks between all pairs of nuclei, even if the nuclei are not directly coupled together. This feature allows one to determine subspectra

8.3 MODIFIED 2D CORRELATION EXPERIMENTS

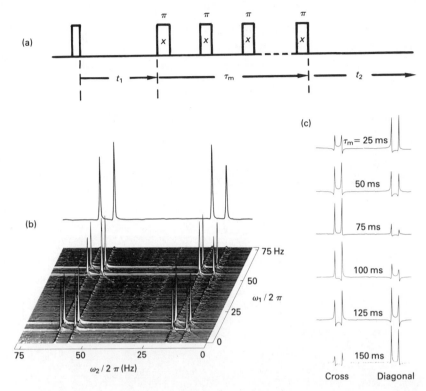

FIG. 8.3.12. (a) Simple sequence for 2D total correlation spectroscopy with a mixing period of duration τ_m consisting of a string of equidistant $(\pi)_x$-pulses. (b) 2D spectrum of the two-spin system of 2,3 dibromothiophene, obtained by co-addition of signals recorded with six different τ_m-intervals. The section shown in (b) illustrates the cancellation of the dispersion components. (c) Phase-sensitive sections taken through six spectra obtained with regular increments of τ_m (pulse interval 500 μs). Note the oscillatory behaviour: for $\tau_m = 75$ ms $\simeq (2J)^{-1}$, the diagonal peaks are almost suppressed, while the amplitude of the cross-peaks is maximized. The inverse situation occurs for $\tau_m = 150$ ms. The dispersion contributions to the peakshapes originate from the $(I_{1y}I_{2z} - I_{1z}I_{2y})$ term in eqn (8.3.15a). These contributions can be cancelled by co-addition of signals obtained with different τ_m. (Adapted from Ref. 8.50.)

originating from subunits in a complex molecule such as a protein. Unlike relayed coherence transfer methods (§ 8.3.4), the approach described here does not require any knowledge of the magnitude of the couplings.

Analogous sequences have also been proposed (8.98) for heteronuclear correlation spectroscopy as explained in § 8.5.6. These experiments are closely related to cross-polarization in the rotating frame that has been applied extensively in solids (8.47, 8.48) as well as in liquids (8.10, 8.49, 8.50) (see § 4.5.3).

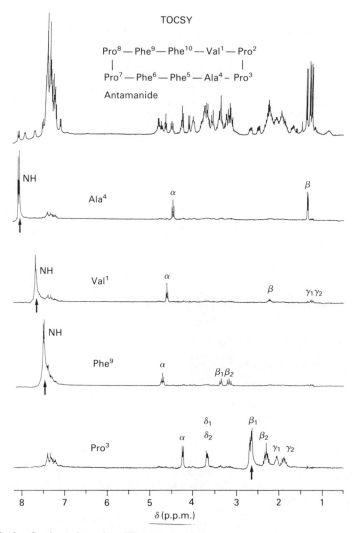

FIG. 8.3.13. Sections through a 2D total correlation spectrum of the cyclic decapeptide antamanide. The top trace shows for comparison the conventional 300 MHz 1D proton resonance spectrum. The following three traces represent sections parallel to the ω_2 axis at ω_1 frequencies corresponding to the NH resonances of alanine[4], valine[1] and phenylalanine[9], respectively. The last trace shows a section at the ω_1 frequency of the β_1 proton of proline[3]. The arrows indicate the crossing of the sections with the diagonal. The spectra can be interpreted as originating from coherence being excited at the position of the arrows and transferred by the extended mixing process to those spins whose resonances appear in the spectra. In effect, subspectra of individual amino acid residues have been obtained. The pulse scheme of Fig. 8.3.12a was applied with offset compensated composite π pulses of the form $336^\circ_0 \ 246^\circ_\pi \ 10^\circ_{\pi/2} \ 74^\circ_{3\pi/2} \ 10^\circ_{\pi/2} \ 246^\circ_\pi \ 336^\circ_0$ (R. Tycko, H. M. Cho, E. Schneider, and A. Pines, *J. mag. Reson.* **61**, 90 (1985)); 90° pulse duration: 8.6 μs, pulse separation: 2 μs, interval between composite π pulses: 200 μs. Six experiments with different mixing times, τ_m = 13.3 ms, 26.5 ms, 39.8 ms, 53.1 ms, 66.4 ms, and 79.6 ms, were coadded (recorded data matrix: 984 × 4096 samples, Fourier-transformed data matrix: 2048 × 8192 data points). From unpublished work by C. Griesinger and O. W. Sørensen.

8.4. Homonuclear two-dimensional multiple-quantum spectroscopy

The 2D experiments discussed in §§ 8.2 and 8.3 are primarily concerned with the precession of single-quantum coherence, although in some cases multiple-quantum coherence is involved in a transient manner for the purpose of selection or filtration (§ 8.3.3). The present section describes experiments where multiple-quantum coherence is allowed to evolve in the t_1-period in order to measure the relevant frequencies and relaxation times. Multiple-quantum spectroscopy may be considered as a generalization of 2D correlation spectroscopy, as illustrated in Fig. 8.4.1: the preparation pulse in the COSY experiment is simply replaced by a more elaborate sequence capable of exciting coherences of various orders. Conversely, one may look at 2D correlation spectroscopy as a special case of p-quantum spectroscopy limited to $p = \pm 1$.

8.4.1. Excitation and detection of multiple-quantum coherence

Various methods for the excitation of multiple-quantum coherence were reviewed in § 5.3.1. Some of these involve selective pulses and therefore presuppose some knowledge of the single-quantum spectrum. Although selective excitation is not incompatible with 2D spectroscopy, it does not

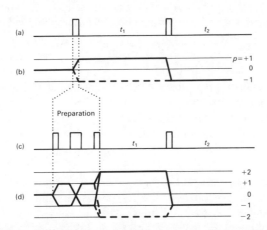

FIG. 8.4.1. Analogy between 2D correlation spectroscopy, (a) and (b), and 2D multiple-quantum spectroscopy, (c) and (d). The two methods are merely distinguished by the preparation and by the selection of coherence transfer pathways. The example shown in (d) refers to the selection of $p = \pm 2$. More sophisticated sequences can be used for the excitation of multiple-quantum coherence and for its reconversion into observable magnetization with $p = -1$. The phase-cycling procedures required for proper pathway selection are described in § 6.3, while techniques for obtaining pure 2D absorption peaks in multiple-quantum spectra are described in § 6.5.3.

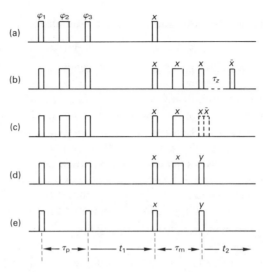

FIG. 8.4.2. Pulse sequences that are commonly employed for 2D multiple-quantum spectroscopy of homonuclear systems, as discussed in the text. In (b), the interval τ_z represents a z-filter (§ 8.3.1).

allow one to exploit all the advantages of 2D methods. Excitation sequences with non-selective pulses make it possible to obtain 2D spectra of unknown compounds with unknown coupling networks.

Some of the techniques that are commonly used in 2D spectroscopy are shown in Fig. 8.4.2. The excitation is usually achieved with a pulse sandwich $[(\pi/2)_{\varphi_1} - \tau_p/2 - (\pi)_{\varphi_2} - \tau_p/2 - (\pi/2)_{\varphi_3}]$, with $\varphi_1 = \varphi_2 = \varphi_3$ for even p and $\varphi_1 = \varphi_2 = \varphi_3 \pm \pi/2$ for odd p. In some cases, the central refocusing pulse may be dropped.

The reconversion into observable magnetization may be achieved either with a single mixing pulse (Fig. 8.4.2(a)) or alternatively with a sequence designed to reverse the effect of the excitation, which produces t_1-modulated longitudinal magnetization that is subsequently 'read' by a $\pi/2$-pulse. This is shown in Fig. 8.4.2(b), which may be compared with the general scheme of Fig. 5.3.1(a) (8.32, 8.33, 8.51, 8.52). In practice, these two approaches are not always clearly distinguishable, as shown in Fig. 8.4.2(c): if the last pulse of the reconversion sandwich and the 'read' pulse cancel each other (dashed lines), the mixing sequence is reduced to a $\pi/2$-pulse followed by a $[\tau_m/2 - (\pi) - \tau_m/2]$ rephasing period. In Fig. 8.4.2(d), a so-called 'purging pulse' (8.32) has been included, which reconverts antiphase magnetization into unobservable multiple-quantum coherence just before the detection period. All schemes with symmetrical excitation and detection sequences with $\tau_p = \tau_m$ (Fig. 8.4.2(b) to (e)) lend themselves to signal-averaging over a range of intervals $\tau_{min} < \tau_p < \tau_{max}$ in order to obtain responses that are largely independent of the magnitude of the couplings. This averaging procedure is most effective if the refocusing pulses are dropped, as in Fig. 8.4.2(e) (8.53).

8.4.2. Double-quantum spectra of two-spin systems

In this section we discuss some basic features of multiple-quantum spectroscopy by means of a two-spin system. At the same time, we shall encounter an application of practical importance for the assignment of carbon-13 spectra which exploits ^{13}C–^{13}C couplings in natural abundance.

Double-quantum spectra of two-spin systems contain in principle the same information as double-quantum filtered single-quantum 2D correlation spectra (§ 8.3.3). However, double-quantum spectra have the advantage that the magnetization is spread over a smaller number of peaks: whereas the COSY spectrum of an AX system contains four multiplets with four components each, a double-quantum spectrum of the same system contains only two multiplets with two components each (assuming a real Fourier transformation with respect to t_1 in both cases). As in ω_1-decoupled COSY (§ 8.3.2), the simplification is obtained at the expense of unpredictable peak intensities, unless the magnitude of the scalar couplings is approximately known before setting up the experiment. If the couplings are not known, averaging procedures may be employed as discussed below.

Consider the pulse sequence in Fig. 8.4.2(a). The excitation of double-quantum coherence by the preparation sandwich $[(\pi/2)_x - \tau_p/2 - (\pi)_x - \tau_p/2 - (\pi/2)_x]$ is described in eqn (5.3.4). If a complex Fourier transformation is calculated with respect to t_1, and if the carrier is positioned outside the spectrum in ω_2, the signals associated with $p = +2$ and $p = -2$ quantum coherences appear in distinct quadrants (Fig. 8.4.3). The mixing pulse with r.f. rotation angle β applied after the evolution time t_1 converts $p = -2$ quantum coherence into observable $p = -1$ quantum coherences according to eqn (2.1.111)

$$I_k^- I_l^- \xrightarrow{\beta(I_{kx}+I_{lx})} -\frac{i}{2}\sin\beta\cos^2\beta/2\,[I_k^- I_l^\alpha - I_k^- I_l^\beta + I_k^\alpha I_l^- - I_k^\beta I_l^-], \quad (8.4.1)$$

while the conversion of $p = +2$ quantum coherence into transverse magnetization is given by

$$I_k^+ I_l^+ \xrightarrow{\beta(I_{kx}+I_{lx})} +\frac{i}{2}\sin\beta\sin^2\beta/2\,[I_k^- I_l^\alpha - I_k^- I_l^\beta + I_k^\alpha I_l^- - I_k^\beta I_l^-]. \quad (8.4.2)$$

The eight coherence transfer processes in eqns (8.4.1) and (8.4.2) lead to the eight signals shown in the right half of Fig. 8.4.3. The intensities associated with the two pathways $p = 0 \rightarrow -2 \rightarrow -1$ and $p = 0 \rightarrow +2 \rightarrow -1$

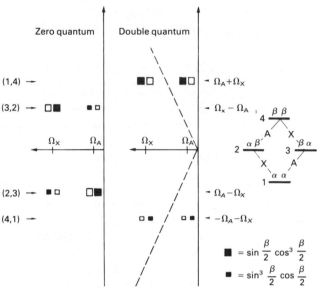

FIG. 8.4.3. Schematic zero- and double-quantum spectra of a weakly coupled two-spin system, assuming a single detection pulse with rotation angle β, a complex Fourier transformation with respect to t_1, and assuming that all coherences are excited uniformly with the same phase. This can be achieved in practice with a preparation sequence $[(\pi/2)_x - \tau_p/2 - (\pi)_x - \tau_p/2 - (\pi/4)_y]$ which leads to a density operator containing, in addition to single-quantum coherence, the term $4I_{kx}I_{lx} = (I_k^+I_l^+ + I_k^+I_l^- + I_k^-I_l^+ + I_k^-I_l^-)$. By contrast, the excitation sequence of Fig. 8.4.2(a) does not yield any zero-quantum terms, and generates coherences of orders $p = \pm 2$ with opposite signs (see eqn (5.3.2)). Note the alternating signs within the doublets (filled and open symbols represent positive and negative peaks respectively), and the symmetry properties of the amplitudes which have a ratio of $\cot^2\beta/2$ (large symbols correspond to dominant signals for $0 < \beta < \pi/2$). All signals have mixed phases described by eqn (6.5.10). The dashed lines in the double-quantum spectrum indicate skew diagonals $\omega_1 = \pm 2\omega_2$. The double-quantum signals appear symmetrically disposed on either side of these skew diagonals. (Reproduced from Ref. 8.8.)

have an amplitude ratio of $\cot^2\beta/2$. As in single-quantum spectra, the two classes of signals are superimposed if a real (cosine) transformation is calculated with respect to t_1, leading to pure 2D absorption peaks if $\beta = \pi/2$ (see § 6.5.3).

The amplitudes of the signals associated with double-quantum coherence stand in contrast to the intensities observed in zero-quantum spectra, shown in the left half of Fig. 8.4.3 (8.8). Although zero-quantum spectra offer some attractive features (most notably the lack of sensitivity of zero-quantum evolution to inhomogeneous static fields, discussed in § 5.4.3), practical applications are hampered by the fact that phase-cycling schemes do not allow the separation of signals stemming from zero-quantum coherence and from longitudinal polarization.

8.4 HOMONUCLEAR 2D MULTIPLE-QUANTUM SPECTROSCOPY

In carbon-13 NMR, the homonuclear $^1J_{CC}$ couplings between bonded carbon nuclei are on the order of 30–45 Hz for saturated carbons, and it is possible to excite double-quantum coherence fairly uniformly with the sequence of Fig. 8.4.2(a), setting $\tau = (2J_{CC}^{\text{average}})^{-1}$. In natural abundance carbon-13 NMR, the intensity of the satellites due to homonuclear couplings is a factor of 200 lower than the intensity of the signals stemming from isolated ^{13}C spins. Under such adverse conditions, the simplicity of double-quantum spectra (i.e. the lack of a complicated multiplet structure) is particularly attractive. The dynamic range problem is less accute than in various forms of difference spectroscopy (including double-quantum filtered COSY), since the excitation sandwich of Fig. 8.4.1(a) acts as a 2π-pulse on the singlet magnetization associated with isolated carbon-13 spins.

The application of double-quantum spectroscopy to the identification of pairs of carbon-13 spins has become known under the acronym of INADEQUATE (Incredible Natural Abundance Double Quantum Transfer Experiment) (8.54–8.64). The pulse scheme that is commonly employed is shown in Fig. 8.4.2(a). The phase-cycling required to reject undesired single-quantum coherence may be derived from the coherence transfer pathways shown in Fig. 8.4.1(d) with the rules set out in § 6.3, leading to a minimum four-step cycle. To improve the suppression, this can be expanded into a 16, 32, or 128-step cycle (8.61). It is possible to retain only the $p = 0 \rightarrow +2 \rightarrow -1$ pathway to reduce the spectral width in the ω_1-domain, either by using z-pulses (8.58), by using a mixing pulse with $\beta = 135°$, which favours the signals represented by small symbols in Fig. 8.4.3 (8.60), or more accurately with a phase-cycle with $N > 4$ steps (8.65). However, it appears preferable to retain both mirror-image pathways in Fig. 8.4.1(d) with $\beta = \pi/2$, and to use time-proportional phase increments to separate peaks with $p = \pm 2$, as shown in Fig. 6.6.4. Because the double-quantum signals of two-spin systems lie within a narrow band along the diagonal (see Fig. 8.4.9), it is possible to reduce the spectral width in ω_1 by a factor of two in conjunction with foldover correction or delayed acquisition (8.62–8.65).

An eloquent experimental illustration of an 'INADEQUATE' spectrum is shown in Fig. 8.4.4, where the connectivities of all neighbouring carbon nuclei in panamine can be derived from a single spectrum obtained from a sample with natural isotopic abundance.

The multiplets with alternating phases illustrated in Fig. 8.4.3 can be converted into in-phase multiplets by appending a sequence $[\tau_m/2 - (\pi) - \tau_m/2]$ with $\tau_m = (2J)^{-1}$ (Fig. 8.4.2(c)). This leads to the idea of symmetrical excitation and detection, which is best described by considering a hypothetical experiment with two $[(\pi/2)_x - \tau/2 - (\pi)_x - \tau/2 - (\pi/2)_x]$ sandwiches for excitation and detection (Fig. 8.4.2(b)). The

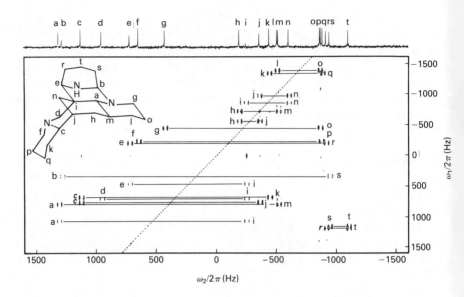

FIG. 8.4.4. Double-quantum spectrum revealing the connectivities through scalar couplings (and hence the vicinity in the molecular skeleton) of pairs of carbon-13 spins in a natural abundance sample of panamine (2D 'INADEQUATE' spectrum), obtained with the sequence $[(\pi/2) - \tau_p/2 - (\pi) - \tau_p/2 - (\pi/2)] - t_1 - (\beta) - t_2$. The 1D spectrum along the top shows the conventional proton-decoupled carbon-13 spectrum; the horizontal bars in the 2D spectrum indicate the individual AX patterns, each with the structure shown in the lower half of Fig. 8.4.3, with pairs of signals symmetrically disposed about the skew diagonal. The pathways with $p = 0 \to +2 \to -1$ were emphasized by setting $\beta = 135°$. (Reproduced from Ref. 8.60.)

overall coherence transfer pathways in this experiment are $p = 0 \to \pm 2 \to 0 \to -1$, i.e. double-quantum coherence is temporarily reconverted into (t_1-modulated) longitudinal polarization. According to eqn (5.3.4), one obtains, immediately after the preparation sandwich of duration τ_p (neglecting residual longitudinal polarization),

$$\sigma(t_1 = 0) = 2\{2QT\}_y \sin \pi J_{kl}\tau_p. \qquad (8.4.3)$$

Free evolution of double-quantum coherence (eqn (5.3.34)) leads to the following state just before the detection sandwich

$$\sigma(t_1) = 2\{2QT\}_y \cos(\Omega_k + \Omega_l)t_1 \sin \pi J_{kl}\tau_p$$
$$- 2\{2QT\}_x \sin(\Omega_k + \Omega_l)t_1 \sin \pi J_{kl}\tau_p. \qquad (8.4.4)$$

Only the first term is reconverted into Zeeman polarization by a

8.4 HOMONUCLEAR 2D MULTIPLE-QUANTUM SPECTROSCOPY

symmetrical mixing sandwich of duration τ_m, described by a propagator $\exp\{-i\pi J_{kl}\tau_m 2I_{ky}I_{ly}\}$

$$\sigma(t_1 + \tau_m) = -(I_{kz} + I_{lz})\cos(\Omega_k + \Omega_l)t_1 \sin \pi J_{kl}\tau_p \sin \pi J_{kl}\tau_m$$
$$+ \{2I_{kx}I_{ly} + 2I_{ky}I_{lx}\}\cos(\Omega_k + \Omega_l)t_1 \sin \pi J_{kl}\tau_p \cos \pi J_{kl}\tau_m$$
$$- (2I_{kx}I_{lx} - 2I_{ky}I_{ly})\sin(\Omega_k + \Omega_l)t_1 \sin \pi J_{kl}\tau_p. \quad (8.4.5)$$

The desirable t_1-modulated I_{kz}-terms may be converted into observable magnetization by another $\pi/2$-pulse (Fig. 8.4.2(b)). The double-quantum coherences remaining in eqn (8.4.5) may be cancelled by cycling the phase of this 'read pulse'. In practice, this is not essential, and one may append a $(\pi/2)_{-x}$-pulse with constant phase to the mixing sequence. The same effect is achieved by dropping the last $(\pi/2)_x$-pulse of the sandwich (Fig. 8.4.2(c)), leading to the observable terms

$$\sigma^{obs}(t_1, t_2 = 0) = -(I_{ky} + I_{ly})\cos(\Omega_k + \Omega_l)t_1 \sin \pi J_{kl}\tau_p \sin \pi J_{kl}\tau_m$$
$$- (2I_{kx}I_{lz} + 2I_{kz}I_{lx})\cos(\Omega_k + \Omega_l)t_1 \sin \pi J_{kl}\tau_p \cos \pi J_{kl}\tau_m. \quad (8.4.6)$$

The undesirable antiphase magnetization terms can be converted back into multiple-quantum coherence with a $(\pi/2)_y$ 'purging' pulse (Fig. 8.4.2(d)), which leaves the in-phase t_1-modulated $(I_{ky} + I_{ly})$-terms invariant. Thus the detection sandwich $[(\pi/2)_x - \tau_m/2 - (\pi)_x - \tau_m/2 - (\pi/2)_y]$ shown in Fig. 8.4.2(d) leads to pure in-phase multiplets. If the durations of the excitation and detection sandwiches are equal ($\tau_p = \tau_m = \tau$), the observable coherence takes the form

$$\sigma^{obs}(t_1, t_2 = 0) = -(I_{ky} + I_{ly})\cos(\Omega_k + \Omega_k)t_1 \sin^2 \pi J_{kl}\tau. \quad (8.4.7)$$

Because of the dependence on the square of the sine, it is possible to average over a range of τ-values, in order to observe a signal that is largely independent of the magnitude of the J-coupling

$$\bar{\sigma}^{obs}(t_1, t_2 = 0) = -\tfrac{1}{2}(I_{ky} + I_{ly})\cos(\Omega_k + \Omega_l)t_1. \quad (8.4.8)$$

The idea of symmetrical excitation and detection is also applicable to higher-order multiple-quantum transitions in larger spin systems (8.32, 8.51–8.53). It can be shown (8.53) that the averaging procedure is more effective if the π-pulses are removed from the preparation and detection sandwiches (sequence in Fig. 8.4.1(d)).

8.4.3. Multiple-quantum spectra of scalar-coupled networks in isotropic phase

Although conventional single-quantum correlation spectroscopy (§ 8.2) is useful for identifying coupled nuclei, it does not allow one to analyse unknown networks of coupled spins without ambiguities. Thus in a linear fragment of the type A – M – X with $J_{AX} = 0$, relayed coherence transfer (see § 8.3.4) is required to verify that the remote nuclei A and X indeed belong to one and the same network, to exclude an accidental superposition of two separate systems A – M and M' – X with degenerate shifts $\Omega_M = \Omega_{M'}$. In addition, there is a need for identifying equivalent spins, since it is often difficult to distinguish subsystems of the type AX, A_2X_3, etc. in complex molecules where multiplets are not fully resolved and integrals cannot be measured reliably. Multiple-quantum NMR may be used to verify the existence of magnetically or chemically equivalent nuclei and to check whether remote nuclei belong to a common coupling network.

In order to appreciate the distinction between single- and multiple-quantum experiments, Fig. 8.4.5 shows a variety of coherence transfer processes that may occur in a weakly-coupled four-spin system. The coupling network chosen for illustration is of the AMX_2 type with $J_{AX} = 0$. Figure 8.4.5(b) describes a typical coherence transfer process in conventional correlation spectroscopy, the symbolic notation being equivalent to

$$I_A^{\pm} I_{Mz} \rightarrow I_{Az} I_M^{\pm} \qquad (8.4.9)$$

which amounts to a transfer of antiphase coherence from spin A to spin M. Relayed transfer from A to X, described in detail in § 8.3.4, is shown in Fig. 8.4.5(c). The last three schemes in Fig. 8.4.5 illustrate three classes of coherence transfer processes that must be distinguished in double-quantum NMR: case (d) corresponds to the transfer processes within a two-spin subsystem, while Fig. 8.4.5(e) and (f) show double-quantum coherences which involve two 'remote' nuclei that are not directly coupled together but possess a common coupling partner. In the case of remote connectivity (Fig. 8.4.5(e)), the relevant transfer under the mixing pulse at the end of the evolution period is

$$I_A^{\pm} I_{Mz} I_X^{\pm} \rightarrow I_{Az} I_M^{\pm} I_{Xz} \qquad (8.4.10)$$

where the double-quantum coherence involving spins I_A and I_X must be in antiphase with respect to the central spin I_M. Note that this type of coherence can be excited even though $J_{AX} = 0$. Since couplings between magnetically equivalent nuclei are ineffective in isotropic phase, the process shown in Fig. 8.4.5(f) is essentially analogous to Fig. 8.4.5(e)

$$I_{Mz} I_X^{\pm} I_{X'}^{\pm} \rightarrow I_M^{\pm} I_{Xz} I_{X'z}. \qquad (8.4.11)$$

8.4 HOMONUCLEAR 2D MULTIPLE-QUANTUM SPECTROSCOPY

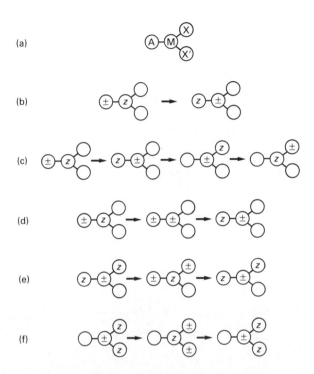

FIG. 8.4.5. Symbolic representation of coherence transfer processess in an AMX$_2$ coupling network (a). The symbols ± and z refer to operators $I^\pm$ and I_z associated with the corresponding spins. (b) Single-quantum transfer between A and M in conventional 2D correlation spectroscopy. (c) Relayed magnetization transfer between A and X. (d) Excitation and reconversion of double-quantum coherence associated with direct connectivity. (e) Same as (d), but for remote connectivity. (f) Same as (d), but for magnetically equivalent spins.

The distinction lies in the type of information that can be derived from the frequency coordinates of the resulting multiple-quantum signals.

The three types of coherence transfer processes shown schematically in Fig. 8.4.5(d), (e), and (f) give rise to characteristic signal patterns in double quantum spectra, that are shown in Fig. 8.4.6. Directly connected pairs of nuclei lead to pairs of signals at $\omega_1 = \pm(\Omega_A + \Omega_X)$ that are symmetrically disposed at $\omega_2 = \Omega_A, \Omega_X$ on either side of the skew diagonals $\omega_1 = \pm 2\omega_2$, as in two-spin systems. Magnetically equivalent nuclei in an A$_2$X subsystem lead to double-quantum signals at $\omega_1 = \pm 2\Omega_A$ and $\omega_2 = \Omega_X$. In the presence of strong coupling or in the case of chemical (as opposed to magnetic) equivalence, i.e. in A$_2$B and in AA'X systems, or in systems with multi-exponential T$_2$ relaxation, additional signals appear that coincide with the skew diagonal in Fig. 8.4.6 at $\omega_1 = \pm 2\Omega_A$ and $\omega_2 = \Omega_A$. Remotely connected nuclei in a linear AMX

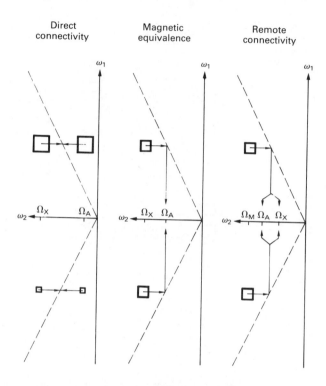

FIG. 8.4.6. Characteristic patterns of double-quantum signals of networks of weakly coupled spins. The symbols represent entire multiplets rather than individual transitions. Large symbols correspond to dominant signals for $0 < \beta < \pi/2$. Direct connectivity signals appear in pairs symmetrically disposed on either side of the skew diagonal $\omega_1 = \pm 2\omega_2$ (dashed lines). Signals associated with remote connectivity and with magnetic equivalence can be distinguished by geometric construction as shown. (Reproduced from Ref. 8.8.)

system with $J_{AX} = 0$ lead to double-quantum signals at $\omega_1 = \pm(\Omega_A + \Omega_X)$ and $\omega_2 = \Omega_M$, which can be identified by geometric construction as indicated in Fig. 8.4.6.

A more detailed picture of the remote connectivity and equivalence signals obtained in double-quantum spectra is given in Fig. 8.4.7, which is based on a calculation of amplitudes $Z_{rs\,tu}$ for transfers from double-quantum coherences $|t\rangle\langle u|$ into observable magnetization $|r\rangle\langle s|$, in analogy to the COSY spectrum in Fig. 8.2.12. It is assumed in Fig. 8.4.7 that all coherences are excited initially with equal amplitudes and equal phases at the beginning of the evolution period, an assumption which cannot easily be fulfilled in practice. Note that the remote connectivity

8.4 HOMONUCLEAR 2D MULTIPLE-QUANTUM SPECTROSCOPY

signals in Fig. 8.4.7 (appearing at $\omega_1 = \pm(\Omega_A + \Omega_X)$ in the AMX system, and at $\omega_1 = \pm 2\Omega_A$ in the A_2X case) have amplitudes that are symmetrical with respect to $\omega_1 = 0$, while the remaining signals (direct connectivity peaks) do not have symmetrical amplitudes.

The coherence transfer phenomena that give rise to these signals can be explained by considering a linear three-spin system. (The symmetrical A_2X spin system can be treated as an $A-X-A'$ system with $J_{AA'} = 0$, in analogy to the $A-M-X$ system with $J_{AX} = 0$.) The system is assumed to be initially in thermal equilibrium, i.e. $\sigma(0) = I_{kz} + I_{lz} + I_{mz}$. For the sake of clarity, we consider the transformations of these terms individually, assuming that $J_{km} = 0$, and neglecting residual longitudinal

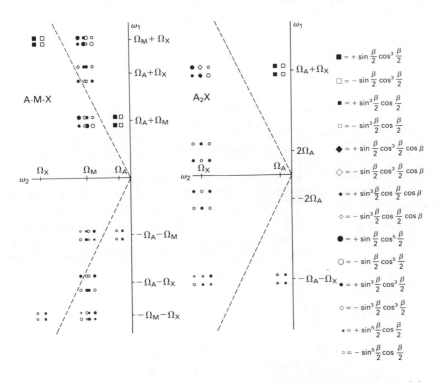

FIG. 8.4.7. Schematic double-quantum spectra of weakly-coupled three-spin systems of the linear AMX type ($J_{AX} = 0$) and of the symmetrical A_2X type, assuming a single mixing pulse with r.f. rotation angle β, a complex Fourier transformation with respect to t_1, and assuming that all double-quantum coherences are excited uniformly with the same phase (in actual fact, this cannot be easily achieved experimentally). Large symbols correspond to dominant signals for $0 < \beta < \pi/2$. The dashed lines indicate skew diagonals $\omega_1 = \pm 2\omega_2$. All signals have composite peakshapes described by eqn (6.5.10). (Reproduced from Ref. 8.8.)

terms that remain after the excitation sandwich of Fig. 8.4.2(a)

$$I_{kz} \xrightarrow{\pi J_{kl}\tau I_{ky}I_{ly}} \xrightarrow{\pi J_{lm}\tau I_{ly}I_{my}} 2I_{kx}I_{ly} \sin \pi J_{kl}\tau,$$

$$I_{lz} \xrightarrow{\pi J_{kl}\tau I_{ky}I_{ly}} \xrightarrow{\pi J_{lm}\tau I_{ly}I_{my}} 2I_{ky}I_{lx} \sin \pi J_{kl}\tau \cos \pi J_{lm}\tau$$
$$+ 2I_{lx}I_{my} \cos \pi J_{kl}\tau \sin \pi J_{lm}\tau$$
$$- 4I_{ky}I_{lz}I_{my} \sin \pi J_{kl}\tau \sin \pi J_{lm}\tau,$$

$$I_{mz} \xrightarrow{\pi J_{kl}\tau I_{ky}I_{ly}} \xrightarrow{\pi J_{lm}\tau I_{ly}I_{my}} 2I_{ly}I_{mx} \sin \pi J_{lm}\tau. \qquad (8.4.12)$$

The resulting terms correspond to superpositions of zero- and double-quantum coherences. For simplicity, we consider $\tau = (2J_{kl})^{-1} = (2J_{lm})^{-1}$, and obtain immediately after the excitation sandwich

$$\sigma(t_1 = 0) = 2I_{kx}I_{ly} - 4I_{ky}I_{lz}I_{my} + 2I_{ly}I_{mx}. \qquad (8.4.13)$$

These terms may be recast in the shorthand notation of eqns (5.3.36) and (5.3.37)

$$\sigma(t_1 = 0) = \{2\text{QT}(k, l)\}_y - \{\text{ZQT}(k, l)\}_y$$
$$+ 2I_{lz}\{2\text{QT}(k, m)\}_x - 2I_{lz}\{\text{ZQT}(k, m)\}_x$$
$$+ \{2\text{QT}(l, m)\}_y - \{\text{ZQT}(l, m)\}_y. \qquad (8.4.14)$$

We concentrate on the double-quantum terms and eliminate the zero-quantum coherences by phase cycling, although in some situations their evolution may be of interest (8.66). The double-quantum terms $\{2\text{QT}(k, l)\}_y$ and $\{2\text{QT}(l, m)\}_y$ are analogous to those found in a two-spin system. The third term in eqn (8.4.14) carries information that cannot be derived from single-quantum methods. The evolution of this term is governed by the sum of the chemical shifts of the two remote spins, and by the couplings to the central spin I_l

$$2I_{lz}\{2\text{QT}(k, m)\}_x \xrightarrow{(\Omega_k I_{kz} + \Omega_m I_{mz})t_1} \xrightarrow{(\pi J_{kl} 2I_{kz}I_{lz} + \pi J_{lm} 2I_{lz}I_{mz})t_1}$$
$$2I_{lz}\{2\text{QT}(k, m)\}_x \cos \Omega_{\text{eff}} t_1 \cos J_{\text{eff}} t_1$$
$$+ 2I_{lz}\{2\text{QT}(k, m)\}_y \sin \Omega_{\text{eff}} t_1 \cos J_{\text{eff}} t_1$$
$$+ \{2\text{QT}(k, m)\}_y \cos \Omega_{\text{eff}} t_1 \sin J_{\text{eff}} t_1$$
$$- \{2\text{QT}(k, m)\}_x \sin \Omega_{\text{eff}} t_1 \sin J_{\text{eff}} t_1 \qquad (8.4.15)$$

with $\Omega_{\text{eff}} = \Omega_k + \Omega_m$ and $J_{\text{eff}} = J_{kl} + J_{lm}$. If we consider a system with magnetically equivalent nuclei I_k and I_m, the effective frequencies are $\Omega_{\text{eff}} = 2\Omega_k$ and $J_{\text{eff}} = 2J_{kl}$. At the end of the evolution period, the $(\pi/2)_x$-mixing pulse can only convert the first term in eqn (8.4.15) into observable single-quantum magnetization. Thus we focus attention on the transformation of this term and retain only the observable part in the resulting density operator after a $(\pi/2)_x$-mixing pulse

$$\sigma^{obs}(t_1, t_2=0) = +\tfrac{1}{2}4I_{kz}I_{ly}I_{mz} \cos(\Omega_k + \Omega_m)t_1 \cos(J_{kl} + J_{lm})t_1. \quad (8.4.16)$$

As usual, this antiphase magnetization may be partly converted into in-phase magnetization by appending a sequence $[\tau/2 - (\pi)_x - \tau/2]$ with $\tau \simeq (2J_{kl})^{-1} \simeq (2J_{lm})^{-1}$, and we obtain

$$\sigma^{obs}(t_1, t_2=0) = +\tfrac{1}{2}I_{ly} \cos(\Omega_k + \Omega_m)t_1 \cos(J_{kl} + J_{lm})t_1$$
$$\times \sin \pi J_{kl}\tau \sin \pi J_{lm}\tau$$
$$+ \text{antiphase terms.} \quad (8.4.17)$$

Residual antiphase terms can be purged by applying a $(\pi/2)_y$-pulse (Fig. 8.4.2(d)). Note that the sign of the I_{ly} term obtained in eqn (8.4.17) is *opposite* with respect to the terms obtained in the two-spin system, eqn (8.4.6). Thus in double-quantum spectra of unknown spin systems, all signals associated with direct connectivity can be made to appear positive, while signals which arise from remote connectivities appear negative (8.53). This allows one to obtain 'edited' spectra, containing either direct connectivity or remote connectivity signals.

The symmetrical excitation and detection scheme allows one to average over different τ-values to obtain broadband excitation that is largely independent of the values of the coupling constants. As mentioned before, this approach is most effective if the π-pulses are removed from the excitation and detection sandwiches (8.53).

An early experimental illustration of a double-quantum spectrum obtained with a single mixing pulse is shown in Fig. 8.4.8. The high-resolution single-quantum spectrum of 3-aminopropanol in isotropic solution shows that the protons in the aliphatic chain $-CH_2-CH_2-CH_2-$ have vanishing long-range couplings and nearly equal vicinal couplings. The system can be treated to a good approximation as an $A_2M_2X_2$ system. If the multiplicities and integrals cannot be determined accurately (as would be the case for an aliphatic fragment in a macromolecule), it would be difficult to ascertain the number of equivalent nuclei, and to prove that all three multiplets stem from nuclei in the same chain. Several erroneous interpretations are compatible with a low-resolution 2D correlation spectrum. For example, one might assume an A_2-M-X_3 system, or a superposition of two fragments A_2-M and $M'-X_2$ with an accidental degeneracy $\Omega_M = \Omega_{M'}$. The double-quantum spectrum of the aliphatic chain makes it possible to resolve these ambiguities: the signals at $2\Omega_A$, $2\Omega_M$, and $2\Omega_X$ prove the existence of at least two equivalent nuclei at each site; the signal at $\Omega_A + \Omega_X$ excludes the possibility of an

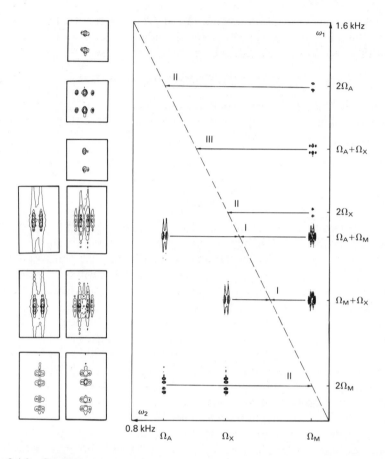

FIG. 8.4.8. Double-quantum spectrum of 3-aminopropanol (DOCH$_2$CH$_2$CH$_2$ND$_2$), obtained with the sequence $[(\pi/2) - \tau/2 - (\pi) - \tau/2 - (\pi/2)] - t_1 - (\pi/2) - t_2$ with $\tau \simeq$ 38 ms $\simeq (4J)^{-1}$. The spin system can be treated to a good approximation as an A$_2$M$_2$X$_2$ system. Signals characteristic for direct connectivity (I), magnetic equivalence (II), and remote connectivity (III) can be readily identified by geometric construction; they are consistent with an A$_a$M$_m$X$_x$ network with a, m, $x \geq 2$ and $J_{AX} \simeq 0$. The spectrum is displayed in absolute-value mode; the multiplets are reproduced on an enlarged scale on the left. (Reproduced from Ref. 8.8.)

accidental superposition of two fragments, while the signals at $\Omega_A + \Omega_M$ and $\Omega_M + \Omega_X$ yield the same information about vicinal couplings as could be obtained from 2D correlation spectroscopy.

In macromolecular systems, the assignment of conventional 2D correlation spectra is associated with a number of difficulties that can be circumvented with double-quantum NMR. In correlation spectroscopy, directly coupled nuclei with small chemical shift differences lead to cross-peaks near to the diagonal that are difficult to detect if the

8.4 HOMONUCLEAR 2D MULTIPLE-QUANTUM SPECTROSCOPY

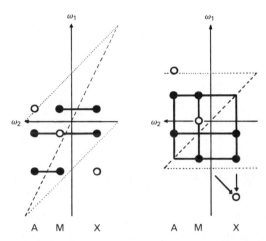

FIG. 8.4.9. Left: schematic double-quantum spectrum of an AMX system obtained by selecting the pathway $p = 0 \to +2 \to -1$. Filled and open symbols correspond to direct and remote connectivity signals. Note that the filled symbols are contained within a frequency band indicated by dotted lines; if there are no systems with more than two spins, as in natural abundance carbon-13 spectra, there can be no signals outside this band. Right: schematic spectrum obtained by 'shearing' the same spectrum according to the procedure of foldover correction (see § 6.6.1). This representation, which is reminiscent of a COSY spectrum (except for the absence of diagonal peaks) can also be obtained by delaying the acquisition. (Reproduced from Ref. 8.65.)

amplitude of the diagonal ridge is dominant. No such problem is encountered in double-quantum NMR, since the skew diagonal does not carry any signals except in the case of strongly coupled or chemically equivalent nuclei (8.67, 8.68) or in systems with multi-exponential T_2 relaxation.

The unusual frequency coordinates of the signals in double-quantum spectra may be a stumbling block to spectroscopists who are used to single-quantum correlation spectra. It is possible to convert double-quantum spectra into a COSY-like representation by shearing the data matrix $S(\omega_1, \omega_2) \to S(\omega_1', \omega_2)$ with $\omega_1' = \omega_1 - \omega_2$ (§ 6.6.1). Alternatively, if the phase-cycle selects the pathway $p = 0 \to +2 \to -1$ and rejects $p = 0 \to -2 \to -1$, a COSY-like spectrum can be obtained by delaying the beginning of signal acquisition by an additional time t_1 (8.62–8.65). In either case, the resulting double-quantum spectrum appears as shown schematically in Fig. 8.4.9 for a three-spin AMX system.

8.4.4. Multiple-quantum spectra of dipole-coupled nuclei in anisotropic phase

The relevance of multiple-quantum transitions in oriented systems, such as solutes in liquid crystalline solvents, lies chiefly in the simplification of

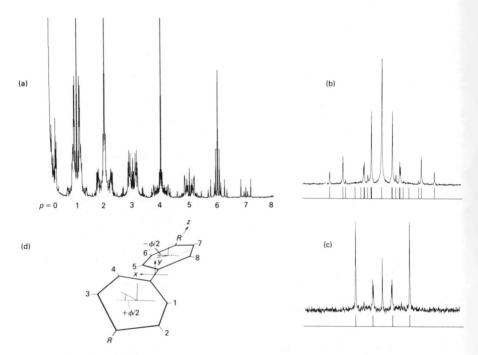

FIG. 8.4.10. (a) Multiple-quantum spectra with transitions of all orders $p = 0, 1, \ldots, 8$ of n-pentyl-cyano-biphenyl-d_{11}. The aliphatic chain is deuterated to restrict the analysis to the eight protons on the biphenyl moiety. (b) and (c) Enlarged views of the experimental six- and seven-quantum spectra, with theoretical stick spectra based on an iterative analysis. All lines in the theoretical spectra have unit height. (d) Schematic conformation of the biphenyl moiety ($R = CN$, $R' = C_5D_{11}$) with a dihedral angle of 32° that is consistent with the frequencies in (b) and (c). (Reproduced from Ref. 8.70.)

spectral information (8.35). The statistics of the decreasing number of transitions with increasing order of coherence has been discussed in § 5.1. While multiple-quantum coherence has seldom been used in dipolar coupled solids (8.69), many elegant applications are concerned with solute molecules in liquid crystalline solvents.

Figure 8.4.10 shows multiple-quantum spectra of n-pentyl-cyano-biphenyl-d_{11} dissolved in a liquid crystal (8.70). If one assumes that the geometry is rigid and that the ring protons have a permutation symmetry isomorphous with the D_2 point group (strictly speaking this is only true for symmetrical substituents $R = R'$), there are ten unique dipolar couplings and two independent order parameters. An iterative analysis of the $p = 5$, 6, and 7 quantum spectra yielded a set of coupling constants consistent with a dihedral angle $\phi \simeq 32°$ between the two rings. The single-quantum spectrum would be more difficult to analyse in this case.

The inherent information content of multiple-quantum spectra is clearly important for structural studies. An additional merit of dipole-coupled systems in liquid crystals is that they provide an ideal testing ground for multiple-quantum methodology. The resolution (narrow lines and large couplings) and the nature of the dipolar Hamiltonian (which lends itself to time-reversal methods (8.71)) make it possible to design and verify experimentally a great variety of sophisticated techniques. Various methods for the separation of different orders of multiple-quantum signals (8.72, 8.73), techniques such as total spin coherence transfer echo spectroscopy (TSCTES) (8.33–8.35), and selective p-quantum excitation (8.51, 8.74, 8.75) have been developed for liquid crystals. A comprehensive review of this work was given by Weitekamp (8.35).

8.4.5. Double-quantum spectra of quadrupolar nuclei with $S = 1$ in anisotropic phase

The principal motivation for recording double-quantum spectra of $S = 1$ nuclei is the elimination of quadrupolar effects. In oriented systems (solids or solutes dissolved in liquid crystals), the quadrupole Hamiltonian of eqn (2.2.20) leads to a splitting of the single-quantum transitions. The resulting powder patterns or inhomogeneously broadened multiplets (the latter occur in liquid crystals where the order parameter is often strongly temperature-dependent) tend to obscure the chemical shifts.

For systems with small quadrupolar couplings (e.g. for ^{2}D), non-selective pulses can be used for excitation and detection of double-quantum coherence. We limit the discussion to high-field NMR where second- and higher-order quadrupole effects can be disregarded. The sequence $[(\pi/2)_x - \tau/2 - (\pi)_x - \tau/2 - (\pi/2)_x]$, which leads to a composite rotation with a propagator $\exp\{-i\omega_Q \tau S_y^2\}$, excites double-quantum coherence (8.15)

$$S_z \xrightarrow{\omega_Q \tau S_y^2} S_z \cos \omega_Q \tau + \{S_x S_y + S_y S_x\} \sin \omega_Q \tau. \qquad (8.4.18)$$

The amplitude of the double-quantum term $\{S_x S_y + S_y S_x\}$ is maximum for $\tau = \pi/(2\omega_Q)$ (analogous to $\tau = 1/(2J)$, since the splitting is $2\omega_Q$ instead of $2\pi J$). The double-quantum coherence is invariant to evolution under the secular part of the quadrupole Hamiltonian of eqn (2.2.24)

$$[\mathcal{H}'_Q, \{S_x S_y + S_y S_x\}] = 0. \qquad (8.4.19)$$

This invariance can be understood by noting that the eigenstates $|M_s = +1\rangle$ and $|M_s = -1\rangle$ are shifted to first order by the same amount under $\mathcal{H}_Q$ (8.76, 8.77). However, under the Zeeman term the double-quantum coherence evolves according to

$$\{S_xS_y + S_yS_x\} \xrightarrow{\Omega t_1 S_z} \{S_xS_y + S_yS_x\}\cos 2\Omega t_1 - \{S_x^2 - S_y^2\}\sin 2\Omega t_1$$
(8.4.20)

where $2\Omega = 2(-\gamma_s B_0 - \omega_{r.f.})$ is equal to twice the single-quantum chemical shift in the rotating frame.

An experimental example of a double-quantum spectrum of a single crystal of 10 per cent deuterated oxalic acid dihydrate is shown in Fig. 8.4.11. The chemical shifts of the carboxyl and hydrate deuterons are well resolved, although they cannot be separated in the single-quantum spectrum (8.77).

In powders, the double-quantum precession under the Zeeman term leads to a powder pattern in the ω_1-domain that allows one to determine the principal values of the chemical shift tensor. The normal quadrupole powder pattern appears in the ω_2-domain and the chemical shielding and quadrupole tensors can be related to each other. Double-quantum spectroscopy can be combined with magic-angle sample spinning (8.80),

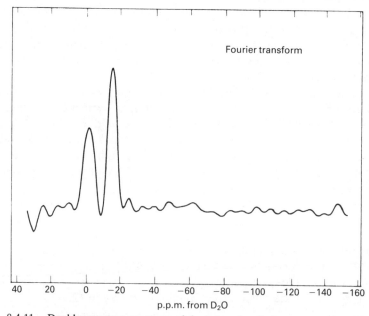

FIG. 8.4.11. Double-quantum spectrum of deuterium (equivalent to a projection of a 2D double-quantum spectrum on to the ω_1-axis) in a single crystal of 10 per cent deuterated oxalic acid dihydrate, obtained in the presence of proton decoupling. The signals from the carboxyl (right) and hydrate deuterons (left) are well-resolved although they cannot be separated in the single-quantum deuterium spectrum. (Reproduced from Ref. 8.77.)

in which case the anisotropic part of the Zeeman interaction is eliminated from the double-quantum ω_1-domain. Similar information can be obtained in conventional (single-quantum) magic-angle spectroscopy (8.78) if the signal is sampled on top of the rotational echoes (8.79), but the double-quantum experiment has the advantage that the spectra are less sensitive to deviations from the magic angle.

For nuclei with larger quadrupole couplings, such as nitrogen-14, non-selective excitation is no longer feasible. In such cases it is easier to excite double-quantum coherence with a selective double-quantum pulse applied in the centre between the two allowed transitions (§ 5.3.1) or to use a cross-polarization procedure to excite and detect the double-quantum coherence via abundant $I = \frac{1}{2}$ spins such as protons (see § 8.5.6).

8.5. Heteronuclear coherence transfer

The phenomenon of coherence transfer is by no means restricted to homonuclear systems: all schemes described in §§ 8.2–8.4 can be extended to systems containing two nuclear species I and S. An even greater variety of heteronuclear experiments can be designed, because the pulses can be applied selectively to either species, and heteronuclear broadband decoupling can be employed at will.

Heteronuclear two-dimensional coherence transfer experiments may be performed for a variety of reasons:

1. Increased sensitivity of indirect detection;
2. Unravelling of overlapping I resonances by exploiting the chemical shift dispersion of S spins, and vice versa;
3. Correlation of the spectra of different nuclear species for assignment purposes.

In many cases, one of the two nuclear species involved is a rare nucleus (S-spin), like carbon-13, nitrogen-15, or silicon-29, while the other nuclei (I-spins) are usually protons, fluorine, or other abundant nuclei. Each molecule contains usually only one single S-spin, and special design considerations are required with regard to optimum sensitivity and information content.

A great variety of heteronuclear coherence transfer schemes in liquids and solids have been proposed, some of which have been mentioned in the context of 1D Fourier spectroscopy (§ 4.5). Most of these techniques are also suitable for 2D experiments. In *liquid phase* we should mention:

1. *Coherence transfer by r.f. pulses* seems to be most versatile in heteronuclear 2D experiments. The first published examples of heteronuclear 2D spectra (8.9) relied on pulsed coherence transfer, and many further schemes have been proposed since (8.10–8.13, 8.81–8.96).

2. *Cross-polarization in the rotating frame* may lead to an oscillatory transfer of coherence between two spin species in liquid phase (8.50), in analogy to Hartmann–Hahn cross-polarization in solids (8.47, 8.48, 8.97). A related mechanism is involved in isotropic mixing in the rotating frame (8.98, 8.99), which is less sensitive to deviations from the Hartmann–Hahn matching condition.

3. *Adiabatic cross-polarization.* A dynamic matching of energy levels can be achieved by a level anticrossing experiment where the r.f. fields applied to the two nuclear species are varied in amplitude through the matching condition (8.49, 8.100). With minor modifications, the same schemes can also be used for heteronuclear 2D spectroscopy in solid phase, although the conditions typical for solids are more in favour of cross-polarization in the rotating frame (8.101–8.104).

8.5.1. Sensitivity considerations

Heteronuclear systems offer several options for designing 2D experiments which differ in their relative sensitivities. The experiment can be started either with I or S spin polarization and may end with the observation of either I or S magnetization, as shown schematically in Fig. 8.5.1. The sensitivity of these experiments is determined by the following factors.

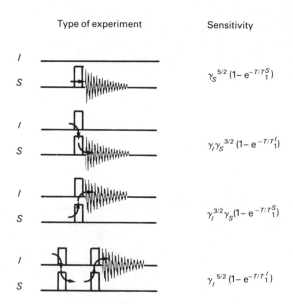

FIG. 8.5.1. Relative sensitivity of various heteronuclear coherence transfer experiments. The formulae are applicable to two-spin systems under ideal conditions.

1. The available polarization is determined by the extent of recovery through T_1 relaxation between subsequent scans separated by an interval T. It may be expressed by a saturation factor $\{1 - \exp(-T/T_1)\}$, and by the equilibrium polarization which is proportional to the gyromagnetic ratio γ_{exc} of the nuclear species excited at the beginning of the sequence.

2. The response of the observed nucleus is proportional to the amplitude and to the frequency of the precessing magnetization. Both are proportional to the gyromagnetic ratio γ_{obs} of the observed nucleus.

3. The detector noise is, according to experience, approximately proportional to $(\omega_{\text{obs}})^{\frac{1}{2}}$ and hence to $(\gamma_{\text{obs}})^{\frac{1}{2}}$.

The overall sensitivity is therefore proportional to

$$S/N \propto \gamma_{\text{exc}} \gamma_{\text{obs}}^{\frac{3}{2}} \{1 - \exp(-T/T_1^{(\text{exc})})\}. \qquad (8.5.1)$$

Some implications of this relationship are shown in Fig. 8.5.1. For the sake of illustration, we may consider coherence transfer between protons and nitrogen-15. In this case, the ratios of the factors in Fig. 8.5.1 are $\gamma_S^{\frac{5}{2}} : \gamma_I \gamma_S^{\frac{3}{2}} : \gamma_I^{\frac{3}{2}} \gamma_S : \gamma_I^{\frac{5}{2}} \simeq 1 : 10 : 30 : 300$. The more favourable relaxation of protons ($T_1^I < T_1^S$) gives an additional advantage of the schemes involving $I \to S$ and $I \to S \to I$ transfer.

In actual fact, several additional factors may significantly reduce the sensitivity.

4. Distribution of the signal intensity over numerous multiplet components (this is particularly important for I-spin observation).

5. Dynamic range problems in the presence of large background signals (important for the indirect detection of rare spins via abundant nuclei).

6. Inefficient coherence transfer due to short transverse relaxation times.

Thus the factors in Fig. 8.5.1 tend to give an optimistic estimate of the sensitivity. Furthermore, all schemes which require the I signal to be detected presuppose that the dominant fraction of I-magnetization, which does not stem from coherence transfer from the dilute S nuclei, is eliminated by saturation or phase cycling schemes, both of which are bound to be incompletely successful. The resulting artefacts ('t_1-noise') tend to offset the theoretical sensitivity advantage.

The initial coherence transfer from I to S used in several schemes to enhance the magnetization can in many cases be replaced by a saturation period to obtain an Overhauser enhancement. In carbon-13 spectroscopy, the Overhauser effect, which is proportional to $(1 + \frac{1}{2}\gamma_I/\gamma_S)$, is nearly as effective as the gain from $I \to S$ coherence transfer, but for other nuclei such as nitrogen-15, the transfer of coherence by r.f. pulses is more efficient.

8.5.2. Coherence transfer pathways

When the coherence transfer pathways discussed in § 6.3 are extended to encompass heteronuclear spin systems, it is convenient to keep track separately of the coherence orders associated with each nucleus (8.105, 8.106), as shown schematically in Fig. 8.5.2. A combination $[p_I = \pm 1, p_S = 0]$ represents single-quantum coherence involving only I spins. A combination $[p_I = 0, p_S = \pm 1]$ represents either pure single-quantum coherence of the dilute S nuclei, or a coherence of the type $I_k^+ I_l^- S_m^\pm$, i.e. a combination of I-spin zero-quantum coherence and S-spin single-quantum coherence. The combinations $[p_I = \pm 1, p_S = \pm 1]$ correspond to heteronuclear double-quantum coherence, while $[p_I = \pm 1, p_S = \mp 1]$ describes heteronuclear zero-quantum coherence. In general, the distinction of zero- and double-quantum coherence turns out to be of lesser significance than for homonuclear systems. In analogy to the selection of pathways in homonuclear systems (§ 6.3), it is possible to derive phase-cycles from first principles. Assuming that an experiment involves n pulses (or n propagators), applied to either I- or S-spins, that

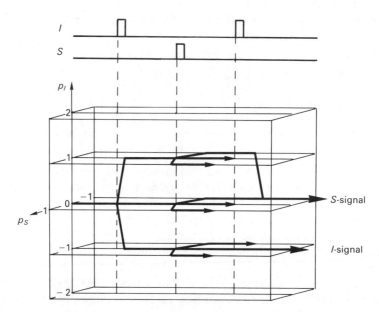

FIG. 8.5.2. Coherence transfer pathways in heteronuclear systems. The figure is appropriate for a system containing two $I = \frac{1}{2}$ spins and one $S = \frac{1}{2}$ spin, with $-2 \leq p_I \leq +2$ and $-1 \leq p_S \leq +1$. The pulse sequence shown on top is an arbitrary example. Note that all pathways start with $p_I = p_S = 0$ (thermal equilibrium), and must end either with $p_I = -1$, $p_S = 0$ (observable transverse I-magnetization), or with $p_I = 0$, $p_S = -1$ (transverse S-magnetization).

bring about n coherence transfer steps, one may number these pulses chronologically (or in arbitrary order if they occur simultaneously). The complete pathway can be specified by a single vector, eqn (6.3.18),

$$\Delta \mathbf{p} = \{\Delta p_1, \Delta p_2, \ldots, \Delta p_n\}.$$

The phases of the n pulses are also specified by a vector, eqn (6.3.20),

$$\boldsymbol{\varphi} = \{\varphi_1, \varphi_2, \ldots, \varphi_n\}.$$

The proper pathway can be selected by shifting the receiver reference phase according to eqn (6.3.26)

$$\varphi^{\text{ref}} = -\Delta \mathbf{p} \cdot \boldsymbol{\varphi}.$$

If I-magnetization must be detected, the vector $\Delta \mathbf{p}$ must be defined (in analogy to eqn (6.3.19)) so as to have

$$\sum_{i} \Delta p_i = -1, \qquad \sum_{i} \Delta p_i = 0,$$
$$\text{\emph{I}-pulses} \qquad \qquad \text{\emph{S}-pulses}$$

If, on the other hand, it is the S-magnetization that one wishes to detect, the first sum must be zero and the second equal to -1. If the phase cycle of the ith pulse extends over N_i increments to select a desired pathway, a manifold of other pathways are selected simultaneously according to eqn (6.3.27)

$$\Delta p_i^{(\text{selected})} = \Delta p_i^{(\text{desired})} \pm nN_i, \qquad n = 1, 2, 3. \ldots$$

If N_i is sufficiently large, these additional pathways can be neglected.

8.5.3. Heteronuclear two-dimensional correlation spectroscopy in isotropic phase

The basic scheme of heteronuclear 2D correlation spectroscopy (Fig. 8.5.3(a)) may be regarded as an extension of homonuclear 2D correlation spectroscopy, where the second pulse in the $\pi/2 - t_1 - \pi/2 - t_2$ sequence is applied simultaneously to both nuclei (8.9). At first sight it may appear impossible to transfer coherence with two pulses with different frequencies that are not coherent in phase. It turns out however that it is sufficient for the two $(\pi/2)^I$ pulses to be phase-coherent (i.e. derived from the same frequency source), while the relative phases of the $(\pi/2)^I$- and $(\pi/2)^S$-pulses do not affect coherence transfer.

To appreciate the mechanism of heteronuclear coherence transfer, it is instructive to assume that the two $\pi/2$-pulses, which are drawn as simultaneous in Fig. 8.5.3(a), follow each other in close succession in the order $(\pi/2)^I$, $(\pi/2)^S$. Only I-magnetization that is in antiphase with

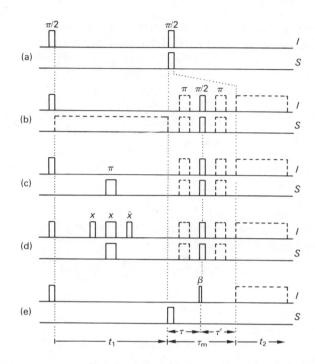

FIG. 8.5.3. Schemes for heteronuclear 2D spectroscopy with coherence transfer from I spins to S spins. (a) Basic scheme with a pair of pulses applied to both species. (b) Scheme with fixed delays τ and τ' before and after coherence transfer, to allow the I-magnetization to dephase and the S-magnetization to rephase. The optional π-pulses in the τ and τ' intervals remove offset-dependent phase errors (§ 8.5.3.1). This may be combined with optional broadband decoupling of the S and I spins in the evolution and detection periods respectively (§ 8.5.3.2). (c) Similar scheme, but with a $(\pi)^S$-pulse in the centre of the evolution period to refocus the $\mathcal{H}_{IS}$ interaction, which is equivalent to continuous S-spin decoupling in t_1 (§ 8.5.3.3). (d) The insertion of a bilinear rotation sandwich at $\tfrac{1}{2}t_1$ allows refocusing of both $\mathcal{H}_{IS}$ and $\mathcal{H}_{II}$ interactions (§ 8.5.3.4). (e) By inserting a delay between the I and S pulses responsible for coherence transfer, it is possible to separate or 'edit' signals from I_nS groups according to the number of equivalent nuclei n (§ 8.5.3.5).

respect to the S spin can be transferred

$$2I_{ky}S_{lz} \xrightarrow{(\pi/2)I_{kx}} 2I_{kz}S_{lz} \xrightarrow{(\pi/2)S_{lx}} -2I_{kz}S_{ly}. \tag{8.5.2}$$

In this case, the coherence transfer involves an intermediate state of longitudinal two-spin order. If the sequence of events is reversed, the final result is equivalent

$$2I_{ky}S_{lz} \xrightarrow{(\pi/2)S_{lx}} -2I_{ky}S_{ly} \xrightarrow{(\pi/2)I_{kx}} -2I_{kz}S_{ly} \tag{8.5.3}$$

but the information is temporarily stored in the form of heteronuclear zero- and double-quantum coherence. The two views are clearly equivalent, since the operators of the two pulses commute. However, the sequence in eqn (8.5.2) has the advantage that the individual steps can be explained in terms of semi-classical magnetization vectors and populations, a feature that has been helpful in developing new experiments with the aid of 'pulse cascades' (8.14). If the pulses are separated by an interval, the sequence of events is no longer immaterial, and different effects can be obtained because of the precession of coherence. The decomposition of a pair of pulses in the order $(\pi/2)^I$, $(\pi/2)^S$ of eqn (8.5.2) shows that the pulses applied to the I and S spins need not be coherent in phase. The intermediate product of z-components does not carry any phase information, and the phase of the observed signal is fully defined by the phase of the S-pulse.

Since the observation is restricted to one species, e.g. the S spins, the heteronuclear 2D spectrum is limited to a window in the ω_2-domain that covers the range of the chemical shifts Ω_S. Since there is no $(\pi/2)^S$ pulse at the beginning of the t_1-period, the appearance of diagonal peaks associated with S-magnetization is avoided.

Phase cycling can be employed to select both pathways $[p_I = \pm 1, p_S = 0] \to [p_I = 0, p_S = -1]$ to obtain pure absorption-mode spectra. This boils down to a simple phase alternation of one of the $(\pi/2)^I$-pulses with alternating addition and subtraction, which leads to the suppression of the native S-spin magnetization. The selection of a single pathway (8.107, 8.108) leads as usual to mixed phase lineshapes.

The basic scheme in Fig. 8.5.3(a) yields 2D spectra that reflect the unperturbed Hamiltonians in the evolution and detection periods

$$\mathcal{H}^{(e)} = \mathcal{H}_{ZI} + \mathcal{H}_{II} + \mathcal{H}_{IS} + [\mathcal{H}_{SS} + \mathcal{H}_{ZS}],$$
$$\mathcal{H}^{(d)} = \mathcal{H}_{ZS} + \mathcal{H}_{SS} + \mathcal{H}_{IS} + [\mathcal{H}_{II} + \mathcal{H}_{ZI}] \qquad (8.5.4)$$

where the terms in brackets are only relevant in strongly coupled systems.

In analogy to cross-peaks in homonuclear correlation spectra, the signals stemming from a transfer of coherence $I \to S$ are in antiphase with respect to the heteronuclear coupling J_{IS} that is the vehicle of coherence transfer. For this reason, it is not possible to decouple the I nuclei immediately after coherence transfer, since the antiphase multiplets would cancel.

8.5.3.1. Transfer of in-phase magnetization

In the modified scheme of Fig. 8.5.3(b), the mixing period has been extended by free precession periods before and after the transfer pulse pair. This allows the coherence to dephase before and to rephase after coherence transfer (8.10). For optimum intervals $\tau = \tau' = (2J_{IS})^{-1}$, the transformation relevant for a two-spin system is

$$I_{kx} \xrightarrow{(\pi/2)2I_{kz}S_{lz}} 2I_{ky}S_{lz} \xrightarrow{(\pi/2)(I_{kx}+S_{lx})} -2I_{kz}S_{ly} \xrightarrow{(\pi/2)2I_{kz}S_{lz}} S_{lx}. \quad (8.5.5)$$

Thus the extended mixing period leads to a net transfer of in-phase magnetization. The four π-pulses in the mixing interval remove offset-dependent phase shifts and can be dropped if absolute-value spectra are acceptable. If the τ- and τ'-intervals are not properly matched to the inverse coupling, the amplitude of the transfer from I_{kx} to S_{lx} is given by the transfer function

$$a_{kl}(\tau, \tau') = \sin(\pi J_{IS}\tau)\sin(\pi J_{IS}\tau') \quad (8.5.6)$$

neglecting relaxation and couplings to further spins. The amplitude of the transfer of in-phase coherence in this extended scheme with rephasing of coherence depends on the multiplicity and on the magnitude of the scalar coupling (8.109). In carbon-13 NMR, it is possible to select τ for efficient in-phase transfer through one-bond couplings, while the weaker long-range couplings normally yield only little in-phase magnetization.

8.5.3.2. Broadband decoupling

The use of a rephasing sequence for transfer of in-phase magnetization opens the way to various decoupling schemes. Broadband I-spin decoupling can be used in the S-spin detection period (Fig. 8.5.3(b)–(d)). The effective Hamiltonian in the detection period is then simplified to

$$\mathcal{H}^{(d)} = \mathcal{H}_{ZS} + \mathcal{H}_{SS}. \quad (8.5.7)$$

In dilute spin systems, only chemical shifts remain in the ω_2-frequency dimension.

In the evolution period, one may choose between a variety of manipulations to achieve selective simplification of the full Hamiltonian. The simplest scheme employs broadband S-spin decoupling throughout the evolution period (8.10, 8.110) (Fig. 8.5.3(b))

$$\mathcal{H}^{(e)} = \mathcal{H}_{ZI} + \mathcal{H}_{II}. \quad (8.5.8)$$

This approach is equally applicable to weakly and strongly coupled systems, but may be difficult to implement technically because of the wide range of chemical shifts of many dilute nuclei. It is recommended to apply offset-compensated decoupling sequences such as WALTZ-16 (8.111) to the S-spectrum, which provide satisfactory decoupling provided that the spectral width does not exceed the amplitude of the r.f. field $\gamma_S B_1^S$ by more than a factor two.

8.5.3.3. Decoupling by refocusing pulses

The problems with broadband decoupling mentioned above may be partly circumvented with the scheme in Fig. 8.5.3(c), which employs a single (possibly composite) $(\pi)^S$-pulse in the centre of the evolution period (8.10, 8.11). This pulse reverses the sign of the I-magnetization that is in antiphase with respect to the S spin, which leads to refocusing of the heteronuclear coupling.

On the other hand, the I-magnetization remains modulated by shifts and homonuclear couplings. In systems with weak couplings, one obtains the effective I-spin Hamiltonian

$$\mathcal{H}^{(e)} = \mathcal{H}_{ZI} + \mathcal{H}_{II}. \tag{8.5.9}$$

Thus the ω_1-domain reveals the chemical shifts and homonuclear couplings of the I nuclei.

The combination of S-spin decoupling in t_1 and I-spin decoupling in t_2, in conjunction with a mixing sequence designed for in-phase coherence transfer through one-bond couplings, leads to so-called 'shift correlation maps', where the coordinates of the signals correspond to the chemical shifts Ω_I and Ω_S of I_nS-subunits. The $\mathcal{H}_{II}$ interaction can often be ignored in practical applications. Such shift correlation spectra have proven useful for the assignment of both proton and carbon spectra.

An experimental example of a heteronuclear 2D shift correlation spectrum is shown in Fig. 8.5.4. The dispersion of the ^{13}C shifts makes it possible to unravel resonances which have nearly overlapping proton shifts. The elongated lineshapes in the proton dimension are due to ill-resolved homonuclear couplings.

In systems with strong coupling between the I nuclei, the $(\pi)^S$-pulse leads to coherence transfer between transitions in the I-spectrum, in analogy to the effects discussed in § 7.2.4. It should be emphasized that strong coupling effects arise whenever the S (e.g. carbon-13) satellites in the I (proton) spectrum are subject to strong I–I coupling. In some cases, the satellite spectrum is weakly-coupled while the normal I-spectrum is strongly coupled, but the reverse situation can also occur. It is often preferable to record heteronuclear correlation spectra without attempting any kind of decoupling in the ω_1-domain. In this case, one obtains an indirect measurement of the undistorted S-spin satellites in the I spectrum, which may be of interest to study $\mathcal{H}_{II}$ couplings in strongly-coupled systems with complicated I-spectra (8.21, 8.112–8.114).

8.5.3.4. Bilinear rotation decoupling

In weakly coupled systems which have one dominant heteronuclear coupling that is much larger than other homo- and heteronuclear

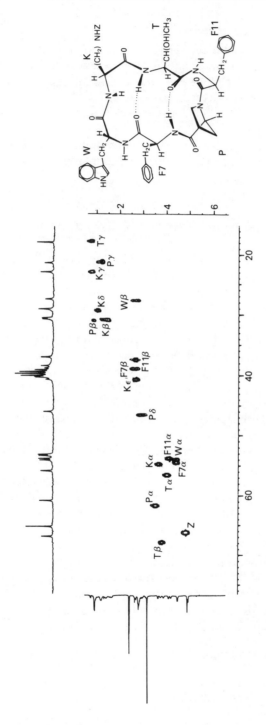

FIG. 8.5.4. Molecular structure of the hexapeptide cyclo [Phe[7]-D-Trp-Lys(Z)-Thr-Phe[11]-Pro] and heteronuclear proton–carbon shift correlation spectrum, obtained with the sequence of Fig. 8.5.3(c). The 1D proton spectrum is shown on the left, and the proton-decoupled carbon-13 spectrum on top (aliphatic region only). Note the peak-shapes that appear elongated in the vertical ω_1-dimension because of ill-resolved homonuclear proton–proton couplings. Many signals that are well-resolved in the 2D spectrum overlap in either of the 1D spectra. (Reproduced from Ref. 8.88.)

couplings, bilinear rotation decoupling (BIRD) (Fig. 8.5.3(d)) can be used to eliminate the effects of all relevant homo- and heteronuclear couplings during the evolution period (8.115–8.117). The bilinear rotation sandwich has been discussed in §§ 7.2.2.6 and 7.2.2.7 for the separation of one-bond and long-range IS couplings. In a general weakly coupled system, the effect of the pulse sandwich is described by the transformations in eqn (7.2.17), with $\tau = (J_{kl})^{-1}$, i.e. provided the width of the BIRD sandwich is matched to the inverse of the one-bond coupling constant.

Consider again a system containing a single spin S_k with one-bond couplings to neighbour spins I_l and with long-range couplings to remote spins I_m. With the full Hamiltonian

$$\mathcal{H} = \mathcal{H}_k + \mathcal{H}_l + \mathcal{H}_m + \mathcal{H}_{kl} + \mathcal{H}_{km} + \mathcal{H}_{lm} + \mathcal{H}_{ll'} + \mathcal{H}_{mm'} \quad (8.5.10)$$

we find, according to eqn (7.2.20), the average Hamiltonian in the evolution period

$$\bar{\mathcal{H}}^{(e)} = \mathcal{H}_l + \mathcal{H}_{km} + \mathcal{H}_{ll'} + \mathcal{H}_{mm'}. \quad (8.5.11)$$

If the mixing period in Fig. 8.5.3(d) is tailored to transfer magnetization through one-bond couplings ($\tau = \tau' = (2J_{kl})^{-1}$), only neighbour spins I_l lead to signals in the 2D spectrum. For these I_l spins, the evolution in the t_1-period is determined only by their chemical shifts and by geminal couplings between non-equivalent neighbours. The homonuclear couplings between neighbour and remote spins I_l and I_m and the one-bond heteronuclear couplings between I_l and S_k are fully eliminated, provided the spin system is weakly coupled. The procedure can be made less sensitive to mismatch with respect to J_{kl} with a compensated bilinear rotation scheme (8.115).

The experimental example in Fig. 8.5.5 shows that the splittings due to $\mathcal{H}_{ll}$ that appear in the conventional proton spectrum (top) are collapsed in the 2D spectrum, a feature that is beneficial both for sensitivity and resolution (8.116).

8.5.3.5. Editing of heteronuclear correlation spectra

In the sequence shown in Fig. 8.5.3(e), the two pulses that lead to the transfer from I to S are separated by an interval τ of the order of $(2J_{IS})^{-1}$. This scheme involves heteronuclear multiple-quantum coherence in the τ-period and allows one to separate and identify I_nS fragments according to the number n of equivalent I spins, in analogy to the DEPT experiment discussed in § 4.5.6. At the end of the evolution period of Fig. 8.5.3(e), antiphase coherence $2I_{kx}S_{lz}$ of one of the equivalent spins is converted by the $(\pi/2)_x^S$-pulse into a superposition of heteronuclear zero- and double-quantum coherences $-2I_{kx}S_{ly}$. At a time

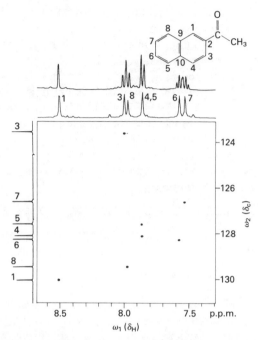

FIG. 8.5.5. Heteronuclear proton–carbon correlation spectrum of the aromatic resonances in 2-acetonaphthalene, obtained with the sequence in Fig. 8.5.3(d). All relevant $\mathcal{H}_{II}$ and $\mathcal{H}_{IS}$ interactions are eliminated in the $F_1 = \omega_1/(2\pi)$-domain, leaving only proton shifts in ω_1 and carbon shifts in ω_2. (Reproduced from Ref. 8.116.)

$\tau = (2J_{IS})^{-1}$ after the $(\pi/2)^S$ pulse, the couplings to the remaining $(n-1)$ equivalent spins that are not actively involved in the two-spin coherence lead to antiphase heteronuclear two-spin coherence

$$\sigma_{IS}(\tau) = -2I_{kx}S_{ly},$$
$$\sigma_{I_2S}(\tau) = +4I_{kx}I_{k'z}S_{lx},$$
$$\sigma_{I_3S}(\tau) = +8I_{kx}I_{k'z}I_{k''z}S_{ly}. \qquad (8.5.12)$$

The reconversion into single-quantum S-magnetization under a $(\beta)_x^I$-pulse is proportional to $\sin\beta$ for IS groups, $\frac{1}{2}\sin 2\beta$ for I_2S groups and $\frac{1}{4}(\sin\beta + \sin 3\beta)$ for I_3S groups.

It is possible to separate the responses from I_nS groups in heteronuclear correlation spectra according to n by one of three methods (8.92).

1. Linear combinations of spectra obtained with different β.
2. With $\beta = 3\pi/4$, the responses of I_2S groups are inverted in sign with respect to those of IS and I_3S groups.

8.5 HETERONUCLEAR COHERENCE TRANSFER

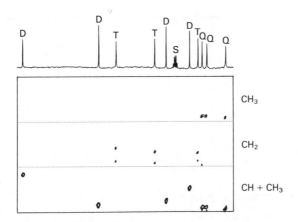

FIG. 8.5.6. Heteronuclear proton–carbon correlation spectrum of menthol, with separation of the signals associated with CH_n groups according to the number n of equivalent protons. The spectrum was obtained with the sequence of Fig. 8.5.3(e) with a t_1-proportional pulse rotation angle according to eqn (8.5.13). The spectrum consists of three regions separated by dotted lines. The vertical ω_1-frequency axis extends over the full range of proton shifts (0.25–3.6 p.p.m.) in each region; the horizontal ω_2-domain contains the carbon-13 shifts (14–73 p.p.m.). The lowest region shows a superposition of signals of CH and CH_3 groups, which can be identified by comparison with the upper CH_3 region. A decoupled one-dimensional carbon-13 spectrum of menthol is shown on top, with indication of multiplicities (D, T, Q) and solvent peaks (S). (Adapted from Ref. 8.92.)

3. By incrementing the r.f. pulse angle in concert with the evolution time

$$\beta = \frac{2\pi}{7} \frac{t_1}{\Delta t_1}, \qquad (8.5.13)$$

the signals are shifted in the ω_1-domain, according to the different trigonometric functions in β, in a manner that is reminiscent of the time-proportional phase incrementation procedure (8.72, 8.73). An experimental correlation spectrum based on this principle is shown in Fig. 8.5.6. Note that the CH_n fragments lead to signals in distinct frequency bands, except for the duplication of the CH_3 subspectrum in the CH region.

8.5.4. Relayed heteronuclear correlation spectroscopy

Conventional heteronuclear correlation spectroscopy leads to spectra where the chemical shifts Ω_I and Ω_S of I_nS subsystems are put into evidence, but the proximity of these fragments in the molecular frame cannot be determined unless the transfer through long-range couplings is exploited (8.95).

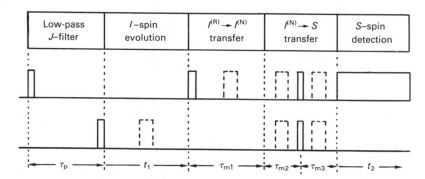

FIG. 8.5.7. Pulse sequence for heteronuclear relayed magnetization transfer from a remote $I_m^{(R)}$ spin to an S_k spin *via* a neighbour $I_l^{(N)}$ spin ($I_m^{(R)} \to I_l^{(N)} \to S_k$), employing a single $(\pi/2)$-pulse for the transfer $I_m^{(R)} \to I_l^{(N)}$ and an INEPT-type sequence for the heteronuclear $I_l^{(N)} \to S_k$ transfer. The optional low-pass J-filter (see text) consists of a period τ_p preceding the evolution period t_1. In this interval, the magnetization of neighbour protons $I_l^{(N)}$ is partly transformed into antiphase magnetization with respect to the S_k spin, which is then 'purged' by a $(\pi/2)_{\pm x}^S$-pulse, i.e. transformed into unobservable heteronuclear zero- and double-quantum coherences ($2I_{ly}S_{kz} \to \mp 2I_{ly}S_{ky}$).

These limitations can be overcome by employing relayed magnetization transfer. In the basic form of this experiment, single-quantum coherence is transferred first from a remote $I_m^{(R)}$ spin to a neighbour $I_l^{(N)}$ spin, and subsequently from $I_l^{(N)}$ to the S_k spin (8.7, 8.42, 8.87, 8.88). This type of experiment, which may be denoted symbolically by $I_m^{(R)} \to I_l^{(N)} \to S_k$, can yield information that is crucial for assignment, since the chemical shift Ω_k of the dilute spin can be correlated not only with that of the immediate neighbour $\Omega_l^{(N)}$, but also with the shifts of remote nuclei $\Omega_m^{(R)}$.

This information can be obtained with the experimental sequence of Fig. 8.5.7. The second $(\pi/2)_x^I$-pulse in the sequence leads to homonuclear coherence transfer (e.g. $2I_{my}I_{lz} \to -2I_{mz}I_{ly}$). The subsequent transfer to the dilute spin (e.g. $-2I_{mz}I_{ly} \to I_{lx} \to 2I_{ly}S_{kz} \to -2I_{lz}S_{ky} \to S_{kx}$) can be achieved with a variety of pulse schemes. The sequence employed in Fig. 8.5.7 is analogous to the INEPT sequence discussed in § 4.5.5.

The experimental spectra of a mixture of α and β glucose in Fig. 8.5.8 show many features that are characteristic of relayed spectroscopy. The relayed peaks (for example $H^{3\alpha} \to C^{2\alpha}$ and $H^{2\alpha} \to C^{3\alpha}$) and the signals stemming from direct transfer through one-bond couplings (e.g. $H^{3\alpha} \to C^{3\alpha}$ and $H^{2\alpha} \to C^{2\alpha}$) appear at the corners of a rectangle in 2D frequency space. This pattern is indicative of the fact that the $H^{2\alpha}-C^{2\alpha}$ and $H^{3\alpha}-C^{3\alpha}$ fragments are neighbours in the molecular framework, or, more accurately, that these two fragments are linked through a homonuclear coupling between $H^{2\alpha}$ and $H^{3\alpha}$.

It is clear that the type of information obtainable from a relayed experiment is related to the information in double-quantum spectra of

8.5 HETERONUCLEAR COHERENCE TRANSFER

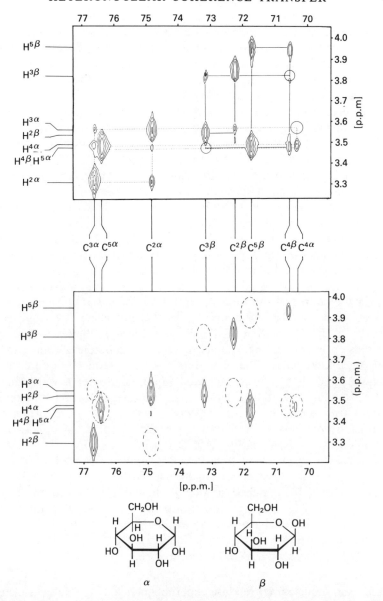

FIG. 8.5.8. Heteronuclear proton–carbon correlation spectra with relayed magnetization transfer $I^{(R)} \rightarrow I^{(N)} \rightarrow S$ of a mixture of the α- and β-anomers of glucose. (a) Spectrum obtained without suppression of neighbour signals. Neighbouring CH_n–CH_m fragments can be identified because the corresponding signals appear at the corners of rectangles in frequency space (dashed lines for α-glucose, solid lines for the β-anomer). Circles indicate the location of expected relayed signals that are missing because of unfavourable coupling constants. (b) Suppression of signals from neighbour protons (dashed ellipses) obtained with a low-pass J filter, as discussed in § 8.5.4. Both spectra are shown in absolute-value mode. (Reproduced from Ref. 8.42.)

carbon-13 in natural abundance (see § 8.4.2). However, relayed-transfer spectra have better inherent sensitivity and a greater information content (since the proton shifts are available at the same time), although these advantages are partly offset by the complexity of the spectra and by misleading long-range proton–proton couplings.

In heteronuclear correlation spectra obtained with relayed transfer, it is often difficult to recognize by inspection which signals stem from direct and which from relayed transfer. A remedy to this problem consists in the suppression of the direct connectivity signals. If a relayed-transfer spectrum is 'cleaned up' in this manner, it may be compared with a conventional correlation spectrum which reveals only direct connectivity signals. The pulse sequence in Fig. 8.5.7 shows one possible strategy for the elimination of redundant signals in relayed-transfer spectra (8.42). This scheme relies on the fact that one-bond heteronuclear couplings between neighbouring IS pairs are usually much larger than long-range couplings (typically by an order of magnitude for $^nJ_{CH}$ couplings). The magnetization of the neighbour proton therefore rapidly dephases under $^1J_{CH}$ in the course of the time τ_p, and the resulting antiphase coherence can be transferred into unobservable heteronuclear two-spin coherence by a $(\pi/2)^S_{\pm x}$-pulse (e.g. $2I_{ly}S_{kz} \to \mp 2I_{ly}S_{ky}$). Phase-alternation of this pulse prevents the heteronuclear two-spin coherence from reappearing after a later pulse. The procedure can be reiterated, preferably with unequal intervals τ_p, to ensure good suppression over a wide range of scalar coupling constants. Note that the magnetization of 'remote' protons with small, long-range $^nJ_{CH}$ couplings is not affected, hence the expression 'low-pass J filter'. An experimental spectrum obtained with this procedure is shown in Fig. 8.5.8(b).

A practical problem in heteronuclear relayed transfer of the type $I_m^{(R)} \to I_l^{(N)} \to S_k$ is the selection of the fixed delay τ_{ml} in Fig. 8.5.7. This delay can only be matched to one particular proton–proton coupling constant J_{ml} and the transfer may be inefficient for other proton pairs in the system. This deficiency can be avoided by incrementing the mixing time τ_{ml} in concert with the evolution period (8.7, 8.41) in analogy to 'accordion spectroscopy' as discussed in § 9.6.

8.5.5. Heteronuclear correlation experiments involving double transfer

As discussed in § 8.5.1, the sensitivity of a heteronuclear 2D experiment is most favourable if the coherence is first transferred from I spins with high γ to S spins with low γ in the preparation period, and reconverted into I-magnetization after the evolution period.

In the scheme of Fig. 8.5.9(a), the S-magnetization is simply enhanced by saturating the I spins. The resulting Overhauser effect is dependent on

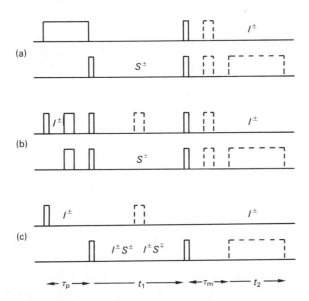

FIG. 8.5.9. Pulse sequences for heteronuclear 2D correlation spectroscopy involving a double transfer between I and S spins. (a) Incoherent transfer of longitudinal polarization ($I_z \to S_z$) by nuclear Overhauser enhancement, followed by the excitation of single-quantum S coherence that is transferred by a pair of r.f. pulses after the evolution period to observable I-magnetization (8.9). (b) Coherent transfer by r.f. pulses of I-magnetization into single-quantum S coherence, evolution, and transfer from S to I (8.12). (c) Transfer of I-magnetization into heteronuclear zero- and multiple-quantum coherence, evolution, and transfer into I-magnetization (8.13, 8.81).

the gyromagnetic ratios and on motional correlation times, but independent of the magnitude of the scalar couplings (8.9, 8.10). In comparison with the scheme of Fig. 8.5.3(a), the Hamiltonians that are effective in the evolution and detection periods of eqn (8.5.4) are interchanged. The basic scheme of Fig. 8.5.9(a) can be extended in analogy to Fig. 8.5.3(b)–(d) to obtain spectra where the $\mathcal{H}_{IS}$ splittings are removed.

In the scheme of Fig. 8.5.9(b), the preparation sequence transfers coherence $I^\pm$ into single-quantum coherence $S^\pm$ by r.f. pulses as discussed in § 4.5.2. In the evolution period, a $(\pi)^I$-pulse refocuses the $\mathcal{H}_{IS}$ interactions, and the coherence is transferred back into observable I-magnetization for detection.

This type of experiment has been employed for the indirect detection of nitrogen-15 and mercury-199 spectra (8.12, 8.89, 8.118, 8.119). The same approach can be used to monitor the evolution of homonuclear S-spin multiple-quantum coherence in systems with quadrupolar spins $S \geq 1$, provided that there is a resolved scalar or dipolar $\mathcal{H}_{IS}$ coupling (8.82, 8.86).

The correlation of I and S spectra can also be obtained by monitoring the evolution of heteronuclear zero- and double-quantum coherence ($I^\pm S^\mp$ and $I^\pm S^\pm$ respectively). The sequence in Fig. 8.5.9(c) shows one possible realization (8.13, 8.120, 8.121), which is a simplified form of a sequence originally proposed by Müller (8.81).

In the preparation interval τ_p of Fig. 8.5.9(c), the magnetization of the I spins dephases under the heteronuclear coupling ($-I_{ky} \to 2I_{kx}S_{mz}$). The $(\pi/2)_x^S$-pulse converts this term into a superposition of zero- and double-quantum coherence

$$-2I_{kx}S_{my} = -\frac{1}{2i}[(I_k^+ S_m^+ - I_k^- S_m^-) - (I_k^+ S_m^- - I_k^- S_m^+)]. \quad (8.5.14)$$

In a two-spin system, the four terms on the right-hand side of eqn (8.5.14) evolve with the effective chemical shifts $(\Omega_I + \Omega_S)$, $-(\Omega_I + \Omega_S)$, $(\Omega_I - \Omega_S)$, and $-(\Omega_I - \Omega_S)$, respectively. The zero- and double-quantum terms can be interconverted by applying a $(\pi)^I$-pulse in the middle of the evolution period, with the result that the effective Hamiltonian depends only on the shift Ω_S of the S-spin (8.81). The heteronuclear two-spin coherence is reconverted into observable I-magnetization by the last $(\pi/2)^S$-pulse in the sequence.

If one focuses attention on, say, the heteronuclear zero-quantum

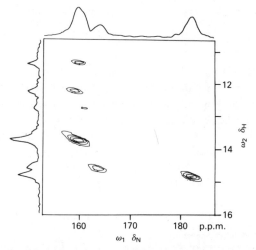

FIG. 8.5.10. Heteronuclear proton–nitrogen-15 shift correlation spectrum of an 0.7 mM solution of tRNA$_f^{Met}$, labelled with 65 per cent ^{15}N at N3 of all uridine-related bases, obtained with the sequence in Fig. 8.5.9(c), with decoupling in t_2 but without $(\pi)^I$-pulse in the evolution period. Only signals associated with heteronuclear double-quantum coherence were retained by phase cycling, and the proton shifts were removed from the ω_1-domain by a procedure which amounts to a shearing of the data matrix (§ 6.6.1). The spectrum was obtained in 6 hours with a sample of 200 μl. (Reproduced form Ref. 8.13.)

coherence by suitable phase cycling and pathway selection, one obtains 2D spectra with signals appearing at $\omega_1 = \pm(\Omega_I - \Omega_S)$ and $\omega_2 = \Omega_I$. Such spectra can be converted by shearing and foldover correction procedures (see § 6.6.1) into shift correlation spectra with $(\omega_1, \omega_2) = (\Omega_I, \Omega_S)$. The same procedure can be applied independently to the heteronuclear double-quantum components (8.13). An example of a proton–nitrogen-15 correlation spectrum obtained with such procedures is shown in Fig. 8.5.10. In larger systems, the heteronuclear coherences are affected by couplings to remote protons, thus allowing one to measure multiplet structures of selected fragments within extensive coupling networks (8.90). For a discussion of similar experiments in liquid crystalline phase, the reader is referred to Refs. 8.35 and 8.99.

Although substantial sensitivity enhancements can be achieved with methods that use a double transfer of the type $I \to S \to I$, the techniques suffer from the need of suppressing the I-magnetization that is not involved in the transfer. These methods are therefore most appropriate for S nuclei with low gyromagnetic ratios and with an isotopic abundance that is not too low.

8.5.6. Heteronuclear correlation in solids

The advantages of heteronuclear 2D correlation spectroscopy can also be exploited for studying solid samples. However, the strength of the homonuclear dipolar interactions severely limits the achievable resolution and leads to fast spin diffusion among the I spins. As a result, it is difficult to attain a selective transfer between neighbouring I and S spins. Special attention has to be paid to the suppression of homo- and heteronuclear dipolar interactions in the evolution and detection periods, and to the suppression of homonuclear interactions in the coherence transfer period. The use of multiple-pulse sequences in all three periods is essential to meet these requirements (8.98, 8.99). The scheme of Fig. 8.5.11 shows one possible realization: the proton magnetization evolves in t_1 under a windowless BLEW-12 sequence (8.122) while the S nuclei are decoupled by a WALTZ-8 sequence (8.111), resulting in the effective evolution Hamiltonian

$$\mathcal{H}^{(e)} = \kappa(\mathcal{H}_{ZI}^{iso} + \mathcal{H}_{ZI}^{aniso}) \qquad (8.5.15)$$

where the scaling factor κ is 0.475 for the BLEW-12 sequence employed in the experiments shown below. In the mixing interval, the magnetization can be transferred by applying a windowless isotropic mixing sequence (WIM-24) simultaneously to both I and S nuclei, as shown in Fig. 8.5.11, leading to the isotropic mixing Hamiltonian

$$\mathcal{H}^{(m)} = D_{IS}\mathbf{I} \cdot \mathbf{S} \qquad (8.5.16)$$

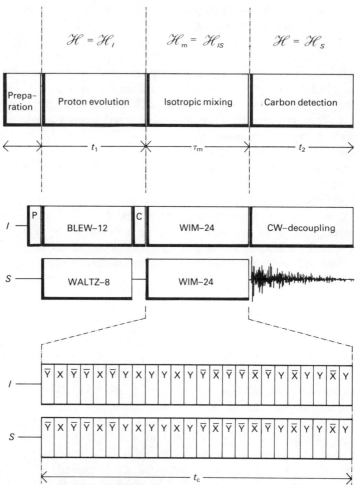

FIG. 8.5.11. Pulse schemes for heteronuclear shift correlation in solids. In the evolution period, the homonuclear dipolar interaction is decoupled by a multiple-pulse method such as the BLEW-12 sequence (8.122), while the carbon-13 spins are decoupled by a broadband method such as WALTZ-8 (8.111). The actual transfer of coherence from I to S occurs under an isotropic mixing Hamiltonian defined by the WIM-24 pulse sequence shown. Finally, the S-signal is detected in the presence of conventional high-power I-spin decoupling. Because the effective axis of rotation under the BLEW sequence is tilted by 63° with respect to the z-axis, a preparation pulse P and a compensation pulse C are inserted as shown. (Reproduced from Ref. 8.98.)

where

$$D_{IS} = \tfrac{1}{3}\kappa b_{IS}(3\cos^2\theta_{IS} - 1).$$

The isotropic mixing sequence is designed to suppress $\mathcal{H}_{II}$ interactions in addition to $\mathcal{H}_{ZI}$ and $\mathcal{H}_{ZS}$ Zeeman terms. A component $I_\alpha (\alpha = x, y, \text{or } z)$

8.5 HETERONUCLEAR COHERENCE TRANSFER

in a two-spin system transforms under the isotropic mixing Hamiltonian

$$I_\alpha \xrightarrow{\mathcal{H}^{(m)}\tau_m} \tfrac{1}{2}I_\alpha[1+\cos(D_{IS}\tau_m)]$$
$$+\tfrac{1}{2}S_\alpha[1-\cos(D_{IS}\tau_m)]$$
$$+(I_\beta S_\gamma - I_\gamma S_\beta)\sin(D_{IS}\tau_m). \qquad (8.5.17)$$

In solids with numerous interactions, the oscillations are rapidly damped, leading to an equilibrium state $\tfrac{1}{2}(I_\alpha + S_\alpha)$ (8.98). Note that this amounts to a transfer of the *complex* magnetization $I^+ = I_x + iI_y \rightarrow S^+ = S_x + iS_y$, i.e. the sense of rotation is preserved without taking recourse to phase-cycling schemes. In the final stage of the pulse scheme of Fig. 8.5.11, the S-magnetization is observed with conventional I decoupling

$$\mathcal{H}^{(d)} = \mathcal{H}^{iso}_{ZS} + \mathcal{H}^{aniso}_{ZS} + [\mathcal{H}_{SS}]. \qquad (8.5.18)$$

In applications to single crystals (8.103), experiments of this type make it possible to correlate the chemical shielding tensors of coupled (i.e. neighbouring) I and S spins.

A more generally applicable experiment combines the averaging in spin space achieved by the pulse sequence with spatial averaging by magic-angle sample spinning. If the spinning is sufficiently fast, the anisotropic terms in $\mathcal{H}^{(e)}$ and $\mathcal{H}^{(d)}$ may be dropped, and the heteronuclear 2D spectrum strongly resembles the spectra that can be obtained in isotropic phase, since they show only isotropic chemical shifts (8.98, 8.104).

The success of this approach may be appreciated in Fig. 8.5.12. The four carbon-13 sites in a polycrystalline sample of threonine give rise to well-resolved isotropic carbon shifts in ω_2, while only two distinct proton shifts can be resolved in ω_1 which belong to CH_3 and to the overlapping $C^\alpha H$ and $C^\beta H$ proton resonances. It is obvious that rather complex spectra could be unravelled and assigned. This is a significant improvement over the direct observation of abundant spins with dipolar decoupling, where it is difficult to resolve more than a few distinct resonances.

In heteronuclear systems with abundant $I = \tfrac{1}{2}$ nuclei and rare $S \geq 1$ nuclei, it is possible to use cross-polarization (§ 4.5.1) to transfer coherence from single-quantum I transitions into (homonuclear) multiple-quantum coherence of spin S. This procedure is attractive for insensitive nuclei such as ^{14}N (8.101, 8.102). The Hartmann–Hahn condition must be modified to account for the effective r.f. nutation frequency

$$[\omega_{1I}^2 + (\Delta\omega_I)^2]^{\tfrac{1}{2}} = [(\omega_{1S}^2/\omega_Q)^2 + (2\Delta\omega_S)^2]^{\tfrac{1}{2}} \qquad (8.5.19)$$

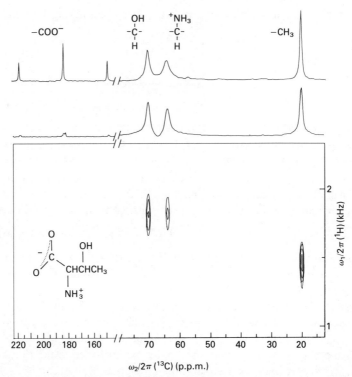

FIG. 8.5.12. Heteronuclear proton–carbon correlation spectrum of a polycrystalline powder sample of threonine, obtained with the sequence of Fig. 8.5.11 combined with magic-angle spinning at 2.6 kHz. Only isotropic shifts of the protons and carbons appear in the ω_1- and ω_2-domains, respectively. The signals indicate that the transfer $I \to S$ is essentially confined to neighbouring spins, and fails in the case of the carboxyl group. The top trace shows the normal 1D carbon-13 spectrum (obtained with cross-polarization and magic-angle spinning; note the spinning sidebands of the COO$^-$ resonance). The second trace shows the projection of the 2D spectrum. (Reproduced from Ref. 8.98.)

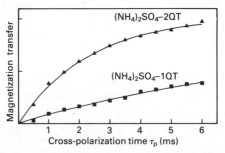

FIG. 8.5.13. Comparison of single- and double-quantum cross-polarization rates. The signal intensities after double cross-polarization $^1\text{H} \to {}^{14}\text{N} \to {}^1\text{H}$ are shown as a function of the cross-polarization time for single- and double-quantum cross-polarization in a single crystal of $(\text{NH}_4)_2\text{SO}_4$. The Hartmann–Hahn condition is nearly fulfilled in both cases. (Reproduced from Ref. 8.101.)

where $\omega_{1I} = -\gamma_I B_{1I}$, and $\omega_{1S} = -\gamma_S B_{1S}$ are the r.f. field strengths, $\Delta\omega_I$, and $\Delta\omega_S$ are offsets, and ω_Q is half the quadrupole splitting in the S spectrum. The rate of the transfer is greatest when the effective fields are matched according to eqn (8.5.19). In multiple-quantum applications, it is important to take into account that both the transfer rate T_{IS}^{-1} and the multiple-quantum decay rate $T_{1\rho}^{-1}$ are increased in comparison to analogous single-quantum experiments. For equal effective r.f. fields and tilt angles, one can show that

$$(T_{IS}^{2QT})^{-1} = 4(T_{IS}^{1QT})^{-1}. \qquad (8.5.20)$$

These predictions have been verified experimentally for single- and double-quantum cross-polarization between protons and nitrogen-14 in single crystals, as shown in Fig. 8.5.13.

9
DYNAMIC PROCESSES STUDIED BY TWO-DIMENSIONAL EXCHANGE SPECTROSCOPY

The investigation of dynamic processes such as chemical exchange, cross-relaxation, nuclear Overhauser effects, spin diffusion, and cross-polarization by 2D spectroscopy has a number of advantages over the 1D techniques discussed in § 4.6.1.4, particularly when the system comprises an extended network of exchange processes that occur simultaneously. The 2D methods are most useful for studying slow dynamic processes with rates that are too low to affect the lineshapes. Two-dimensional exchange spectroscopy is therefore particularly well suited for studies of cross-relaxation (transient Overhauser effects) and of spin diffusion in solids. In applications to chemical exchange, the information content of 2D exchange spectra is greatest if the temperature is chosen such that the exchange rate is fast compared to longitudinal relaxation and slow compared to the spectral parameters affected by the exchange.

9.1. Polarization transfer in one- and two-dimensional methods

The fundamental idea of 2D exchange spectroscopy (9.1, 9.2) is the 'frequency-labelling' of the longitudinal magnetization of various sites before exchange takes place, such that after exchange the pathways of the magnetization can be traced back to their origins. While the magnetization is put into a non-equilibrium state, the concentrations of the chemical species remain in dynamic equilibrium throughout the experiment.

Consider the basic sequence in Fig. 9.1.1(a). A pair of non-selective $\pi/2$-pulses, separated by the evolution period t_1, is used to prepare non-equilibrium populations at the beginning of the mixing time τ_m. For the sake of clarity, we consider a symmetrical two-site *chemical* exchange case with equal concentrations ($k_{AB} = k_{BA} = k$), equal spin–lattice relaxation rates ($R_1^A = R_1^B = R_1$) and equal transverse relaxation ($T_2^A = T_2^B = T_2$). The transverse magnetization, excited by the initial $(\pi/2)_y$-pulse, precesses freely in the t_1-interval. If the exchange is slow, we may neglect its effect on the lineshape in this period and obtain two complex magnetization components (see eqn (4.2.16))

$$M_A^+(t_1) = M_{A0} \exp\{i\Omega_A t_1 - t_1/T_2\},$$
$$M_B^+(t_1) = M_{B0} \exp\{i\Omega_B t_1 - t_1/T_2\}. \qquad (9.1.1)$$

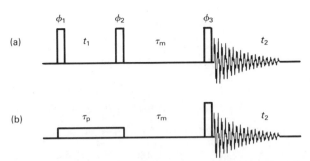

FIG. 9.1.1. (a) Basic scheme for 2D exchange spectroscopy, where t_1 and t_2 indicate the evolution and detection periods. The exchange time τ_m is normally kept constant in a 2D experiment. (b) Related scheme used in 1D spectroscopy to study slow exchange with selective inversion and subsequent recovery (see also § 4.6.1.4).

If the second pulse in Fig. 9.1.1(a) is applied along the y-axis, the real components of the transverse magnetization are converted into longitudinal magnetization

$$M_{Az}(\tau_m = 0) = -M_{A0}\cos\Omega_A t_1 \exp\{-t_1/T_2\},$$
$$M_{Bz}(\tau_m = 0) = -M_{B0}\cos\Omega_B t_1 \exp\{-t_1/T_2\}. \qquad (9.1.2)$$

The y-components remain unaffected by the pulse and are normally destroyed by an inhomogeneous magnetic field or cancelled by phase-cycling. In case the second pulse is not exactly $\pi/2$, it is advisable that contributions from magnetization that is recovering towards M_{A0} and M_{B0} in t_1 are cancelled by phase alternation (suppression of the axial peaks, as discussed in § 9.2).

The t_1-modulated longitudinal components in eqn (9.1.2) migrate from one site to another because of chemical exchange or cross-relaxation, while spin-lattice relaxation attenuates the memory of the initial labelling, as will be discussed in § 9.3

$$M_{Az}(\tau_m) = M_{Az}(\tau_m = 0)\tfrac{1}{2}[1 + \exp\{-2k\tau_m\}] \exp\{-\tau_m/T_1\}$$
$$\quad + M_{Bz}(\tau_m = 0)\tfrac{1}{2}[1 - \exp\{-2k\tau_m\}] \exp\{-\tau_m/T_1\},$$
$$M_{Bz}(\tau_m) = M_{Az}(\tau_m = 0)\tfrac{1}{2}[1 - \exp\{-2k\tau_m\}] \exp\{-\tau_m/T_1\}$$
$$\quad + M_{Bz}(\tau_m = 0)\tfrac{1}{2}[1 + \exp\{-2k\tau_m\}] \exp\{-\tau_m/T_1\} \qquad (9.1.3)$$

where k is the rate constant of the exchange process. The final $(\pi/2)_y$-pulse converts these longitudinal components into observable transverse magnetization. After 2D Fourier transformation, a cross-peak at $(\omega_1, \omega_2) = (\Omega_A, \Omega_B)$ with an integrated amplitude $I_{BA}(\tau_m)$ appears if a magnetization component that precessed at Ω_A in t_1 resumes precession at Ω_B in t_2. The amplitudes $I_{kl}(\tau_m)$ of the diagonal and cross-peaks depend on the equilibrium magnetization M_{l0} and on mixing coefficients

$a_{kl}(\tau_m)$

$$I_{AA}(\tau_m) = a_{AA}(\tau_m)M_{A0},$$
$$I_{BB}(\tau_m) = a_{BB}(\tau_m)M_{B0},$$
$$I_{AB}(\tau_m) = a_{AB}(\tau_m)M_{B0},$$
$$I_{BA}(\tau_m) = a_{BA}(\tau_m)M_{A0}. \quad (9.1.4)$$

The mixing coefficients correspond to the factors in eqn (9.1.3)

$$a_{AA}(\tau_m) = a_{BB}(\tau_m) = \tfrac{1}{2}[1 + \exp\{-2k\tau_m\}]\exp\{-\tau_m/T_1\},$$
$$a_{AB}(\tau_m) = a_{BA}(\tau_m) = \tfrac{1}{2}[1 - \exp\{-2k\tau_m\}]\exp\{-\tau_m/T_1\}. \quad (9.1.5)$$

The pathways that lead to diagonal and cross-peaks are represented schematically in Fig. 9.1.2. Note that for systems without resolved couplings, the very appearance of a cross-peak is sufficient proof that exchange is taking place.

In the symmetrical two-site case discussed here for $M_{A0} = M_{B0}$, the exchange rate can be determined from the ratio of the peak intensities

$$\frac{I_{AA}}{I_{AB}} = \frac{a_{AA}}{a_{AB}} = \frac{1 + \exp\{-2k\tau_m\}}{1 - \exp\{-2k\tau_m\}} \simeq \frac{1 - k\tau_m}{k\tau_m}. \quad (9.1.6)$$

The latter equation holds within the limitations of the initial rate approximation (9.3).

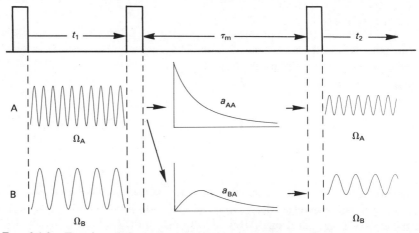

FIG. 9.1.2. Transfer of frequency-labelled longitudinal polarization in 2D exchange spectroscopy of a symmetrical two-site system. The amplitudes of the diagonal peaks, proportional to $a_{AA}(\tau_m) = a_{BB}(\tau_m)$ defined in eqn (9.1.5), decay biexponentially, while the cross-peak amplitudes, proportional to $a_{BA}(\tau_m) = a_{AB}(\tau_m)$, first increase due to exchange before decaying because of spin–lattice relaxation.

We shall see in § 9.3 that these expressions can be generalized to N-site exchange problems which simultaneously involve chemical exchange and cross-relaxation.

It is instructive to compare the 2D exchange method in Fig. 9.1.1(a) with the conventional 1D polarization transfer method shown schematically in Fig. 9.1.1(b) and discussed in § 4.6.1.4. In the 1D case, a selective π-pulse with a frequency Ω_A is used to invert the polarization M_{Az} of one particular site. The migration and recovery are monitored in the same manner as in 2D exchange spectroscopy by a non-selective pulse. Both schemes are applicable to the same type of slow-exchange cases.

The most obvious advantage of the 2D scheme lies in the ability to study networks involving a large number of sites: whereas in the 1D approach, the magnetization of each site must be inverted in turn, a single 2D experiment allows one to impose a modulation on the polarization of all sites. Although a large number of experiments with different t_1-values are required in the 2D method, a significant sensitivity advantage results from the fact that all processes are investigated simultaneously.

The selectivity of the excitation is determined in both 1D and 2D experiments by the duration of the preparation, i.e. by t_1^{max} in Fig. 9.1.1(a), or by τ_p in Fig. 9.1.1(b). In the 2D method however, it is possible to study partially overlapping spectra, where selective excitation is not possible. Furthermore, if the selective pulse in the 1D method is long, it is necessary to take account of exchange processes during the pulse, making it difficult to separate excitation and recovery. In the 2D experiment on the other hand, the longitudinal magnetization in the t_1-interval is immaterial, while the exchange of the transverse components evolving in t_1 does not affect the integrated cross-peak intensities, since it merely leads to line broadening (see § 9.3). The second $\pi/2$-pulse generates non-equilibrium populations almost instantaneously, and the relevant mixing process is initiated at this point. Hence the observed migration of Zeeman polarization starts from a clear-cut initial condition, making it possible to determine the rates of the dynamic processes with improved accuracy.

A third advantage of the 2D method becomes apparent if the dynamic processes are to be monitored as a function of mixing time τ_m (9.3). Strictly speaking, systematic variation of τ_m turns both experiments into 3D techniques: the selective experiment leads to a set of spectra $S(\Omega_k, \tau_m, \omega)$ with parametric variation of Ω_k and τ_m, whereas the 2D approach leads to an array of 2D spectra $S(\omega_1, \tau_m, \omega_2)$. In the latter case, the experimental time and the data volume can be greatly reduced by converting a 3D experiment into a 2D experiment by simultaneous variation of t_1 and τ_m, as discussed in § 9.6. Thus, at least in favourable cases, the systematic variation of τ_m does not greatly enhance the complexity of the 2D experiment, while the same variation makes the 1D approach more cumbersome.

9.2. Selection of coherence transfer pathways

The basic three-pulse sequence employed in 2D exchange spectroscopy can also lead to undesirable coherent phenomena such as relayed magnetization transfer (§ 8.3.4) and multiple-quantum excitation (§ 8.4). In order to focus attention on the migrating longitudinal magnetization, it is essential to select the proper coherence transfer pathways (9.4), as shown in Fig. 9.2.1.

In the evolution period, contributions from longitudinal magnetization (which would lead to unmodulated axial peaks at $\omega_1 = 0$) must be suppressed. Pure 2D absorption peakshapes can be obtained if the two pathways with $p = \pm 1$ quantum coherence in t_1 (solid and dashed lines in Fig. 9.2.1) are retained.

In the mixing period, it is essential to cancel all pathways involving $p = \pm 1$ quantum coherence. In systems with resolved scalar or dipolar couplings, additional care must be taken to remove $p = \pm 2$, $p = \pm 3, \ldots$ quantum coherences. In coupled spin systems, the order $p = 0$ that must be selected comprises, in addition to the desired longitudinal Zeeman polarization (represented by density operator components proportional to I_{kz}), undesirable zero-quantum coherence (e.g. $I_k^+ I_l^-$) and longitudinal

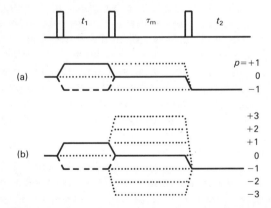

FIG. 9.2.1. Coherence transfer pathways in 2D exchange spectroscopy: (a) in systems without resolved couplings; (b) in systems with up to three coupled $I = \frac{1}{2}$ nuclei. Solid lines correspond to pathways that must be retained. For pure 2D absorption lineshapes, the dashed 'mirror-image' pathways must be retained as well. The dotted pathways (longitudinal terms in t_1, single- and multiple-quantum pathways in τ_m) must be suppressed.

scalar or dipolar order (e.g. $2I_{kz}I_{lz}$). These components may lead to spurious effects which are discussed in §§ 9.4.2–9.4.4.

The pathway selection for pure 2D absorption lineshapes can be achieved by selecting the following changes in magnetic quantum numbers under the three pulses (see § 6.4)

$$\Delta p_1: \mathbf{-1}, (0), \mathbf{+1}, \tag{9.2.1a}$$

$$\Delta p_2: \text{no need for selection, since } \sum_{i=1}^{3} \Delta p_i = -1, \tag{9.2.1b}$$

$$\Delta p_3: (-p^{\max} - 1), \ldots, (-2), \mathbf{-1}, (0), \ldots, (p^{\max} - 1) \tag{9.2.1c}$$

where all Δp-values in eqn (9.2.1a) and in eqn (9.2.1c) which are set in brackets must be cancelled, while the Δp-values set in boldface are retained. This can be achieved with a cycle of $N_1 \cdot N_3 = 2(p^{\max} + 1)$ steps. The cycles suitable for systems without couplings ($p^{\max} = 1$) and for coupled systems with $p^{\max} = 3$ are indicated in Table 9.2.1(a) and (b).

If mixed-phase or absolute value peakshapes are acceptable, the selection may be narrowed down to the pathways indicated by solid lines in Fig. 9.2.1

$$\Delta p_1 = (-1)(0)\mathbf{+1}.$$

This can be achieved by cycling the first pulse in three increments of $2\pi/3$ or, if such increments are not available, in four steps of $\pi/2$. In either case, the net effect of the phase cycle is to allow the transfer of the *complex* magnetization precessing in the t_1-interval to the detection period. This may appear paradoxical, since the longitudinal magnetization in the τ_m-period is necessarily real. The M_x- and M_y-components are in fact transferred in successive experiments of the phase cycle. A cycle for systems without couplings ($p^{\max} = 1$) is given in Table 9.2.1(c).

9.3. Cross-relaxation and exchange in systems without resolved couplings

Consider a system with N sites, which may be contained within one molecule or in different chemical species (9.1, 9.5). These sites can be interconverted through chemical exchange or communicate through intra- or intermolecular cross-relaxation. Each site l may contain n_l magnetically equivalent nuclei, but we assume for the time being that there are no resolved couplings. Since multiple-quantum effects are precluded in this case, the dynamics can be described in terms of modified classical Bloch equations (§ 2.4.1).

The equilibrium z-magnetizations of the N sites are collected in the

Table 9.2.1

Phase cycles for 2D exchange spectroscopy. Time-proportional phase increments (TPPI) are only required if the carrier is positioned within the spectrum

(a) In spin systems without resolved couplings ($p^{\max} = 1$): selection of $p = 0 \to \pm 1 \to 0 \to -1$ pathways to obtain pure 2D absorption lineshapes (see Fig. 9.2.1(a))

$\Delta p_1 = \pm 1$	$\Delta p_2 =$ free	$\Delta p_3 = \pm 1$	
$\varphi_1 = $ TPPI	$\varphi_2 = 0$	$\varphi_3 = 0$	$\varphi^{\text{ref}} = 0$
$= \pi + $ TPPI	$= 0$	$= 0$	$= \pi$
$= $ TPPI	$= 0$	$= \pi$	$= \pi$
$= \pi + $ TPPI	$= 0$	$= \pi$	$= 0$

(b) In coupled spin systems with $p^{\max} = 3$: selection of $p = 0 \to \pm 1 \to 0 \to -1$ pathways for pure 2D absorption (see Fig. 9.2.1(b))

$\Delta p_1 = \pm 1$	$\Delta p_2 =$ free	$\Delta p_3 = -1$	
$\varphi_1 = $ TPPI	$\varphi_2 = 0$	$\varphi_3 = 0$	$\varphi^{\text{ref}} = 0$
$= \pi + $ TPPI	$= 0$	$= 0$	$= \pi$
$= $ TPPI	$= 0$	$= \pi/2$	$= \pi/2$
$= \pi + $ TPPI	$= 0$	$= \pi/2$	$= 3\pi/2$
$= $ TPPI	$= 0$	$= \pi$	$= \pi$
$= \pi + $ TPPI	$= 0$	$= \pi$	$= 0$
$= $ TPPI	$= 0$	$= 3\pi/2$	$= 3\pi/2$
$= \pi + $ TPPI	$= 0$	$= 3\pi/2$	$= \pi/2$

(c) In systems with $p^{\max} = 1$: selection of $p = 0 \to +1 \to 0 \to -1$ pathway to transfer complex magnetization (yields lineshapes with mixed phases)

$\Delta p_1 = +1$	$\Delta p_2 =$ free	$\Delta p_3 = -1$	
$\varphi_1 = 0$	$\varphi_2 = 0$	$\varphi_3 = 0$	$\varphi^{\text{ref}} = 0$
$= 2\pi/3$	$= 0$	$= 0$	$= 4\pi/3$
$= 4\pi/3$	$= 0$	$= 0$	$= 2\pi/3$
$= 0$	$= 0$	$= \pi$	$= \pi$
$= 2\pi/3$	$= 0$	$= \pi$	$= \pi/3$
$= 4\pi/3$	$= 0$	$= \pi$	$= 5\pi/3$

vector $\mathbf{M}_0$ with elements $M_{l0} = n_l x_l M_0$ that are proportional to the number of equivalent nuclei n_l in the site l, to the mole fraction x_l in chemical equilibrium, and to the equilibrium magnetization M_0 per mole of spins.

If we consider the pulse sequence in Fig. 9.1.1(a), the initial $(\pi/2)_y$-preparation pulse generates transverse magnetization along the (real) x-axis

$$\mathbf{M}^+(t_1 = 0) = \mathbf{M}_x(t_1 = 0) + i\mathbf{M}_y(t_1 = 0) = \mathbf{M}_0. \tag{9.3.1}$$

with $\mathbf{M}_y(t_1 = 0) = \mathbf{0}$. Note that these quantities are vectors, i.e. ordered linear arrays of N components, but that they are *not* associated with a

9.3 SYSTEMS WITHOUT RESOLVED COUPLINGS

direction in physical space. The evolution of the complex transverse magnetization is governed by a set of N coupled differential equations (see eqn (2.4.20))

$$\frac{d}{dt}\mathbf{M}^+ = \mathbf{L}^+\mathbf{M}^+ \qquad (9.3.2)$$

where

$$\mathbf{L}^+ = i\mathbf{\Omega} - \mathbf{\Lambda} + \mathbf{K}. \qquad (9.3.3)$$

The diagonal matrices $\mathbf{\Omega}$ and $\mathbf{\Lambda}$ contain the chemical shifts Ω_l and the transverse relaxation rates $\lambda_l = 1/T_{2l}$, and the kinetic matrix $\mathbf{K}$ represents the effects of chemical exchange. The precession in the t_1-period is described by

$$\mathbf{M}^+(t_1) = \exp\{\mathbf{L}^+ t_1\}\mathbf{M}^+(t_1 = 0). \qquad (9.3.4)$$

This equation describes chemical shift precession, line-broadening, and coalescence in intermediate exchange regimes.

The second $(\pi/2)_y$-pulse in the sequence of Fig. 9.1.1(a) simply converts the x-component into longitudinal magnetization

$$\mathbf{M}_z(\tau_m = 0) = -\mathrm{Re}\{\mathbf{M}^+(t_1)\}. \qquad (9.3.5)$$

The evolution of the longitudinal magnetization is described by eqn (2.4.21)

$$\frac{d}{d\tau_m}\Delta\mathbf{M}_z(\tau_m) = \mathbf{L}\,\Delta\mathbf{M}_z(\tau_m) \qquad (9.3.6)$$

with the deviations from thermal equilibrium $\Delta\mathbf{M}_z(\tau_m) = \mathbf{M}_z(\tau_m) - \mathbf{M}_0$ and the dynamic matrix

$$\mathbf{L} = -\mathbf{R} + \mathbf{K} \qquad (9.3.7)$$

which expresses the effect of spin-lattice relaxation (the diagonal elements of $\mathbf{R}$ are equal to $1/T_{1l}$), cross-relaxation (off-diagonal elements of $\mathbf{R}$), and chemical exchange (kinetic matrix $\mathbf{K}$). The relationship between the elements of $\mathbf{R}$ and transition probabilities W_{ij} is shown in eqns (9.7.1)–(9.7.12). This general treatment is possible because of the formal similarity between cross-relaxation and exchange. In practice, we will often encounter situations where one of these mechanisms may be neglected. Thus in natural abundance carbon-13 NMR cross-relaxation will be absent due to the isotopic dilution and it becomes possible to measure pure exchange.

The solution of eqn (9.3.6)

$$\mathbf{M}_z(\tau_m) = \mathbf{M}_0 + \exp\{\mathbf{L}\tau_m\}\Delta\mathbf{M}_z(\tau_m = 0) \qquad (9.3.8)$$

implies that the magnetization components recover in the course of τ_m towards equilibrium $\mathbf{M}_0$. The last $(\pi/2)_y$-pulse in the sequence generates transverse magnetization

$$\mathbf{M}^+(t_2 = 0) = \mathbf{M}_z(\tau_m) \tag{9.3.9}$$

which is real (i.e. along the x-axis). The overall time dependence can be summed up in the expression

$$\mathbf{M}^+(t_1, \tau_m, t_2) = \exp\{\mathbf{L}^+ t_2\}\left[1 - \exp\{\mathbf{L}\tau_m\}(\mathrm{Re}[\exp\{\mathbf{L}^+ t_1\}] + 1)\right]\mathbf{M}_0 \tag{9.3.10}$$

Terms which do not depend on t_1 give rise to so-called axial peaks at $\omega_1 = 0$. They are not informative and are usually eliminated by alternation of the phase of the first pulse and of the receiver reference phase (which amounts to a form of difference spectroscopy, see Table 9.2.1). The remainder takes the form

$$\mathbf{M}^+(t_1, \tau_m, t_2) = -\exp\{\mathbf{L}^+ t_2\}\exp\{\mathbf{L}\tau_m\}\mathrm{Re}[\exp\{\mathbf{L}^+ t_1\}]\mathbf{M}_0 \tag{9.3.11}$$

If the single experiment considered so far is expanded into a cycle of phase-shifted experiments, the net effect can amount to a transfer of both real and imaginary components of $\mathbf{M}^+(t_1)$ (see §9.2), and we arrive finally at the simple result

$$\mathbf{M}^+(t_1, \tau_m, t_2) = -\exp\{\mathbf{L}^+ t_2\}\exp\{\mathbf{L}\tau_m\}\exp\{\mathbf{L}^+ t_1\}\mathbf{M}_0. \tag{9.3.12}$$

9.3.1. Slow exchange

If the exchange is slow, the lineshapes are not noticeably affected by the transport of transverse magnetization from one site to another, and the contributions of the exchange matrix $\mathbf{K}$ to the dynamic matrix $\mathbf{L}^+$ in eqn (9.3.3) may be neglected in the evolution and detection periods

$$\mathbf{L}^{+(\mathrm{slow})} = i\mathbf{\Omega} - \mathbf{\Lambda}. \tag{9.3.13}$$

In this case, each transverse term develops independently

$$M_l^+(t_1) = \exp(i\Omega_l t_1 - \lambda_l t_1) M_{l0} \tag{9.3.14}$$

as assumed in eqn (9.1.1). As a result, the time-domain signal simplifies to

$$s(t_1, \tau_m, t_2) = -\sum_k \sum_l \exp\{i\Omega_k t_2 - \lambda_k t_2\}[\exp\{\mathbf{L}\tau_m\}]_{kl}$$
$$\times \exp\{i\Omega_l t_1 - \lambda_l t_1\} M_{l0}. \tag{9.3.15}$$

After 2D Fourier transformation, the integrated amplitude of a signal with frequency coordinates $(\omega_1, \omega_2) = (\Omega_l, \Omega_k)$ is

$$I_{kl}(\tau_m) = a_{kl}(\tau_m) M_{l0} \tag{9.3.16}$$

and

$$a_{kl}(\tau_m) = [\exp\{\mathbf{L}\tau_m\}]_{kl}. \tag{9.3.17}$$

Hence the 2D spectrum amounts to a pictorial representation of the exponential mixing operator.

9.3.2. Two-site systems

In two-site systems, eqn (9.3.17) may be evaluated analytically. The dynamic matrix

$$\mathbf{L} = \begin{pmatrix} L_{AA} & L_{AB} \\ L_{BA} & L_{BB} \end{pmatrix} = \begin{pmatrix} -K_{BA} - R_1^A & K_{AB} - R_{AB} \\ K_{BA} - R_{BA} & -K_{AB} - R_1^B \end{pmatrix} \tag{9.3.18}$$

has the eigenvalues

$$\lambda_{\pm} = -\sigma \pm D \tag{9.3.19}$$

where

$$\sigma = -\tfrac{1}{2}(L_{AA} + L_{BB}) = \tfrac{1}{2}(K_{BA} + K_{AB} + R_1^A + R_1^B),$$
$$D = [\delta^2 + L_{AB} L_{BA}]^{\tfrac{1}{2}},$$
$$\delta = -\tfrac{1}{2}(L_{AA} - L_{BB}) = \tfrac{1}{2}(K_{BA} - K_{AB} + R_1^A - R_1^B),$$

and K_{AB} is the rate constant of the reaction B→A (see eqn (2.4.12)). The equilibrium magnetization $M_{l0} = n_l x_l M_0$ is proportional to the number n_l of magnetically equivalent nuclei in site l, and to the mole fraction x_l ($x_l = 1$ for pure cross-relaxation between two nuclei within one molecule; $x_l = 0.5$ for symmetrical two-site chemical exchange). The integrated amplitudes of the diagonal and cross-peaks are

$$I_{AA}(\tau_m) = \frac{1}{2}\left[\left(1 - \frac{\delta}{D}\right) \exp\{(-\sigma + D)\tau_m\} + \left(1 + \frac{\delta}{D}\right) \exp\{(-\sigma - D)\tau_m\}\right] M_{A0}, \tag{9.3.20a}$$

$$I_{BB}(\tau_m) = \frac{1}{2}\left[\left(1 + \frac{\delta}{D}\right) \exp\{(-\sigma + D)\tau_m\} + \left(1 - \frac{\delta}{D}\right) \exp\{(-\sigma - D)\tau_m\}\right] M_{B0}, \tag{9.3.20b}$$

$$I_{AB}(\tau_m) = I_{BA}(\tau_m)$$
$$= \tfrac{1}{2}[\exp(-\sigma + D)\tau_m - \exp(-\sigma - D)\tau_m]\bar{M}, \tag{9.3.20c}$$

where

$$\bar{M} = M_{B0} \frac{L_{AB}}{D} = M_{A0} \frac{L_{BA}}{D}.$$

Note that the integrated amplitudes of the cross-peaks are always equal, regardless of the populations of the two sites. This symmetry of 2D exchange spectra with respect to the diagonal is preserved in N-site exchange processes.

The peak amplitudes can be conveniently expressed by introducing a *cross rate constant* R_C and a *leakage rate constant* R_L

$$R_C = \lambda_+ - \lambda_- = 2D,$$
$$R_L = -\lambda_+ = \sigma - D. \tag{9.3.21}$$

The amplitudes can now be written in the form

$$I_{AA}(\tau_m) = \frac{1}{2}\left[\left(1 - \frac{\delta}{D}\right) + \left(1 + \frac{\delta}{D}\right)\exp\{-R_C\tau_m\}\right]\exp\{-R_L\tau_m\}M_{A0}, \tag{9.3.22a}$$

$$I_{BB}(\tau_m) = \frac{1}{2}\left[\left(1 + \frac{\delta}{D}\right) + \left(1 - \frac{\delta}{D}\right)\exp\{-R_C\tau_m\}\right]\exp\{-R_L\tau_m\}M_{B0}, \tag{9.3.22b}$$

$$I_{AB}(\tau_m) = I_{BA}(\tau_m) = \tfrac{1}{2}[1 - \exp\{-R_C\tau_m\}]\exp\{-R_L\tau_m\}\bar{M}. \tag{9.3.22c}$$

For a symmetrical case without cross-relaxation, we obtain $\delta = 0$, $R_L = R_1 = 1/T_1$, and $R_C = 2k$, and eqns (9.3.22) reduce to eqn (9.1.4).

9.3.3. Multiple-site exchange

In larger systems, eqn (9.3.17) cannot be solved analytically. However, for sufficiently short mixing periods τ_m, approximate expressions for the amplitude can be obtained by invoking the initial rate approximation. Since

$$\exp(\mathbf{L}\tau_m) \simeq \mathbf{1} + \mathbf{L}\tau_m, \tag{9.3.23}$$

we obtain

$$a_{kk}(\tau_m) = 1 + L_{kk}\tau_m, \tag{9.3.24a}$$
$$a_{kl}(\tau_m) = L_{kl}\tau_m. \tag{9.3.24b}$$

Thus the cross-peak amplitudes are directly proportional to the corresponding matrix elements. In the absence of cross-relaxation (pure chemical exchange), we obtain a direct measure of the exchange rates

$$\frac{a_{kl}(\tau_m)}{\tau_m} = \frac{I_{kl}(\tau_m)}{M_{l0}\tau_m} = K_{kl}, \tag{9.3.25}$$

whereas in the absence of exchange (e.g. Overhauser effects between non-labile protons) the cross-peak amplitudes reflect the cross-relaxation rates

$$\frac{a_{kl}(\tau_m)}{\tau_m} = \frac{I_{kl}(\tau_m)}{M_{l0}\tau_m} = R_{kl}. \quad (9.3.26)$$

If the mixing interval τ_m is increased beyond the range of the initial rate approximation, it is necessary to include higher terms of the expansion of $\exp(\mathbf{L}\tau_m)$

$$I_{kl} = \left[L_{kl}\tau_m + \tfrac{1}{2}\sum_j L_{kj}L_{jl}\tau_m^2 + \tfrac{1}{6}\sum_j\sum_i L_{kj}L_{ji}L_{il}\tau_m^3 + \ldots \right]M_{l0}. \quad (9.3.27)$$

The quadratic terms correspond to a two-step migration of longitudinal magnetization, first from M_l to M_j, and then from M_j to M_k. The triple term corresponds to a transfer $M_l \to M_i \to M_j \to M_k$. Long mixing times no longer give a straightforward picture of individual pathways of polarization transfer, but rather a measure of the sum of all itineraries that the polarization can follow on its way from M_l to M_k.

9.4. Two-dimensional exchange spectroscopy in coupled spin systems

The basic principles of 2D exchange spectroscopy have been discussed so far in terms of classical magnetization. This treatment is only valid in the absence of resolved homonuclear spin–spin couplings. In systems with resolved scalar or dipolar couplings, coherence transfer phenomena can occur, involving single-, zero-, and multiple-quantum coherence (9.6, 9.7). In addition to the incoherent transfer of longitudinal magnetization brought about by exchange or cross-relaxation, the transverse magnetization can be transferred between different spins within a network of coupled nuclei, as in the experiments described in Chapter 8. These effects may be understood in a qualitative manner in terms of coherence transfer pathways, as discussed in § 9.2. It is always possible to design pathway selection procedures (phase-cycling, field gradient pulses, etc.) to remove all terms in τ_m with orders $p \neq 0$, that is to say, it is possible to suppress all single- and multiple-quantum coherence transfer processes. To measure the migration of longitudinal magnetization (Zeeman polarization, represented by density operator terms I_{kz}), the order $p = 0$ must be retained in the mixing interval τ_m. This implies that zero-quantum coherence (i.e. terms of the type $I_k^\pm I_l^\mp$, $I_k^\pm I_l^\mp I_{mz}$, $I_k^\pm I_l^\mp I_m^\pm I_n^\mp$, ...) and longitudinal scalar or dipolar order (terms of the type $2I_{kz}I_{lz}$, $4I_{kz}I_{lz}I_{mz}$, ...) are also retained. In 2D exchange spectroscopy, these terms are usually not desirable, since they lead to additional signals, often referred to as 'J cross-peaks' (9.6), which interfere with genuine exchange cross-peaks. Several schemes for the elimination of such parasitic signals associated with coherence transfer will be discussed in § 9.4.4.

9.4.1. Density operator treatment

The reader is referred to § 2.4 for a discussion of the density operator of systems with chemical exchange. We are concerned here with systems in dynamic chemical equilibrium, and assume weakly coupled spin systems for clarity (9.6). The slow exchange limit is considered where chemical transformations are important only during the extended mixing period τ_m.

Consider the effect of the sequence $(\pi/2)_y - t_1 - (\pi/2)_y - \tau_m - (\pi/2)_x - t_2$ in Fig. 9.1.1(a) on a two-spin system. It is easy to show (9.8) that the density operator immediately after the second pulse can be described in terms of Cartesian operator products (see § 2.1.5)

$$\begin{aligned}\sigma(t_1, \tau_m = 0) = &-[I_{kz} \cos \Omega_k t_1 + I_{lz} \cos \Omega_l t_1]\cos \pi J_{kl}t_1 & ① \\ &+[I_{ky} \sin \Omega_k t_1 + I_{ly} \sin \Omega_l t_1]\cos \pi J_{kl}t_1 & ② \\ &+[2I_{kz}I_{lx} \sin \Omega_k t_1 + 2I_{kx}I_{lz} \sin \Omega_l t_1]\sin \pi J_{kl}t_1 & ③ \\ &+\tfrac{1}{2}(2I_{kx}I_{ly} + 2I_{ky}I_{lx})(\cos \Omega_k t_1 + \cos \Omega_l t_1)\sin \pi J_{kl}t_1 & ④ \\ &-\tfrac{1}{2}(2I_{kx}I_{ly} - 2I_{ky}I_{lx})(\cos \Omega_k t_1 - \cos \Omega_l t_1)\sin \pi J_{kl}t_1 & ⑤\end{aligned}$$
(9.4.1)

In addition to the t_1-labelled polarizations of term ①, which carry the relevant information in 2D exchange spectroscopy, we recognize in-phase single-quantum coherence in term ② (which also occurs in systems without resolved couplings), antiphase single-quantum coherence in term ③, pure double-quantum coherence $\{2QT\}_y$ in term ④, and, most important of all, pure zero-quantum coherence $\{ZQT\}_y$ in term ⑤. Suitable phase-cycling schemes may cancel all terms except ① and ⑤.

The migration of the longitudinal terms I_{kz} through chemical exchange, cross-relaxation, or spin diffusion proceeds in exact analogy to the classical components M_{kz}. By analogy to eqn (2.4.25), the N longitudinal components of the density operator of the N-spin system

$$\sigma^{\text{Zeeman}} = \sum_{k=1}^{N} b_k I_{kz} \qquad (9.4.2)$$

can be collected in a vector which obeys the differential equation

$$\frac{d}{dt}\mathbf{b}(t) = \mathbf{L}[\mathbf{b}(t) - \mathbf{b}_0]. \qquad (9.4.3)$$

described by eqn (9.3.16). Their significance is not altered in any way by the presence of resolved couplings, although the signals are split into in-phase multiplets in both frequency dimensions.

9.4.2. Zero-quantum interference

The zero-quantum term ⑤ in eqn (9.4.1) must be considered carefully, because it can persist in the mixing time τ_m without being affected by the static field inhomogeneity or by phase-cycling, and its transverse relaxation is in some cases inefficient. Consider the effect of the chemical shifts on the zero-quantum coherence in a two-spin system

$$\tfrac{1}{2}(2I_{kx}I_{ly} - 2I_{ky}I_{lx}) \xrightarrow{\Sigma \Omega_k I_{kz}\tau_m} \tfrac{1}{2}(2I_{kx}I_{ly} - 2I_{ky}I_{lx})\cos(\Omega_k - \Omega_l)\tau_m$$
$$+ \tfrac{1}{2}(2I_{kx}I_{lx} + 2I_{ky}I_{ly})\sin(\Omega_k - \Omega_l)\tau_m. \quad (9.4.4)$$

If the third pulse in the sequence has a rotation angle $\beta_x = \pi/2$, only the cosine-modulated component is converted into observable antiphase single-quantum coherence

$$\sigma^{ZQT \to 1QT} = -\tfrac{1}{2}(2I_{kx}I_{lz} - 2I_{kz}I_{lx})$$
$$\times (\cos \Omega_k t_1 - \cos \Omega_l t_1)\sin \pi J_{kl}t_1 \cos(\Omega_k - \Omega_l)\tau_m. \quad (9.4.5)$$

These terms give rise to four antiphase multiplets, two centred on the diagonal, and two in the form of cross-peaks at $(\omega_1, \omega_2) = (\Omega_k, \Omega_l)$ and (Ω_l, Ω_k). These are known as 'J cross-peaks', since they originate from a coherent transfer via J couplings, and should not be confused with exchange cross-peaks, which consist of in-phase multiplets. If the true exchange peaks appear in pure 2D absorption, the J cross-peaks originating from zero-quantum coherence have pure 2D dispersion lineshapes.

In systems with more than two spins the evolution of zero-quantum coherence is affected by couplings to passive nuclei (9.8). Thus in a three-spin system (k, l, n) one may obtain antiphase zero-quantum coherence

$$\tfrac{1}{2}(2I_{kx}I_{ly} - 2I_{ky}I_{lx}) \xrightarrow{\pi J_{kn}\tau_m 2I_{kz}I_{nz}} \xrightarrow{\pi J_{ln}\tau_m 2I_{lz}I_{nz}}$$
$$\tfrac{1}{2}(2I_{kx}I_{ly} - 2I_{ky}I_{lx})\cos \pi(J_{kn} - J_{ln})\tau_m$$
$$+ \tfrac{1}{2}(2I_{kx}I_{lx} + 2I_{ky}I_{ly})2I_{nz} \sin \pi(J_{kn} - J_{ln})\tau_m. \quad (9.4.6)$$

The operator product $4I_{ky}I_{ly}I_{nz}$ may be converted into observable magnetization by a $(\pi/2)_x$-pulse, and leads to J-cross-peaks at

$(\omega_1, \omega_2) = (\Omega_k, \Omega_n)$ and (Ω_l, Ω_n), although the nucleus n is not actively involved in the zero-quantum coherence. This phenomenon must be borne in mind when designing schemes for zero-quantum suppression.

The evolution of zero-quantum coherence is also damped by transverse relaxation (9.6). In a two-spin system with pure intramolecular *dipolar* relaxation caused by an isotropic random motion, one finds the zero-, single-, and double-quantum relaxation rates in the presence of four non-overlapping single quantum transitions[1]

$$1/T_2^{(ZQT)} = \tfrac{1}{4}q_{kl}[2J_{kl}(\omega_{0k} - \omega_{0l}) + 3J_{kl}(\omega_{0k}) + 3J_{kl}(\omega_{0l})]$$

$$1/T_2^{(1QT)} = \tfrac{1}{4}q_{kl}[4J_{kl}(0) + J_{kl}(\omega_{0k} - \omega_{0l}) + 3J_{kl}(\omega_{0k}) + 3J_{kl}(\omega_{0l})$$
$$+ 6J_{kl}(\omega_{0k} + \omega_{0l})],$$

$$1/T_2^{(2QT)} = \tfrac{1}{4}q_{kl}[3J_{kl}(\omega_{0k}) + 3J_{kl}(\omega_{0l}) + 12J_{kl}(\omega_{0k} + \omega_{0l})] \quad (9.4.7)$$

with the constant

$$q_{kl} = \tfrac{1}{10}\gamma_k^2\gamma_l^2\hbar^2 r_{kl}^{-6}\left[\frac{\mu_0}{4\pi}\right]^2 \quad (9.4.8)$$

and the spectral density functions

$$J_{kl}(\omega_{0k}) = \frac{2\tau_c^{kl}}{1 + \omega_{0k}^2(\tau_c^{kl})^2} \quad (9.4.9)$$

where ω_{0k} is the Zeeman frequency of spin k in the laboratory frame and where τ_c^{kl} is the correlation time of the isotropic random process which modulates the orientation of the internuclear vector $\mathbf{r}_{kl}$. In the fast-motion limit, $(\tau_c \ll 1/\omega_{0k})$ the three transverse relaxation rates in eqns (9.4.7) have ratios of $8:17:18$, while in the slow-motion limit $(\tau_c \gg 1/\omega_{0k})$, the ratios are $2:5:0$. In either case, the zero-quantum decay is slower than single-quantum relaxation.

If, on the other hand, we consider pure *external random field* relaxation (9.6), we find the relaxation rates

$$1/T_2^{(ZQT)} = \tfrac{1}{6}[\overline{(\gamma_k B_k - \gamma_l B_l)^2}J(0) + \overline{(\gamma_k B_k)^2}J(\omega_{0k}) + \overline{(\gamma_l B_l)^2}J(\omega_{0l})],$$
$$1/T_2^{(1QT)} = \tfrac{1}{6}[\overline{(\gamma_k B_k)^2}J(0) + \overline{(\gamma_k B_k)^2}J(\omega_{0k}) + \overline{(\gamma_l B_l)^2}J(\omega_{0l})],$$
$$1/T_2^{(2QT)} = \tfrac{1}{6}[\overline{(\gamma_k B_k + \gamma_l B_l)^2}J(0) + \overline{(\gamma_k B_k)^2}J(\omega_{0k}) + \overline{(\gamma_l B_l)^2}J(\omega_{0l})]$$

$$(9.4.10)$$

where $B_k(t)$ and $B_l(t)$ are isotropic random magnetic fields at the sites of the two nuclei k and l with the spectral density function $J(\omega)$. If the

---

[1] The single-quantum relaxation rate in the *absence* of scalar coupling, but for different chemical shifts, $\omega_{0k} \neq \omega_{0l}$, takes the form (cf. eqn (161), chapter VIII of Ref. 9.57)

$$1/T_{2k}^{(1QT)} = \tfrac{1}{4}q_{kl}[4J_{kl}(0) + J_{kl}(\omega_{0k} - \omega_{0l}) + 3J_{kl}(\omega_{0k}) + 6J_{kl}(\omega_{0l}) + 6J_{kl}(\omega_{0k} + \omega_{0l})].$$

For degenerate Larmor frequencies, $\omega_{0k} = \omega_{0l}$, one finds the expression

$$1/T_2^{(1QT)} = \tfrac{1}{4}q_{kl}[9J_{kl}(0) + 15J_{kl}(\omega_0) + 6J_{kl}(2\omega_0)].$$

fluctuations are completely correlated, and if the amplitudes are equal at both sites, the three rates have ratios of $2:3:6$ in the fast-motion limit and $0:1:4$ for slow motion. In the latter case (which is relevant for Overhauser studies in macromolecules), the zero-quantum coherence may persist for extensive mixing intervals. If the fluctuations have equal amplitudes but vanishing correlation coefficients, i.e. if (cf. eqn (2.3.33))

$$C_{kl} = \frac{\overline{B_k B_l}}{[\overline{B_k^2}\,\overline{B_l^2}]^{\frac{1}{2}}} \qquad (9.4.11)$$

is zero, the ratios of the three relaxation rates in eqn (9.4.10) are $4:3:4$ in the fast-motion limit and $2:1:2$ in the slow-motion limit. Note that the correlation times for dipolar and random-field relaxation can be different.

In order to appreciate the practical implications, Fig. 9.4.1 shows the theoretical τ_m-dependence of the amplitudes of the exchange cross-peaks (smooth curves) and of the J cross-peaks (oscillating curves) stemming from zero- and double-quantum coherence (the latter has been included on the assumption that it would not be cancelled by phase-cycling).

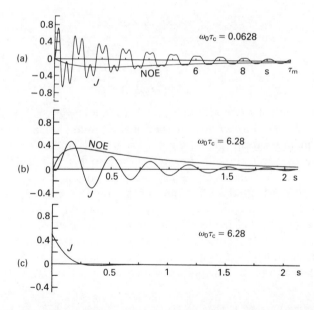

FIG. 9.4.1. Contributions to cross-peaks in 2D exchange spectra: smooth curves represent true exchange, oscillating curves correspond to J cross-peaks. (a) Two-spin system in the fast-motion limit with pure dipolar relaxation. Nuclear Overhauser Effect (NOE) cross-peaks appear together with J cross-peaks due to zero- and double-quantum coherence. (b) Two-spin system in the slow-motion limit with dipolar and external random-field relaxation with J cross-peaks due to zero- and double-quantum coherence. (c) Same conditions as (b), J cross-peak contribution due to zero-quantum coherence. (Adapted from Ref. 9.6.)

Figure 9.4.1(a) shows the amplitudes in a two-spin system with pure dipole relaxation in the fast-motion limit, in a homogeneous static field (assumption: $q\tau_c = 0.06\,\text{s}^{-1}$). Note that the undesired coherent contributions exceed the relevant cross-peaks (which are negative in this case, see § 9.7) by as much as a factor of five. Figure 9.4.1(b) and (c) show the situation in the slow-motion regime, where it has been assumed that uncorrelated external fields are acting in addition to the dipolar mechanism. In this case, the zero-quantum interference decays in a time of the order of the cross-relaxation rate (Fig. 9.4.1(c)), while the double-quantum coherence persists longer if the static field inhomogeneity and the external fluctuating fields are not effective.

9.4.3. Longitudinal scalar or dipolar order

The basic 2D experiment may be modified, either on purpose or inadvertently, by using r.f. pulses with rotation angles $\beta \neq \pi/2$. In this case, the second pulse in Fig. 9.1.1(a) converts antiphase single-quantum coherence into four terms

$$
\begin{aligned}
2I_{ky}I_{lz} \xrightarrow{(\beta)_x} \; & 2I_{ky}I_{lz}\cos^2\beta \\
& +2I_{kz}I_{lz}\sin\beta\cos\beta \\
& -2I_{ky}I_{ly}\sin\beta\cos\beta \\
& -2I_{kz}I_{ly}\sin^2\beta.
\end{aligned}
\qquad (9.4.12)
$$

The second term, known as longitudinal two-spin order, will persist in the mixing time *in addition* to the various Zeeman, zero-, single-, and multiple-quantum terms given in eqn (9.4.1), and can be reconverted into observable magnetization if the last pulse also has $\beta \neq \pi/2$. Measuring the decay and diffusion of $2I_{kz}I_{lz}$ amounts to a 2D version of the Jeener–Broekaert method for measuring dipolar relaxation in solids (9.9). The $2I_{kz}I_{lz}$ term does not oscillate as a function of τ_m, and gives rise to cross- and diagonal peaks with antiphase multiplet structure with the same 2D absorption characteristics as the exchange signals associated with I_{kz} terms. The $2I_{kz}I_{lz}$ signals can be minimized by careful calibration of $\pi/2$ rotation angles, and suppressed by using composite pulses (9.10).

9.4.4. Suppression of J cross-peaks

In contrast to single- and multiple-quantum interference, which can be cancelled by phase-cycling techniques (§ 9.2), zero-quantum coherence and longitudinal spin order cannot easily be separated from Zeeman magnetization. To suppress J cross-peaks originating from zero-quantum

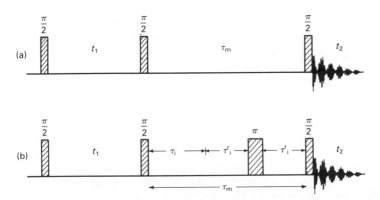

FIG. 9.4.2. (a) Basic pulse sequence for 2D exchange spectroscopy and (b) sequence with a π-pulse inserted in the mixing period τ_m. The effective zero-quantum precession interval is restricted to τ_i, since the chemical shifts are refocused in the remaining intervals $\tau_i^r = (\tau_m - \tau_i)/2$. (Reproduced from Ref. 9.12.)

coherence, one can exploit the oscillatory evolution of these terms during the mixing time of eqn (9.4.4). Various strategies can be used.

1. Repetition of the full 2D experiment for several values of τ_m and co-addition of the signals. Often a random variation of τ_m is used (9.6) although a set of well-chosen values can lead to better results (9.12). Components with low zero-quantum frequencies $(\Omega_k - \Omega_l)$ in eqn (9.4.5), which lead to J cross-peaks near to the diagonal, are not suppressed, since the co-addition over different τ_m values acts as a low-pass filter, the cut-off frequency being determined by the total variation of τ_m. This variation has to be kept sufficiently small to avoid errors in the amplitudes of the exchange cross-peaks.

2. Repetition of the full 2D experiment with an additional π-pulse inserted at a variable position within the mixing time τ_m (9.6, 9.12), as shown in Fig. 9.4.2. Except for a sign change, this π-pulse does not affect the time evolution of the z-components that are relevant for the exchange peaks. However, the π-pulse inverts the precession phase of the zero-quantum coherence

$$\{ZQT\}_x = \tfrac{1}{2}(2I_{kx}I_{lx} + 2I_{ky}I_{ly}) \xrightarrow{\pi F_x \text{ or } \pi F_y} \{ZQT\}_x,$$

$$\{ZQT\}_y = \tfrac{1}{2}(2I_{ky}I_{lx} - 2I_{kx}I_{ly}) \xrightarrow{\pi F_x \text{ or } \pi F_y} -\{ZQT\}_y. \quad (9.4.13)$$

Variation of the position of the π-pulse through the mixing interval τ_m (see Fig. 9.4.2) and co-addition leads to efficient suppression of J cross-peaks.

3. By a concerted incrementation of the position of the π-pulse (or of the duration of mixing time τ_m in the absence of a π-pulse) and the evolution time t_1, i.e. by adding a time increment

$$\tau_i = \chi t_1 \tag{9.4.14a}$$

or

$$\tau_m = \tau_m^0 + \chi t_1, \tag{9.4.14b}$$

it is possible to shift the ω_1-frequency of the zero-quantum peaks (9.11). In this case, the diagonal and cross-peaks are both flanked by a pair of zero-quantum 'satellite' signals which are displaced in ω_1 by $\pm \chi(\Omega_k - \Omega_l)$. However, there are no zero-quantum contributions that coincide with the exact positions of the diagonal and exchange cross-peaks. By symmetrization of the 2D spectrum (see § 6.6.4) it is then possible to eliminate the J cross-peaks.

4. In some practical applications τ_i (or τ_m) is varied in a random manner with t_1. This leads to a 'smearing' of the J cross-peaks along ω_1 and causes a corresponding increase in t_1-noise.

9.5. Two-dimensional exchange difference spectroscopy

A quantitative evaluation of exchange rates is particularly straightforward in the initial rate regime, by recording a set of 2D spectra $S(\omega_1, \omega_2, \tau_m)$ with $\tau_m < L_{kl}^{-1}$. Unfortunately, exchange spectra obtained for short mixing times tend to be obscured by dominant diagonal peaks.

Several schemes have been proposed to reduce the dominant diagonal signals (9.13, 9.14). We shall restrict our discussion to a technique which yields diagonal peak amplitudes opposite to the sum of the amplitudes of the cross-peaks in the same row

$$I_{kk}(\tau_m) = -\sum_{l \neq k} I_{kl}. \tag{9.5.1}$$

If this condition is fulfilled, the 2D spectrum corresponds to an exact map of the exchange matrix, including the diagonal elements.

Consider a network with N sites, where the integrated signal intensities are given by eqn (9.3.16). If exchange or cross-relaxation is much more efficient than spin–lattice relaxation ($L_{kl} \gg R_1$), it is sufficient to subtract a 2D spectrum obtained with $\tau_m = 0$ from a conventional 2D exchange spectrum to cancel diagonal peaks that stem from magnetization that is not involved in exchange.

If spin–lattice relaxation cannot be neglected, more sophisticated experiments are necessary. Consider the four schemes in Fig. 9.5.1. Only scheme I gives rise to exchange cross-peaks; in scheme II one observes neither exchange nor leakage relaxation, while schemes III and IV

9.5 2D EXCHANGE DIFFERENCE SPECTROSCOPY

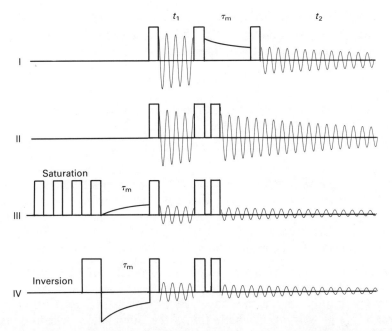

FIG. 9.5.1. Four experimental schemes used in 2D difference spectroscopy. (I) Three-pulse scheme used for conventional 2D exchange spectroscopy. Both leakage and exchange occur in the mixing time τ_m. (II) The same sequence with vanishing mixing time. (III) Three-pulse scheme with an additional pre-relaxation period τ_m starting from a saturated state. (IV) Similar experiment but starting from a state with inverted magnetization. In the pre-relaxation periods, leakage relaxation is effective but not the actual exchange processes, if only initial rates are considered. (Adapted from Ref. 9.14.)

involve spin–lattice relaxation but no exchange. In the initial rate approximation, the amplitudes in the four experiments are obtained by combining eqns (9.3.16) and (9.3.24)

$$I_{kl}^{(I)} = L_{kl}\tau_m M_{l0} = (K_{kl} - R_{kl})\tau_m M_{l0} \quad \text{for} \quad k \neq l,$$

$$I_{kl}^{(II)} = I_{kl}^{(III)} = I_{kl}^{(IV)} = 0 \quad \text{for} \quad k \neq l,$$

$$I_{kk}^{(I)} = \left(1 - R_1^k \tau_m - \sum_{l \neq k} K_{lk}\tau_m\right)M_{k0},$$

$$I_{kk}^{(II)} = M_{k0},$$

$$I_{kk}^{(III)} = R_1^k \tau_m M_{k0},$$

$$I_{kk}^{(IV)} = (2R_1^k \tau_m - 1)M_{k0}. \tag{9.5.2}$$

Two distinct procedures lead to the desired amplitudes of diagonal peaks (9.14).

1. In 'saturation-recovery difference spectroscopy' (SRD), the signals of the first three experiments in Fig. 9.5.1 are combined in such a way that the initial rate diagonal peak intensities are

$$I_{kk}^{(\text{SRD})} = I_{kk}^{(\text{I})} - I_{kk}^{(\text{II})} + I_{kk}^{(\text{III})} = -M_{k0} \sum_{l \neq k} K_{lk} \tau_m. \tag{9.5.3}$$

2. In 'inversion-recovery difference spectroscopy' (IRD), the following weighted linear combination is used

$$I_{kk}^{(\text{IRD})} = I_{kk}^{(\text{I})} - \tfrac{1}{2} I_{kk}^{(\text{II})} + \tfrac{1}{2} I_{kk}^{(\text{IV})} = -M_{k0} \sum_{l \neq k} K_{lk} \tau_m. \tag{9.5.4}$$

In this case, the sensitivity is improved by a factor $(3/2)^{\frac{1}{2}}$ compared to saturation-recovery difference spectroscopy, since one-half rather than one-third of the experiments contribute to cross-peaks. Experimental examples of this procedure are shown in § 9.9. The basic ideas are equally applicable to chemical exchange and nuclear Overhauser studies.

The initial rate approximation has been invoked only to simplify the discussion, disregarding the interplay between relaxation and exchange. For magnetization that is not involved in either exchange or cross-relaxation (e.g. solvent magnetization), the subtraction schemes discussed here lead to complete elimination of the diagonal peak even if the initial rate condition is violated.

9.6. Determination of rate constants by 'accordion' spectroscopy

For a quantitative measurement of exchange rates, it is normally not sufficient to record a 2D exchange spectrum with a single mixing time τ_m. If the build-up and decay is monitored in a series of 2D spectra $S(\omega_1, \tau_m, \omega_2)$ with different τ_m-delays (9.15), one may speak of an extension from two- to three-dimensional NMR. Indeed, a Fourier transformation with respect to the third time variable τ_m leads to a new frequency domain ω_m (9.3). The resulting three-dimensional spectrum is represented schematically in Fig. 9.6.1. Provided the chemical shifts are well resolved in the ω_1-domain, it is possible to reduce the dimension from three to two by a skew projection, in such a way that the lineshape information contained in the (scaled) ω_m-domain is combined with the shift information of the ω_1-domain. A genuine 3D experiment tends to be exceedingly time-consuming. Fortunately, the skew projection shown in Fig. 9.6.1 can be obtained more efficiently by a concerted variation of two intervals of the basic 2D exchange experiment shown in Fig. 9.1.1(a)

$$\tau_m = \chi t_1. \tag{9.6.1}$$

This proportionality is the central feature of the so-called 'accordion' experiment (9.3, 9.16). The coupling of two parameters in effect reduces

9.6 'ACCORDION' SPECTROSCOPY

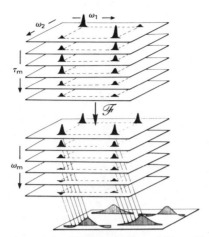

FIG. 9.6.1. Schematic representation of the reduction of 3D to 2D spectroscopy achieved in the accordion method. Top: A true 3D exchange spectrum can be visualized as a stack of 2D spectra $S(\omega_1, \omega_2)$ recorded with systematic increments of τ_m. The diagonal peaks decay monotonically, while cross-peaks first increase and later decay as a function of τ_m. Middle: By Fourier transformation with respect to τ_m, a three-dimensional frequency domain $S(\omega_1, \omega_m, \omega_2)$ is obtained. Provided the spectrum is well-resolved in ω_1, a skew projection (bottom) can be made without entailing any loss of information. The same spectrum can be obtained directly and much more efficiently by the accordion method, where τ_m and t_1 are incremented together. (Reproduced from Ref. 9.3.)

a 3D experiment to a special form of 2D experiment. The proportionality factor χ is chosen in such a way that τ_m^{max} is on the order of three times the longest time constant in the mixing process (usually the longitudinal spin–lattice time T_1) to avoid truncation errors. Typically this is achieved with $10 \leq \chi \leq 100$.

Because t_1 and τ_m are varied together, a Fourier transformation with respect to t_1 is at the same time a transformation with respect to τ_m. In the 2D frequency domain, the ω_1- and ω_m-axes run in parallel, but the spectral width spanned in ω_1 is χ times the spectral width in ω_m. The peak *positions* along the ω_1, ω_m-axis correspond to the chemical shifts in the ω_1-domain of the conventional 2D exchange experiment. The *lineshapes* on the other hand reflect the dynamic processes in the τ_m-period. The ω_m-lineshapes of diagonal and cross-peaks in a symmetrical two-site exchange case with equal populations and equal spin–lattice relaxation rates are obtained by Fourier transformation of eqn (9.1.4) with respect to τ_m

$$S_{AA}(\omega_m) = S_{BB}(\omega_m) = \tfrac{1}{4}M_0\left[\frac{R_1}{R_1^2 + \omega_m^2} + \frac{2k + R_1}{(2k + R_1)^2 + \omega_m^2}\right], \quad (9.6.2a)$$

$$S_{AB}(\omega_m) = S_{BA}(\omega_m) = \tfrac{1}{4}M_0\left[\frac{R_1}{R_1^2 + \omega_m^2} - \frac{2k + R_1}{(2k + R_1)^2 + \omega_m^2}\right]. \quad (9.6.2b)$$

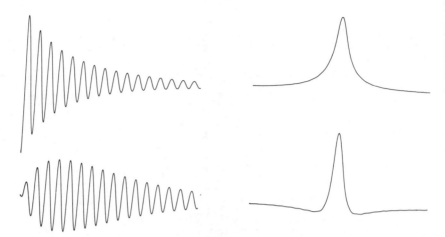

FIG. 9.6.2. Simulated accordion lineshapes for diagonal peaks (top) and cross-peaks (below) for a symmetrical two-site exchange case. The time-domain signals (left) show oscillating components $\cos \Omega_A t_1$ with envelopes $a_{AA}(\tau_m)$ (top left) and $a_{AB}(\tau_m)$ (bottom left). The Fourier transforms reveal characteristic lineshapes in the new ω_m-domain. The diagonal peak $S_{AA}(\omega_m)$ (top right) consists of the sum of two Lorentzians with line-widths R_1 and $2k + R_1$ (eqn (9.6.2a)), while the cross-peak $S_{AB}(\omega_m)$ (bottom right) is composed of the difference of the same pair of Lorentzians according to eqn (9.6.2b). In these illustrations, $2k + R_1 = 3R_1$. (Reproduced from Ref. 9.3.)

These lineshapes apply to sections taken parallel to the ω_1/ω_m axis through absorption-mode phase-sensitive spectra. The origin $\omega_m = 0$ for each function is centred at the chemical shift (Ω_A or Ω_B) in the ω_1-domain. Both the time-domain functions (eqns (9.1.4)) and their Fourier transformations (eqns (9.6.2)) are shown in Fig. 9.6.2. The frequency-domain signals consist of sums and differences of Lorentzian lines with equal integrated intensities but different widths. In practice, these ideal lineshapes may be affected by transverse relaxation and inhomogeneous broadening in t_1, expressed by the apparent relaxation rate $R_1^{app} = R_1 + \chi^{-1} R_2^*$ that must be substituted instead of R_1 in eqns (9.6.2).

In a complex exchange network with N sites, the evolution of the longitudinal magnetization is governed by eqn (9.3.6)

$$\frac{d}{d\tau_m} \Delta \mathbf{M}_z(\tau_m) = \mathbf{L} \, \Delta \mathbf{M}_z(\tau_m). \quad (9.6.3)$$

The dynamic matrix $\mathbf{L} = -\mathbf{R} + \mathbf{K}$ can be diagonalized by a transformation $\mathbf{T}$

$$\mathbf{T}^{-1} \mathbf{L} \mathbf{T} = \mathbf{D} \quad (9.6.4)$$

which makes it possible to reformulate the solution in eqn (9.3.8)

$$\Delta \mathbf{M}_z(\tau_m) = \mathbf{T}\exp\{\mathbf{D}\tau_m\}\mathbf{T}^{-1}\Delta \mathbf{M}_z(\tau_m = 0). \tag{9.6.5}$$

In accordion spectroscopy, it is possible to measure the eigenvalues (diagonal elements of $\mathbf{D}$) by monitoring the evolution of normal modes defined by

$$\Delta \mathbf{N}(\tau_m) = \mathbf{T}^{-1}\Delta \mathbf{M}_z(\tau_m). \tag{9.6.6}$$

Each normal mode ΔN_k is a linear combination of deviations from the equilibrium magnetization and corresponds to a particular preparation of the spin system. These normal modes have a simple exponential time dependence

$$\Delta \mathbf{N}(\tau_m) = \exp\{\mathbf{D}\tau_m\}\Delta \mathbf{N}(\tau_m = 0). \tag{9.6.7}$$

The corresponding frequency-domain lineshapes have pure Lorentzian characteristics, and the observable lineshapes due to the magnetization $\Delta \mathbf{M}_z = \mathbf{T}\Delta \mathbf{N}$ consist of a superposition of N Lorentzians.

There are three distinct approaches to the analysis of the ω_m-lineshapes in view of extracting the exchange rates and spin–lattice relaxation rates.

1. *Direct lineshape analysis.* The superposition of N Lorentzian line contributions in a system with N sites can be separated by least-squares analysis if N is not too large.

2. *Reverse Fourier transformation.* The lineshape $S(\omega_m)$ taken from a 2D accordion spectrum may be converted by reverse Fourier transformation into a signal $s(\tau_m)$, where the build-up and decay of first- and higher-order exchange processes can be readily identified. The oscillating part of the time domain (see Fig. 9.6.2) can be removed by taking the absolute value of the complex signal in the τ_m time domain, provided the envelopes are real and positive.

3. *Normal-mode analysis.* If the diagonalizing transformation $\mathbf{T}$ is known, (or if it can be identified experimentally), the Lorentzian lineshapes associated with the normal modes in eqn (9.6.6) can be separated by linear combinations of sections taken from 2D accordion spectra (9.3). In the symmetrical two-site case, the *sum* of the diagonal and cross-peak lineshapes in eqn (9.6.2) yields a narrow Lorentzian

$$S^\Sigma(\omega_m) = \tfrac{1}{2}M_0 \frac{R_1}{R_1^2 + \omega_m^2} \tag{9.6.8}$$

while the *difference* leads to a broader Lorentzian

$$S^\Delta(\omega_m) = \pm\tfrac{1}{2}M_0 \frac{2k + R_1}{(2k + R_1)^2 + \omega_m^2}. \tag{9.6.9}$$

Note that $S^\Sigma(\omega_m)$ is not affected by chemical exchange, since the exchange process preserves the total magnetization. The effect of chemical exchange can be observed in the difference $S^\Delta(\omega_m)$. The exchange rate k is readily obtained by subtracting the line-widths (expressed in Hz) of eqns (9.6.8) and (9.6.9)

$$k = \left(\frac{\pi}{2}\right)(\Delta\nu_m^\Delta - \Delta\nu_m^\Sigma). \tag{9.6.10}$$

A typical case of a symmetrical two-site exchange is the ring inversion of *cis*-decalin, which has been investigated by conventional methods (9.17–9.19) and 2D NMR (9.3, 9.16, 9.20). The 2D exchange accordion spectrum shown in Fig. 9.6.3 reveals five diagonal peaks: the resonance

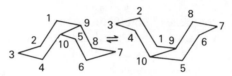

of $C_9 + C_{10}$ (bottom left) is invariant to ring inversion, while four pairs of sites interchange their chemical shifts: $C_1 \rightleftarrows C_4$, $C_2 \rightleftarrows C_3$, $C_5 \rightleftarrows C_8$, and $C_6 \rightleftarrows C_7$. This qualitative information can be obtained at a glance from the contour plot in Fig. 9.6.3. The exchange rate can be derived from the phase-sensitive cross-sections in Fig. 9.6.3, by least-squares analysis, reverse Fourier transformation (9.16), or normal-mode analysis (9.3).

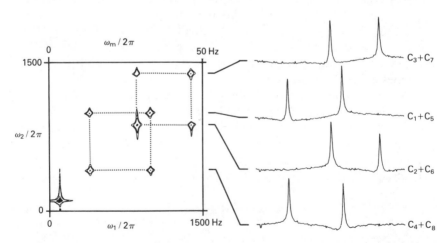

FIG. 9.6.3. Proton-decoupled carbon-13 accordion spectrum of the ring-puckering process in *cis*-decalin at 240 K. (Left) 2D survey spectrum. (Right) Four phase-sensitive cross-sections showing the characteristic accordion lineshapes. (Reproduced from Ref. 9.16.)

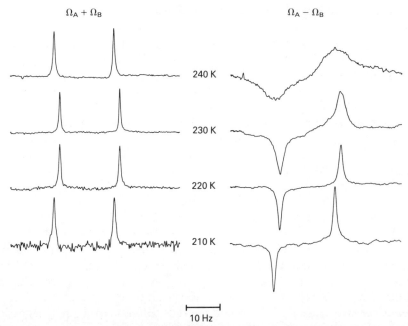

FIG. 9.6.4. Linear combinations of cross-sections taken at $\omega_2 = \Omega_A$ and $\omega_2 = \Omega_B$ from carbon-13 accordion spectra of *cis*-decalin at four different temperatures. The sums $\Omega_A + \Omega_B$ (left) show essentially temperature-independent line-widths determined by R_1. The differences $\Omega_A - \Omega_B$ show line-widths $2k + R_1$. Comparison of the line-widths left and right yields a straightforward measure of the temperature dependence of the exchange rate k. (Reproduced from Ref. 9.3.)

The linear combinations appropriate for a symmetrical two-site case (i.e. simple sums and differences) are shown as a function of temperature in Fig. 9.6.4. (Site A corresponds to $C_1 + C_5$ in *cis*-decalin, site B to $C_4 + C_8$.) Note that the line-width of the sum is nearly temperature-independent in this range, while the difference mode contains two Lorentzians with opposite sign that become broader as the temperature is increased. The activation energy of the exchange process can be readily derived from such a series of experiments (9.3).

It should be emphasized that accurate exchange rates can only be obtained over a relatively narrow temperature interval. The accuracy of the lineshape analysis suffers if the spectra are crowded, and the normal-mode analysis breaks down if the diagonal in the 2D spectrum carries signals from numerous superimposed sites. Finally, ill-resolved scalar couplings must be considered carefully, and interference with zero-quantum coherence must be suppressed in spin systems with homonuclear couplings.

9.7. Cross-relaxation and nuclear Overhauser effects

Nuclear cross-relaxation in liquids is caused by mutual spin flips in pairs of dipolar-coupled spins which are induced by motional processes. Cross-relaxation leads to a transfer of magnetization between the spins and hence to intensity changes, known as nuclear Overhauser effects (NOE) (9.21–9.27).

Cross-relaxation depends both on the character of the motional processes and on the separation of the interacting spins. Nuclear Overhauser effects can provide unique information on molecular structure in solution which cannot be obtained with any other known technique. Such measurements have become of central importance in molecular biology where they allow the complete determination of the 3D structure of large biomolecules (9.26–9.30).

Cross-relaxation and nuclear Overhauser effects can be studied either by 1D or by 2D techniques. Early 1D measurements were mostly concerned with *steady-state* Overhauser effects obtained by selectively saturating one spin while observing the intensity changes occurring on other spins (9.22). More specific information can be obtained from *transient* Overhauser effects where the redistribution of the magnetization is studied as a function of time after a selective inversion of one spin (9.22, 9.24, 9.26, 9.27). Two-dimensional NOE experiments (9.1, 9.5, 9.28) are closely related to transient NOE measurements, with the advantage that all transfer pathways can be traced out at once.

The usefulness of these techniques strongly depends on the time-scale of the motional processes causing cross-relaxation. We have to distinguish the fast-motion limit (extreme narrowing limit) with a short correlation time $\tau_c \ll \omega_0^{-1}$, which applies for small molecules in non-viscous solutions, and the slow-motion limit (spin diffusion limit) with a long correlation time $\tau_c \gg \omega_0^{-1}$ which applies to macromolecules at high magnetic fields. Table 9.7.1 gives a survey on the effects to be expected for the three techniques.

It is apparent from Table 9.7.1 that Overhauser effects in small molecules in the fast-motion limit are best studied by steady-state saturation methods where the largest effect can be expected. On the other hand, large molecules with slow motions are best investigated by transient NOE measurements or by 2D NOE spectroscopy.

The advantages of the 2D approach have been discussed in § 9.1: one obtains information from all exchanging sites in one single experiment, and the initial conditions of the exchange process are well defined. In coupled systems, one must however take care to suppress coherent interference (§ 9.4).

Table 9.7.1

Nuclear Overhauser effect measured by	Fast-motion limit $\tau_c \ll \omega_0^{-1}$	Slow-motion limit $\tau_c \gg \omega_0^{-1}$
Steady-state saturation	Positive enhancement of relaxation partner	Rapid saturation, lack of selectivity due to extended spin diffusion
Recovery after selective inversion	Weak positive transient enhancement	Negative effect, rapidly washed out by extended spin diffusion
2D NOE spectroscopy (with positive diagonal peaks)	Weak negative cross-peaks	Strong positive cross-peaks

9.7.1. Intramolecular cross-relaxation

We refer to the general discussion in § 9.3 and focus attention on cross-relaxation within a molecule or a molecular fragment containing n_A magnetically equivalent A spins and n_B magnetically equivalent B spins with spin quantum numbers $I = \frac{1}{2}$. Thus eqns (9.3.18)–(9.3.22) can be simplified with the reduced dynamic matrix

$$\mathbf{L} = -\mathbf{R} = \begin{pmatrix} -R_1^A & -R_{AB} \\ -R_{BA} & -R_1^B \end{pmatrix}. \quad (9.7.1a)$$

Several authors, following Solomon (9.21), have used the symbol σ for the cross-relaxation rate and ρ for the spin–lattice relaxation rate (9.22). The correspondence between our notation and Solomon's is straightforward

$$R = \begin{pmatrix} R_1^A & R_{AB} \\ R_{BA} & R_1^B \end{pmatrix} = \begin{pmatrix} \rho_A & \sigma_{AB} \\ \sigma_{BA} & \rho_B \end{pmatrix}. \quad (9.7.1b)$$

The elements of the cross-relaxation matrix can be expressed by the transition probabilities W resulting from AA, BB, and AB interactions and by the external relaxation rates R_{1A}^{ext} and R_{1B}^{ext} which take into account possible interactions with further spins (9.5)

$$R_1^A = 2(n_A - 1)(W_1^{AA} + W_2^{AA}) + n_B(W_0^{AB} + 2W_{1A}^{AB} + W_2^{AB}) + R_{1A}^{\text{ext}},$$
$$R_1^B = 2(n_B - 1)(W_1^{BB} + W_2^{BB}) + n_A(W_0^{AB} + 2W_{1B}^{AB} + W_2^{AB}) + R_{1B}^{\text{ext}},$$
$$R_{AB} = n_A(W_2^{AB} - W_0^{AB}),$$
$$R_{BA} = n_B(W_2^{AB} - W_0^{AB}). \quad (9.7.2)$$

W_1^{AA} and W_2^{AA} represent the single- and double-quantum transition probabilities caused by A–A dipolar interactions

$$W_1^{AA} = \tfrac{3}{4} q_{AA} J_{AA}(\omega_{0A}),$$
$$W_2^{AA} = 3 q_{AA} J_{AA}(2\omega_{0A}), \qquad (9.7.3)$$

which contribute only to the rate constant R_1^A. Similarly, relaxation due to B–B dipolar interactions affects R_1^B alone. Relaxation due to the dipolar interaction of A and B spins, on the other hand, contributes to all four rate constants in eqn (9.7.2)

$$W_{1A}^{AB} = \tfrac{3}{4} q_{AB} J_{AB}(\omega_{0A}),$$
$$W_{1B}^{AB} = \tfrac{3}{4} q_{AB} J_{AB}(\omega_{0B}),$$
$$W_0^{AB} = \tfrac{1}{2} q_{AB} J_{AB}(\omega_{0A} - \omega_{0B}),$$
$$W_2^{AB} = 3 q_{AB} J_{AB}(\omega_{0A} + \omega_{0B}) \qquad (9.7.4a)$$

with the constants q_{kl} and the spectral densities $J_{kl}(\omega_{0k})$ defined in eqns (9.4.8) and (9.4.9). The dependence on the motional correlation time τ_c of the transition probabilities W_0^{AB}, W_1^{AB}, and W_2^{AB} due to the dipolar interaction is shown in Fig. 9.7.1.

The corresponding rate constants for external isotropic random-field (RF) relaxation (see also eqn (9.4.10)) are given by

$$W_{1A}^{RF} = \tfrac{1}{6} \overline{(\gamma_A B_A)^2} J(\omega_{0A}),$$
$$W_{1B}^{RF} = \tfrac{1}{6} \overline{(\gamma_B B_B)^2} J(\omega_{0B}),$$

and

$$W_0^{RF} = W_2^{RF} = 0. \qquad (9.7.4b)$$

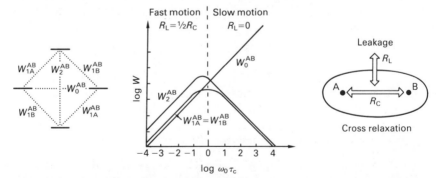

FIG. 9.7.1. Transition probabilities W_0^{AB}, W_1^{AB}, and W_2^{AB} between states separated by a difference in quantum numbers $\Delta M = 0$, 1, and 2 (zero-, single-, and double-quantum relaxation rates), induced by fluctuations of the dipolar AB interaction, as a function of the motional correlation time τ_c. In the fast-motion limit ('extreme narrowing'), the leakage and cross-relaxation rates for $R_1^{ext} = 0$ are $R_L = \tfrac{1}{2} R_C$, while in the slow-motion limit ('spin diffusion limit'), $R_L = 0$.

9.7.2. Intramolecular cross-relaxation in a two-spin system

We consider the special case of a homonuclear two-spin system with $\Omega_A \simeq \Omega_B \simeq \Omega$ and assume equal external relaxation for both nuclei. We can therefore simplify eqns (9.7.2)–(9.7.4) by setting

$$n_A = n_B = 1,$$
$$W_{1A}^{AB} = W_{1B}^{AB} = W_1^{AB},$$
$$R_{1A}^{ext} = R_{1B}^{ext} = R_1^{ext}. \quad (9.7.5)$$

The integrated intensities of the diagonal and cross-peaks, eqn (9.3.22), are in this case

$$I_{AA}(\tau_m) = I_{BB}(\tau_m) = \frac{M_0}{2}[1 + e^{-R_C \tau_m}]e^{-R_L \tau_m},$$

$$I_{AB}(\tau_m) = I_{BA}(\tau_m) = -\frac{M_0}{2}\frac{W_2^{AB} - W_0^{AB}}{|W_2^{AB} - W_0^{AB}|}[1 - e^{-R_C \tau_m}]e^{-R_L \tau_m} \quad (9.7.6)$$

with the cross-relaxation and leakage rate constants

$$R_C = 2|W_2^{AB} - W_0^{AB}|,$$
$$R_L = R_1^{ext} + 2W_1^{AB} + W_0^{AB} + W_2^{AB} - |W_2^{AB} - W_0^{AB}|. \quad (9.7.7)$$

The peak intensities are plotted in Fig. 9.7.2 as a function of the mixing time τ_m for three characteristic values of the correlation time τ_c. When the mixing time τ_m is optimized for maximum cross-peak amplitude, the ratio of the cross- to diagonal-peak intensities is

$$\frac{I_{AB}(\tau_m^{opt})}{I_{AA}(\tau_m^{opt})} = \pm \frac{R_C}{2R_L + R_C}. \quad (9.7.8)$$

For a critical correlation time $\omega_0 \tau_{crit} = 5^{\frac{1}{2}}/2$, the cross-relaxation rate in eqn (9.7.7) is zero and the cross-peaks vanish irrespective of the mixing time τ_m.

In the *fast-motion limit* ($\tau_c \ll \omega_0^{-1}$), the extreme narrowing approximation applies ($J(0) = J(\omega_0) = J(2\omega_0) = 2\tau_c$) and one obtains

$$W_0^{AB} = q\tau_c,$$
$$W_{1A}^{AB} = W_{1B}^{AB} = \tfrac{3}{2}q\tau_c,$$
$$W_2^{AB} = 6q\tau_c,$$
$$R_C = 10q\tau_c,$$
$$R_L = R_1^{ext} + 5q\tau_c \quad (9.7.10)$$

with

$$q = \tfrac{1}{10}\gamma^4 \hbar^2 r_{AB}^{-6}\left(\frac{\mu_0}{4\pi}\right)^2.$$

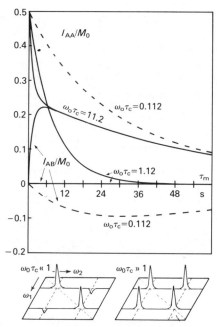

FIG. 9.7.2. Dependence of the diagonal and cross-peak intensities $I_{AA} = I_{BB}$ and $I_{AB} = I_{BA}$ on the mixing time τ_m for cross-relaxation in an AB spin system. Three typical correlation times τ_c have been assumed: $\omega_0\tau_c = 0.112$ corresponds to a short correlation time (extreme narrowing, negative cross-peaks), while $\omega_0\tau_c = 11.2$ represents a case of long correlation time (slow motion, positive cross-peaks). The critical case $\omega_0\tau_c = 1.12$ leads to vanishing cross-peaks irrespective of the mixing time τ_m. The indicated time-scale assumes a Larmor frequency $\omega_0/2\pi = 100$ MHz and $q = 3.33 \times 10^6\,\text{s}^{-2}$. (Reproduced from Ref. 9.5.)

The cross-peaks are negative in this case (see eqn (9.7.6)), as shown in Fig. 9.7.2. In the extreme narrowing limit, the AB dipolar relaxation contributes to the leakage rate R_L. For this reason, the cross-peaks are always weak. If $R_1^{\text{ext}} = 0$, one obtains

$$\frac{I_{AB}(\tau_m^{\text{opt}})}{I_{AA}(\tau_m = 0)} = -0.19.$$

This ratio drops to -0.09 if $R_1^{\text{ext}} = 10q\tau_c$.

In the *slow-motion limit* $(\tau_c \gg \omega_0^{-1})$, only the spectral density $J(0)$ contributes, and one obtains

$$\begin{aligned}
W_0^{AB} &= q\tau_c, \\
W_{1A}^{AB} &= W_{1B}^{AB} = W_2^{AB} = 0, \\
R_C &= 2q\tau_c, \\
R_L &= R_1^{\text{ext}}.
\end{aligned} \qquad (9.7.11)$$

The cross-peaks are positive in this case (see eqn (9.7.6)), and the dipolar AB interaction does not contribute to leakage relaxation. In the absence of external relaxation, the cross-peaks can achieve the same intensity as the diagonal peaks. This case is formally equivalent to pure chemical exchange.

The different behaviour for short and long correlation times can easily be rationalized. For long correlation times, the transition probability W_0^{AB} dominates. It is responsible for energy-conserving flip-flop transitions $\alpha\beta \rightleftarrows \beta\alpha$. These transitions lead to an exchange of energy between the two spins. Cross-peaks will therefore become positive. For short correlation times, on the other hand, the transition probability W_2^{AB}, which leads to transitions $\alpha\alpha \rightleftarrows \beta\beta$, dominates. The negative cross-peaks can be explained by the fact that a spin may lose an energy quantum preferably when a second spin also loses a quantum. As a result, one observes a mutual enhancement of relaxation, rather than an exchange of magnetization.

It should be noted that signal intensities in conventional Overhauser saturation experiments follow an opposite trend (see Table 9.7.1). For long correlation times, negative Overhauser effects (reduced signal intensities) are observed, while for short correlation times positive Overhauser effects (increased signal intensities) are obtained. The apparent discrepancy between steady-state saturation and 2D NOE experiments is due to the fact that in steady-state saturation experiments, negative magnetization (i.e. saturation) is transferred, whereas in a 2D NOE experiment positive magnetization is exchanging.

9.7.3. Intramolecular cross-relaxation in a system with equivalent spins

If there are several magnetically equivalent nuclei in each site, we have to consider the AA and BB dipolar interactions within the groups. These provide competitive leakage mechanisms which tend to quench cross-relaxation effects (9.5).

Consider for example a molecule with two non-equivalent methyl groups ($n_A = n_B = n = 3$). For simplicity, we assume that $W^{AA} = W^{BB}$. We find with eqns (9.3.18)–(9.3.22) and (9.7.2)–(9.7.4) the cross-relaxation and leakage rates

$$R_C = 2n(W_2^{AB} - W_0^{AB}),$$
$$R_L = R_1^{ext} + 2(n-1)(W_1^{AA} + W_2^{AA})$$
$$+ n(2W_1^{AB} + W_0^{AB} + W_2^{AB} - |W_2^{AB} - W_0^{AB}|). \quad (9.7.12)$$

In the fast-motion limit, these expressions reduce to

$$R_C = 10nq_{AB}\tau_c^{AB},$$
$$R_L = R_1^{ext} + 5[3(n-1)\lambda + n]q_{AB}\tau_c^{AB}, \quad (9.7.13)$$

with the efficiency ratio

$$\lambda = \frac{q_{AA}\tau_c^{AA}}{q_{AB}\tau_c^{AB}}. \qquad (9.7.14)$$

For short correlation times the cross-peaks will be negative and of extremely small amplitude due to the strongly enhanced leakage relaxation. For example, for cross-relaxation between two interacting methyl groups with a carbon–carbon separation of 2.5 Å, an efficiency ratio $\lambda = 10$ results, and we obtain a maximum cross-peak intensity for $R_1^{\text{ext}} = 0$

$$\frac{I_{AB}(\tau_m^{\text{opt}})}{I_{AA}(\tau_m = 0)} = -0.0012. \qquad (9.7.15)$$

In the slow-motion limit, on the other hand, eqns (9.7.12) reduce to

$$R_L = R_1^{\text{ext}}. \qquad (9.7.16)$$

In this case, the intragroup relaxation causes merely a redistribution among the equivalent spins and does not contribute at all to the leakage, and strong cross-peaks can be expected irrespective of the number of magnetically equivalent nuclei.

9.7.4. Intermolecular cross-relaxation

Consider a mixture of two chemical species A and B each containing an isolated spin (9.5). If we assume equal concentrations and equal interaction strengths for AB, AA, and BB pairs, we find the optimum cross-peak intensity in the extreme narrowing limit

$$\frac{I_{AB}(\tau_m^{\text{opt}})}{I_{AA}(\tau_m = 0)} = -0.074. \qquad (9.7.17)$$

In this case, cross-relaxation between different species is partly quenched by cross-relaxation between nuclei of the same species, in analogy to the intragroup relaxation discussed in § 9.7.3.

For a very dilute solution of species A in a solvent B, one obtains the ratio

$$\frac{I_{AB}(\tau_m^{\text{opt}})}{I_{AA}(\tau_m = 0)} = -0.148. \qquad (9.7.18)$$

The 2D exchange spectra in Fig. 9.7.3 show evidence of intermolecular cross-relaxation in a 20 per cent (by volume) solution of chloroform in cyclohexane (9.25). Only one cross-peak I_{BA} is clearly visible, its symmetrical counterpart being hidden in t_1-noise associated with the I_{BB} signal.

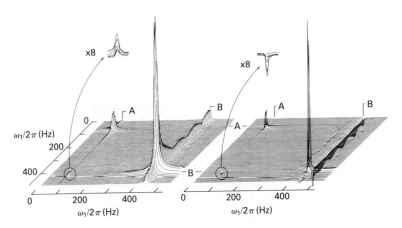

FIG. 9.7.3. Two-dimensional exchange spectrum showing intermolecular cross-relaxation in a degassed mixture of 20 per cent chloroform (A) and 80 per cent cyclohexane (B), obtained with the sequence in Fig. 9.1.1(a) with $\tau_m = 12$ s. Left: Absolute-value spectrum; right: phase-sensitive spectrum showing a negative cross-peak. (Reproduced from Ref. 9.5.)

9.7.5. Cross-relaxation in the slow-motion limit: applications to macromolecules

Two-dimensional NOE spectroscopy is particularly useful for the study of macromolecules (9.28–9.30). Thus the 2D approach is attractive both because it is suitable for investigating many sites at once, and because it circumvents problems with selective pulses (9.28, 9.29).

In the slow-motion limit, eqns (9.7.11) are applicable, and the integrated intensities of diagonal and cross-peaks are derived from eqn (9.7.6) for a pair of interacting spins

$$I_{AA}(\tau_m) = I_{BB}(\tau_m) = \frac{M_0}{2}[1 + \exp\{-2q_{AB}\tau_c\tau_m\}]\exp\{-R_1^{ext}\tau_m\},$$

$$I_{AB}(\tau_m) = I_{BA}(\tau_m) = \frac{M_0}{2}[1 - \exp\{-2q_{AB}\tau_c\tau_m\}]\exp\{-R_1^{ext}\tau_m\}. \qquad (9.7.19)$$

Because longer mixing times τ_m lead to complicated cross-peaks due to a transfer of polarization within networks of dipolar-coupled nuclei, eqn (9.3.27), 2D NOE spectra are most straightforward to interpret in the initial rate regime, where cross-peak intensities yield

$$I_{AB}(\tau_m) \simeq M_0 q_{AB}\tau_c\tau_m(1 - R_1^{ext}\tau_m). \qquad (9.7.20)$$

If external relaxation can be neglected, we obtain with the constants in

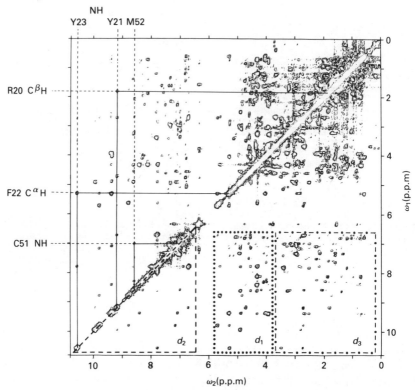

FIG. 9.7.4. Contour plot of a symmetrized, absolute-value 500-MHz ^{1}H NOESY spectrum of a 0.02 M solution of basic pancreatic trypsin inhibitor (BPTI) in D_2O, p^2H 4.6, $T = 36°C$. The spectrum was recorded in ~6 h, immediately after dissolving the protein in D_2O, so that, in addition to the non-labile protons, the resonances of ~30 backbone amide protons are seen between 7 and 10.6 p.p.m. In the lower right triangle, three spectral regions of interest for sequential resonance assignments are outlined, i.e. the regions where NOE connectivities between different amide protons (- - -), between amide protons and C^α protons (. . . .) and between amide protons and C^β protons (- · - · -) are usually observed. In the upper left triangle, the assignment of one of each of these types of connectivity is shown (C = cysteine, F = phenylalanine, M = methionine, R = arginine, Y = tyrosine). (From Ref. 9.30.)

eqn (9.4.8)

$$I_{AB}(\tau_m) \propto \frac{\tau_c}{r_{AB}^6} \tau_m. \qquad (9.7.21)$$

The correlation time τ_c refers to the reorientation of the internuclear vector $\mathbf{r}_{AB}$. In systems with isotropic reorientation, such as globular proteins, τ_c is usually assumed to be common to all AB pairs, although this assumption certainly breaks down for mobile side-chains (which, incidentally, may not fulfil the slow-motion approximation).

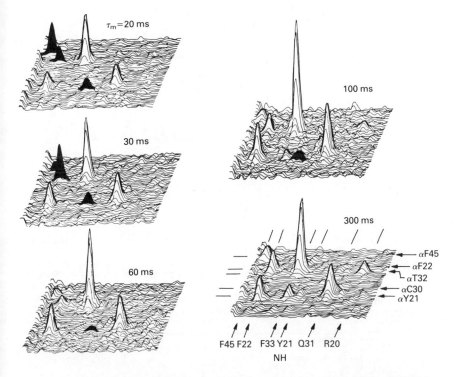

FIG. 9.7.5. Two-dimensional NOE spectra of basic pancreatic trypsin inhibitor (BPTI) for five different mixing times τ_m. A blow-up of the region $5 \leq \omega_1 \leq 6$ p.p.m. and $8 \leq \omega_2 \leq 10$ p.p.m. is shown. Abbreviations: C = cysteine, F = phenylalanine, Q = glutamine, R = arginine, T = threonine, Y = tyrosine. The blacks peaks are due to zero-quantum coherence (so-called 'J-peaks', discussed in § 9.4.2). (Adapted from Ref. 9.15.)

Many features of 2D NOE spectroscopy ('NOESY') have been explored with experiments on basic pancreatic trypsin inhibitor (BPTI), a small globular protein with 58 amino acid residues and a molecular weight of 6500 (9.7, 9.10, 9.11, 9.15, 9.28–9.31). As shown in the lower right triangle of the 2D NOE spectrum in Fig. 9.7.4, there are three regions of particular interest: NOEs between different amide protons (---), between amide and $C^\alpha H$-protons (···) and between amide and $C^\beta H$-protons (–·–·–·). Some assignments are shown in the top left triangle.

In order to measure the cross-relaxation rates, a series of 2D NOE spectra must be recorded with parametric variation of τ_m, as shown in Fig. 9.7.5. The NOE cross-peaks in this region stem from cross-relaxation between NH protons and $C^\alpha H$ protons in the β-sheet of BPTI, shown schematically in Fig. 9.7.6. The J cross-peaks due to zero-quantum interference (see § 9.4.2) appear in black in Fig. 9.7.5. The NOE

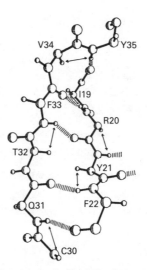

FIG. 9.7.6. Perspective view of the central part of the β-sheet of BPTI. The 10 amino-acid residues are labelled by the one-letter codes for amino-acids (C = cysteine, F = phenylalanine, I = isoleucine, Q = glutamine, R = arginine, T = threonine, V = valine, Y = tyrosine). Hydrogen bonds between NH and CO groups are indicated by hatching. Note the close distances between the NH protons of residues n and the $C^\alpha H$ protons of residues $(n-1)$, indicated by arrows. The resulting NOEs allow one to establish sequential resonance assignments. (Reproduced from Ref. 9.31.)

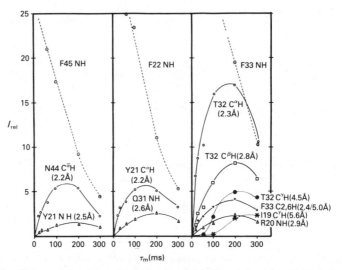

FIG. 9.7.7. The τ_m-dependence of a few selected diagonal peaks (dashed lines) and cross-peaks (solid lines) in 2D NOE spectra of BPTI (see Fig. 9.7.5). Abbreviations: F = phenylalanine, I = isoleucine, N = asparagine, R = arginine, T = threonine, Q = glutamine, Y = tyrosine. The curves in the left, centre, and right diagrams indicate cross-relaxation between NH protons in F45, F22, and F33, respectively and various protons on other residues as indicated. In parenthesis, the proton–proton distances known from X-ray studies are indicated. Two cases of second-order Overhauser effects are apparent in the diagram on the right, with vanishing initial rates (Reproduced from Ref. 9.15.)

9.7 NUCLEAR OVERHAUSER EFFECTS

cross-peaks observed in this region increase monotonically with τ_m until they are damped by external relaxation. This τ_m-dependence is shown in more detail in Fig. 9.7.7. Note that second-order processes involving magnetization transfer through three sites $A \to B \to C$ can be readily distinguished because of the vanishing initial slope at $\tau_m = 0$.

Since 2D NOE spectra are symmetrical with respect to the diagonal, it is convenient to compare one-half of the spectrum (e.g. the upper left triangle in Fig. 9.7.8) with a 2D correlation spectrum (see § 8.1), which is also symmetrical (lower right triangle in Fig. 9.7.8). This representation simplifies assignment, since one can see at a glance which signals are coupled by resolved scalar couplings, and which sites are connected by cross-relaxation.

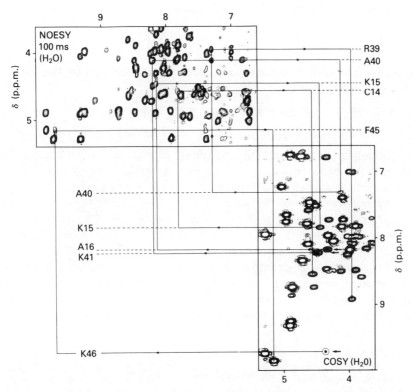

FIG. 9.7.8. Combined 2D NOE and 2D correlation spectra (NOESY: upper triangle, COSY: lower triangle) of basic pancreatic trypsin inhibitor (BPTI). The spectra were obtained separately, symmetrized, and presented in absolute-value mode. Note the sequential resonance assignment, indicated by a helicoidal series of arrows ('snail') for the segments 46 to 45, 41 to 39, and 16 to 14. The starting points are indicated by arrows in the COSY spectrum. (Reproduced from Ref. 9.30.)

9.8. Chemical exchange

Nuclear magnetic resonance has proven to be a powerful and versatile technique for the investigation of chemical exchange processes. Much of the present detailed knowledge on chemical and biological rate processes is due to NMR investigations (9.32, 9.33). Depending on the rates, various approaches can be employed, ranging from relaxation studies to lineshape analysis and polarization transfer studies. Two-dimensional exchange spectroscopy is closely related to 1D polarization transfer studies (see § 4.6.1.4), and is most useful in slow exchange, i.e. when the lineshapes are not affected by the dynamic processes. Two-dimensional exchange spectra give a particularly vivid representation of exchange networks.

Slow chemical exchange in two-site systems is described by excluding cross-relaxation in eqns (9.3.18)–(9.3.22)

$$\mathbf{L} = \begin{pmatrix} -K_{BA} - R_1^A & K_{AB} \\ K_{BA} & -K_{AB} - R_1^B \end{pmatrix}. \quad (9.8.1)$$

The diagonal and cross-peak amplitudes are determined by the leakage and cross-relaxation rates

$$R_L = \sigma - D = \tfrac{1}{2}(K_{AB} + K_{BA} + R_1^A + R_1^B) - R_C/2$$
$$R_C = 2D = 2[\tfrac{1}{4}(K_{BA} - K_{AB} + R_1^A - R_1^B) + K_{AB}K_{BA}]^{1/2}. \quad (9.8.2)$$

In symmetrical two-site systems we obtain $R_L = R_1$ and $R_C = 2k$, and the diagonal and cross-peak intensities are given in eqns (9.1.4) and (9.1.5).

If chemical exchange is fast enough to affect the lineshapes, the 2D spectra feature exchange broadening, coalescence, and exchange narrowing, described by eqn (9.3.12). These effects are demonstrated experimentally in Fig. 9.8.1. They are in complete analogy to the lineshape effects known in 1D spectroscopy.

A demonstration example is the dynamic rearrangement of the heptamethylbenzenonium ion (9.34–9.35). This ion undergoes an alkide shift which makes all seven methyl groups equivalent at sufficiently high temperature. There has been some discussion whether this shift is

intramolecular, involving either a 1–2-shift or a random shift with jumps between all possible positions, or whether it is intermolecular. Nuclear magnetic resonance lineshape analysis (9.36) indicated in agreement with many other systems of similar structure that the dynamics is governed by

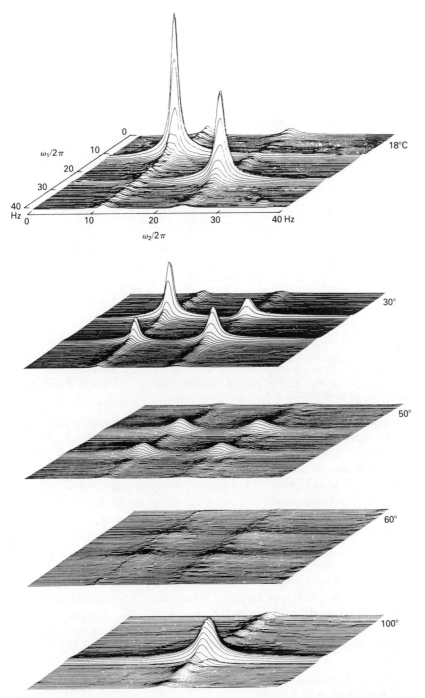

FIG. 9.8.1. Two-dimensional exchange spectra showing the exchange of the two methyl groups in N,N-dimethylacetamide for five different temperatures, demonstrating the lineshape effects for rapid chemical exchange (9.1).

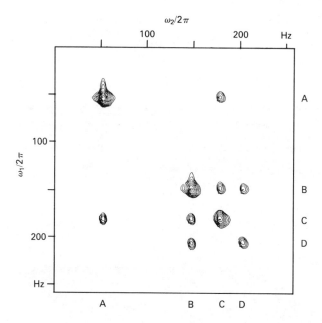

FIG. 9.8.2. Two-dimensional exchange spectrum of the protons in heptamethylbenzenonium ion in 9.4 M H_2SO_4 obtained with the sequence in Fig. 9.1.1(a) with $\tau_m = 280$ ms. The cross-peak amplitudes are consistent with a 1–2-alkide shift mechanism. (Reproduced from Ref. 9.2.)

an intramolecular 1–2-methyl shift. Although the least-squares fit gives an unambiguous result, the visual evidence for this conclusion is just a slight line distortion for intermediate exchange rates. The mechanism has been verified also by a saturation transfer study (9.35).

A 2D exchange spectrum of heptamethylbenzenonium ion is shown in Fig. 9.8.2. The cross-peaks indicate that exchange occurs between the sites $A \leftrightarrows C \leftrightarrows B \leftrightarrows D$, a network that is consistent with a 1–2-alkide shift mechanism and rules out random jumps (9.2).

The proton 2D exchange spectra in Figs. 9.8.1 and 9.8.2 are particularly simple because of the absence of resolved scalar couplings. In many systems, J couplings lead to zero- and multiple-quantum interference (see § 9.4). These complications can be circumvented by studying exchange phenomena by carbon-13 spectroscopy (9.20). An example of ^{13}C chemical exchange was encountered in § 9.6: the ring inversion of cis-decalin leads to pairwise interconversion of eight sites (Fig. 9.6.3).

9.9. Indirect detection of longitudinal relaxation in multilevel spin systems

In a system with N coupled nuclei, the longitudinal relaxation between the $(2I+1)^N$ energy levels can be described in terms of the master

9.9 INDIRECT DETECTION OF LONGITUDINAL RELAXATION

equation (eqn (2.3.3))

$$\frac{d}{dt}\mathbf{P}(t) = \mathbf{W} \Delta \mathbf{P}(t). \tag{9.9.1}$$

In conventional 1D or 2D investigations, one invariably measures combinations of the elements of the $\mathbf{W}$ matrix, and extensive data-processing is required to determine the individual elements W_{kl} (9.37, 9.38).

It is however possible to measure the elements W_{kl} separately if the relaxation within a manifold of I spins is observed indirectly via a so-called 'spy nucleus' S. This spin should not significantly contribute to the I-spin relaxation, but it should have a resolved scalar coupling to all I spins under investigation (9.39). In this case, each transition in the S-spin multiplet corresponds to one particular state of the I manifold, and a 2D exchange spectrum of the S resonances leads to cross-peaks with amplitudes I_{kl} which are simply proportional to the transition probability W_{kl}, provided the relaxation of the S nucleus can be neglected, and provided the initial rate approximation is valid ($\tau_m < W_{kl}^{-1}$).

An experimental 2D exchange spectrum of the proton-coupled multiplet of carbon C_X in imidazole (I) is shown in Fig. 9.9.1(a). The eight

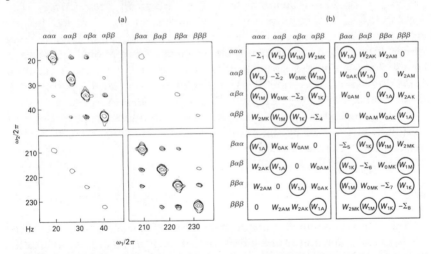

FIG. 9.9.1. (a) Two-dimensional exchange spectrum, obtained with the pulse sequence of Fig. 9.1.1(a) with $\tau_m = 2.5$ s, of the proton-coupled eight-line multiplet of C_X in imidazole ($J_{AX} = 189$ Hz, $J_{MX} = 13$ Hz, $J_{KX} = 8$ Hz). The external random-field relaxation was enhanced by addition of $5 \cdot 10^{-5}$ M Gd(fod)$_3$. In the initial rate regime, the cross-peak amplitudes are proportional to the transition probabilities W_{kl} between the proton states, provided the dipolar relaxation of the protons by the carbon-13 nuclei can be neglected. (b) The theoretical $\mathbf{W}$ matrix contains 24 single-quantum, 12 zero-quantum, and 12 double-quantum elements. In the presence of Gd(fod)$_3$, the single-quantum pathways (in circles) dominate, in agreement with the amplitudes of the cross-peaks in the experimental spectrum. (Reproduced from Ref. 9.39.)

lines along the diagonal correspond to the proton states $\alpha\alpha\alpha$, $\alpha\alpha\beta$, etc. of a weakly-coupled AMK system (the coupling to the NH proton is not observable because of rapid chemical exchange). If the relaxation of the

I [chemical structure showing tautomeric equilibrium of an imidazole-like ring with H_A, H_M, $^{13}C_X$, H_K, and N–H protons]

S nucleus is disregarded, there is a one-to-one correspondence between the cross-peaks in the 2D spectrum and the elements W_{kl} of the 8×8 matrix of transition probabilities shown in Fig. 9.9.1(b). For the sake of illustration, the single-quantum proton relaxation processes W_{1A}, W_{1M}, and W_{1K} (in circles) were enhanced by the external random field mechanism through addition of a gadolinium complex.

The elements of the type W_0^{AM} and W_2^{AM} (which are not effective in this case because of the dominant external field relaxation) correspond to concerted flips of two protons A and M, in the form of zero-quantum ($\alpha\beta \rightleftharpoons \beta\alpha$) or double-quantum transitions ($\alpha\alpha \rightleftharpoons \beta\beta$), respectively. The amplitudes of the corresponding cross-peaks are not affected by the relaxation of the S nucleus (in the initial rate approximation), and deliver direct information on internuclear distances or motional correlation times. The amplitudes of cross-peaks corresponding to single-quantum transitions (W_{1A}, W_{1M}, and W_{1K}) on the other hand can be perturbed if dipolar relaxation between the spy X and the protons A, M, and K opens additional relaxation pathways. The apparent transition probability of proton A is reduced by the zero- and double-quantum relaxation rates W_0^{AX} and W_2^{AX} but enhanced by W_{1A}^{AX}

$$W_{1A}^{app} = W_{1A}^{external} + W_{1A}^{AM} + W_{1A}^{AK} + W_{1A}^{AX} - \tfrac{1}{2}(W_0^{AX} + W_2^{AX}) \quad (9.9.2)$$

where the first three terms correspond to the total transition probability W_{1A} in the absence of the X spin.

The method is equally applicable to systems containing $I > \tfrac{1}{2}$ nuclei (9.14). The transition probabilities due to nuclear quadrupole relaxation between the three levels $|+\rangle$, $|0\rangle$, and $|-\rangle$ of an isolated deuterium nucleus in the extreme narrowing regime are described by a simple relaxation matrix

$$\mathbf{W} = c \begin{pmatrix} -3J & J & 2J \\ J & -2J & J \\ 2J & J & -3J \end{pmatrix} \quad (9.9.3)$$

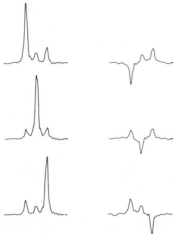

FIG. 9.9.2. Cross-sections through 2D exchange spectra of carbon-13 coupled to deuterium in benzene-d_6. The contour plot reveals a matrix of 3×3 peaks due to scalar coupling with $J_{CD} = 24$ Hz. Left; conventional 2D exchange spectrum, where the cross-peak amplitudes have the 2:1 ratio expected in the extreme narrowing regime, but the diagonal peak amplitudes do not agree with the diagonal elements of the **W**-matrix in eqn (9.9.3). Right: in the inversion-recovery 2D exchange difference spectrum (see § 9.5), all peak amplitudes are consistent with the elements of the **W** matrix. (Reproduced from Ref. 9.14.)

where $J = J(\omega_0) = J(2\omega_0)$ is the spectral density of the molecular reorientation processes. Note that the double-quantum relaxation processes are twice as likely as single-quantum relaxation transitions ($W_{13} = 2W_{12} = 2W_{23}$). This may be verified by recording a 2D exchange spectrum of the triplet of an S nucleus (carbon-13) with a resolved scalar coupling to the deuteron under investigation. Figure 9.9.2 shows cross-sections through the nine resonances in the 2D spectrum of deuterobenzene (9.14). In the normal 2D exchange experiment (Fig. 9.9.2, left), the cross-peaks show 2:1 intensity ratios in agreement with the elements of the **W**-matrix. In the right half of Fig. 9.9.2, an inversion-recovery 2D exchange difference spectrum of the same compound is shown (see § 9.5), where the diagonal and cross-peaks have the $-3:1:2$ intensity ratios of the theoretical **W**-matrix.

9.10. Dynamic processes in solids

The 2D experiments described in the preceding sections can also be applied to solids, provided sufficient spectral resolution can be obtained. In most cases, this requires a combination of 2D spectroscopy with magic-angle spinning and/or multiple-pulse sequences to remove line-broadening due to dipolar interactions.

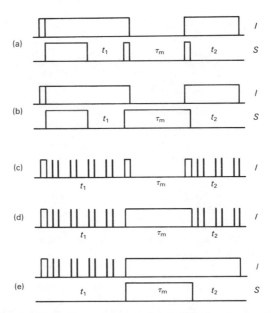

FIG. 9.10.1. Schemes for the measurement of spin diffusion in the solid state by 2D spectroscopy. (a) Diffusion between dilute nuclei in the laboratory frame. The transverse S-magnetization is enhanced by cross-polarization; the remainder of the sequence is analogous to Fig. 9.1.1(a). (b) Diffusion between dilute nuclei in the rotating frame (i.e. during spin-locking). (c) Diffusion between abundant nuclei in the laboratory frame. This sequence is identical to Fig. 9.1.1(a), except for the insertion of a multiple-pulse sequence for decoupling of homonuclear dipolar interactions, e.g. MREV-8. (d) Diffusion between abundant spins in the rotating frame (i.e. during spin-locking). (e) Diffusion between two nuclear species in the rotating frame (cross-polarization).

Chemical exchange in the solid state has been studied by carbon-13 2D exchange spectroscopy (9.49), by combining the basic 2D sequence of Fig. 9.1.1(a) with cross-polarization and decoupling of the abundant spins, as shown in Fig. 9.10.1(a). Figure 9.10.2 shows an example of a carbon-13 exchange spectrum of solid tropolone (I) at 40°C.

Slow exchange processes in solids between sites with resolved chemical shifts are rather rare. Of greater practical importance are 2D studies of *spin diffusion* in solids (9.50–9.56). The static dipolar interaction between nuclear spins causes the propagation of Zeeman order and of dipolar order through the crystal lattice by mutual spin flips (9.40–9.43). Spin

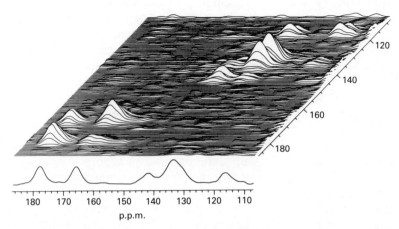

FIG. 9.10.2. Two-dimensional exchange spectrum of solid tropolone (I), recorded with the sequence of Fig. 9.10.1(a) with $\tau_m = 3$ s in combination with magic-angle spinning at 40°C. The exchange process in the solid is slow, as evidenced by the relatively narrow lines of the hydroxyl-bearing carbon (at 166 p.p.m.) and the carbonyl resonance (at 178 p.p.m.). The cross-peaks between these resonances indicate that the interconversion between the two forms I(a) and I(b) occurs in the mixing period. The other carbon pairs (with the exception of the C-5 resonance, which is unaffected by the exchange process) also give rise to cross-peaks. (Reproduced from Ref. 9.49.)

diffusion is responsible for the fact that a uniform spin temperature is obtained throughout a solid. The propagation of spin order under the influence of a static interaction can be seen as the evolution of a non-equilibrium state under the Hamiltonian with which it does not commute. In particular, the Zeeman energy of an individual spin is not a constant of motion if the dipolar coupling is strong, and will evolve into Zeeman energy of other spins. Spin diffusion is of direct importance for many phenomena in solid-state magnetic resonance, including relaxation by paramagnetic impurities (9.40), dynamic nuclear polarization (9.44), distant ENDOR (9.45), and cross-polarization (9.46–9.48).

The spin diffusion rate is dependent on the internuclear dipolar interaction and on the energy mismatch of mutual spin flips. The diffusion rate is maximum when the resonance lines of the coupled spins overlap. Due to its dipolar origin, the spin diffusion rate is proportional to r_{kl}^{-6} (9.40–9.42, 9.54, 9.55). The measurement of spin diffusion rates therefore provides information on the spatial proximity of nuclei in a solid. Such measurements can be used most fruitfully to study heterogeneity in solids.

In 2D experiments designed to measure spin-diffusion, the dipolar couplings must be removed in the evolution and detection periods in order to obtain resolved resonances, while the dipolar couplings must be allowed to act in the mixing interval τ_m. These requirements can be fulfilled with various experimental schemes shown in Fig. 9.10.1. Schemes (a) and (c) are suitable for measuring spin diffusion in the laboratory frame for dilute and abundant spins, respectively. With schemes (b) and (d), it is possible to measure spin diffusion in the rotating frame for dilute and abundant spins. Finally, scheme (e) allows one to study the dynamics of cross-polarization between two different spin species in the rotating frame.

Spin diffusion among isotopically dilute spins such as carbon-13 is rather slow due to the large average separation between spins (9.49–9.55). The statistical distribution makes it difficult to derive structural information from dilute spin diffusion measurements. However, in favourable cases, it is possible to obtain a qualitative picture of heterogeneity based on the spatial proximity of molecules that carry the carbon-13 nuclei. A simple example is shown in Fig. 9.10.3(a): in a heterogeneous mixture of adamantane and hexamethyl ethane, spin

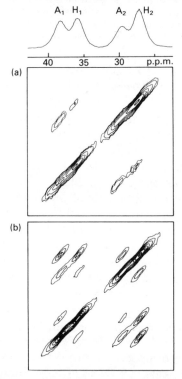

FIG. 9.10.3. Two-dimensional spin diffusion carbon-13 spectra obtained with the scheme of Fig. 9.10.1(a) of mixtures of solid adamantane and hexamethylethane: (a) heterogeneous mixture of powders; (b) homogeneous, glassy mixture by melt. Note the absence of cross-peaks between signals belonging to different chemical species in the heterogenous sample. (Reproduced from Ref. 9.51.)

diffusion is confined to nuclei within the same chemical species. As a result, the only cross-peaks that are observed in the figure connect the resonances $A_1 \rightleftarrows A_2$ (CH and CH_2 carbons of adamantane) and $H_1 \rightleftarrows H_2$ (quaternary and methyl carbons of hexamethyl ethane). In a homogeneous mixture, prepared by melting and cooling equal amounts of the two materials, additional cross-peaks appear (Fig. 9.10.3(b)), indicating that spin diffusion proceeds with similar rates between all four carbon sites. This can be regarded as evidence that the two chemical species form a homogeneous glass. The ratio of the amplitudes of cross-peaks

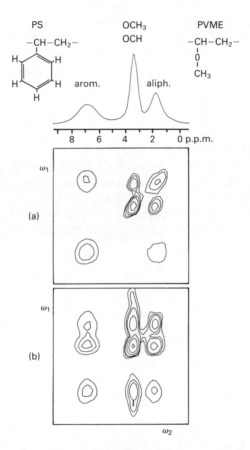

FIG. 9.10.4. Spin diffusion spectrum of protons in blends of polystyrene (PS) and poly(vinyl methyl ether) (PVME), obtained with the sequence of Fig. 9.10.1(c) ($\tau_m = 100$ ms) in conjunction with magic-angle spinning ($v_r = 2.8$ kHz). (a) Heterogeneous blend cast from chloroform and (b) homogeneous blend cast from toluene. Note the absence of cross-peaks between signals belonging to different polymers in (a), while strong cross-peaks appear in (b) between the aromatic resonances of PS and the $OCH_3 + OCH$ resonances of PVME. (Reproduced from Ref. 9.56.)

connecting different species to the amplitudes of cross-peaks relating signals of the same species provides a quantitative measure of microscopic homogeneity in disordered solids (9.51). This type of information is comparatively difficult to obtain from diffraction measurements.

The elongated lineshapes in Fig. 9.10.3 indicates inhomogeneous broadening due to susceptibility effects. Similar shapes are obtained in rigid solids where the shielding anisotropy is not averaged by molecular motion. These line-widths can be greatly reduced by combining 2D exchange spectroscopy with magic-angle sample spinning. Although dipolar interactions are reduced, this procedure does not lead to a complete quenching of the spin diffusion rates. A general limitation of natural abundance ^{13}C spin diffusion lies in the low diffusion rates, which make it necessary to use long mixing times of the order of $1-10^2$ s.

Much faster spin diffusion can be expected for abundant spins such as protons, even in combination with magic-angle sample spinning. The scheme in Fig. 9.10.1(c) combines the basic three-pulse 2D exchange experiment with multiple-pulse dipolar decoupling techniques (typically MREV-8).

Two examples of differently prepared polymer blends are shown in Fig. 9.10.4. The absence of cross-peaks between signals belonging to different components in Fig. 9.10.4(a) shows that the blend cast from chloroform is heterogeneous, while the strong cross-peaks in Fig. 9.10.4(b) demonstrate the presence of homogeneous domains in the blend cast from toluene. This clearly shows the potential for the study of heterogeneity in polymers.

10
NUCLEAR MAGNETIC RESONANCE IMAGING

After its successful application in physics, chemistry, and molecular biology, NMR is now beginning to play a vital role in the field of clinical medicine. Many fruitful applications have emerged, like investigations of whole tissue *in vitro* and *in vivo* to trace out biological processes. But the most promising application of NMR in medicine is the non-invasive imaging of entire biological organisms. The first experimental demonstration of the feasibility of macroscopic imaging by NMR was given by Lauterbur in 1972 (10.1, 10.2). The use of NMR for investigations of the human body, particularly for cancer detection, was advocated first by Damadian (10.3, 10.4).

At first sight, it may not be obvious at all why macroscopic imaging should employ just nuclear magnetic resonance and how an imaging concept could be realized. Consider first the attenuation of radiation by human tissue, schematically indicated in Fig. 10.0.1, taking into account both electromagnetic and acoustical radiation. It is obvious from this figure that nature provides three windows which permit us to look inside the human body. The X-ray window has been exploited since the basic experiments by Röntgen in 1895 and has completely revolutionized medical diagnosis. In more recent years, X-ray computer tomography has had a further significant impact on medicine (10.5–10.8), despite potential dangers by the non-negligible radiation doses.

The second window which has been taken advantage of is the low-frequency ultrasonic window. It led to the development of ultrasonic image scanners (10.9) which permit one to obtain images of acceptable quality in very fast sequence.

The radio-frequency window, on the other hand, was not exploited until 1972. This is not astonishing considering the achievable resolution, which is usually limited by the wavelength of the applied radiation through the uncertainty relation. The maximum radio-frequency useful for imaging is about 100 MHz, leading to a resolution of 3 m which is not sufficient even for imaging elephants.

The crucial idea, as first proposed by Lauterbur (10.1, 10.2), is to utilize a magnetic field *gradient* to disperse the NMR resonance frequencies of the various volume elements. The basic principle of NMR imaging is visualized in Fig. 10.0.2. The NMR spectrum of an object recorded in the presence of a strong linear magnetic field gradient can be considered as a 1D projection of the 3D proton density on to the direction of the field gradient. All nuclei in a plane perpendicular to the

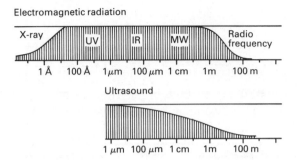

FIG. 10.0.1. Attenuation of radiation by human tissue. All electromagnetic radiation is absorbed except in the X-ray and radio-frequency ranges. Acoustical radiation is strongly absorbed for wavelengths below 1 mm.

field gradient will experience the same field and contribute to the signal amplitude at the same frequency.

For a complete representation of an object, each volume element in physical space must have its correspondence in frequency domain. A full NMR image therefore takes the form of a 3D spectrum in which the signal intensities represent the local spin density. Obviously, there is a direct connection with 2D spectroscopy. Correspondingly, much technology is common to both NMR imaging and conventional 2D spectroscopy. In particular, the data treatment procedures are quite similar. This gives us the motivation for a brief discussion of the basic approaches to NMR imaging.

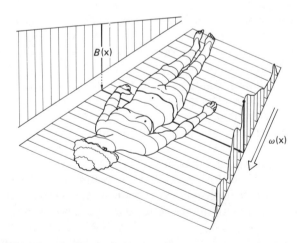

FIG. 10.0.2. Basic idea of NMR imaging: the NMR signal is recorded in the presence of a linear magnetic field gradient $B(x)$, such that the resonance frequency $\omega(x)$ is a linear function of the spatial coordinate.

10.1. Classification of imaging techniques

The major handicap of NMR is its low inherent sensitivity (10.10, 10.11). In conventional NMR, this deficiency can be overcome by longer measuring times and by signal averaging (10.12). In medical applications, however, the available time is often limited by the object under investigation. Thus, the sensitivity and performance time of NMR imaging are of major concern, and it is appropriate to use these two characteristics to classify imaging techniques (10.10).

In complete analogy to 1D Fourier spectroscopy (§ 4.3), the sensitivity can be optimized by making use of the multiplex advantage: The more volume elements are observed simultaneously, the higher will be the achievable sensitivity.

Four possible types of experiments are illustrated in Fig. 10.1.1.

(a) *Sequential point techniques*. A single volume element is selectively excited and observed at a time (10.13–10.17). For the construction of a complete image, one volume element after the other must be scanned through sequentially. Sequential point techniques produce a direct image without any intermediate image processing. However, the sensitivity will necessarily be low and the measurement time correspondingly long. Sequential point techniques are of particular merit when a localized area of the object must be investigated in more detail.

(b) *Sequential line techniques*. To increase sensitivity, an entire line of volume elements is excited and observed simultaneously (10.18–10.37). A static magnetic field gradient is applied along this line to obtain the required frequency dispersion for distinguishing the volume elements that are simultaneously observed. Sequential line techniques normally utilize a 1D Fourier experiment for simultaneous excitation and observation of the entire line. The reconstruction of the line image merely requires a simple 1D Fourier transformation.

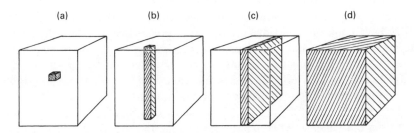

FIG. 10.1.1. Four types of imaging experiments: (a) sequential point; (b) sequential line; (c) sequential plane; (d) simultaneous techniques. (Reproduced from Ref. 10.50.)

(c) *Sequential plane techniques.* The sensitivity can be further increased when an entire plane of volume elements is excited and observed at once (10.38–10.49). 2D image reconstruction can, in some procedures, be achieved by a 2D Fourier transformation (10.47, 10.48). In other cases, alternative reconstruction procedures, for example filtered backprojection, are required (10.7).

(d) *Simultaneous techniques.* The most sensitive techniques involve a simultaneous excitation and observation of the entire 3D object. Such an experiment, however, is quite demanding with regard to the amount of data to be acquired and the necessary performance time.

We shall describe in the following sections the most prominent techniques for NMR imaging, classified according to the scheme discussed above.

10.2. Sequential point techniques

The selection of a single volume element in an extended object can be achieved by three different approaches, which may also be used in combination.

1. Selective excitation of a single volume element;
2. Non-selective excitation and selective destruction of unwanted magnetization;
3. Selective observation of a single volume element.

Two such techniques will briefly be discussed.

10.2.1. Sensitive point technique

A particularly successful variant of the sequential point techniques is the sensitive point technique proposed by Hinshaw (10.13–10.15). It utilizes non-selective excitation pulses and defocuses the unwanted magnetization by means of time-dependent field gradients.

The principle of the sensitive point technique is demonstrated in Fig. 10.2.1. A continuous string of strong r.f. pulses is applied to the sample to create a steady-state transverse magnetization of *all* volume elements. Steady-state free precession was originally described by Bradford et al. (10.51) and by Carr (10.52) and is discussed in § 4.2.5. At maximum, half the equilibrium magnetization can be maintained in the steady state.

For the selection of the sensitive point at coordinates (x_0, y_0, z_0), three time-dependent gradients are necessary. A sinusoidally modulated magnetic field gradient $g_x(t)$ is applied along the x-axis such that its nodal plane passes through the point at $x = x_0$ (Fig. 10.2.1). This gradient

10.2 SEQUENTIAL POINT TECHNIQUES

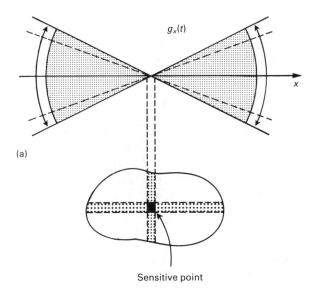

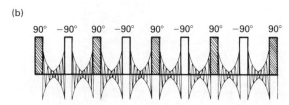

FIG. 10.2.1. Sensitive point technique: a sequence of r.f. pulses with alternating phases (b) generates a steady-state magnetization. Three time-dependent gradients (one of which is shown in (a)) modulate the resonance frequencies of all volume elements except for the sensitive point (intersection of the nodal planes of the alternating gradients). The modulation interferes with the formation of a steady-state magnetization, which is destroyed except for that of the sensitive point.

modulates the resonance frequencies of all volume elements lying outside the nodal plane. The modulation inhibits the formation of a steady-state transverse magnetization and leads to destruction of all magnetization outside the plane $x = x_0$. Two additional time-varying gradients $g_y(t)$ and $g_z(t)$ are applied simultaneously with nodal planes at $y = y_0$ and $z = z_0$, respectively. When three incommensurate modulation frequencies are used, only the magnetization of the sensitive point remains and determines the observed signal.

By moving the three nodal planes, it is possible to move the sensitive point systematically through the object. A point-by-point image is then obtained. This extremely simple technique produces good image quality. However, it is slow: in particular, it should be noticed that after each measured point the destroyed magnetization must be allowed to recover before the next point can be measured. The data acquisition rate is limited to about one point per T_1 relaxation time.

10.2.2. Field focusing NMR (FONAR) and topical NMR

Another scheme for the selection of a 'sensitive point' was suggested by Damadian (10.16). In this technique, called FONAR, the static magnetic field is shaped in such a way that good homogeneity is obtained only in one single area. The homogeneous region is normally near a saddle-point of the field as shown in Fig. 10.2.2.

Most of the object will give rise to a broad background signal, while the region around the saddle-point gives a dominant contribution at one particular frequency. Additional contributions to the same frequency may be negligible under suitable circumstances. The sensitive point is fixed in space, and it is necessary to move the object to measure a complete image.

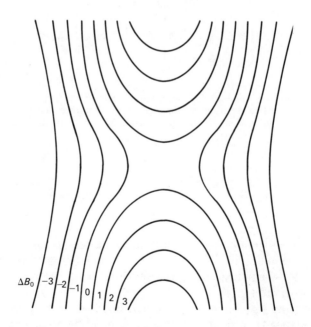

FIG. 10.2.2. Shaping of the static magnetic field used in FONAR and topical NMR. Only the region around the saddle-point is sufficiently homogeneous to give rise to narrow signals, while the remainder of the object contributes only broad resonances.

In the context of imaging, this scheme often does not lead to adequate spatial resolution, and it may be necessary to enhance selectivity further by suitably shaped radio-frequency fields. This approach is useful, however, for acquiring spectral information (chemical shifts) of a localized area within an object. This approach is known as 'topical NMR' (10.53).

With this technique it is possible to record resolved NMR spectra of a localized organ in a living being. This is the basis for studying physiological processes with a non-destructive and non-invasive technique. Particularly ^{31}P NMR proved revealing in metabolic investigations (10.54–10.56). It allows the measurement of concentrations and pH in living tissue. For experiments of this type, the requirements for spatial resolution are less severe than for true imaging and a field-focusing method is often adequate, although many other approaches have been proposed in the mean time. Surface coils (10.57a) and volume-selective pulse sequences (10.57b) have become important tools for achieving spatial selectivity.

10.3. Sequential line techniques

In sequential line techniques, a column of volume elements is selected. By means of a linear field gradient applied along this line, the necessary frequency dispersion can be obtained. A single Fourier experiment delivers simultaneously information on the entire line. In comparison to sensitive point methods, a substantial time saving can be achieved in this way, exploiting the multiplex advantage of Fourier spectroscopy. The various sequential line techniques described in this section differ by the scheme used for selective excitation or detection of the 'sensitive line'.

10.3.1. Sensitive line or multiple sensitive point method

The sensitive line or multiple sensitive point (MSP) technique, suggested by Hinshaw (10.15, 10.23), is a straightforward extension of the sensitive point method. Instead of three time-varying gradients, only two time-varying gradients are used and a static field gradient is applied along the z-axis as shown in Fig. 10.3.1. Again steady-state free precession is generated by a repetitive pulse sequence. The frequencies of precession between two successive pulses are analysed to determine the spin density along the selected line.

The sensitive line technique is simple and leads to fair sensitivity. Some high-quality images have been obtained using this technique (10.18–10.24). A relatively early image of a human wrist is shown in Fig. 10.3.2.

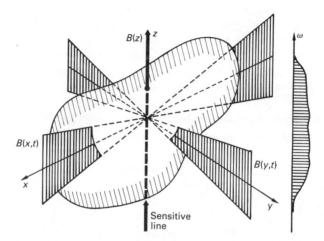

Fig. 10.3.1. In the sensitive line technique, only two time-dependent gradients are applied, while a static gradient is applied along the z-axis. The steady-state magnetization, generated by a sequence of pulses (see Fig. 10.2.1(b)), is destroyed by the modulation, except for volume elements in one column.

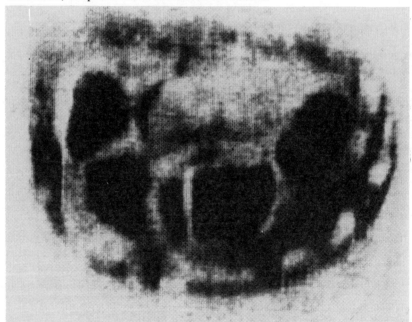

Fig. 10.3.2. Image of the left wrist of one of the authors of Ref. 10.18. The image is oriented as though the patient were facing the reader with palm upward, and was taken at the level of the distal tip of the anterior horn of the lunate. Dark areas indicate regions that contain high concentrations of mobile protons such as the marrow and subcutaneous fat. Light areas correspond to tissue such as tendons, nerves, and solid bone. Blood in the veins and arteries appears light because of its motion during the imaging process. (Reproduced from Ref. 10.18.)

10.3.2. Line scan technique

One of the inherent disadvantages of the sensitive line technique is the complete saturation of all volume elements outside of the investigated sensitive line. Therefore, after each measured line, a waiting time has to be inserted before the next line can be excited.

The line scan technique, proposed by Mansfield (10.28, 10.30) overcomes this deficiency (Fig. 10.3.3). At first, a single plane of volume elements perpendicular to the x-axis is selected by applying a magnetic field gradient along the x-axis and selectively saturating all volume

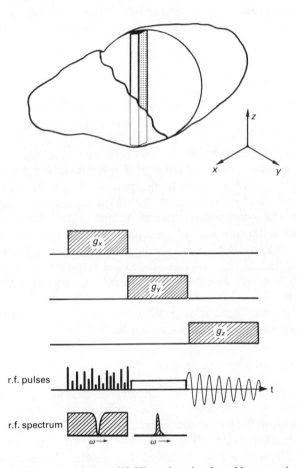

FIG. 10.3.3. In the line scan technique (10.28), a plane is selected by saturating all volume elements except for those in a plane perpendicular to the x-axis by tailored excitation (the r.f. spectrum should be white except for a 'dip'). A selective r.f. irradiation in the presence of a y-gradient excites the magnetization associated with a column of volume elements, and the signal is recorded in the presence of a z-gradient. (Compare with Fig. 10.4.8.)

elements outside this plane by means of a suitably tailored excitation (10.58). Within this plane, a line perpendicular to the y-axis is then selected by a selective r.f. pulse in the presence of a y-gradient. Finally, the free induction decay of this line is observed in the presence of a z-gradient to disperse the responses of the volume elements along this line. Because a selective pulse has been used for excitation, it is possible to repeat the experiment without delay on another line within the same plane without any adverse saturation effects.

There has been some controversy on the use of selective pulses (10.59, 10.60). In particular, it should be noted that after a selective pulse the various magnetization components show significant phase dispersion. The dispersion can be eliminated by the application of a reversed field gradient for a short time (10.36, 10.37, 10.59, 10.60).

10.3.3. Echo line imaging

A modified procedure which also leads to the selection of a single line has been indicated by Hutchison and co-workers (10.35–10.37). Figure 10.3.4 shows that by means of a selective 180° pulse in the presence of a g_x-gradient, an entire plane of spins is inverted. A subsequent selective 90° pulse applied in the presence of a g_z-gradient rotates the spins of a selected plane perpendicular to the z-axis. Therefore, a negative signal will be produced by the selected line which is parallel to the y-axis while the remaining plane perpendicular to the z-axis produces a positive signal. The difference of two free induction signals, one obtained with and one without 180° pulse, yields a signal that originates exclusively from the selected line. In this scheme, a refocusing gradient has been included to produce an echo before the FID is sampled in the presence of a g_y-gradient.

With this technique, it is possible to repeat the experiment without delay on another line as no saturation has been employed.

10.4. Sequential plane techniques

In most applications of medical imaging, it is sufficient to select planar sections through the object to be investigated. From the sensitivity standpoint, it is best to excite simultaneously an entire plane of volume elements, leading to the planar techniques to be treated in this section. In rare cases, it is necessary to record a complete 3D image which can be done by a sequential plane measurement or by a fully 3D excitation technique. We will not treat the 3D techniques separately, but we include in this section a few remarks on extensions of the 2D techniques to three dimensions.

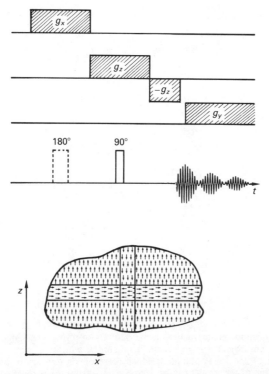

FIG. 10.3.4. Echo line imaging: the transverse magnetization is excited by a 90°-pulse in the presence of a z-gradient, which selects a slice perpendicular to the z-axis. In alternate scans, a 180° pulse applied in the presence of an x-gradient inverts the magnetization in a slice perpendicular to the x-axis. The difference of the free induction decays obtained in the two experiments yields a signal stemming from a column parallel to the y-axis. The arrows in the lower part of the figure indicate the distribution of the spin orientation after the 180°–90°-sequence.

10.4.1. Projection–reconstruction technique

The projection–reconstruction technique was first introduced into NMR by Lauterbur (10.1, 10.2, 10.38–10.43). The inspiration came from X-ray tomography (10.5–10.8) where an analogous scheme is routinely used. A signal measured in the presence of a strong linear field gradient corresponds to a 1D projection of the object on to the axis of the gradient. It is well known that it is not sufficient to obtain three orthogonal projections in order to reconstruct an image of the object. In fact, a large number of projections is required to acquire sufficient information for a well-resolved image (Fig. 10.4.1). The number of projections recorded must be of the order of the number of resolution elements to be distinguished in a line in the case of 2D imaging, or equal to the number of resolution elements in a plane for 3D imaging.

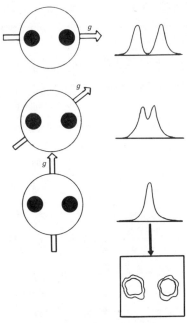

FIG. 10.4.1. Projection-reconstruction technique: by applying gradients with a variable orientation to an object (in this case a phantom with two water-filled cylindrical volumes), a set of spectra is obtained, which corresponds to a set of projections of the object. The image of the object can be reconstructed from the projections.

Let us restrict the discussion here to imaging in two dimensions. By means of a selective pulse applied in the presence of a magnetic field gradient, e.g. along the z-axis, one entire plane perpendicular to the z-axis is excited. The free induction decay (FID) is then observed in the presence of a magnetic field gradient applied along a line in the xy-plane subtending an angle φ with the x-axis. A whole set of FIDs for different angles φ covering the range from $\varphi = 0°$ to $\varphi = 180°$ is recorded. The signals are Fourier-transformed and yield the required projections.

A number of reconstruction techniques have been developed for X-ray tomography (10.7). The same procedures can be used equally well for the reconstruction of NMR images.

The simplest method of reconstruction is the *back-projection technique*. The intensity of each projection is back-projected into the image plane along the direction of projection. If the projections are used directly without filtering, a blurred image will be obtained. However, by suitable prefiltering of the projections, it is possible to obtain a faithful image. This leads to the *filtered back-projection* procedure.

Another possibility is *iterative reconstruction*. which also employs back-projection. However, after each cycle of back-projection, new projections of the obtained image are computed and compared with the real projections of the object. The differences are again back-projected to improve the image in an iterative fashion. The recursive procedure leads to a faithful image, without prefiltering.

Of particular interest in the context of Fourier spectroscopy is the *Fourier reconstruction technique*. It is based on the projection cross-section theorem (eqns (6.4.25)–(6.4.28) in § 6.4.1.5). Let $S(\omega_1, \omega_2)$ be the desired image of the object and $P(\omega, \phi)$ be a projection obtained by applying a gradient in the direction ϕ. The projection cross-section theorem then states that the 1D Fourier transform $c(t, \phi)$ of the projection $P(\omega, \phi)$ represents a central cross-section through the 2D Fourier transform $s(t_1, t_2)$ of the image $S(\omega_1, \omega_2)$. The measured frequencies ω_1 and ω_2 are related to the spatial coordinates x_1 and x_2 through the relations

$$x_i = -\omega_i/(\gamma g), \qquad i = 1, 2, \ldots . \tag{10.4.1}$$

where g is the magnetic field gradient applied to obtain the projection. This theorem is visualized in Fig. 10.4.2.

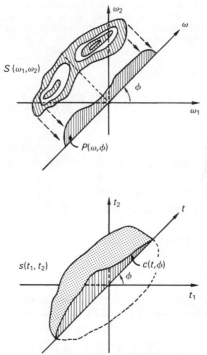

FIG. 10.4.2. Fourier reconstruction: the projection $P(\omega, \phi)$ of the desired image $S(\omega_1, \omega_2) = S(-\gamma g x_1, -\gamma g x_2)$ on to an axis that subtends an angle ϕ with respect to the ω_1-axis is obtained by applying a gradient in the direction ϕ. The 1D Fourier transform of $P(\omega, \phi)$ is equal to a cross-section $c(t, \phi)$ through the origin $t_1 = t_2 = 0$ (central cross-section) of the Fourier transform $s(t_1, t_2)$ of the image.

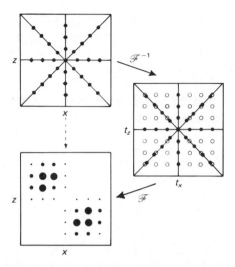

FIG. 10.4.3. Fourier reconstruction technique: the spectra are observed in the presence of gradients with different directions ϕ, as in Fig. 10.4.1. These signals correspond to projections $P(\omega, \phi)$ (top left), which can be Fourier-transformed to give cross-sections $c(t, \phi)$, with grid-points indicated by filled circles in the right-hand figure. These cross-sections actually correspond to time-domain signals (FIDs) obtained in the presence of gradients with angles ϕ. A regular grid (open circles) is obtained by interpolation, and a reverse 2D Fourier transformation leads to the desired image (bottom left).

We can use this theorem to reconstruct an image from projections (Fig. 10.4.3).

1. The measured projections $P(\omega, \phi)$ for different directions ϕ of projection are individually Fourier-transformed.

2. The Fourier transforms represent central cross-sections $c(t, \phi)$ through the Fourier-transformed object $s(t_1, t_2)$. By means of an interpolation procedure, a regular grid of sample points is computed in the (t_1, t_2) plane.

3. A reverse 2D Fourier transformation of $s(t_1, t_2)$ produces the desired image $S(\omega_1, \omega_2)$.

It should be noted that a Fourier experiment permits a direct measurement of the central cross-sections $c(t, \phi)$, and Step 1 in the reconstruction sequence can be omitted. The only approximation involved in this image reconstruction procedure is in the interpolation required to obtain a regular grid of sampling points. It can be appreciated in Fig. 10.4.3 that the measured sampling points are more densely distributed around the central point, $t_1 = t_2 = 0$, than in the outer parts of the (t_1, t_2)-plane. This implies that the 'low-frequency components' or the coarse features of the image are better represented than the finer details contained in the high-frequency components.

The question therefore arises whether it is not possible to obtain directly an equally spaced grid of sampling points $s(t_1, t_2)$. This is indeed possible by the Fourier imaging technique.

10.4.2. Fourier imaging

Fourier imaging (10.47, 10.48) of slices or volumes can be considered as a typical 2D spectroscopy technique. It belongs to the class of separation techniques (Chapter 7). The two (or three) frequency coordinates which determine the location of a volume element are measured sequentially in an experiment consisting of one (or two) evolution periods and a detection period. The experiment is sketched in Fig. 10.4.4. In the 2D (planar) version, a selective pulse is applied in the presence of a gradient g_x to select a plane parallel to the yz-plane. Evolution takes place under the influence of a g_y-gradient, while the detection is done in the presence of a g_z-gradient. In accordance with the basic recipe of 2D spectroscopy, it is necessary to repeat this experiment for a complete set of increments of the evolution period. In the case of a 3D experiment, two evolution periods with g_x and g_y gradients must be inserted, and the lengths of both must be varied systematically from experiment to experiment.

The set of measured signals forms an equidistant grid of sampling points of the 2D (or 3D) Fourier transform $s(t_1, t_2)$ of the object (open circles in Fig. 10.4.3). The signal contribution originating from a volume

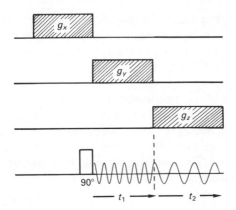

FIG. 10.4.4. Fourier imaging: in the 2D version of the experiment, a plane is selected by applying a selective pulse in the presence of an x-gradient. The precession frequencies in the evolution and detection periods are determined by the position of the volume element in the presence of y- and z-gradients respectively.

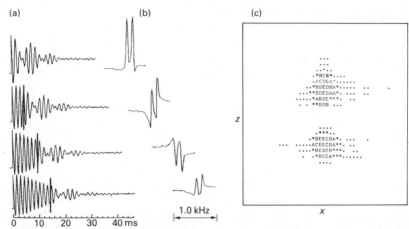

FIG. 10.4.5. Fourier imaging of a phantom consisting of two water-filled capillaries with an inner diameter of 1 mm and a centre-to-centre separation of 2.2 mm, immersed in D_2O. The tubes are parallel to the y-axis. (a) Time-domain signals: four typical signals were recorded throughout evolution and detection periods. In the t_1- and t_2-periods, gradients were applied along the x- and z-axes, respectively (in the t_1-period, the signals from both tubes have the same frequency). (b) Signals $S(t_1, \omega_2)$ obtained by 1D Fourier transformation over the t_2-interval. (c) Signal $S(\omega_1, \omega_2)$ obtained after the second Fourier transformation (absolute-value display). (Adapted from Ref. 10.48.)

element at coordinates (x, y, z) is given for a 2D experiment by

$$s(t_1, t_2) = s(0, 0)\exp\{-i\gamma y g_y t_1 - i\gamma z g_z t_2\} \qquad (10.4.2)$$

and for a 3D experiment by

$$s(t_1, t_2, t_3) = s(0, 0, 0)\exp\{-i\gamma x g_x t_1 - i\gamma y g_y t_2 - i\gamma z g_z t_3\}. \qquad (10.4.3)$$

To obtain the desired image, it is sufficient to perform a 2D (or 3D) Fourier transformation of the acquired data. An example of the recorded signals with the two successive stages of Fourier transformation is shown in Fig. 10.4.5.

The close relation to imaging by a Fourier projection–reconstruction technique is obvious. The two techniques differ merely in the distribution of the sampling points in the 2D or 3D time domain. Fourier imaging yields an equally spaced grid of sampling points and leads therefore to equal accuracy of high and low frequencies. Thus the finer details will be better represented by Fourier imaging than in an image obtained with a projection–reconstruction technique.

10.4.3. Spin-warp imaging

Fourier imaging proved to be the most reliable method in commercial realizations of NMR tomography. In a modified form, known under the

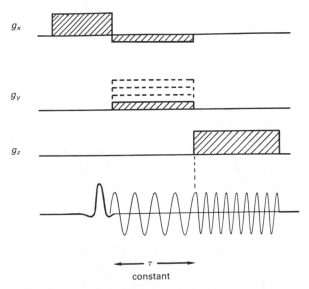

FIG. 10.4.6. Experimental scheme for spin-warp imaging. In contrast to Fourier imaging (Fig. 10.4.4), the phase of free precession at the end of the evolution period is modified by increasing the amplitude of the gradient g_y rather than the duration of the evolution period.

term spin-warp imaging, it is nowadays the most frequently applied method.

Spin-warp imaging, proposed by Hutchison and co-workers (10.61, 10.62), differs from conventional Fourier imaging in that the evolution period is fixed in length and that the *amplitude* of the applied magnetic field gradient(s) is incremented from experiment to experiment. This has the advantage that relaxation effects during the evolution time t_1 remain the same through all experiments. It is a constant-time experiment in the sense of §8.3.2 and the achievable resolution is independent of relaxation.

The experimental scheme of spin-warp imaging shown in Fig. 10.4.6 incorporates some further improvements with respect to the basic scheme of Fourier imaging. The excitation is effected by a selective r.f. pulse in the presence of a g_x-gradient. During the evolution period a reversed gradient $-g_x$ is applied to refocus the excited magnetization. At the same time a gradient g_y of variable amplitude serves to obtain the differentiation of the volume elements in the y-direction. Finally during observation a g_z-gradient is used to disperse the volume elements in the z-direction. The gradients may be turned on and off in a smooth manner without adverse effects provided that the gradient shape is the same in all experiments of a series.

Spin-warp imaging can easily be extended to three dimensions (10.62). It can also be combined with measurements of chemical shifts.

10.4.4. Rotating-frame imaging

A variant of Fourier imaging, proposed by Hoult (10.49), combines the preparation and evolution periods into one single interval. The transverse magnetization is excited by a linearly inhomogeneous r.f. field. Thus with an r.f. field gradient g_x, different planes will experience a different pulse rotation angle $\beta(x)$ (Fig. 10.4.7). A systematic variation of the pulse length from experiment to experiment creates a characteristic amplitude modulation of the resulting signal which carries the x-coordinate information. Detection takes place in the presence of a static g_y-field gradient.

Except for the replacement of a static field gradient by an r.f. field gradient, rotating-frame imaging is entirely equivalent to Fourier imaging. The same data-processing is required. An advantage of rotating-frame imaging is that no switched static field gradients are required.

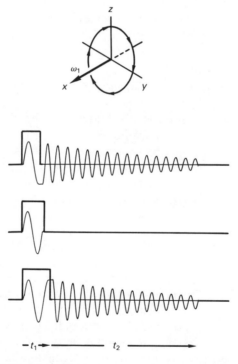

FIG. 10.4.7. Rotating frame imaging: the evolution period consists of a pulse with an inhomogeneous r.f. field with a gradient g_x. The pulse width is incremented systematically, leading to an amplitude modulation of the signal which depends on the x-coordinate of the volume element.

There has been some concern about possible adverse effects of switched field gradients on human beings. On the other hand, it is more difficult to create a clean linear r.f. field gradient than it is to generate a linear static field gradient.

10.4.5. Planar and multiplanar imaging

Two further sequential plane techniques were proposed by Mansfield and co-workers (10.44). Both are extensions of the line scan technique (§ 10.3.2).

The principle of planar imaging is visualized in Fig. 10.4.8. At first, all

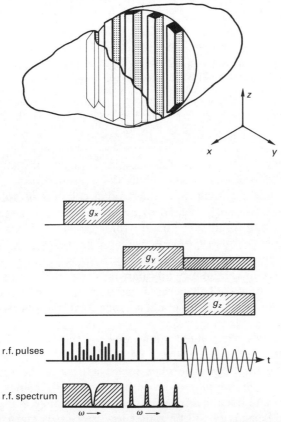

FIG. 10.4.8. In planar imaging, a tailored excitation sequence is applied in the presence of an x-gradient, such that the r.f. spectrum is essentially white except for a dip (compare Fig. 10.3.3). Except for a plane perpendicular to the x-axis, all volume elements are therefore saturated. A different tailored excitation sequence (e.g. a sequence of regularly-spaced pulses) with a spectrum consisting of discrete sidebands is then applied in the presence of a y-gradient. Finally, the signal is observed in the presence of two weighted gradients along the y- and z-axes.

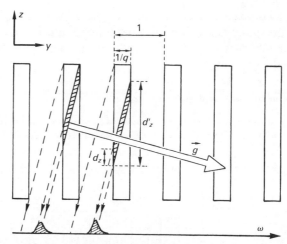

FIG. 10.4.9. In the planar imaging method (see Fig. 10.4.8), the y- and z-gradients applied in the detection period must be adjusted in such a way that the skew projections of the individual columns do not overlap. Good spatial resolution in the z-direction can only be obtained if the columns are narrow ($1/q \ll 1$), i.e. if the excitation sequence applied in the presence of the y-gradient consists of well-separated sidebands. (Reproduced from Ref. 10.50.)

parts of the object, except for one single plane, are saturated by the application of tailored excitation in the presence of a g_x-gradient. In contrast to line scanning, a set of parallel columns of volume elements are then simultaneously excited by a suitable multifrequency pulse in the presence of a g_y-gradient. Finally, the FID is observed in the simultaneous presence of two properly weighted gradients g_y and g_z, leading to an inclined gradient in the xy-plane.

It is important that sufficiently narrow strips of the selected plane are excited (Fig. 10.4.9). This permits one to project the spin density by the application of an inclined gradient such that the spectra of different columns do not overlap.

Thus, a single experiment allows one to image an entire plane (or at least a family of narrow strips of this plane). Planar imaging is an exceptionally fast technique. The inherent trick is the reduction of the dimension by the mapping of an entire plane on to a single straight line.

Although the idea behind this technique is ingenious, it has some drawbacks. Sensitivity will be rather low, since very narrow strips of the plane have to be observed. In addition, resolution is severely restricted because of the elongated shape of the distinguishable volume elements (Fig. 10.4.8).

In a further extension of this technique, Mansfield and Maudsley (10.44) proposed to image simultaneously an entire 3D object by suitable selective excitation of narrow columns evenly distributed through the object. A twofold reduction of dimension can be achieved in this manner. This technique, known as multiplanar imaging, has the same advantages and disadvantages as planar imaging but in an even more pronounced form.

10.4.6. Echo planar imaging

Echo planar imaging, also proposed by Mansfield (10.46), can be considered as another modification of Fourier imaging in which all experiments necessary to reconstruct the image of an entire plane are performed sequentially within a single FID. It is also related to planar imaging but does not require selective excitation of narrow strips.

At first, transverse magnetization of an entire plane of volume elements is excited, e.g. by a selective pulse in the presence of a magnetic field gradient g_z. The signal is observed in the presence of a weak static field gradient g_x and a strong switched gradient g_y as shown in Fig. 10.4.10. This leads to a sequence of echoes due to the refocusing effect of the reversal of the field gradient.

To understand the principle and to grasp the relation to Fourier imaging, we can consider this procedure as a sequence of experiments, each experiment lasting from one echo peak to the next one. The signal originating from the volume element at coordinates (x, y) in the nth

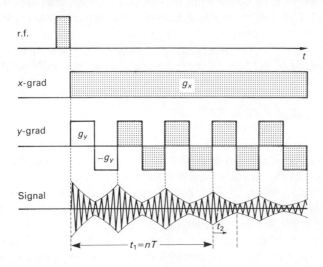

FIG. 10.4.10. Echo planar imaging: the magnetization of a plane is excited selectively in the presence of a z-gradient, and observed in the presence of a weak static x-gradient with a switched y-gradient.

experiment (following the nth echo) is then given by

$$s(nT, t_2) = s(0, 0)\exp\{-i\gamma x g_x nT - i\gamma y g_y t_2\} \quad (10.4.4)$$

where $t_2 = 0$ at the top of the nth echo. The effects caused by g_y in the previous periods have all been refocused, and the phase at the time of the nth echo is entirely determined by the local field yg_x. The signal in eqn (10.4.4) is exactly the same as in a Fourier imaging experiment, eqn (10.4.2).

The reconstruction in echo planar imaging also requires a 2D Fourier transformation of the signal $s(nT, t_2)$. The method is one of the fastest and most sensitive techniques known today, as sufficient information about an entire plane is acquired during a single FID.

10.5. Comparison of sensitivity and performance time of various imaging techniques

A comparative study of the various techniques discussed so far has been published in Ref. (10.50). This section recapitulates some of the main results, considering both 2D and 3D imaging. The differentiation of the various techniques is more pronounced in the 3D case and clearly shows their characteristics.

To present numerical results, we select a ratio of relaxation times $T_1/T_2 = 3$. We assume that equal numbers of $n = 32$ volume elements must be resolved in each of the two or three dimensions. The sensitivity of the different techniques is arbitrarily normalized with respect to the most sensitive technique, the projection–reconstruction method, by setting its sensitivity for minimum performance time equal to unity. The characteristics of 2D and 3D imaging are summarized in Figs. 10.5.1 and 10.5.2.

10.5.1. Sensitivity

In agreement with expectations, Figs. 10.5.1 and 10.5.2 show that for long performance times the sensitivity will increase proportionally to the square root of the available time. Near the minimum performance time, however, the sensitivity increases faster for some of the techniques, particularly for the sensitive point and the planar imaging methods. This behaviour is particularly pronounced for large T_1/T_2 ratios.

It is apparent that the total spread of sensitivity of the different techniques is quite large. It strongly depends on the number of volume elements that must be resolved. For $n = 32$, the sensitivity of 3D techniques covers a range of more than four orders of magnitude, while 2D techniques can differ by two orders of magnitude in sensitivity.

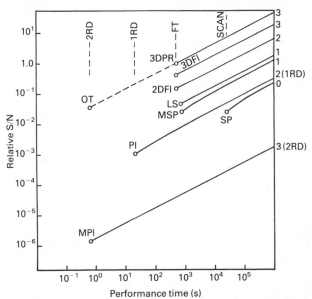

FIG. 10.5.1. Relative sensitivity versus performance time for 3D imaging of a cubic object with $n = 32$ volume elements in each of the three directions. The following techniques are included: OT = optimum technique; 3DPR = 3D projection reconstruction; 3DFI = 3D Fourier imaging; 2DFI = 2D Fourier imaging used for imaging of the 3D objects; LS = line scan; PI = planar imaging; MPI = multiplanar imaging; SP = sensitive point; and MSP = multiple sensitive point. The following parameter values have been used: $T_1 = 0.5$ sec, $T_1/T_2 = 3$, observation time $T = 3T_2$. The numbers on the right-hand side refer to the following classification: 0 = sequential point measurement, 1 = sequential line measurement, 2 = sequential plane measurement, and 3 = simultaneous measurement. 1RD and 2RD indicate techniques with onefold and twofold reduction of dimension respectively. The minimum possible performance times for the different techniques are indicated by circles. A classification in four groups with regard to minimum performance time is apparent: SCAN (scan methods); FT (Fourier techniques); and 1RD and 2RD techniques. (Reproduced from Ref. 10.50.)

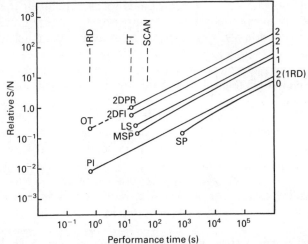

FIG. 10.5.2. Relative sensitivity versus performance time for the imaging of a 2D slice of a cubic object with $n = 32$ volume elements in each of the three directions. The parameter values are identical to those employed for Fig. 10.5.1. (Reproduced from Ref. 10.50.)

In both 3D and 2D experiments, projection–reconstruction (3DPR, 2DPR) and Fourier imaging techniques (3DFI, 2DFI) are the most sensitive methods, since they are simultaneous measurements. The sensitivity of the projection–reconstruction technique is always the best. The slight sensitivity loss of Fourier imaging methods is caused by the extensive observation time, assumed to be $T = 3T_2$, and by the fact that the signal decays in the evolution period. By using spin-warp imaging as explained in § 10.4.3, the same sensitivity can be achieved as for the projection–reconstruction technique, if the evolution time is shortened and the applied field gradients correspondingly increased.

Sequential line measurements, like line scan (LS) and multiple sensitive point (MSP), are usually less sensitive by about one order of magnitude than simultaneous measurements. The line scan technique has a somewhat higher sensitivity than the multiple sensitive point technique. This can be explained by the advantage of selective excitation in the line scan method, which avoids saturation of the magnetization of unobserved volume elements. This permits a faster repetition rate, and consequently higher sensitivity. The sensitivity of the line scan method may even surpass the sensitivity of 2D Fourier imaging if the T_1/T_2 ratio is large.

An advantage of line scan as well as of multiple sensitive point and sensitive line techniques over projection-reconstruction and Fourier imaging methods is the simplicity of data-processing: in particular, a single line can be processed at a time without the need for storing a complete 3D data array. Slow physical motion of a living object will severely affect the resolution of 2D or 3D Fourier methods because the entire time-domain data set contributes to each point in the spectrum. The time to image one line is comparatively shorter and thus less sensitive to motion. In this respect the sensitive point technique is an ideal method, as the local spin density is directly measured and no data-processing, except perhaps for some matched filtering, is required. For obtaining a complete image however, the sensitivity of the sensitive point technique is markedly lower than that of all other methods.

Special cases are planar imaging (PI) and multiplanar imaging (MPI). They involve sequential plane or simultaneous 3D measurements, respectively. Nevertheless, they are low in sensitivity. This is due to the need of restricting the active volume. A onefold reduction of dimension as used in planar imaging, reduces the sensitivity below that of a sequential line measurement. Twofold reduction as used in multiplanar imaging leads to a sensitivity considerably lower than that of the sensitive point method. In addition, these two techniques have a disadvantage in resolution in comparison with other techniques, which could only be overcome by sacrificing more sensitivity.

We have only discussed the relative sensitivities. The absolute sensitivity is of course also of great practical interest, but strongly depends on the available instrumentation. A critical evaluation of some realistic cases was published by Hoult and Lauterbur (10.11).

10.5.2. Performance time

The various techniques may also be classified according to the minimum required performance time for a complete 2D or 3D experiment. Minimum performance times, indicated by circles in Figs. 10.5.1 and 10.5.2, cover a range of four orders of magnitude for 3D experiments and three orders of magnitude for 2D experiments for the assumed number of volume elements. The longest performance time is required by sequential point measurements. An example of this class is the sensitive point method (SP), which is a typical scan method.

Data accumulation can be speeded up by about a factor n (number of volume elements to be resolved in each dimension) by employing Fourier techniques. In this case it is not important whether a sequential line, sequential plane, or simultaneous 3D measurement is involved: roughly the same minimum performance time is required although the various methods have different sensitivities. The same total number of experiments is required for all these Fourier techniques. The methods differ only in the number of experiments which contain information on one particular volume element.

Line scan and multiple sensitive point techniques require somewhat longer minimum performance times than the other Fourier methods. In line scan, it is the additional time necessary for selective saturation and the waiting time before moving to a new plane, which lengthen the performance time, particularly for long T_1 relaxation times.

A further reduction of performance time is possible by a one- or twofold reduction of dimension, as used in planar imaging and in multiplanar imaging, respectively. Here it is possible to record an image of an entire plane or of a complete 3D object in the time required for a single FID. In summary, the four classes of techniques may be classified in order of decreasing performance time:

1. scan methods;
2. Fourier techniques;
3. techniques using onefold reduction of dimension;
4. methods employing twofold reduction of dimension.

At the time of writing, it appears that Fourier methods are most promising for obtaining high-quality images for diagnostic medical purposes. It is not impossible however that the fashion in imaging technology will change, favouring other techniques which have particular merits for more specialized applications, such as the combination of imaging and high-resolution NMR in order to monitor metabolic processes *in vivo*.

More extensive treatments of NMR imaging and further references can be found in Refs. 10.63–10.68.

REFERENCES

Chapter 1

1.1. F. Bloch, W. W. Hansen, and M. Packard, *Phys. Rev.* **69,** 127 (1946).
1.2. F. Bloch, *Phys. Rev.* **70,** 460 (1946).
1.3. F. Bloch, W. W Hansen, and M. Packard, *Phys. Rev.* **70,** 474 (1946).
1.4. E. M. Purcell, H. C. Torrey, and R. V. Pound, *Phys. Rev.* **69,** 37 (1946).
1.5. E. M. Purcell, *Phys. Rev.* **69,** 681 (1946).
1.6. A. Abragam, *Principles of nuclear magnetism,* Oxford University Press, London (1961).
1.7. L. A. Zadeh and C. A. Desoer, *Linear system theory, the state space approach,* McGraw-Hill, New York (1963).
1.8. E. A. Guillemin, *Theory of linear physical systems,* Wiley, New York (1963).
1.9. B. M. Brown, *The mathematical theory of linear systems,* Science Paperbacks, Chapman and Hall, London (1965).
1.10. T. F. Bogart, *Basic concepts in linear systems: theory and experiments,* J. Wiley, New York (1984).
1.11. F. Bitter et al., *MIT Res. Lab. Electron., Quart. Progr. Rep.,* p. 26 (15 July, 1947).
1.12. B. A. Jacobsohn and R. K. Wangsness, *Phys. Rev.* **73,** 942 (1948).
1.13. M. Weger, *Bell System Tech. J.* **39,** 1013 (1960).
1.14. R. R. Ernst, *Adv. mag. Reson.* **2,** 1 (1966).
1.15. F. A. Nelson and H. E. Weaver, *Science* **146,** 223 (1964).
1.16. M. Sausade and S. Kan, *Adv. electronics and electron physics,* Vol. 34, Academic Press, New York (1973).
1.17. D. I. Hoult and R. E. Richards, *J. magn. Reson.* **24,** 71 (1976).
1.18. E. G. Paul and D. M. Grant, *J. Am. chem. Soc.* **86,** 2977 (1964).
1.19. J. H. Noggle and R. E. Schirmer, *The nuclear Overhauser effect, chemical applications,* Academic Press, New York (1971).
1.20. S. R. Hartmann and E. L. Hahn, *Phys. Rev.* **128,** 2042 (1962).
1.21. A. Pines, M. G. Gibby, and J. S. Waugh, *J. Chem. Phys.* **59,** 569 (1973).
1.22. M. Mehring, *High resolution NMR spectroscopy in solids,* Springer, Berlin, 2nd edn (1983).
1.23. A. A. Maudsley, L. Müller, and R. R. Ernst, *J. mag. Reson.* **28,** 463 (1977).
1.24. L. Müller and R. R. Ernst, *Mol. Phys.* **38,** 963 (1979).
1.25. R. D. Bertrand, W. B. Moniz, A. N. Garroway, and G. C. Chingas, *J. Am. chem. Soc.* **100,** 5227 (1978).
1.26. R. D. Bertrand, W. B. Moniz, A. N. Garroway, and G. C. Chingas, *J. mag. Reson.* **32,** 465 (1978).
1.27. G. C. Chingas, A. N. Garroway, R. D. Bertrand, and W. B. Moniz, *J. Chem. Phys.* **74,** 127 (1981).
1.28. A. A. Maudsley and R. R. Ernst, *Chem. Phys. Lett.* **50,** 368 (1977).

1.29. G. Bodenhausen and D. J. Ruben, *Chem. Phys. Lett.* **69,** 185 (1980).
1.30. A. Bax, R. G. Griffey, and B. L. Hawkins, *J. mag. Reson.* **55,** 301 (1983).
1.31. A. W. Overhauser, *Phys. Rev.* **92,** 411 (1953).
1.32. C. D. Jeffries, *Dynamic nuclear orientation,* Wiley, New York (1963).
1.33. G. Feher, *Phys. Rev.* **103,** 834 (1956).
1.34. E. B. Baker, L. W. Burch, and G. N. Root, *Rev. sci. Instr.* **34,** 243 (1963).
1.35. P. Mansfield and P. K. Grannell, *J. Phys.* **C4,** L197 (1971).
1.36. P. Brunner, M. Reinhold, and R. R. Ernst, *J. Chem. Phys.* **73,** 1086 (1980).
1.37. M. Reinhold, P. Brunner, and R. R. Ernst, *J. Chem. Phys.* **74,** 184 (1981).
1.38. H. S. Gutowsky and J. C. Tai, *J. Chem. Phys.* **39,** 208 (1963).
1.39. G. Suryan, *Proc. Ind. Acad. Sci.* **A33,** 107 (1951).
1.40. A. I. Zhernovoi and G. D. Latyshev, *Nuclear magnetic resonance in a flowing liquid,* Consultants Bureau, New York (1965).
1.41. M. P. Klein and G. W. Barton, *Rev. sci. Instr.* **34,** 754 (1963).
1.42. P. Laszlo and P. R. Schleyer, *J. Am. chem. Soc.* **85,** 2017 (1963).
1.43. L. C. Allen and L. F. Johnson, *J. Am. chem. Soc.* **85,** 2668 (1963).
1.44. O. Jardetsky, N. G. Wade, and J. J. Fisher, *Nature* **197,** 183 (1963).
1.45. R. R. Ernst, *Rev. sci. Instr.* **36,** 1689 (1965).
1.46. R. R. Ernst and W. A. Anderson, *Rev. sci. Instr.* **36,** 1696 (1965).
1.47. J. Dadok and R. F. Sprecher, *J. mag. Reson.* **13,** 243 (1974).
1.48. R. K. Gupta, J. A. Ferretti, and E. D. Becker, *J. mag. Reson.* **13,** 275 (1974).
1.49. J. A. Ferretti and R. R. Ernst, *J. Chem. Phys.* **65,** 4283 (1976).
1.50. J. Delayre, Europ. Exp. NMR Conference, Univ. of Kent, April 1974.
1.51. I. J. Lowe and R. E. Norberg, *Phys. Rev.* **107,** 46 (1957).
1.52. R. R. Ernst and W. A. Anderson, *Rev. sci. Instr.* **37,** 93 (1966).
1.53. T. C. Farrar and E. D. Becker, *Pulse and Fourier transform NMR,* Academic Press, New York (1971).
1.54. D. Shaw, *Fourier transform NMR spectroscopy,* Elsevier, Amsterdam, 2nd edn. (1984).
1.55. M. L. Martin, J.-J. Delpuech and G. J. Martin, *Practical NMR spectroscopy,* Heyden, London (1980).
1.56. T. G. Schmalz and W. H. Flygare, In *Laser and coherence spectroscopy* (ed. E. I. Steinfeld), Plenum Press, New York (1977).
1.57. S. M. Klainer, T. B. Hirschfeld, and R. A. Marino, In *Fourier, Hadamard and Hilbert transforms in chemistry* (ed. A. G. Marshall). Plenum Press, New York, 1982.
1.58. M. B. Comisarow, In *Transform techniques in chemistry* (ed. P. R. Griffiths), Plenum Press, New York, 1978; M. B. Comisarow, In *Fourier, Hadamard and Hilbert transforms in chemistry* (ed. A. G. Marshall), Plenum Press, New York, (1982).
1.59. P. Fellgett, Thesis, Cambridge University, 1951.
1.60. G. A. Vanasse and H. Sakai, *Progr. Optics* **6,** 261 (1967); R. J. Bell, *Introductory Fourier transform spectroscopy,* Academic Press, New York (1972); G. A. Vanasse (Ed.) *Spectrometric techniques,* Academic Press, New York (1977).

1.61. P. Connes, *Infrared Phys.* **24,** 69 (1984).
1.62. A. G. Marshall (Ed.), *Fourier, Hadamard and Hilbert transforms in chemistry,* Plenum Press, New York (1982).
1.63. E. L. Hahn, *Phys. Rev.* **80,** 580 (1950).
1.64. H. Y. Carr and E. M. Purcell, *Phys. Rev.* **94,** 630 (1954).
1.65. R. L. Vold, J. S. Waugh, M. P. Klein, and D. E. Phelps, *J. Chem. Phys.* **48,** 3831 (1968).
1.66. R. Freeman and H. D. W. Hill, In *Dynamic NMR spectroscopy* (ed. L. M. Jackman and F. A. Cotton), Academic Press, New York (1975).
1.67. R. L. Vold and R. R. Vold. *Prog. NMR Spectrosc.* **12,** 79 (1978).
1.68. E. O. Stejskal and J. E. Tanner, *J. Chem. Phys.* **42,** 288 (1965).
1.69. H. S. Gutowsky, R. L. Vold, and E. J. Wells, *J. Chem. Phys.* **43,** 4107 (1965).
1.70. L. M. Jackman and F. A. Cotton, *Dynamic NMR spectroscopy,* Academic Press, New York (1975).
1.71. J. I. Kaplan and G. Fraenkel, *NMR of chemically exchanging systems,* Academic Press, New York (1980).
1.72. N. Wiener, *Nonlinear problems in random theory,* Wiley, New York (1958); MIT Press (1966).
1.73. L. Amorcho and A. Brandsletter, *Water Reson. Res.* **7,** 1087 (1971).
1.74. G. H. Canavan, *J. fluid Mech.* **41,** 405 (1970).
1.75. P. Marmarelis and K. J. Naka, *Science* **175,** 1276 (1972).
1.76. R. R. Ernst and H. Primas, *Helv. Phys. Acta* **36,** 583 (1963).
1.77. R. R. Ernst, *J. mag. Reson.* **3,** 10 (1970).
1.78. R. Kaiser, *J. mag. Reson.* **3,** 28 (1970).
1.79. E. Bartholdi, A. Wokaun, and R. R. Ernst, *Chem. Phys.* **18,** 57 (1976).
1.80. R. Kaiser, *J. magn. Reson.* **15,** 44 (1974).
1.81. D. Ziessow and B. Blümich, *Ber. Bunsenges. Phys. Chem.* **78,** 1169 (1974).
1.82. B. Blümich, *Bull. mag. Reson.* **7,** 5 (1985).
1.83. A. L. Bloom and J. N. Shoolery, *Phys. Rev.* **97,** 1261 (1955).
1.84. R. A. Hoffman and S. Forsen, *Prog. NMR Spectrosc.* **1,** 15 (1966).
1.85. R. R. Ernst, *J. Chem. Phys.* **45,** 3845 (1966).
1.86. R. Freeman and W. A. Anderson, *J. Chem. Phys.* **37,** 2053 (1962).
1.87. U. Haeberlen, High Resolution NMR in Solids, *Adv. mag. Reson.,* Suppl. 1 (1976).
1.88. J. S. Waugh, L. M. Huber, and U. Haeberlen, *Phys. Rev. Lett.* **20,** 180 (1968).
1.89. P. Mansfield, M. J. Orchard, D. C. Stalker, and K. H. B. Richards, *Phys. Rev.* **B7,** 90 (1973).
1.90. W. K. Rhim, D. D. Elleman, and R. W. Vaughan, *J. Chem. Phys.* **59,** 3740 (1973).
1.91. E. R. Andrew, A. Bradbury, and R. G. Eades, *Nature* **182,** 1659 (1958).
1.92. I. J. Lowe, *Phys. Rev. Lett.* **2,** 285 (1959).
1.93. J. Schaefer and E. O. Stejskal, *J. Am. chem. Soc.* **98,** 1031 (1976).
1.94. M. Maricq and J. S. Waugh, *J. Chem. Phys.* **70,** 3300 (1979).
1.95. J. Jeener, Ampère Summer School, Basko Polje, Yugoslavia (1971).
1.96. W. P. Aue, E. Bartholdi, and R. R. Ernst, *J. Chem. Phys.* **64,** 2229 (1976).

1.97. S. Yatsiv, *Phys. Rev.* **113,** 1522 (1952).
1.98. W. A. Anderson, R. Freeman, and C. A. Reilly, *J. Chem. Phys.* **39,** 1518 (1963).
1.99. J. I. Musher, *J. Chem. Phys.* **40,** 983 (1964).
1.100. G. Bodenhausen, *Prog. NMR Spectrosc.* **14,** 137 (1983).
1.101. D. P. Weitekamp, *Adv. mag. Reson.* **11,** 111 (1983).
1.102. B. H. Meier and R. R. Ernst, *J. Am. chem. Soc.* **101,** 6441 (1979).
1.103. J. Jeener, B. H. Meier, and R. R. Ernst, *J. Chem. Phys.* **71,** 4546 (1979).
1.104. Anil Kumar, R. R. Ernst, and K. Wüthrich, *Biochem. Biophys. Res. Commun.* **95,** 1 (1980).

Chapter 2

2.1. U. Fano, *Rev. mod. Phys.* **29,** 74 (1957).
2.2. M. Weissbluth, *Atoms and molecules,* Academic Press, New York (1978).
2.3. A. Böhm, *Quantum mechanics,* Springer, New York (1979).
2.4. K. Blum, *Density matrix theory and applications,* Plenum Press, New York (1981).
2.5. C. P. Slichter, *Principles of magnetic resonance,* 2nd edn, Springer, Berlin (1978).
2.6. F. Dyson, *Phys. Rev.* **75,** 486 (1949); *Phys. Rev.* **75,** 1736 (1949).
2.7. J. A. Crawford, *Nuovo Cim.* **10,** 698 (1958).
2.8. C. N. Banwell and H. Primas, *Mol. Phys.* **6,** 225 (1963); H. Primas, *Helv. Phys. Acta* **34,** 331 (1961).
2.9. J. Jeener, *Adv. mag. Reson.* **10,** 2 (1982).
2.10. O. Platz, In *Electron spin relaxation in liquids* (ed. L. T. Muus and P. W. Atkins), Plenum Press, New York (1972).
2.11. R. Zwanzig, In *Lectures in theoretical physics III,* Interscience Publ., New York (1961).
2.12. K. O. Friedrichs, *Spectral theory of operators in Hilbert space,* Springer, New York (1973).
2.13. O. W. Sørensen, G. W. Eich, M. H. Levitt, G. Bodenhausen, and R. R. Ernst, *Prog. NMR Spectrosc.* **16,** 163 (1983).
2.14. D. P. Weitekamp, J. R. Garbow, and A. Pines, *J. Chem. Phys.* **77,** 2870 (1982).
2.15. M. H. Levitt, In *Two-dimensional NMR and related techniques* (ed. W. S. Brey), Academic Press, New York, to be published.
2.16. A. Wokaun and R. R. Ernst, *J. Chem. Phys.* **67,** 1752 (1977).
2.17. S. Vega, *J. Chem. Phys.* **68,** 5518 (1978).
2.18. B. C. Sanctuary, *J. Chem. Phys.* **64,** 4352 (1976); B. C. Sanctuary, *Mol. Phys.* **48,** 1155 (1983); B. C. Sanctuary, T. K. Halstead, and P. A. Osment, *Mol. Phys.* **49,** 753 (1983); B. C. Sanctuary, *Mol. Phys.* **49,** 785 (1983); B. C. Sanctuary and T. K. Halstead, *J. mag. Reson.* **53,** 187 (1983).
2.19. A. R. Edmonds, *Angular momentum in quantum mechanics,* 2nd edn, Princeton University Press, Princeton (1974).
2.20. M. E. Rose, *Elementary theory of angular momentum,* Wiley, New York (1957).

2.21. D. M. Brink and C. R. Satchler, *Angular momentum*, Clarendon, Oxford (1962).
2.22. S. Schäublin, A. Höhener, and R. R. Ernst, *J. mag. Reson.* **13**, 196 (1974).
2.23. J. D. Memory, *Quantum theory of magnetic resonance parameters*, McGraw-Hill, New York (1968); I. Ando and G. A. Webb, *Theory of NMR parameters*, Academic Press, New York (1983).
2.24. U. Haeberlen, *High resolution NMR in solids, Adv. mag. Reson.* Suppl. 1 (1976).
2.25. M. Mehring, *High resolution NMR spectroscopy in solids*, 2nd edn, Springer, Berlin (1983).
2.26. H. W. Spiess, *NMR Basic Principles Prog.* **15**, 55 (1978).
2.27. A. Abragam, *Principles of nuclear magnetism*, Oxford University Press, London (1961).
2.28. A. G. Redfield, *Adv. mag. Reson.* **1**, 1 (1965).
2.29. D. Wolf, *Spin temperature and nuclear spin relaxation in matter*, Clarendon Press, Oxford (1979).
2.30. R. K. Wangsness and F. Bloch, *Phys. Rev.* **89**, 728 (1953).
2.31. P. S. Hubbard, *Rev. mod. Phys.* **33**, 249 (1961).
2.32. A. G. Redfield, *IBM J. Res. Develop.* **1**, 19 (1957).
2.33. P. N. Argyres and P. N. Kelley, *Phys. Rev.* **134A**, 98 (1964).
2.34. F. Bloch, *Phys. Rev.* **70**, 460 (1946).
2.35. H. S. Gutowsky, D. M. McCall, and C. P. Slichter, *J. Chem. Phys.* **21**, 279 (1953).
2.36. J. A. Pople, W. G. Schneider, and H. J. Bernstein, *High-resolution NMR*, McGraw-Hill, New York (1959).
2.37. A. Loewenstein and T. M. Connor, *Ber. Bunsenges. Physik. Chem.* **67**, 280 (1963).
2.38. J. Delpuech, *Bull. Soc. Chim. France*, p. 2697 (1964).
2.39. J. W. Emsley, J. Feeney, and L. H. Sutcliffe, *High resolution NMR spectroscopy*, Pergamon Press, Oxford (1965).
2.40. L. M. Jackman and F. A. Cotton (Eds.) *Dynamic NMR spectroscopy*, Academic Press, New York (1975).
2.41. A. Steigel, *NMR Basic Principles Progr.* **15**, 1 (1978).
2.42. M. L. Martin, G. J. Martin, and J. J. Delpuech, *Practical NMR spectroscopy*, Heyden, London (1980).
2.43. J. I. Kaplan and G. Fraenkel, *NMR of chemically exchanging systems*, Academic Press, New York (1980).
2.44. J. Sandström, *Dynamic NMR spectroscopy*, Academic Press, London (1982).
2.45. K. Schaumburg, *Dan. Kemi.* **47**, 177 (1966).
2.46. J. L. Sudmeier and J. J. Pesek, *Inorg. Chem.* **10**, 860 (1971).
2.47. J. J. Grimaldi, J. Baldo, C. McMurray, and B. D. Sykes, *J. Am. chem. Soc.* **94**, 7641 (1972).
2.48. C. A. Fyfe, M. Cocivera, and S. W. H. Damij, *J. chem. Soc., Chem. Commun.* 743 (1973).
2.49. M. Cocivera, C. A. Fyfe, S. P. Vaish, and H. E. Chen, *J. Am. chem. Soc.* **96**, 1611 (1974).
2.50. R. G. Lawler and M. Halfon, *Rev. sci. Instr.* **45**, 84 (1974).

2.51. J. J. Grimaldi and B. D. Sykes, *J. Biol. Chem.* **250,** 1618 (1975); *J. Am. chem. Soc.* **97,** 273 (1975); *Rev. sci. Instr.* **46,** 1201 (1975).
2.52. D. W. Jones and T. F. Child, *Adv. mag. Reson.* **8,** 123 (1976).
2.53. J. Bargon, H. Fischer, and U. Johnsen, *Z. Naturforsch.* **A22,** 1551 (1967).
2.54. R. G. Lawler, *Prog. NMR Spectrosc.* **9,** 145 (1973); A. R. Lepley and G. L. Closs (Eds).) *Chemically induced magnetic polarization,* Wiley, New York (1973); C. Richard and P. Granger, In *NMR—Basic principles and progress,* Vol. 8 (ed. P. Diehl *et al.*), Springer, Berlin (1974); R. Kaptein In *Advances in free radical chemistry,* Vol. 5 (ed. G. H. Williams), Elek Science, London (1975).
2.55. S. Schäublin, A. Wokaun, and R. R. Ernst, *Chem. Phys.* **14,** 285 (1976).
2.56. S. Schäublin, A. Wokaun, and R. R. Ernst, *J. mag. Reson.* **27,** 273 (1977).
2.57. R. O. Kühne, T. Schaffhauser, A. Wokaun, and R. R. Ernst, *J. mag. Reson.* **35,** 39 (1979).
2.58. S. W. Benson, *The foundation of chemical kinetics,* McGraw-Hill, New York (1960).
2.59. K. J. Laidler, *Chemical kinetics,* McGraw-Hill, New York (1965).
2.60. A. Bauder and Hs. H. Günthard, *Helv. Chim. Acta* **55,** 2263 (1972).
2.61. L. W. Reeves and W. G. Schneider, *Can. J. Chem.* **36,** 793 (1958).
2.62. W. G. Schneider and L. W. Reeves, *Ann. NY Acad. Sci.* **70,** 858 (1958).
2.63. E. L. Hahn and D. E. Maxwell, *Phys. Rev.* **88,** 1070 (1952).
2.64. J. T. Arnold, *Phys. Rev.* **102,** 136 (1956).
2.65. H. M. McConnell, *J. Chem. Phys.* **28,** 430 (1958).
2.66. S. Alexander, *J. Chem. Phys.* **37,** 974 (1962).
2.67. R. R. Ernst, *J. Chem. Phys.* **59,** 989 (1973).
2.68. G. Binsch, *J. Am. chem. Soc.* **91,** 1304 (1969).
2.69. R. Freeman, S. Wittekoek, and R. R. Ernst, *J. Chem. Phys.* **52,** 1529 (1970).
2.70. M. Goldmann, *Quantum Description of High-Resolution NMR in Liquids,* Clarendon Press, Oxford (1988).

Chapter 3

3.1. N. F. Ramsey and R. V. Pound, *Phys. Rev.* **81,** 278 (1951).
3.2. D. P. Weitekamp, A. Bielecki, D. Zax, K. W. Zilm, and A. Pines, *Phys. Rev. Lett.* **50,** 1807 (1983).
3.3. A. Bielecki, J. B. Murdoch, D. P. Weitekamp, D. B. Zax, K. W. Zilm, H. Zimmermann, and A. Pines, *J. Chem. Phys.* **80,** 2232 (1984).
3.4. J. H. Shirley, *Phys. Rev.* **B138,** 979 (1965).
3.5. S. R. Barone, M. A. Narcowich, and F. J. Narcowich, *Phys. Rev.* **A15,** 1109 (1977).
3.6. D. Suwelack and J. S. Waugh, *Phys. Rev.* **B22,** 5110 (1980).
3.7. U. Haeberlen and J. S. Waugh, *Phys. Rev.* **175,** 453 (1968).
3.8. U. Haeberlen, *High resolution NMR in solids, Adv. mag. Reson.*, Suppl. 1 (1976).
3.9. M. Mehring, *High resolution NMR spectroscopy in solids,* 2nd edn, Springer, Berlin (1983).
3.10. P. Mansfield, M. J. Orchard, D. C. Stalker, and K. H. B. Richards, *Phys. Rev.* **B7,** 90 (1973).

REFERENCES

3.11. W. K. Rhim, D. D. Elleman, and R. W. Vaughan, *J. Chem. Phys.* **59**, 3740 (1973).
3.12. W. K. Rhim, D. D. Elleman, L. B. Schreiber, and R. W. Vaughan, *J. Chem. Phys.* **60**, 4595 (1974).
3.13. D. P. Burum and W. K. Rhim, *J. Chem. Phys.* **71**, 944 (1979).
3.14. D. P. Burum, M. Linder, and R. R. Ernst, *J. mag. Reson.* **44**, 173 (1981).
3.15. M. H. Levitt and R. Freeman, *J. mag. Reson.* **43**, 502 (1981).
3.16. M. H. Levitt, R. Freeman, and T. Frenkiel, *J. mag. Reson.* **47**, 328 (1982).
3.17. M. H. Levitt, R. Freeman, and T. Frenkiel, *J. mag. Reson.* **50**, 157 (1982).
3.18. M. H. Levitt, R. Freeman, and T. Frenkiel, *Adv. mag. Reson.* **11**, 47 (1983).
3.19. A. J. Shaka, J. Keeler, T. Frenkiel, and R. Freeman, *J. mag. Reson.* **52**, 335 (1983).
3.20. A. J. Shaka, J. Keeler, and R. Freeman, *J. mag. Reson.* **53**, 313 (1983).
3.21. J. D. Ellett and J. S. Waugh, *J. Chem. Phys.* **51**, 2851 (1969).
3.22. W. P. Aue and R. R. Ernst, *J. mag. Reson.* **31**, 533 (1978).
3.23. M. Munowitz, W. P. Aue, and R. G. Griffin, *J. Chem. Phys.* **77**, 1686 (1982).
3.24. W. P. Aue, D. J. Ruben, and R. G. Griffin, *J. mag. Reson.* **46**, 354 (1982).
3.25. H. Y. Carr and E. M. Purcell, *Phys. Rev.* **94**, 630 (1954).
3.26. J. S. Waugh, *Proc. 4th Ampère Int. Summer School,* Pula, Yugoslavia, 1976.
3.27. F. J. Dyson, *Phys. Rev.* **75**, 486 (1949); *Phys. Rev.* **75**, 1736 (1949).
3.28. R. P. Feynman, *Phys. Rev.* **84**, 108 (1951).
3.29. W. Magnus, *Commun. pure appl. Math.* **7**, 649 (1954).
3.30. R. M. Wilcox, *J. Math. Phys.* **8**, 962 (1967).
3.31. J. S. Waugh, L. M. Huber, and U. Haeberlen, *Phys. Rev. Lett.* **20**, 180 (1968).
3.32. C. H. Wang and J. D. Ramshaw, *Phys. Rev.* **B6**, 3253 (1972).
3.33. A. Abragam and M. Goldman, *Nuclear magnetism: order and disorder,* Clarendon Press, Oxford (1982).
3.34. Anil Kumar and R. R. Ernst, *J. mag. Reson.* **24**, 425 (1976).
3.35. M. M. Maricq, *Phys. Rev.* **B25**, 6622 (1982).
3.36. Y. Zur and S. Vega, *J. Chem. Phys.* **79**, 548 (1983).
3.37. Y. Zur, M. H. Levitt, and S. Vega, *J. Chem. Phys.* **78**, 5293 (1983).
3.38. M. H. Levitt, G. Bodenhausen, and R. R. Ernst, *J. mag. Reson.* **53**, 443 (1983):

Chapter 4

4.1. R. R. Ernst and W. A. Anderson, *Rev. sci. Instrum.* **37**, 93 (1966).
4.2. R. R. Ernst, *Adv. mag. Reson.* **2**, 1 (1966).
4.3. T. C. Farrar and E. D. Becker, *Pulse and Fourier transform NMR,* Academic Press, New York (1971).
4.4. D. Shaw, *Fourier transform NMR spectroscopy,* 2nd edn, Elsevier, Amsterdam (1984).

REFERENCES

4.5. M. Mehring, *High resolution NMR spectroscopy in solids*, 2nd edn. Springer, Berlin (1983).
4.6. U. Haeberlen, *High resolution NMR in solids*, Academic Press, New York (1976).
4.7. S. Goldman, *Information theory*, Prentice Hall, Englewood Cliffs, New Jersey (1953).
4.8. W. B. Davenport and W. L. Root, *An introduction to the theory of random signals and noise*, McGraw-Hill, New York (1958).
4.9. L. A. Zadeh and C. A. Desoer, *Linear system theory, the state space approach*, McGraw-Hill, New York (1963).
4.10. E. A. Guillemin, *Theory of linear physical systems*, Wiley, New York (1963).
4.11. B. M. Brown, *The mathematical theory of linear systems*, Science Paperbacks, Chapman and Hall, London (1965).
4.12. R. Deutsch, *System analysis techniques*, Prentice Hall, Englewood Cliffs, New Jersey (1969).
4.13. T. F. Bogart, *Basic concepts in linear systems: theory and experiments*, J. Wiley, New York (1984).
4.14. N. Wiener, *Nonlinear problems in random theory*, J. Wiley, New York (1958).
4.15. R. Deutsch, *Nonlinear transformations of random processes*, Prentice Hall, Englewood Cliffs, New Jersey (1962).
4.16. M. Schetzen, *The Volterra and Wiener theories of nonlinear systems*, J. Wiley, New York (1980).
4.17. W. J. Rugh, *Nonlinear system theory: the Volterra/Wiener approach*, Johns Hopkins Univ. Press, Baltimore (1981).
4.18. R. M. Bracewell, *The Fourier transform and its applications*, McGraw-Hill, New York (1965).
4.19. A. Abragam, *The principles of nuclear magnetism*, Oxford University Press, Oxford (1960).
4.20. C. P. Slichter, *Principles of magnetic resonance*, 2nd edn, Springer, Berlin (1978).
4.21. E. Bartholdi and R. R. Ernst, *J. mag. Reson.* **11,** 9 (1973).
4.22. D. C. Champeney, *Fourier transforms and their physical applications*, Academic Press, New York (1973).
4.23. D. Achilles, *Die Fourier-Transformation in der Signal-verarbeitung*, Springer, Berlin (1978).
4.24. D. Ziessow, *On-line Rechner in der Chemie, Grundlagen und Anwendung in der Fourierspektroskopie*, de Gruyter, Berlin (1973).
4.25. J. Max, *Méthodes et techniques de traitement du signal*, Masson et Cie., Paris (1972).
4.26. J. C. Lindon and A. G. Ferrige, *Prog. NMR Spectrosc.* **14,** 27 (1980).
4.27. R. Freeman, unpublished calculations.
4.28. L. R. Rabiner and B. Gold, *Theory and application of digital signal processing*, Prentice Hall, Englewood Cliffs, New Jersey (1975).
4.29. A. V. Oppenheim and R. W. Schafer, *Digital signal processing*, Prentice Hall, Englewood Cliffs, New Jersey (1975).
4.30. A. Papoulis, *Signal analysis*, McGraw-Hill, New York (1977).
4.31. W. D. Stanley, *Digital signal processing*, Reston Publ. Reston, Va (1975).

4.32. J. W. Brault and O. R. White, *Astron. Astrophys.* **13**, 169 (1971).
4.33. P. Jacquinot and B. Roizen-Dossier, *Progress in optics* (ed. E. Wolf), Vol. III, p. 31, North-Holland, Amsterdam (1964).
4.34. G. A. Vanasse and H. Sakai, *Progress in optics* (ed. E. Wolf), Vol. VI, p. 261, North-Holland, Amsterdam (1967).
4.35. R. H. Norton and R. Beer, *J. opt. Soc. Am.* **66**, 259 (1976).
4.36. H. C. Schau, *Infrared Phys.* **19**, 65 (1979).
4.37. A. G. Marshall (Ed.), *Fourier, Hadamard and Hilbert transforms in chemistry*, Plenum Press, New York (1982).
4.38. C. L. Dolph, *Proc. IRE* **35**, 335 (1946).
4.39. H. D. Helms, *IEEE Trans. Audio and Electroacoustics* **AU 16**, 336 (1968).
4.40. F. F. Kuo and J. F. Kaiser (Ed.), *System analysis by digital computer*, J. Wiley, New York (1967).
4.41. J. Makhoul, *Proc. IEEE* **63**, 561 (1975); J. Markhoul, in *Modern spectral analysis* (ed. D. G. Childers) pp. 99–118, IEEE Press, Wiley, New York, (1978).
4.42. R. Kumaresan and D. W. Tufts, *IEEE Trans.* **ASSP-30**, 833 (1982).
4.43. H. Barkhuijsen, R. de Beer, W. M. M. J. Bovée and D. van Ormondt, *J. mag. Reson.* **61**, 465 (1985).
4.44. S. F. Gull and G. J. Daniell, *Nature* **272**, 686 (1978); S. F. Gull and J. Skilling, *Proc. IEE* (F) **131**, 646 (1984).
4.45 S. Sibisi, *Nature* **301**, 134 (1983); S. Sibisi, J. Skilling, R. G. Brereton, E. D. Laue and J. Staunton, *Nature* **311**, 466 (1984); E. D. Laue, J. Skilling, J. Staunton, S. Sibisi and R. G. Brereton, *J. mag. Reson.* **62**, 437 (1985).
4.46. I. D. Campbell, C. D. Dobson, R. J. P. Williams, and A. V. Xavier, *J. mag. Reson.* **11**, 172 (1973).
4.47. A. G. Ferrige and J. C. Lindon, *J. mag. Reson.* **31**, 337 (1978).
4.48. A. DeMarco and K. Wüthrich, *J. mag. Reson.* **24**, 201 (1976).
4.49. M. Guéron, *J. mag. Reson.* **30**, 515 (1978).
4.50. W. M. Wittbold, A. J. Fischman, C. Ogle, and D. Cowburn, *J. mag. Reson.* **39**, 127 (1980).
4.51. V. Volterra, *Theory of functionals*, Dover, New York (1959).
4.52. N. Wiener, *Nonlinear problems in random theory*, J. Wiley, New York (1958).
4.53. M. Schetzen, *Int. J. Contr.* **1**, 251 (1965).
4.54. Y. W. Lee and M. Schetzen, *Int. J. Contr.* **2**, 237 (1965).
4.55. R. Deutsch, *Nonlinear transformations of random processes*, Prentice Hall, Englewood Cliffs, New Jersey (1962).
4.56. E. Bedrosian and S. O. Rice, *Proc. IEEE* **59**, 1688 (1971).
4.57. M. Schetzen, *The Volterra and Wiener theories of non-linear systems*, J. Wiley, New York (1980).
4.58. R. R. Ernst, *Chimia* **26**, 53 (1972).
4.59. R. R. Ernst, *J. mag. Reson.* **3**, 10 (1970).
4.60. R. Kaiser and W. R. Knight, *J. mag. Reson.* **50**, 467 (1982).
4.61. R. Kubo and K. Tomita, *J. Phys. Soc. Japan* **9**, 888 (1954).
4.62. R. Kubo, *J. Phys. Soc. Japan* **12**, 570 (1957).
4.63. M. Asdente, M. C. Pascucci, and A. M. Ricca, *Nuovo Cim.* **32B**, 369 (1976).
4.64. R. R. Ernst and H. Primas, *Helv. Phys. Acta* **36**, 583 (1963).

4.65. R. R. Ernst, *J. Chem. Phys.* **45,** 3845 (1966).
4.66. R. Kaiser, *J. mag. Reson.* **3,** 28 (1970).
4.67. E. Bartholdi, A. Wokaun, and R. R. Ernst, *Chem. Phys.* **18,** 57 (1976).
4.68. B. Blümich and D. Ziessow, *Bull. mag. Reson.* **2,** 299 (1981).
4.69. B. Blümich and D. Ziessow, *J. mag. Reson.* **46,** 385 (1982).
4.70. B. Blümich and D. Ziessow, *Ber. Bunsenges. Phys. Chem.* **84,** 1090 (1980).
4.71. R. Kaiser and W. R. Knight, *J. mag. Reson.* **50,** 467 (1982).
4.72. B. Blümich and D. Ziessow, *Mol. Phys.* **48,** 955 (1983).
4.73. B. Blümich and D. Ziessow, *J. Chem. Phys.* **78,** 1059 (1983).
4.74. B. Blümich and D. Ziessow, *J. mag. Reson.* **52,** 42 (1983).
4.75. B. Blümich and R. Kaiser, *J. mag. Reson.* **54,** 486 (1983).
4.76. B. Blümich, *Mol. Phys.* **51,** 1283 (1984).
4.77. B. Blümich and R. Kaiser, *J. mag. Reson.* **58,** 149 (1984).
4.78. B. Blümich, *J. mag. Reson.* **60,** 37 (1984).
4.79. B. Blümich, *Bull. mag. Reson.* **7,** 5 (1985).
4.80. W. P. Aue, E. Bartholdi, and R. R. Ernst, *J. Chem. Phys.* **64,** 2229 (1976).
4.81. H. Grad, *Commun. pure appl. Math.* **2,** 325 (1949).
4.82. L. A. Zadeh, *IRE Wescon Conv. Record,* Part 2, 105 (1957).
4.83. H. Primas, *Helv. Phys. Acta* **34,** 36 (1961).
4.84. P. Meakin and J. P. Jesson, *J. mag. Reson.* **10,** 296 (1973).
4.85. M. H. Levitt and R. Freeman, *J. mag. Reson.* **33,** 473 (1979).
4.86. R. Freeman, S. P. Kempsell, and M. H. Levitt, *J. mag. Reson.* **38,** 453 (1980).
4.87. M. H. Levitt and R. R. Ernst, *J. mag. Reson.* **55,** 247 (1983).
4.88. A. G. Redfield and R. K. Gupta, *J. Chem. Phys.* **54,** 1418 (1971).
4.89. P. Plateau, C. Dumas, and M. Guéron, *J. mag. Reson.* **54,** 46 (1983).
4.90. A. G. Redfield, S. D. Kunz, and E. K. Ralph, *J. mag. Reson.* **19,** 114 (1975).
4.91. P. Plateau and M. Guéron, *J. Am. chem. Soc.* **104,** 7310 (1982).
4.92. P. Hore, *J. mag. Reson.* **54,** 539 (1983); **55,** 283 (1983).
4.93. D. E. Jones and H. Sternlicht, *J. mag. Reson.* **6,** 167 (1972).
4.94. R. Freeman and H. D. W. Hill, *J. mag. Reson.* **4,** 366 (1971).
4.95. R. R. Ernst and W. A. Anderson, *Rev. sci. Instr.* **37,** 93 (1966).
4.96. R. Bradford, C. Clay, and E. Strick, *Phys. Rev.* **84,** 157 (1951).
4.97. H. Y. Carr, *Phys. Rev.* **112,** 1693 (1958).
4.98. J. Kaufmann and A. Schwenk, *Phys. Lett.* **24A,** 115 (1967).
4.99. P. Waldstein and W. E. Wallace, *Rev. sci. Instr.* **42,** 437 (1971).
4.100. W. S. Hinshaw, *Phys. Lett.* **48A,** 87 (1984).
4.101. W. S. Hinshaw, *J. appl. Phys.* **47,** 3709 (1976).
4.102. R. Kaiser, E. Bartholdi, and R. R. Ernst, *J. Chem. Phys.* **60,** 2966 (1974).
4.103. A. Schwenk, *J. mag. Reson.* **5,** 376 (1971).
4.104. R. R. Ernst, US patent 3968424.
4.105. I. Hainmüller, Wiss. Arbeit, Phys. Inst. der Universität Tübingen (1976).
4.106. M. H. Levitt, *J. mag. Reson.* **48,** 234 (1982).
4.107. M. H. Levitt, *J. mag. Reson.* **50,** 95 (1982).
4.108. M. H. Levitt and R. Freeman, *J. mag. Reson.* **43,** 65 (1981).
4.109. M. H. Levitt and R. Freeman, *J. mag. Reson.* **43,** 502 (1981).

4.110. M. H. Levitt, R. Freeman, and T. Frenkiel, *J. mag. Reson.* **47,** 328 (1982).
4.111. M. H. Levitt, R. Freeman, and T. Frenkiel, *J. mag. Reson.* **50,** 157 (1982).
4.112. M. H. Levitt, R. Freeman, and T. Frenkiel, *Adv. mag. Reson.* **11,** 47 (1983).
4.113. J. S. Waugh, *J. mag. Reson.* **49,** 517 (1982).
4.114. J. S. Waugh, *J. mag. Reson.* **50,** 30 (1982).
4.115. A. J. Shaka, J. Keeler, T. Frenkiel, and R. Freeman, *J. mag. Reson.* **52,** 335 (1983).
4.116. A. J. Shaka, J. Keeler, and R. Freeman, *J. mag. Reson.* **53,** 313 (1983).
4.117. M. H. Levitt and R. R. Ernst, *Mol. Phys.* **50,** 1109 (1983).
4.118. M. H. Levitt, D. Suter, and R. R. Ernst, *J. Chem. Phys.* **80,** 3064 (1984).
4.119. J. Baum, R. Tycko, and A. Pines, *J. Chem. Phys.* **79,** 4643 (1983).
4.120. R. Tycko, *Phys. Rev. Lett.* **51,** 775 (1983).
4.121. A. J. Shaka and R. Freeman, *J. mag. Reson.* **55,** 487 (1983).
4.122. S. Meiboom and D. Gill, *Rev. sci. Instrum.* **29,** 688 (1958).
4.123. R. Freeman, T. A. Frenkiel, and M. H. Levitt, *J. mag. Reson.* **44,** 409 (1981).
4.124. E. S. Pearson, *Biometrica* **24,** 404 (1932).
4.125. R. Courant and D. Hilbert, *Methods of mathematical physics,* Interscience Publ., New York (1953).
4.126. J. S. Waugh, *J. mol. Spectrosc.* **35,** 298 (1970).
4.127. R. R. Ernst and W. A. Anderson, *Rev. sci. instrum.* **36,** 1696 (1965).
4.128. W. A. Anderson, *Rev. sci. Instrum.* **33,** 1160 (1962).
4.129. O. Haworth and R. E. Richards, *Prog. NMR Spectrosc.* **1,** 1 (1966).
4.130. R. R. Ernst, in *The application of computer techniques in chemical research,* p. 61, Inst. of Petroleum, London, (1972).
4.131. S. Schäublin, A. Höhener, and R. R. Ernst, *J. mag. Reson.* **13,** 196 (1974).
4.132. O. W. Sørensen, G. W. Eich, M. H. Levitt, G. Bodenhausen, and R. R. Ernst, *Prog. NMR Spectrosc.* **16,** 163 (1983).
4.133. G. Bodenhausen and R. Freeman, *J. mag. Reson.* **36,** 221 (1979).
4.134. F. J. Adrian, *J. Chem. Phys.* **53,** 3374 (1970); *J. Chem. Phys.* **54,** 3912 (1970).
4.135. H. Hatanaka and C. S. Yannoni, *J. mag. Reson.* **42,** 330 (1981).
4.136. D. P. Burum, unpublished calculations.
4.137. A. Schwenk, *Z. Phys.* **213,** 482 (1968).
4.138. A. Schwenk, *Phys. Lett.* **31A,** 513 (1970).
4.139. H. Y. Carr and E. M. Purcell, *Phys. Rev.* **94,** 630 (1954).
4.140. A. Allerhand and D. W. Cochran, *J. Am. chem. Soc.* **92,** 4482 (1970).
4.141. E. D. Becker, J. A. Ferretti, and T. C. Farrar, *J. Am. chem. Soc.* **91,** 7784 (1969).
4.142. R. R. Shoup, E. D. Becker, and T. C. Farrar, *J. mag. Reson.* **8,** 298 (1972).
4.143. S. R. Hartmann and E. L. Hahn, *Phys. Rev.* **128,** 2042 (1962).
4.144. A. Pines, M. G. Gibby, and J. S. Waugh, *J. Chem. Phys.* **59,** 569 (1973).
4.145. A. G. Anderson and S. R. Hartmann, *Phys. Rev.* **128,** 2023 (1962).

4.146. M. Goldman, *Spin temperature and NMR in solids*, Oxford University Press, London (1970).
4.147. J. H. Noggle and R. E. Schirmer, *The nuclear Overhauser effect, chemical applications*, Academic Press, New York (1971).
4.148. S. Sørensen, R. S. Hansen, and H. J. Jakobsen, *J. mag. Reson.* **14,** 243 (1974).
4.149. H. J. Jakobsen, S. A. Linde, and S. Sørensen, *J. mag. Reson.* **15,** 385 (1974).
4.150. G. A. Morris and R. Freeman, *J. Am. chem. Soc.* **101,** 760 (1979).
4.151. D. P. Burum and R. R. Ernst, *J. mag. Reson.* **39,** 163 (1980).
4.152. M. H. Levitt and R. Freeman, *J. mag. Reson.* **39,** 533 (1980).
4.153. C. Le Cocq and J.-Y. Lallemand, *J. chem. Soc., Chem. Commun.* p. 150 (1981).
4.154. D. J. Cookson and B. E. Smith, *Org. mag. Reson.* **16,** 111 (1981).
4.155. D. W. Brown, T. T. Nakashima, and D. L. Rabenstein, *J. mag. Reson.* **45,** 302 (1981).
4.156. M. R. Bendall, D. M. Doddrell, and D. T. Pegg, *J. Am. chem. Soc.* **103,** 4603 (1981).
4.157. S. L. Patt and J. N. Shoolery, *J. mag. Reson.* **46,** 535 (1982).
4.158. Feng-Kui Pei and R. Freeman, *J. mag. Reson.* **48,** 318 (1982).
4.159. H. J. Jakobsen, O. W. Sørensen, W. S. Brey, and P. Kanyha, *J. mag. Reson.* **48,** 328 (1982).
4.160. H. Bildsøe, S. Dønstrup, H. J. Jakobsen, and O. W. Sørensen, *J. mag. Reson.* **53,** 154 (1983).
4.161. O. W. Sørensen, S. Dønstrup, H. Bildsøe, and H. J. Jakobsen, *J. mag. Reson.* **55,** 347 (1983).
4.162. D. M. Doddrell, D. T. Pegg, and M. R. Bendall, *J. mag. Reson.* **48,** 323 (1982).
4.163. M. R. Bendall and D. T. Pegg, *J. mag. Reson.* **53,** 272 (1983).
4.164. O. W. Sørensen and R. R. Ernst, *J. mag. Reson.* **51,** 477 (1983).
4.165. A. Bax, R. Freeman, and S. P. Kempsell, *J. Am. chem. Soc.* **102,** 4849 (1980).
4.166. G. Bodenhausen and C. M. Dobson, *J. mag. Reson.* **44,** 212 (1981).
4.167. P. J. Hore, E. R. P. Zuiderweg, K. Nicolay, K. Dijkstra, and R. Kaptein, *J. Am. chem. Soc.* **104,** 4286 (1982).
4.168. P. J. Hore, R. M. Scheek, A. Volbeda, and R. Kaptein, *J. mag. Reson.* **50,** 328 (1982).
4.169. P. J. Hore, R. M. Scheek, A. Volbeda, R. Kaptein, and J. H. van Boom, *J. mag. Reson.* **50,** 328 (1982).
4.170. P. J. Hore, R. M. Scheek, and R. Kaptein, *J. mag. Reson.* **52,** 339 (1983).
4.171. U. Piantini, O. W. Sørensen, and R. R. Ernst, *J. Am. chem. Soc.* **104,** 6800 (1982).
4.172. A. J. Shaka and R. Freeman, *J. mag. Reson.* **51,** 169 (1983).
4.173. O. W. Sørensen, M. H. Levitt, and R. R. Ernst, *J. mag. Reson.* **55,** 104 (1983).
4.174. M. H. Levitt and R. R. Ernst, *Chem. Phys. Lett.* **100,** 119 (1983).
4.175. M. H. Levitt and R. R. Ernst, *J. Chem. Phys.* **83,** 3297 (1985).
4.176. P. Mansfield and P. K. Grannell, *J. Phys. C* **4,** L197 (1971).

4.177. H. E. Bleich and A. G. Redfield, *J. Chem. Phys.* **67,** 5040 (1977).
4.178. A. A. Maudsley, L. Müller, and R. R. Ernst, *J. mag. Reson.* **28,** 463 (1977).
4.179. L. Müller and R. R. Ernst, *Mol. Phys.* **38,** 963 (1979).
4.180. R. D. Bertrand, W. B. Moniz, A. N. Garroway, and G. C. Chingas, *J. Am. chem. Soc.* **100,** 5227 (1978).
4.181. G. C. Chingas, A. N. Garroway, R. D. Bertrand, and W. B. Moniz, *J. Chem. Phys.* **74,** 127 (1981).
4.182. D. A. McArthur, E. L. Hahn, and R. E. Walstedt, *Phys. Rev.* **188,** 609 (1969).
4.183. D. E. Demco, J. Tegenfeldt, and J. S. Waugh, *Phys. Rev.* **B11,** 4133 (1975).
4.184. J. Tegenfeldt and U. Haeberlen, *J. mag. Reson.* **36,** 453 (1979).
4.185. L. Müller, Anil Kumar, T. Baumann, and R. R. Ernst, *Phys. Rev. Lett.* **32,** 1402 (1974).
4.186. P. Caravatti, G. Bodenhausen, and R. R. Ernst, *Chem. Phys. Lett.* **89,** 363 (1982).
4.187. P. Caravatti, L. Braunschweiler, and R. R. Ernst, *Chem. Phys. Lett.* **100,** 305 (1983).
4.188. R. L. Vold and R. R. Vold, *Prog. NMR Spectrosc.* **12,** 79 (1978).
4.189. R. Freeman and H. D. W. Hill, In *Dynamic NMR spectroscopy* (ed. L. M. Jackman and F. A. Cotton), p. 131, Academic Press, New York (1975).
4.190. D. A. Wright, D. E. Axelson, and G. C. Levy, *Top. ^{13}C NMR Spectrosc.* **3,** 104 (1979).
4.191. M. L. Martin, J.-J. Delpuech, and G. J. Martin, *Practical NMR spectroscopy,* Heyden, London (1980).
4.192. R. L. Vold, J. S. Waugh, M. P. Klein, and D. E. Phelps, *J. Chem. Phys.* **48,** 3831 (1968).
4.193. R. Freeman and H. D. W. Hill, *J. Chem. Phys.* **53,** 4103 (1970).
4.194. R. Freeman and H. D. W. Hill, *J. Chem. Phys.* **54,** 3367 (1971).
4.195. D. C. Look and D. R. Locker, *Rev. sci. Instrum.* **41,** 250 (1970).
4.196. R. Kaptein, K. Dijkstra, and C. E. Tarr, *J. mag. Reson.* **24,** 295 (1976).
4.197. D. E. Demco, P. van Hecke, and J. S. Waugh, *J. mag. Reson.* **16,** 467 (1974).
4.198. R. R. Vold and G. Bodenhausen, *J. mag. Reson.* **39,** 363 (1980).
4.199. G. C. Levy and I. R. Peat, *J. mag. Reson.* **18,** 500 (1975).
4.200. D. Canet, G. C. Levy, and I. R. Peat, *J. mag. Reson.* **18,** 199 (1975).
4.201. H. Hanssum, *J. mag. Reson.* **45,** 461 (1981).
4.202. R. Gupta, J. Ferretti, E. D. Becker, and G. Weis, *J. mag. Reson.* **38,** 447 (1980).
4.203. J. L. Markley, W. J. Horsley, and M. P. Klein, *J. Chem. Phys.* **55,** 3604 (1971).
4.204. R. Freeman and H. D. W. Hill, *J. Chem. Phys.* **54,** 3367 (1971).
4.205. R. K. Gupta, *J. mag. Reson.* **25,** 231 (1977).
4.206. K. A. Christensen, D. M. Grant, E. M. Schulman, and C. Walling, *J. Chem. Phys.* **78,** 1971 (1974).
4.207. Y. N. Luzikov, N. M. Sergeyev, and M. G. Levkovitch, *J. mag. Reson.* **21,** 359 (1976).

4.208. R. Freeman, S. Wittekoek, and R. R. Ernst, *J. Chem. Phys.* **52,** 1529 (1970).
4.209. S. H. Forsén and R. A. Hoffman, *J. Chem. Phys.* **39,** 2892 (1963).
4.210. S. H. Forsén and R. A. Hoffman, *J. Chem. Phys.* **40,** 1189 (1964).
4.211. S. H. Forsén and R. A. Hoffman, *J. Chem. Phys.* **45,** 2049 (1966).
4.212. T. R. Brown and S. Ogawa, *Proc. nat. Acad. Sci., USA* **74,** 3627 (1977).
4.213. A. A. Bothner-By, In *Magnetic resonance studies in biology* (ed. R. G. Shulman), p. 177, Academic Press, New York (1979).
4.214. A. Kalk and H. J. C. Berendsen, *J. mag. Reson.* **24,** 343 (1976).
4.215. A. Dubs, G. Wagner, and K. Wüthrich, *Biochem. Biophys. Acta* **577,** 177 (1979).
4.216. G. Wagner and K. Wüthrich, *J. mag. Reson.* **33,** 675 (1979).
4.217. E. L. Hahn, *Phys. Rev.* **80,** 580 (1950).
4.218. E. O. Stejskal and J. E. Tanner, *J. Chem. Phys.* **42,** 288 (1965).
4.219. R. L. Vold and S. O. Chan, *J. Chem. Phys.* **53,** 449 (1970).
4.220. R. Freeman and H. D. W. Hill, *J. Chem. Phys.* **54,** 301 (1971).
4.221. A. Allerhand, *J. Chem. Phys.* **44,** 1 (1966).
4.222. Anil Kumar and R. R. Ernst, *Chem. Phys. Lett.* **37,** 162 (1976).
4.223. R. R. Vold and R. L. Vold, *J. Chem. Phys.* **64,** 320 (1976).
4.224. Anil Kumar and R. R. Ernst, *J. mag. Reson.* **24,** 425 (1976).
4.225. A. Loewenstein and T. M. Connor, *Ber. Bunsenges. Physik. Chem.* **67,** 280 (1963).
4.226. J. Delpuech, *Bull. Soc. Chim. France* p. 2697 (1964).
4.227. L. M. Jackman and F. A. Cotton (Eds.), *Dynamic NMR spectroscopy,* Academic Press, New York (1975).
4.228. A. Steigel, In *Dynamic NMR spectroscopy, NMR basic principles and progress,* **15,** 1 (1978).
4.229. H. W. Spiess, *NMR basic principles and progress* **15,** 55 (1978).
4.230. J. I. Kaplan and G. Fraenkel, *NMR of chemically exchanging systems,* Academic Press, New York (1980).
4.231. J. Sandström, *Dynamic NMR spectroscopy,* Academic Press, London (1982).
4.232. J. I. Kaplan, *J. Chem. Phys.* **28,** 278 (1958).
4.233. R. K. Harris, N. C. Pyper, R. E. Richards, and G. W. Schultz, *Mol. Phys.* **19,** 145 (1970).
4.234. R. K. Harris and N. C. Pyper, *Mol. Phys.* **20,** 467 (1971); *Mol. Phys.* **23,** 277 (1971).
4.235. R. K. Harris and K. M. Worvill, *J. mag. Reson.* **9,** 383 (1973); *J. mag. Reson.* **9,** 394 (1973).
4.236. G. Fraenkel, J. I. Kaplan, and P. P. Yang, *J. Chem. Phys.* **60,** 2574 (1974).
4.237. R. O. Kühne, T. Schaffhauser, A. Wokaun, and R. R. Ernst, *J. Mag. Reson.* **35,** 39 (1979).
4.238. S. Schäublin, A. Wokaun, and R. R. Ernst, *J. mag. Reson.* **27,** 273 (1977).
4.239. M. Eigen and L. De Maeyer, In *Techniques of chemistry,* Vol. VI, Part II, (ed. G. G. Hammes), p. 63, Wiley, New York (1974); G. G. Hammes (Ed.), In *Techniques of chemistry,* Vol. VI, Part II, p. 147, Wiley, New York (1974).

4.240. E. F. Greene and J. P. Toennies, *Chemical reactions in shock waves*, Academic Press, New York (1964); W. Knoche, In *Techniques of chemistry*, Vol. VI, Part II, (ed. G. G. Hammes), p. 187, Wiley, New York (1974).
4.241. G. Porter and M. A. West, In *Techniques of chemistry*, Vol. VI, Part II, (ed. G. G. Hammes), p. 367, Wiley, New York (1974).
4.242. N. C. Verma and R. W. Fessenden, *J. Chem. Phys.* **65**, 2139 (1976).
4.243. A. D. Trifunac, K. W. Johnson, and R. H. Lowers, *J. Am. chem. Soc.* **98**, 6067 (1976).
4.244. H. Hartridge and F. J. W. Roughton, *Proc. R. Soc.* **A104**, 376, 395 (1923).
4.245. B. Chance, *J. Franklin Inst.* **229**, 455, 613, 737 (1940).
4.246. B. Chance, R. H. Eisenhardt, Q. H. Gibson, and K. K. Lonberg-Holm (Eds.), *Rapid Mixing and sampling techniques in biochemistry*, Academic Press, New York (1964).
4.247. D. W. Jones and T. F. Child, *Adv. mag. Reson.* **8**, 123 (1976).
4.248. K. Schaumburg, *Dan. Kemi.* **47**, 177 (1966).
4.249. J. L. Sudmeier and J. J. Pesek, *Inorg. Chem.* **10**, 860 (1971).
4.250. J. J. Grimaldi, J. Baldo, C. McMurray, and B. D. Sykes, *J. Am. chem. Soc.* **94**, 7641 (1972).
4.251. C. A. Fyfe, M. Cocivera, and S. W. H. Damij, *J. chem. Soc., Chem. Commun.* 743 (1973).
4.252. M. Cocivera, C. A. Fyfe, S. P. Vaish, and H. E. Chen, *J. Am. chem. Soc.* **96**, 1611 (1974).
4.253. R. G. Lawler, *Prog. NMR Spectrosc.* **9**, 145 (1975).
4.254. J. J. Grimaldi and B. D. Sykes, *J. Biol. Chem.* **250**, 1618 (1975); *J. Am. chem. Soc.* **97**, 273 (1975); *Rev. sci. Instr.* **46**, 1201 (1975).
4.255. J. Bargon, H. Fischer, and U. Johnsen, *Z. Naturforsch.* **A22**, 1551 (1967).
4.256. R. G. Lawler and M. Halfon, *Rev. sci. Instr.* **45**, 84 (1974).
4.257. S. Schäublin, A. Wokaun, and R. R. Ernst, *Chem. Phys.* **14**, 285 (1976).
4.258. A. W. Overhauser, *Phys. Rev.* **92**, 411 (1953).
4.259. R. V. Pound, *Phys. Rev.* **79**, 685 (1950).
4.260. R. Freeman and W. A. Anderson, *J. Chem. Phys.* **37**, 2053 (1962).
4.261. R. A. Hoffman and S. Forsén, *Prog. NMR Spectrosc.* **1**, 15 (1966).
4.262. W. A. Anderson and R. Freeman, *J. Chem. Phys.* **37**, 85 (1963); W. A. Anderson and F. A. Nelson, *J. Chem. Phys.* **39**, 183 (1963).
4.263. N. R. Krishna and S. L. Gordon, *Phys. Rev.* **A6**, 2059 (1972).
4.264. R. R. Ernst, W. P. Aue, E. Bartholdi, A. Höhener, and S. Schäublin, *Pure appl. Chem.* **37**, 47 (1974).
4.265. N. R. Krishna, *J. Chem. Phys.* **63**, 4329 (1975).
4.266. R. E. D. McClung and N. R. Krishna, *J. mag. Reson.* **29**, 573 (1978).
4.267. R. Freeman, H. D. W. Hill, and R. Kaptein, *J. mag. Reson.* **7**, 327 (1972).
4.268. A. L. Bloom and J. N. Shoolery, *Phys. Rev.* **97**, 1261 (1955).
4.269. A. Wokaun and R. R. Ernst, *Mol. Phys.* **38**, 1579 (1979).
4.270. H. J. Reich, M. Jautelat, M. T. Messe, F. J. Weigert, and J. D. Roberts, *J. Am. chem. Soc.* **91**, 7445 (1969).

4.271. L. F. Johnson, Tenth Experimental NMR Conference, Pittsburgh, Pa. (1969).
4.272. B. Birdsall, N. J. M. Birdsall, and J. Feeney, *J. chem. Soc., Chem. Commun.* 316 (1972).
4.273. E. Breitmaier and W. Voelter, ^{13}C *NMR Spectroscopy*, Verlag Chemie, Weinheim (1974).
4.274. M. Tanabe, T. Hamasaki, D. Thomas, and L. F. Johnson, *J. Am. chem. Soc.* **93**, 273 (1971).
4.275. W. P. Aue and R. R. Ernst, *J. mag. Reson.* **31**, 533 (1978).
4.276. G. A. Morris, G. L. Nayler, A. J. Shaka, J. Keeler, and R. Freeman, *J. mag. Reson.* **58**, 155 (1984).
4.277. J. P. Jesson, P. Meakin, and G. Kneissel, *J. Am. chem. Soc.* **95**, 618 (1973).
4.278. W.-K. Rhim, D. P. Burum, and D. D. Elleman, *Phys. Rev. Lett.* **37**, 1764 (1976).
4.279. W.-K. Rhim, D. P. Burum, and D. D. Elleman, *J. chem. Phys.* **68**, 692 (1978).
4.280. W.-K. Rhim, D. P. Burum, and D. D. Elleman, *Phys. Lett.* **62A**, 507 (1977).
4.281. D. P. Burum, D. D. Elleman, and W.-K. Rhim, *J. Chem. Phys.* **68**, 1164 (1978).
4.282. P. Mansfield, K. H. B. Richards, and D. Ware, *Phys. Rev.* **B1**, 2048 (1970).
4.283. M. E. Stoll, W.-K. Rhim, and R. W. Vaughan, *J. Chem. Phys.* **64**, 4808 (1976).
4.284. J. B. Grutzner and R. E. Santini, *J. mag. Reson.* **19**, 173 (1975).
4.285. V. J. Basus, P. D. Ellis, H. D. W. Hill, and J. S. Waugh, *J. mag. Reson.* **35**, 19 (1979).
4.286. J. D. Ellett and J. S. Waugh, *J. Chem. Phys.* **51**, 2851 (1969).
4.287. W. P. Aue, D. J. Ruben, and R. G. Griffin, *J. mag. Reson.* **46**, 354 (1982).
4.288. W. P. Aue, D. P. Burum, and R. R. Ernst, *J. mag. Reson.* **38**, 376 (1980).
4.289. D. P. Burum, unpublished.
4.290. M. H. Levitt, G. Bodenhausen, and R. R. Ernst, *J. mag. Reson.* **53**, 443 (1983).
4.291. D. L. Turner, D.Phil thesis, Oxford University (1977).
4.292. D. T. Pegg, M. R. Bendall, and D. M. Doddrell, *J. mag. Reson.* **49**, 32 (1982).
4.293. W. R. Hamilton, *Proc. R. Irish Acad.* **2**, 424 (1944), In W. R. Hamilton, *Mathematical papers*, Vol. 3, Cambridge University Press (1967).
4.294. B. Blümich and H. Spiess, *J. mag. Reson.* **61**, 356 (1985).
4.295. C. Counsell, M. H. Levitt, and R. R. Ernst, *J. mag. Reson.*, **63**, 133 (1985).
4.296. J. P. Elliott and P. G. Dawber, *Symmetry in physics*, Vol. 2, pp. 480–3, Macmillan, London (1979).
4.297. W. K. Rhim, D. P. Burum, and D. D. Elleman, *Phys. Rev. Lett.* **37**, 1764 (1976).
4.298. J. Jeener and P. Broekaert, *Phys. Rev.* **157**, 232 (1967).

Chapter 5

5.1. S. Yatsiv, *Phys. Rev.* **113,** 1522 (1952).
5.2. W. A. Anderson, R. Freeman, and C. A. Reilly, *J. Chem. Phys.* **39,** 1518 (1963).
5.3. H. Hatanaka, T. Terao, and T. Hashi, *J. Phys. Soc. Japan* **39,** 835 (1975).
5.4. H. Hatanaka and T. Hashi, *J. Phys. Soc. Japan* **39,** 1139 (1975).
5.5. H. Hatanaka, T. Ozawa, and T. Hashi, *J. Phys. Soc. Japan* **42,** 2069 (1977).
5.6. H. Hatanaka and T. Hashi, *Phys. Lett.* **67A,** 183 (1978).
5.7. S. Vega, T. W. Shattuck, and A. Pines, *Phys. Rev. Lett.* **37,** 43 (1976).
5.8. S. Vega and A. Pines, *J. Chem. Phys.* **66,** 5624 (1977).
5.9. A. Pines, D. Wemmer, J. Tang, and S. Sinton, *Bull. Am. Phys. Soc.* **23,** 21 (1978).
5.10. G. Drobny, A. Pines, S. Sinton, D. Weitekamp, and D. Wemmer, *Faraday Div. Chem. Soc. Symp.* **13,** 49 (1979).
5.11. W. S. Warren, S. Sinton, D. P. Weitekamp, and A. Pines, *Phys. Rev. Lett.* **43,** 1791 (1979).
5.12. S. Sinton and A. Pines, *Chem. Phys. Lett.* **76,** 263 (1980).
5.13. J. Tang and A. Pines, *J. Chem. Phys.* **72,** 3290 (1980).
5.14. W. S. Warren, D. P. Weitekamp, and A. Pines, *J. Chem. Phys.* **73,** 2084 (1980).
5.15. W. S. Warren and A. Pines, *J. Chem. Phys.* **74,** 2808 (1981).
5.16. W. S. Warren and A. Pines, *Chem. Phys. Lett.* **88,** 441 (1982).
5.17. J. B. Murdoch and A. Pines, *J. Am. chem. Soc.* **103,** 3578 (1981).
5.18. J. R. Garbow, D. P. Weitekamp, and A. Pines, *J. Chem. Phys.* **79,** 5301 (1983).
5.19. D. P. Weitekamp, J. R. Garbow, and A. Pines, *J. mag. Reson.* **46,** 529 (1982).
5.20. D. P. Weitekamp, J. R. Garbow, and A. Pines, *J. Chem. Phys.* **77,** 2870 (1982).
5.21. D. Zax and A. Pines, *J. Chem. Phys.* **78,** 6333 (1983).
5.22. W. P. Aue, E. Bartholdi, and R. R. Ernst, *J. Chem. Phys.* **64,** 2229 (1976).
5.23. A. Wokaun and R. R. Ernst, *Chem. Phys. Lett.* **52,** 407 (1977).
5.24. A. Wokaun and R. R. Ernst, *J. Chem. Phys.* **67,** 1752 (1977).
5.25. A. Wokaun and R. R. Ernst, *Mol. Phys.* **36,** 317 (1978).
5.26. A. Wokaun and R. R. Ernst, *Mol. Phys.* **38,** 1579 (1979).
5.27. P. Brunner, M. Reinhold, and R. R. Ernst, *J. Chem. Phys.* **73,** 1086 (1980).
5.28. A. Minoretti, W. P. Aue, M. Reinhold, and R. R. Ernst, *J. mag. Reson.* **40,** 175 (1980).
5.29. S. Macura, Y. Huang, D. Suter, and R. R. Ernst, *J. mag. Reson.* **43,** 259 (1981).
5.30. M. Reinhold, P. Brunner, and R. R. Ernst, *J. Chem. Phys.* **74,** 184 (1981).
5.31. M. Hintermann, L. Braunschweiler, G. Bodenhausen, and R. R. Ernst, *J. mag. Reson.* **50,** 316 (1982).
5.32. S. Macura, K. Wüthrich, and R. R. Ernst, *J. mag. Reson.* **46,** 269 (1982).

5.33. S. Macura, K. Wüthrich, and R. R. Ernst, *J. mag. Reson.* **47,** 351 (1982).
5.34. U. Piantini, O. W. Sørensen, and R. R. Ernst, *J. Am. chem. Soc.* **104,** 6800 (1982).
5.35. O. W. Sørensen, M. H. Levitt, and R. R. Ernst, *J. mag. Reson.* **55,** 104 (1983).
5.36. H. Kessler, H. Oschkinat, O. W. Sørensen, H. Kogler, and R. R. Ernst, *J. mag. Reson.* **55,** 329 (1983).
5.37. L. Braunschweiler, G. Bodenhausen, and R. R. Ernst, *Mol. Phys.* **48,** 535 (1983).
5.38. O. W. Sørensen, G. W. Eich, M. H. Levitt, G. Bodenhausen, and R. R. Ernst, *Prog. NMR Spectrosc.* **16,** 163 (1983).
5.39. M. H. Levitt and R. R. Ernst, *Chem. Phys. Lett.* **100,** 119 (1983).
5.40. G. Bodenhausen, H. Kogler, and R. R. Ernst, *J. mag. Reson.* **58,** 370 (1984).
5.41. M. Rance, O. W. Sørensen, G. Bodenhausen, G. Wagner, R. R. Ernst, and K. Wüthrich, *Biochem. Biophys. Res. Commun.* **117,** 479 (1983).
5.42. R. Poupko, R. L. Vold, and R. R. Vold, *J. mag. Reson.* **34,** 67 (1979).
5.43. G. Bodenhausen, N. M. Szeverenyi, R. L. Vold, and R. R. Vold, *J. Am. chem. Soc.* **100,** 6265 (1978).
5.44. G. Bodenhausen, R. L. Vold, and R. R. Vold, *J. mag. Reson.* **37,** 93 (1980).
5.45. R. L. Vold, R. R. Vold, R. Poupko, and G. Bodenhausen, *J. mag. Reson.* **38,** 141 (1980).
5.46. G. Bodenhausen, *J. mag. Reson.* **34,** 357 (1979).
5.47. D. Jaffe, R. L. Vold, and R. R. Vold, *J. Chem. Phys.* **78,** 4852 (1983).
5.48. A. Bax, R. Freeman, and S. P. Kempsell, *J. Am. chem. Soc.* **102,** 4849 (1980).
5.49. A. Bax, R. Freeman, and S. P. Kempsell, *J. mag. Reson.* **41,** 349 (1980).
5.50. A. Bax and R. Freeman, *J. mag. Reson.* **41,** 507 (1980).
5.51. A. Bax, R. Freeman, and T. Frenkiel, *J. Am. chem. Soc.* **103,** 2102 (1981).
5.52. A. Bax, R. Freeman, T. Frenkiel, and M. H. Levitt, *J. mag. Reson.* **43,** 478 (1981).
5.53. R. Freeman, T. Frenkiel, and M. B. Rubin, *J. Am. chem. Soc.* **104,** 5545 (1982).
5.54. T. H. Mareci and R. Freeman, *J. mag. Reson.* **48,** 158 (1982).
5.55. T. H. Mareci and R. Freeman, *J. mag. Reson.* **51,** 531 (1983).
5.56. A. J. Shaka and R. Freeman, *J. mag. Reson.* **51,** 169 (1983).
5.57. S. Vega, *J. Chem. Phys.* **68,** 5518 (1978).
5.58. S. Vega and Y. Naor, *J. Chem. Phys.* **75,** 75 (1981).
5.59. Y. Zur and S. Vega, *J. Chem. Phys.* **79,** 548 (1983).
5.60. G. Bodenhausen, *Prog. NMR Spectrosc.* **14,** 137 (1981).
5.61. D. P. Weitekamp, *Adv. mag. Reson.* **11,** 111 (1983).
5.62. R. C. Hewitt, S. Meiboom, and L. C. Snyder, *J. Chem. Phys.* **58,** 5089 (1973); L. C. Snyder and S. Meiboom, *J. Chem. Phys.* **58,** 5096 (1973).
5.63. A. Pines, D. J. Ruben, S. Vega, and M. Mehring, *Phys. Rev. Lett.* **36,** 110 (1976).
5.64. A. Pines, S. Vega, and M. Mehring, *Phys. Rev.* **B18,** 112 (1978).

5.65. P. L. Corio, *Structure of high-resolution NMR spectra*, Academic Press, New York (1966).
5.66. R. G. Jones, *NMR Basic Principles Prog.* **1**, 100 (1969).
5.67. P. Bucci, G. Ceccarelli, and C. A. Veracini, *J. Chem. Phys.* **50**, 1510 (1969).
5.68. J. I. Kaplan and S. Meiboom, *Phys. Rev.* **106**, 499 (1957).
5.69. A. D. Cohen and D. H. Whiffen, *Mol. Phys.* **7**, 449 (1964).
5.70. J. I. Musher, *J. Chem. Phys.* **40**, 983 (1964).
5.71. M. L. Martin, G. J. Martin, and R. Couffignal, *J. Chem. Phys.* **49**, 1985 (1968).
5.72. A. Abragam, *Principles of nuclear magnetism*, Oxford University Press, London (1961).
5.73. M. Rance, O. W. Sørensen, W. Leupin, H. Kogler, K. Wüthrich, and R. R. Ernst, *J. mag. Reson.* **61**, 67 (1985).
5.74. A. Wokaun, Ph.D. Thesis, ETH, Zurich (1978).
5.75. M. H. Levitt and R. R. Ernst, *J. Chem. Phys.* **83**, 3297 (1985).
5.76. W. K. Rhim, A. Pines, and J. S. Waugh, *Phys. Rev. Lett.* **25**, 218 (1970).
5.77. W. K. Rhim, A. Pines, and J. S. Waugh, *Phys. Rev.* **B3**, 684 (1971).
5.78. B. C. Sanctuary, *J. Chem. Phys.* **64**, 4352 (1976).
5.79. B. C. Sanctuary, *J. Chem. Phys.* **73**, 1048 (1980).
5.80. B. C. Sanctuary and F. Garisto, *J. Chem. Phys.* **73**, 2927 (1980).
5.81. B. C. Sanctuary, *Mol. Phys.* **48**, 1155 (1983).
5.82. B. C. Sanctuary, T. K. Halstead, and P. A. Osment, *Mol. Phys.* **49**, 753 (1983).
5.83. A. D. Bain, *Chem. Phys. Lett.* **57**, 281 (1978).
5.84. A. D. Bain, *J. mag. Reson.* **39**, 335 (1980).
5.85. A. D. Bain and S. Brownstein, *J. mag. Reson.* **47**, 409 (1982).
5.86. A. D. Bain, *J. mag. Reson.* **56**, 418 (1984).
5.87. H. W. Spiess, In *NMR Basic Principles Prog.* **15**, (1980).
5.88. S. Hsi, H. Zimmermann, and Z. Luz, *J. Chem. Phys.* **69**, 4126 (1978).
5.89. K. M. Worvill, *J. mag. Reson.* **18**, 217 (1975).
5.90. S. Vega and A. Pines, In *Proc. 19th Ampère Congress*, Heidelberg, 1976, (ed. H. Brunner, K. H. Hausser and D. Schweizer) p. 395 (1976).
5.91. A. A. Maudsley, A. Wokaun, and R. R. Ernst, *Chem. Phys. Lett.* **55**, 9 (1978).

Chapter 6

6.1. J. Jeener, Ampère International Summer School, Basko Polje, Yugoslavia (1971).
6.2. R. R. Ernst, VIth International Conference on Magnetic Resonance in Biological Systems, Kandersteg, Switzerland (1974).
6.3. R. R. Ernst, *Chimia* **29**, 179 (1975).
6.4. L. Müller, Anil Kumar, and R. R. Ernst, *J. Chem. Phys.* **63**, 5490 (1975).
6.5. W. P. Aue, E. Bartholdi, and R. R. Ernst, *J. Chem. Phys.* **64**, 2229 (1976).
6.6. B. Blümich and D. Ziessow, *Ber. Bunsenges. Phys. Chem.* **84**, 1090 (1980).

6.7. B. Blümich and D. Ziessow, *J. Chem. Phys.* **78,** 1059 (1983).
6.8. A. D. Bain, *J. mag. Reson.* **56,** 418 (1984).
6.9. G. Bodenhausen, H. Kogler, and R. R. Ernst, *J. mag. Reson.* **58,** 370 (1984).
6.10. A. Wokaun and R. R. Ernst, *Chem. Phys. Lett.* **52,** 407 (1977).
6.11. G. Bodenhausen, R. Freeman, and D. L. Turner, *J. mag. Reson.* **27,** 511 (1977).
6.12. K. Nagayama, Anil Kumar, K. Wüthrich, and R. R. Ernst, *J. mag. Reson.* **40,** 321 (1980).
6.13. A. Bax and G. A. Morris, *J. mag. Reson.* **42,** 501 (1981).
6.14. P. H. Bolton and G. Bodenhausen, *J. mag. Reson.* **46,** 306 (1982).
6.15. R. Bracewell, *The Fourier transform and its applications,* McGraw-Hill, New York (1965).
6.16. D. C. Champeney, *Fourier transforms and their physical application,* Academic Press, New York (1973).
6.17. K. Nagayama, P. Bachmann, K. Wüthrich, and R. R. Ernst, *J. mag. Reson.* **31,** 133 (1978).
6.18. E. Bartholdi and R. R. Ernst, *J. mag. Reson.* **11,** 9 (1973).
6.19. R. N. Bracewell, *Aust. J. Phys.* **9,** 198 (1956).
6.20. R. M. Merserau and A. V. Oppenheim, *Proc. IEEE* **62,** 1319 (1974).
6.21. P. Bachmann, W. P. Aue, L. Müller, and R. R. Ernst, *J. mag. Reson.* **28,** 29 (1977).
6.22. G. Bodenhausen, R. Freeman, R. Niedermeyer, and D. L. Turner, *J. mag. Reson.* **26,** 133 (1977).
6.23. M. H. Levitt, G. Bodenhausen, and R. R. Ernst, *J. mag. Reson.* **58,** 462 (1984).
6.24. A. A. Maudsley, A. Wokaun, and R. R. Ernst, *Chem. Phys. Lett.* **55,** 9 (1978).
6.25. A. Bax and R. Freeman, *J. mag. Reson.* **44,** 542 (1981).
6.26. L. Braunschweiler, G. Bodenhausen, and R. R. Ernst, *Mol. Phys.* **48,** 535 (1983).
6.27. R. Freeman, S. P. Kempsell, and M. H. Levitt, *J. mag. Reson.* **34,** 663 (1979).
6.28. D. J. States, R. A. Haberkorn, and D. J. Ruben, *J. mag. Reson.* **48,** 286 (1982).
6.29. P. H. Bolton, In *Biological magnetic resonance* (ed. L. J. Berliner and J. Reuben), Plenum, New York (1984).
6.30. A. Bax, A. F. Mehlkopf, and J. Smidt, *J. mag. Reson.* **35,** 373 (1979).
6.31. A. Bax, R. Freeman, and G. A. Morris, *J. mag. Reson.* **43,** 333 (1981).
6.32. A. J. Shaka, J. Keeler, and R. Freeman, *J. mag. Reson.* **56,** 294 (1984).
6.33. B. Blümich and D. Ziessow, *J. mag. Reson.* **49,** 151 (1982).
6.34. R. R. Ernst, *Adv. mag. Reson.* **2,** 1 (1966).
6.35. W. P. Aue, P. Bachmann, A. Wokaun, and R. R. Ernst, *J. mag. Reson.* **29,** 523 (1978).
6.36. J. Max, *Traitement du signal,* Masson, Paris (1972).
6.37. A. G. Ferrige and J. C. Lindon, *J. mag. Reson.* **31,** 337 (1978).
6.38. J. C. Lindon and A. G. Ferrige, *Prog. NMR Spectrosc.* **14,** 27 (1982).
6.39. A. Bax, *Two-dimensional nuclear magnetic resonance in liquids,* Delft Univ. Press, Dordrecht (1982).

6.40. A. de Marco and K. Wüthrich, *J. mag. Reson.* **24**, 201 (1976).
6.41. G. Wagner, K. Wüthrich, and H. Tschesche, *Eur. J. Biochem.* **86**, 67 (1978).
6.42. B. Clin, J. de Bony, P. Lalanne, J. Biais, and B. Lemanceau, *J. mag. Reson.* **33**, 457 (1979).
6.43. G. Wider, S. Macura, Anil Kumar, R. R. Ernst, and K. Wüthrich, *J. mag. Reson.* **56**, 207 (1984).
6.44. A. Bax and T. H. Mareci, *J. mag. Reson.* **53**, 360 (1983).
6.45. A. Bax, R. H. Griffey, and B. L. Hawkins, *J. mag. Reson.* **55**, 301 (1983).
6.46. W. P. Aue, J. Karhan, and R. R. Ernst, *J. Chem. Phys.* **64**, 4226 (1976),
6.47. A. G. Redfield and S. D. Kunz, *J. mag. Reson.* **19**, 250 (1975).
6.48. G. Bodenhausen, R. Freeman, G. A. Morris, R. Niedermeyer, and D. L. Turner, *J. mag. Reson.* **25**, 559 (1977).
6.49. G. Drobny, A. Pines, S. Sinton, D. Weitekamp, and D. Wemmer, *Faraday Div. Chem. Soc. Symp.* **13**, 49 (1979).
6.50. G. Bodenhausen, R. L. Vold, and R. R. Vold, *J. mag. Reson.* **37**, 93 (1980).
6.51. D. P. Weitekamp, *Adv. mag. Reson.* **11**, 111 (1983).
6.52. R. Baumann, Anil Kumar, R. R. Ernst, and K. Wüthrich, *J. mag. Reson.* **44**, 76 (1981).
6.53. R. Baumann, G. Wider, R. R. Ernst, and K. Wüthrich, *J. mag. Reson.* **44**, 402 (1981).
6.54. B. U. Meier, G. Bodenhausen, and R. R. Ernst, *J. mag. Reson.* **60**, 161 (1984); P. Pfändler, G. Bodenhausen, B. U. Meier, and R. R. Ernst, *Analyt. Chem.* **57**, 2510 (1985).
6.55. D. L. Turner, *J. mag. Reson.* **49**, 175 (1982).
6.56. D. L. Turner, *J. mag. Reson.* **53**, 259 (1983).
6.57. O. W. Sørensen, G. W. Eich, M. H. Levitt, G. Bodenhausen, and R. R. Ernst, *Prog. NMR Spectrosc.* **16**, 163 (1983).
6.58. G. Bodenhausen and P. H. Bolton, *J. mag. Reson.* **39**, 399 (1980).
6.59. G. Bodenhausen and R. R. Ernst, *J. Am. chem. Soc.* **104**, 1304 (1982).
6.60. A. Bax, P. G. de Jong, A. F. Mehlkopf, and J. Smidt, *Chem. Phys. Lett.* **69**, 567 (1980).
6.61. J. E. Bertie, In *Analytical applications of Fourier transform infrared spectroscopy to molecular and biological systems* (ed. J. R. Durig), p. 25 Reidel, Dordrecht (1980).
6.62. O. W. Sørensen, M. Rance, and R. R. Ernst, *J. mag. Reson.* **56**, 527 (1984).
6.63. C. J. R. Counsell, M. H. Levitt, and R. R. Ernst, *J. mag. Reson.* **64**, 470 (1985).
6.64. R. R. Ernst, W. P. Aue, P. Bachmann, J. Karhan, Anil Kumar, and L. Müller, *Proc. 4th Ampère Int. Summer School*, Pula, Yugoslavia, 89 (1976).

Chapter 7

7.1. L. Müller, Anil Kumar, and R. R. Ernst, *J. Chem. Phys.* **63**, 5490 (1975).
7.2. W. P. Aue, J. Karhan, and R. R. Ernst, *J. Chem. Phys.* **64**, 4226 (1976).

7.3. K. Nagayama, P. Bachmann, K. Wüthrich, and R. R. Ernst, *J. mag. Reson.* **31,** 133 (1978).
7.4. W. P. Aue and R. R. Ernst, *J. mag. Reson.* **31,** 533 (1978).
7.5. A. Bax, A. F. Mehlkopf, and J. Smidt, *J. mag. Reson.* **35,** 167 (1979).
7.6. H. Y. Carr and E. M. Purcell, *Phys. Rev.* **94,** 630 (1954).
7.7. A. Allerhand, *J. Chem. Phys.* **44,** 1 (1966).
7.8. A. Bax, A. F. Mehlkopf, and J. Smidt, *J. mag. Reson.* **40,** 213 (1980).
7.9. K. Nagayama, *J. Chem. Phys.* **71,** 4404 (1979).
7.10. G. Bodenhausen, R. Freeman, R. Niedermeyer, and D. L. Turner, *J. mag. Reson.* **26,** 133 (1977).
7.11. G. Bodenhausen, R. Freeman, and D. L. Turner, *J. mag. Reson.* **27,** 511 (1977).
7.12. G. Bodenhausen, H. Kogler, and R. R. Ernst, *J. mag. Reson.* **58,** 370 (1984).
7.13. L. Müller, *J. mag. Reson.* **36,** 301 (1979).
7.14. K. Nagayama, Anil Kumar, K. Wüthrich, and R. R. Ernst, *J. mag. Reson.* **40,** 321 (1980).
7.15. G. Bodenhausen, R. Freeman, and D. L. Turner, *J. Chem. Phys.* **65,** 839 (1976).
7.16. G. Bodenhausen, R. Freeman, R. Niedermeyer, and D. L. Turner, *J. mag. Reson.* **24,** 291 (1976).
7.17. L. Müller, Anil Kumar, and R. R. Ernst, *J. mag. Reson.* **25,** 383 (1977).
7.18. R. Freeman, G. A. Morris, and D. L. Turner, *J. mag. Reson.* **26,** 373 (1977).
7.19. R. Freeman, S. P. Kempsell, and M. H. Levitt, *J. mag. Reson.* **34,** 663 (1979).
7.20. G. Bodenhausen, R. Freeman, G. A. Morris, and D. L. Turner, *J. mag. Reson.* **28,** 17 (1977).
7.21. A. Bax and R. Freeman, *J. Am. chem. Soc.* **104,** 1099 (1982).
7.22. J. R. Garbow, D. P. Weitekamp, and A. Pines, *Chem. Phys. Lett.* **93,** 504 (1982).
7.23. A. Bax, *J. mag. Reson.* **53,** 517 (1983).
7.24. C. Bauer, R. Freeman, and S. Wimperis, *J. mag. Reson.* **58,** 526 (1984).
7.25. O. W. Sørensen, G. W. Eich, M. H. Levitt, G. Bodenhausen, and R. R. Ernst, *Prog. NMR Spectrosc.* **16,** 163 (1983).
7.26. V. Rutar, *J. mag. Reson.* **56,** 87 (1984).
7.27. G. Bodenhausen and D. L. Turner, *J. mag. Reson.* **41,** 200 (1980).
7.28. R. Freeman and J. Keeler, *J. mag. Reson.* **43,** 484 (1981).
7.29. P. Bachmann, W. P. Aue, L. Müller, and R. R. Ernst, *J. mag. Reson.* **28,** 29 (1977).
7.30. G. Bodenhausen, R. Freeman, G. A. Morris, and D. L. Turner, *J. mag. Reson.* **31,** 75 (1978).
7.31. G. Wider, R. Baumann, K. Nagayama, R. R. Ernst, and K. Wüthrich, *J. mag. Reson.* **42,** 73 (1981).
7.32. Anil Kumar and C. L. Khetrapal, *J. mag. Reson.* **30,** 137 (1978).
7.33. C. L. Khetrapal, Anil Kumar, A. C. Kunwar, P. C. Mathias, and K. V. Ramanathan, *J. mag. Reson.* **37,** 349 (1980).
7.34. J. W. Emsley, 7th European Experimental NMR Conference, Palermo (1984).

REFERENCES

7.35. Anil Kumar and R. R. Ernst, *Chem. Phys. Lett.* **37**, 162 (1976).
7.36. R. R. Vold and R. L. Vold, *J. Chem. Phys.* **64**, 320 (1976).
7.37. Anil Kumar and R. R. Ernst, *J. mag. Reson.* **24**, 425 (1976).
7.38. R. K. Hester, J. L. Ackerman, B. L. Neff, and J. S. Waugh, *Phys. Rev. Lett.* **36**, 1081 (1976).
7.39. E. F. Rybaczewski, B. L. Neff, J. S. Waugh, and J. S. Sherfinski, *J. Chem. Phys.* **67**, 1231 (1977).
7.40. M. E. Stoll, A. J. Vega, and R. W. Vaughan, *J. Chem. Phys.* **65**, 4093 (1976).
7.41. U. Haeberlen, *High resolution NMR in solids, Adv. mag. Reson.* Suppl. 1 (1976).
7.42. M. Mehring, *High resolution NMR spectroscopy in solids*, 2nd edn, Springer, Berlin (1983).
7.43. N. Schuff and U. Haeberlen, *J. mag. Reson.* **52**, 267 (1983).
7.44. D. P. Burum, M. Linder, and R. R. Ernst, *J. mag. Reson.* **44**, 173 (1981).
7.45. P. Caravatti, G. Bodenhausen, and R. R. Ernst, *Chem. Phys. Lett.* **89**, 363 (1982).
7.46. P. Caravatti, L. Braunschweiler, and R. R. Ernst, *Chem. Phys. Lett.* **100**, 305 (1983).
7.47. G. Bodenhausen, R. E. Stark, D. J. Ruben, and R. G. Griffin, *Chem. Phys. Lett.* **67**, 424 (1979).
7.48. M. Lee and W. I. Goldburg, *Phys. Rev.* **A140**, 1261 (1965).
7.49. M. Linder, A. Höhener, and R. R. Ernst, *J. Chem. Phys.* **73**, 4959 (1980).
7.50. M. G. Munowitz, R. G. Griffin, G. Bodenhausen, and T. H. Huang, *J. Am. chem. Soc.* **103**, 2529 (1981).
7.51. R. G. Griffin, G. Bodenhausen, R. A. Haberkorn, T. H. Huang, M. G. Munowitz, R. Osredkar, D. J. Ruben, R. E. Stark, and H. van Willigen, *Phil. Trans. R. Soc., London* **A299**, 547 (1981).
7.52. M. G. Munowitz and R. G. Griffin, *J. Chem. Phys.* **76**, 2848 (1982).
7.53. M. G. Munowitz and R. G. Griffin, *J. Chem. Phys.* **78**, 613 (1983).
7.54. M. G. Munowitz, W. P. Aue, and R. G. Griffin, *J. Chem. Phys.* **77**, 1686 (1982).
7.55. W. P. Aue, D. J. Ruben, and R. G. Griffin, *J. Chem. Phys.* **80**, 1729 (1984).
7.56. J. Schaefer, R. A. McKay, E. O. Stejskal, and W. T. Dixon, *J. mag. Reson.* **52**, 123 (1983).
7.57. M. M. Maricq and J. S. Waugh, *J. Chem. Phys.* **70**, 3300 (1979).
7.58. J. Herzfeld and A. M. Berger, *J. Chem. Phys.* **73**, 6021 (1980).
7.59. K. W. Zilm and D. M. Grant, *J. mag. Reson.* **48**, 524 (1982).
7.60. T. Terao, H. Miura and A. K. Saika, *J. Am. chem. Soc.* **104**, 5228 (1982).
7.61. C. L. Mayne, R. J. Pugmire, and D. M. Grant, *J. mag. Reson.* **56**, 151 (1984).
7.62. E. Lippmaa, M. Alla, and T. Tuherm, in *Proc. 19th Ampère Congress*, Heidelberg, 1976.
7.63. Y. Yarim-Agaev, P. N. Tutunjian, and J. S. Waugh, *J. mag. Reson.* **47**, 51 (1982).
7.64. A. Bax, N. M. Szeverenyi, and G. E. Maciel, *J. mag. Reson.* **51**, 400 (1983).

7.65. W. P. Aue, D. J. Ruben, and R. G. Griffin, *J. mag. Reson.* **43,** 472 (1981).
7.66. A. Bax, N. M. Szeverenyi, and G. E. Maciel, *J. mag. Reson.* **55,** 494 (1983).
7.67. A. Bax, N. M. Szeverenyi, and G. E. Maciel, *J. mag. Reson.* **52,** 147 (1983).
7.68. B. H. Meier, F. Graf, and R. R. Ernst, *J. Chem. Phys.* **76,** 767 (1982).
7.69. Anil Kumar, W. P. Aue, P. Bachmann, J. Karhan, L. Müller, and R. R. Ernst, *Proc. 19th Ampère Congress,* Heidelberg, p. 473 (1976).

Chapter 8

8.1. J. Jeener, Ampère International Summer School, Basko Polje, Yugoslavia (1971).
8.2. W. P. Aue, E. Bartholdi, and R. R. Ernst, *J. Chem. Phys.* **64,** 2229 (1976).
8.3. K. Nagayama, K. Wüthrich, and R. R. Ernst, *Biochem. Biophys. Res. Commun.* **90,** 305 (1979).
8.4. G. Wagner, Anil Kumar, and K. Wüthrich, *Eur. J. Biochem.* **114,** 375 (1981).
8.5. A. Bax and R. Freeman, *J. mag. Reson.* **44,** 542 (1981).
8.6. G. W. Eich, G. Bodenhausen, and R. R. Ernst, *J. Am. chem. Soc.* **104,** 3731 (1982).
8.7. P. H. Bolton and G. Bodenhausen, *Chem. Phys. Lett.* **89,** 139 (1982).
8.8. L. Braunschweiler, G. Bodenhausen, and R. R. Ernst, *Mol. Phys.* **48,** 535 (1983).
8.9. A. A. Maudsley and R. R. Ernst, *Chem. Phys. Lett.* **50,** 368 (1977).
8.10. A. A. Maudsley, L. Müller, and R. R. Ernst, *J. mag. Reson.* **28,** 463 (1977).
8.11. G. Bodenhausen and R. Freeman, *J. mag. Reson.* **28,** 471 (1977).
8.12. G. Bodenhausen and D. J. Ruben, *Chem. Phys. Lett.* **69,** 185 (1980).
8.13. A. Bax, R. H. Griffey, and B. L. Hawkins, *J. mag. Reson.* **55,** 301 (1983).
8.14. G. Bodenhausen and R. Freeman, *J. mag. Reson.* **36,** 221 (1979).
8.15. O. W. Sørensen, G. W. Eich, M. H. Levitt, G. Bodenhausen, and R. R. Ernst, *Prog. NMR Spectrosc.* **16,** 163 (1983).
8.16. S. Schäublin, A. Höhener, and R. R. Ernst, *J. mag. Reson.* **13,** 196 (1974).
8.17. G. Wagner and K. Wüthrich, *J. mol. Biol.* **155,** 347 (1982).
8.18. D. Marion and K. Wüthrich, *Biochem. Biophys. Res. Commun.* **113,** 967 (1983).
8.19. C. Brévard, R. Schimpf, G. Tourne, and C. M. Tourne, *J. Am. chem. Soc.* **105,** 7059 (1983).
8.20. T. L. Venable, W. C. Hutton, and R. N. Grimes, *J. Am. chem. Soc.* **104,** 4716 (1982); **106,** 29 (1984).
8.21. G. Bodenhausen and P. H. Bolton, *J. mag. Reson.* **39,** 399 (1980).
8.22. K. Nagayama, Anil Kumar, K. Wüthrich, and R. R. Ernst, *J. mag. Reson.* **40,** 321 (1980).
8.23. K. Nagayama, K. Wüthrich, and R. R. Ernst, *Biochem. Biophys. Res. Commun.* **90,** 305 (1979).

8.24. G. Wider, S. Macura, Anil Kumar, R. R. Ernst, and K. Wüthrich, *J. mag. Reson.* **56,** 207 (1984).
8.25. O. W. Sørensen, M. Rance, and R. R. Ernst, *J. mag. Reson.* **56,** 527 (1984).
8.26. A. Bax and R. Freeman, *J. mag. Reson.* **44,** 542 (1981).
8.27. M. Rance, G. Wagner, O. W. Sørensen, K. Wüthrich, and R. R. Ernst, *J. mag. Reson.* **59,** 250 (1984).
8.28. U. Piantini, O. W. Sørensen, and R. R. Ernst, *J. Am. chem. Soc.* **104,** 6800 (1982).
8.29. A. J. Shaka and R. Freeman, *J. mag. Reson.* **51,** 169 (1983).
8.30. M. Rance, O. W. Sørensen, G. Bodenhausen, G. Wagner, R. R. Ernst, and K. Wüthrich, *Biochem. Biophys. Res. Commun.* **117,** 479 (1983).
8.31. B. U. Meier, G. Bodenhausen, and R. R. Ernst, *J. mag. Reson.* **60,** 161 (1984).
8.32. O. W. Sørensen, M. H. Levitt, and R. R. Ernst, *J. mag. Reson.* **55,** 104 (1983).
8.33. D. P. Weitekamp, J. R. Garbow, J. B. Murdoch, and A. Pines, *J. Am. chem. Soc.* **103,** 3578 (1981).
8.34. J. R. Garbow, D. P. Weitekamp, and A. Pines, *J. Chem. Phys.* **79,** 5301 (1983).
8.35. D. P. Weitekamp, *Adv. mag. Reson.* **11,** 111 (1983).
8.36. M. H. Levitt and R. R. Ernst, *Chem. Phys. Lett.* **100,** 119 (1983).
8.37. M. H. Levitt and R. R. Ernst, *J. Chem. Phys.* **83,** 3297 (1985).
8.38. G. Wagner, *J. mag. Reson.* **55,** 151 (1983).
8.39. G. Wagner, *J. mag. Reson.* **57,** 497 (1984).
8.40. G. King and P. E. Wright, *J. mag. Reson.* **54,** 328 (1983).
8.41. O. W. Sørensen and R. R. Ernst, *J. mag. Reson.* **55,** 338 (1983).
8.42. H. Kogler, O. W. Sørensen, G. Bodenhausen, and R. R. Ernst, *J. mag. Reson.* **55,** 157 (1983).
8.43. A. D. Cohen, R. Freeman, K. A. McLauchlan, and D. H. Whiffen, *Mol. Phys.* **7,** 45 (1963).
8.44. R. A. Hoffman, B. Gestblom, and S. Forsén, *J. mol. Spectrosc.* **13,** 221 (1964).
8.45. L. Braunschweiler and R. R. Ernst, *J. mag. Reson.* **53,** 521 (1983).
8.46. A. Bax, D. G. Davies, and S. K. Sarkar, *J. mag. Reson.* **63,** 230 (1985).
8.47. S. R. Hartmann and E. L. Hahn, *Phys. Rev.* **128,** 2042 (1962).
8.48. A. Pines, M. G. Gibby, and J. S. Waugh, *J. Chem. Phys.* **59,** 569 (1973).
8.49. R. D. Bertrand, W. B. Moniz, A. N. Garroway, and G. C. Chingas, *J. Am. chem. Soc.* **100,** 5227 (1978).
8.50. L. Müller and R. R. Ernst, *Mol. Phys.* **38,** 963 (1979).
8.51. W. S. Warren, D. P. Weitekamp, and A. Pines, *J. Chem. Phys.* **73,** 2084 (1980).
8.52. D. P. Weitekamp, J. R. Garbow, and A. Pines, *J. mag. Reson.* **46,** 529 (1982).
8.53. M. Rance, O. W. Sørensen, W. Leupin, H. Kogler, K. Wüthrich, and R. R. Ernst, *J. mag. Reson.* **61,** 67 (1985).
8.54. A. Bax, R. Freeman, and S. P. Kempsell, *J. Am. chem. Soc.* **102,** 4849 (1980).
8.55. A. Bax, R. Freeman, and S. P. Kempsell, *J. mag. Reson.* **41,** 349 (1980).

8.56. A. Bax and R. Freeman, *J. mag. Reson.* **41,** 507 (1980).
8.57. A. Bax, R. Freeman and T. Frenkiel, *J. Am. chem. Soc.* **103,** 2102 (1981).
8.58. A. Bax, R. Freeman, T. Frenkiel, and M. H. Levitt, *J. mag. Reson.* **43,** 478 (1981).
8.59. R. Freeman, T. Frenkiel, and M. B. Rubin, *J. Am. chem. Soc.* **104,** 5545 (1982).
8.60. T. H. Mareci and R. Freeman, *J. mag. Reson.* **48,** 158 (1982).
8.61. A. Bax, *Two-dimensional nuclear magnetic resonance in liquids,* Delft University Press/Reidel, Dordrecht (1982).
8.62. D. L. Turner, *J. mag. Reson.* **49,** 175 (1982).
8.63. D. L. Turner, *J. mag. Reson.* **53,** 259 (1983).
8.64. A. Bax and T. H. Mareci, *J. mag. Reson.* **53,** 360 (1983).
8.65. G. Bodenhausen, H. Kogler, and R. R. Ernst, *J. mag. Reson.* **58,** 370 (1984).
8.66. G. Pouzard, S. Sukumar, and L. D. Hall, *J. Am. chem. Soc.* **103,** 4209 (1981).
8.67. G. Wagner and E. R. P. Zuiderweg, *Biochem. Biophys. Res. Commun.* **113,** 854 (1983).
8.68. J. B. Boyd, C. M. Dobson, and C. Redfield, *J. mag. Reson.* **55,** 170 (1983).
8.69. S. Emid, A. Bax, J. Konijnendijk, J. Smidt, and A. Pines, *Physica* **B96,** 333 (1979); S. Emid, J. Smidt, and A. Pines, *Chem. Phys. Lett.* **73,** 496 (1980); Y.-S. Yen and A. Pines, *J. Chem. Phys.* **78,** 3579 (1983).
8.70. S. Sinton and A. Pines, *Chem. Phys. Lett.* **76,** 263 (1980).
8.71. W. K. Rhim, A. Pines, and J. S. Waugh, *Phys. Rev. Lett.* **25,** 218 (1970); W. K. Rhim, A. Pines, and J. S. Waugh, *Phys. Rev.* **B3,** 684 (1971).
8.72. G. Drobny, A. Pines, S. Sinton, D. Weitekamp, and D. Wemmer, *Faraday Div. chem. Soc. Symp.* **13,** 49 (1979).
8.73. G. Bodenhausen, R. L. Vold, and R. R. Vold, *J. mag. Reson.* **37,** 93 (1980).
8.74. W. S. Warren, S. Sinton, D. P. Weitekamp, and A. Pines, *Phys. Rev. Lett.* **43,** 1791 (1979).
8.75. W. S. Warren and A. Pines, *J. Chem. Phys.* **74,** 2808 (1981).
8.76. S. Vega and A. Pines, *J. Chem. Phys.* **66,** 5624 (1977).
8.77. S. Vega, T. W. Shattuck, and A. Pines, *Phys. Rev. Lett.* **37,** 43 (1976).
8.78. M. M. Maricq and J. S. Waugh, *J. Chem. Phys.* **70,** 3300 (1979).
8.79. R. Eckman, M. Alla, and A. Pines, *J. mag. Reson.* **41,** 440 (1980).
8.80. R. Eckman, L. Müller, and A. Pines, *Chem. Phys. Lett.* **74,** 376 (1980).
8.81. L. Müller, *J. Am. chem. Soc.* **101,** 4481 (1979).
8.82. A. Minoretti, W. P. Aue, M. Reinhold, and R. R. Ernst, *J. mag. Reson.* **40,** 175 (1980).
8.83. G. Bodenhausen, *J. mag. Reson.* **39,** 175 (1980).
8.84. G. A. Morris, *J. mag. Reson.* **44,** 277 (1981).
8.85. L. Müller, *Chem. Phys. Lett.* **91,** 303 (1982).
8.86. Y. S. Yen and D. P. Weitekamp, *J. mag. Reson.* **47,** 476 (1982).
8.87. P. H. Bolton, *J. mag. Reson.* **48,** 336 (1982).
8.88. H. Kessler, M. Bernd, H. Kogler, J. Zarbock, O. W. Sørensen, G. Bodenhausen, and R. R. Ernst, *J. Am. chem. Soc.* **105,** 6944 (1983).
8.89. A. G. Redfield, *Chem. Phys. Lett.* **96,** 537 (1983).

8.90. P. H. Bolton, *J. mag. Reson.* **52,** 326 (1983).
8.91. P. H. Bolton, *J. mag. Reson.* **54,** 333 (1983).
8.92. M. H. Levitt, O. W. Sørensen, and R. R. Ernst, *Chem. Phys. Lett.* **94,** 540 (1983).
8.93. M. A. Delsuc, E. Guittet, N. Trotin, and J. Y. Lallemand, *J. mag. Reson.* **56,** 163 (1984).
8.94. D. Neuhaus, G. Wider, G. Wagner, and K. Wüthrich, *J. mag. Reson.* **57,** 164 (1984).
8.95. S. Wimperis and R. Freeman, *J. mag. Reson.* **58,** 348 (1984).
8.96. H. Kessler, C. Griesinger, J. Zarbock, and H. R. Loosli, *J. mag. Reson.* **57,** 331 (1984).
8.97. L. Müller, Anil Kumar, T. Baumann, and R. R. Ernst, *Phys. Rev. Lett.* **32,** 1402 (1974).
8.98. P. Caravatti, L. Braunschweiler, and R. R. Ernst, *Chem. Phys. Lett.* **100,** 305 (1983).
8.99. D. P. Weitekamp, J. R. Garbow, and A. Pines, *J. Chem. Phys.* **77,** 2870 (1982).
8.100. G. C. Chingas, A. N. Garroway, R. D. Bertrand, and W. B. Moniz, *J. Chem. Phys.* **74,** 127 (1981).
8.101. P. Brunner, M. Reinhold, and R. R. Ernst, *J. Chem. Phys.* **73,** 1086 (1980).
8.102. M. Reinhold, P. Brunner, and R. R. Ernst, *J. Chem. Phys.* **74,** 184 (1981).
8.103. P. Caravatti, G. Bodenhausen, and R. R. Ernst, *Chem. Phys. Lett.* **89,** 363 (1982).
8.104. J. E. Roberts, S. Vega, and R. G. Griffin, *J. Am. chem. Soc.* **106,** 2506 (1984).
8.105. A. D. Bain, *J. mag. Reson.* **56,** 418 (1984).
8.106. H. Kogler, Ph.D. Thesis, Frankfurt University (1984).
8.107. A. Bax and G. A. Morris, *J. mag. Reson.* **42,** 501 (1981).
8.108. P. H. Bolton and G. Bodenhausen, *J. mag. Reson.* **46,** 306 (1982).
8.109. D. P. Burum and R. R. Ernst, *J. mag. Reson.* **39,** 163 (1980).
8.110. P. H. Bolton, *J. mag. Reson.* **51,** 134 (1983).
8.111. A. J. Shaka, J. Keeler, and R. Freeman, *J. mag. Reson.* **53,** 313 (1983).
8.112. P. H. Bolton and G. Bodenhausen, *J. Am. chem. Soc.* **101,** 1080 (1979).
8.113. P. H. Bolton, In *NMR of newly accessible nuclei* (ed. P. Laszlo), Vol. I, Chapter 2, Academic Press (1983).
8.114. P. H. Bolton, *J. mag. Reson.* **45,** 239 (1981).
8.115. J. R. Garbow, D. P. Weitekamp, and A. Pines, *Chem. Phys. Lett.* **93,** 504 (1982).
8.116. A. Bax, *J. mag. Reson.* **53,** 517 (1983).
8.117. V. Rutar, *J. mag. Reson.* **56,** 87 (1984).
8.118. M. F. Roberts, D. A. Vidusek, and G. Bodenhausen, *FEBS Lett.* **117,** 311 (1980).
8.119. D. A. Vidusek, M. F. Roberts, and G. Bodenhausen, *J. Am. chem. Soc.* **104,** 5452 (1982).
8.120. A. Bax, R. H. Griffey, and B. L. Hawkins, *J. Am. chem. Soc.* **105,** 7188 (1983).

8.121. B. L. Hawkins, Z. Yamaizumi, and S. Nishimura, *Proc. nat. Acad. Sci., USA,* **80,** 5895 (1983).
8.122. D. P. Burum, M. Linder, and R. R. Ernst, *J. mag. Reson.* **44,** 173 (1981).
8.123. R. R. Ernst, W. P. Aue, P. Bachmann, J. Karhan, Anil Kumar, and L. Müller, *Proc. 4th Ampère Int. Summer School,* Pula, Yugoslavia, 89 (1976).

Chapter 9

9.1. J. Jeener, B. H. Meier, P. Bachmann, and R. R. Ernst, *J. Chem. Phys.* **71,** 4546 (1979).
9.2. B. H. Meier and R. R. Ernst, *J. Am. chem. Soc.* **101,** 6441 (1979).
9.3. G. Bodenhausen and R. R. Ernst, *J. Am. chem. Soc.* **104,** 1304 (1982).
9.4. G. Bodenhausen, H. Kogler, and R. R. Ernst, *J. mag. Reson.* **58,** 370 (1984).
9.5. S. Macura and R. R. Ernst, *Mol. Phys.* **41,** 95 (1980).
9.6. S. Macura, Y. Huang, D. Suter, and R. R. Ernst, *J. mag. Reson.* **43,** 259 (1981).
9.7. S. Macura, K. Wüthrich, and R. R. Ernst, *J. mag. Reson.* **47,** 351 (1982).
9.8. O. W. Sørensen, G. W. Eich, M. H. Levitt, G. Bodenhausen, and R. R. Ernst, *Prog. NMR Spectrosc.* **16,** 163 (1983).
9.9. J. Jeener and P. Broekaert, *Phys. Rev.* **157,** 232 (1967).
9.10. G. Bodenhausen, G. Wagner, M. Rance, O. W. Sørensen, K. Wüthrich, and R. R. Ernst, *J. mag. Reson.* **59,** 542 (1984).
9.11. S. Macura, K. Wüthrich, and R. R. Ernst, *J. mag. Reson.* **46,** 269 (1982).
9.12. M. Rance, G. Bodenhausen, G. Wagner, K. Wüthrich, and R. R. Ernst, *J. mag. Reson.,* **62,** 497 (1985).
9.13. K. Nagayama, *J. mag. Reson.,* **51,** 84 (1983).
9.14. G. Bodenhausen and R. R. Ernst, *Mol. Phys.* **47,** 319 (1982).
9.15. Anil Kumar, G. Wagner, R. R. Ernst, and K. Wüthrich, *J. Am. chem. Soc.* **103,** 3654 (1981).
9.16. G. Bodenhausen and R. R. Ernst, *J. mag. Reson.* **45,** 367 (1981).
9.17. F. R. Jensen and B. H. Beck, *Tetrahedron Lett.* **1966,** 4523 (1966).
9.18. D. K. Dalling, D. M. Grant, and L. F. Johnson, *J. Am. chem. Soc.* **93,** 3678 (1971).
9.19. J. Dale, *Topics Stereochem.* **9,** 258 (1976).
9.20. Y. Huang, S. Macura, and R. R. Ernst, *J. Am. chem. Soc.* **103,** 5327 (1981).
9.21. I. Solomon, *Phys. Rev.* **99,** 559 (1955).
9.22. J. H. Noggle and R. E. Schirmer, *The nuclear Overhauser effect, chemical applications,* Academic Press, New York (1971).
9.23. I. D. Campbell and R. Freeman, *J. mag. Reson.* **11,** 143 (1973).
9.24. A. A. Bothner-By, In *Magnetic resonance studies in biology* (ed. R. G. Shulman), p. 177, Academic Press, New York (1979).
9.25. R. Kaiser, *J. Chem. Phys.* **42,** 1838 (1965).
9.26. A. Kalk and H. J. C. Berendsen, *J. mag. Reson.* **24,** 343 (1976).
9.27. R. Richarz and K. Wüthrich, *J. mag. Reson.* **30,** 147 (1978).
9.28. Anil Kumar, R. R. Ernst, and K. Wüthrich, *Biochem. Biophys. Res. Commun.* **95,** 1 (1980).

9.29. Anil Kumar, G. Wagner, R. R. Ernst, and K. Wüthrich, *Biochem. Biophys. Res. Commun.* **96,** 1156 (1980).
9.30. G. Wagner and K. Wüthrich, *J. mol. Biol.* **155,** 347 (1982).
9.31. A. Dubs, G. Wagner, and K. Wüthrich, *Biochem. Biophys. Acta* **577,** 177 (1979).
9.32. L. M. Jackman and F. A. Cotton (Eds.), *Dynamic NMR spectroscopy,* Academic Press, New York (1975).
9.33. J. I. Kaplan and G. Fraenkel, *NMR of chemically exchanging systems,* Academic Press, New York (1980).
9.34. V. A. Koptyug, V. G. Shubin, A. I. Rezbukhin, D. V. Korchogina, V. P. Tretyakov, and E. S. Rudakov, *Dokl. Chem.* **171,** 1109 (1966).
9.35. B. D. Derendyaev, V. I. Mamatyuk, and V. A. Koptyug, *Bull. Acad. Sci. USSR, Chem. Sci.* **5,** 972 (1971).
9.36. M. Saunders, In *Magnetic resonance in biological systems* (ed. A. Ehrenberg, B. G. Malmström, and T. Vänngård), p. 85, Pergamon Press, Oxford (1967).
9.37. R. Freeman, S. Wittekoek, and R. R. Ernst, *J. Chem. Phys.* **52,** 1529 (1970).
9.38. R. L. Vold and R. R. Vold, *Prog. NMR Spectrosc.* **12,** 79 (1978).
9.39. Y. Huang, G. Bodenhausen, and R. R. Ernst, *J. Am. chem. Soc.* **103,** 6988 (1981).
9.40. N. Bloembergen, S. Shapiro, P. S. Pershan, and J. O. Artman, *Phys. Rev.* **114,** 445 (1959).
9.41. I. J. Lowe and S. Gade, *Phys. Rev.* **156,** 817 (1967).
9.42. A. G. Redfield and W. N. Yu, *Phys. Rev.* **169,** 443 (1968).
9.43. A. M. Portis, *Phys. Rev.* **104,** 584 (1956).
9.44. C. D. Jeffries, *Dynamic nuclear orientation,* Wiley, New York (1963).
9.45. L. Kevan and L. D. Kispert, *Electron spin double resonance spectroscopy,* Wiley, New York (1976).
9.46. S. R. Hartmann and E. L. Hahn, *Phys. Rev.* **128,** 2042 (1962).
9.47. D. A. McArthur, E. L. Hahn, and R. E. Walstedt, *Phys. Rev.* **188,** 609 (1969).
9.48. D. E. Demco, J. Tegenfeldt, and J. S. Waugh, *Phys. Rev.* **B11,** 4133 (1975).
9.49. N. M. Szeverenyi, M. J. Sullivan, and G. E. Maciel, *J. mag. Reson.* **47,** 462 (1982).
9.50. D. Suter and R. R. Ernst, *Phys. Rev.* **B25,** 6038 (1982).
9.51. P. Caravatti, J. A. Deli, G. Bodenhausen, and R. R. Ernst, *J. Am. chem. Soc.* **104,** 5506 (1982).
9.52. P. Caravatti, G. Bodenhausen, and R. R. Ernst, *J. mag. Reson.* **55,** 88 (1983).
9.53. C. E. Bronniman, N. M. Szeverenyi, and G. E. Maciel, *J. Chem. Phys.* **79,** 3694 (1983).
9.54. D. L. VanderHart and A. N. Garroway, *J. Chem. Phys.* **71,** 2773 (1979).
9.55. J. Virlet and D. Ghesquieres, *Chem. Phys. Lett.* **73,** 323 (1980).
9.56. P. Caravatti, P. Neuenschwander, and R. R. Ernst, *Macromolecules,* **18,** 119 (1985).
9.57. A. Abragam, *Principles of nuclear magnetism,* Oxford University Press, London (1961).

Chapter 10

10.1. P. C. Lauterbur, *Bull. Am. phys. Soc.* **18**, 86 (1972).
10.2. P. C. Lauterbur, *Nature* **242**, 190 (1973).
10.3. R. Damadian, *Science* **171**, 1151 (1971).
10.4. R. Damadian, U.S. Patent 3.789.832.
10.5. R. S. Ledley, G. Di Chiro, A. J. Luessenhop, and H. L. Twigg, *Science* **186**, 207 (1974).
10.6. A. L. Robinson, *Science* **190**, 542, 647 (1975).
10.7. R. A. Brooks and G. Di Chiro, *Phys. Med. Biol.* **21**, 689 (1976).
10.8. M. M. Ter-Pogossian, *Sem. Nucl. Med.* **7**, 109 (1977).
10.9. K. R. Erikson, F. J. Fry, and J. P. Jones, *IEEE Trans. Sonics Ultrasonics* **SU-21**, 144 (1974).
10.10. P. Brunner and R. R. Ernst, *J. mag. Reson.* **33**, 83 (1979).
10.11. D. I. Hoult and P. C. Lauterbur, *J. mag. Reson.* **34**, 425 (1979).
10.12. R. R. Ernst, *Adv. mag. Reson.* **2**, 1 (1966).
10.13. W. S. Hinshaw, *Phys. Lett.* **A48**, 87 (1974).
10.14. W. S. Hinshaw, *Proc. 18th Ampère Congress,* Nottingham, p. 433 (1974).
10.15. W. S. Hinshaw, *J. appl. Phys.* **47**, 3709 (1976).
10.16. R. Damadian, M. Goldsmith, and L. Minkoff, *Physiol. Chem. Phys.* **9**, 97 (1977).
10.17. R. Damadian, L. Minkoff, M. Goldsmith, and J. A. Koutcher, *Naturwiss.* **65**, 250 (1978).
10.18. W. S. Hinshaw, P. A. Bottomley, and G. N. Holland, *Nature* **270**, 722 (1977).
10.19. W. S. Hinshaw, E. R. Andrew, P. A. Bottomley, G. N. Holland, W. S. Moore, and B. S. Worthington, *Br. J. Radiol.* **51**, 273 (1978).
10.20. G. N. Holland, P. A. Bottomley, and W. S. Hinshaw, *J. mag. Reson.* **28**, 133 (1977).
10.21. H. R. Brooker and W. S. Hinshaw, *J. mag. Reson.* **30**, 129 (1978).
10.22. W. S. Hinshaw, *Proc. IEEE* **71**, 338 (1983).
10.23. E. R. Andrew, P. A. Bottomley, W. S. Hinshaw, G. N. Holland, W. S. Moore, and C. Simaroj, *Phys. Med. Biol.* **22**, 971 (1977).
10.24. E. R. Andrew, P. A. Bottomley, W. S. Hinshaw, G. N. Holland, W. S. Moore, and C. Simaroj, *Proc. 20th Ampère Congress,* Tallinn (1978).
10.25. P. Mansfield and P. K. Grannell, *J. Phys.* **C6**, L422 (1973).
10.26. P. Mansfield and P. K. Grannell, *Phys. Rev.* **B12**, 3618 (1975).
10.27. P. Mansfield, P. K. Grannell, and A. A. Maudsley, *Proc. 18th Ampère Congress,* Nottingham, p. 431 (1974).
10.28. P. Mansfield, A. A. Maudsley, and T. Baines, *J. Phys.* **E9**, 271 (1976).
10.29. P. Mansfield and A. A. Maudsley, *Proc. 19th Ampère Congress,* Heidelberg, p. 247 (1976).
10.30. A. N. Garroway, P. K. Grannell, and P. Mansfield, *J. Phys. C: Solid State Phys.* **7**, L457 (1974).
10.31. P. Mansfield and A. A. Maudsley, *Phys. Med. Biol.* **21**, 847 (1976).
10.32. P. Mansfield and A. A. Maudsley, *Br. J. Radiol.* **50**, 188 (1977).
10.33. P. Mansfield, *Contemp. Phys.* **17**, 553 (1976).
10.34. P. Mansfield and I. L. Pykett, *J. mag. Reson.* **29**, 355 (1978).
10.35. J. M. S. Hutchison, C. C. Goll, and J. R. Mallard, *Proc. 18th Ampère Congress,* Nottingham, p. 283 (1974).

10.36. R. J. Sutherland and J. M. S. Hutchison, *J. Phys. E: Sci. Instrum.* **11,** 79 (1978).
10.37. J. M. S. Hutchison, R. J. Sutherland, and J. R. Mallard, *J. Phys. E: Sci. Instrum.* **11,** 217 (1978).
10.38. P. C. Lauterbur, *Proc. First Int. Conf. on Stable Isotopes in Chem., Biol., and Medicine,* May 9–11, 1973.
10.39. P. C. Lauterbur, *Pure appl. Chem.* **40,** 149 (1974).
10.40. P. C. Lauterbur, *Proc. 18th Ampère Congress,* Nottingham, p. 27 (1974).
10.41. P. C. Lauterbur, In *NMR in biology* (eds. R. A. Dwek, I. D. Campbell, R. E. Richards, and R. J. P. Williams), p. 323, Academic Press, London (1977).
10.42. P. C. Lauterbur, D. M. Kramer, W. V. House, and C.-N. Chen, *J. Am. chem. Soc.* **97,** 6866 (1975).
10.43. P. C. Lauterbur, *IEEE Trans. nucl. Sci.* **NS26,** 2808 (1979).
10.44. P. Mansfield and A. A. Maudsley, *J. Phys. C: Solid State Phys.* **9,** L409 (1976).
10.45. P. Mansfield and A. A. Maudsley, *J. mag. Reson.* **27,** 101 (1977).
10.46. P. Mansfield, *J. Phys. C: Solid State Phys.* **10,** L55 (1977).
10.47. Anil Kumar, D. Welti, and R. R. Ernst, *Naturwiss.* **62,** 34 (1975).
10.48. Anil Kumar, D. Welti, and R. R. Ernst, *J. mag. Reson.* **18,** 69 (1975).
10.49. D. I. Hoult, *J. mag. Reson.* **33,** 183 (1979).
10.50. P. Brunner and R. R. Ernst, *J. mag. Reson.* **33,** 83 (1979).
10.51. R. Bradford, C. Clay, and E. Strick, *Phys. Rev.* **84,** 157 (1951).
10.52. H. Y. Carr, *Phys. Rev.* **112,** 1693 (1958).
10.53. R. E. Gordon, P. E. Hanley, and D. Shaw, *Prog. NMR Spectrosc.* **15,** 1 (1982).
10.54. J. H. Ackerman, T. H. Grove, G. G. Wong, D. G. Gadian, and G. K. Radda, *Nature* **283,** 167 (1980).
10.55. R. E. Gordon, P. E. Hanley, D. Shaw, D. G. Gadian, G. K. Radda, P. Styles, P. J. Bore, and L. Chan, *Nature* **287,** 736 (1980).
10.56. D. G. Gadian, *Nuclear magnetic resonance and its applications to living systems,* Oxford University Press, Oxford (1982).
10.57a. J. J. H. Ackerman, T. H. Grove, G. G. Wong, D. G. Gadian and G. K. Radda, *Nature* **283,** 167 (1980); J. L. Evelhoch, M. G. Crowley, and J. J. H. Ackerman, *J. mag. Reson.* **56,** 110 (1984).
10.57b. W. P. Aue, S. Müller, T. A. Cross, and J. Seelig, *J. mag. Reson.* **56,** 350 (1984); M. R. Bendall, *J. mag. Reson.* **59,** 406 (1984); A. J. Shaka, J. Keeler, M. B. Smith, and R. Freeman, *J. mag. Reson.* **61,** 175 (1985).
10.58. B. L. Tomlinson and H. D. W. Hill, *J. Chem. Phys.* **59,** 1775 (1973).
10.59. D. I. Hoult, *J. mag. Reson.* **26,** 165 (1977).
10.60. P. Mansfield, A. A. Maudsley, P. G. Morris, and I. L. Pykett, *J. mag. Reson.* **33,** 261 (1979).
10.61. W. A. Edelstein, J. M. S. Hutchison, G. Johnson, and T. W. Redpath, *Phys. Med. Biol.* **25,** 751 (1980).
10.62. G. Johnson, J. M. S. Hutchison, T. W. Redpath, and L. M. Eastwood, *J. mag. Reson.* **54,** 374 (1983).
10.63. P. Mansfield and P. G. Morris, NMR imaging in biomedicine. *Adv. mag. Reson.,* Suppl. 2 (1982).
10.64. J. Jaklovsky, *NMR imaging, a comprehensive bibliography,* Addison Wesley, Reading, Mass. (1983).

10.65. L. Kaufman, L. E. Crooks, and A. R. Margulis, *Nuclear magnetic resonance imaging in medicine,* Igakun-Shoin, Tokyo (1981).
10.66. S. Wende and M. Thelen (Eds.), *Kernresonanz-Tomographie in der Medizin,* Springer, Berlin (1983).
10.67. K. Roth, *NMR—Tomographie und Spektroskopie in der Medizin,* Springer, Berlin (1984).
10.68. P. G. Morris, *Nuclear magnetic resonance imaging in medicine and biology,* Clarendon Press, Oxford (1986).

INDEX

absolute-value, 2D spectra 326, 364, 411
  representation, non-linearity of 327
absorption lineshape 119
absorption, pure 2D 317, 374
accordion spectroscopy
  direct lineshape analysis 513
  lineshape 511
  normal-mode analysis 513
  reduction of dimension 510
  reverse Fourier transformation 513
activation energy 515
active coupling 420
actively involved spins 244, 271
adiabatic
  demagnetization 191
  remagnetization 191
amplitude modulation 318, 323
anisotropic
  chemical shifts 45, 383
  molecular rotation 278
anticommutator 32
antiechoes 316
antiphase
  coherence 402, 407
  square patterns 345
aperiodic perturbations 72, 83
apodization 101
  function 355
arrow notation 21, 27, 268, 295
artefacts 365, 373
asymmetric spectra 380
asymmetry parameter 49
attached proton test 181
attenuation of radiation by tissue 539
auto-correlation function 51, 278
average Hamiltonian 72, 232, 265, 383, 444
  cancellation of irrelevant terms 88
  condition for the existence of 86
  in multiple pulse experiments 77
  in spin-echo experiments 86
average time-domain signal amplitude 351
averaging by time-dependent
       perturbations 75
axial peaks 288

Baker–Campbell–Hausdorff expansion 72
band structure of 2D matrix 336
bandwidth
  of receiver 352
  of spectrometer 150

bilinear rotation, 178, 370
  decoupling (BIRD) 477
  sandwich 370,472
binary random process 112
Bloch equation 115
  in laboratory frame 116
  in rotating frame 116
  modified 60, 211, 495
  non-linearity of 96
Bloch–Siegert shift 252
Boltzmann distribution 14
bra 10
broadband decoupling 237

cancellation effects 328
captive volume 350
Carr–Purcell sequence 365
Cartesian operator products 25, 173, 176, 273, 347
cascade
  of pulses 257
  of semi-selective pulses 169
causality 94, 110, 305, 306
central cross-section 304, 551
chemical compounds
  2-acetonaphthalene 478
  N,N-dimethylacetamide 529
  antamanide 448
  basic pancreatic trypsin inhibitor 265, 326, 362, 411, 431, 439, 442, 524
  benzene 533
  bull seminal inhibitor 344, 412
  cis-decalin 514
  glucose 481
  heptamethylbenzenonium ion 528
  imidazole 531
  menthol 479
  methyl groups 386, 437
  oxalic acid dihydrate 466
  panamine 453
  poly(vinyl methyl ether) 537
  polystyrene 537
  threonine 488
chemical equilibrium 210
chemical exchange 57, 201, 209, 495, 528
  see also chemical reaction
  in the solid state 534
chemical non-equilibrium systems 68, 219

INDEX

chemical reaction 57, 201, 209, 495, 528
  density operator description 64, 68
  first-order 60, 68, 211, 215
  higher-order 63
  in chemical equilibrium, 57, 68
  investigation by 2D spectroscopy 528
  involving coupled spin systems 64, 217
  networks 58
  non-equilibrium 69
  reaction rate 59
  reaction rate constant 60
  stopped flow 57
  transient 57, 210, 215
chemical shielding 44
  anisotropy 393
  tensor 45, 387
chemical shift 45
  concertina 239
chemically induced dynamic nuclear polarization (CIDNP) 57, 170, 215
chromatography 358
Clebsch–Gordon coefficient 41
coalescence 528
coherence 11; *see also* transition
  antiphase 28, 173, 176, 182
  double-quantum, 273
  heteronuclear zero- and double-quantum 470, 484
  in-phase 28, 173, 182
  in-phase multiplet 273
  in-phase p-quantum 174
  multiple-quantum 145, 175, 182
  multiplet 292
  order of 38, 43, 145
  p-quantum 43, 434
  q-spin-p-quantum 271, 273, 438
  single-quantum 43
  spin inversion 271
  three-spin 174
  total spin 271
  transfer 16, 43
  two-spin 173
  vector representation 321
  zero-quantum 43, 494
coherence order 43, 293
coherence transfer 16, 43, 270, 378, 400, 440
  amplitude 290, 320, 360, 402, 416, 422
  by isotropic mixing 445
  by r.f. pulses 467
  consecutive steps 298
  diagram 401
  echo 315, 331, 356
  echo in transfer of multiple-quantum coherence 282
  echo in heteronuclear coherence transfer 282, 315
  heteronuclear 467
  in multiple-quantum spectroscopy 457, 459
  in strongly coupled systems, 39, 280, 374
  in-phase components 446
  in-phase magnetization 473
  induced by π-pulses 370, 374
  map 401
  multiple 298, 442
  pathway 268, 292, 298;
    fanning out 294
    for multiple-quantum spectroscopy 449
    in 2D exchange spectroscopy 494
    in heteronuclear systems 470
    in relayed transfer 441
    mirror image 342
  pathway selection 294, 494
    aliasing in 296
  relayed 440, 456, 482
  selection rules 402, 434
  single-quantum 457
  through one-bond couplings 474
coherent
  state 43
  superposition of eigenfunctions 43
collective spin modes 445
combination lines 174, 244, 260
complex amplitude, in 2D spectra 289, 292, 320
complex signal
  in 1D spectra 119, 160
  in 2D spectra 292, 346, 347
complex transverse magnetization 160, 497
composite Liouville space 66
composite pulse 133, 238, 448
  MLEV 146
  WALTZ 140, 146
  cyclic 146
  for accurate inversion 137
  for accurate refocusing 144
  for minimum phase dispersion 137
  for minimum residual z component 136
  recursive expansion procedures 140
  z-pulses 145
composite rotations 177
connectivity 414
  direct 322, 456, 461
  parallel 414, 418, 425
  progressive 414, 425
  progressive to order q 418
  regressive 414, 425
  regressive to order q 418
  remote 322, 456, 461
constant-time
  correlation spectroscopy 429
  J-spectroscopy 363

INDEX

continuous-wave
  detection of multiple-quantum transitions 247, 250
  sensitivity in one dimension 155
convolution 94
  filtering 304
  integral 99
  theorem 303
correlated state 192
correlation
  coefficient 276, 505
  function 53
  of chemical shielding and dipolar coupling 387
  spectroscopy
    homonuclear (COSY) 405
    heteronuclear 467
  time 516, 524
COSY, 2D correlation spectroscopy 400
counter-rotating component of the r.f. field 117
coupling
  invariance to r.f. pulses 360
  degeneracy, effect on 2D spectra 405
  residual, in scaling 236
  topology 438
cross-correlation function 278
cross-peak 345, 400, 410
  amplitude of 419
  antiphase 405
  due to strong coupling 380
  in-phase 445
  multiplet 408, 418
  phase of 419
  vanishing 404
cross-polarization 185, 468, 535
  adiabatic 468
  double-quantum 488
  in liquids 189
  in solids 186
  in the rotating frame 185, 468
  in two-dimensional spectroscopy 191
  instrumental point of view 191
  measurement of 187
  multiple contact 187, 188
  rate constant 188
  sensitivity enhancement 188
cross-relaxation 495, 500, 516; *see also* nuclear Overhauser effect
  fast-motion limit 519, 521
  in a system with equivalent spins 521
  intermolecular 522
  intramolecular 517
  matrix 517
  rate constants 500, 519
  slow-motion limit 520

cross-section 305
cross-section projection theorem 304, 364, 551
cumulant expansion of the propagator 73
cut-off frequency of audio filter 150
CW, *see* continuous wave

decoupling 146, 232; *see also* double resonance, spin decoupling;
  broadband 237, 367, 474
  by bilinear rotation 475
  by refocusing pulses 475
  dipolar 77
  homonuclear broadband 364
  illusions of 239
  in ω_1 429
  off-resonance 234
delayed acquisition 98, 338, 428, 463
density matrix 15, 268
density of spin states 245
density operator 9, 10
  composite 66
  concentration-dependent 69, 217
  concentration-independent 68
  description of Fourier experiments 159
  direct product 65
  direct summation 66
  equation 12
  expansion in base operators 18
  expansion in single transition operators 291
  idempotent 12
  matrix representation 268, 289
  of systems with chemical exchange 502
DEPT (distortionless enhancement by polarization transfer) 143, 181, 198, 368, 477
detection
  indirect 180, 185, 187, 469, 483
  quadrature phase 95, 150, 159, 175, 293, 308
  single channel 346
detection period 286
diagonal multiplet 318, 407
diagonal peak 410, 425
difference spectroscopy 498, 508
diffusion 158
  effects on echo formation 362
dipolar
  coupling 47, 383
  decoupling in solids 71
  order 506
  tensor 387, 396
Dirac delta function 93
dispersion signal 119, 309
dispersion relations 95, 306

distortionless enhancement by polarization transfer (DEPT) 143, 181, 198, 368, 477
double resonance 220, 365; *see also* decoupling, spin decoupling
  continuous Fourier 225
  during detection 223
  in strongly coupled systems 226
  in weakly coupled systems 227
  phase and intensity anomalies 226
  signal averaging 223
  spin decoupling 232
  theoretical formulation 222
double-quantum
  coherence 11, 43, 245, 408, 494
  cross polarization 488
  excitation 465
  filters 298, 435
  relaxation 275
  signal patterns 458
  spectra 278, 294, 451
  of quadrupolar nuclei 465
double-resonance 71
down-conversion of high-frequency noise 150, 357
dressed spin states 83
driven equilibrium Fourier transform (DEFT) 158
dummy pulses 260
duty ratio of the receiver 151
dynamic chemical equilibrium 490, 502
dynamic matrix 62, 497, 517
dynamic nuclear polarization 535
dynamic processes 201, 286, 490
  in solids 533
  see also chemical reaction, cross-polarization, cross-relaxation, nuclear Overhauser effect
Dyson time-ordering operator 12, 74

echo
  amplitude 87
  line imaging 548
  modulation 144
    by non-resonant spins 381, 382
  planar imaging 559
  primary 209
  secondary 209
  signal 411, 316
  stimulated 209
editing 196, 198, 431, 477
effective magnetic field 120, 227
effective nutation angle 120
effective r.f. field for excitation of p-quantum coherence 262
eigenbasis 39, 422

eigenfunctions of a linear system 94
eigenoperator of a superoperator 80
eigenstates of weakly-coupled spin systems 168
electric field gradient tensor 48
ENDOR (Electron nuclear double resonance) 535
energy levels
  assignment 254
  crossing and avoided crossing 248
ensemble average 10
envelope function, weighted 350
envelope of the free induction signal 107
equilibrium magnetization 184
equivalence of slow-passage and Fourier spectroscopy 162, 210
Euler angles 41
evolution period 285, 288
exchange 60, 201, 495, 499; *see also* chemical exchange, chemical reaction
  broadened lineshapes 210
  broadening 528
  cross-peaks 501
  difference spectroscopy 508
  in coupled spin systems 501
  multiple-site 500
  maps 528
  narrowing 528
  networks 512
  processes 209
  rates 514
excitation *see also* pulse
  of multiple-quantum transitions 257–65
exorcycle 297, 366
expectation value of operator 13, 18
extreme narrowing (fast motion limit) 52, 516

fast motion limit 516, 518
field gradient 132
field-cycling 71
filtering
  convolution 149
  linear 99
  matched 152, 155, 331, 355
  multiple-quantum 328, 434
  p-spin 438
  spin-pattern 438
  spin-system-selective 438
  two-dimensional 329
flip angle, *see also* pulse rotation angle
flip angle effect in 2D spectra 419
flip-back pulse 188
flip–flop transitions, energy conserving 521
Floquet Hamiltonian 82
Floquet theory 71, 81
fluctuating fields 56, 274, 275

INDEX

FOCSY (foldover-corrected spectroscopy) 428
folding 337, see also aliasing
foldover correction 98, 337, 368, 428
Fourier analysis 299
  in pathway selection 296
Fourier imaging 553
  reconstruction 551
Fourier spectroscopy
  advantages 91
  density operator formalism 159
  double resonance 220
  drawbacks 91
  four-phase 133
  of non-equilibrium systems 165
  pulse rotation angle dependence 170
  quadriga 132
  quantum-mechanical description 158
  sensitivity 148
  two-dimensional 283
Fourier transform pair 303
Fourier transformation 97
  complex 301, 409
  convolution theorem 98
  derivative theorem 98
  discrete 105
  power theorem 99
  real 317, 409
  shift theorem 98
  similarity theorem 98
Fourier transformation, two-dimensional 301
  convolution theorem 303
  hypercomplex 307
  power theorem 304
  projection cross-section theorem 304
  real transforms 302
  similarity theorem 303
  vector notation 302
free induction decay 118, 160
free precession 118
frequency response function 94, 99
frequency vector 27
frequency-dependent phase shift 98, 128

gated decoupling 369
Gaussian envelope 332
Gaussian random process 112
generating operators 26
grid of sampling points 554
group spin quantum numbers 247

Hahn echo 208
Hamiltonian 44; see also average Hamiltonian
  antisymmetric part 86
  average 72, 444
  bilinear 28, 46
  chain of non-commuting terms 88
  dipolar 47
  effective 358
  isotropic mixing 445
  linear 44
  mixing 444
  periodic 394
  periodic time-dependent 81
  quadratic 48
  relaxation 55
  symmetric part 86
Hartmann–Hahn condition 186
  matching 190
heat capacity of the dipolar reservoir 191
Heisenberg operator 13
  representation 14
Hermitian operators 18
heterogeneity in solids 536
heteronuclear
  coherence transfer 180, 467
  correlation by 2D spectroscopy 313, 471
  in solids 485
  involving double transfer 482
heteronuclear separated local field spectra 384
high-resolution spectra in inhomogeneous fields 280
high-temperature approximation 65, 161, 182
Hilbert space 10, 44
Hilbert transformation 96, 306
HOMCOR (homonuclear correlation spectroscopy) 400
homonuclear J-spectroscopy 328
hypercomplex Fourier transformation 307

idempotent operator 22
illusions of decoupling 90, 239
imaging 539
  classification 541
  echo line 548
  echo planar 559
  field focusing NMR (FONAR) 544
  Fourier 553
  line scan 547
  multiplanar 559
  performance time 560
  planar 557
  projection-reconstruction 549
  rotating frame 556
  sensitive line 545
  sensitive point 542
  sensitivity 560

imaging (cont.)
  sequential line 545
  sequential plane 548
  sequential point 542
  spin-warp 554
impulse response 93, 99, 112
  higher-order 110
INADEQUATE experiment 453
incoherent non-equilibrium 162, 165
INEPT (insensitive nuclei enhanced by polarization transfer) 181, 193, 368, 480
  refocused 195
inhomogeneous
  r.f. fields 133
  broadening 207, 312
    of p-quantum transitions 279
  decay 404
  fields 280
initial rate approximation 492, 501, 523
input/output relations 92
insensitive nuclei enhanced by polarization transfer (INEPT) 181, 193, 368, 480
intensity anomalies 129, 131
intensity in CW-NMR 250
interactions
  dipolar 47
  electric quadrupole 48
  indirect scalar 46
  linear 44
  quadratic 48
  Zeeman 44
interference
  constructive 313
  destructive 312, 314
  in 2D spectra 313
  longitudinal 124
  of neighbouring peaks 312
  transverse 125, 131
  zero-quantum 503
inverse spin temperature 184
inverse temperature of lattice 183
inversion of magnetization 122, 137, 203
inversion-recovery 138, 203
  difference spectroscopy 510
in vivo NMR 545
irreducible tensor operator 40, 56
  rank of 269
  transformation properties 269
isochromats 207
isotropic mixing Hamiltonian 445
isotropic mixing sequence 486

J cross-peak 501, 505, 525
  suppression 506
J couplings to passive nuclei 503

Jeener–Broekaert method 506
J-filter 480
  low-pass 507
J-spectroscopy (J-resolved spectroscopy) 208, 359–82

kernel function 110, 114
ket 10
kinetics, chemical 57, 201, 495
kinetic matrix 61, 497
  pseudo first-order 64
Kramers–Kronig relations 95, 306

Larmor frequency 45, 118
lattice, degrees of freedom 14
leakage rate constant 500, 519
level anti-crossing 192
level shift in CW-NMR 252, 262
line scan technique 547
line-width, inhomogeneous 280
line-narrowing, *see* decoupling
linear dependence among operators 36
linear response theory 124
lineshape 102, 119; *see also* peakshape
  absolute-value, 333
  analysis 528
  in equilibrium chemical exchange 210
  in one-sided chemical reactions 213
  in two-sided chemical reactions 216
Liouville space 17, 39
  composite 66
  interaction 65
  molecular 66
Liouville–von Neuman equation 12, 111
liquid crystals 276, 464
long-range coupling 313, 370
longitudinal relaxation 49, 202, 530; *see also* relaxation
longitudinal scalar order 506
  three-spin order 174
  two-spin order 166, 173, 182, 196, 408, 472
Lorentz–Gauss transformation 330, 332
Lorentzian peak 309, 332
low-pass J filter 482

magic angle 386
  flipping 398
  hopping 398
  sample spinning 393, 396, 466
magnetic equivalence 404, 426, 458
magnetic field gradient 539
  gradient pulse 131
  sinusoidally modulated 542

INDEX

magnetic field, effective 120, 227
magnetic quantum number 244, 247, 293
magnetically equivalent nuclei, 247, 456
magnetization
  complex notation 119
  transverse 43, 131
  vector 62
  see also coherence
Magnus expansion 72, 75, 233
manipulation of nuclear spin
    Hamiltonians 70
manipulation of two-dimensional
    spectra 336
master equation 530
  for the composite density operator 67
  for populations 50, 221
  quantum mechanical 15, 51
  with saturation 221
matched filtering 152, 331, 355
matched preparation delay 259
matrix
  kinetic 61
  representation 11, 15, 269, 289
  representation, of relaxation
    superoperator 53
  stoichiometric 58
maximum enhancement of polarization 184
maximum entropy method (MEM) 106
McConnell equation 60
microscopic reversibility 63
mirror image pathway 293, 298, 322, 342, 442
mixed phase peakshape 310
mixed states 375
mixing effect of an r.f. pulse 222
mixing operator 499
mixing period 285, 288, 450
  in 2D correlation 290
  in 2D exchange 491
  in 2D separation 290
  optimized for maximum cross-peak
    amplitude 519
mobile side-chains 524
modified Bloch equation 211, 495
molecular structure in solution 516
motional processes 396
multichannel spectrometer 4
multiple refocusing 144
multiple-channel experiment 156
multiple-pulse dipolar decoupling 383, 386
  BLEW-12, 485
  WALTZ-8, 485
  WIM-24 (windowless isotropic mixing) 485
multiple-quantum, CW detection 254, 247
  cross-polarization, 487
  filters 434

multiple-quantum coherence 11, 204, 403
  creation of 256
  detection 257, 449
  excitation 257, 449
  excitation by non-selective pulses 258
  excitation by selective pulses 261
  excitation of even and odd orders 260
  excitation of specific orders 265
  for oriented spins 262
  heteronuclear 483
  in an inhomogeneous static field 272
  number of transitions 244
  phase-shift dependence 268
  precession frequency 271, 265
  relaxation 274
  selective excitation 266
  separation of orders 265
  spin-locking 281
  tailored excitation 263
  transformation properties 268
  uniform excitation 259
multiple-quantum
  double resonance 273
  frequency 265, 271
  pulse 258
  refocusing pulse 281
  relaxation 274
    measurement of decay rate 278
  signal separation of different orders 267
multiple-quantum spectroscopy 256, 270, 328, 456
  in liquid crystalline solvents 464
  of dipole-coupled spins 463
  of scalar-coupled networks 456
  time-domain 256
  two-dimensional 449
multiple-quantum transition 242
  effective resolution 254
  intensity 250
  level shift 252
  line-width 249, 254
  offset-dependence 256
  saturation 251
  spin–lattice relaxation 249
multiplet coherence 292
multiplet operators 271
multiplet patterns 176
multiplet structures 347
  for magnetically equivalent spins 426
  of cross-peaks 420
multiplets, antiphase 434
multiplex advantage of Fourier
    spectroscopy 545

nematic solvent 276
nodal plane of field gradient 542
NOE (nuclear Overhauser effect) 516

INDEX

NOE cross-peaks 521
NOESY, 2D NOE spectroscopy 516
noise
  Gaussian 111
  power per unit bandwidth 152
  pseudo-random 115
  r.m.s. amplitude 450, 352
  white 113
noise decoupling 238
non-equilibrium state
  coherent 165
  incoherent 165, 170
  of the first kind 162, 165
  of the second kind 165
  preparation of 219
N peaks 316, 339
nuclear Overhauser effect (NOE) 181, 184, 221, 469, 516; see also cross-relaxation
  build-up curves 526
  sign of 517, 521
  steady-state 516
  transient 516
Nyquist frequency 104, 150, 337, 340, 352

observable 13, 177
off-resonance effects 119, 227, 234
one-dimensional rotation group, irreducible representation of 293
operator
  adjoint 18
  algebra 18, 26
  angular momentum 25
  Cartesian single-transition 34
  exponential 73
  Hermitian 11
  idempotent 22
  lowering 32
  multiplet 271
  non-Hermitian 32
  permutation 375
  pictorial representation of products 175
  polarization 30, 33, 166
  products of Cartesian 173
  projection 21
  raising 32
  shift 32
  single-element 33, 271
  single-spin 172
  single-transition 172, 291
  single-transition shift 18, 38
  spectral resolution 21
  spectral set 22
  spin 25
optimum pulse rotation angle 125, 127, 153
optimum r.f. field strength for maximum signal intensity in CW–NMR 252

order see dipolar, scalar order
order of coherence 244, 265, 360
oscillatory signal ('ripple') 101
Overhauser effect see nuclear Overhauser effect

parallel transition 416
partition function 14
passive J coupling 420
passive spin 403
pattern recognition 343
Pauli matrix 34
peakshape in two-dimensional spectra 310
  2D absorption 310, 321, 431, 446
  2D dispersion 310, 321
  in 2D correlation spectroscopy 425, 409
  mixed phase 309, 321, 431
  phase-twist 310
performance time of imaging techniques 563
perturbation
  aperiodic 83, 285
  cyclic 76
  expansion 247
  operator 247
  periodic 76
  theory 81, 250
  time-dependent 71
  time-independent 70
phase
  adjustment 307
  alternation 133, 491
  anomalies 129, 131
  cycle 133, 179, 204, 257, 268, 280, 295, 297, 470, 496
  encoding 554
  error 122
  in relayed coherence transfer 442
  in ω_1- and ω_2-domain 347, 348
  manipulations in the mixed time-frequency domain 338
  modulation 317, 323
  of peakshape 128, 161, 321, 419
  selection pulse 374
  shift 179, 268
  shifter 145
  synchronization 230
phase-sensitive
  detector 265
  spectra 412
  2D correlation spectra 437
phase-shifted
  propagator 295
  r.f. pulse 32; see also phase cycle
phase-twist 311
  peakshape 321, 361
phenomenological Bloch equation 49

polarization 173
  transfer 180, 368, 490, 493
    adiabatic 191
    by cross-polarization in the rotating frame 185
    by nuclear Overhauser effect 184
    by radio-frequency pulses 192, 400
    heteronuclear 180
polymer blends 537
populations 11, 34
  non-equilibrium 261
powder spectrum 390
power spectral density 50, 56
power theorem 304
P peak 316, 339, 411
precession frequency in the rotating frame 118
preparation period 285
principal axes, values of the shielding tensor 45, 389
principle of linear prediction 106
product basis 39, 422
product function 404
product operator 25, 32, 173
progressive saturation 136
projection
  of 2D spectra 256, 327, 328
  of absolute-value spectra 329
  orthogonal 363
  skew 305, 328, 363
  skyline 329
projection cross-section theorem 304, 363, 551
projection operator 12
projection-reconstruction technique 549
propagation of spin order 535
propagator 13, 73
  mixing 257
  preparation 257
proton flip method 370
pseudo double-quantum spectra 444
pseudo-echo envelope 334
  transformation 333
pulse
  cascade 169, 402, 473
  composite, see composite pulse
  dummy 178
  modulation sidebands 237
  non-selective 171
  operator matrix representation 422
  rotation angle 118, 171
    nominal 135
    optimum 127, 153
  rotation-synchronized 397
  sandwich 259

  selective 171, 261, 370
  selective p-quantum 262
  semi-selective 171
pulse sequence
  2D correlation spectroscopy, COSY 406, 447
  2D difference spectroscopy 509
  2D exchange and NOE spectroscopy 491, 507
  BR–24, 383
  DEPT, 199
  INEPT, 180
  MLEV, 238
  MREV-8, 383
  SEMUT-GL, 200
  WALTZ, 140, 238
  WHH-4, 78, 383
  Carr–Purcell 158
  constant-time correlation spectroscopy 430
  Fourier imaging 553
  Jeener–Broekaert 192
  echo line imaging 549
  echo planar imaging 559
  heteronuclear 2D spectroscopy 472
  heteronuclear relayed magnetization transfer 480
  heteronuclear shift correlation in solids 486
  line scan technique 547
  measurement of spin diffusion 534
  modified SECSY 428
  multiple-quantum filtered COSY 433
  multiple-quantum spectroscopy 258, 449, 450
  phase alternated 133
  planar imaging 557
  relayed 2D correlation spectroscopy 441
  rotating frame imaging 556
  'sandwich' symmetry 178
  sensitive point technique 543
  spin–echo correlation spectroscopy 428
  spin–warp imaging 555
  windowless 384
pure phase in two-dimensional spectra 310, 317, 324, 341
purging pulse 196, 455

quadratic response 110, 112
quadrature phase detection 95, 150, 159, 175, 293, 308, 346
quadrupolar relaxation 57
quadrupolar splitting, elimination of 465
quadrupole coupling 48
  powder pattern 466
  rotations induced by 31

quantization axis, tilted 235
quantum numbers 11
  'good' quantum numbers 293
  group spin 247
quaternion formalism 135

r.m.s. noise amplitude 150, 352
random field
  correlation 275
  model 56
  relaxation 56
random process 51
  ergodic 150
rank of irreducible tensor 42, 269
rate constant 64
reaction, see chemical reaction
reaction number 59
reconstruction techniques
  Fourier 551
  back-projection 550
  iterative 550
recovery of the longitudinal
      magnetization 124
Redfield
  matrix 160
  relaxation supermatrix 53
reduction of dimension 511, 558
redundancy of multiplet patterns 345
refocusing 144, 280; see also echo
  of chemical shifts 177, 207, 371
  of heteronuclear couplings 371
  of the IS couplings 369
  strong coupling effects 374
relaxation 49, 201, 495
  by external random fields 274, 504, 518, 531
  by paramagnetic impurities 535
  dipolar 56
  double-quantum 275
  in multi-level spin systems 530
  indirect detection 530
  intermolecular 276
  longitudinal 49, 202, 530
  matrix 49, 53, 497
  mechanisms 55
  multi-exponential 404
  of multiple-quantum coherence 274
  paramagnetic 276
  quadrupolar 57, 276
  single quantum 275
  semi-classical theory 50
  superoperator 15, 49, 51
  transverse 49, 206
  zero-quantum 275
relaxation measurements
  Carr–Purcell 208

Hoffman–Forsen method 205
indirect detection 530
inversion-recovery 203
progressive saturation 205
saturation-recovery methods 204
single-scan T_1-recovery 203
steady-state Overhauser
    measurement 206
transient NOE method 205
transverse relaxation 206
truncated driven NOE method
    (TOE) 206
relaxation rate,
  adiabatic 54
  double-quantum 275, 504
  fast motion limit 518
  non-adiabatic 55
  single-quantum 275, 504
  slow motion limit 518
  transverse 289
  zero-quantum 275, 504
relaxation reagent 531
relaxation theory
  quantum-mechanical 50
  semi-classical 50
relayed coherence transfer 440, 456, 482
relayed heteronuclear correlation
    spectroscopy 479
remote connectivity 458
remote spin 373, 456
resolution in the ω_1-domain 355
resolution and sensitivity
    compromise 154
resolution enhancement 330
  convolution-difference technique 107
  Lorentz–Gauss transformation 108
  sine-bell function 108
  ultimate ripple-free 109
response theory 92
  Kubo 111
  linear 93
  non-linear 109
  quantum-mechanical 111
  stochastic 112
ridge plots 391
rotary echo 282
rotating frame 45, 116, 265
rotating frame imaging 556
rotation
  bilinear 30, 259
  in operator subspaces 28, 37
  sense of 27
rotation angle 118, 171
  nominal 135
  optimum 153, 430
  self-compensating effect of the
      effective 137

INDEX

rotation matrix 121
rotation operator 41, 135
   matrix representation 121, 41, 422
rotation-synchronized pulses 39
rotational echo 393, 398
rotational sideband 395

sampling, synchronous 384, 393
sampling theorem 104
satellite signals 273, 508
saturation 220
   broadening 255
   in CW NMR 251
   matrix 221
   parameter 155, 251
saturation-recovery 136, 204, 510
scalar order 495, 506
scalar product
   of operators 18
   of state functions 10
scaling
   factor 383
   of heteronuclear couplings 238
   of homonuclear interactions 71
   of multiplet splittings 363
   of the chemical shift 398
   uniform 236
Schroedinger equation 9
Schroedinger representation 13
second order shifts 51
second rank tensor 383
SECSY (spin echo correlated spectroscopy) 338, 428
section, see cross-section
secular contributions 52
segmentation of the time axis 284
selection rules of coherence transfer 402
selective inversion 181, 517
   population transfer 181
   pulse 171, 370, 548
selectivity of the excitation 493
SEMUT (subspectral editing using a multiple-quantum trap) 181
sensitive line method 545
   point method 129, 542, 544
sensitivity 148, 151, 349
   advantage of INEPT 194
   comparison of Fourier and slow-passage 156
   comparison of one- and two-dimensional spectra 353
   in heteronuclear coherence transfer 468
   maximum 156
   of imaging techniques 560
   optimization of a 2D experiment 357
sensitivity enhancement 330
   by double transfer 485
   by recycling of magnetization 157
sensitivity of two-dimensional spectra 349
sensitivity ratio of 2D and 1D spectroscopy 353
separated local field spectroscopy 383
   heteronuclear 384
   homonuclear 383
   magic angle spinning 394
separation of interactions 358
sequential resonance assignment 527
shearing transformation 336, 362, 368, 376, 386, 463
shift correlation maps 475
sideband frequencies of the r.f. pulse sequence 127
sign alternation of cross-peak multiplets 345
sign of J couplings 421
signal
   average power 152, 353
   complex 119
   energy 153
   envelope function 149, 316, 331, 349
   envelope function in 2D time domain 315
   phase 128
signal-to-noise ratio 353
   in 1D spectroscopy 151
   in 2D spectroscopy 353
   maximum achievable 155
   per unit time 151, 353
   slow passage spectra 155
similarity theorem 303, 336
simple lines 249
$\text{sinc}(x)$ function 101
single-channel detection 346
single-transition operators 18, 38, 172, 291
skew diagonal 452, 463
skew projections 304
slow-exchange limit 493, 498
slow-motion limit 516
slow-passage spectroscopy 148, 283
Solomon's notation 517
Son of Laocoon 380
spectral density function 53, 278
spectral editing 181
spectral elements, number 156
spectrum, complex 119, 163
spherical harmonics 278
spin
   active 244, 271
   passive 271
spin decoupling 221, 232, 366; see also decoupling, double resonance.
   broadband 237
   by continuous irradiation 240
   in liquids 71
   multiple-pulse 238

spin decoupling (*cont.*)
  noise 238
  off-resonance 234
  time-shared 236
spin diffusion
  among isotopically dilute spins 536
  between abundant spins 534
  between two nuclear species 534
  in the laboratory frame 534
  in the rotating frame 534
spin diffusion limit 516
spin entropy 182
spin filter 431
  bandpass 432, 438
  building blocks 433
  high-pass 432
  multiple quantum 434
  spin-topology selective 432
spin Hamiltonian 44
spin order 182
spin pattern recognition 181
spin system
  AA'X 457
  AB 410, 422, 457
  ABC 381
  ABX 377
    AB subspectra 378
    2D spin echo spectrum 379
    eigenstates 377
  AMX 168, 364, 456, 461
  AX 168, 451, 459
  strongly coupled 39
spin temperature 167, 535
  inverse 161
spin-echo 86; *see also* echo
spin-echo correlation spectroscopy
    (SECSY) 338
spin-echo Fourier transform method
    (SEFT) 158
spin-echo spectroscopy 208, 360, 428
spin-flip number 168
spin-knotting 137
spin-lattice relaxation, *see* longitudinal
    relaxation
spin-locking 137
  by r.f. pulses 129
spin-connectivity selective excitation 263,
    438
spin-connectivity selective pulse sequence
    257
spin–rotation interaction 55
spin–spin coupling, indirect 46
spin-tickling 225, 228, 230
spin-warp imaging 554
spinor transformation properties 375
SPT (selective population transfer) 181
spy nucleus 531
star effect 311, 327, 330, 332

state
  coherent 11
  function 9
  incoherent non-equilibrium 33
  mixed 10
  non-stationary 11
  of maximum information 12
  pure 10
statistical ensemble 43
steady-state free precession 129, 542
  magnetization 127, 130, 157, 543
  saturation 517
stimulated echo 208
stochastic Hermite polynomial 113
stochastic multidimensional
    spectroscopy 111
stoichiometric matrix 59
stopped-flow 219
stroboscopic observation 72, 76, 81, 398
strong coupling 374, 403, 422
  2D *J*-spectra 377
  2D correlation 422.
subspectral analysis 419, 446
supercycle 146
supermatrix 19, 40
superoperator 19
  adjoint 19
  algebra 25
  commutator 16, 20
  derivative 20
  eigenoperator 24
  eigenvalue 24
  exchange 67
  general representation 23
  Hermitian 20
  inversion 163
  left-, right-translation 20
  linear 19
  Liouville 20, 159, 162
  matrix representation 23
  of relaxation 287
  of rotation 287
  p-quantum projection 22
  projection 21
  relaxation 15, 46
  unitary transformation 20
superposition of signals 312
superposition principle 93
suppression
  of ripple 103
  of sidebands 397
surface coils 545
susceptibility effects 538
symmetric cycles 79
symmetric excitation and detection 257, 453
symmetrization 341
symmetry-adapted base function 404
symmetry about $\omega_1 = 0$, 323, 374

restrictions on coherence transfer 403
 of 2D spectra 341
 of operator 247
symmetry-adapted base functions 426
system operator 92

tailored excitation 548
tensor
 chemical shielding 45, 387
 dipolar coupling 45, 387
 quadrupole coupling 48
tensor operator
 irreducible 40, 47
 rank of 41
thermal equilibrium 14
thermal noise 351
three j symbols 42
three-dimensional rotation group,
  irredicible representation 41
three-dimensional spectroscopy 365, 510
 projection 286
tickling effects 221, 230
tilt pulses 384
tilted effective field 119, 126, 133
tilted frame 189
time reversal 265
time-dependent Schroedinger equation 9
time-proportional phase incrementation
  (TPPI) 268, 322, 340, 479, 496
time-reversal 265, 324, 374, 465
time-shared decoupling 236
TOCSY (total correlation
  spectroscopy) 445
toggling frame 76, 208
topical NMR 544
topology of coupling network 263
topology of the energy level scheme 414
total correlation spectroscopy
  (TOCSY) 444
total spin coherence 438
total spin coherence transfer echo
  spectroscopy 465
total spin operator 244
TPPI (time proportional phase
  incrementation) 340
trace metric 18
transfer
 adiabatic 183
 see also coherence transfer; polarization,
  spin order
transfer function 94
transfer of the complex magnetization 487,
  495
transformation
 into rotating frame 117
 Lorentz–Gauss 330, 332
transition; see also coherence

connected 232, 247
connectivity 414
 double-quantum 244
 ladder of connected 261
 number of transitions 244
 order of 244
 parallel 168, 414
 progressive connectivity 232, 250, 254
  to order q 414, 417
 regressive connectivity 232, 250
  to order q 414
 remotely connected 414
transition probability 49, 201, 275, 517
 double-quantum relaxation 518
 single-quantum relaxation 518
translational diffusion 132, 208, 365
transverse relaxation
 adiabatic contributions 274
 non-adiabatic contributions 274
triangular multiplication 343
trigonometric interpolation 106
truncated signal 101
truncation of internal Hamiltonians 79
TSCTES (total spin coherence transfer echo
  spectroscopy) 465
two-dimensional
 absorption peakshape 310, 317, 386
 chemical exchange spectroscopy 528
  indirect detection 530
  in solids 533
 correlation spectroscopy (COSY) 294,
  298, 400
  homonuclear 405
  heteronuclear 471
  modified experiments 427
  relayed 440
  relayed heteronuclear 479
  small flip angle 421
  strong coupling 422
  with constant time evolution 429
 dispersion peakshape 310
 double quantum spectroscopy 298, 449
 exchange difference spectroscopy 508
 exchange spectroscopy 298, 490
  in coupled spin systems 501
 foldover-corrected spectroscopy
  (FOCSY) 337, 428
 Fourier transformation 301
 J-spectroscopy 181, 360, 428
 INADEQUATE 454
 NOE spectroscopy (NOESY) 516
  for macromolecules 523
  intermolecular 522
  intramolecular 517
 powder spectra 316, 387
 separation 358
  in heteronuclear systems 366
  in homonuclear systems 360

two-dimensional (cont.)
  separation in oriented phase 383
  signal patterns 348
  spectroscopy
    basic principles 283
    double resonance experiments 283
    formal theory 287
    stochastic excitation 113, 284
  spectrum
    coordinate system 311
    multiplet structure 347
    symmetry 318
  spin echo correlated spectroscopy
    (SECSY) 338, 428
  total correlation spectroscopy
    (TOCSY) 444
t_1-noise 351, 343
two-spin system
  2D correlation 406
  relaxation 275, 504
  strongly coupled 422

ultrasonic scanner 539
uniform excitation 259
unit step response 94
unitary transformation 12, 16, 20, 73

variable nutation angle method 205
virtual coupling 426
Volterra functional expansion 109
volume element 541
volume-selective pulse sequence 545

WALTZ-16, 474
WHH-4 multiple-pulse sequence 78
weighting function
  for enhancing the resolution 107
  matched 152, 353
  optimum 108
  rectangular 101
white noise 113
Wigner rotation matrix 269
window
  cosine 103
  Dolph--Chebycheff 103, 109
  Hamming 103, 109
  Hanning 103
  Kaiser 103
  matched 152, 331, 355
  sine-bell 108
windowless sequences 384

X-ray tomography 539, 549

zero-quantum coherence 43, 408, 494, 245
zero-field magnetic resonance 71
zero-filling 104
zero-quantum suppression 504

z-filter 340, 429, 433
z-pulse 141, 145
z-rotation 179